T0180088

Lecture Notes in Artificial Intelligence   10870

Subseries of Lecture Notes in Computer Science

LNAI Series Editors

Randy Goebel
  University of Alberta, Edmonton, Canada
Yuzuru Tanaka
  Hokkaido University, Sapporo, Japan
Wolfgang Wahlster
  DFKI and Saarland University, Saarbrücken, Germany

LNAI Founding Series Editor

Joerg Siekmann
  DFKI and Saarland University, Saarbrücken, Germany

More information about this series at http://www.springer.com/series/1244

Francisco Javier de Cos Juez · José Ramón Villar
Enrique A. de la Cal · Álvaro Herrero
Héctor Quintián · José António Sáez
Emilio Corchado (Eds.)

Hybrid Artificial Intelligent Systems

13th International Conference, HAIS 2018
Oviedo, Spain, June 20–22, 2018
Proceedings

 Springer

Editors
Francisco Javier de Cos Juez
Department of Mine Operating
and Prospection
University of Oviedo
Oviedo
Spain

José Ramón Villar
Department of Computer Science
University of Oviedo
Oviedo
Spain

Enrique A. de la Cal
Department of Computer Science
University of Oviedo
Oviedo
Spain

Álvaro Herrero
Department of Civil Engineering
University of Burgos
Burgos
Spain

Héctor Quintián
University of A Coruña
A Coruña
Spain

José António Sáez
University of Salamanca
Salamanca
Spain

Emilio Corchado 🅸🅳
University of Salamanca
Salamanca
Spain

ISSN 0302-9743   ISSN 1611-3349 (electronic)
Lecture Notes in Artificial Intelligence
ISBN 978-3-319-92638-4   ISBN 978-3-319-92639-1 (eBook)
https://doi.org/10.1007/978-3-319-92639-1

Library of Congress Control Number: 2018944405

LNCS Sublibrary: SL7 – Artificial Intelligence

Printed on acid-free paper

This Springer imprint is published by the registered company Springer International Publishing AG
part of Springer Nature
The registered company address is: Gewerbestrasse 11, 6330 Cham, Switzerland

Preface

This volume of *Lecture Notes in Artificial Intelligence* (LNAI) includes accepted papers presented at HAIS 2018 held in the beautiful city of Oviedo (Asturias), Spain, June 2018.

The International Conference on Hybrid Artificial Intelligence Systems HAIS has become a unique, established, and broad interdisciplinary forum for researchers and practitioners who are involved in developing and applying symbolic and sub-symbolic techniques aimed at the construction of highly robust and reliable problem-solving techniques and being responsible the most relevant achievements in this field.

Hybridization of intelligent techniques, coming from different computational intelligence areas, has become popular because of the growing awareness that such combinations frequently perform better than the individual techniques such as neurocomputing, fuzzy systems, rough sets, evolutionary algorithms, agents and multiagent systems, etc.

Practical experience has indicated that hybrid intelligence techniques might be helpful for solving some of the challenging real-world problems. In a hybrid intelligence system, a synergistic combination of multiple techniques is used to build an efficient solution to deal with a particular problem. This is, thus, the setting of the HAIS conference series, and its increasing success is the proof of the vitality of this exciting field.

The HAIS 2018 International Program Committee selected 62 papers that are published in this conference proceedings, with a percentage of acceptance about the 60% of the submissions.

The selection of papers was extremely rigorous in order to maintain the high quality of the conference and we would like to thank the Program Committee for their hard work in the reviewing process. This process is very important to the creation of a conference of high standard and the HAIS conference would not exist without their help.

The large number of submissions is certainly not only to testimony to the vitality and attractiveness of the field but an indicator of the interest in the HAIS conferences themselves.

HAIS 2018 enjoyed outstanding keynote speeches by distinguished guest speakers: Prof. Antony Bagnall, Department of Computer Science, University of East Anglia, UK, and Prof. Luciano Sánchez, Computer Science Department, University of Oviedo, Spain.

HAIS 2018 teamed up with *Sensors* (MDPI) and the *Logic Journal of the IGPL* Oxford Journals for a suite of special issues including selected papers from HAIS 2018.

Particular thanks go as well to the main sponsors of the conference, Startup OLE, Government of Principado de Asturias, Government of the Local Council of Oviedo, University of Oviedo, Computer Science Department at University of Oviedo,

University of Salamanca, who jointly contributed in an active and constructive manner to the success of this initiative.

We would like to thank Alfred Hofmann and Anna Kramer from Springer for their help and collaboration during this demanding publication project.

June 2018

Francisco Javier de Cos Juez
José Ramón Villar
Enrique A. de la Cal
Álvaro Herrero
Héctor Quintián
José António Sáez
Emilio Corchado

Organization

General Chair

Emilio Corchado   University of Salamanca, Spain

Local Chairs

Francisco Javier   University of Oviedo, Spain
de Cos Juez
Jose Ramón Villar   University of Oviedo, Spain
Enrique A. de la Cal   University of Oviedo, Spain

Honorary Committee

Javier Fernández Fernández   President of the Government of Asturias
Santiago García Granda   Chancellor of the University of Oviedo
Francisco Wenceslao   Mayor of the City of Oviedo
López Martínez

International Advisory Committee

Ajith Abraham   Machine Intelligence Research Labs, Europe
Antonio Bahamonde   University of Oviedo, Spain
Andre de Carvalho   University of São Paulo, Brazil
Sung-Bae Cho   Yonsei University, South Korea
Juan M. Corchado   University of Salamanca, Spain
José R. Dorronsoro   Autonomous University of Madrid, Spain
Michael Gabbay   Kings College London, UK
Ali A. Ghorbani   UNB, Canada
Mark A. Girolami   University of Glasgow, Scotland
Manuel Graña   University of País Vasco, Spain
Petro Gopych   Universal Power Systems USA-Ukraine LLC, Ukraine
Jon G. Hall   The Open University, UK
Francisco Herrera   University of Granada, Spain
César Hervás-Martínez   University of Córdoba, Spain
Tom Heskes   Radboud University Nijmegen, The Netherlands
Dusan Husek   Academy of Sciences of the Czech Republic,
     Czech Republic
Lakhmi Jain   University of South Australia, Australia
Samuel Kaski   Helsinki University of Technology, Finland
Daniel A. Keim   University Konstanz, Germany
Marios Polycarpou   University of Cyprus, Cyprus

Witold Pedrycz	University of Alberta, Canada
Xin Yao	University of Birmingham, UK
Hujun Yin	University of Manchester, UK
Michał Woźniak	Wroclaw University of Technology, Poland
Aditya Ghose	University of Wollongong, Australia
Ashraf Saad	Armstrong Atlantic State University, USA
Fanny Klett	German Workforce Advanced Distributed Learning Partnership Laboratory, Germany
Paulo Novais	Universidade do Minho, Portugal
Rajkumar Roy	The EPSRC Centre for Innovative Manufacturing in Through-life Engineering Services, UK
Amy Neustein	Linguistic Technology Systems, USA
Jaydip Sen	Innovation Lab, Tata Consultancy Services Ltd., India

Program Committee

Enrique de la Cal (PC Co-chair)	University of Oviedo, Spain
Francisco Javier de Cos Juez (PC Co-chair)	University of Oviedo, Spain
Héctor Quintián (PC Co-chair)	University of A Coruña, Spain
José Ramón Villar (PC Co-chair)	University of Oviedo, Spain
Emilio Corchado (PC Chair)	University of Salamanca, Spain
Abdel-Badeeh Salem	Ain Shams University, Egypt
Alberto Cano	Virginia Commonwealth University, USA
Alfredo Cuzzocrea	ICAR-CNR and University of Calabria, Italy
Alicia Troncoso	Pablo de Olavide University, Spain
Álvaro Herrero	University of Burgos, Spain
Amelia Zafra Gómez	University of Córdoba, Spain
Ana M. Bernardos	Polytechnic University of Madrid, Spain
Ana Madureira	Instituto Superior de Engenharia do Porto, Portugal
Anca Andreica	Babes-Bolyai University, Romania
Andreea Vescan	Babes-Bolyai University, Romania
Andrés Enrique	University of Oviedo, Spain
Andrés Pinón	University of A Coruña, Spain
Ángel Arroyo	University of Burgos, Spain
Antonio de Jesús Díez	University of Oviedo, Spain
Antonio D. Masegosa	University of Deusto/IKERBASQUE, Spain
Antonio Dourado	University of Coimbra, Portugal
Antonio Morales-Esteban	University of Seville, Spain
Arkadiusz Kowalski	Wrocław University of Technology, Poland
Barna Laszlo Iantovics	Petru Maior University of Tg. Mures, Romania
Beatriz Remeseiro	University of Oviedo, Spain
Bogdan Trawinski	Wrocław University of Science and Technology, Poland

Bruno Baruque	University of Burgos, Spain
Camelia Pintea	Technical University of Cluj-Napoca, North University Center at Baia Mare, Romania
Carlos Carrascosa	GTI-IA DSIC Universidad Politecnica de Valencia, Spain
Carlos Mencía	University of Oviedo, Spain
Carlos Pereira	ISEC, Portugal
Cezary Grabowik	Silesian Technical University, Poland
Damian Krenczyk	Silesian University of Technology, Poland
Dario Landa-Silva	The University of Nottingham, UK
David Iclanzan	Sapientia - Hungarian Science University of Transylvania, Romania
Diego P. Ruiz	University of Granada, Spain
Dragan Simic	University of Novi Sad, Serbia
Edward R. Nuñez	University of Oviedo, Spain
Eiji Uchino	Yamaguchi University, Japan
Eneko Osaba	University of Deusto, Spain
Esteban Jove Pérez	University of A Coruña, Spain
Eva Volna	University of Ostrava, Czech Republic
Federico Divina	Pablo de Olavide University, Spain
Fermin Segovia	University of Granada, Spain
Fidel Aznar	University of Alicante, Spain
Francisco Javier Martínez de Pisón Ascacíbar	University of La Rioja, Spain
Francisco Martínez-Álvarez	Pablo de Olavide University, Spain
George Papakostas	EMT Institute of Technology, Greece
Georgios Dounias	University of the Aegean, Greece
Giancarlo Mauri	University of Milano-Bicocca, Italy
Giorgio Fumera	University of Cagliari, Italy
Gloria Cerasela Crisan	Vasile Alecsandri University of Bacau, Romania
Gonzalo A. Aranda-Corral	University of Huelva, Spain
Gualberto Asencio-Cortés	Pablo de Olavide University, Spain
Guiomar Corral	La Salle University, Spain
Héctor Aláiz	University of León, Spain
Henrietta Toman	University of Debrecen, Hungary
Ignacio Turias	University of Cádiz, Spain
Ioannis Hatzilygeroudis	University of Patras, Greece
Irene Diaz	University of Oviedo, Spain
Isabel Barbancho	University of Málaga, Spain
Iskander Sánchez-Rola	University of Deusto, Spain
Javier Bajo	Polytechnic University of Madrid, Spain
Javier De Lope	Polytechnic University of Madrid, Spain
Javier Sedano	ITCL, Spain
Jorge García-Gutiérrez	University of Seville, Spain
Jorge Reyes	NT2 Labs, Chile
José Alfredo Ferreira Costa	Federal University, UFRN, Brazil

José Antonio Sáez	University of Salamanca, Spain
José Dorronsoro	Universidad Autónoma de Madrid, Spain
José García-Rodriguez	University of Alicante, Spain
José Luis Calvo-Rolle	University of A Coruña, Spain
José Luis Casteleiro-Roca	University of A Coruña, Spain
José Luis Verdegay	University of Granada, Spain
José M. Molina	Carlos III University of Madrid, Spain
José Manuel Lopez-Guede	Basque Country University, Spain
José María Armingol	Carlos III University of Madrid, Spain
Jose-Ramón Cano De Amo	University of Jaen, Spain
Juan Humberto Sossa Azuela	National Polytechnic Institute, México
Juan J. Flores	Universidad Michoacana de San Nicolás de Hidalgo, Mexico
Juan Pavón	Complutense University of Madrid, Spain
Julio Ponce	Universidad Autónoma de Aguascalientes, México
Khawaja Asim	PIEAS, Pakistan
Krzysztof Kalinowski	Silesian University of Technology, Poland
Lauro Snidaro	University of Udine, Italy
Lenka Lhotska	Czech Technical University in Prague, Czech Republic
Leocadio G. Casado	University of Almeria, Spain
Luis Alfonso Fernández Serantes	FH Joanneum, University of Applied Sciences, Austria
M. Chadli	University of Picardie Jules Verne, France
Manuel Graña	University of Basque Country, Spain
María Sierra	University of Oviedo, Spain
Mario Koeppen	Kyushu Institute of Technology, Japan
Oscar Fontenla-Romero	University of A Coruña, Spain
Oscar Mata-Carballeira	University of A Coruña, Spain
Ozgur Koray Sahingoz	Turkish Air Force Academy, Turkey
Paula M. Castro	University of A Coruña, Spain
Paulo Novais	University of Minho, Portugal
Pavel Brandstetter	VSB-Technical University of Ostrava, Czech Republic
Pedro López	University of Deusto, Spain
Peter Rockett	University of Sheffield, UK
Ramon Rizo	University of Alicante, Spain
Ricardo Del Olmo	University of Burgos, Spain
Ricardo Leon Talavera Llames	Pablo de Olavide University, Spain
Robert Burduk	Wroclaw University of Technology, Poland
Rodolfo Zunino	University of Genoa, Italy
Roman Senkerik	TBU in Zlin, Czech Republic
Rubén Fuentes-Fernández	Complutense University of Madrid, Spain
Sean Holden	University of Cambridge, UK
Sebastián Ventura	University of Córdoba, Spain
Theodore Pachidis	Kavala Institute of Technology, Greece

Urszula Stanczyk	Silesian University of Technology, Poland
Wiesław Chmielnicki	Jagiellonian University, Poland
Yannis Marinakis	Technical University of Crete, Greece
Zuzana Kominkova Oplatkova	Tomas Bata University in Zlin, Czech Republic

Organizing Committee

Enrique de la Cal	University of Oviedo, Spain
José R. Villar	University of Oviedo, Spain
Francisco Javier de Cos Juez	University of Oviedo, Spain
Noelia Rico	University of Oviedo, Spain
Mirko Fáñez	University of Oviedo, Spain
Carmen Peñalver	University of Oviedo, Spain
Juan Manuel Marín	University of Oviedo, Spain
Marta Blanco	University of Oviedo, Spain
Héctor Quintian	University of A Coruña, Spain
José Antonio Sáez	University of Salamanca, Spain
Emilio Corchado	University of Salamanca, Spain

Contents

Bio-inspired Models and Evolutionary Computation

Learning Algorithms

Visual Analysis and Advanced Data Processing Techniques

Hybrid Intelligent Applications

Data Mining, Knowledge Discovery and Big Data

A Deep Learning-Based Recommendation System to Enable End User Access to Financial Linked Knowledge

Luis Omar Colombo-Mendoza[1], José Antonio García-Díaz[1] ⓘ,
Juan Miguel Gómez-Berbís[2], and Rafael Valencia-García[1](✉) ⓘ

[1] Dpto. Informática y Sistemas, Facultad de Informática,
Universidad de Murcia, Murcia, Spain
{luisomar.colombo,joseantonio.garcia8,valencia}@um.es
[2] Departamento de Informática,
Universidad Carlos III de Madrid, Madrid, Spain
juanmiguel.gomez@uc3m.es

Abstract. Motivated by the assumption that Semantic Web technologies, especially those underlying the Linked Data paradigm, are not sufficiently exploited in the field of financial information management towards the automatic discovery and synthesis of knowledge, an architecture for a knowledge base for the financial domain in the Linked Open Data (LOD) cloud is presented in this paper. Furthermore, from the assumption that recommendation systems can be used to make consumption of the huge amounts of financial data in the LOD cloud more efficient and effective, we propose a deep learning-based hybrid recommendation system to enable end user access to the knowledge base. We implemented a prototype of a knowledge base for financial news as a proof of concept. Results from an Information Systems-oriented validation confirm our assumptions.

Keywords: Linked Open Data · Knowledge base · Ontology · Deep learning
Collaborative filtering · Content-based recommendation

1 Introduction

As the management of financial data moved from traditional desktop-based solutions towards Web-based solutions diverse data sets became available across departments in financial companies and even across different financial entities. A high price has been paid, however: the emergence of several heterogeneous formats that make it difficult to systematically manage an ever-increasing amount of financial data.

Thanks to their ability to enable computer systems to integrate, share, process and interpret the information in the Web of documents, i.e., information formerly readable by humans [1], Semantic Web technologies have gained momentum in the development of software systems among different domains over the last decade.

Motivated by the assumption that Semantic Web technologies, especially those underlying the Linked Data paradigm, are not sufficiently exploited in the research field of financial information management towards the automatic discovery and synthesis of

© Springer International Publishing AG, part of Springer Nature 2018
F. J. de Cos Juez et al. (Eds.): HAIS 2018, LNAI 10870, pp. 3–14, 2018.
https://doi.org/10.1007/978-3-319-92639-1_1

knowledge, an architecture for a knowledge base for the financial domain in the Linked Open Data (LOD) cloud is presented in this paper. As stated by the DBPedia project, knowledge bases are playing an increasingly important role in enhancing the intelligence of Web and in supporting information integration.

Furthermore, recommendation systems have over the years proved to be effective in overcoming the challenges related to the incredible growth of the information on the Web. In this context, deep learning techniques, which can exploit massive amounts of data, have recently gained much attention from recommender systems researchers after companies such as Yahoo, YouTube and Google bet on these techniques in an attempt to improve their recommendation quality [2].

From the assumption that these facts are true also in the case of the huge amounts of financial knowledge in the LOD cloud, we proposed a deep learning-based hybrid recommendation system as the means to enable end user access to the knowledge base.

One of the advantages of our knowledge base compared with others in the literature is the integration of a recommendation system, which eliminates the need for the users to known the underlying ontologies and use both an ontology language and a graph query language. At the same time, this integration makes the proposed recommendation system different from others in the literature. In fact, the results of the state-of-the-art analysis carried out in this research suggest that recommendation systems have not yet been sufficiently studied as a means to support knowledge bases in the Linked Data cloud.

An Information Systems-oriented evaluation based on traditional Information Retrieval metrics is used for the validation of the presented architecture.

The remainder of this paper is structured as follows: the fundamentals of the topics of this research are described in Sect. 2. The architecture that is the salient contribution of this paper is outlined in Sect. 3. Section 4 presents the evaluation method used for validation purposes. Finally, conclusions are discussed in Sect. 5.

2 Background

2.1 Ontologies and Linked Data for the Financial Domain

Semantic Web technologies, specifically ontologies, have been deemed as a promising way for the large-scale integration and access of financial data from disparate sources.

Ontologies, which can be defined as "formal explicit specifications of shared conceptualizations" [3], provide a common vocabulary for a domain and define the meaning of the terms and the relations between them. They also facilitate the retrieval of contents and information. Moreover, Linked Data is a paradigm to expose, share and link pieces of structured data on the Semantic Web (using ontologies) by relying on the URI, XML and RDF technologies. There have been, however, few efforts toward the automatic discovery and synthesis of financial knowledge by means of Linked Data.

For instance, HIKAKU [4] is an effort to enrich financial reporting practice by linking heterogeneous financial data and tracing their provenance using the Linked Data paradigm. It utilizes three main data sources: (1) XBRL standard-based data sources, (2) Linked Open Data (LOD) datasets from DBPedia (http://wiki.dbpedia.org/) and crunchbase (https://www.crunchbase.com/) and (3) news media.

Furthermore, a method for publishing and linking financial data, which is called "Financial Linked Data (FLD)", was presented in [5]. This method uses the XBRL standard to link facts to corresponding business reports and then publish these facts using Linked Data principles. FLD also allows interlinking resulting financial data with relevant external LOD datasets such as the Corporate and Individual Ownership dataset.

The building of ontologies is key to the materialization of the vision of the Semantic Web that is enabled by the Linked Data principles. Despite many proposals for the automatic construction and maintenance of ontologies have arisen in the last few years [6–8], these tasks are considered a pending issue in ontology engineering.

In this context, it is possible to distinguish three major categories of automatic ontology building tasks: ontology learning and ontology evolution, which are beyond the scope of this work, and ontology population. A particular case of ontology population that specifically concerns this work is ontology population from semi-structured documents, such as XML, RSS and HTML documents.

Ontology population from semi-structured HTML documents has traditionally received much attention from ontology engineering researches because most of the content on the Web is provided by means of HTML-based Web pages in which tables can be seen as the natural means to structure information about entities [9].

Various approaches for populating ontologies from semi-structured HTML documents that focus on HTML tables have been proposed to date. Nonetheless, a general methodology for this task can involve the following phases: (1) crawl the Web in search of HTML documents relevant to the domain of interest; (2) use Web wrappers to process the tables within the HTML documents and extract information about occurrences of domain entities by regarding each table entry as an occurrence of an unknown entity given by a set of pairs of a possible property name and the corresponding value; (3) interpret each property name in a set as a predicate of an RDF triple, and each value as an object of this predicate in the same triple; (3.1) match the domain entity represented by the table as a whole with the most proper concept in the ontology; interpret the resulting concept as the subject common to all the outlined RDF triples [10].

2.2 Recommender Systems for Knowledge Bases in the Linked Data Cloud

In order to extract information from knowledge bases in the Linked Data cloud, users are first required to known the underlying ontologies and some ontology language, such as RDF and OWL. They then need to use a graph query language, especially an RDF graph query language, such as SPARQL. This coupled with the problem that knowledge bases typically manage huge amounts of information cause users to be overwhelmed, highlighting the need for end-user tools to facilitate the access and consumption of these huge amounts of knowledge in the Linked Data cloud [11].

Since their conception, recommender systems have proved to be a natural solution to the problem of the information overload on the Web [12], but they have not yet been sufficiently studied as a means to support knowledge bases in the Linked Data cloud.

As proved by companies like Yahoo, YouTube and Google, deep learning techniques significantly improved traditional model-based CF techniques [13–15], which are probably the most widely-used techniques in recommender systems at present [16]. Moreover, deep learning techniques can be applied to problems involving massive amounts of data, which is the case of problems related to access and consumption of knowledge from knowledge bases in the Linked Data cloud.

In this context, an approach to learning top-N recommendations from knowledge graphs, which is called "entity2Rec", was presented in [17]. In that work, graphs are feed with linked open data, and a global user-item relatedness model, which is computed from vector representations of the graph nodes, is proposed by relying on the "Adarank" and "LambdaMART" learning to rank algorithms. Node vector representations are computed using "node2vec", a deep feature learning framework for networks, which is considered an improvement over the "DeepWalk" algorithm [18].

Furthermore, in [19] a unified framework for collaborative filtering and knowledge base embedding, which is called "Collaborative Knowledge Base Embedding (KBE)" is proposed to simultaneously learn items' semantic representations from the knowledge base and capture the implicit relations between users and items. Semantic representations are automatically extracted by using deep learning embedding techniques.

Nonetheless, unlike these proposals, in this work the deep learning-based recommendation techniques are not the end but the means to support a knowledge base.

3   Proposed Architecture

At a high level of abstraction, the architecture for the knowledge base for the financial domain, which is the salient contribution of this paper, is composed of five major components: (1) Domain Ontology, (2) Ontology Population Subsystem, (3) Knowledge Repository, (4) Semantic Annotation Subsystem and (5) Recommendation Subsystem. (see Fig. 1). The roles and associations between all these components are explained below.

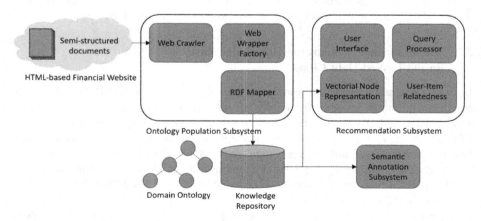

Fig. 1. Proposed architecture.

3.1 Domain Ontology

We have adapted a previous ontology of our research group in order to obtain an ontology for the financial domain in the Linked Open Data (LOD) cloud [20]. This ontology was manually designed using the OWL ontology language, and it covers four top-level concepts: financial market, financial intermediary, asset and legislation. For a more complete reference about this ontology, please refer to the aforementioned work.

In this adaptation process we basically have redesigned the ontology to conform to the Linked Data principles. We reused vocabularies from the LOD cloud, specifically, the DBPedia ontology and the Linked Chrunchbase ontology, as much as we could. DBPedia provides a large dataset comprising around 50000 companies. It is, in the words of Tim Berners-Lee, one of the more famous pieces of Linked Data as a project. Linked Chrunchbase (http://km.aifb.kit.edu/services/crunchbase/) is a Linked Data server for Chrunchbase, an open database about innovative companies and people that is an ideal source of data about funding rounds, investors and acquisitions.

In particular, we reused the "Company" class defined in the DBPedia ontology, whose IRI reference is "http://dbpedia.org/ontology/Company". Some other classes from the same ontology were also reused to cover bottom-level concepts in our Domain Ontology, such as currencies (the "Currency" class) and laws (the "Law" class), where currencies, as well as companies, are considered to be (types of) investment assets (the Investment_Asset own class).

We accordingly reused the "Investment" class defined in the Linked Chrunchbase ontology. From the same ontology we also reused the News class, whose IRI reference is "http://ontologycentral.com/2010/05/cb/vocab#News", to cover a concept that will be crucial to the validation of the proposed architecture.

3.2 Ontology Population Subsystem and Knowledge Repository

The Ontology Population Subsystem extracts some information from semi-structured HTML documents gathered from a financial website, such as the Yahoo! Finance website, and creates RDF triples according to the schema represented by the Domain Ontology. The populated ontology, or rather, the populated Knowledge Repository is the foundation of an up-to-date knowledge base for the financial domain in the LOD cloud.

We have implemented the Ontology Population Subsystem in a series of modules based on the general methodology for populating ontologies that was outlined in Sect. 2.1 of this paper: (1) a Web Crawler, (2) a Web Wrapper Factory and (3) an RDF Mapper as well on a previous work of our research group [21].

The Web Crawler is in charge of constantly gathering HTML-based documents from a number of different sources, e.g., the World Indices Page and the Index Components Page of the Yahoo! Finance website. One concrete Web Wrapper needs to be used to process one kind of collected documents and extract information of possible interest about different individuals of a predefined class in the Domain Ontology from a target HTML table. This information is internally represented using an XML-based format.

The elements in an XML document object generated by a concrete Web Wrapper are finally used by the RDF Mapper to create RDF triples. The Apache Jena framework has been used for that purpose. Each possible property name represented by an attribute name in an XML element must be specifically mapped to the name of an actual property in the Domain Ontology. We have proposed to identify the name of the matching ontology property for each XML attribute name by finding the pair of names between which the Levenshtein distance is minimum. Because the class that individuals represented in an XML document object belong to is known beforehand, an RDF triple can be created for each XML element by interpreting the matching property name as the predicate, the corresponding value as the object and the predefined class as the subject.

Moreover, we have used the full text search capability from the SPARQL implementation of the RDF store behind the DBPedia dataset to interlink our dataset with the aforementioned one. We have specifically searched for semantically similar individuals of the "Company" class and created owl:sameAs statements accordingly, thus conforming to the principle of Linked Data that the value of the data increases as it is interlinked.

By relying on the families of XML-based Web Wrappers, the Ontology Population Subsystem is intended to be easily adapted for use with different HTML-based websites providing semi-structured financial data in table structures in a seamless manner.

The Knowledge Repository, which is based on the OpenLink Virtuoso's RDF quadstore, the open source edition of the Virtuoso Universal Server's RDF quadstore, is responsible for storing the RDF triples created by the Ontology Population Subsystem.

In detail, the RDF triples are stored in a named graph whereas the schema (the Domain Ontology) is serialized in RDF/XML format and stored in a second named graph. Furthermore, thanks to the RDFS and OWL reasoning capabilities of the SPARQL implementation of the OpenLink Vrtuoso's RDF quadstore, any third-party user and application is allowed to "on the fly" infer, through SPARQL, additional RDF triples from the triples and the associated axioms (the schema) that are physically stored.

3.3 Semantic Annotation Subsystem

The Semantic Annotation Subsystem is in charge of automatically creating semantic annotations for financial news using concepts from the Domain Ontology. A prerequisite for this task is to continuously crawl the Web in search for news as in the case with the Ontology Population Subsystem's Web crawler. Although the functionality of the Semantic Annotation Subsystem is described here in the context of semantic annotation of news, which is crucial to the proof of concept proposed in this paper, it can be exploited for semantic annotation of any kind of financial document.

In general, we have used the General Architecture for Text Engineering (GATE) framework to implement an ontology-based semantic annotation process comprising two major phases: linguistic preprocessing and Named Entity Recognition (NER) [22].

With respect to NER we have used a gazetteer-based approach. In particular, a gazetteer, which is dynamically generated from the RDF triples in the Knowledge

Repository, is matched against the processed documents to find mentions in text matching knowledge entities (individuals of classes of the ontology and classes themselves). It is essential for this to preprocess the RDF dataset as with the corpus of documents. The aim in this case is to extract human-understandable lexicalizations from knowledge entity names as well as from datatype property values including values of the "rdfs:label" property.

3.4   Recommendation Subsystem

The Knowledge Repository can be queried by users using keyword-based queries. A query preprocessing phase is first required to generate a candidate set of financial news. Its particularity lies in the search for synonyms for the lexemes that are obtained from the user query as a result of a morphological analysis. Some lexical databases such as Wordnet are exploited for that purpose. They were also used at design time to obtain synonyms for the terms assigned to the classes in the Domain Ontology. Both synonyms and terms were included as labels of the classes (values of the "rdfs:label" property).

The candidate set of news is generated by matching the synonyms and lexemes against the labels in the Knowledge Repository to detect knowledge entities and selecting all the news that are annotated with these entities.

The relatedness between a user querying the knowledge base (the active user) and each of the news (items) in the candidate set needs to be then computed. The aim is to filter out such items preferred by other users who have preferred less items of the active user and such items sharing less features with items preferred by the active user in the past. This recommendation approach is based on the one proposed in [17].

In order for the Recommendation Subsystem to compute active user-item relatedness (see Formula 1) it is necessary to learn vectorial node representations from the graph in the Knowledge Repository storing individuals of classes of the Domain Ontology (the data graph). This task is feature-specific. It must be therefore carried out over a subgraph of the data graph by considering one property at a time. There are features encoding collaborative filtering information and features encoding content information, the former are represented by the "likes" object property, whose domain is the "User" class of the Domain Ontology. The vectorial representations are in fact computed by simulating random walks on the subgraphs using the "node2vec" deep feature learning algorithmic framework [23].

According to the paragraph above, users interact with the Recommendation Subsystem to implicitly provide positive feedback on the recommended news by actually viewing them.

$$p_p(u, i) = \begin{cases} s\big(x_p(u), x_p(i)\big) & \text{if } p = \text{'likes'} \\ \frac{1}{|R_+(u)|} \sum_{i' \in R_+(u)} s\big(x_p(i), x_p(i')\big) & \text{otherwise} \end{cases} \quad (1)$$

Where $R+(u)$ are the items (i') liked by the active user u in the past, s is the adjusted cosine similarity metric, $xp(u)$, $xp(i)$ and $xp(i')$ are the property-specific vector representations.

We have proposed to then adjust each property-specific relatedness score by a normalized weighting factor as follows: if the knowledge entities identified from the expanded user query are classes of the Domain Ontology, then a greater weighting factor is given to the object properties in which the domain is the "News" class and the range is represented by these classes. Otherwise a greater weighting factor is given to the "likes" object property. The weighting factors need to be experimentally established.

The actual relatedness scores are finally calculated by using the adjusted property-specific relatedness scores as the features of a global active user-item relatedness model that needs to be learnt to provide recommendations. This is done by finding the parameters θ of a function f of the property-specific scores $\vec{p}(u, i)$ and of a set of parameters θ that optimize the top-N item recommendation as a supervised learning to rank problem. We have used the "LambdaMART" algorithm [24] as the ranking algorithm.

The subset of the N financial news in the candidate set that have the top-N best global relatedness scores is presented to the user in descending order as the result of the query.

4 Evaluation

We conducted a user study in which ten undergraduate students of the University of Murcia interacted with a prototype of a knowledge base for the financial domain implemented from the proposed architecture. Each student set up their preferences by freely searching and reading financial news. At this point in the study the Recommendation Subsystem was configured to not carry out any filtering process during the training phase of the ranking algorithm.

Once the Knowledge Repository was populated with preferences information, a group of three volunteers from the department of Informatics and Systems at the University of Murcia selected 20 news unknown to the students, and they also wrote a keyword-based query for the students to interact with the knowledge base prototype once again. The students manually classified each of the news selected as either relevant or non-relevant to their preferences in the context of the predefined user query. At this point in the study the Recommendation Subsystem was able to actually filter the set of selected news from the preferences information.

The Ontology Population Subsystem was configured to extract financial information from semi-structured HTML documents gathered from the website of the "Bolsa de Madrid", which allowed the knowledge base prototype to deal with news about the Spanish Stock Market.

It is worth mentioning that this user study was actually repeated 10 times as an offline experiment with the aim of establishing the normalized weighting factors for the calculation of the property-specific user-item relatedness scores.

During a second stage of this evaluation, we implemented three alternative approaches of our recommendation approach (called baseline approaches). For that purpose, we alternately used "DeepWalk" as an alternative network-based feature learning algorithm and the cosine similarity metric as an alternative vector similarity metric in calculating the property-specific relatedness scores. We then recreated the user study. We specifically carried out a new offline experiment by leveraging the already generated dataset (user preferences). We also leveraged the classifications (relevant/not relevant) provided by the participants for the news of the sample already selected. Finally, we were able to generate three additional lists of recommendation results for each simulated student using the baseline approaches. These recommendation results were compared with the results already produced by our recommendation approach.

Table 1 depicts the results of the calculation of the Recall, Precision and F1 score measures and the input parameters used in each case for the corresponding formulas. These results actually correspond to the best ranked iteration of the offline experiment carried out during the first stage of the evaluation.

Table 1. Results of Recall, Precision and F1 score calculation for the proposed approach.

Relevant		Non-relevant		Recom.	Recall	Prec.	F1 score
User	System-hits	User	System-errors				
14	10	6	2	12	0.833	0.714	0.769
12	9	8	2	11	0.818	0.750	0.783
13	11	7	3	14	0.786	0.846	0.815
14	10	6	4	14	0.714	0.714	0.714
13	11	7	4	15	0.733	0.846	0.786
14	10	6	4	14	0.714	0.714	0.714
13	10	7	3	13	0.769	0.769	0.769
12	9	8	2	12	0.818	0.750	0.783
12	10	8	3	13	0.769	0.833	0.800
11	9	9	3	12	0.750	0.818	0.783
Average					0.771	0.776	0.772

Table 2 depicts the results of the comparative analysis carried out during the second stage of the evaluation. These results are given in terms of average Recall, Precision and F1 score values.

Table 2. Results of the comparative analysis.

Approach	Avg. Recall	Avg. Prec.	Avg. F1 score
DeepWalk + cosine sim.	0.710	0.729	0.719
DeepWalk + adjusted cosine sim.	0.755	0.760	0.756
node2vec + cosine sim.	0.732	0.744	0.738
node2vec + adjusted cos. sim. (our approach)	0.771	0.776	0.772

According to the results from Table 1, the average Recall value of the proposed recommendation approach is 0.771 (77.1%), whereas its average Precision value is 0.776 (77.6%). In particular, in the best-case scenario in terms of Precision, 11 out of

13 news recommended by the proposed recommendation approach were actually news judged as relevant by students, which is equivalent to a Precision value of 0.846 (84.6%), whereas in the worst-case scenario in the same terms, only 10 out of 14 news recommended were actually relevant news, which is equivalent to a Precision value of 0.714 (71.4%). According to the results from Table 2, our recommendation approach achieved slightly higher average Precision and Recall measures than the three baseline approaches. In particular, Precision was improved by 6.45% with respect to the "DeepWalk+cosine sim." baseline approach (the best-case scenario), and it was improved by 2.11% with respect to the "DeepWalk+adjusted cosine sim." baseline approach (the worst-case scenario).

5 Conclusions and Future Work

Starting from the realization that the exploitation of the Semantic Web technologies, especially ontologies and those underlying the Linked Data paradigm, represents an area of opportunity for the automatic discovery and synthesis of knowledge in the financial domain, in this paper we presented an architecture for a knowledge base for the financial domain in the Linked Open Data (LOD) cloud. The architecture relies on a domain ontology, which was manually designed by reusing existing vocabularies from the LOD cloud as well as on the functionality of a semantic annotation subsystem that can be used for semantically annotating any kind of financial document. Furthermore, from the realization that the use of recommendation systems to support knowledge bases has not yet been sufficiently studied, we proposed a deep learning-based hybrid recommendation system to enable end user access to the knowledge base.

We implemented a prototype of a knowledge base for financial news based on the presented architecture. We used an Information Systems-oriented evaluation for validation purposes. According to the results obtained, recommendation systems can be used to make access to LOD-based knowledge bases more transparent by eliminating the need for knowing underlying schemas and using ontology and query languages. Moreover, recommendation systems can be used to actually make consumption of the huge amounts of knowledge in the LOD cloud more efficient and effective in terms of the quality of the retrieval output.

For future work, we have planned to repeat the evaluation presented in this paper using larger training and testing datasets to assess the extent to which our conclusions on Recall and Precision are valid in such conditions. We believe, however, that the behavior of these measures will remain more or less stable thanks to the query preprocessing operation that is performed by the recommendation subsystem. Likewise, we have planned to validate our proposal using a complementary Decision Support Systems-oriented evaluation. This will allow us to determine the quality of the support for accessing and consuming financial knowledge in the LOD cloud as it is perceived by end users.

Acknowledgements. This work has been supported by the Spanish National Research Agency (AEI) and the European Regional Development Fund (FEDER/ERDF) through project (TIN2016-76323-R) and by the Fundación Séneca through grant 19371/PI/14.

References

1. Berners-Lee, T.: The semantic web. Sci. Am. **284**, 34–43 (2001)
2. Zhang, S., Yao, L., Sun, A.: Deep Learning based Recommender System: A Survey and New Perspectives (2017). ArXiv170707435 Cs
3. Gruber, T.R.: A translation approach to portable ontology specifications. Knowl. Acquis. **5**, 199–220 (1993)
4. Lee, V., Goto, M., Hu, B., Naseer, A., Vandenbussche, P.-Y., Shakair, G., Rodrigues, E.M.: Exploiting linked data in financial engineering. In: Liu, K., Gulliver, S.R., Li, W., Yu, C. (eds.) ICISO 2014. IFIP AICT, vol. 426, pp. 116–125. Springer, Heidelberg (2014). https://doi.org/10.1007/978-3-642-55355-4_12
5. Ashraf, J., Hussain, O.K.: Integrating financial data using semantic web for improved visibility. In: 2012 Eighth International Conference on Semantics, Knowledge and Grids, pp. 265–268 (2012)
6. Wong, W., Liu, W., Bennamoun, M.: Ontology learning from text: a look back and into the future. ACM Comput. Surv. **44**, 20:1–20:36 (2012)
7. Petasis, G., Karkaletsis, V., Paliouras, G., Krithara, A., Zavitsanos, E.: Ontology population and enrichment: state of the art. In: Paliouras, G., Spyropoulos, C.D., Tsatsaronis, G. (eds.) Knowledge-Driven Multimedia Information Extraction and Ontology Evolution. LNCS (LNAI), vol. 6050, pp. 134–166. Springer, Heidelberg (2011). https://doi.org/10.1007/978-3-642-20795-2_6
8. Liu, K., Hogan, W.R., Crowley, R.S.: Natural language processing methods and systems for biomedical ontology learning. J. Biomed. Inform. **44**, 163–179 (2011)
9. Sugibuchi, T., Tanaka, Y.: Interactive web-wrapper construction for extracting relational information from web documents. In: Special Interest Tracks and Posters of the 14th International Conference on World Wide Web, pp. 968–969. ACM, New York (2005)
10. Park, S.-B., Kim, S.-S., Oh, S., Zeong, Z., Lee, H., Park, S.R.: Target concept selection by property overlap in ontology population. Int. J. Comput. Electr. Autom. Control Inf. Eng. **2**, 50–54 (2008)
11. Kaufmann, E., Bernstein, A.: Evaluating the usability of natural language query languages and interfaces to semantic web knowledge bases. Web Semant. Sci. Serv. Agents World Wide Web **8**, 377–393 (2010)
12. Goldberg, D., Nichols, D., Oki, B.M., Terry, D.: Using collaborative filtering to weave an information Tapestry. Commun. ACM **35**, 61–70 (1992)
13. Cheng, H.-T., Koc, L., Harmsen, J., Shaked, T., Chandra, T., Aradhye, H., Anderson, G., Corrado, G., Chai, W., Ispir, M., Anil, R., Haque, Z., Hong, L., Jain, V., Liu, X., Shah, H.: Wide & deep learning for recommender systems. In: Proceedings of the 1st Workshop on Deep Learning for Recommender Systems, pp. 7–10. ACM, New York (2016)
14. Covington, P., Adams, J., Sargin, E.: Deep neural networks for YouTube recommendations. In: Proceedings of the 10th ACM Conference on Recommender Systems, pp. 191–198. ACM, New York (2016)
15. Okura, S., Tagami, Y., Ono, S., Tajima, A.: Embedding-based news recommendation for millions of users. In: Proceedings of the 23rd ACM SIGKDD International Conference on Knowledge Discovery and Data Mining, pp. 1933–1942. ACM, New York (2017)
16. Ricci, F., Rokach, L., Shapira, B.: Introduction to recommender systems handbook. In: Ricci, F., Rokach, L., Shapira, B., Kantor, P.B. (eds.) Recommender Systems Handbook, pp. 1–35. Springer, Boston (2011). https://doi.org/10.1007/978-0-387-85820-3_1

17. Palumbo, E., Rizzo, G., Troncy, R.: Entity2Rec: learning user-item relatedness from knowledge graphs for Top-N item recommendation. In: Proceedings of the Eleventh ACM Conference on Recommender Systems, pp. 32–36. ACM, New York (2017)
18. Perozzi, B., Al-Rfou, R., Skiena, S.: DeepWalk: online learning of social representations. In: Proceedings of the 20th ACM SIGKDD International Conference on Knowledge Discovery and Data Mining, pp. 701–710. ACM, New York (2014)
19. Zhang, F., Yuan, N.J., Lian, D., Xie, X., Ma, W.-Y.: Collaborative knowledge base embedding for recommender systems. In: Proceedings of the 22nd ACM SIGKDD International Conference on Knowledge Discovery and Data Mining, pp. 353–362. ACM, New York (2016)
20. Lupiani-Ruiz, E., García-Manotas, I., Valencia-García, R., García-Sánchez, F., Castellanos-Nieves, D., Fernández-Breis, J.T., Camón-Herrero, J.B.: Financial news semantic search engine. Expert Syst. Appl. 38, 15565–15572 (2011)
21. García-Manotas, I., Lupiani, E., García-Sánchez, F., Valencia-García, R.: Populating knowledge based decision support systems. Int. J. Decis. Support Syst. Technol. 2, 1–20 (2010)
22. Rodríguez-García, M.Á., Valencia-García, R., García-Sánchez, F., Samper-Zapater, J.J.: Ontology-based annotation and retrieval of services in the cloud. Knowl. Based Syst. 56, 15–25 (2014)
23. Grover, A., Leskovec, J.: Node2Vec: scalable feature learning for networks. In: Proceedings of the 22nd ACM SIGKDD International Conference on Knowledge Discovery and Data Mining, pp. 855–864. ACM, New York (2016)
24. Burges, C.J.C.: From RankNet to LambdaRank to LambdaMART: An Overview. Microsoft Research (2010)

On the Use of Random Discretization
and Dimensionality Reduction
in Ensembles for Big Data

Diego García-Gil$^{(\boxtimes)}$, Sergio Ramírez-Gallego, Salvador García,
and Francisco Herrera

Department of Computer Science and Artificial Intelligence,
University of Granada, CITIC-UGR, 18071 Granada, Spain
{djgarcia,sramirez,salvagl,herrera}@decsai.ugr.es
http://sci2s.ugr.es

Abstract. Massive data growth in recent years has made data reduction techniques to gain a special popularity because of their ability to reduce this enormous amount of data, also called Big Data. Random Projection Random Discretization is an innovative ensemble method. It uses two data reduction techniques to create more informative data, their proposed Random Discretization, and Random Projections (RP). However, RP has some shortcomings that can be solved by more powerful methods such as Principal Components Analysis (PCA). Aiming to tackle this problem, we propose a new ensemble method using the Apache Spark framework and PCA for dimensionality reduction, named Random Discretization Dimensionality Reduction Ensemble. In our experiments on five Big Data datasets, we show that our proposal achieves better prediction performance than the original algorithm and Random Forest.

Keywords: Big Data · Ensemble · Discretization · Apache Spark
PCA · Data reduction

1 Introduction

Nowadays everything is constantly creating and storing data. In 2014 IDC predicted that by 2020, the digital universe will be 10 times as big as it was in 2013, totaling an astonishing 44 zettabytes[1]. Big Data is not only a huge amount of data, but a new paradigm and set of technologies that can store and process this data. This scenario becomes particularly important using data reduction techniques [10]. These techniques are frequently applied to reduce the size of the original data and to clean some errors that it may contain [8,9].

Ensembles are methods that combine a set of base classifiers to make predictions [5]. These classifiers have been proven to be *accurate* and *diverse*. Ensembles of decision trees like Random Forest [2] are well known for creating diverse

---

[1] IDC: The Digital Universe of Opportunities. 2018 [Online] Available: http://www.emc.com/infographics/digital-universe-2014.htm.

© Springer International Publishing AG, part of Springer Nature 2018
F. J. de Cos Juez et al. (Eds.): HAIS 2018, LNAI 10870, pp. 15–26, 2018.
https://doi.org/10.1007/978-3-319-92639-1_2

decision trees. This diversity is usually introduced via randomization. Through small changes in input data, diverse decision trees are created and better ensembles are obtained.

This principle is followed by Ahmad and Brown in [1]. Random Projection Random Discretization (RPRD) is an ensemble method that applies two data reduction techniques, Random Discretization (RD) and Random Projection (RP) [12], to the input data and joins the results to create a more informative dataset. However, despite its good performance against other popular ensemble methods, it still has three main drawbacks: (1) As the projected dimension is decreased, as it drops below $\log k$, random projection suffers a gradual degradation in performance [3]. (2) RP is highly unstable, different random projections may lead to radically different results [6]. (3) RPRD is not prepared for working with Big Data.

In order to fill this gap and inspired by the RPRD ensemble algorithm, we propose a new ensemble method under Apache Spark using PCA, called Random Discretization Dimensionality Reduction Ensemble (RD^2R) for Big Data. In our design we use PCA instead of RP for improving the dimensionality reduction step.

To show the effectiveness of our approach, we have carried out an experimental evaluation with five large datasets, namely *poker, SUSY, HIGGS, epsilon* and *ECBDL14*. These datasets have very different properties and allow us to test all aspects of our implementation. Finally, we show a comparative study of the performance of RD^2R, RPRD and Random Forest. Spark's implementation of the algorithm can be downloaded from the Spark's community repository[2].

The remainder of this contribution is organized as follows: Sect. 2 outlines the main concepts of the RPRD Ensemble. Section 3 explains the new ensemble design based on PCA. Section 4 describes the experiments carried out to check the effectiveness of this proposal. Finally, Sect. 5 concludes the contribution.

2   Background

In this section we first introduce the RPRD Ensemble algorithm used as reference in our ensemble interpretation and its two components, RD and RP. Finally we describe the MapReduce Model.

2.1   Random Discretization

Discretization is the process of partitioning a set of continuous attributes into discrete attributes by associating categorical values to the intervals [7]. To create s categories we need $s - 1$ different intervals. There are different methods to create these intervals, some of them based on evolutionary optimization [14], implemented in Apache Spark. The main problem is that they create the same discretized dataset after different executions. In an ensemble some randomization is necessary in order to introduce diversity to the decision trees.

---

[2] https://spark-packages.org/package/djgarcia/RD2R.

In RD randomization is introduced to the discretization process. First $s - 1$ data points are randomly selected from the training data to create s categories. Then for each feature, every $s - 1$ data points are sorted. Finally the dataset is discretized into s categories using these $s - 1$ sorted data points. These thresholds are selected randomly each iteration of the ensemble.

2.2   Random Projection

The objective of dimensionality reduction techniques is to produce a compact low-dimensional encoding of a given high dimensional dataset. In RP, the original m-dimensional data is projected to a d-dimensional $(d << m)$ subspace through the origin, using a random $d \times m$ matrix $\boldsymbol{R}$ whose columns have unit lengths, and whose elements $r_{i,j}$ are often Gaussian distributed. Using matrix notation where $\boldsymbol{X}_{m \times N}$ is the original set of N m-dimensional observations,

$$\boldsymbol{X}_{d \times N}^{RP} = \boldsymbol{R}_{d \times m} \boldsymbol{X}_{m \times N}$$

is the projection of the data onto a lower d-dimensional subspace. The key idea of random mapping arises from the Johnson-Lindenstrauss lemma [12]: if points in a vector space are projected onto a randomly selected subspace of suitably high dimension, then the distances between the points are approximately preserved.

2.3   Random Projection Random Discretization Ensembles and Classification

RD and RP perform data reduction, but have different mechanisms; RD performs random discretization whereas RP creates new features that are the linear combinations of the original features. RPRD ensemble was based on the idea that both RP and RD can be combined to create a better ensemble method. In each iteration RD and RP are performed on the input data, then the results are fused. Finally a decision tree is trained using this new data. This results in better trees compared to the original data as they have more features to select at each node.

In the prediction phase a data point is converted into a $m + d$ dimensional data point using the corresponding values of RD and RP for the iteration. Then the probabilities of each class are calculated by the decision tree. Finally the confidence value for each class is calculated. The class with the highest confidence value will be the class of the data point.

2.4   MapReduce Model

MapReduce is a framework designed by Google in 2003 [4]. This model is composed of a Map procedure that performs a transformation, and a Reduce method that performs a summary operation. The workflow of a MapReduce program is as follows: first the master node splits the dataset and distributes the results across the cluster. Then each node applies the Map function to the local data.

After that process is finished the data is redistributed based on the key-value pairs generated in the Map phase. Once the data has been redistributed so that all pairs belonging to one key are in the same node, it is processed in parallel [15].

Apache Hadoop[3] is the most popular open-source framework for large-scale data storing and processing based on the MapReduce model. The framework is designed to handle hardware errors automatically. In spite of its popularity and performance, Hadoop presents some important limitations [13]: poor performance on online and iterative computing, low inter-communication capacity and insufficiency for in-memory computation.

Apache Spark[4] is an open-source framework built around speed, ease of use and in-memory computation [11]. Spark's core concepts are Resilient Distributed Datasets (RDDs) [17]. RDDs are a distributed and immutable memory abstraction, they can be described as a collection of data partitioned across the clusters. RDDs support two types of operations: transformations, which are not evaluated when defined and will produce a new RDD. And actions, which evaluate and return a new value. When an action is called on a RDD, all the previous transformations are applied in parallel to each partition of the RDD.

3 Random Discretization Dimensionality Reduction Ensemble

In this section, we present the design of the ensemble by using a more powerful method like PCA, proving its performance over big real-world problems.

For the implementation of the algorithm, we have used some basic Spark primitives. Here, we outline those more relevant for the algorithm:

- *map*: Applies a transformation to each element of a RDD and returns a new RDD representing the results.
- *zip*: Joins one RDD with another one.
- *zipWithIndex*: Zips a RDD with its element indices.
- *lookup*: Returns the list of values in the RDD for a given key.

RD^2R has two phases, learning and prediction. In the learning phase we train a model from the input data and in the prediction phase, we apply this model to the test data in order to obtain a prediction. In the learning phase we discretize the training data using RD. As RDDs are unsorted by nature, to select a specific instance for the RD method it performs the *zipWithIndex* operation to the RDD in order to add an index to each instance. With the added index we can get the values of the features using the *lookup* operation. For iterating through every instance and to discretize them, Spark's *map* function is used. Once RD has been performed, PCA is also applied to the training data with a random value of principal components in the interval $[1, m-1]$ (m number of features). Finally we join the results with the *zip* function and learn a decision tree using this new dataset. We repeat this process L times, L the size of the ensemble.

---

[3] Apache Hadoop Project 2018 [Online] Available: https://hadoop.apache.org/.
[4] Apache Spark Project 2018 [Online] Available: https://spark.apache.org/.

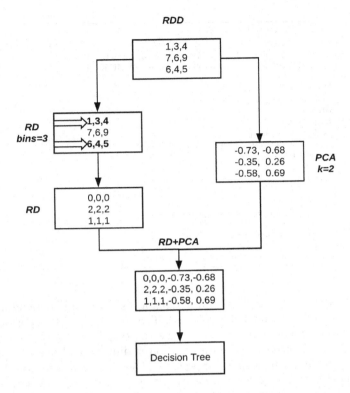

Fig. 1. RD²R learning phase flowchart

In Fig. 1 we can see a flowchart of the RD²R learning phase process.

In the prediction phase we discretize the data point with the cut points obtained in the learning phase and perform PCA with the model obtained previously. Then we predict the probability of the data point belonging to each class. We repeat this process L times. Finally we add all probabilities. The class with the largest probability will be the class of the data point.

Our algorithm is divided into two procedures explained in two sections as follows: Sect. 3.1 describes the learning phase. And Sect. 3.2 provides details of the prediction phase.

3.1   Learning Models Phase

Algorithm 1 explains the learning phase in the ensemble. The algorithm discretizes using RD method and performs PCA, both with the training data, then joins the result of both methods to create a new dataset. Then a decision tree is learned with this new data. It requires the following as input parameters: the dataset, the size of the ensemble and the number of bins for the discretization.

The first step is to perform RD. First we calculate the thresholds for the discretization. With these thresholds the data is discretized through a Map function.

Algorithm 1. Main RD^{2}R algorithm

1: **Input:** *data:* an RDD of type LabeledPoint (features, label), *L:* the size of the ensemble and *s:* the number of categories for the discretization.
2: **Output:** The model created, an object of class RD2RModel
3: **for** $i = 0...L$ **do**
4:     **Random Discretization**
5:     $cutPoints(i) \leftarrow get\_cut\_points(data, s)$
6:     $rdData \leftarrow$
7:     **map** $l \in data$
8:         **for** $c = 0...size(l) - 1$ **do**
9:             $l \leftarrow discretize(l(c), cutPoints(i)(c))$
10:         **end for**
11:     **end map**
12:     **PCA**
13:     $d \leftarrow random(1, size(data) - 1)$
14:     $pcaModels(i) \leftarrow PCA(data, d)$
15:     $pcaData \leftarrow transform(data, pcaModels(i))$
16:     $rd2rData \leftarrow zip(rdData, pcaData)$
17:     $trees(i) \leftarrow decisionTree(rd2rData)$
18: **end for**
19: $return(RD2RModel(L, cutPoints, pcaModels, trees))$

It iterates through every feature and assigns a discrete value depending on the feature's value and the thresholds selected. The second step is to perform PCA. First we select a random number d in the interval $[1, m - 1]$ (m number of features). Then we project the data to a lower dimensional space using PCA, keeping only the first d principal components. The final step is to fuse the results from RD and PCA, and to learn a decision tree with it. We repeat this process L times, saving the cut points, the PCA models and the trees created at each iteration. Once all the trees have been trained, the model is created and returned.

Algorithm 2 describes the process of selecting the thresholds for the discretization. First we select $s - 1$ random numbers in the interval $[0, n]$ (n the number of instances in the data). Then we add an index to each instance in the dataset in order to get the values of the features we have selected in the previous step. The next step is to get those values, sort them and check if they are all equal. This process is described as follows: first we transpose the thresholds array in order to access to the cut points of each feature through a Map function, and add an index to each threshold. Then we iterate through the thresholds using a Map function. The cut points are sorted and checked if they are different. If they are all equal, we get all the different values for that feature, then take $s - 1$ points randomly and finally sort them. The result is a list of the thresholds for each feature, which is returned to the main algorithm as *cutPoints*.

3.2   Prediction Phase

Algorithm 3 explains the prediction phase in the ensemble. The algorithm discretizes using RD and performs PCA on the test data point using the cut points and the models for PCA provided the same way as was described in the main procedure. Then the results of the two methods are joined and then predicted

---

Algorithm 2. Function to select the cut points for a given dataset (get_cut_points)

1: **Input:** $data$: an RDD of type LabeledPoint (features, label) and s: the number of thresholds.
2: **Output:** An array with thresholds for each feature
3: $instances \leftarrow get\_random\_array(s - 1, size(data))$
4: $indexData \leftarrow zipWithIndex(data)$
5: **for** $i = 0...s - 1$ **do**
6:     $values \leftarrow lookup(indexData, instances(i))$
7: **end for**
8: $thresholds \leftarrow zipWithIndex(transpose(values))$
9: $cutPoints \leftarrow$
10: **map** $(l, i) \in thresholds$
11:     $feature \leftarrow sorted(distinct(l))$
12:     **if** $size(feature) = 1$ **then**
13:         $col \leftarrow distinct(get\_feature\_values(data, i))$
14:         $sorted(take\_random(s - 1, col))$
15:     **else**
16:         $feature$
17:     **end if**
18: **end map**
19: $return(cutPoints)$

Algorithm 3. RD2RModel algorithm

1: **Input:** L: the size of the ensemble, $cutPoints$: the thresholds, $pcaModels$: the models for performing PCA and $trees$: the models of the trained trees.
2: **Output:** The class of the data point.
3: **function** TEST($test : LabeledPoint$)
4:     $rawPredictions \leftarrow 0$
5:     **for** $i = 0...L$ **do**
6:         $rd \leftarrow 0$
7:         **for** $c = 0...size(test) - 1$ **do**
8:             $rd(c) \leftarrow discretize(test(c), cutPoints(i)(c))$
9:         **end for**
10:        $pcaTest \leftarrow transform(test, pcaModels(i))$
11:        $rd2rData \leftarrow zip(rd, pcaTest)$
12:        $rawPredictions \leftarrow rawPredictions + predict(trees(i), rd2rData)$
13:    **end for**
14:    $label \leftarrow max\_index(rawPredictions)$
15:    $return(label)$
16: **end function**

---

with the corresponding tree. The tree gives the probabilities for each class. These probabilities are added to a list of predictions at each iteration. The class with the greatest probability is selected. It selects the index of the maximum probability for an instance as a decision.

4   Experimental Results

This section describes the experiments carried out to show the performance of RD²R for Big Data algorithm in five huge problems. We carried out the comparative study of RD²R method facing the original proposal, and MLlib's implementation of Random Forest. RPRD ensemble algorithm have been also implemented in Apache Spark.

4.1   Experimental Framework

Five huge classification datasets are used in our experiments:

- Poker hand dataset, which has 1,025,000 instances with 11 attributes. Each record is an example of a hand consisting of five playing cards drawn from a standard deck of 52.
- SUSY dataset, which consists of 5,000,000 instances and 18 attributes. The task is to distinguish between a signal process which produces supersymmetric (SUSY) particles and a background process which does not.
- HIGGS dataset, which has 11,000,000 instances and 28 attributes. This dataset is a classification problem to distinguish between a signal process which produces Higgs bosons and a background process which does not.
- Epsilon dataset, which consists of 500,000 instances with 2,000 numerical features. This dataset was artificially created for the Pascal Large Scale Learning Challenge in 2008.
- ECBDL14 dataset. This dataset was used as a reference at the ML competition of the Evolutionary Computation for Big Data and Big Learning under the international conference GECCO-2014. It consists of 631 characteristics and 32 million instances. It is a binary classification problem where the class distribution is highly imbalanced: 2% of positive instances. For this problem, the Random OverSampling (ROS) algorithm used in [16] was applied in order to replicate the minority class instances until the number of instances for both classes was equalized, summing a total of 65 million instances.

The experiments were conducted following 5 fold cross-validation and three different sizes of ensembles: 10, 50 and 100 iterations.

We have established 5 intervals for RD, the same for RPRD. For RPRD method, recommended values are used (5 bins for RD, the number of new features created by using RP d as $2(\log_2 c)$ where c is the number of features, and the elements r_{ij} of the Random Matrix $\boldsymbol{R}$ are Gaussian distributed.) For Random Forest, default values are used (featureSubsetStrategy = "auto", impurity = "gini", maxDepth = 5 and maxBins = 32).

As evaluation criteria, prediction accuracy is used to evaluate the accuracy produced by the predictors (number of examples correctly labeled as belonging to a given class divided by the total number of elements). As ECBDL14 dataset is highly unbalanced, we have used the True Positive Rate (TPR) and True Negative Rate (TNR) TPR·TNR metric.

For all experiments we have used a cluster composed of 14 computing nodes. The nodes hold the following characteristics: 2 x Intel Core i7-4930K, 6 cores each, 3.40 GHz, 12 MB cache, 4 TB HDD, 64 GB RAM. Regarding software, we have used the following configuration: Hadoop 2.6.0-cdh5.10.0 from Cloudera's opensource Apache Hadoop distribution, Apache Spark and MLlib 2.2.0, 252 cores (18 cores/node), 728 RAM GB (52 GB/node).

4.2   Experimental Results and Analysis

Along this section we show the test accuracy values for 5 fold cross-validation.

Table 1. RD vs PCA vs RD^{2}R vs RPRD Test Accuracy

Dataset	Trees	RD	PCA	RD^{2}R	RPRD
Poker	10	54.73($\pm$0.43)	54.68($\pm$0.24)	55.07($\pm$0.19)	53.84($\pm$0.26)
	50	54.76($\pm$0.49)	54.81($\pm$0.13)	54.92($\pm$0.20)	53.82($\pm$0.25)
	100	54.73($\pm$0.28)	54.76($\pm$0.28)	54.97($\pm$0.23)	53.82($\pm$0.07)
SUSY	10	78.00($\pm$0.09)	75.30($\pm$0.20)	78.31($\pm$0.07)	78.19($\pm$0.05)
	50	78.26($\pm$0.04)	74.97($\pm$0.08)	78.47($\pm$0.09)	78.28($\pm$0.09)
	100	78.31($\pm$0.07)	75.31($\pm$0.34)	78.49($\pm$0.03)	78.35($\pm$0.08)
HIGGS	10	68.64($\pm$0.25)	60.10($\pm$2.01)	68.75($\pm$0.56)	68.36($\pm$0.09)
	50	68.98($\pm$0.15)	60.44($\pm$0.76)	69.28($\pm$0.18)	69.01($\pm$0.13)
	100	69.17($\pm$0.12)	60.81($\pm$0.25)	69.35($\pm$0.10)	69.22($\pm$0.17)
Epsilon	10	68.78($\pm$0.39)	78.14($\pm$0.09)	78.57($\pm$0.37)	68.64($\pm$0.33)
	50	69.04($\pm$0.19)	78.14($\pm$0.09)	78.57($\pm$0.25)	69.10($\pm$0.27)
	100	69.22($\pm$0.25)	78.14($\pm$0.09)	78.58($\pm$0.27)	69.31($\pm$0.29)
ECBDL14[a]	10	0.1884	0.2400	0.4742	0.4735
	50	0.1885	0.2410	0.4717	0.4775
	100	0.1880	0.2415	0.4742	0.4757

[a]For this dataset TPR·TNR metric is being used.

Table 1 compares the accuracy values in prediction obtained by RD^{2}R and RPRD for the five datasets using a Decision Tree as a classifier. To prove that the combination of RD and PCA produces a better ensemble method, we also show the prediction accuracy for both RD and PCA independently. According to these results, it is demonstrated that the combination of RD and PCA produces a better ensemble method than RPRD. This improvement is especially important in the Epsilon dataset, where there is a difference of 10% more accuracy. We can assert that ensembles of 10 trees are the best choice as the improvement in prediction with 5 and 10 times more trees is minimal. The ensemble also proves to be very stable, as there is little or no improvement in bigger ensemble sizes.

Table 2 compares the accuracy values in prediction obtained by RD^{2}R with 10 trees and Random Forest with 200 and 500 trees. We show that our algorithm also outperforms Random Forest. Even with big ensembles with up to 500 trees, Random Forest can not match or outperform RD^{2}R with 10 trees.

In Figs. 2 and 3 we can see a graphic representation of the test accuracy of RD, PCA, RD^{2}R, RPRD and Random Forest.

Table 3 shows learning runtime values obtained by RD^{2}R, RPRD (all two with 10 trees) and Random Forest. As we can see, for datasets with a small number of features RD^{2}R performs faster than RPRD algorithm. Epsilon and

Table 2. RD^2R vs Random Forest Test Accuracy

Dataset	RD^2R 10	RF 200	RF 500
Poker	55.07($\pm$0.19)	51.56($\pm$0.98)	51.61 ($\pm$0.97)
SUSY	78.31($\pm$0.07)	77.73($\pm$0.04)	77.76($\pm$0.07)
HIGGS	68.75($\pm$0.56)	67.98($\pm$0.12)	67.94($\pm$0.13)
Epsilon	78.57($\pm$0.37)	73.24($\pm$0.32)	73.41($\pm$0.22)
ECBDL14[5]	0.4742	0.4642	0.4634

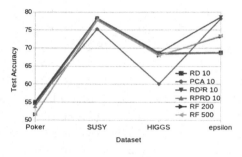

Fig. 2. Figure test accuracy in poker, HIGGS, SUSY and epsilon datasets

Fig. 3. Figure test accuracy in ECBDL14 dataset

Table 3. Learning Time Values in Seconds

Dataset	RD^2R 10	RPRD 10	RF 200	RF 500
Poker	159	169	112	294
SUSY	193	328	161	351
HIGGS	248	604	143	325
Epsilon	2,048	338	79	124
ECBDL14	22,093	12,607	1,664	4,460

ECBDL14 datasets represent a challenge for PCA, as they have a very large number of features to compute. RP performs a matrix multiplication whilst PCA has to compute the principal components of 2,000 features in the case of the epsilon dataset. However the performance improvement obtained by RD^2R over RPRD and Random Forest justifies this result.

Table 4 shows prediction runtime values for one test example obtained by RD^2R, RPRD (all two with 10 trees) and Random Forest. As we can see, RD^2R is more competitive in prediction. RPRD only performs better than RD^2R in the Epsilon dataset.

In view of the results we can conclude that:

– The performance of PCA against RP has proven to be better for every tested dataset, achieving up to 10% more accuracy.

Table 4. Prediction Time Values in Microseconds

Dataset	RD^{2}R 10	RPRD 10	RF 200	RF 500
Poker	63.41	78.05	68.31	82.93
SUSY	34.00	41.00	37.41	44.00
HIGGS	16.36	23.18	13.15	14.55
Epsilon	2,350	325.00	38.49	50.00
ECBDL14	148.97	214.14	11.16	13.10

– The RD^{2}R algorithm has shown to be able to work with huge datasets in a short amount of time.
– It is a very stable method for 10 trees. It shows little or no improvement with bigger ensemble sizes.
– It outperforms the original proposal as well as Random Forest for most of the tested datasets. This difference is more noticeable in the Epsilon dataset.
– PCA can perform faster than the other methods for datasets with small number of features.

5 Conclusions

In this contribution, a new ensemble method is proposed inspired by RPRD Ensemble. It replaces the inconsistency of Random Projections by using a more informative dimensionality reduction method such as PCA.

Thereby we proposed the RD^{2}R algorithm, a new ensemble method based on PCA for the dimensionality reduction step and Random Discretization, capable of working with Big Data and integrated in Spark's MLlib Library as a third-party package.

The experimental results have demonstrated the stability and improvement in prediction accuracy when using our ensemble solution for the five datasets used. RD^{2}R learning times have shown to be faster than RPRD and Random Forest for datasets with a small number of features. Additionally, results suggest that RD^{2}R is very effective for ensembles with a small number of trees.

Acknowledgments. This contribution is supported by FEDER, the Spanish National Research Projects TIN2014-57251-P and TIN2017-89517-P, and the Project BigDaP-TOOLS - Ayudas Fundación BBVA a Equipos de Investigación Científica 2016.

References

1. Ahmad, A., Brown, G.: Random projection random discretization ensembles - ensembles of linear multivariate decision trees. IEEE Trans. Knowl. Data Eng. **26**(5), 1225–1239 (2014)
2. Breiman, L.: Random forests. Mach. Learn. **45**(1), 5–32 (2001)

3. Dasgupta, S.: Experiments with random projection. In: Proceedings of the Sixteenth Conference on Uncertainty in Artificial Intelligence, UAI 2000, pp. 143–151. Morgan Kaufmann Publishers Inc., San Francisco (2000)
4. Dean, J., Ghemawat, S.: Mapreduce: simplified data processing on large clusters. Commun. ACM **51**(1), 107–113 (2008)
5. Dietterich, T.G.: Ensemble methods in machine learning. In: Kittler, J., Roli, F. (eds.) MCS 2000. LNCS, vol. 1857, pp. 1–15. Springer, Heidelberg (2000). https://doi.org/10.1007/3-540-45014-9_1
6. Fradkin, D., Madigan, D.: Experiments with random projections for machine learning. In: Proceedings of the Ninth ACM SIGKDD International Conference on Knowledge Discovery and Data Mining, KDD 2003, pp. 517–522. ACM, New York (2003)
7. García, S., Luengo, J., Sáez, J., López, V., Herrera, F.: A survey of discretization techniques: taxonomy and empirical analysis in supervised learning. IEEE Trans. Knowl. Data Eng. **25**(4), 734–750 (2013)
8. García, S., Luengo, J., Herrera, F.: Tutorial on practical tips of the most influential data preprocessing algorithms in data mining. Knowl. Syst. **98**, 1–29 (2016)
9. García, S., Luengo, J., Herrera, F.: Data Preprocessing in Data Mining. Springer, Heidelberg (2015). https://doi.org/10.1007/978-3-319-10247-4
10. García, S., Ramírez-Gallego, S., Luengo, J., Benítez, J.M., Herrera, F.: Big data preprocessing: methods and prospects. Big Data Anal. **1**(1), 9 (2016)
11. García-Gil, D., Ramírez-Gallego, S., García, S., Herrera, F.: A comparison on scalability for batch big data processing on Apache Spark and Apache Flink. Big Data Anal. **2**(1), 11 (2017)
12. Johnson, W.B., Lindenstrauss, J.: Extensions of Lipschitz mappings into a Hilbert space. Contemp. Math. **26**(189–206), 1 (1984)
13. Lin, J.: Mapreduce is good enough? If all you have is a hammer, throw away everything that's not a nail!. Big Data **1**(1), 28–37 (2013)
14. Ramírez-Gallego, S., García, S., Benítez, J., Herrera, F.: A distributed evolutionary multivariate discretizer for big data processing on apache spark. Swarm Evolut. Comput. **38**, 240–250 (2018)
15. Ramírez-Gallego, S., Fernández, A., García, S., Chen, M., Herrera, F.: Big data: tutorial and guidelines on information and process fusion for analytics algorithms with mapreduce. Inf. Fusion **42**, 51–61 (2018)
16. del Río, S., López, V., Benítez, J.M., Herrera, F.: On the use of mapreduce for imbalanced big data using random forest. Inf. Sci. **285**, 112–137 (2014)
17. Zaharia, M., Chowdhury, M., Das, T., Dave, A., Ma, J., McCauley, M., Franklin, M.J., Shenker, S., Stoica, I.: Resilient distributed datasets: A fault-tolerant abstraction for in-memory cluster computing. In: Proceedings of the 9th USENIX Conference on Networked Systems Design and Implementation, NSDI 2012, pp. 15–28. USENIX Association, Berkeley (2012)

Hybrid Deep Learning Based on GAN for Classifying BSR Noises from Invehicle Sensors

Jin-Young Kim, Seok-Jun Bu, and Sung-Bae Cho$^{(\boxtimes)}$

Department of Computer Science, Yonsei University, Seoul, South Korea
{seago0828, sjbuhan, sbcho}@yonsei.ac.kr

Abstract. BSR (Buzz, squeak, and rattle) noises are essential criteria for the quality of a vehicle. It is necessary to classify them to handle them appropriately. Although many studies have been conducted to classify noise, they suffered some problems: the difficulty in extracting features, a small amount of data to train a classifier, and less robustness to background noise. This paper proposes a method called transferred encoder-decoder generative adversarial networks (tedGAN) which solves the problems. Deep auto-encoder (DAE) compresses and reconstructs the audio data for capturing the features of them. The decoder network is transferred to the generator of GAN so as to make the process of training generator more stable. Because the generator and the discriminator of GAN are trained at the same time, the capacity of extracting features is enhanced, and a knowledge space of the data is expanded with a small amount of data. The discriminator to classify whether the input is the real or fake BSR noises is transferred again to the classifier; then it is finally trained to classify the BSR noises. The classifier yields the accuracy of 95.15%, which outperforms other machine learning models. We analyze the model with t-SNE algorithm to investigate the misclassified data. The proposed model achieves the accuracy of 92.05% for the data including background noise.

1 Introduction

BSR (Buzz, squeak, and rattle) noises are a severe problem to deteriorate the quality of a vehicle. The decline in vehicle quality leads to customer complaints and vehicle repairs, and at least 50% of motor vehicle repairs are associated with noise in the interior of a vehicle [1]. Although there are many approaches to reducing the noise for improving the quality, the factors which produce noises increase, for example, electronic devices and light bodywork and materials. However, because it can be improved immediately when a noise source is identified [2], studies about classifying noise have been conducted.

Some problems exist in classifying noise: the difficulty in extracting features, a small amount of data to train a classifier, and reduction in robustness to background noise. The first problem is challenging to extract features. Figure 1 shows sound data of each type: normal, retractor, seat rattle, and weatherstrip, which are the types of BSR noises. As data have complicated and messy features, we need a preprocessing method.

© Springer International Publishing AG, part of Springer Nature 2018
F. J. de Cos Juez et al. (Eds.): HAIS 2018, LNAI 10870, pp. 27–38, 2018.
https://doi.org/10.1007/978-3-319-92639-1_3

Since the sound data have spatial (frequency) and temporal (time) features, we convert the data to sound map image with short-time Fourier transform (STFT) to maintain temporal features as well as spatial features, as illustrated in Fig. 2. The features that appear are relatively clear, but because they are still complex to be classified, we need a more sophisticated method to extract features. The second problem is that there is a small amount of data to train a classifier. If there is lots of data, structuring them is expensive. Classifier needs lots of data to get better performance. We can collect BSR noise easily without background noise in a laboratory, but not quickly natural data. The third problem is that a real BSR noise has background noise such as wind sound or chat sound. We can divide BSR noises into two types: data with background noise and data without it. The latter can be collected relatively easily but not practical. Therefore, it is needed to make classifier which can classify previous data, not just latter data.

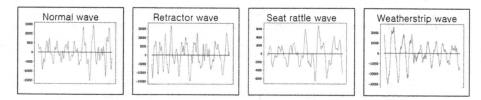

Fig. 1. Sound data of each type. The blue line is a data and orange line is just 0 line. (Color figure online)

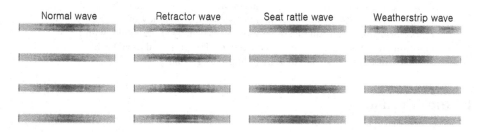

Fig. 2. Sound data which are converted to sound map image with STFT. Height means time and width means frequency. The color of each pixel means amplitude on that time and frequency. Red color means higher amplitude than blue. (Color figure online)

Noises raised in the vehicle are shown in Fig. 3. The noise generated in the vehicle can be roughly divided into two types: noise from the vehicle itself and noise from external factors (background noise). BSR classification is to distinguish yellow star in Fig. 3 from the normal sound. Because there are background noises, not just BSR noises, the third problem occurs.

The rest of the paper is organized as follows. Section 2 reviews the related works and the hybrid deep learning model is proposed in Sect. 3. In Sect. 4, we show the performance of the proposed method and compare it with the conventional methods. Some conclusions and discussion are presented in Sect. 5.

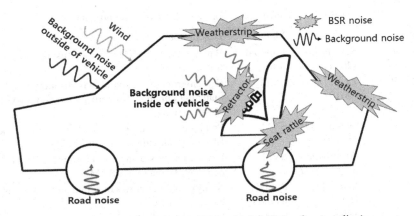

Fig. 3. Types of noises raised in vehicle. (Color figure online)

2 Related Works

Many research works have been conducted to classify various types of noise. The related works are summarized in Table 1. Machine learning techniques were used to classify noise. Saki and Kehtarnavaz classified background noise using random forest [3]. Li preprocessed eco-environmental sound based on matching pursuit and classified them with support vector machine (SVM) [4]. Many researchers used SVM with different preprocessing methods. For example, Wang et al. preprocessed data with principal component analysis, linear discriminant, and Gabor dictionary [5]. Amiriparian et al. used spectrograms of data and extracted features with convolutional neural networks (CNN) [6]. Lee et al. analyzed features of sound using Prony's method and classified the data based on it [7]. Salamon and Bello used other methods that extract features based on spherical k-means algorithm [8].

These research tried finding efficient or superior feature extraction methods. Because most of them extracted characteristics of data based on existing data, they may not be robust to new data. Some studies overcame this problem with the fact that deep learning can discover representations that are stable with respect to variations in data [9]. Tanweer et al. classified environmental noise using LDA, quadratic discriminant analysis, and artificial neural network [10]. Rahim et al. made homogeneous multi-classifier system for classifying moving vehicles noise with multilayer perceptron [11]. Piczak used CNN architecture and classified environmental sound [12]. Medgat et al. classified sound with masked conditional neural networks which can learn exploration of different feature combinations [13].

These methods can be robust to new data thanks to deep learning, but they have also disadvantage that it needs lots of data, since deep learning with a small amount of

Table 1. Related words for classification of sound data.

Category	Authors	Method	Description
Focus on feature extraction	Saki F. [3] (2014)	Random forest	Background noise classification using random forest tree classifier
	Li Y. [4] (2010)	SVM	Eco-environmental sound classification based on matching pursuit for preprocessing and SVM for classification
	Wang J.-C. [5] (2014)	SVM	Preprocessing with Gabor dictionary, PCA and LDA, and classification with SVM
	Amiriparian S. [6] (2017)	SVM	Extract features using spectrograms and CNN, and classify data with SVM
	Lee J. [7] (2015)	Prony's method	Analyze characteristics of sound and classify data based on them
	Salamon J. [8] (2015)	Spherical k-means	Use method which extracts features based on spherical k-means algorithm
Deep learning to classify data	Tanweer S. [10] (2016)	LDA, QDA, and ANN	Extract features using mel frequency cepstral coefficient (MFCC) and classify data with LDA, QDA, and ANN
	Rahim N. A. [11] (2015)	MLP	Classify sound with homogeneous multi-classifier system
	Piczak K. J. [12] (2015)	CNN	Construct CNN architecture and classify data with it
	Medhat F. [13] (2017)	MCLNN	Propose MCLNN model which can learn exploration of different feature combinations

data can be overfitting easily [14]. To solve this problem, we propose a model that extracts features with deep auto-encoder (DAE) and generates new data. This model finally helps classifier to be trained with more data.

3   The Proposed Method

We divide the proposed model into three parts: extracting feature, generating, and classifying parts. In the first part, DAE compresses and reconstructs the sound data. Besides, the decoder of DAE is used in the generator (G) of GAN. G and discriminator (D) are trained at the same time so that G can generate data as similar to the real as possible and D can distinguish the real and the fake as precise as possible. The architecture of the whole process is shown in Fig. 4.

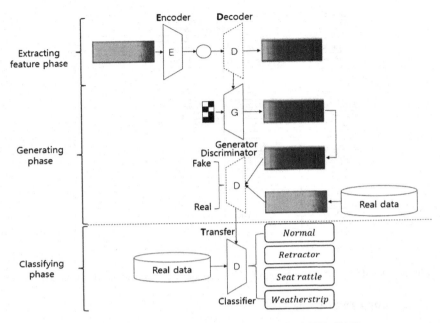

Fig. 4. The architecture of whole process of TED-GAN.

3.1 Feature Extraction with DAE

Since sound data have not only spatial-temporal feature but also lots of combinations of a feature, extracting a feature of sound is an important part of the classification. The DAE, which is widely used for capturing features [15, 16], can extract characteristics while it compresses and reconstructs data. A loss function to train DAE is following:

$$\mathcal{L}_{\text{DAE}} = H(X, Dec(Enc(X))) + KL(Enc(X)\|N(0, 1)) \tag{1}$$

The first term of Eq. (1) indicates the degree of reconstruction. H means information entropy. Enc and Dec are encoder and decoder functions, respectively. The smaller the first term, the more precise the reconstruction. The second term is Kullback-Leibler divergence between $Enc(X)$ and a normal distribution with mean of 0 and standard deviation of 1. It improves G's performance because G generates data from normal distribution, so it can be called a process of pre-training of G. This will be explained more in the next section.

Details of DAE construction is illustrated in Fig. 5. We use long-term recurrent convolutional networks (LRCN) [17] for encoder, and deconvolution layers [18] for decoder. LeakyReLU and dropout layer are followed all of the conv1D, conv2D, and deconv2D except the second conv1D of ConvBlock in Fig. 6.

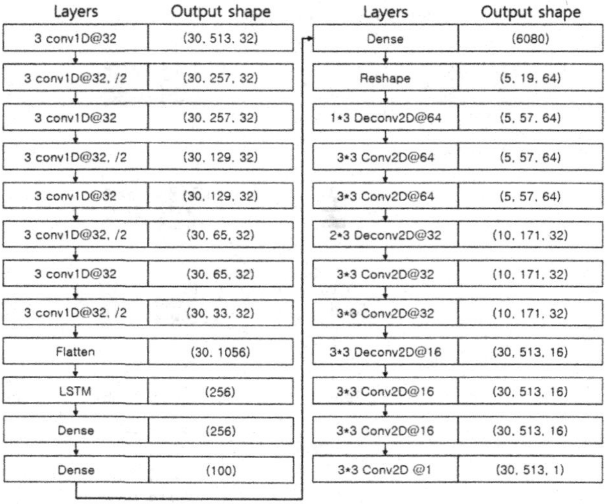

Layers	Output shape	Layers	Output shape
3 conv1D@32	(30, 513, 32)	Dense	(6080)
3 conv1D@32, /2	(30, 257, 32)	Reshape	(5, 19, 64)
3 conv1D@32	(30, 257, 32)	1*3 Deconv2D@64	(5, 57, 64)
3 conv1D@32, /2	(30, 129, 32)	3*3 Conv2D@64	(5, 57, 64)
3 conv1D@32	(30, 129, 32)	3*3 Conv2D@64	(5, 57, 64)
3 conv1D@32, /2	(30, 65, 32)	2*3 Deconv2D@32	(10, 171, 32)
3 conv1D@32	(30, 65, 32)	3*3 Conv2D@32	(10, 171, 32)
3 conv1D@32, /2	(30, 33, 32)	3*3 Conv2D@32	(10, 171, 32)
Flatten	(30, 1056)	3*3 Deconv2D@16	(30, 513, 16)
LSTM	(256)	3*3 Conv2D@16	(30, 513, 16)
Dense	(256)	3*3 Conv2D@16	(30, 513, 16)
Dense	(100)	3*3 Conv2D @1	(30, 513, 1)

Fig. 5. Details of DAE construction. Left: encoder of DAE. Right: decoder of DAE

3.2   Generating Data Using GAN

GAN has led to significant improvements in data generation [19]. The basic training process of GAN is to adversely interact and simultaneously train G and D. Equation (2) shows the objective function of a GAN. p_{data} is the probability distribution of the real data. $G(z)$ is generated from a probability distribution p_z by the G, and it is distinguished from the real data by the discriminator D. The discriminator is trained such that $D(x)$ of the first term is 1 and $D(G(z))$ of the second term is 0, to maximize $V(D, G)$, and G is trained such that $D(G(z))$ of the second term is 1, to minimize $V(D, G)$.

$$
\min_G \max_D V(D, G) = E_{x \sim p_{data}(x)}[\log D(x)] \\
+ E_{z \sim p_z(z)}[\log(1 - D(G(z)))]
\tag{2}
$$

Since original GAN has a disadvantage that the generated data are insensible because of the unstable learning process of the generator, we pre-train G with the decoder of DAE as discussed in the previous section. To overcome the difference related to the fact that the G generates fake data from a random variable z but the decoder generates it from $Enc(x)$, we add Kullback-Leibler divergence to loss of DAE as shown in Eq. (1). The result of DAE is that $\left| p_{data}^{DAE} - p_G^{DAE} \right| \leq |p_{data} - p_G|$ so that it can reach a goal of GAN ($p_{data} \approx p_G$) stably.

The reason for using GAN is to classify new data, including background noise, into the model learned using existing data. Because GAN generates data from a random distribution, new data has some variants compared to existing data, resulting in expanding a knowledge space of data. Therefore, D is robust to deformation [20], and we can train D with more data.

Details of GAN is shown in Fig. 6. The structure of G is same to that of decoder of DAE, and we use LRCN in D which is transferred to classifier later. A ConvBlock is a concatenation of two convolutional layers which is shown at the right in Fig. 6.

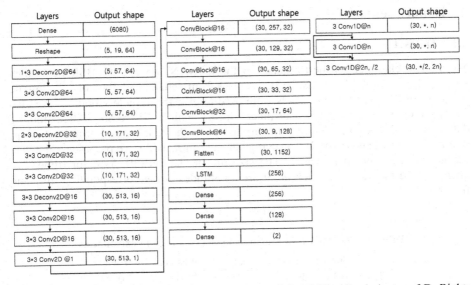

Layers	Output shape		Layers	Output shape		Layers	Output shape
Dense	(6080)		ConvBlock@16	(30, 257, 32)		3 Conv1D@n	(30, *, n)
Reshape	(5, 19, 64)		ConvBlock@16	(30, 129, 32)		3 Conv1D@n	(30, *, n)
1*3 Deconv2D@64	(5, 57, 64)		ConvBlock@16	(30, 65, 32)		3 Conv1D@2n, /2	(30, */2, 2n)
3*3 Conv2D@64	(5, 57, 64)		ConvBlock@16	(30, 33, 32)			
3*3 Conv2D@64	(5, 57, 64)		ConvBlock@32	(30, 17, 64)			
2*3 Deconv2D@32	(10, 171, 32)		ConvBlock@64	(30, 9, 128)			
3*3 Conv2D@32	(10, 171, 32)		Flatten	(30, 1152)			
3*3 Conv2D@32	(10, 171, 32)		LSTM	(256)			
3*3 Deconv2D@16	(30, 513, 16)		Dense	(256)			
3*3 Conv2D@16	(30, 513, 16)		Dense	(128)			
3*3 Conv2D@16	(30, 513, 16)		Dense	(2)			
3*3 Conv2D @1	(30, 513, 1)						

Fig. 6. Details of GAN construction. Left: generator of G. Middle: discriminator of D. Right: ConvBlock layer in discriminator.

3.3 Classifying Data with Transfer Learning

In the previous process, we extract features with DAE and generates data for expanding knowledge space with GAN so that D can classify data which are not in existing data. The last process is a classification of data with transfer learning [21]. We transfer D to a classifier so that the ability of D is inherited to a classifier. Since the goals of D and classifier are different, we transfer ConvBlock and long short-term memory (LSTM) to the classifier and train fully connected layers of it newly. The detail of a classifier is shown in Fig. 7.

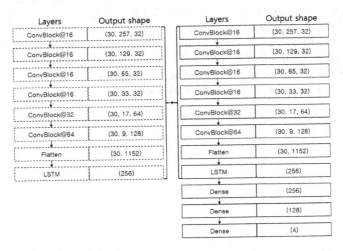

Layers	Output shape		Layers	Output shape
ConvBlock@16	(30, 257, 32)		ConvBlock@16	(30, 257, 32)
ConvBlock@16	(30, 129, 32)		ConvBlock@16	(30, 129, 32)
ConvBlock@16	(30, 65, 32)		ConvBlock@16	(30, 65, 32)
ConvBlock@16	(30, 33, 32)		ConvBlock@16	(30, 33, 32)
ConvBlock@32	(30, 17, 64)		ConvBlock@32	(30, 17, 64)
ConvBlock@64	(30, 9, 128)		ConvBlock@64	(30, 9, 128)
Flatten	(30, 1152)		Flatten	(30, 1152)
LSTM	(256)		LSTM	(256)
			Dense	(256)
			Dense	(128)
			Dense	(4)

Fig. 7. Details of classifier construction. Dashed line means transfer learning.

4   Experiments

4.1   Dataset

To validate classification performance of the tedGAN, we use the BSR noise data described in Table 2. We collected BSR noises from Sedan in a laboratory. To imitate the noise of the road, 2 shaker axes were used to give the vehicle vibration. We also collected sound data with a sensor placed in the vehicle. There are 10,022 training data and 3,112 test data. The data belong to one of the four types: normal, retractor, seat rattle, and weatherstrip. Because raw data are intractable, we preprocess the data using STFT, as shown in Fig. 2. The size of one data is (30, 513) for (time, frequency), and the values of each point are amplitudes at the corresponding time and frequency.

Table 2.   The number of training and test data used in experiments.

	Normal	Retractor	Seat rattle	Weatherstrip	Total
Train	3885	567	3273	2297	10022
Test	978	223	928	983	3112

4.2   Result of Classification

In this section, we show a result of classification and compare it with those of other classification methods, such as the k nearest neighbors (NN), naïve Bayes (NB), random forest (RF), decision trees (DT), AdaBoost(AB), support vector machine (SVM) with a polynomial (Poly) kernel and radial basis function (RBF) kernel, which are provided in the scikit-learn library. For all of these algorithms, we used default values, except maximal depth = 5 in the decision tree and random forest methods. We compared the proposed method not only to those methods provided by the scikit-learn library but also to the multi-layer perceptron (MLP) and the CNN which has the same structure as the discriminator.

We use 10-fold cross-validation for all of the algorithms. Results of these experiments are summarized in Fig. 8. The average correct classification achieved by the proposed model is 95.15%. The accuracy difference between tedGAN and CNN is small, but it has statistical significance with 2.44e-05 p-value in a t-test.

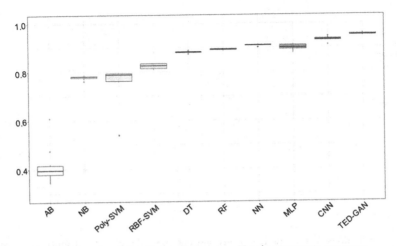

Fig. 8. Result of classifying BSR noise. Y-axis represents accuracy.

4.3 Analysis of Result

The confusion matrix for BSR classification for validating the performance of the proposed classifier is shown in Fig. 9. The matrix confirms that the classifier cannot distinguish well normal and retractor. However, the proposed model can classify seat rattle and weatherstrip better than others.

Figure 10 reveals clustering patterns in the BSR data by applying the t-SNE algorithm [22], which groups data points according to their similarity. It can be seen that data are sufficiently modified for the clustering pattern to become apparent to the classifier.

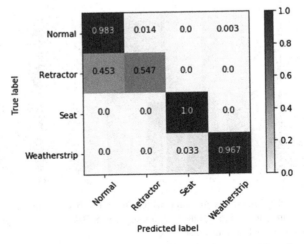

Fig. 9. Confusion matrix for result of classifying BSR noise.

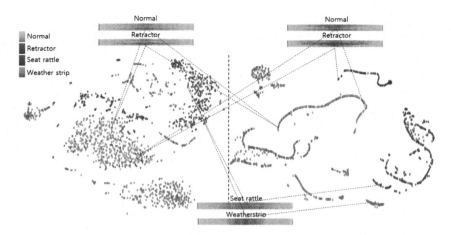

Fig. 10. Distributions of the raw data (left) and output of the classifier (right). The classifier changes the values so that similar data points are clustered together.

4.4   Robust to Noise

To verify the effect of expanding knowledge space, we add background noise to data. We use motorized noise (such as taxi, private, police, and ambulance) and non-motorized noise (such as street, road, bicycle, construction, and vehicle air conditioner) among Urban Sound Dataset[1]. We train model with data which do not include background noise and test the model with data which have background noise. The result of the experiment is shown in Fig. 11. The proposed model gets 92.05% performance which is less than before but still higher than others.

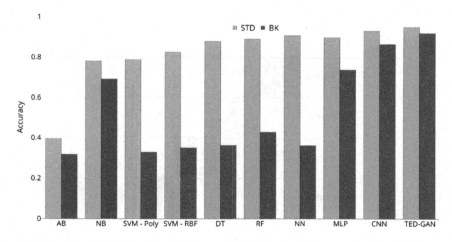

Fig. 11. Result of classification with data which includes background noise. STD means 'standard', i.e., they are data which has no surrounding noise. BK means 'background', i.e., they are data which includes background noise.

---

[1] https://serv.cusp.nyu.edu/projects/urbansounddataset/.

5 Conclusion

In this paper, we introduce the necessity of classifying BSR noises. We attempt to classify BSR noises using the information extracted from in-vehicle sensors. Since there are three problems: the difficulty in extracting features, a small amount of data, and less robustness to surrounding noise, we propose tedGAN to extract features and generate more data. Knowledge space of data is expanded using the generator G which is initialized by the decoder of DAE so that the discriminator D can learn more characteristics of data. Finally, the capacity of D is transferred to a classifier. The proposed classifier achieves the accuracy of 95.15%, thereby outperforming other conventional models. Moreover, it can classify new data which includes surrounding noise well, resulting in more robustness to noise.

In the future work, we plan to generate more various data that make the model more robust to noise. To solve the problem of not distinguishing between normal and rattle, we will construct a model that generates a specific class so that it can solve class imbalance problem. We will investigate the generated data with inverse-STFT to check how close it is to the sound from the vehicle.

Acknowledgement. This work has been supported by a grant from Hyundai motors, Inc.

References

1. Mattingly-Hannigan, E.: Vibration Testing on the Job, pp. 55–57. Mattingly Publishing Co. Inc. (2008)
2. Mog, M.G., Min, B.H., Chio, S.W., Lee, H.J.: Development of the reproduction test method of automobile buzz, squeak, rattle noise and the noise tracking system. In: Korean Society of Automotive Engineers (KSAE), pp. 1475–1481 (2010)
3. Saki, F., Kehtarnavaz, N.: Background noise classification using random forest tree classifier for cochlear implant applications. In: 2014 IEEE International Conference on Acoustics, Speech and Signal Processing (ICASSP), pp. 3591–3595. IEEE (2014)
4. Li, Y., Li, Y.: Eco-environmental sound classification based on matching pursuit and support vector machine. In: 2010 2nd International Conference on Information Engineering and Computer Science (ICIECS), pp. 1–4. IEEE (2010)
5. Wang, J.C., Lin, C.H., Chen, B.W., Tsai, M.K.: Gabor-based nonuniform scale-frequency map for environmental sound classification in home automation. IEEE Trans. Autom. Sci. Eng. **11**(2), 607–613 (2014)
6. Amiriparian, S., Gerczuk, M., Ottl, S., Cummins, N., Freitag, M., Pugachevskiy, S., Baird, A., Schuller, B.: Snore sound classification using image-based deep spectrum features. In: Proceedings of Interspeech, vol. 17, pp. 3512–3516 (2017)
7. Lee, J., Lee, S., Kwak, Y., Kim, B., Park, J.: Temporal and spectral characteristics of BSR noises and influence on auditory perception. J. Mech. Sci. Technol. **29**(12), 5199–5204 (2015)
8. Salamon, J., Bello, J.P.: Unsupervised feature learning for urban sound classification. In: 2015 IEEE International Conference on Acoustics, Speech and Signal Processing (ICASSP), pp. 171–175. IEEE (2015)

9. Deng, L., Li, J., Huang, J.T., Yao, K., Yu, D., Seide, F., Seltzer, M.L., Zweig, G., He, X., Williams, J., Gong, Y., Acero, A.: Recent advances in deep learning for speech research at Microsoft. In: 2013 IEEE International Conference on Acoustics, Speech and Signal Processing (ICASSP), pp. 8604–8608. IEEE (2013)
10. Tanweer, S., Mobin, A., Alam, A.: Environmental noise classification using LDA, QDA and ANN methods. Indian J. Sci. Technol. 9(33), 1–8 (2016)
11. Abdul Rahim, N., Paulraj, M.P., Adom, A.H., Shukor, S.A.A., Masnan, M.J.: Homogeneous multi-classifier system for moving vehicles noise classification based on multilayer perceptron. J. Intell. Fuzzy Syst. 29(1), 149–157 (2015)
12. Piczak, K.J.: Environmental sound classification with convolutional neural networks. In: 2015 IEEE 25th International Workshop on Machine Learning for Signal Processing (MLSP), pp. 1–6. IEEE (2015)
13. Medhat, F., Chesmore, D., Robinson, J.: Masked conditional neural networks for environmental sound classification. In: Bramer, M., Petridis, M. (eds.) SGAI-AI 2017. LNCS (LNAI), vol. 10630, pp. 21–33. Springer, Cham (2017). https://doi.org/10.1007/978-3-319-71078-5_2
14. LeCun, Y., Bengio, Y., Hinton, G.: Deep learning. Nature 521(7553), 436–444 (2015)
15. Krizhevsky, A., Hinton, G.E.: Using very deep autoencoders for content-based image retrieval. In: European Symposium on Artificial Neural Networks, Computational Intelligence and Machine Learning, pp. 489–494 (2011)
16. Vincent, P., Larochelle, H., Bengio, Y., Manzagol, P.A.: Extracting and composing robust features with denoising autoencoders. In: International Conference on Machine Learning, pp. 1096–1103 (2008)
17. Donahue, J., Anne Hendricks, L., Guadarrama, S., Rohrbach, M., Venugopalan, S., Saenko, K., Darrell, T.: Long-term recurrent convolutional networks for visual recognition and description. In: Proceedings of the IEEE Conference on Computer Vision and Pattern Recognition, pp. 2625–2634 (2015)
18. Zeiler, M.D., Fergus, R.: Visualizing and understanding convolutional networks. In: Fleet, D., Pajdla, T., Schiele, B., Tuytelaars, T. (eds.) ECCV 2014, Part I. LNCS, vol. 8689, pp. 818–833. Springer, Cham (2014). https://doi.org/10.1007/978-3-319-10590-1_53
19. Goodfellow, I., Pouget-Abadie, J., Mirze, M., Xu, B., Warde-Farley, D., Ozair, S., Courville, A., Bengio, Y.: Generative adversarial nets. In: Advances in Neural Information Processing Systems, pp. 2672–2680 (2014)
20. Radford, A., Metz, L., Chintala, S.: Unsupervised representation learning with deep convolutional generative adversarial networks (2015). arXiv preprint: arXiv:1511.06434
21. Arnold, A., Nallapati, R., Cohen, W.: A comparative study of methods for transductive transfer learning. In: IEEE International Conference on Data Mining, pp. 77–82 (2007)
22. Maaten, L., Hinton, G.: Visualizing data using t-SNE. J. Mach. Learn. Res. 9, 2579–2605 (2008)

Inferring User Expertise from Social Tagging in Music Recommender Systems for Streaming Services

Diego Sánchez-Moreno[1], María N. Moreno-García[1]([✉]),
Nasim Sonboli[2], Bamshad Mobasher[2], and Robin Burke[2]

[1] Department of Computing and Automation, University of Salamanca, Plaza de
los Caídos s/n, 37008 Salamanca, Spain
sanchezhh@gmail.com, mmg@usal.es
[2] Center for Web Intelligence, DePaul University, 243 S Wabash Ave, Chicago,
IL 60604, USA
nsonboli@depaul.edu, {mobasher,rburke}@cs.depaul.edu

Abstract. Suppliers of music streaming services are showing an increasing interest for providing users with reliable personalized recommendations since their practically unlimited offerings make it difficult for users to find the music they like. In this work, we take advantage of social tags that users give to music through streaming platforms for improving recommendations. Most of the works in the literature use the tags in the context of content based methods for finding similarities between songs and artists, but we use them for characterizing users, instead of characterizing music, aiming at improving user-based collaborative filtering algorithms. The expertise level of users is inferred from the frequency analysis of their tags by using TF-IDF (Term Frequency-Inverse Document Frequency), which is an indicator of the quantity and relevance of the tags that users provide to items. User expertise has been studied in the context of recommender systems and other domains, but, as far as we know, it has not been studied in the context of music recommendations.

Keywords: Music recommender systems · User expertise
Collaborative filtering · Streaming services

1 Introduction

The way people consume digital music contents has drastically changed in recent years. Getting music on-demand is the dominant tendency in digital downloading. Current streaming services facilitate access to almost all existing music from anywhere. Hence, there is increasing interest in developing recommendation algorithms that help users to filter the huge amount of musical content available in the digital space and discover the music that fits their preferences.

Most streaming platforms have search services and some of them have some recommendation mechanism. The methods used are diverse, from those based simply on the popularity or genre of the songs, to those based on measures of similarity between songs or users. Some companies such as Spotify, Apple and Pandora have developed

F. J. de Cos Juez et al. (Eds.): HAIS 2018, LNAI 10870, pp. 39–49, 2018.
https://doi.org/10.1007/978-3-319-92639-1_4

playlist generation algorithms based on content analysis performed by human experts or collaborative filtering. However, the details of these algorithms are unknown and there are not available data about their reliability.

Moreover, streaming services allow their users to share songs, artists and playlists, which makes them authentic social networks that are also endowed with facilities such as tagging, friendship relations and so on. This social information can be exploited in order to provide more reliable recommendations. In this work, social tagging information is used for improving the results of collaborative filtering methods. The idea is to infer user expertise from social tags and then use this information for giving a higher level of influence to the expert users when generating the recommendations. In prior research, social tags have been used for obtaining music similarity, nevertheless, they have not been used for characterizing users. The proposed method is a hybrid approach that makes use of user features in addition to ratings, aiming at improving traditional collaborative filtering methods.

User expertise has been studied in the context of recommender systems and other domains. In some of these studies, the main objective is modeling long-term temporal effects on user preferences as consequence of user personal development [1]. In other areas, such as question answer communities (CQA), the purpose is the identification of experts in specific topics who would provide more reliable answers. In the field of music recommendations, we are not aware of any paper where user expertise has been studied.

Our work also addresses the sparsity problem, described in the next section, by computing ratings from number of plays. Instead of using simple frequency functions, we applied a method indicated for play frequencies that have a power law distribution.

The rest of the paper is organized as follows: In Sect. 2 the basis and classification of collaborative filtering techniques are introduced. Section 3 includes a short survey of related works. The proposed methodology is described in Sects. 4, 5 and 6. Section 7 encloses the study conducted for its validation. Finally, the conclusions are given in Sect. 8.

2   Background

Most existing recommender systems use some type of collaborative filtering (CF) based approach. The aim of CF is to predict the rating that a target user would give to an item considering users having similar preferences regarding previously rated items.

In memory-based (user-based or user-user) CF methods the predictions for the active user are based on his nearest neighbors [2]. There are users who have similar preferences to the active user since they have rated items in common with a similar score. Different measures for similarity/distance computing can be used: Pearson correlation coefficient, cosine, Chebyshev, Jaccard, etc.

Item-based (or item-item) CF was proposed in [3] with the aim of avoiding the scalability problems associated with memory-based methods by precomputing the similarities between items. This can be done since it is expected that new ratings given to items in large rating databases do not significantly change the similarity between items, especially for frequently rated items. However, recommendations provided by item-based methods usually have less quality than those provided by user-based approaches and are thus used in large-scale systems where scalability is a serious problem. They have been used in popular systems like Amazon [4].

Collaborative filtering requires explicit expression of user personal preferences for items in the form of ratings, which are usually difficult to obtain. This fact is the cause of one of the main drawbacks of this approach, the sparsity problem, which arises when the number of ratings needed for prediction is greater than the number of the ratings obtained from the users. This is the main drawback that prevents the application of CF approach in many systems. Time that users spend examining the items is an alternative way to obtain implicit user preferences, but it requires to process log files and this implicit information about user preferences is not as reliable as the explicit ratings.

Currently, hybrid techniques are the most extensively implemented in recommender systems, in an attempt to address the limitations of CF and content-based approaches. These methods combine either different categories of CF methods or CF with other recommendation techniques such as content-based schemes [5]. The main drawbacks they present are their complexity and the information needed for inducing the models.

3 Related Work

Collaborative filtering (CF) methods are widely used in recommender systems. The GroupLens research system for Usenet news [6] was the first recommender system using CF and Ringo [3] was one of the first and most popular music recommender systems based on CF. In the music recommender area several ways of dealing with the sparsity problem presented by these methods have been proposed. The access history of users is often used to implicitly obtain user interests in a music recommendation system [7]. In several works where the last.fm database is used, the times that the users play the songs (play counts) are converted to ratings by means of various functions [8, 9].

Regarding content-based methods, in the music field, metadata of items, such as title, artist, genre and lyrics, can be exploited as content attributes, but also audio features like timbre, melody, rhythm or harmony. In [10] music similarity was determined from chord structure (spectrum, rhythm and harmony). Melody style is the music feature used in [11] for music recommendation. A content-based method is proposed where a classification of music objects in melody styles is performed and users' music preferences are learned by mining the melody patterns from the music access behavior of the users. Clustering of similar songs according to different features of audio content is performed in [12] to provide users with recommendations of music from the appropriate clusters.

The combination of memory-based and model-based CF methods is a common way of building hybrid CF approaches that usually yields better recommendations than the single methods applied separately. Hybrid strategies have been adopted in the development of music recommendation systems. In [13], the authors associate rating and content data with latent variables that directly describe the unobserved user preferences. For music recommendation, they adapt a Bayesian network which was originally designed for recommendation of documents. Unobservable user preferences are represented as a set of latent variables that are statistically estimated and introduced into the Bayesian network. Another hybrid music recommendation system is presented in [14]. Its authors propose a content-based scheme to recommend music without ratings, a collaborative algorithm to make recommendations based on the suggestions of other users and a recommendation procedure based on emotions that determines the music of interest to users by calculating the differences between their interests and musical emotions. The combination of the three methods is done through a weighting system based on the user's listening behavior. This method requires users to complete a questionnaire that allows the system to discover their interests, which is not always possible. There are other hybrid proposals for music recommendation, but most of them enclose very complex procedures and require information about music and users that often is unavailable.

Concerning the topic of user expertise, as mentioned before, the main application in the field of recommender systems is modeling the evolution of users' preferences as they gain knowledge. In domains such as CQA and online reviews, most of the works make use of voting scores that users give to posts, answers and so on [15, 16]. In some of them, directed relationships between asker and answerer users are also considered [17]. There are some works in which domain specific ontologies are defined for inducing expertise. For example, in the e-commerce domain the authors of [18] define an ontology of products, then they compute expertise in a category considering the quantity and diversity of products of different subcategories rated by the users. In our proposal, expertise is directly obtained from the tags without need of defining any ontology.

Recently, many hybrid recommendation systems exploit social networks and other web sources in order to gather information that may be useful in the recommendation process [19, 20]. In [21] social tagging is used to derive the latent themes associated with songs from the most frequent tags given to them. In [22] a recommendation system is proposed that also uses social media tags to establish the similarity between the songs. In addition, tag information is used to capture user preferences. The results reveal that social labeling is a valid means of predicting musical preferences.

4   Inferring User Expertise from Social Tagging

Music datasets do not contain voting scores for users that could be used to obtain expertise levels. However, tags that users give to music items can provide an indication of their expertise degree.

Social tags express different music features such as genre, feelings, mood, instruments, periods of time or more subjective aspects. In general, social tagging information, provided by thousands of users of social networks, has given rise to a broad domain-specific body of knowledge that is called folksonomy [23]. This term, unlike taxonomy, represents a form of indexing information that does not follow a hierarchical organization or any other relationship and is performed by non-experts.

Social tagging allows the creation of a very rich user-driven description of musical items in multiple dimensions as well as a dynamic and not pre-established classification.

This kind of indexation obtained through social tags is much more appropriate for music than the one based on classical taxonomies since its categories, such as musical genres, have fuzzy and unstable boundaries, new types are regularly introduced and the existing ones change.

Although social tagging is usually used for music characterization, we propose to use it for characterizing users aiming at taking advantage of this information for improving user-based collaborative filtering methods. Our proposal involves the inference of the expertise degree of the user by means of the analysis of the tags they give to items.

To understand the idea in an intuitive way, let's consider tags like *rock* and *pop* that correspond to well-known and easily identifiable genres, thus they could be given by a user with a low level of expertise. However, assigning tags as *darkwave* or *trip hop*, associated with less identifiable or more specialized genres, requires a higher level of music expertise. Therefore, the expertise of users can be determined by analyzing their tags.

Based on the fact that the tags requiring more knowledge are less common than the others, we have resorted to tag frequency analysis to establish the degree of user expertise. We have used *tf–idf* (term frequency–inverse document frequency), a measure widely used for document retrieval and classification, to characterize users according to the quantity and relevance of the tags they provide to items. In this context, terms are replaced by tags and documents by users.

When *tf-idf* is used for classifying documents, the weights of terms that occur very frequently in the document set are decreased while the weights of terms that occur rarely are increased. When finding user expertise profiles from tags, *tf-idf* provides an indication of the tags' frequency but giving more relevance to tags that are less frequently used by many users. This metric has been used in some works to identify relevant tags for items (i.e. songs or artists), however, our aim is to identify relevant tags for users. Tags from expert users would have high values of *tf-idf* (high level of specialization) while tags from non-expert users would have low values of this metric. *Tf-idf* for a user $u \in U$ and a tag t is defined as follows:

$$tf\text{-}idf(U, u, t) = tf(u, t) \times idf(U, t) \qquad (1)$$

$$idf(U, t) = 1 + \log\left(\frac{|U|}{df(U, t)}\right) \qquad (2)$$

Where $tf(u, t)$ is the frequency of the tag t for the user u, $|U|$ is the total number of users and $df(U, t)$ is the number of users that have the tag t.

In our approach, it is necessary to compute $tf\text{-}idf(U, u, t)$ for each pair (u, t). Then, the degree of expertise of a user u is given by *tf-idf* average of all tags of u.

5 Incorporation of User Expertise to Collaborative Filtering

User expertise computed from social tags is used for improving the quality of the recommendations provided by user-based collaborative filtering methods.

These recommendations are predictions of items that a user could like based on the ratings given by other users with similar preferences. Let's consider a set of m users. $U = \{u_1, u_2, ..., u_m\}$ and a set of n items $I = \{i_1, i_2, ..., i_n\}$. Each user u_i have a list of k ratings that he has given to a set of items I_{ui}, where $I_{ui} \subseteq I$. In this context, a recommendation for the active user $u_a \in U$ involves a set of items $I_r \subset I$ that fulfill the condition $I_r \cap I_{ua} = \varnothing$, since only items not rated by him can be recommended. Ratings are stored in a $m \times n$ matrix called the rating matrix, where each element is the rating that a user u_i gives to an item i_j. Usually, this matrix has many empty elements since each user only rates a small percentage of available items.

Similarity between users is computed from the rating matrix. To do that, there are different distance based measures such as cosine, Chebyshev and Jaccard; or correlations coefficients such as Pearson, Kendall and Spearman. Regardless of the method used, the similarity between the active user u_a and another user u_i is denoted as $sim(u_a, u_i)$.

At this point, we introduce user expertise to increase the influence of expert users in the recommendations against that of non-experts. Since we assume that the degree of expertise of a user is given by his $tf\text{-}idf$ average, we use this value as a weight in the similarity measure. Then, a weighted similarity $w\text{-}sim(u_a, u_i)$ is computed as follows.

$$\overline{tf\text{-}idf}(u_i) = \sum_t \frac{tf\text{-}idf(u_i, t)}{N(u_i)} \tag{3}$$

where $N(u_i)$ is the number of tags given by user u_i.

$$w\text{-}sim(u_a, u_i) = sim(u_a, u_i) \times \overline{tf\text{-}idf}(u_i) \tag{4}$$

After having the weighted similarity, we obtain the list of the nearest neighbors, that is, the n users $U = \{u_1, u_2, ..., u_n\}$ most similar to the active user u_a, where u_1 is the closest user to u_a, u_2 the second and so on.

Once the nearest neighbors to u_a have been obtained, the prediction of the rating pr_{aj} that the active user would give to a certain item j is computed from the weighted sum of other users' ratings using the following equation:

$$pr_{aj} = \bar{r}_a + \frac{\sum_{i=1}^{n} w\text{-}sim(u_a, u_i)(r_{ij} - \bar{r}_i)}{\sum_{i=1}^{n} |w\text{-}sim(u_a, u_i)|} \tag{5}$$

where

$$\bar{r}_i = \frac{1}{|I_i|} \sum_{j \in I_i} r_{ij} \tag{6}$$

6   Implicit Ratings

Most of the music datasets lack of explicit rating information, so that it is usually estimated from the number of plays of the tracks, which is often the only available indication of user preferences. There have been proposed few methods for computing ratings from plays and most of them are based on simple frequency functions [8, 9]. However, these methods are not indicated in the context of artist recommendation where most of the artists have low number of plays and there are few highly played artists. As consequence, play frequencies have a clear power law distribution, also known as the "long tail" distribution. For that situation the method proposed by Pacula [24] is proved to be more suitable. Below we describe this approach, which is adopted in our work.

The play frequency for a given artist i and a user j is defined as follow:

$$Freq_{i,j} = \frac{p_{i,j}}{\sum_{i'} p_{i',j}} \tag{7}$$

Where $p_{i,j}$ is the number of times that a user j plays an artist i.

On the other hand, $Freq_k(j)$ denote the k-th most listened artist for user j. Then, a rating for an artist with rank k is computed as a linear function of the frequency percentile:

$$r_{i,j} = 4\left(1 - \sum_{k'=1}^{k-1} Freq_{k'}(j)\right) \tag{8}$$

Once the ratings are calculated, collaborative filtering methods can be applied in the way it is done for dataset containing explicit user preferences.

7 Validation of the Proposed Method

7.1 Dataset

The study has been carried out using a sample with 1000 users, 5604 tags and 11680 artists from the dataset Hetrec2011-lastfm [25]. This dataset contains social network-ing, tagging, and music artist listening information from last.fm online music system. It is composed by the following files:

- artists.dat: Contains information about artists listened and tagged by the users.
- tags.dat: Set of tags available in the dataset.
- user_artists.dat: Contains the artists listened by each user. It also provides a lis-tening count for each [user, artist] pair.
- user_taggedartists.dat and user_taggedartists-timestamps.dat: These files contain the tag assignments of artists provided by each particular user. They also contain the timestamps when the tag assignments were done.
- user_friends.dat: Contains the friend relations between users in the database.

7.2 Preprocessing and Data Analysis

The first step was to compute tf-idf for the tags of each user in the dataset. Then, ratings were obtained from the count of plays in order to obtain user preferences.

Aiming at comparing the information that can be representative of the user profiles, average and standard deviation of TF-IDF and rating were computed for each user. The distribution of number of tags, number of plays, TF-IDF average and rating average per user in the dataset is showed in Fig. 1. We can observe a long-tail distribution for most of the variables.

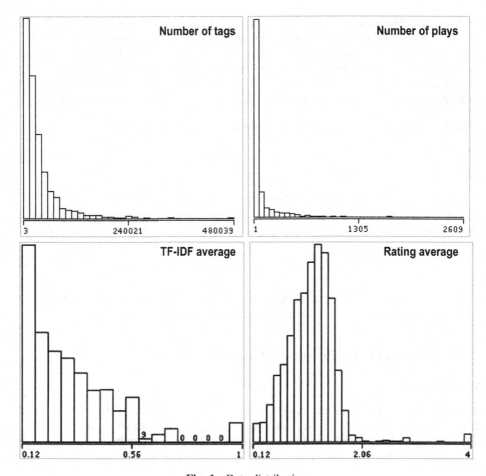

Fig. 1. Data distribution

7.3   Results

In order to validate the proposal, it is necessary to confirm that the introduction of user expertise in user-based CF by means of the weighted similarity *w-sim* improves the results against using CF with traditional similarity measures. The most used metrics are Pearson and cosine similarity coefficients, but there are other distance-based measures such as Chebyshev distance, Jaccard similarity, or Euclidean distance, all of which have been analysed in this comparative study. Euclidean and Chebyshev distances have been normalized to have values between 0 and 1 and the corresponding similarity metrics are obtained subtracting this value from 1.

Table 1 shows the values of NRMSE (Normalized Root Mean Square Error) obtained when using CF with some well-known and widely used similarity measures based on distances (first column), and when using CF with these similarity metrics weighted according to the expertise level of the user (second column). We can observe the significant improvement of the results for the last approach for all the tested measures. This important reduction in the error rate can also be visualized in Fig. 2.

Table 1. NRMSE for user-based CF and user-based CF with weighted similarity

	User-CF	Weighted user-CF
Cosine	0.2992	0.0254
Jaccard	0.3013	0.0168
Chebyshev	0.2527	0.0303
Euclidean	0.3622	0.2418

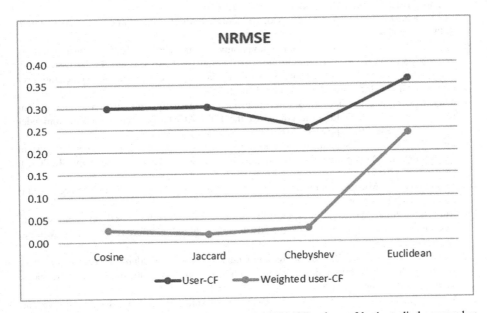

Fig. 2. Visualization of the difference between the NRMSE values of both studied approaches using common similarity metrics

8   Conclusions

Social tagging is usually used in the context of recommender system in content-based CF methods for finding similarities between items. However, in this work we take advantage of social tags for inducing the expertise degree of the users. This information is used for improving user-based CF methods by giving more relevance in the computation of the similarity between users to the opinions of more expert users.

We have conducted a study with a dataset containing information obtained from last.fm music system. The results prove that the introduction of a weight proportional to user expertise produces significantly better predictions regardless of the similarity metrics applied.

References

1. McAuley, J.J., Leskovec, J.: From amateurs to connoisseurs: modeling the evolution of user expertise through online reviews. In: Proceedings of the 22nd International Conference on World Wide Web, Rio de Janeiro, Brazil, pp. 897–908 (2013)
2. Breese, J.S., Heckerman, D., Kadie C.: Empirical analysis of predictive algorithms for collaborative filtering. In: Proceedings of the Fourteenth Conference on Uncertainty in Artificial Intelligence, Madison, pp. 43–52 (1998)
3. Sarwar, B., Karypis, G., Konstan, J., Riedl, J.: Item-based collaborative filtering recommendation algorithm. In: Proceedings of the Tenth International World Wide Web Conference, pp. 285–295 (2001)
4. Linden, G., Smith, B., York, J.: Amazon.com recommendations: item to item collaborative filtering. IEEE Internet Comput. **7**(1), 76–80 (2003)
5. Su, X., Khoshgoftaar, T.M.: A survey of collaborative filtering techniques. Adv. Artif. Intell. **2009**, 1–19 (2009)
6. Resnik, P.: Semantic similarity in a taxonomy: an information based measure and its application to problems of ambiguity in natural language. J. Artif. Intell. **11**, 94–130 (1999)
7. Chen, H.C., Chen, A.L.P.: A music recommendation system based on music and user grouping. Intell. Inf. Syst. **24**(2/3), 113–132 (2005)
8. Vargas, S., Castells, P.: Rank and relevance in novelty and diversity metrics for recommender systems. In: Proceedings of the Fifth ACM Conference on Recommender Systems, RecSys 2011, pp. 109–116. ACM, New York (2011)
9. Lee, K., Lee, K.: Escaping your comfort zone: a graph-based recommender system for finding novel recommendations among relevant items. Expert Syst. Appl. **42**(2015), 4851–4858 (2015)
10. Tzanetakis, G.: Musical genre classification of audio signals. IEEE Trans. Speech Audio Process. **10**(5), 293–302 (2002)
11. Kuo, F.F., Shan, M.K.: A personalized music filtering system based on melody style classification. In: Proceedings of the IEEE International Conference on Data Mining, pp. 649–652 (2002)
12. Cataltepe, Z., Altinel, B.: Music recommendation based on adaptive feature and user grouping. In: 22nd International Symposium on Computer and Information Sciences, Ankara, Turkey, pp. 1–6 (2007)
13. Yoshii, K., Goto, M., Komatani, K., Ogata, T., Okuno, H.G.: Hybrid collaborative and content-based music recommendation using probabilistic model with latent user preferences. In: Proceedings of the 7th International Conference on Music Information Retrieval, pp. 296–301 (2006)
14. Lu, C.C., Tseng, V.S.: A novel method for personalized music recommendation. Expert Syst. Appl. **36**, 10035–10044 (2009)
15. Yang, L., Qiu, M., Gottopati, S., Zhu, F., Jiang, J.: CQARank: jointly model topics and expertise in Community Question Answering. In: Proceedings of the 22nd ACM International Conference on Information and Knowledge Management, CIKM 2013, San Francisco, CA, USA, p. 108 (2013)
16. Yang, B., Manandhar, S.: Tag-based expert recommendation in Community Question Answering. In: 2014 IEEE/ACM International Conference on Advances in Social Networks Analysis and Mining (ASONAM 2014), pp. 960–963 (2014)

17. Zhou, G., Lai, S., Liu, K., Zhao, J.: Topic-sensitive probabilistic model for expert finding in question answer communities. In: Proceedings of the 21st ACM International Conference on Information and Knowledge Management, CIKM 2012, Maui, Hawaii, USA, pp. 1662–1666 (2012)
18. Martín-Vicente, M.I., Gil-Solla, A., Ramos-Cabrer, M., Blanco-Fernández, Y., López-Nores, M.: Semantic inference of user's reputation and expertise to improve collaborative recommendations. Expert Syst. Appl. **39**(2012), 8248–8258 (2012)
19. Hyung, Z., Lee, K., Lee, K.: Music recommendation using text analysis on song requests to radio stations. Expert Syst. Appl. **41**(5), 2608–2618 (2014)
20. Deng, S., Wang, D., Li, X., Xu, G.: Exploring user emotion in microblogs for music recommendation. Expert Syst. Appl. **42**(23), 9284–9293 (2015)
21. Hariri, N., Mobasher, B., Burke, R.: Context-aware music recommendation based on latent topic sequential patterns. In: Proceedings of the Sixth ACM Conference on Recommender Systems, Dublin, Ireland, pp. 131–138 (2012)
22. Su, J.H., Chang, W.Y., Tseng, V.S.: Personalized music recommendation by mining social media tags. Procedia Comput. Sci. **22**, 303–312 (2013)
23. Schedl, M., Sordo, M., Koenigstein, N., Weinsberg, U.: Mining user generated data for music information retrieval. In: Moens, M.F., Li, J., Chua T.S. (eds.) Mining User Generated Content, pp. 67–96, Chapman and Hall/CRC Press (2013)
24. Pacula, M.: A matrix factorization algorithm for music recommendation using implicit user feedback. http://www.mpacula.com/publications/lastfm.pdf. Accessed 6 Mar 2018
25. Cantador, I., Brusilovsky, P., Kuflik, T.: 2nd workshop on information heterogeneity and fusion in recommender systems (HETREC 2011). In: Proceedings of the 5th ACM Conference on Recommender Systems, RecSys 2011. ACM, New York (2011)

Learning Logical Definitions of n-Ary Relations in Graph Databases

Furkan Goz and Alev Mutlu$^{(\boxtimes)}$

Department of Computer Engineering, Kocaeli University, İzmit, Kocaeli, Turkey
{furkan.goz,alev.mutlu}@kocaeli.edu.tr

Abstract. Given a set of facts and related background knowledge, it has always been a challenging task to learn theories that define the facts in terms of background knowledge. In this study, we focus on graph databases and propose a method to learn definitions of n-ary relations stored in such mediums. The proposed method distinguishes from state-of-the-art methods as it employs hypergraphs to represent relational data and follows substructure matching approach to discover concept descriptors. Moreover, the proposed method provides mechanisms to handle inexact substructure matching, incorporate numerical attributes into concept discovery process, avoid target instance ordering problem and concept descriptors suppress each other. Experiments conducted on two benchmark biochemical datasets show that the proposed method is capable of inducing concept descriptors that cover all the target instances and are similar to those induced by state-of-the-art methods.

Keywords: Concept discovery · n-ary relation · Hypergraph
Graph database · Neo4j

1 Introduction

Given a set of facts and related background knowledge, concept discovery is concerned with inducing logical definitions of the facts in terms of background knowledge [1]. The problem has extensively been studied from Inductive Logic Programming (ILP) perspective where data is represented within first order logic framework and logic operators are used to induce concept descriptors [2]. The problem has also been addressed within graph mining framework where relational data is represented as graphs and graph algorithms are used to mine concept descriptors [3,4].

In this study, we propose a method to find definitions of n-ary relations stored in graph databases. The proposed method inputs a graph database and a number of relation names to guide the search. The proposed method queries the graph database to find frequently appearing substructures that involve the relations names provided. The user guided search enables discovery of concept descriptors of particular interest as well as limiting the search space.

© Springer International Publishing AG, part of Springer Nature 2018
F. J. de Cos Juez et al. (Eds.): HAIS 2018, LNAI 10870, pp. 50–61, 2018.
https://doi.org/10.1007/978-3-319-92639-1_5

The contributions of this study are two folds. The first contribution is about the representation of data: in this study we represent the data using hypergraphs rather than ordinary graphs. The second contribution is related to the concept descriptor discovery process: in this study we follow a substructure matching approach instead of pathfinding. Moreover, the proposed method performs inexact subgraph matching, handles numeric attributes, avoids target instance ordering problem. The proposed method also avoids certain concept descriptors to suppress discovery of other concept descriptors.

To evaluate the performance of the proposed method, experiments are conducted on two biochemical datasets, namely Predictive Toxicology Evaluation (PTE) and Mutagenesis. The experimental results show that the proposed method is capable of inducing concept descriptors reported in literature and achieves 100% coverage.

The rest of the paper is organized as follows. Section 2 presents the related work and highlights differences of the proposed method from state-of-the-art methods. Section 3 presents the data representation model, advantages of hypergraph representation over graph representation, and introduces the proposed method. Section 4 presents the experimental findings and the last section concludes the paper.

2 Background

Given a set of positive and negative instances that belong a *target relation* and related facts, called *background knowledge*, concept discovery is concerned with inducing relational definitions of the target relation in terms of background knowledge relations and possibly the relation itself, such that the induced concept descriptors explain all of the positive target instances and none of the negative target instances [1]. As an example, given a kinship dataset and *father* relation as the target relation, a typical concept discovery system would induce concept descriptors such as *father(A,B):-mother(C,A), wife(C,D), daughter(B,C)*. The concept discovery problem is formulated in Eq. 1, where B is background knowledge, $E = E^+ \cup E^-$ is set of target instances, and H is a concept descriptor.

$$
\begin{aligned}
&\text{Prior Satisfiability. } B \wedge E^- \not\models \Box \\
&\text{Posterior Satisfiability. } B \wedge H \wedge E^- \not\models \Box \\
&\text{Prior Necessity. } B \not\models E^+ \\
&\text{Posterior Sufficiency. } B \wedge H \models E^+
\end{aligned}
\tag{1}
$$

The problem has primarily been investigated by ILP research where data is represented within first order logic and logical operators are used to induce concept descriptors. Although ILP-based concept discovery systems have applications in several domains [5], they are reported to suffer from large search space [6] and the local plateau problem [4].

More recently, the problem is attacked from graph perspective and methods for concept discovery based on pathfinding [4,7,8] and substructure discovery [3,9] have been proposed. Pathfinding-based approaches assume that concept descriptors should be represented by fixed length paths that connect some arguments of the target relation. Substructure-based approaches, on the other hand, assume that frequently appearing substructures that involve the target relation should be concept descriptors. Pathfinding-based approaches are more suitable for learning descriptive concept descriptors while substructure discovery-based approaches can learn both descriptive and predictive concept descriptors. Although graph-based approaches are promising in concept discovery they suffer from complexity of graph algorithms.

With the increasing popularity of graph databases, studies dealing with concept discovery in graph databases have also been published. Our previous works [10,11] focused on concept discovery in graph databases for binary relations and followed pathfinding-based approach. [12], another previous work of our group, proposed a pathfinding-based approach for learning concept descriptors of n-ary relations where concept descriptors of length l are joined to generate candidate concept descriptors of length *(l + 1)* in Apriori manner.

The method proposed in this study distinguishes from state-of-the-art methods by two means: (a) data representation and (b) concept descriptor inference mechanism. In this study we represent data using hypergraphs rather than graphs. Hypergraphs are generalization of ordinary graphs and are more powerful in representing n-ary relations [13]. Although not all graph databases provide direct implementation of hypergraphs, graph rewriting techniques are available for transforming hypergraphs into graphs [14]. Concept descriptor inference is based on user-guided substructure discovery. Although user guidance may require domain expertise, it enables the discovery of concept descriptors of particular interest and limits the search space.

3   The Proposed Method

3.1   Data Representation

In this study, directed, labeled hypergraphs are used to represent relational data. Hypergraphs, $G = (V, E)$, are generalization of graphs where V is a set of nodes and edges connect any number of nodes, $E \in 2^V$. Relation between relational data model and hypergraphs are discussed in [14,15]. We use the sample dataset provided in Table 1 to illustrate our data model.

Data presented in Table 1 is a subset of Mutagenesis dataset and the target instance indicates that chemical *d1* is mutagenically active. The background knowledge *muta_atom(d1,d1_3,c,22,-0.117)* states that *d1* has an atom namely *d1_3* which is described using three features: element *c*, type *22*, and charge -0.117. The relation *muta_bond(d1_3,d1_4,7)* indicates that there is bond between *d1_3* and *d1_4* of type *7*. The predicates *indA(d1,1)* and *indV(d1,0)* indicate *d1* has values *1* and *0* for properties, respectively, *indA* and *indV*. The remaining predicates can be interpreted in a similar way.

Table 1. Sample dataset

Target instance	Background knowledge
Mutagenic(d1, active)	muta_atom(d1,d1_3,h,22,−0.117)
	muta_atom(d1,d1_4,c,195,−0.087)
	muta_bond(d1,d1_3,d1_4,7)
	muta_bond(d1,d1_4,d1_5,7)
	indA(d1,1)
	indV(d1,0)

To represent the data as a graph, we represent each distinct argument value as a vertex and place an edge between vertices that are related and label the edge after the relation/property name. Ordinary graphs are suitable to represent relations such as $indA(d1,1)$: two vertices representing $d1$ and 1 connected by an edge labeled $indA$, Fig. 1(a) illustrates this mapping. $muta\_atom(d1,d1\_4,c,195, -0.087)$ can be represented with 5 vertices labeled with c, 195, and -0.087 connected to a node labeled $d1\_1$ which is connected to a vertex labeled $d1$, Fig. 1(b) illustrates this representation.

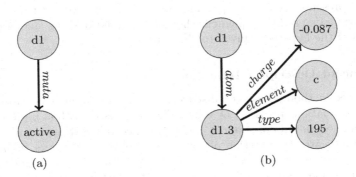

(a)   (b)

Fig. 1. Graphs corresponding to some relations

However, it is not possible to represent relations $muta\_bond(d1\_3,d1\_4,7)$ and $muta\_bond(d1\_4,d1\_5,7)$ in a similar fashion. If we do represent these relations via binary relations as indicated in Fig. 2(a), the semantics can not be retrieved properly. From Fig. 2 one may infer there is a bond between $d1\_3$ and $d1\_5$ with property 7 which is not the case. In Fig. 2(b) we illustrate hypergraph representation of these two relations which captures all semantics without any ambiguity.

As graph databases may not directly support hypergraphs, such structures need to be transformed into graph. In literature there are several transformations of hypergraphs into graphs. One such transformation proposes to represent

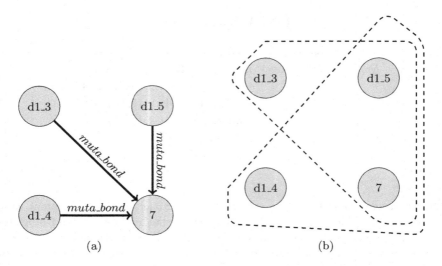

Fig. 2. Graph vs. Hypergraph representation

a hyperedge via a new vertex and creating pairwise relationships among end-nodes of the hyperedge [16]. In this study we follow this mechanism and Fig. 3 partially represents the dataset provided in Table 1. The *bond* vertex in Fig. 3 is an auxiliary node used to implement the pairwise relationships among the three end-nodes of the hyperedge given in Fig. 2(b).

3.2   Concept Descriptor Induction

The proposed method is based on discovering frequently appearing substructures that include relations provided by a user. The proposed method inputs a graph, relations names, minimum support and confidence values. Support of a concept descriptor indicates the fraction of the number of positive target instances the induced concept descriptor holds for over the total number of target instances. Confidence of a concept descriptor indicates the fraction of the number of positive target instances the induced concept descriptor holds for over the total number of positive target instances. The ultimate goal of the proposed method is to discover as much specific concept descriptors as possible that satisfy the minimum support and minimum confidence values and describe every target instance.

Algorithm 1 outlines the proposed method. The proposed method takes a target instance and retrieves the substructures that are connected to it via the edges named after the relation names provided. Support and confidence values of each such retrieved substructure are calculated, if the substructure is not evaluated before. If the support and confidence values are above the thresholds, it is added into a solution set, i.e. *S*. Once all target instances are handled, concept descriptors in solution set are simplified and the simplified versions of the concept descriptors constitute the final solution set.

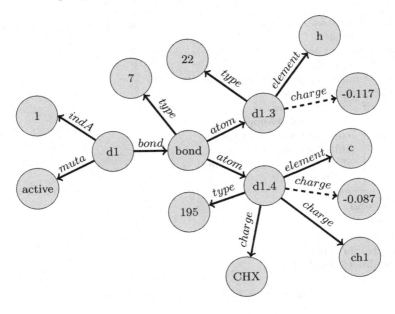

Fig. 3. Graph for sample *mutagenesis* dataset

In graph-based concept discovery systems, a substructure is generally associated with more than one target instance. Hence the same substructure will be discovered multiple times. In order to avoid evaluation of the same substructures more than once, we store each discovered substructure in a hash table, *H* in Algorithm 1, and evaluate those substructures that are not in the hash table only. If a substructure is discovered for the first time, its support and confidence values are calculated and inserted into hash table along with the substructure itself.

Once all concept descriptors are discovered a concept descriptor simplification processes is executed. In this step if each substructure that corresponds to a relation is removed if it does not enhance coverage of the concept descriptor.

In concept discovery systems there are some issues that need special attention. These are handling numeric attributes, inexact subgraph matching, avoiding target instance ordering problem, and avoiding one discovery of a concept descriptor to suppress discovery of another concept descriptor. In the subsection below we explain these problems in more detail and explain our approach to handle them.

Handling Numerical Values: In order for a value to be present as a constant in a concept descriptor it should appear at least $ms \times \#target\_instance$ times in the dataset. As some numeric values such as *charge* in *muta_atom* relation do not appear that many times, they can not be represented in concept descriptors as they are. For this reason we employ equal width discretization where width

Data: D: database, R: set of relations, ms: minimum support, mc: minimum
          confidence
Result: S: a set of concept descriptors
foreach t *in* D **do**
   | C_c = buildCandidateConceptDes(t, R);
   | **foreach** c *in* C_c **do**
   | | **if** c *not in* H **then**
   | | | sp = calculateSupport(c);
   | | | cn= calculateConfidence(c);
   | | | H.insert(c, sp, cn);
   | | | **if** $sp \geq ms \wedge cn \geq mc$ **then**
   | | | | S.insert(c);
   | | | **end**
   | | **end**
   | **end**
end
foreach c *in* S **do**
   | Sol.insert(simplify(c));
end

Algorithm 1: The proposed method

of an interval is determined by $ms \times \#target\_instance$. Hence the graph is
modified by replacing continuous values with their corresponding discrete values
and adjusting edges accordingly. In Fig. 3 the node labeled with *ch1* is a such
node, and nodes with dashed edges and labeled with *–0.117* and *–0,087*, are
indeed not used in concept discovery process.

Inexact Subgraph Matching: Suppose that more than $ms \times \#target\_$
instance target instances agree on (n − 1) nodes that represent a relation p/n.
In such a case this substructure can not participate in a concept descriptor as
a substructure corresponding a relation should fully be present in the concept
descriptor. In order to handle such situations for each distinct attribute value
we create and add a node that represents a *don't care* value and adjust edges
respectively. In Fig. 3 node labeled with *CHX* is such a node. By this way we
allow the proposed method to handle inexact subgraph matchings.

Avoiding Target Instance Ordering Problem: Suppose that concept
descriptor c_k is discovered while target instance t_i is handled and c_k also covers
target instance t_j - which is not examined yet. In concept discovery systems
that follow covering principle [17], target instance t_j will not be handled in subsequent iterations and any concept descriptor due to t_j will not be discovered.
In order to overcome this problem, in the proposed method we do not employ
covering mechanisms but examine every target instance even if it is covered by
some previously discovered concept descriptors.

Avoiding Suppression of Concept Descriptors: In predicate logic g subsumes f if there is substitution θ such that $g\theta = f$. As an example $p(X, Y, Z)$ subsumes $p(a, Y, b)$. Similarly, a substructure s_1 subsumes substructure s_2 if some constant nodes of s_2 are *don't care* nodes in s_1. In traditional concept discovery systems, s_2 will not be considered as a valid concept descriptor as s_1 will probably have higher support and confidence value even though s_2 provides valuable information. Such a problem is also valid for graph-based concept discovery systems such as [18] that aim to find a substructure that best compresses the original graph. Any subgraph that does not compress the original graph best but is good enough will be missed. In order to overcome this situation, we add substructures that satisfy the minimum support and minimum confidence criteria to the solution set even if they are subsumed by some other concept descriptors.

4 Experiments

4.1 Datasets and Experimental Settings

To evaluate the performance of the proposed method, we conducted experiments on two biochemical datasets: Mutagenesis and extended version of Predictive Toxicology Evaluation (PTE) [19] namely PTE-5 [20]. These are benchmark datasets for concept discovery [21,22], and the problem is to predict class labels of drugs as carcinogenic and mutagenic, respectively. Table 2 lists the properties of the datasets and the corresponding graphs. The first column indicates the dataset name, the second column indicates the number of relations in the dataset, the third column indicates the number of instances that belong to the target relation and the fourth column indicates the number of instances provided as background knowledge. The last two columns indicate the number of the vertices and the number of edges of graph representation of the datasets.

Table 2. Properties of datasets and the corresponding graphs

Dataset	Dataset properties			Graph properties	
	# Pred.	# T.I.	# B.I.	# Vertices	# Edges
PTE	37	298	28969	19363	27132
Mutagenesis	8	188	13123	12650	35776

Hence, the graph representation of the dataset stores two vertices for each distinct argument value and the auxiliary vertices required for hypergraph representation. In the experiments, minimum support and minimum confidence values are set 0.1 and 0.7 [20].

Neo4j [23] graph database engine is used to store the data. Neo4j's Java API is used to retrieve the substructures and query the database to calculate support and confidence of the candidate concept descriptors.

4.2   Experimental Results

Due to space limitation in Fig. 4 we provide graph representation of only one of the discovered concept descriptors and list the remaining ones in Table 3.

An unbounded capital letter in a concept descriptor represents a *don't care* value while a bounded capital letter enforces integrity among relations. A small letter represents a constant value. As an example *atom(A,B,h,3,ch14)* can be interpreted as any drug connected to an atom with element *h*, type *3* and charge *ch14* is mutagenic. Similarly, *atom(A,B,h,3,ch16)*, *atom(A,C,c,22,ch8)*,

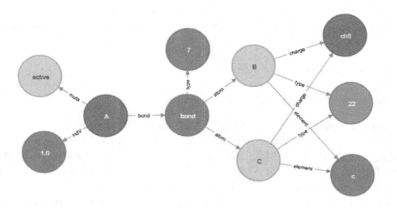

Fig. 4. Graph representation of concept descriptor *atom(A,B,c,22,ch8)*, *atom(A,C,c, 22,ch8)*, *bond(A,C,B,7)*, *indV(A,1.0)*

Table 3. Concept descriptors discovered for the Mutagenesis dataset

Concept descriptor	Support	Confidence
atom(A,B,h,3,ch14)	0.16	0.77
atom(A,B,c,22,ch8)	0.83	0.74
atom(A,B,h,3,ch15), lumo(A,lumo2)	0.1	1.0
atom(A,B,h,3,ch15), indA(A,0.0), indV(A,1.0)	0.42	0.98
atom(A,B,h,3,ch15), atom(A,D,c,22,ch8), bond(A,D,B,1), indA(A,0.0), indV(A,1.0)	0.41	1.0
atom(A,B,c,22,ch8), atom(A,C,c,22,ch8), bond(A,C,B,7), indV(A,1.0)	0.68	1.0
atom(A,B,o,40,ch4), atom(A,C,n,38,ch22), bond(A,C,B,2)	0.69	0.70
atom(A,B,h,3,ch16), atom(A,C,c,22,ch8), bond(A,C,B,1)	0.35	0.83
atom(A,B,c,29,ch11), atom(A,C,c,29,ch11), bond(A,C,B,1), indA(A,0.0), indV(A,1.0)	0.16	0.95
atom(A,B,o,40,ch4), atom(A,C,n,38,ch23), bond(A,C,B,2), indA(A,0.0), indV(A,1.0)	0.13	1.0
atom(A,B,element,atomType,ch4), bond(A,D,E,2), bond(A,B,E,1), ($-0,398 \leq ch4 \leq 0,376$)	0.87	0.74

Table 4. Concept descriptors discovered for the PTE-5 dataset

Concept descriptor	Support	Confidence
atom(A,B,h,C,D), ind(A,di10,1), has_property(A,cytogen_ca,E)	0.14	0.70
atom(A,B,c,C,D), ind(A,E,1), has_property(A,salmonella,F)	0.11	0.83
atom(A,B,C,D,ch12), ind(A,E,1), has_property(A,cytogen_sce,F), $(-0.159 \leq ch12 \leq -0.126)$	0.10	0.87
atom(A,B,C,22,D), ind(A,di10,1), has_property(A,salmonella,E)	0.13	0.74
atom(A,B,C,D,ch22), atom(A,E,F,32,G), $(0.191 \leq ch22 \leq 0.227)$	0.245	0.702

bond(A,C,B,1) can be read as a drug connected to an atom with element *h*, type *3* and charge *ch16* and connected to another atom with element *c*, type *22* and charge *ch8* and there is bond between these two atoms of type *1* is mutagenic.

When compared to ILP-based concept discovery system Progol [24] the proposed method is able to find similar concept descriptors. The proposed method discovered similar concept descriptors reported in a recent ILP-based concept discovery system CRIS [20]. The proposed method is also able to find the concept descriptor that is reported the best by [18].

In Table 4, we list some of the concept descriptors discovered for the PTE-5 dataset as well as the support and confidence values of the concept descriptors. The obtained results are compatible with those reported in ILP-based studies such as [20,24] and graph-based approaches such as [25,26].

Running time is another important criteria in comparing performance of concept discovery systems. CRIS [20] is reported to run about 5 h for PTE-5 and around 3 h and 40 min for Mutagenesis dataset. The proposed method terminates around 10 min for each datasets. When compared to SUBDUECL and gSpan, the proposed method has longer running time.

5 Conclusion

In this study we introduced a method to learn logical definitions of n-ary relations stored in graph databases. The proposed method represents relational data within hypergraph framework, incorporates numeric values into concept discovery process after a discretization step, can perform inexact substructure matching, overcome target instance ordering problem, and avoid concept discovery suppression problem. Experimental results show that the proposed method is capable of discovering concept descriptors that are reported in the literature in much shorter time when compared to state-of-the-art ILP-based concept discovery system, however it is slower when compared to other graph-based concept discovery systems.

Future research directions include analysis of impact of indexing on the running time performance.

References

1. Dzeroski, S.: Multi-relational data mining: an introduction. SIGKDD Explor. **5**(1), 1–16 (2003)
2. Muggleton, S.: Inductive logic programming. New Gener. Comput. **8**(4), 295–318 (1991)
3. Yan, X., Han, J.: gSpan: graph-based substructure pattern mining. In: Proceedings of the 2002 IEEE International Conference on Data Mining (ICDM 2002), 9–12 December 2002, Maebashi City, Japan, pp. 721–724 (2002)
4. Richards, B.L., Mooney, R.J.: Learning relations by pathfinding. In: Proceedings of the 10th National Conference on Artificial Intelligence, San Jose, 12–16 July 1992, pp. 50–55 (1992)
5. De Raedt, L.: Inductive logic programming. In: Encyclopedia of Machine Learning, pp. 529–537. Springer (2011)
6. Zeng, Q., Patel, J.M., Page, D.: QuickFOIL: scalable inductive logic programming. Proc. VLDB Endow. **8**(3), 197–208 (2014)
7. Gao, Z., Zhang, Z., Huang, Z.: Learning relations by path finding and simultaneous covering. In: WRI World Congress on Computer Science and Information Engineering, CSIE 2009, 31 March–2 April 2009, Los Angeles, vol. 7, pp. 539–543 (2009)
8. Gao, Z., Zhang, Z., Huang, Z.: Extensions to the relational paths based learning approach RPBL. In: ACIIDS, pp. 214–219. IEEE Computer Society (2009)
9. Gonzalez, J.A., Holder, L.B., Cook, D.J.: Graph based concept learning. AAAI/IAAI 1072 (2000)
10. Goz, F., Mutlu, A.: Concept discovery in graph databases. In: Martínez de Pisón, F.J., Urraca, R., Quintián, H., Corchado, E. (eds.) HAIS 2017. LNCS (LNAI), vol. 10334, pp. 63–74. Springer, Cham (2017). https://doi.org/10.1007/978-3-319-59650-1_6
11. Abay, N.C., Mutlu, A., Karagoz, P.: A path-finding based method for concept discovery in graphs. In: 6th International Conference on Information, Intelligence, Systems and Applications, IISA 2015, Corfu, 6–8 July 2015, pp. 1–6 (2015)
12. Abay, N.C., Mutlu, A., Karagoz, P.: A graph-based concept discovery method for n-ary relations. In: Madria, S., Hara, T. (eds.) DaWaK 2015. LNCS, vol. 9263, pp. 391–402. Springer, Cham (2015). https://doi.org/10.1007/978-3-319-22729-0_30
13. Li, L., Li, T.: News recommendation via hypergraph learning: encapsulation of user behavior and news content. In: Proceedings of the Sixth ACM International Conference on Web Search and Data Mining, pp. 305–314. ACM (2013)
14. Blockeel, H., Witsenburg, T., Kok, J.: Graphs, hypergraphs and inductive logic programming. In: Proceedings of the 5th International Workshop on Mining and Learning with Graphs, pp. 93–96 (2007)
15. Gallo, G., Longo, G., Pallottino, S., Nguyen, S.: Directed hypergraphs and applications. Discrete Appl. Math. **42**(2), 177–201 (1993)
16. Zien, J.Y., Schlag, M.D., Chan, P.K.: Multilevel spectral hypergraph partitioning with arbitrary vertex sizes. IEEE Trans. Comput. Aided Des. Integr. Circuits Syst. **18**(9), 1389–1399 (1999)
17. Muggleton, S.: Inverse entailment and Progol. New Gener. Comput. **13**(3–4), 245–286 (1995)
18. Ketkar, N.S., Holder, L.B., Cook, D.J.: Subdue: compression-based frequent pattern discovery in graph data. In: Proceedings of the 1st International Workshop on Open Source Data Mining: Frequent Pattern Mining Implementations, pp. 71–76. ACM (2005)

19. Srinivasan, A., King, R.D., Muggleton, S.H., Sternberg, M.J.: The predictive toxicology evaluation challenge. In: IJCAI, vol. 1, pp. 4–9. Citeseer (1997)
20. Kavurucu, Y., Senkul, P., Toroslu, I.H.: Concept discovery on relational databases: new techniques for search space pruning and rule quality improvement. Knowl. Based Syst. **23**(8), 743–756 (2010)
21. Srinivasan, A., King, R.D., Bristol, D.W.: An assessment of submissions made to the predictive toxicology evaluation challenge (1999)
22. Lodhi, H., Muggleton, S.: Is mutagenesis still challenging. In: Proceedings of the 15th International Conference on Inductive Logic Programming, ILP, pp. 35–40. Citeseer (2005)
23. Neo4j: The Neo4j Graph Platform. https://neo4j.com. Accessed 5 Feb 2018
24. Srinivasan, A., Muggleton, S., King, R.D., Sternberg, M.J.: Mutagenesis: ILP experiments in a non-determinate biological domain. In: Proceedings of the 4th International Workshop on Inductive Logic Programming, vol. 237, pp. 217–232. Citeseer (1994)
25. Gonzalez, J., Holder, L., Cook, D.J.: Application of graph-based concept learning to the predictive toxicology domain. In: Proceedings of the Predictive Toxicology Challenge Workshop (2001)
26. Chittimoori, R.N., Holder, L.B., Cook, D.J.: Applying the subdue substructure discovery system to the chemical toxicity domain. In: FLAIRS Conference, pp. 90–94 (1999)

GAparsimony: An R Package
for Searching Parsimonious Models
by Combining Hyperparameter
Optimization and Feature Selection

F. J. Martinez-de-Pison[(✉)], R. Gonzalez-Sendino, J. Ferreiro, E. Fraile,
and A. Pernia-Espinoza

EDMANS Group, University of La Rioja, Logroño, Spain
fjmartin@unirioja.es, edmans@dim.unirioja.es
http://www.mineriadatos.com

Abstract. Nowadays, there is an increasing interest in automating KDD processes. Thanks to the increasing power and costs reduction of computation devices, the search of best features and model parameters can be solved with different meta-heuristics. Thus, researchers can be focused in other important tasks like data wrangling or feature engineering. In this contribution, `GAparsimony` R package is presented. This library implements GA-PARSIMONY methodology that has been published in previous journals and HAIS conferences. The objective of this paper is to show how to use `GAparsimony` for searching accurate parsimonious models by combining feature selection, hyperparameter optimization, and parsimonious model search. Therefore, this paper covers the cautions and considerations required for finding a robust parsimonious model by using this package and with a regression example that can be easily adapted for another problem, database or algorithm.

Keywords: GA-PARSIMONY · Hyperparameter optimization
Feature selection · Parsimonious model · Genetic algorithms

1 Introduction

In the last years, companies have been demanding new methodologies to automatize tedious machine learning tasks such as hyperparameter optimization (HO) or feature selection (FS). Therefore, the effort can be focused in other important processes as feature engineering or data munging that are harder to automatize [6]. Besides, these tools can be useful for many scientific or engineers that are not expert in machine learning but who need to easily obtain useful and robust models.

To facilitate these tasks, new libraries are emerging to perform HO when the number of model parameters is high. For example, Bayesian Optimization (BO) is implemented in `Auto-WEKA` [12] for *Weka* suite, in *Python* `Hyperopt` [2] and

© Springer International Publishing AG, part of Springer Nature 2018
F. J. de Cos Juez et al. (Eds.): HAIS 2018, LNAI 10870, pp. 62–73, 2018.
https://doi.org/10.1007/978-3-319-92639-1_6

`bayes_opt`, or `rBayesianOptimization` and `mlr` [3] in *R*. However, new tools try also to optimize the model structure with Soft Computing (SC) strategies. For example, `TPOT` [9] in *python* automatically optimizes machine learning pipelines with genetic programming (GP), *SUMO-Toolbox* [5] in *MATLAB* uses different plug-ins for optimizing each KDD stage or *DEvol* uses GP to automatically modify the layers structure of a deep neural network.

In this context, we present `GAparsimony` [7], a public R package for searching robust and parsimonious models. This library implements previously published GA-PARSIMONY methodology [10,13] that uses genetic algorithms (GA) to search robust and parsimonious models with FS, HO. and parsimonious model selection (PMS). The objective of this paper is to describe the use of this new package for searching parsimonious models. Therefore, paper covers an explanation, with a regression example, of how to create the fitness function which measure the errors and model complexity, how to correctly initialize the settings of the GA optimization and how to tune the *rerank_error* parameter for searching parsimonious solutions, and so on.

The rest of the paper is organized as follows: Sect. 2 presents a brief description of `GAparsimony` R Package. Section 3 describes the use of the new package to obtain a robust Artificial Neuronal Network (ANN) model for the *Boston* UCI dataset. Finally, Sect. 4 presents the conclusions and suggestions for further works.

2 The `GAparsimony` R Package

2.1 Package Description

`GAparsimony` [7] is a *R* package for implementing GA-PARSIMONY methodology [10,13] to search accurate parsimonious models (PM) by combining feature selection (FS), hyperparameter optimization (HO), and parsimonious model selection (PMS).

This R package, that has been written in S4, provides a flexible tools for automatically searching parsimonious models within a pre-established error margin. It can be run sequentially or in parallel, using an explicit master-slave parallelization or a grid of computers.

`GAparsimony` has successfully been used with Random Forest (RF), Artificial Neural Networks (ANNs), Extreme Gradient Boosting Machines (XGBoost) or Support Vector Regression (SVR) in many fields, such as solar radiation estimation [1], mechanical design [4], industrial processes [11], or hotel booking forecasting [14].

The released version can be installed directly from CRAN repository with:

```
install.packages("GAparsimony")
```

or the development version from GitHub with `devtools` package:

```
devtools::install_github("jpison/GAparsimony")
```

2.2   The Search of Parsimonious Solutions

GAparsimony [10,13] uses a GA-based optimization method with a similar flowchart to other classical GA methods. The main difference is that method selects best individuals in two separated cost and complexity evaluations. Cost (J) measures the model's error or accuracy and is usually obtained with cross-validation, hold-out or bootstrapping. On the other hand, model complexity depends on the model structure and it usually defines the model "flatness" which is related to its robustness. For example, the sum of the squared coefficients in *ridge regression* or the sum of the squared weights in *ANNs with weight decay*, are two popular complexity metrics in regularization methods. Also, other metrics as *Vapnik-Chervonenkis* (VC) dimension, *degrees of freedom* (GDF) [15], the number of selected input features, N_{SF}, or a combination of them, can be used.

Table 1 shows, with a little example how GAparsimony works in the selection process of four individuals. In the first step (left part of the table), models are sorted by their J. However, in a second step (right part of the table), individuals of the three top positions are rearranged by their complexity because the absolute difference between their J are lower than a predefined error margin of 0.01. As a consequence, the best parsimonious models, with lower complexity, are promoted to be elitist in the next GA population.

Table 1. First step: individuals sorted by their J. Second step: individuals are rearranged by *Complexity* if their absolute difference of J is < 0.01 (*ReRank* = 0.010).

First step			Second step		
Position	J	*Complexity*	Position	J	*Complexity*
1st	0.455	120	1st	0.461	80
2nd	0.460	100	2nd	0.460	100
3rd	0.461	80	3rd	0.455	120
4th	0.480	75	4th	0.480	75

2.3   Objective Function and GA Methods

In GAparsimony, users have to write their own fitness function to evaluate each i individual which has to be defined with a chromosome λ_g^i:

$$\lambda_g^i = [Param, \ Q] \tag{1}$$

where *Param* corresponds with the model's parameters and Q is a vector with 0 and 1s values for selecting the input features. Fitness function must return J and *Complexity* for each i individual in order to be evaluated.

By defect, a non-linear selection method based on the rank is used [8]. Crossover function uses *heuristic blending* [8] with $\alpha = 0.1$ for *Param*, and *random swapping* for Q. Finally, mutation procedure also treats *Param* and Q in a separated way with two different thresholds: *pmutation*, the percentage of parameters to be mutated, and $feat\_muth\_thres$, the probability of a value from Q to be changed. However other selection, crossover or mutation methods can be provided by the user.

In order to start with an homogeneously distributed first population, the package uses a *Random Latin Hypercube Sampling (LHS)*. However, other configurations can be selected such as *geneticLHS*, *improvedLHS*, *maxminLHS*, *optimumLHS*, or *random*.

Next section presents a deeper explanation of how to configure and use GAparsimony in a regression example.

3 Example: Searching a Parsimonious Regression Model for *Boston* database

This example shows how to obtain, for the *Boston* dataset, a robust parsimony model with GAparsimony and caret R packages. For this purpose, a artificial neuronal network (ANN) algorithm is employed. The main objective is to seek, with GAparsimony, a robust and parsimony ANN model by using FS, HO and PMS.

All experiments were implemented in dual 28-core servers from *Beronia* cluster of the Universidad de La Rioja.

3.1 Data Preprocessing

In order to check the generalization capability of each model, we use *createDataPartition()* command to split the database in a 90% for training/validation and the other 10% for testing. The test database is only used for checking the model generalization capability. Training database is composed of 13 input features and 458 rows, and test database with 48 instances.

```
library(MASS)
library(caret)
library(GAparsimony)
library(data.table)
library(nnet)

# Preprocess data
set.seed(1234)
trainIndex <- createDataPartition(Boston[,"medv"], p=0.90, list=FALSE)
data_train <- Boston[trainIndex,]
label_train <- data_train[,ncol(data_train)]
data_test <- Boston[-trainIndex,]
label_test <- data_test[,ncol(data_test)]

# Z-score
mean_train <- apply(data_train,2,mean)
sd_train <- apply(data_train,2,sd)
data_train <- data.frame(scale(data_train, center = mean_train, scale = sd_train))
data_test <- data.frame(scale(data_test, center = mean_train, scale = sd_train))
# Add a little noise to avoid sd=0 in columns
data_train <- data_train+ matrix(runif(prod(dim(data_train)))/1000,
    nrow = nrow(data_train), ncol=ncol(data_train))

# Restore original target
data_train[,ncol(data_train)] <- label_train
data_test[,ncol(data_test)] <- label_test
print(dim(data_train))
print(dim(data_test))
## [1] 458  14
## [1] 48  14
```

3.2   Fitness Function Description

Although **GAparsimony** usually obtains useful solutions, it is highly recommended to execute it repeatedly with different random seeds. The objective is to ensure a correct estimation of the best model parameters and more important features. For this purpose, a reliable validation process will be necessary. For example, with small databases, n-repeated k-fold cross validation it is usually a good choice.

The following code shows the R function *fitness_NNET()* that has been designed to evaluate with a 25-repeated 10-fold cross validation each ANN. Selected function *nnet()* uses *Broyden-Fletcher-Goldfarb-Shanno* (BFGS) for the weight optimization process with *decay* as regularization term.

```r
# --------------------------------------------------
# Function to evaluate each ANN individual
# --------------------------------------------------
fitness_NNET <- function(chromosome, ...)
{
    # Extract parameters and select features from chromosome
    # --------------------------------------------------
    # First two values in chromosome are 'size' &
    # 'decay' of 'nnet' method
    tuneGrid <- expand.grid(size=round(chromosome[1]), decay=chromosome[2])

    # Next values of chromosome are the selected features
    # (Selected if > 0.50)
    selec_feat <- chromosome[3:length(chromosome)]>0.50

    # Return -Inf if there is not selected features
    if (sum(selec_feat)<1) return(c(mse_val=-Inf, rmse_test=-Inf, complexity=-Inf))

    # Extract features from the original DB + the response
    data_train_model <- data_train[,c(selec_feat,TRUE)]
    data_test_model  <- data_test[,c(selec_feat,TRUE)]

    # Use a 25-repeated 10-fold CV for validating each individual
    train_control <- trainControl(method = "repeatedcv", number = 10, repeats = 25)
    # Train the model
    # ----------------
    set.seed(1234)
    model <- train(medv ~ ., data=data_train_model, trControl=train_control,
            method="nnet", tuneGrid=tuneGrid, trace=F, linout = 1,
            MaxNWts=10000)

    # Extract validation & test metrics
    # --------------------------------------------------
    # RMSE repeated k-fold CV
    rmse_val <- model$results$RMSE

    # RMSE with test DB
    rmse_test <- sqrt(mean((unlist(predict(model, data_test_model))-
            data_test_model$medv)^2))

    # Model Complexity
    # --------------------------------------------------
    # Complexity = sum(ann_weights^2)
    weights <- model$finalModel$wts
    complexity <- sum(weights*weights)

    # Return errors and complexity. Errors are negative
    # GA-PARSIMONY tries to maximize them
    vect_errors <- c(rmse_val=-rmse_val,rmse_test=-rmse_test, complexity=complexity)
    return(vect_errors)
}
```

ANN parameters and selected input features are extracted from *chromosome* parameter. Fitness function requires a λ_g^i configuration for each individual i of the generation g:

$$\lambda_g^i = [size,\ decay,\ Q] \tag{2}$$

where the first two values correspond with the ANN parameters: number of hidden neurons (*size*) and the weight decay parameter (*decay*). The second part of *chromosome*, Q, is a vector with 13 real numbers between 0 and 1 which are binarized with a threshold of 0.50 for selecting the input features.

Finally, the function returns a vector with the mean of the Root Mean Squared Error (*RMSE*) of a 25-repeated 10-fold CV, $RMSE_{val}$, the RMSE measured with the test database, $RMSE_{tst}$, and the *complexity* that is calculated as the sum of ANN squared weights.

3.3   GAparsimony Settings

GAparsimony is executed 10 times with different random seeds. Each optimization process begins defining the range for searching the ANN parameters and their names (*min_param*, *max_param* and *names_param* parameters). The GA optimization process is defined with 40 individuals per generation (*popSize*) and a maximum number (*maxiter*) of 100 iterations with an early stopping criteria (*early_stop*) of 20 generations. In addition, results of each iteration are saved by setting *keep_history* to TRUE with the aim of using *plot()* and *parsimony_importance()* methods. Default value of crossover probability between pairs of chromosomes (*pcrossover*) is set to 0.8. Percentage of elitists is 20% (*elitism* parameter). Also, in order to start with a high percentage of features in the first population, *feat_thres* is established with a value of 0.90. Therefore, optimization process starts with a 90% of the features selected.

Mutation configuration has three important parameters: the number of top elitists that are not muted in each generation (in this example is *not_muted* = 2), the probability of mutation in a parent chromosome (in this case *pmutation* = 0.10), and the probability to be one when a feature is selected to be muted (*feat_mut_thres* = 0.10). This last parameter is established to a low value of 10% to ease the parsimony search (reduction of the input features in the following generations).

The *rerank_error* parameter is the maximum difference of J between two models to be considered similar. Individuals with similar J but lower complexity are promoted to the top positions into the GA selection process. In this first example, *rerank_error* has been set to a very low value, 0.0001, to disable the re-ranking process. In Sect. 3.5 other values of *rerank_error* are used with the objective of improving the trade-off between model complexity and accuracy, and obtaining models with less number of features.

```
# ------------------------------------------------------------
# Search the best parsimonious model with GA-PARSIMONY by
# using FS, Parameter Tuning and Parsimonious Model Selection
# ------------------------------------------------------------
library(GAparsimony)

# GA optimization process with 40 individuals per population,
# 100 max generations with an early stopping of 20 generations

GAparsimony_model <- vector(mode = "list", 10)
# GA-Parsimony 10 times
for (n_iter in 1:10)
{
print("########################################")
print("########################################")
print(n_iter)
print("########################################")
print("########################################")
GAparsimony_model[[n_iter]] <- ga_parsimony(
fitness=fitness_NNET, # Fitness function
min_param=c(1, 0.0001),   # min size and decay
max_param=c(30 , 0.9999), # max size and decay
names_param=c("size","decay"), # Parameters name
nFeatures=ncol(data_train)-1, #Num of input features
names_features=colnames(data_train)[-ncol(data_train)],
keep_history = TRUE, # Save all generations
rerank_error = 0.0001, # Max diff of scores to be similar
elitism=round(40*0.20), # Number of Elitists
popSize = 40,  # Population size of each generation
maxiter = 100,  # Max number of generations
early_stop=20, # Num of generations for early stopping
feat_thres=0.90, # Perc selected features in first gen
feat_mut_thres=0.10, # Prob in feature to be mutated
not_muted=2, # Not to mute first top 2 individual
parallel = TRUE, seed_ini = 1234*n_iter)
gc()
}
save(GAparsimony_model, file="GAparsimony_model_ANN_tipo.RData")
```

Finally, *GAparsimony* permits parallelization via *foreach* method in multi-core Windows, Mac OSX, or Unix/Linux systems. With *parallel = TRUE* uses algorithm all cores. Also, a specific number of cores (n) can be provided with *parallel = n*. Besides, a cluster parallelization function can be provided.

3.4 Extracting the Best Solutions

At the end of the for-loop, a list with 10 *ga_parsimony* objects are provided. Each object keeps the initial configuration, the historical data, the elapsed time, the last population and the final results.

The following code presents the method to extract the best solution of each iteration and, for the iteration 9, the best individual, the elapsed time in minutes, and the final number of generations. Finally, the code shows how to extract the population of generation 2 of the third iteration.

Table 2 shows the best individual of each iteration. The best model corresponds with the iteration 9 which has the lowest $RMSE_{val}$. Final solution is a *ANN* with 10 input features, 50 hidden neurons and a decay rate of 2.760. Elapsed time was 222 min.

```
# Extract best solution of each interation
resultados <- NULL
for (n_iter in 1:10)
{
  resultados <- cbind(resultados,
  GAparsimony_model[[n_iter]]@bestsolution)
}
write.csv(resultados, file="resultados.csv")

# Best individual of iter 9
GAparsimony_object <- GAparsimony_model[[9]]
print(summary(GAparsimony_object))
...
print(paste0("BEST_SOLUTION:"))
print(GAparsimony_object@bestsolution)
  fitnessVal fitnessTst complexity    size  decay crim
   -2.878759  -3.380322 888.456352      50 2.760325  0.0
    zn  indus   chas   nox    rm   age
   1.0    0.0    0.0   1.0   1.0   1.0
   dis  rad  tax ptratio black lstat
   1.0  1.0  1.0     1.0   1.0   1.0

print("ELAPSED_TIME_(min):")
print(GAparsimony_object@minutes_total)
  222.0121
print("NUMBER_OF_GENERATIONS:")
print(GAparsimony_object@iter)
  40
print("POPULATION_GENERATION")
pop_mat <- cbind(GAparsimony_object@population, GAparsimony_object@fitnessval,
         GAparsimony_object@fitnesstst, GAparsimony_object@complexity)
names_mat <- c(GAparsimony_object@names_param, GAparsimony_object@names_features,
         "RMSEval","RMSEtst","complexity")
colnames(pop_mat) <- names_mat
print(head(pop_mat,2))
        size    decay crim zn indus chas nox rm age dis rad tax
[1,] 49.51992 2.76032    0  1     0    0   1  1   1   1   1   1
[2,] 49.51992 2.76032    0  1     0    0   1  1   1   1   1   1
     ptratio black lstat     RMSEval   RMSEtst  complexity
[1,]       1     1     1   -2.878759 -3.380322    888.4564
[2,]       1     1     1   -2.878759 -3.380322    888.4564

# Best results of generation 2 in iter 9
GAparsimony_object <- GAparsimony_model[[9]]@history[[2]]
pop_mat <- cbind(GAparsimony_object$population, GAparsimony_object$fitnessval,
GAparsimony_object$fitnesstst, GAparsimony_object$complexity)
colnames(pop_mat) <- names_mat
print(head(pop_mat,3))
         size    decay crim zn indus chas nox rm age dis rad tax
[1,] 12.56129  2.76032    1  1     1    1   1  1   1   1   1   0
[2,] 12.56129  2.76032    1  1     1    1   1  1   1   1   1   0
[3,] 11.05759 11.72260    1  0     1    1   1  1   1   0   1   1
     ptratio black lstat     RMSEval   RMSEtst complexity
[1,]       1     1     1   -3.368904 -3.644329   958.8236
[2,]       0     1     1   -3.482480 -3.445078   743.7078
[3,]       1     1     1   -3.664144 -4.075636   339.5751

# Plot GA evolution ('keep_history' must be TRUE)
elitists <- plot(GAparsimony_model[[9]], window=FALSE, general_cex=0.6,
          pos_cost_num=-1, pos_feat_num=-1.5, digits_plot=3,min_ylim=-5.0)
```

The *plot()* command displays $RMSE_{val}$ and $RMSE_{tst}$ evolution of elitist individuals for iteration nine. *general_cex* defines the letter size, and *pos_cost* and *pos_feat_num* set the relative position of the axes text. Also, y-axis limits can be modified with *min_ylim* and *max_ylim*. In the Fig. 1, white and gray box-plots represent respectively the evolution of $RMSE_{val}$ and $RMSE_{tst}$ for the elitists. Shaded area delimits the maximum and minimum number of features, N_{SF}. Continuous line, dot and stripped line, and discontinuous line represent, respectively, $RMSE_{val}$, $RMSE_{tst}$ and N_{SF} of the best individual.

Table 2. Best individual for each iteration.

Var	Iter1	Iter2	Iter3	Iter4	Iter5	Iter6	Iter7	Iter8	**Iter9**	Iter10
$RMSE_{val}$	3.094	3.314	3.432	2.914	3.466	3.341	2.927	2.949	**2.879**	3.550
$RMSE_{tst}$	3.524	3.767	3.701	3.185	3.538	3.546	3.453	3.382	**3.380**	3.654
$Time(min)$	215.7	126.15	232.3	332.6	190.9	172.3	301.9	291.1	**222.0**	97.8
$complexity$	595.6	459.8	361.7	1016.4	350.5	516.9	748.3	787.2	**888.5**	299.0
$size$	49	16	15	46	15	17	50	49	**50**	15
$decay$	5.564	10.158	13.879	2.864	15.032	10.794	3.573	3.650	**2.760**	17.698
crim	0	1	1	1	1	1	0	0	**0**	1
zn	1	1	1	1	1	1	1	1	**1**	1
indus	0	1	1	1	1	1	0	0	**0**	1
chas	0	0	0	0	0	0	0	0	**0**	0
nox	1	1	1	1	1	1	1	1	**1**	1
rm	1	1	1	1	1	1	1	1	**1**	1
age	1	1	1	1	1	1	1	1	**1**	1
dis	1	1	1	1	1	1	1	1	**1**	1
rad	1	1	1	1	1	1	1	1	**1**	1
tax	1	1	1	1	1	1	1	1	**1**	1
ptratio	1	1	1	1	1	1	1	1	**1**	1
black	1	1	1	0	1	1	1	1	**1**	1
lstat	1	1	1	1	1	1	1	1	**1**	1

3.5   Searching Parsimonious Solutions

In many real applications, researchers are not so worried about improving small differences of J. For example, the decimal digits of the temperature set point of an industrial furnace with a range of 700–$900°C$ is usually superfluous. However, the priority can be to find solutions with a reduced input set, more robust when dealing with noise. Besides, these parsimonious models will be easier and cheaper to build, update and maintain.

The search of parsimonious solutions with a smaller quantity of features can be approached by tuning the $rerank\_error$ parameter and including the number of features, N_{SF}, in the complexity model calculation.

For example,

```
complexity <-   1E6*sum(selec_feat)+sum(weights*weights)
```

considers $selec\_feat$ (N_{SF}) to be more important than the internal model complexity by multiplying the first by 1.000.000. Between two models with the same N_{SF}, the model complexity will play in the PMS process.

Table 3 shows the best individuals obtained with the new complexity metric and for four values of $rerank\_error$. With $rerank\_error = 0.05$ the solution has a difference of 0.111 in the $RMSE_{val}$ but with a 30% of reduction in N_{SF} (7 vs 10 features). With smaller values of $rerank\_error$ (0.01 or 0.0001) solutions

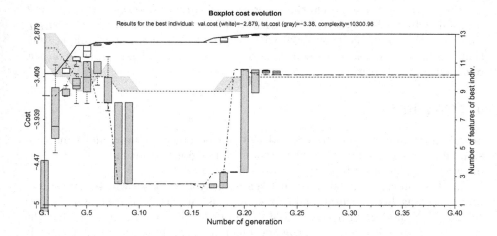

Fig. 1. Evolution of elitist individuals in iteration 9.

Table 3. Best individual with the new complexity metric and different values of *rerank_error*.

Var	Iter8	**Iter9**	Iter9	Iter9
rerank_error	0.1	**0.05**	0.01	0.0001
$RMSE_{val}$	3.341	**2.993**	2.882	2.879
$RMSE_{tst}$	4.519	**4.509**	3.334	3.380
$Time(min)$	85.9	**186.5**	213.6	221.6
complexity	6000492.0	**7000862.5**	10000902.3	10000888.5
size	12	**48**	48	50
decay	4.313	**2.760**	2.760	2.760
crim	0	**0**	0	0
zn	0	**0**	1	1
indus	0	**0**	0	0
chas	0	**0**	0	0
nox	1	**1**	1	1
rm	1	**1**	1	1
age	0	**1**	1	1
dis	1	**1**	1	1
rad	1	**0**	1	1
tax	0	**1**	1	1
ptratio	1	**1**	1	1
black	0	**0**	1	1
lstat	1	**1**	1	1

are similar to the Sect. 3.4. On the other side, with $rerank\_error = 0.1$, the best solution reduces the N_{SF} to 6 but with a difference of 0.462 in the $RMSE_{val}$. However, $RMSE_{tst}$ increase 1.129 with respect to the non-parsimonious solution, probably because some solutions of the testing database depends on the removed inputs features.

4 Conclusions

Searching the best model by FS and HO can become a tedious and time consuming process. Besides, many researchers and engineers that are not experts in machine learning need tools that can facilitate these tasks.

In this contribution, we present GAparsimony R package which can be useful for seeking high accuracy and parsimonious models. The process to search robust ANN models with *Boston* database has been exposed, showing that the strategy of adjusting $rerank\_error$ and including N_{SF} in the complexity metric can help to obtain accurate parsimonious solutions. This example code can be easily adapted to another problem and algorithm.

Future works will be focused in improving the optimization process by including other Meta-heuristics.

Acknowledgements. We are greatly indebted to *Banco Santander* for the APPI17/04 fellowship and to the University of La Rioja for the EGI16/19 fellowship. Also, A. Pernia wants to express her gratitude with the Instituto de Estudios Riojanos (IER) for the fellowship. This work used the Beronia cluster (Universidad de La Rioja), which is supported by FEDER-MINECO grant number UNLR-094E-2C-225.

References

1. Antonanzas-Torres, F., Urraca, R., Antonanzas, J., Fernandez-Ceniceros, J., de Pison, F.M.: Generation of daily global solar irradiation with support vector machines for regression. Energy Convers. Manag. **96**, 277–286 (2015)
2. Bergstra, J., Komer, B., Eliasmith, C., Yamins, D., Cox, D.D.: Hyperopt: a python library for model selection and hyperparameter optimization. Comput. Sci. Discov. **8**(1), 014008 (2015)
3. Bischl, B., Lang, M., Kotthoff, L., Schiffner, J., Richter, J., Studerus, E., Casalicchio, G., Jones, Z.M.: mlr: Machine learning in R. J. Mach. Learn. Res. **17**(170), 1–5 (2016)
4. Fernandez-Ceniceros, J., Sanz-Garcia, A., Antonanzas-Torres, F., de Pison, F.M.: A numerical-informational approach for characterising the ductile behaviour of the t-stub component. part 2: parsimonious soft-computing-based metamodel. Eng. Struct. **82**, 249–260 (2015)
5. Gorissen, D., Couckuyt, I., Demeester, P., Dhaene, T., Crombecq, K.: A surrogate modeling and adaptive sampling toolbox for computer based design. J. Mach. Learn. Res. **11**, 2051–2055 (2010)
6. Hashem, I.A., Yaqoob, I., Anuar, N.B., Mokhtar, S., Gani, A., Ullah Khan, S.: The rise of big data on cloud computing: review and open research issues. Inf. Syst. **47**, 98–115 (2015)

7. Martinez-de-Pison, F.: GAparsimony package for R (2017). https://github.com/jpison/GAparsimony

8. Michalewicz, Z., Janikow, C.Z.: Handling constraints in genetic algorithms. In: ICGA, pp. 151–157 (1991)

9. Olson, R.S., Bartley, N., Urbanowicz, R.J., Moore, J.H.: Evaluation of a tree-based pipeline optimization tool for automating data science. In: Proceedings of the Genetic and Evolutionary Computation Conference 2016, GECCO '16, pp. 485–492. ACM, New York (2016)

10. Sanz-Garcia, A., Fernandez-Ceniceros, J., Antonanzas-Torres, F., Pernia-Espinoza, A., Martinez-de Pison, F.J.: GA-PARSIMONY: A GA-SVR approach with feature selection and parameter optimization to obtain parsimonious solutions for predicting temperature settings in a continuous annealing furnace. Appl. Soft Comput. **35**, 13–28 (2015)

11. Sanz-García, A., Fernández-Ceniceros, J., Antoñanzas-Torres, F., Martínez-de Pisón, F.J.: Parsimonious support vector machines modeling for set points in industrial processes based on genetic algorithm optimization. In: International Joint Conference SOCO13-CISIS13-ICEUTE13, Advances in Intelligent Systems and Computing, vol. 239, pp. 1–10. Springer. Cham (2014)

12. Thornton, C., Hutter, F., Hoos, H.H., Leyton-Brown, K.: Auto-WEKA: Combined selection and hyperparameter optimization of classification algorithms. In: Proceedings of the 19th ACM SIGKDD International Conference on Knowledge Discovery and Data Mining, KDD '13, pp. 847–855. ACM, New York (2013)

13. Urraca, R., Sodupe-Ortega, E., Antonanzas, J., Antonanzas-Torres, F., de Pison, F.M.: Evaluation of a novel ga-based methodology for model structure selection: the ga-parsimony. Neurocomputing **271**, 9–17 (2018)

14. Urraca, R., Sanz-Garcia, A., Fernandez-Ceniceros, J., Sodupe-Ortega, E., Martinez-de-Pison, F.J.: Improving hotel room demand forecasting with a hybrid GA-SVR methodology based on skewed data transformation, feature selection and parsimony tuning. In: Onieva, E., Santos, I., Osaba, E., Quintián, H., Corchado, E. (eds.) HAIS 2015. LNCS (LNAI), vol. 9121, pp. 632–643. Springer, Cham (2015)

15. Ye, J.: On measuring and correcting the effects of data mining and model selection. J. Am. Stat. Assoc. **93**(441), 120–131 (1998)

Improving Adaptive Optics Reconstructions with a Deep Learning Approach

Sergio Luis Suárez Gómez[1]([⊠]), Carlos González-Gutiérrez[2],
Enrique Díez Alonso[2], Jesús Daniel Santos Rodríguez[1],
Maria Luisa Sánchez Rodríguez[1], Jorge Carballido Landeira[1],
Alastair Basden[3], and James Osborn[3]

[1] Department of Physics, University of Oviedo, Oviedo, Spain
suarezsergio@uniovi.es
[2] Prospecting and Exploitation of Mines Department,
University of Oviedo, Oviedo, Spain
[3] Department of Physics, Centre for Advanced Instrumentation,
University of Durham, Durham, UK

Abstract. The use of techniques such as adaptive optics is mandatory when performing astronomical observation from ground based telescopes, due to the atmospheric turbulence effects. In the latest years, artificial intelligence methods were applied in this topic, with artificial neural networks becoming one of the reconstruction algorithms with better performance. These algorithms are developed to work with Shack-Hartmann wavefront sensors, which measures the turbulent profiles in terms of centroid coordinates of their subapertures and the algorithms calculate the correction over them. In this work is presented a Convolutional Neural Network (CNN) as an alternative, based on the idea of calculating the correction with all the information recorded by the Shack-Hartmann, for avoiding any possible loss of information. With the support of the Durham Adaptive optics Simulation Platform (DASP), simulations were performed for the training and posterior testing of the networks. This new CNN reconstructor is compared with the previous models of neural networks in tests varying the altitude of the turbulence layer and the strength of the turbulent profiles. The CNN reconstructor shows promising improvements in all the tested scenarios.

Keywords: Adaptive optics · Artificial neural networks
Convolutional neural networks

1 Introduction

In the latest years, the growing interest in artificial intelligence lead the development of these techniques to search its application in a broad range of science branches; in astronomical imaging, neural networks were already used, for example, for object detection and star/galaxy classification [1]. Nowadays, Adaptive Optics (AO) are a series of techniques, used for astronomical observation, that are mandatory for the search the correction of astronomical images taken from ground based telescopes [2]. AO techniques are based on the measurement of the turbulent profiles of the

© Springer International Publishing AG, part of Springer Nature 2018
F. J. de Cos Juez et al. (Eds.): HAIS 2018, LNAI 10870, pp. 74–83, 2018.
https://doi.org/10.1007/978-3-319-92639-1_7

atmosphere, to estimate a phase correction on the upcoming light that will form the astronomical image. A wide variety of algorithms and methods had been developed to perform the estimation of the corrections to the images; between them, artificial intelligence techniques, in particular Artificial Neural Networks (ANNs) have been proven to be a useful alternative as reconstructor to the traditional algorithms, for it good performance and for its improvements in computational time [3].

ANNs and other artificial intelligence techniques are based on the learning of characteristics and patterns that lie within a set of data [4], [5], to replicate the modelled system or to infer a response when new data is presented. In particular, ANNs are formed by neurons, which are its processing units, and the interconnections between them are adapted to fit the training set. This idea showed excellent results when applied to the adaptive optics as the Complex Atmospheric Reconstructor based on Machine lEarNing (CARMEN) [6], both in simulations [7], and in its posterior live testing on-sky [8]. These remarkable results were obtained when CARMEN was contrasted with other usual techniques in AO reconstruction, such as Learn and Apply [9] or Least Squares [10].

This network worked by means of learning from the measurements of a Shack-Hartmann WaveFront Sensor (SH-WFS), which estimates the turbulence of the atmosphere considering sections of the sky and the effect of the turbulence in the light that comes to each of the subapertures of the sensor. The calculated centroids from the image of each subaperture constitutes the inputs of CARMEN and the corrected for the scientific object are the outputs of the network.

Based on previous works [11] and the nature of the problem, information from the turbulence is likely to be lost once the calculation of the centroid from the SH-WFS images is performed. We propose a reconstructor based in a more complex architecture of ANNs, the Convolutional Neural Networks (CNNs), which allow images as inputs and consequently, avoiding the centroid calculations and its implicit loss of information.

CNNs are commonly used in document recognition [12], image classification [13, 14] or speech recognition [15]. In our case, the ability of CNNs to extract the most relevant features from the image, allows to process the full image of the SH-WFS measurements and improve the results that CARMEN achieves.

This work presents an introduction about SH-WFS and Durham Adaptive optics Simulation Platform (DASP) [16], the AO simulation platform used, in Sect. 2. The next section details how CARMEN solved this problem, and what is intended to achieve with a new reconstructor based on CNNs that will be referred as Convolutional CARMEN. In Sect. 4, the architecture of the CNN proposed as reconstructor is specified, as well as other details of the experiment, such as the performed simulations. In Sect. 5, results from Convolutional CARMEN are presented and discussed. Finally, in the last section, conclusions and future work are stated.

2 Adaptive Optics

The principal usage of AO is in the correction of the optical aberrations induced in the astronomical images by the atmospheric turbulent profiles. To do that, AO consist on different procedures; first, the turbulence has to be measured, usually with sensors such as the SH-WFS sensor.

The sensing is performed with the support of known sources of light that are close to the scientific object and allow to estimate the turbulence that affects the upcoming light from the scientific object, these are called Natural Guide Stars (NGS). When available, Laser Guide Stars (NGS) are usually used for gaining more information in the near field of the scientific object.

After the measurements are obtained, the turbulent profile is estimated with a reconstruction algorithm, and the correction shape of the deformable mirrors is calculated. When the deformable mirror fits the correct shape to modify the phase of the wavefront from the scientific object image, it compensates the effects of the aberrations induced by the atmosphere to form the final image.

2.1   Shack-Hartmann Wavefront Sensor

The SH-WFS sensor consists on a set of lenses with the same focal length, called subapertures, which allows to divide the incoming wavefront. The turbulence is estimated in the space corresponding to each subaperture. In conditions of no turbulence, in each lense the focused spot should be centred in the middle of the subaperture, however, for the light that passes through the atmosphere, the focused spot becomes a blurry image.

The gravity centre of that image can be calculated in terms of the coordinates of each subaperture, being called centroids, as Fig. 1 shows. By using the information obtained with the centroids, it is possible to reconstruct the incoming wavefront, by means of Zernike polynomials. Once this information is obtained from the SH-WFS, it is possible to compensate the turbulence aberration, by adapting the deformable mirror.

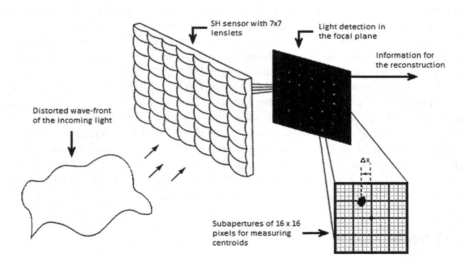

Fig. 1. Diagram of the working of a SH-WFS sensor. Light wave-fronts are measured locally in each subaperture. Each lens forms a blurred image from where centroids are estimated.

2.2 Simulations and DASP Platform

DASP is a simulation platform developed by the University of Durham [16] that allows to produce AO simulations at all stages of an AO system.

In order to perform a comparison of the performance between CARMEN reconstructor and our proposal, the simulations for this work were done as a replica of the ones performed in the original articles where the reconstructor CARMEN was presented and contrasted with computational simulations [6, 7].

For this work, among all the options that DASP offers, the simulations required different values of r0, different heights for the turbulent layers, as different number of turbulent layers. With these parameters, DASP is able to simulate the images of off-axis SH-WFS sensors with guide stars, including the focused points in each subaperture, or its centroids as it was originally required for CARMEN. Also, DASP allows the simulation of the centroids from the on-axis SH-WFS for the scientific object, which are used as the outputs in both the original reconstructor and our newest proposal.

3 CARMEN

3.1 CARMEN as Multi-Layer Perceptron

The first AO reconstructor based in ANNs presented, CARMEN, was a Multi-Layer Perceptron (MLP) [17].

MLPs consist on series of processing units, called neurons, grouped in sets called layers [18]. All the neurons of one layer are connected to the neurons of the adjacent layers. These connections are characterized by a weight. Each neuron processes all the information it had as input applying an activation function [3].

All techniques of artificial intelligence have as trait the capability for learning form the data and approximate nonlinear systems [19]; this is performed by algorithms that measure the error and adjust the weights of the connections between neurons, such as the backpropagation algorithm [20, 21].

A graphical representation of the topology of a MLP is shown in Fig. 2.

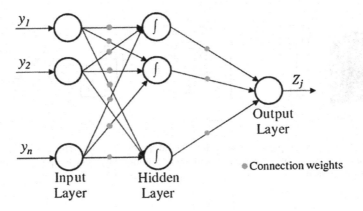

Fig. 2. Topology of a MLP neural network. Reproduced from [11].

In the AO field, ANNs were introduced with a first approximation of the reconstructor CARMEN to simulation [6, 7], trained with the centroids of off-axis SH-WFS with NGS as inputs, and the centroids of the scientific object as outputs; this allowed to estimate, when applied to new data, the turbulence affecting the image of the scientific object with the data of the off-axis SH-WFS.

For comparison with the new reconstructor presented in this work we use, for CARMEN, the topology presented in [8], where excellent results were achieved in on-sky testing.

3.2 CARMEN as Convolutional Neural Network

CNNs models emerged to solve the issues that MLPs could not deal with. When the information which is needed to process is too high, for example in an image, the number of connection weights between neurons in an MLP may increase enough to be impractical for a computer to handle. The improvements of CNNs are based in the implementation of convolutional layers that work as filters, extracting the main features from the input data. These models provide versatility to the systems of study and can be applied in most scenarios such as document recognition [12], image classification [13, 14] or speech recognition [15].

These layers allow processing of their outputs with activation functions, as for example the Rectified Linear Unit, known as ReLU [22]. Frequently, the outputs of these layers are also adjusted with a pooling layer [23], reducing the size of the resultant images, by extracting the maximum or mean value from a certain region of pixels. Nesting several of these layers, image size can be lowered significantly, although the number of images will increase. When the desired sizes of the data are reached, a MLP can be implemented, using as inputs the features that the convolutional layers computed as outputs [19]. An example of the topology and implementation of a CNN is shown in Fig. 3.

Training process for CNNs involves the adjustment of the weights in the connections between the neurons of adjacent layers of the MLP, as well as the weights of the filters from the convolutional layers.

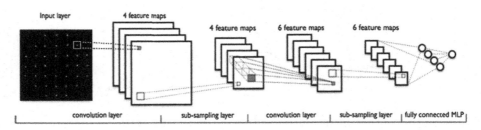

Fig. 3. Example of the topology of a convolutional neural network. The original image is the input of the sequences of convolution and sub-sampling layers, which are followed by a multi-layer perceptron.

In this work, we present a reconstructor based on the convolutional approach, developed to improve the performance of the actual reconstructors, being CARMEN MLP between them. Our proposal relies on the use of the full image as input, instead using only the centroid calculations, as CARMEN nowadays does. This allow us to input all the information recorded by the SH-WFS to the reconstructor, as well as avoiding one computational step, the centroid calculations.

4   Experimental Setup

Simulations used for the training of the networks were performed with DASP. Data sets for training and testing were created for reproducing the original tests presented in [7], with the following properties:

All the SH-WFS simulated had 36 subapertures, being a total of 72 coordinates of centroids per sensor.

Training set was simulated with layers varying the height of the turbulence from 0 to 15500 m with steps of 100 m each. The turbulence strength was simulated with the parameter r0, which was chosen ranging from 8 cm to 20 cm, with steps of 1 cm each. Each of the combinations of heights and r0 values was sampled 100 times, being a total of 202800 samples.

Test set have two layers, the first a ground layer at 0 m and the second with a fixed altitude. The r0 values ranged from 5 cm to 20 cm, and the relative strengths of the layer were 0.5 for both. There were simulated three scenarios, with the second layer at altitudes of 5000, 10000 and 15000 m. Each of the combinations of heights and r0 values was sampled 1000 times, being a total of 16000 samples each set.

Neural networks, both the original CARMEN and the convolutional CARMEN, were trained and tested, in terms of normalized error, with the above sets.

The original CARMEN had 216 neurons as inputs, with a hidden layer of 216 neurons and 72 for the outputs. The inputs corresponded with the centroids of 3 SH-WFS, and the outputs are the centroids of the SH-WFS of the scientific object.

The convolutional CARMEN inputs used the full image of the SH-WFS with 3 channels, since 3 SH-WFS were simulated. It has 4 convolutional layers, each one with 4 kernels of 5×5 size. After the convolutional layers, ReLU was applied as activation function and pooling of sizes 2×2 for the first two convolutional layers, and sizes 4×4 for the las two convolutional layers. This resulted in 768 images of size 2×2 that become the input of the fully-connected layers; consequently, it has 3072 neurons as input. The hidden layer was set with 216 neurons and 72 for the output layer.

The network was implemented in TensorFlow [24] version r1.3.

5   Results and Discussion

The trained networks were compared in terms of normalized error. The error is calculated as the average of the absolute value of the difference between all the output network centroids and the simulated true centroids. In each of the three scenarios defined for the test, the comparison has been made for each value of r0.

In Fig. 4, the convolutional network shows better performance than the MLP for all the values of r0. There is a clear trend of improving results for the weaker turbulence profiles, as it is expected.

The comparison at 10000 m is shown in Fig. 5. In this case, the convolutional network still improves the results provided by the MLP, however with closer values of error.

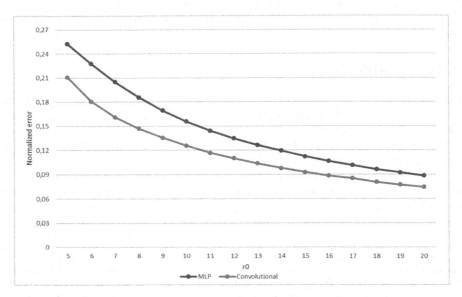

Fig. 4. CARMEN MLP and Convolutional CARMEN performance in terms of normalized error for the test with turbulence layer at 5000 m.

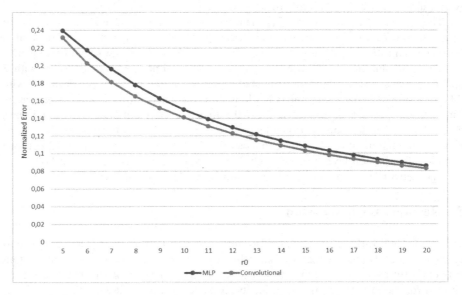

Fig. 5. CARMEN MLP and Convolutional CARMEN performance in terms of normalized error for the test with turbulence layer at 10000 m.

As in the previous cases, the convolutional network improved the results for all the values of r0 when considering the turbulence at 15000 m. The results are shown in Fig. 6. The trend is of higher decay of the error as the turbulence weakens than in the other turbulence altitudes.

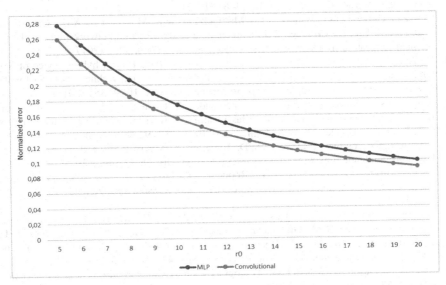

Fig. 6. CARMEN MLP and Convolutional CARMEN performance in terms of normalized error for the test with turbulence layer at 15000 m.

6 Conclusions and Future Lines

Convolutional networks were presented as an alternative for the actual neural networks used as reconstructors, showing great performance in the presented tests. The convolutional approach was considered since it seemed that some information may be lost in other reconstruction algorithms as consequence of the calculation of centroids. Convolutional networks, allowing the usage of full images from the SH-WFS, proved this argument as it is shown in the comparison of errors.

These promising results open possibilities in further work in the topic, improving the topology of the network, setting more solid testing with sets of multilayer turbulence profiles, using optical measurements for the comparison of errors, and testing the tomographic reconstructor in a real telescope.

Other alternatives that can be addressed are different types of artificial intelligence techniques, such as those based on on-sky learning or recurrent neural networks.

References

1. Andreon, S., Gargiulo, G., Longo, G., Tagliaferri, R., Capuano, N.: Wide field imaging—I. Applications of neural networks to object detection and star/galaxy classification. Mon. Not. R. Astron. Soc. **319**(3), 700–716 (2000)
2. Beckers, J.M.: Adaptive optics for astronomy: principles, performance, and applications. Annu. Rev. Astron. Astrophys. **31**(1), 13–62 (1993)
3. González-Gutiérrez, C., Santos, J.D., Martínez-Zarzuela, M., Basden, A.G., Osborn, J., Díaz-Pernas, F.J., De Cos Juez, F.J.: Comparative study of neural network frameworks for the next generation of adaptive optics systems. Sensors **17**(6), 1263 (2017)
4. De Andrés, J., Sánchez-Lasheras, F., Lorca, P., de Cos Juez, F.J.: A hybrid device of Self Organizing Maps (SOM) and Multivariate Adaptive Regression Splines (MARS) for the forecasting of firms' bankruptcy. Account. Manag. Inf. Syst. **10**(3), 351 (2011)
5. Sánchez, A.S., Iglesias-Rodríguez, F.J., Fernández, P.R., de Cos Juez, F.J.: Applying the K-nearest neighbor technique to the classification of workers according to their risk of suffering musculoskeletal disorders. Int. J. Ind. Ergon. **52**, 92–99 (2016)
6. Osborn, J., De Cos Juez, F.J., Guzman, D., Butterley, T., Myers, R., Guesalaga, A., Laine, J.: Using artificial neural networks for open-loop tomography. Opt. Express **20**(3), 2420 (2012)
7. de Cos Juez, F.J., Lasheras, F.S., Roqueñí, N., Osborn, J.: An ANN-based smart tomographic reconstructor in a dynamic environment. Sensors (Switzerland) **12**(7), 8895–8911 (2012)
8. Osborn, J., Guzman, D., de Cos Juez, F.J., Basden, A.G., Morris, T.J., Gendron, E., Butterley, T., Myers, R.M., Guesalaga, A., Lasheras, F.S., Victoria, M.G., Rodríguez, M.L. S., Gratadour, D., Rousset, G.: Open-loop tomography with artificial neural networks on CANARY: on-sky results. Mon. Not. R. Astron. Soc. **441**(3), 2508–2514 (2014)
9. Vidal, F., Gendron, E., Rousset, G.: Tomography approach for multi-object adaptive optics. JOSA A **27**(11), A253–A264 (2010)
10. Ellerbroek, B.L.: First-order performance evaluation of adaptive-optics systems for atmospheric-turbulence compensation in extended-field-of-view astronomical telescopes. JOSA A **11**(2), 783–805 (1994)
11. Suárez Gómez, S.L., Santos Rodríguez, J.D., Iglesias Rodríguez, F.J., de Cos Juez, F.J.: Analysis of the temporal structure evolution of physical systems with the self-organising tree algorithm (SOTA): application for validating neural network systems on adaptive optics data before on-sky implementation. Entropy **19**(3), 103 (2017)
12. Lecun, Y., Bottou, L., Bengio, Y., Haffner, P.: Gradient-based learning applied to document recognition. Proc. IEEE **86**(11), 2278–2323 (1998)
13. Krizhevsky, A., Sutskever, I., Hinton, G.E.: ImageNet classification with deep convolutional neural networks. In: Advances in Neural Information Processing Systems, pp. 1097–1105 (2012)
14. Giusti, A., Ciresan, D.C., Masci, J., Gambardella, L.M., Schmidhuber, J.: Fast image scanning with deep max-pooling convolutional neural networks. In: 2013 20th IEEE International Conference on Image Processing (ICIP), pp. 4034–4038 (2013)
15. Graves, A., Mohamed, A., Hinton, G.: Speech recognition with deep recurrent neural networks. In: ICASSP, vol. 3, pp. 6645–6649, May 2013
16. Basden, A.: DASP the Durham adaptive optics simulation platform: modelling and simulation of adaptive optics systems. https://github.com/agb32/dasp
17. Gardner, M.W., Dorling, S.R.: Artificial neural networks (the multilayer perceptron)—a review of applications in the atmospheric sciences. Atmos. Environ. **32**(14), 2627–2636 (1998)

18. Suárez Gómez, S.L., Gutiérrez, C.G., Rodríguez, J.D.S., Rodríguez, M.L.S., Lasheras, F.S., de Cos Juez, F.J.: Analysing the performance of a tomographic reconstructor with different neural networks frameworks. In: International Conference on Intelligent Systems Design and Applications, pp. 1051–1060 (2016)
19. Hornik, K., Stinchcombe, M., White, H.: Multilayer feedforward networks are universal approximators. Neural Netw. **2**(5), 359–366 (1989)
20. Lasheras, J.E.S., Tardón, A., Tardón, G.G., Gómez, S.L.S., Sánchez, V.M., Donquiles, C.G., de Cos Juez, F.J.: A methodology for the detection of relevant single nucleotide polymorphism in prostate cancer by means of multivariate adaptive regression splines and backpropagation artificial neural networks. In: Proceedings of the International Joint Conference SOCO 2017-CISIS 2017-ICEUTE 2017 León, Spain, 6–8 September 2017, pp. 391–399 (2017)
21. Rumelhart, D.E., Hinton, G.E., Williams, R.J.: Learning representations by back-propagating errors. Cogn. Model. **5**(3), 1 (1988)
22. Nair, V., Hinton, G.E.: Rectified linear units improve restricted Boltzmann machines. In: Proceedings of the 27th International Conference on Machine Learning (ICML-10), pp. 807–814 (2010)
23. Nagi, J., Ducatelle, F., Di Caro, G.A., Cireçsan, D., Meier, U., Giusti, A., Nagi, F., Schmidhuber, J., Gambardella, L.M.: Max-pooling convolutional neural networks for vision-based hand gesture recognition. In: 2011 IEEE International Conference on Signal and Image Processing Applications (ICSIPA), pp. 342–347 (2011)
24. Abadi, M., Barham, P., Chen, J., Chen, Z., Davis, A., Dean, J., Devin, M., Ghemawat, S., Irving, G., Isard, M., et al.: TensorFlow: a system for large-scale machine learning. In: OSDI, vol. 16, pp. 265–283 (2016)

Complexity of Rule Sets in Mining Incomplete Data Using Characteristic Sets and Generalized Maximal Consistent Blocks

Patrick G. Clark[1], Cheng Gao[1], Jerzy W. Grzymala-Busse[1,2(✉)],
Teresa Mroczek[2], and Rafal Niemiec[2]

[1] Department of Electrical Engineering and Computer Science,
University of Kansas, Lawrence, KS 66045, USA
patrick.g.clark@gmail.com, {cheng.gao,jerzy}@ku.edu
[2] Department of Expert Systems and Artificial Intelligence,
University of Information Technology and Management, 35-225 Rzeszow, Poland
{tmroczek,rniemiec}@wsiz.rzeszow.pl

Abstract. In this paper, missing attribute values in incomplete data sets have two possible interpretations, lost values and "do not care" conditions. For rule induction we use characteristic sets and generalized maximal consistent blocks. Therefore we apply four different approaches for data mining. As follows from our previous experiments, where we used an error rate evaluated by ten-fold cross validation as the main criterion of quality, no approach is universally the best. Therefore we decided to compare our four approaches using complexity of rule sets induced from incomplete data sets. We show that the cardinality of rule sets is always smaller for incomplete data sets with "do not care" conditions. Thus the choice between interpretations of missing attribute values is more important than the choice between characteristic sets and generalized maximal consistent blocks.

Keywords: Incomplete data · Lost values · "Do not care" conditions
Characteristic sets · Maximal consistent blocks
MLEM2 rule induction algorithm · Probabilistic approximations

1 Introduction

In this paper, missing attribute values in incomplete data sets have two possible interpretations, lost values and "do not care" conditions. A lost value is the missing attribute value that initially existed but currently is unavailable, for example it is forgot or erased. For incomplete data sets with lost values we use a cautious approach, inducing rules only from existing, specified attribute values. A "do not care" condition is interpreted differently, we are assuming that the missing attribute value may be replaced by any value from the attribute domain. A

© Springer International Publishing AG, part of Springer Nature 2018
F. J. de Cos Juez et al. (Eds.): HAIS 2018, LNAI 10870, pp. 84–94, 2018.
https://doi.org/10.1007/978-3-319-92639-1_8

frequent reason for the "do not care" conditions is a refusal to answer a question during the interview. Our assumption is that an expert should provide the most appropriate interpretation of missing attribute values for a specific data set.

In our experiments we use probabilistic approximations, a generalization of the lower and upper approximations, well-known in rough set theory. A probabilistic approximation is associated with a parameter α, interpreted as a probability. When $\alpha = 1$, a probabilistic approximation becomes the lower approximation; if α is a small positive number, e.g., 0.001, a probabilistic approximation is the upper approximation. Initially, probabilistic approximations were applied to completely specified data sets [9,12–19]. Probabilistic approximations were generalized to incomplete data sets in [8].

Characteristic sets, for incomplete data sets with any interpretation of missing attribute values, were introduced in [7]. Maximal consistent blocks, restricted only to data sets with "do not care" conditions, were introduced in [11]. Additionally, in [11] maximal consistent blocks were used as basic granules to define only ordinary lower and upper approximations. A definition of the maximal consistent block was generalized to cover lost values and probabilistic approximations in [1]. The applicability of characteristic sets and maximal consistent blocks for mining incomplete data, from the view point of an error rate, was studied in [1]. The main result of [1] is that there is a small difference in quality of rule sets, in terms of an error rate computed as the result of ten-fold cross validation, induced either way. Thus, we decided to compare characteristic sets with generalized maximal consistent blocks in terms of complexity of induced rule sets. In our experiments, the Modified Learning from Examples Module, version 2 (MLEM2) was used for rule induction [6].

Our main objective is to compare four approaches to mining incomplete data sets, combining characteristic sets and generalized maximal consistent blocks with two interpretations of missing attribute values, lost values and "do not care" conditions. Our conclusion is that the choice between characteristic sets and generalized maximal consistent blocks is not as important as the choice between two interpretations of missing attribute values. The simplest rule sets are induced from incomplete data sets with an interpretation of "do not care" conditions for missing attribute values.

2   Incomplete Data

In this paper, input data sets are presented in a form of a decision table. An example of the decision table is shown in Table 1. Rows of the decision table represent cases, while columns are labeled by variables. The set of all cases will be denoted by U. In Table 1, $U = \{1, 2, 3, 4, 5, 6, 7, 8\}$. Independent variables are called *attributes* and a dependent variable is called a *decision* and is denoted by d. The set of all attributes is denoted by A. In Table 1, $A = \{$ *Temperature, Headache, Cough* $\}$. The value for a case x and an attribute a is denoted by $a(x)$.

We distinguish between two interpretations of missing attribute values: lost values, denoted by "?" and "do not care" conditions, denoted by "$*$". Table 1 presents an incomplete data set with both lost values and "do not care" conditions.

Table 1. A decision table

Case	Attributes			Decision
	Temperature	Headache	Cough	Flu
1	high	yes	?	yes
2	*	*	yes	yes
3	very-high	*	no	yes
4	normal	*	yes	yes
5	?	yes	?	no
6	normal	yes	*	no
7	high	yes	?	no
8	normal	*	no	no

The set X of all cases defined by the same value of the decision d is called a *concept*. For example, a concept associated with the value *yes* of the decision *Flu* is the set $\{1, 2, 3, 4\}$.

For a completely specified data set, let a be an attribute and let v be a value of a. A *block* of (a, v), denoted by $[(a, v)]$, is the set $\{x \in U \mid a(x) = v\}$ [4].

For incomplete decision tables the definition of a block of an attribute-value pair (a, v) is modified in the following way.

- If for an attribute a and a case x we have $a(x) = ?$, the case x should not be included in any blocks $[(a, v)]$ for all values v of attribute a,
- If for an attribute a and a case x we have $a(x) = *$, the case x should be included in blocks $[(a, v)]$ for all specified values v of attribute a.

For the data set from Table 1 the blocks of attribute-value pairs are:
[(Temperature, normal)] = $\{2, 4, 6, 8\}$,
[(Temperature, high)] = $\{1, 2, 7\}$,
[(Temperature, very-high)] = $\{2, 3\}$,
[(Headache, yes)] = U,
[(Cough, no)] = $\{3, 6, 8\}$,
[(Cough, yes)] = $\{2, 4, 6\}$.

For a case $x \in U$ and $B \subseteq A$, the *characteristic set* $K_B(x)$ is defined as the intersection of the sets $K(x, a)$, for all $a \in B$, where the set $K(x, a)$ is defined in the following way:

- If $a(x)$ is specified, then $K(x, a)$ is the block $[(a, a(x))]$ of attribute a and its value $a(x)$,
- If $a(x) = ?$ or $a(x) = *$, then $K(x, a) = U$.

For Table 1 and $B = A$,

$K_A(1) = \{1, 2, 7\}$,
$K_A(2) = \{2, 4, 6\}$,

$K_A(3) = \{3\}$,
$K_A(4) = \{2, 4, 6\}$,
$K_A(5) = U$,
$K_A(6) = \{2, 4, 6, 8\}$,
$K_A(7) = \{1, 2, 7\}$,
$K_A(8) = \{6, 8\}$.

A binary relation $R(B)$ on U, defined for $x, y \in U$ in the following way

$$(x, y) \in R(B) \; if \; and \; only \; if \; y \in K_B(x)$$

will be called the *characteristic relation*. In our example $R(A) = \{(1, 1), (1, 2),$
$(1, 7), (2, 2), (2, 4), (2, 6), (3, 3), (4, 2), (4, 4), (4, 6), (5, 1), (5, 2), (5, 3), (5,$
$4), (5, 5), (5, 6), (5, 7), (5, 8), (6, 2), (6, 4), (6, 6), (6, 8), (7, 1), (7, 2), (7, 7),$
$(8, 6), (8, 8)\}$.

For Table 1 and both concepts, all conditional probabilities $P(X|K_A(x))$ are
presented in Tables 2 and 3.

Table 2. Conditional probabilities $Pr([(Flu, yes)]|K_A(x))$

x	1	2	3	4
$K_A(x)$	$\{1, 2, 7\}$	$\{2, 4, 6\}$	$\{3\}$	$\{2, 4, 6\}$
$P(\{1, 2, 3, 4\} \mid K_A(x))$	0.667	0.667	1	0.667

Table 3. Conditional probabilities $Pr([(Flu, no)]|K_A(x))$

x	5	6	7	8
$K_A(x)$	U	$\{2, 4, 6, 8\}$	$\{1, 2, 7\}$	$\{6, 8\}$
$P(\{5, 6, 7, 8\} \mid K_A(x))$	0.5	0.5	0.333	1

We quote some definitions from [1]. Let X be a subset of U. The set X is
B-consistent if $(x, y) \in R(B)$ for any $x, y \in X$. If there does not exist a B-
consistent subset Y of U such that X is a proper subset of Y, the set X is called
a *generalized maximal B-consistent block*. The set of all generalized maximal B-
consistent blocks will be denoted by $\mathscr{C}(B)$. In our example, $\mathscr{C}(A) = \{\{1, 7\}, \{2,$
$4, 6\}, \{3\}, \{5\}, \{6, 8\}\}$.

Let $B \subseteq A$ and $Y \in \mathscr{C}(B)$. The set of all generalized maximal B-consistent
blocks which include an element x of the set U, i.e. the set

$$\{Y|Y \in \mathscr{C}(B), x \in Y\}$$

will be denoted by $\mathscr{C}_B(x)$.

For data sets in which all missing attribute values are "do not care" conditions, an idea of a maximal consistent block of B was defined in [10]. Note that
in our definition, the generalized maximal consistent blocks of B are defined for
arbitrary interpretations of missing attribute values. For Table 1, the generalized
maximal A-consistent blocks $\mathscr{C}_A(x)$ are

$$\mathscr{C}_A(1) = \{\{1, 7\}\},$$
$$\mathscr{C}_A(2) = \{\{2, 4, 6\}\},$$
$$\mathscr{C}_A(3) = \{\{3\}\},$$
$$\mathscr{C}_A(4) = \{\{2, 4, 6\}\},$$
$$\mathscr{C}_A(5) = \{\{5\}\},$$
$$\mathscr{C}_A(6) = \{\{2, 4, 6\}, \{6, 8\}\},$$
$$\mathscr{C}_A(7) = \{\{1, 7\}\},$$
$$\mathscr{C}_A(8) = \{\{6, 8\}\}.$$

For Table 1 and the concepts $[(Flu, yes)]$ and $[(Flu, no)]$, all conditional probabilities.

$Pr([(Flu, yes)] | Y)$ and $Pr([(Flu, no)] | Y)$, where $Y \in \mathscr{C}(A)$, are presented in Table 4.

Table 4. Conditional probabilities $Pr([(Flu, yes)] | Y)$ and $Pr([(Flu, no)] | Y)$

Y	$\{1, 7\}$	$\{2, 4, 6\}$	$\{3\}$	$\{5\}$	$\{6, 8\}$
$Pr(\{1, 2, 3, 4\} \mid Y)$	0.5	0.667	1	0	0
$Pr(\{5, 6, 7, 8\} \mid Y)$	0.5	0.333	0	1	1

3   Probabilistic Approximations

In this section, we will discuss two types of probabilistic approximations: based on characteristic sets and on generalized maximal consistent blocks.

3.1   Probabilistic Approximations Based on Characteristic Sets

In general, probabilistic approximations based on characteristic sets may be categorized as singleton, subset and concept [3,7]. In this paper we restrict our attention only to concept probabilistic approximations, for simplicity calling them probabilistic approximations based on characteristic sets.

A *probabilistic approximation based on characteristic sets* of the set X with the threshold α, $0 < \alpha \leq 1$, denoted by $appr_\alpha^{CS}(X)$, is defined as follows

$$\cup \{K_A(x) \mid x \in X, \ Pr(X | K_A(x)) \geq \alpha\}.$$

For Table 1 and both concepts $\{1, 2, 3, 4\}$ and $\{5, 6, 7, 8\}$, all distinct probabilistic approximations, based on characteristic sets, are

$$appr_{0.667}^{CS}(\{1, 2, 3, 4\}) = \{1, 2, 3, 4, 6, 7\},$$
$$appr_1^{CS}(\{1, 2, 3, 4\}) = \{3\},$$
$$appr_{0.5}^{CS}(\{5, 6, 7, 8\}) = U,$$
$$appr_1^{CS}(\{5, 6, 7, 8\}) = \{6, 8\}.$$

If for some β, $0 < \beta \leq 1$, a probabilistic approximation $appr_\beta^{CS}(X)$ is not listed above, it is equal to the probabilistic approximation $appr_\alpha^{CS}(X)$ with the closest α to β, $\alpha \geq \beta$. For example, $appr_{0.2}^{CS}(\{1, 2, 3, 4\}) = appr_{0.667}^{CS}(\{1, 2, 3, 4\})$.

3.2   Probabilistic Approximations Based on Generalized Maximal Consistent Blocks

By analogy with the definition of a probabilistic approximation based on characteristic sets, we may define a probabilistic approximation based on generalized maximal consistent blocks as follows:

A *probabilistic approximation* based on generalized maximal consistent blocks of the set X with the threshold α, $0 < \alpha \leq 1$, and denoted by $appr_\alpha^{MCB}(X)$, is defined as follows

$$\cup\{Y \mid Y \in \mathscr{C}_x(A),\ x \in X,\ Pr(X|Y) \geq \alpha\}.$$

All distinct probabilistic approximations based on generalized maximal consistent blocks are

$$appr_{0.5}^{MCB}(\{1,2,3,4\}) = \{1,2,3,4,6,7\},$$
$$appr_{0.667}^{MCB}(\{1,2,3,4\}) = \{2,3,4,6\},$$
$$appr_1^{MCB}(\{1,2,3,4\}) = \{3\},$$
$$appr_{0.333}^{MCB}(\{5,6,7,8\}) = \{1,2,4,5,6,7,8\},$$
$$appr_{0.5}^{MCB}(\{5,6,7,8\}) = \{1,5,6,7,8\},$$
$$appr_1^{MCB}(\{5,6,7,8\}) = \{5,6,8\}.$$

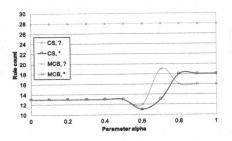

Fig. 1. Number of rules for the *bankruptcy* data set

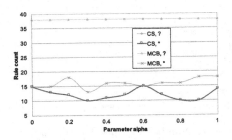

Fig. 2. Number of rules for the *breast cancer* data set

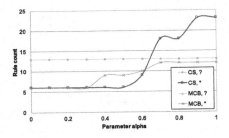

Fig. 3. Number of rules for the *echocardiogram* data set

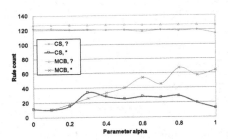

Fig. 4. Number of rules for the *hepatitis* data set

4   Experiments

We conducted our experiments on eight data sets from the University of California at Irvine *Machine Learning Repository*. These data sets were completely specified. We randomly replaced 35% of the existing, specified attribute values by question marks, indicating lost values. After that, all question marks were replaced by asterisks, indicating "do not care" conditions, so additional eight data sets were created. Thus, we ended up with 16 data sets.

In our experiments we used the MLEM2 rule induction algorithm of the LERS (Learning from Examples using Rough Sets) data mining system [2,5,6]. We used characteristic sets and generalized maximal consistent blocks for mining incomplete datasets. Additionally, we used incomplete data sets with lost values and with "do not care" conditions. Thus our experiments were conducted on four different approaches to mining incomplete data sets. These four approaches, denoted by (CS, ?), (CS, *), (MCB, ?), and (MCB, *), where "CS" denotes a characteristic set, "MCB" denotes a generalized maximal consistent block, "?" denotes a lost value and "*" denotes a "do not care" condition, were compared by applying the Friedman rank sum test combined with multiple comparisons, with a 5% level of significance. We applied this test to all pairs of incomplete data sets, with lost values and "do not care" conditions.

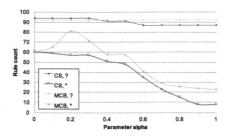

Fig. 5. Number of rules for the *image segmentation* data set

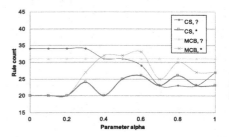

Fig. 6. Number of rules for the *iris* data set

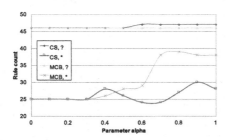

Fig. 7. Number of rules for the *lymphography* data set

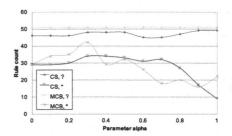

Fig. 8. Number of rules for the *wine recognition* data set

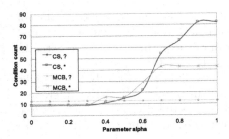

Fig. 9. Total number of conditions for the *bankruptcy* data set

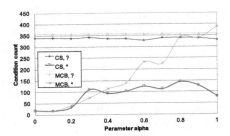

Fig. 10. Total number of conditions for the *breast cancer* data set

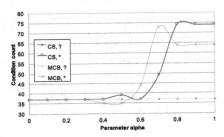

Fig. 11. Total number of conditions for the *echocardiogram* data set

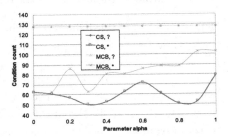

Fig. 12. Total number of conditions for the *hepatitis* data set

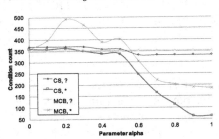

Fig. 13. Total number of conditions for the *image segmentation* data set

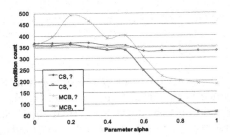

Fig. 14. Total number of conditions for the *iris* data set

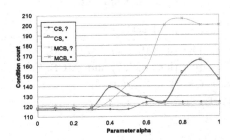

Fig. 15. Total number of conditions for the *lymphography* data set

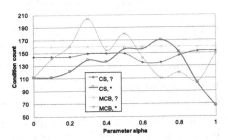

Fig. 16. Total number of conditions for the *wine recognition* data set

In our experiments we recorded the number of rules and the total number of conditions in rule sets induced using characteristic sets and generalized maximal consistent blocks from all data sets. Results of our experiments are presented in Figs. 1, 2, 3, 4, 5, 6, 7, 8, 9, 10, 11, 12, 13, 14, 15 and 16.

For our four approaches, the number of rules differ significantly for all eight data sets, i.e., the null hypotheses H_0 of the Friedman test saying that differences between these approaches are insignificant was rejected. Results of the post-hoc test indicated precisely the differences between the four approaches. Results are presented in Table 5.

Table 5. Results of statistical analysis for the number of rules

Data set	Friedman test results (5% significance level)
Bankruptcy	(MCB, *) is simpler than both (CS, ?) and (MCB, ?)
Breast cancer	(CS, *) is simpler than both (CS, ?) and (MCB, ?) (MCB, *) is simpler than (MCB, ?)
Echocardiogram	(CS, *) and (MCB, *) are simpler than (CS, ?) and (MCB, ?)
Hepatitis	(CS, *) and (MCB, *) are simpler than (CS, ?) and (MCB, ?)
Image recognition	(CS, *) is simpler than both (CS, ?) and (MCB, ?) (MCB, *) is simpler than (MCB, ?)
Iris	(CS, *) is simpler than (MCB, ?)
Lymphography	(CS, *) and (MCB, *) are simpler than (CS, ?) and (MCB, ?)
Wine recognition	(MCB, *) is simpler than (CS, ?) and (MCB, ?) (CS, *) is simpler than (MCB, ?)

For one data set (*bankruptcy*) the Friedman test shows statistical insignificance between all four approaches. For another data set (*echocardiogram*), the Friedman test shows statistical significance for some approaches, but the post-hoc test proves that the differences between the four approaches are statistically insignificant. For remaining six data sets the results are presented in Table 6.

Table 6. Results of statistical analysis for the total number of conditions

Data set	Friedman test results (5% significance level)
Breast cancer	(CS, *) and (MCB, *) are simpler than (CS, ?) and (MCB, ?)
Hepatitis	(CS, *) and (MCB, *) are simpler than (CS, ?) and (MCB, ?)
Image recognition	(CS, *) is simpler than both (CS, ?) and (MCB, *)
Iris	(CS, ?) is simpler than (MCB, *)
Lymphography	(CS, ?) is simpler than (MCB, *)
Wine recognition	(CS, ?) and (CS, *) are simpler than (MCB, ?)

5   Conclusions

Our objective was to compare four approaches to mining incomplete data sets (combining characteristic sets and generalized maximal consistent blocks with two interpretations of missing attribute values, lost values and "do not care" conditions). Our conclusion is that for the number of rules in a rule set, the choice for an interpretation of missing attribute values is more important than the choice between characteristic sets and generalized maximal consistent blocks. The number of rules is always smaller for incomplete data sets with "do not care" conditions. Surprisingly, the total number of conditions does not show a clear preference, here all depends on the choice of the data set. There is no universally best approach.

References

1. Clark, P.G., Gao, C., Grzymala-Busse, J.W., Mroczek, T.: Characteristic sets and generalized maximal consistent blocks in mining incomplete data. In: Polkowski, L., Yao, Y., Artiemjew, P., Ciucci, D., Liu, D., Ślęzak, D., Zielosko, B. (eds.) IJCRS 2017. LNCS (LNAI), vol. 10313, pp. 477–486. Springer, Cham (2017). https://doi.org/10.1007/978-3-319-60837-2_39
2. Clark, P.G., Grzymala-Busse, J.W.: Experiments on probabilistic approximations. In: Proceedings of the 2011 IEEE International Conference on Granular Computing, pp. 144–149 (2011)
3. Clark, P.G., Grzymala-Busse, J.W.: Experiments using three probabilistic approximations for rule induction from incomplete data sets. In: Proceedings of the MCC-SIS 2012, IADIS European Conference on Data Mining ECDM 2012, pp. 72–78 (2012)
4. Grzymala-Busse, J.W.: LERS-a system for learning from examples based on rough sets. In: Slowinski, R. (ed.) Intelligent Decision Support Theory and Decision Library, vol. 11, pp. 3–18. Kluwer Academic Publishers, Dordrecht (1992). https://doi.org/10.1007/978-94-015-7975-9_1
5. Grzymala-Busse, J.W.: A new version of the rule induction system LERS. Fundam. Inf. **31**, 27–39 (1997)
6. Grzymala-Busse, J.W.: MLEM2: a new algorithm for rule induction from imperfect data. In: Proceedings of the 9th International Conference on Information Processing and Management of Uncertainty in Knowledge-Based Systems, pp. 243–250 (2002)
7. Grzymala-Busse, J.W.: Rough set strategies to data with missing attribute values. In: Notes of the Workshop on Foundations and New Directions of Data Mining, in Conjunction with the Third International Conference on Data Mining, pp. 56–63 (2003)
8. Grzymała-Busse, J.W.: Generalized parameterized approximations. In: Yao, J.T., Ramanna, S., Wang, G., Suraj, Z. (eds.) RSKT 2011. LNCS (LNAI), vol. 6954, pp. 136–145. Springer, Heidelberg (2011). https://doi.org/10.1007/978-3-642-24425-4_20
9. Grzymala-Busse, J.W., Ziarko, W.: Data mining based on rough sets. In: Wang, J. (ed.) Data Mining: Opportunities and Challenges, pp. 142–173. Idea Group Publishing, Hershey (2003)

10. Leung, Y., Li, D.: Maximal consistent block technique for rule acquisition in incomplete information systems. Inf. Sci. **153**, 85–106 (2003)
11. Leung, Y., Wu, W., Zhang, W.: Knowledge acquisition in incomplete information systems: a rough set approach. Eur. J. Oper. Res. **168**, 164–180 (2006)
12. Pawlak, Z., Skowron, A.: Rough sets: some extensions. Inf. Sci. **177**, 28–40 (2007)
13. Pawlak, Z., Wong, S.K.M., Ziarko, W.: Rough sets: probabilistic versus deterministic approach. Int. J. Man-Mach. Stud. **29**, 81–95 (1988)
14. Ślęzak, D., Ziarko, W.: The investigation of the Bayesian rough set model. Int. J. Approx. Reason. **40**, 81–91 (2005)
15. Wong, S.K.M., Ziarko, W.: INFER–an adaptive decision support system based on the probabilistic approximate classification. In: Proceedings of the 6th International Workshop on Expert Systems and their Applications, pp. 713–726 (1986)
16. Yao, Y.Y.: Probabilistic rough set approximations. Int. J. Approx. Reason. **49**, 255–271 (2008)
17. Yao, Y.Y., Wong, S.K.M.: A decision theoretic framework for approximate concepts. Int. J. Man-Mach. Stud. **37**, 793–809 (1992)
18. Ziarko, W.: Variable precision rough set model. J. Comput. Syst. Sci. **46**(1), 39–59 (1993)
19. Ziarko, W.: Probabilistic approach to rough sets. Int. J. Approx. Reason. **49**, 272–284 (2008)

Optimization of the University Transportation by Contraction Hierarchies Method and Clustering Algorithms

Israel D. Herrera-Granda[1], Leandro L. Lorente-Leyva[1(✉)],
Diego H. Peluffo-Ordóñez[2], Robert M. Valencia-Chapi[1],
Yakcleem Montero-Santos[1], Jorge L. Chicaiza-Vaca[3],
and Andrés E. Castro-Ospina[4]

[1] Facultad de Ingeniería en Ciencias Aplicadas, Universidad Técnica del Norte,
Av. 17 de Julio, 5-21, y Gral. José María Cordova, Ibarra, Ecuador
{idherrera, lllorente, rmvalencia, ymontero}@utn.edu.ec
[2] Escuela de Ciencias Matemáticas y Tecnología Informática, Yachay Tech,
Hacienda San José s/n, San Miguel de Urcuquí, Ecuador
dpeluffo@yachaytech.edu.ec
[3] Department of Production and Logistics, Technische Universität Dortmund,
44221 Dortmund, Germany
jorge.chicaiza@tu-dortmund.de
[4] Instituto Tecnológico Metropolitano, Medellín, Colombia
andrescastro@itm.edu.co

Abstract. This research work focuses on the study of different models of solution reflected in the literature, which treat the optimization of the routing of vehicles by nodes and the optimal route for the university transport service. With the recent expansion of the facilities of a university institution, the allocation of the routes for the transport of its students, became more complex. As a result, geographic information systems (GIS) tools and operations research methodologies are applied, such as graph theory and vehicular routing problems, to facilitate mobilization and improve the students transport service, as well as optimizing the transfer time and utilization of the available transport units. An optimal route management procedure has been implemented to maximize the level of service of student transport using the K-means clustering algorithm and the method of node contraction hierarchies, with low cost due to the use of free software.

Keywords: Optimization · Vehicle routing · University transportation
K-means · Clustering algorithms · Contraction hierarchies · Free software

1 Introduction

Different methods, techniques and mathematical models have been developed to treat vehicle routing by nodes also known as VRP. The traveling salesman problem (TSP) is the first VRP-type problem, was proposed in 1959 [1], and it is one of the most complex problems (NP-hard) and more studied in Operational Research, especially in Combinatorial Optimization [2]. Exact treatment and heuristic methods have been

© Springer International Publishing AG, part of Springer Nature 2018
F. J. de Cos Juez et al. (Eds.): HAIS 2018, LNAI 10870, pp. 95–107, 2018.
https://doi.org/10.1007/978-3-319-92639-1_9

applied [3], however to the date, an optimal solution to this problem has not been achieved, especially in cases with huge numbers of nodes [2]. The VRP is the result of the intersection of the TSP traveling salesman problem and the Bin Packing Problem (BPP) [2]. The use and development of these methods in this field are the ones that dealt this problem significantly, because of having new and continuous advances in information technology as a fundamental basis for its general operation.

In 2012, Khalid et al. [4] presented the modeling of the school bus routing problem (SBRP), considering bus capacities and time windows in the school in which the bus operates, in order to obtain the information to use in the model, geographic information system (GIS) tools, grouping techniques based on available capacity, and network cutting techniques were used. To try to find a solution to the model (SBRP) they used the hybrid metaheuristics of ant colony optimization with the local improvement heuristic Lin Kernighan iterated.

Another proposal was made by Huo et al. in 2014, when presenting the construction of a mathematical model for the problem of routing school buses for a single center and a single vehicle [5]. To do this, they used location strategy methodologies - allocation - routing (LAR), and a clustering of students to determine optimal stops within a service radius of 500 m. The solution of the proposed model was through the ant colony optimization algorithm (ACO). Finally, it was shown that the model is applicable in real life, and that it can reduce the cost of school bus service.

Authors such as Nezam et al. [6] in 2016 proposed in Iran a complex mathematical model that pursues various objectives, such as the optimal allocation of multiple nodes of the population with their respective demands towards one or more server nodes. Also, find the optimal location of the server nodes and the optimal allocation of the population to different routes that allow reaching the assigned server nodes, so that the total transport time is minimized.

In the year 2017, Hashi et al. [7] established a model for the optimization of student transport that considers the capacity of the vehicle fleet, using the methodologies of the Geographic Information System (GIS), the formulation of the problem was through the mathematical model of the routing and programming of school buses (SBRS) with time windows. Then, the resolution of the proposed mathematical model was through the algorithm Clarke & Wright, obtaining as multiple results, optimized routes, it was also possible to identify the pick-up stops of the students according to their concentration, and the visualization of the routes and the stops in (SIG).

The problem of vehicle routing by nodes has also been treated by means of graph theory, one of the main algorithms used in this field is the Dijkstra algorithm [8], which allows the calculation of the shortest route through a graph consisting of a set of n client nodes joined by branches or arcs, which can even consider directions of the roads and their costs [9]. However, recent advances have been made in this field as the method of contraction hierarchies of nodes-CH which have demonstrated better yields in the time calculation of the shorter route than the Dijkstra algorithm [8, 10, 11], and that additionally allow to simulate vehicle routes in computational environments at low cost [12].

Within the field of vehicular routing, when modifications occur in the locations of the installations, it is necessary to determine a route that can link the different locations of both facilities and customers that require transportation in an optimal way [13]. In that sense, the investigation was initiated considering each of the clients or students to keep a record of the positions of the current residences. In order to compile a database that will be used to perform a clustering using the K-means algorithm [14], through its centroids allowed to determine the stops in which the vehicles must pick up the students and optimize the route which should continue to be allocated by the university for the students transportation.

2 K-means Algorithm

The zoning of the assigned areas to the transportation vehicles is an important aspect of vehicle fleet management, especially for the balancing of the workload between different fleet vehicles and even to determine the coverage of a transportation service [15–17], for that reason in this work we use the K-means algorithm whose performance has been demonstrated in the clustering of spatial databases [14].

Let $X = X_{i,i} = 1,2,...., n$; The set of n d-dimensional points to be grouped into a set of K groupings, $C = C_k, k = 1,...., K$. The K-means algorithm finds such a partition that the quadratic error between the empirical mean of a clustering and the points in the clustering is minimized [14, 18]. If we define μ_k as the mean of C_k, then we can define the square error J (C_k) between μ_k and the points in the cluster C_k can be defined by the following expression:

$$J(G_k) = \sum_{X_i \in C_k} ||X_i - \mu_k||^2 \tag{1}$$

The objective of the K-means algorithm is to minimize the sum of the square errors of the K clusters, which is a NP-hard problem [14].

$$Min\, J(C) = \sum_{k=1}^{k} \sum_{X_i \in C_k} ||X_i - \mu_k||^2 \tag{2}$$

To do this, K-means starts with an initial partition with K-clusters and assigns patterns to clusters to reduce the squared error.

Since the quadratic error always decreases with an increase in the number of K groups (with $J\,(C) = 0$ when $K = n$), it can be minimized only for a fixed number of groups [14].

2.1 Parameters of the K-means Algorithm

The user must indicate the following parameters:

1. K clusters number.
2. Clustering initialization.
3. Euclidean distance metric to calculate the distance between the points and the centers of grouping.

As a result of the execution of this algorithm, we obtain spherical groupings in the data [14].

Operation of the K-means Algorithm

(a) Two-dimensional input data with three groups.
(b) Three seed points selected as clustering centers and initial assignment of data points to groups.
(c) and (d) Intermediate iterations updating the labels of the groups and their centers.
(d) Final grouping obtained by the K-means algorithm in convergence.

The operation of the K-means algorithm is schematized in Fig. 1.

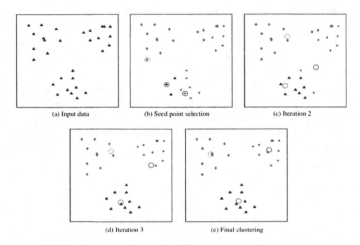

(a) Input data   (b) Seed point selection   (c) Iteration 2

(d) Iteration 3   (e) Final clustering

Fig. 1. Illustration of K-means algorithm [14].

2.2 Open Source Routing Machine (OSRM) and the Method of Node Contraction Hierarchies (CH)

Robert Geisberger in 2008 [8] developed the CH in his doctoral thesis, a method to find the shortest route improving the Dijkstra algorithm, linked it to the free SIG Open-StreetMap platform [19] and made it available to the public in a free way by creating the open source online platform called OSRM - Open Source Routing Machine [8].

CH is a technique to accelerate the calculation by nodes of the shortest vehicular route on a network road using routing by first creation or precomputed, contracting versions of the connection graph according to the importance or hierarchy of its nodes, the forward search uses only the leading arcs to more important nodes and the backward search uses only the coming arcs from more important nodes [8, 11].

The contraction hierarchy method makes the calculation of the shortest route more efficient in comparison with the Dijkstra algorithm, because in CH only the arcs are restored that join nodes with higher hierarchy and it is not necessary to restore all the shorter paths [8].

CH Definition

Let a trained and directed graph $G = (V, E)$ composed by client nodes V and connecting arcs E, in which the set of nodes is ordered by importance through a heuristic method, it is not necessary to know the order of all nodes before beginning to contract them but it is sufficient to know the next node to start contracting, then the selection of the next node is made by choosing the first one in a line generated by the linear combination of several terms of priority of the Candidate nodes. Thus, if $u < v$ then node v is more important than u, which is a method for hierarchizing nodes [8, 10].

Considering any pair of nodes of the graph (v, w), if the shortest route from v to w passes through node u is smaller in said graph, a new arc must be added to guarantee that the procedure is determining the shortest route. The set of the added arcs in the total-CH graph is treated as shortcuts with their respective weights [8]. See Figs. 2, 3 and 4.

Algorithm 1: Simplified Construction Procedure $(G = (V,E), <)$

```
1   foreach u ∈ V ordered by < ascending do
2       foreach (v,u) ∈ E with v>u do
3           foreach (u,w) ∈ E with w>u do
4               if (v,u,w) "may be" the only shortestway from v to w then
5                   E := E U{(v,w)}(use weigth w(v,w):= w(v,u)+w(u,w))
```

Fig. 2. Pseudocode of the node-contraction hierarchy method-CH [8]

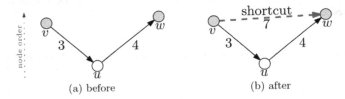

(a) before   (b) after

Fig. 3. Adding a new shortcut. Node order: $u < v < w$ [8, 10].

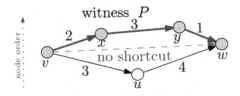

Fig. 4. Inclusion of the witness path [8, 10].

The preprocessing of the initial graph generates a CH graph which contains the data of the initial graph plus the ordering of nodes and the direct access arcs introduced [8].

2.3   Use and Storage of Witness Path in the CH

The CH uses a witness path or also called witness P of lower cost than the initial route, which is stored to determine the necessary shortcuts during the hiring of node u. Figure 5 shows a witness path "witness P" for the route (v, u, w), depending on the weights of the changed arcs, the witness path may remain valid or not [8, 11].

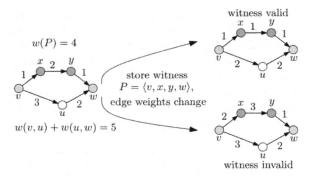

Fig. 5. Method of CH and use of witness paths [8, 10, 11].

2.4 Collection of Data Information

The information rising from all students was made through the institutional databases and personal surveys of the student population. Then they were placed on a scale-map of Ibarra city by means of AutoCAD software to obtain Cartesian coordinates in the X and Y axes of each student of the study population, 314 students. The center of this coordinate system coincided with the Pedro Moncayo Square located in the center of the mentioned city, Fig. 6.

Fig. 6. Spatial distribution of student homes in Ibarra city.

Figure 7 shows the initial route of the buses for the students transfer, the distance traveled was 8.2 km in 18 min and the total time to complete the tour was 30 min, so, the one-way trip to the destination called "UTN stadium" and the return to Universidad Técnica del Norte. It should be mentioned that the route offered a low level of service to the students, about 22.61%, the same that will be evaluated in the analysis of results.

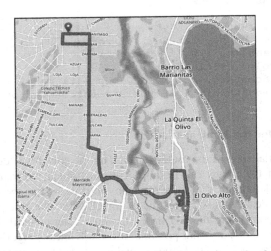

Fig. 7. Initial route traveled by university transportation buses

2.5 Data Processing

To expose the problem, routing with the OSRM application was performed by collecting a significant sample of 50 student nodes from the total of 314 directly in their homes and taking them through the three fixed installations to which they had to arrive, it is say, FICA, UTN stadium, and Old Hospital San Vicente de Paul. As a result, from the first route, a long and costly route was obtained in terms of distance and time (56.8 km to be traveled in more than 2 h and 8 min), as it exceeded the students' 30-minute travel expectation according to surveys Performed on all of them.

It should be mentioned that, if an attempt is made to create a route that collects the 314 students directly in their homes, the duration of such route would rise exponentially in distances. This indicated that it was necessary to facilitate the problem so that the route is feasible in terms of travel times.

Through a literature review, it was determined to smooth the problem using the clustering of the student population and its centroids by means of the K-means algorithm so that vehicular routing is feasible [14–16, 20, 21].

2.6   Grouping of Student Nodes

To perform the homogeneous clustering of the student nodes, the R free software was used, in which the database of the 314 students was called "cluster3", which contains the location X and Y of their homes, as you can see in the next code:

```
library(readxl)
> cluster3
# a 2-dimensional example (x,y)
x <- cluster3
colnames(x) <- c ("x", "y")
#clustering the population in 7 zones
(cl <- kmeans (x, 7))
plot (x, col = cl$cluster)
points (cl$centers, col = 1:2, pch = 8, cex=2)
kmeans(x,1)$withinss # if you are interested in that [22]
```

2.7   Guidelines for the Implementation

First of all, the initial state of the routes for pick up the students is determined, then it must be defined if the routes can be improved, for which the facilities are located geographically and through the survey to the students, it can be determined their directions and the desired maximum time of travel. Next, the students' door-to-door collection is simulated using the CH method, which determines if the routes can be improved. In the case of not being possible, use the K-means algorithm to determine the number and location of optimal stops to pick up the students and simulated optimal route again using the CH method, but this time through the optimal stops. Finally, the amount of improvement is determined by means of the proximity indicator of a transport service, Fig. 8.

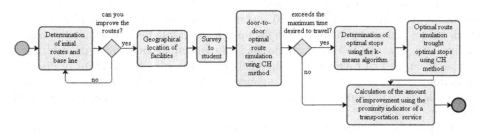

Fig. 8. Flow chart for the implementation process

3   Results and Discussion

Through various executions of K-means algorithm, the population of 314 students was divided into several k-groupings and the data of the locations of the stops were tabulated, in which the students were picked up. Each K-cluster has a K-centroid which was interpreted as the optimal location from the student transportation vehicle stop, Fig. 9.

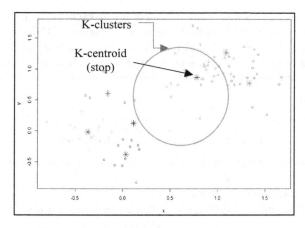

Fig. 9. Interpretation of K-means clustering.

3.1   Simulation of Routes with Multiple Stops

Using the free OSRM software, we simulated the different routes that the vehicle could take when picking up students. This was achieved by varying the number and position of their stops (K-centroids) from R-Project by executing the algorithm K-means as shown in Table 1. It was determined that the execution with 4 stops (K-centroids) and the three fixed installations would be the optimal because it meets the requirements of the students, to complete a tour in approximately 30 min at the lowest possible cost.

Table 1. Simulated routes in OSRM.

		K-means clustering	OSRM routing	
		Stops	Travel time (min)	Distance traveled (km)
Fixed installations	FICA	27	11
	Estadio UTN	18	7,8
	Old HSVP	23	8,8
K-clustering	3	34	16,8
	4	33	16,3
	5	39	20,2
	6	47	24,1
	7	52	28,5
	8	50	27,2
	9	56	29,7
	10	50	25,5

Considering that the project wants to determine an optimal route of collection and transfer of students from the Industrial Engineering Career - CINDU from the closest possible to their homes, it was found that the optimum route would be the grouping with 4 stops and three fixed installations, illustrated in Table 2, which would fulfill the required time by the students to the full course, with a total travel time of 16.3 km in 33 min.

Table 2. Tour with 4 stops (optimal route).

K-groupings	Data of grouping in R		Address located on the map type grid
4	X coordinate	Y coordinate	
	848,2500	1846,0250	North Pan American Customs Detention Center
	809,0452	1464,5400	Olivo Panamericana Norte Street and Dr. Luis Madera Street
	−904,8806	1206,2410	Jose Hidalgo Street and Alfredo Gómez Avenue
	−783,4566	−1241,6970	Street Eduardo Almeida and Street Carlos E. Grijalva (Yacucalle Sector)

The objective of this project is to facilitate the mobilization of students, to optimize the time of transfer, using free software, geographic information systems (GIS) tools, such as QGIS and the application of operations and logistics research methodologies such as algorithms previously reviewed.

Using the route proposed in the present project, the transportation times of CINDU students of the Technical University of North would be optimized, with a single transportation unit being necessary. To provide a transportation coverage level of 100% of the population of 314 students, with a duration of 30 min, the implementation of another unit is suggested. It should be noted that the Faculty of Engineering in Applied Sciences has a unit assigned, so a single route is proposed.

3.2   Comparison of the Routes by the Proximity Indicator of Transportation Stops

By means of the Free Quantum GIS software 2.18.2, it was possible to calculate the areas of influence to the transportation routes by foot in 5 min, that is, the proximity of the residences of the students within a radius of 300 m of proximity with respect to the routes covered was calculated, which is one of the main indicators of the level of urban transportation service [23]. As shown in Fig. 10.

Fig. 10. Proximity to the transportation route in the initial and optimized route

It was proved that the proximity indicator of transportation stops, in the initial route was of 22, 61% which causes a low level of service of the offered transportation. This indicator is improved with the optimal route proposed which offers a coverage level of 88%, Table 3.

Table 3. Route with 4 stops (optimal route)

	Students located 300 m from the route	Indicator of access to transportation stops (%)
Initial route	71	22,61
Optimized route	277	88,22

4   Conclusions

With the development of this project, it takes into consideration the realization and implementation of an optimal route management tool to minimize times and raise the level of service. It is an affordable procedure that uses models and algorithms for road optimization and fleet frequency.

During the grouping of the population of students for the generation of the optimal route that must traverse the transport vehicle it was determined that each K-centroid is a possible stop that the vehicle must carry out to guarantee the maximum coverage of the service of Student transportation, as demonstrated.

Free access software RStudio and OSRM, allowed the implementation of a routing pattern to establish shorter paths, and visualize the shortest route to go from a stop to the fixed installations where you want to arrive, and is to say, to the central campus of the Technical University of the North, old Hospital San Vicente de Paul and University Stadium.

The application OSRM simulated graphically the route and the distance to be traversed by the user, establishing 16.3 km and travel of 33 min, with an indicator of access to stops of the transport of 88.22%, optimizing to the maximum the available resources.

It is suggested, in future research to work on an implementation that includes all the methods and algorithms described in this work in the same database of the client nodes, to develop a software or application that reduces the times in the Processing of the data that in this project are carried out in multiple software's independently, which can be applied in the institutional and business area helping to optimize the routes of transport.

References

1. Dantzig, G.B., Ramser, J.H.: The truck dispatching problem. Manage. Sci. **6**(1), 80–91 (1959)
2. Toth, P., Vigo, D.: Vehicle Routing Problem, Methods, and Application (2014)
3. Martí, R.: Procedimientos Metaheurísticos en Optimización Combinatoria. Dep. d'Estadística i Investig. Oper. pp. 1–60 (2001)
4. Eldrandaly, K.A., Abdallah, A.M.F.: A novel GIS-based decision-making framework for the school bus routing problem. Geo-Spatial Inf. Sci. **15**(1), 51–59 (2012)
5. Huo, L., Yan, G., Fan, B., Wang, H., Gao, W.: School bus routing problem based on ant colony optimization algorithm. In: 2014 IEEE Conference of Expo and Transportation Electrification, Asia-Pacific (ITEC Asia-Pacific), no. 1, pp. 1–5 (2014)
6. Shiripour, S., Mahdavi-Amiri, N., Mahdavi, I.: Optimal location-multi-allocation-routing in capacitated transportation networks under population-dependent travel times. Int. J. Comput. Integr. Manuf. **29**(6), 652–676 (2016)
7. Hashi, E.K., Hasan, M.R., Zaman, M.S.U.: GIS based heuristic solution of the vehicle routing problem to optimize the school bus routing and scheduling. In: 9th International Conference on Computer and Information Technology, ICCIT 2016, pp. 56–60 (2016)
8. Geisberger, R., Sanders, P., Schultes, D., Delling, D.: Contraction hierarchies: faster and simpler hierarchical routing in road networks. In: McGeoch, C.C. (ed.) WEA 2008. LNCS, vol. 5038, pp. 319–333. Springer, Heidelberg (2008). https://doi.org/10.1007/978-3-540-68552-4_24
9. Dijkstra, E.W.: A note on two problems in connexion with graphs. Numer. Math. **1**(1), 269–271 (1959)
10. Geisberger, R., Sanders, P., Schultes, D., Vetter, C.: Exact routing in large road networks using contraction hierarchies. Transp. Sci. **46**(3), 388–404 (2012)
11. Batz, G.V., Geisberger, R., Sanders, P., Vetter, C.: Minimum time-dependent travel times with contraction hierarchies. J. Exp. Algorithm. **18**, 1.1–1.43 (2013)
12. Open Source Routing Machine DEMO (2017). http://map.project-osrm.org/. Accessed 23 Mar 2017
13. Fu, L., Sun, D., Rilett, L.R.: Heuristic shortest path algorithms for transportation applications: state of the art. Comput. Oper. Res. **33**(11), 3324–3343 (2006)
14. Jain, A.K.: Data clustering: 50 years beyond K-means. Pattern Recognit. Lett. **31**(8), 651–666 (2010)
15. Shen, C.-W., Quadrifoglio, L.: Evaluation of zoning design with transfers for paratransit services, no. 2277 (2012)

16. Ng, W.L., Leung, S.C.H., Lam, J.K.P., Pan, S.W.: Petrol delivery tanker assignment and routing: a case study in Hong Kong. J. Oper. Res. Soc. **59**(9), 1191–1200 (2008)
17. Yossi, S.: Logistics Clusters. Mit Press, Cambridge (2012)
18. Kanungo, T., Mount, D.M., Netanyahu, N.S., Piatko, C.D., Silverman, R., Wu, A.Y.: An efficient k-means clustering algorithms: analysis and implementation. IEEE Trans. Pattern Anal. Mach. Intell. **24**(7), 881–892 (2002)
19. OSMF, OpenStreetMap (2017). https://www.openstreetmap.org/#map=5/51.500/-0.100. Accessed 07 Apr 2017
20. Kim, B.-I., Kim, S., Sahoo, S.: Balanced clustering algorithms for improving shapes on vehicle routing problems. In: IIE Annual Conference and Exhibition 2004 (2004)
21. Herrera, I.: Diseño y evaluación de un algoritmo genético para ruteo vehicular que permita optimizar la distribución en una empresa comercializadora de autopartes en quito, Escuela Superior Politécnica del Litoral (2015)
22. R Project: R: K-Means Clustering (2016). https://stat.ethz.ch/R-manual/R-devel/library/stats/html/kmeans.html
23. Rueda, S.: Plan Especial de Indicadores de Sostenibilidad Ambiental de la Actividad Urbanística de Sevilla. Agencia Ecol. Urbana Barcelona, Barcelona (2007)

Identification of Patterns in Blogosphere Considering Social Positions of Users and Reciprocity of Relations

Krzysztof Rudek[✉] and Jarosław Koźlak

Faculty of Computer Science, Electronics and Telecommunications,
AGH University of Science and Technology, Kraków, Poland
{rudek, kozlak}@agh.edu.pl

Abstract. The aim of the paper is to identify and categorize frequent patterns describing interactions between users in social networks. We consider a social network with already identified relationships between users which evolves in time. The social network is based on the Polish blog website pertaining on socio-political issues salon24.pl. It consists of bloggers and links between them, which result from the intensity and characteristic features of posting comments. In our research, we discover patterns based on frequent and fast interactions between pairs of users. The patterns are described by the characteristics of these interactions, such as their reciprocity, the relative difference between estimates of global influence in the pairs of users participating in the discussions and time of day of the conversation. In addition, we consider the roles of system users, determined by the number of interactions initiating discussions, their frequency and the number of strong interactions in which users are involved. We take into account how many such intense conversations individual users participate in.

Keywords: Social network analysis · Social relations
Identification of social patterns · Blogosphere

1 Introduction

The aim of the work is to identify and analyze patterns frequently occurring in the social network evolving in time. Finding patterns allows to determine which relationships in the social network are particularly important and how to better understand their nature, which in turn may lead to a better understanding of the behavior of the entire network and how individual users affect each other and the entire community. Such information may help to understand ways of propagation ideas, which can be used to promote products or political views, and also to identify, for example, unlawful activities, their sources and ways of propagating them.

The analysis of blogosphere can be a useful area for testing solutions to such problems, due to open access to this data, a large scope of available data and the ability to observe the interactions between individual users in a long (multi-year) time horizon.

The graph evolving in time is described by a set of subsequent graphs represented its structure in subsequent time intervals. This allows us to analyze many graphs with a similar set of vertices and extract such an element in the graph (both from the point of

© Springer International Publishing AG, part of Springer Nature 2018
F. J. de Cos Juez et al. (Eds.): HAIS 2018, LNAI 10870, pp. 108–119, 2018.
https://doi.org/10.1007/978-3-319-92639-1_10

view of the existence of individual relationships, sets of relationships and attributes assigned to vertices and edges) which often occurs.

2 State of the Art

In this section we will discuss issues related to the states of a social networks (with data coming usually from networking portals or phone calls), characteristics of relations between users in such portals and types of frequent patterns which may be identified in such graph.

Different names are used for frequently present elements in networks: patterns [10], cascades [5], persistent cascades [12], graphlets [6], fundamental structures [9], motifs [8] or sub-graphs [11]. In this paper we will use the name "patterns". Considered patterns are usually small bricks building networks containing several (usually 3–4) nodes.

The authors of [6] analyze patterns in both statics and dynamics networks. In [8] are analyzed temporal patterns with events which do not overlap in time and considered patterns allow to describe a sub-network topology and a sequence of events. Considered data sets contains information about phone calls. In [10] are studied patterns in data coming from blogosphere and they describe relations between posts or blogs. In [12] the authors analyze patterns containing few nodes in networks created from the data about phone calls. In [14] are considered casual paths of events, temporal distances between events are calculated as well as their distribution and correlations. Measures of closeness centrality are calculated for nodes in analyzed graphs. The dataset concerns phone calls, air transport networks and random graphs. In [7] are identified motifs and dynamic motifs of the network which appear in the network with highest frequency. The authors analyze dynamic motifs mostly having 3 nodes and use the measure of the statistical significance comparing results for *Yahoo Answers* and *Flickr* with the results obtained for random graphs.

In [5] the patterns are considered for datasets gathered from Twitter and containing posts related to crisis situations.

Another approach is an identification of heavy subgraphs [1], and their analysis for time evolving networks. They are the sub—graphs with highest scoring which means having the highest weights of the connections.

Fundamental structures [9] are identified for temporal networks describing frequent meetings in graphs representing meeting of the persons. The authors consider different kinds of networks concerning meeting in various special locations and time periods.

In [4] dynamic patterns are used for the identification of contributions of nodes and paths in the information flow.

In the presented works the patterns were identified in different ways. What distinguishes our approach is to include in the patterns not only links between users and their topologies, but also additional attributes describing users and character of the relationship (absolute and relative social significances, timely distribution of interactions, their intensity, duration, etc.).

3 Methodology

In order to build relations between users in a community portal, all interactions between users are usually considered. In the case of the blogosphere, both the relationship between the author of a given comment and the author of the post in the context of which the comment is written, or between the author of the comment and the author of the commentary, which is commented, may be considered [3].

In our work, we focus on specific types of connections, which we call *strong relationships,* as we consider comments written quickly after the post or comment they respond to. We consider two different time intervals: of 5 min and 15 min. Another particularly important feature describing relationships is the reciprocity - if both users comment on their posts and comments.

The aim of the work is to distinguish patterns describing such strong and frequent connections occurring on the social portal. The patterns include the following parts:

- describing topology of relations between users (RT),
- containing attributes describing relationships (RA),
- social and mutual relations of users (SP).

Therefore, the analyzed pattern can be written as the following n-k

$$P = (RT, RA, SP) \tag{1}$$

Topology of relations between users
The topology is associated with the edges in the network, which are included in the considered pattern. The simplest pattern consists of two users/nodes and one edge, and on those we focus in this work. However, more complex patterns covering more users may also be considered.

The next main issue is whether the relationship is mutual and to what extent, i.e. how many posts and comments of each party meets the appropriate commentary of another member of the pattern.

Features of relations between users (RA)
Relations in the patterns can also be characterized by additional, detailed attributes describing the temporal characteristics of the interactions taking place between the users in question. In particular, the duration of their interactions and their volume are considered. Another important feature describing the relationship is the intensity of interactions at particular periods of the day.

Social and mutual position of users (SP)
An important feature characterizing the relations between users are the social positions of the users participating in them, as well as the mutual relationship between these social positions. Different classes of relationships are the ones between users of high social importance, between users with insignificant social positions, and different between a user with a high social importance and a small social importance.

4 Description of Experiments

4.1 Characteristics of Interactions in Network

The social network that we used for our research is based on data from the www. salon24.pl portal [3, 15]. Dataset was based on objects representing portal users, which are nodes in the network we built, as well as on objects representing posts and comments written by portal users. All objects are described with a set of attributes, which allowed to create an edge between the vertices of the graph and study how these edges change over time. The aim of our research is to find attributes of relations between portal users that will allow us to classify these relationships and find behavioral patterns. To better understand data and to find patterns based on interactions' attributes, we first need to look at amount of data and it's type. We want to know what is the quantitative dependence between the interactions in a given time interval for quick reactions in this range. Another feature we are considering at this stage is the same relation to fast interactions that were reciprocated equally quickly. The average number of interactions in the studied time intervals is 100000. The graphs below (Figs. 1 and 2) show how many fast interactions and mutual quick interactions have occurred, with the assumed delays of 5 min and 15 min.

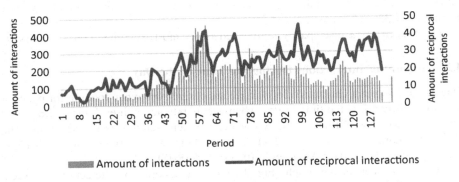

Fig. 1. Amount of interactions and reciprocal interactions (5-min slot).

At the beginning of the graph analysis, it should be noted that the test data was collected from the moment when there was a large amount of interaction between agents in the social network, however it was still the initial stage of network development. Therefore, the increase in both the number of quick interactions between agents and mutual relationships is natural. Looking at Fig. 1. Presenting amount of interactions and reciprocal interactions in 5-min slot, three main peaks are noticeable - they all depend on important socio-political events. Bearing in mind the average number of interactions in the studied time intervals, we note that the number of fast interactions is low, between 20 and 460, with an average value of 149. On average, we report interrelations 22 over a given time interval, with a minimum value of 2 and a maximum of 46.

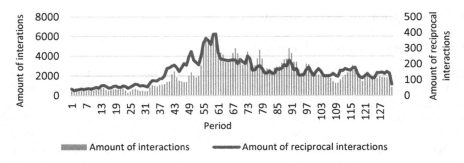

Fig. 2. Amount of interactions and reciprocal interactions (15-min slot).

At a chart presenting 15-min time slot (Fig. 2) we again notice a gradual increase in both values. In contrast, only one summit is dominant, for which an important socio-political event is again responsible. The lack of very visible other peaks, which we can see in the case of the 5-min slot, can be explained by the lower overtones of these events in the country and in the world. We can easily see, however, that we are dealing here with many times greater representation of data. The number of interactions, classified as fast, increased to an average of 2151, with a minimum of 296 and a maximum of 5787. The number of interactions increased on average to 147, with a minimum of 33 and a maximum of 391.

Due to such a large discrepancy, we decided to base most experiments on a 15-min slot.

4.2 Amount of Interactions During the Time of the Day

One of the important characteristics that we are considering when looking for patterns in a social network is the time of day when agents interact with each other. Based on the test set, we chose 4 times of the day, which we then used to classify the interaction. They are: morning (06.00 AM–10.00 AM), working hours (10.00 AM–06.00 PM), evening (06.00 PM–00.00 AM), night (00.00 AM–06.00 AM).

As can be seen in the graph (Fig. 3), the vast majority of interactions between agents takes place during business hours and in the evenings. The increases in the moments of the highest peaks are distributed proportionally. This proves a certain regularity in interactions, which can be used as an important factor to search for patterns in the social network.

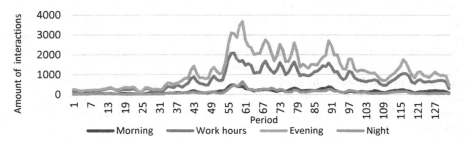

Fig. 3. Amount of interactions during the time of day (15-min slot).

Based on the knowledge presented in the previous graphs (Figs. 2 and 3), it can be said that the distribution of the amount of rapid and mutual interactions between agents looks predictable (see Fig. 4). The increases are proportional and, as in the case of ordinary interactions, most often occur during working hours and in the evenings. It is worth mentioning that the distribution for this 5-min interval looks very similar. However for this characteristics as well as some in further research presented in next paragraphs, distribution of interactions for 5-min slot has not been presented, as, due to the small amount of interaction, in further research we focus on the 15-min interval. This decision was caused by the knowledge of the portal at once, where serious socio-political topics are raised, and often it takes more than 5 min to read the content published by the user.

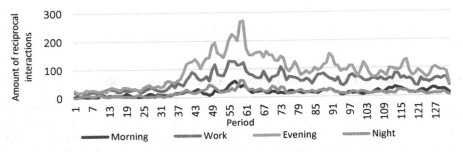

Fig. 4. Amount of reciprocal interactions during the time of day (15-min slot).

4.3   Cardinality of the Sets in Terms of the Number of Interacting Agents

Having knowledge about how fast and mutual interactions are and how the day the agents interact with each other, we want to check with how many agents they usually interact with. For this purpose, we have divided into groups of agents who interact with one, two, three etc. with other agents. Then, for ease of presentation, we made a grouping of results, which is shown in the graphs below.

In the case of a 5-min slot (Fig. 5), we received two groups. The first one includes those agents who interact with up to 5 other agents. This is the vast majority. The second group are agents who interact with more than 5 agents. It is a very small group, but from the point of view of looking for patterns, it is very important, allowing for the bolding of the characteristics of individual interactions.

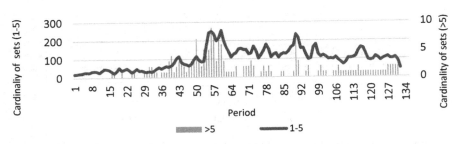

Fig. 5. Cardinality of the sets in terms of the number of interacting agents (5-min slot).

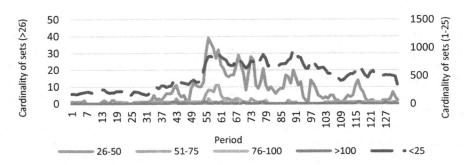

Fig. 6. Cardinality of the sets in terms of the number of interacting agents (15-min slot).

The characteristics for 15-min slot (Fig. 6) looks similar. The vast majority of agents interact with a group of several to 25 other agents.

4.4   Cardinality of the Sets in Terms of Agent's In-Degree Ratio

Very important factor influencing the establishment of interaction between agents is their own characteristics. Agents are characterized by values such as in-degree and out-degree, describing how much the agent brings to the network and how the other members of the network are focused on his actions. We made an analysis that grouped together agent pairs, taking the ratio of the in-degree ratio of each pair as a factor. It turned out that we are dealing with six groups. Markings on the graph indicate the agent's in-degree coefficient in relation to which the executed action takes place.

In the case of quick interactions, it turned out that the agents most often interact with those who have the in-degree parameter higher several times (see Fig. 7). The next in terms of population size was the one containing agents affecting over a hundred times more influential agents. It is worth noting that at the moments of the highest stitches, it is even the most numerous group. The groups of agents interacting with those whose in-degree parameter is higher several dozen times or a few times lower are not much smaller. On the other hand, agents rarely interact with those much less influential.

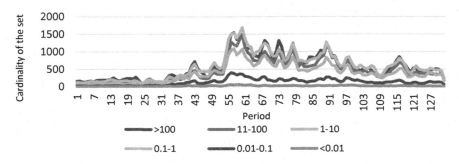

Fig. 7. Cardinality of the sets in terms of agent's in-degree ratio (15-min slot).

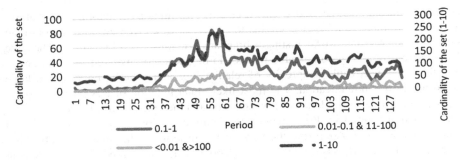

Fig. 8. Cardinality of the sets in terms of agent's in-degree ratio for reciprocal interactions (15-min slot).

The same division made for quick and mutual interactions has emerged four groups of agents (two overlap, because they are symmetrical) (see Fig. 8). It is very clearly visible that agents who interact quickly and interact with each other have very similar characteristics in terms of social network (in-line several times higher, or several times lower). Such situations are almost unnoticeable in the case of very large differences with the pair.

4.5 Studied Relationships

In order to check how the surveyed characteristics translate into the characteristics of the existing relationships, we examined the relationships between agents, characterized by the highest number of interactions. It turned out that in a small collection of 14 accounts, there were different types of relationships, both one-sided and mutual, lasting for a long time, as well as those with very frequent intervals, based on agents with similar characteristics as well as very diverse ones. The relations are described by the following characteristics: (A) in-degree ratio at the beginning of relationship, (B) in-degree ratio at the end of the relationship, (C) in-degree ratio amplitude, (D) – in-degree ratio noted when relationship was mutual, (E) Durability of unilateral relationship (periods when interaction was present), (F) Reciprocity [%], (G) – Reciprocity (periods when relationship was mutual), (H) Longest sequence of periods with interactions, (I) Longest sequence of periods without interactions, (J) Longest sequence of periods with mutual interactions, (K) Longest sequence of periods without mutual interactions, (L) Percentage of interactions taken in the morning [%], (M) Percentage of interactions taken during business hours [%], (N) Percentage of interactions taken in the evening [%], (O) Percentage of interactions taken in the night [%].

Looking at the results presented in Table 1, we can specify several groups to which the given relationships fit. The first of them is defined by relationships 4 and 5. It is characterized by the lack of reciprocity, longevity and a very large difference in in-degree parameters describing two vertices in a pair. Moreover, it is worth noting that the vast majority of interactions take place in the evening and at night. So here we have an example of relationships, when one group of individuals who does not get enough in the social network by writing their own posts, communicates with very influential individuals during the night hours. This is certainly a common behavior pattern that can be easily determined and used to predict the behavior of a given group of users in the future. Another group worth distinguishing are relationships 1, 6 and 7. They are characterized by a very comparable ratio of in-degree users involved in communication, there is a noticeable reciprocity of relationships and, what is important, a large single break in communication. The most likely interpretation of this phenomenon is the late establishment of relations between users, and then the relationship stabilization. This is also supported by high coefficients defining the longest sequences in which interactions occurred. Analyzing the relationship in terms of the time of establishing interaction, it is clearly visible that these relations, which are based on interactions at night, are most often one-sided relations. This thesis is supported by the relations 3.4 and 5. An interesting and very characteristic is the relations 12. This is a highlight of the results we have obtained, proving that the mutual relationship is favored by the situation when both sides share a similar social network. It is clearly visible that the bilateral relationship was established at the moment when the commenting person began to be more involved in the network and ceased to be anonymous.

Table 1. Characteristics of the studied relationships, with the most frequent interactions.

	A	B	C	D	E	F	G	H	I	J	K	L	M	N	O
1	2	0.5	0.5–2	0.5–2	104	26.92	28	38	20	2	42	75	25	0	0
2	1.5	0.1	0.1–1.5	0.1–0.5	96	3.13	3	31	16	1	70	25	50	25	0
3	1.5	0.15	0.15–1.5	X	84	0	0	36	34	0	132	0	0	100	0
4	1500	500	500–1500	X	84	0	0	26	29	0	132	0	0	100	0
5	381	1500	100–1500	X	83	0	0	28	39	0	132	0	20	80	0
6	565	10	2–565	2–5	80	7.5	6	24	30	2	59	30	40	30	0
7	1	0.1	0.1–2	1	73	11.2	9	8	24	5	62	0	35	65	0
8	1.5	0.3	0.3–1.5	X	69	0.00	0	11	18	0	132	0	45	55	0
9	2	0.25	0.25–1	X	67	0.00	0	29	34	0	132	25	45	30	0
10	0.4	0.4	0.35–0.4	0.35–0.4	62	1.61	1	25	52	1	79	0	25	75	0
11	2	20	2–20	2–12	59	52.54	31	17	36	17	35	0	0	65	35
12	5000	10	10–5000	10–20	56	69.23	18	46	63	7	67	0	15	85	0
13	385	0.3	0.3–385	X	55	0	0	23	29	0	132	20	20	60	0
14	1	0.5	20	0.5–40	52	25	13	22	58	4	58	0	25	75	0

4.6   The Influence of In-Degree Ration on Relationship

It should also be noted that the dominant factor that affects the development of the relationship between users is the ratio of their impact on the development of the network - the in-degree ratio. In Fig. 9 we show the ratio of the amount of interaction to mutual interactions depending on the in-degree ratio.

We note here that the highest chances of getting reciprocity in a relationship are those users whose in-degree parameter in the network is up to 10 times smaller than the commented person.

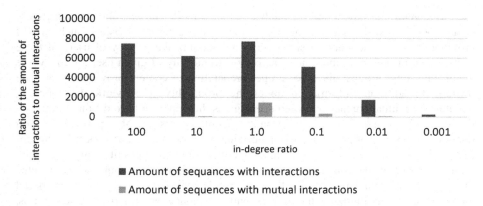

Fig. 9. Ratio of the amount of interaction to mutual interactions depending on the in-degree ratio (15-min slot).

4.7 Experiments Conclusions

The goal set at the beginning of the work was to find relation patterns based on given attributes. At work we were able to analyze the relationships that occurred in a given social network. We presented the distribution of fast relations depending on the time window that the other side of the relationship had on reacting. As it turned out, increasing this window three times allowed to determine a 15-fold larger set of relationships. Similarly, we have done categorizing interactions depending on the time of the day, getting the answer to the information when there are the most, and therefore, when establishing communication gives the highest chance of response. An important piece of information for further analysis is that the vast majority of mutual relations occur where two parties with a similar social position in the network interact. We additionally presented specific relations, along with their characteristics, showing that with their help we are able to categorize a given relationship, and more importantly, and what is the subject of further work, predict how relations will behave in successive intervals. Our analyzes allowed to identify characteristics such as the frequency of interaction, the speed of interaction, their reciprocity and the time when they took place, which are key to determining patterns of relationships in the social network. Nowadays, with the increasing amount of data, it is increasingly difficult to build long-term relationships with many users and, as described in the article "The influence of their strength of importance in blogosphere", these relationships are very often based on trust in the network. By using information about relationship patterns in a given social network, we are able to appropriately profile the activities of specific agents/users in order to achieve a specific goal.

5   Conclusion

In the paper we focus on the strong relationships between users in the blogosphere. The characteristics of such relations are analyzed considering on the social positions of the users, their duration and the periods of day with their highest intensity. The techniques used to develop the model and the analysis of such modeled data, allow to effectively create social networks without the use of tools from external websites (such as API provided by social networking sites) while maintaining the privacy of users of these websites. Information on the time of activity and intensity of given user groups, while identifying these groups, allow for proper preparation of content provided by website owners. In addition, information about the probability of network development, maintaining mutual contacts between users, gives the opportunity to estimate traffic on the site. What's more, having information about the history of a particular user's relationship with other portal users, categorizing it, allows to present a specific offer for this particular user. Increasingly, social media offers different types of membership with more or less extensive functions delivered to users. It is worth noting that the benefits of using the mechanisms presented in this article may have the users of the portal themselves, by familiarizing themselves with the characteristics of specific groups, cliques, and profiling for a specific purpose of their own activity. In further work, we will be using data mining techniques to predict the duration of social relationships.

Acknowledgments. This work is partially funded by the Dean's Grant of the Faculty of Computer Science, Electronics and Telecommunications AGH UST.

References

1. Bogdanov, P. Mongiovì, M., Singh, A.K.: Mining heavy subgraphs in time-evolving networks. In: 2011 IEEE 11th International Conference on Data Mining, pp. 81–90 (2011)
2. Borge-Holthoefer, J., Baños, R.A., González-Bailón, S., Moreno, Y.: Cascading behaviour in complex socio-technical networks. J. Complex Netw. 1(1), 3–24 (2013)
3. Gliwa, B., Koźlak, J., Zygmunt, A., Cetnarowicz, K.: Models of social groups in blogosphere based on information about comment addressees and sentiments. In: Aberer, K., Flache, A., Jager, W., Liu, L., Tang, J., Guéret, C. (eds.) SocInfo 2012. LNCS, vol. 7710, pp. 475–488. Springer, Heidelberg (2012). https://doi.org/10.1007/978-3-642-35386-4_35
4. Harush, U., Barzel, B.: Dynamic patterns of information flow in complex networks. Nat. Commun. 8, 2181 (2017)
5. Hui, C., Tyshchuk, Y., Wallace, W.A., Magdon-Ismail, M., Goldberg, M.: Information cascades in social media in response to a crisis: a preliminary model and a case study. In: Proceedings of the 21st International Conference on World Wide Web (WWW 2012 Companion), pp. 653–656. ACM, New York (2012)
6. Hulovatyy, H., Chen, T., Milenkovic, T.: Exploring the structure and function of temporal networks with dynamic graphlets. Bioinformatics 31 (2014). https://doi.org/10.1093/bioinformatics/btv227
7. Kabutoya, Y., Nishida, K., Fujimura, K.: Dynamic network motifs: evolutionary patterns of substructures in complex networks. In: Du, X., Fan, W., Wang, J., Peng, Z., Sharaf, Mohamed A. (eds.) APWeb 2011. LNCS, vol. 6612, pp. 321–326. Springer, Heidelberg (2011). https://doi.org/10.1007/978-3-642-20291-9_33
8. Kovanen, L., Karsai, M., Kaski, K., Kertész, J., Saramäki, J.: IOP Publishing Lt, J. Stat. Mech. Theory Exp. 2011, November 2011
9. Sekara, V., Stopczynski, A., Lehmann, S.: Fundamental structures of dynamic social networks. PNAS 113, 9977–9982 (2016)

10. Leskovec, J., McGlohon, M., Faloutsos, C., Glance, N., Hurst, M.: Patterns of cascading behavior in large blog graphs. In: Proceedings of the 2007 SIAM International Conference on Data Mining, pp. 551–556 (2007)
11. Milardo, R., Johnson, M., Huston, T.: Developing close relationships: changing patterns of interaction between pair members and social networks. J. Pers. Soc. Psychol. **44**(05), 964–976 (1983)
12. Morse, S., Gonzalez, M., Markuzon, N.: Persistent cascades: measuring fundamental communication structure in social net-works. In: 2016 IEEE International Conference on Big Data, BigData 2016, Washington DC, USA, pp. 969–975, 5–8 December 2016
13. Oliwa, L., Kozlak, J.: Anomaly detection in dynamic social networks for identifying key events. In: 2017 International Conference on Behavioral, Economic, Socio-cultural Computing, BESC 2017, Krakow, Poland, 16–18 October 2017. IEEE (2017)
14. Pan, R.K., Saramäki, J.: Path lengths, correlations, and centrality in temporal networks. Phys. Rev. E **84**, 016105 (2011)
15. Rudek, K., Kozlak, J.: The influence of relationships strength on their duration in blogosphere. In: 2017 International Conference on Behavioral, Economic, Socio-cultural Computing, BESC 2017, Krakow, Poland, 16–18 October 2017. IEEE (2017)
16. Song, D., Wang, Y., Gao, X., Qu, S.X., Lai, Y.C., Wang, X.: Pattern formation and transition in complex networks, March 2017

SmartFD: A Real Big Data Application for Electrical Fraud Detection

D. Gutiérrez-Avilés[1]([⊠]), J. A. Fábregas[2]([⊠]), J. Tejedor[3]([⊠]),
F. Martínez-Álvarez[1], A. Troncoso[1], A. Arcos[4], and J. C. Riquelme[2]

[1] Division of Computer Science, Pablo de Olavide University, Seville, Spain
{dgutavi,fmaralv,atrolor}@upo.es
[2] Department of Computer Science, University of Seville, Seville, Spain
{jfabregas,riquelme}@us.es
[3] Endesa SA, Madrid, Spain
javier.tejedor@enel.com
[4] Department of Industrial Organization and Business Management,
University of Seville, Seville, Spain
aarcos@us.es

Abstract. The main objective of this paper is the application of big data analytics to a real case in the field of smart electric networks. Smart meters are not only elements to measure consumption, but they also constitute a network of millions of sensors in the electricity network. These sensors provide a huge amount of data that, once analyzed, can lead to significant advances for the society. In this way, tools are being developed in order to reach certain goals, such as obtaining a better consumption estimation (which would imply a better production planning), finding better rates based on the time discrimination or the contracted power, or minimizing the non-technical losses in the network, whose actual costs are eventually paid by end-consumers, among others. In this work, real data from Spanish consumers have been analyzed to detect fraud in consumption. First, 1 TB of raw data was preprocessed in a HDFS-Spark infrastructure. Second, data duplication and outliers were removed, and missing values handled with specific big data algorithms. Third, customers were characterized by means of clustering techniques in different scenarios. Finally, several key factors in fraud consumption were found. Very promising results were achieved, verging on 80% accuracy.

Keywords: Big data · Sensors · Classification · Fraud detection

1 Introduction

During most of the 20th century, the interrelationship between electricity users and distribution companies remained unchanged. Suppliers were not chosen and, therefore, there was no need to treat consumers as customers. However, deregulation, the green agenda and the continuous technological leap have changed this relationship. New constraints such as security of supply, competitiveness and

© Springer International Publishing AG, part of Springer Nature 2018
F. J. de Cos Juez et al. (Eds.): HAIS 2018, LNAI 10870, pp. 120–130, 2018.
https://doi.org/10.1007/978-3-319-92639-1_11

sustainability are the three priority axes towards changing the energy model that is currently demanded, which is materialized in objectives such as reducing emissions, renewable energy generation and improving energy efficiency.

An essential tool in this new model is the so-called smart meters that should not be understood only as devices that measure consumption but act as true sensors of the electrical network. These sensors configure a highly flexible and adaptable network that intelligently integrates the actions of the users that are connected to it, in order to achieve an efficient, safe and sustainable supply. The volume of information available from these networks is so huge that it can only be handled with Big Data techniques.

This proposal aims to provide a pioneering solution in the field of electrical distribution oriented to the analysis of the data provided by smart meters using big data techniques. The main objective of the paper is to develop a methodology aimed at the intelligent detection of non-technical losses in the field of electrical distribution, but it is not the only possibility. The data infrastructure and algorithms resulting from this paper may serve for a better understanding of the consumption patterns of customers. This study has the endorsement of the Endesa company to be able to access the data of its network.

With the aim of building a complete big data system for effective fraud detection, the authors have been accomplishing several tasks: A big data infrastructure based on HDFS and Spark has been built. Then, a knowledge discovery in databases (KDD) strategy has been followed. Raw data, which consisted of many duplicates, missing values and, even outliers, needed intensive preprocessing. Later, minable views were created, identifying labels and fraud targets. Finally, classification algorithms have been applied to different scenarios, reporting accuracies higher than 80%. Currently, site visits are being carried out, confirming the effectiveness of the proposed approach.

The rest of the paper is structured as follows. Section 2 reviews the most relevant papers related to this work. The applied methodology is described in Sect. 3. Reported results can be found in Sect. 4. Finally, the conclusions drawn from this study are summarized in Sect. 5.

2 State of the Art

This section reviews the most relevant works published in the field of fraud detection during the last decade. A novel method for fraud detection in high voltage electrical energy consumers using data mining was introduced in [3]. The use of Self-Organizing Maps was proposed in order to identify consumption profiles. The authors mainly compared usage patterns in historical data with current behaviors, detecting anomalies in cases of fraud.

Monedero et al. [14] studied users with anomalous drops in energy in 2012. For this purpose, they used Bayesian networks and decision trees, also finding other types of non-technical loss patterns. The proposed methodology was tested with real customers, also from the Endesa company (hereon the partner).

The authors in [7] addressed the fraud detection problem in electric power distribution networks (low-voltage consumers). Namely, artificial neural networks were applied to discover fraud in Brazilian costumers. The authors claimed an improvement of over 50% when compared to other existing approaches. One year later, in 2014, the use of artificial neural networks was proposed again for smart grid energy fraud detection [9].

The work in [16] analyzed time series without the seasonal component of consumers' power consumption at low voltage for the purpose of detect fraud and illogical consumption by customers. The authors drawn two main conclusions: energy drop in the last series is dominant sign in suspicious customer's detection and in series of suspicious customers it is noticed alternating positive autocorrelation.

Decision tree learning for fraud detection in consumer energy consumption can be found in [5]. In fact, this work proposed this kind of learning to profile normal energy consumption behavior, thus allowing for the detection of potentially fraudulent activity.

In 2016, a supervised approach for fraud detection in energy consumption was introduced in [6]. The model found anomalies in meter readings thanks to the application of machine learning techniques to historical data. Furthermore, the model is updated with incremental learning strategies and was tested on real Spanish data.

Finally, an approach based on machine learning to detect abnormalities is customer behaviors was proposed in [12]. They assessed linear discriminant analysis and logistic regression performances. Reported results by logistic regression reached higher accuracy since it was able to forecast irregularities accurately.

3 Methodology

The main objective of *SmartFD* methodology is the application of big data analytics to smart electric networks and the construction of classification models in order to determine the customers with high probability of fraud in their electricity consumption.

The global process to achieve this objective is shown in Fig. 1. We can see how the methodology has five phases framed in the *KDD* process [8] that are described in the following sections.

3.1 Data Retrieval

The first phase consists in the data retrieval processing wherein the raw files and the data model documentation is provided by the Partner. Data from the smart electric network of the Partner is contained in the raw files. In addition to being elements to measure consumption, smart meters also constitute a network of millions of sensors in a power network. These sensors provide very large amounts of data that, once analyzed, can lead to significant advances in society.

These raw files had to be unzipped and stored our storage system. This task would be trivial in common *KDD* environment, in contrast, it is hard and critical in big data environments since issues related to data transport, network bandwidth and computational cost emerge.

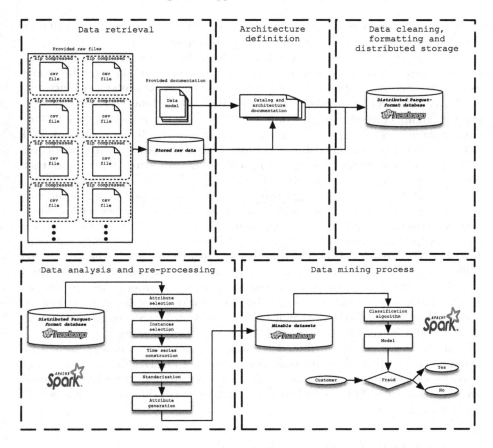

Fig. 1. *SmartDF* methodology

3.2 Architecture Definition

The second phase consists of the study of data model study, the inference of the database architecture and the inventory of the schemes, tables, and relations that contains. The importance of this process lies in the necessity of produce a consistent, precise and self-contained documentation of the catalog and architecture of the database from a disorganized and inaccurate documentation and unstructured source data. Again, the effort and time invested, as well as the difficulty of this task, increases considerably in our big data environment.

3.3 Data Cleaning, Formatting and Distributed Storage

This achievement of this third phase is essential and critical in big data environments. These task enable us to process and handle large amount of data. With the support of the catalog and architecture documentation, the stored raw data is processed in order to drop the inconsistencies, wrong formats and duplicate values for the purpose of obtain database with consistency and integrity.

Then, to be able to work in a big data context, the tables of the database have to be converted form CSV format to Parquet format as well as must be stored in a distributed file system implemented in a cluster of computer machines.

Apache Parquet [1] is an open-source column-oriented data store designed to support highly efficient compression and coding schemes, which allows for lower data storage costs and greater efficiency of one's query. In a column-oriented database, the information is stored in order of registration, so that a particular entry for different columns belongs to the same entry record. This means that one can access individual data elements (the name of a customer, a consumption date, a postal code, etc.) through columns as a group rather than reading row by row. In addition, this compression makes it easier for queries with column operations such as sum, average, minimum, etc. to be carried out much faster, so that when a query is made, it is only executed on the necessary columns. It should be added that, in addition to the considerable size reduction of the tables, they can be compressed in Snappy format [11], compatible with HDFS and Spark. If we also consider the option of using machine learning services in the cloud, such as those provided by Amazon Web Service (where services are billed by runtime and/or size of the scanned data), the savings in time and money would be enormous compared to using the CSV format.

For this purpose of distributed storage, a Hadoop environment has to be installed and configured, which will allow us to have an HDFS architecture [15]; thus, in addition to ensuring greater speed of access to data, we also implement greater tolerance to failures and crashes of cluster nodes due to the replicas that this file system generates.

3.4   Data Analysis and Pre-processing

The fourth stage aims to produce minable datasets to apply classification algorithms to them. From our target distributed Parquet-format database, we use Spark SQL [2] framework to carry out the following five steps:

1. Attribute selection: which attributes may have the greatest influence in identifying possible fraud have been studied. Along with the consumption of each customer, attributes such as postal code, business activity (in the case of non-residential customers), the model and status of smart meters, the power contracted by customers, or certain events that have occurred over the life of a contract have been some of the most interesting when constructing a first set of data.
2. Instances selection: in a second step, it has been necessary to check which of the instances in our set are optimal, so that in the classification stages we can obtain more interpretable sets of rules providing more information. Aspects such as the number of null values or number of correct readings have been some of the most important aspects to take into account in this stage of

data pre-preparation. In addition to this, the data set have to be balanced, a fundamental requirement for a dataset with which to generate a classification model.

3. Time series construction: once a set of quality data is generated. Our objective have been to build time series based on customer consumption. To do this, we have consumption readings with a quarterly frequency, so these time series encompass a wide range of frequencies, from low (quarterly, monthly) to very high (hourly). As a result, different datasets have been built with which to generate different classification models in later stages of the project.

4. Standardization: due to the difference in consumption amongst customers, the possibility of carrying out different standardizations of consumption data has been assessed. Calculating all consumptions with respect to a customer's average or maximum consumption are two possible options when generating new datasets.

5. Attribute generation: in addition to the different time series (standardized or not) and the attributes that we previously considered most relevant, we have also generated attributes that can provide additional information such as the number of estimated readings associated to a customer.

3.5 Data Mining Process

Finally, the fifth stage uses the minable datasets, that were produced in the previous stage, to extract hidden, useful, and valid information from them by means of machine learning algorithms. Specifically, we have used Spark MLlib framework [13] to apply classification algorithms which produce models that detect fraudulent behaviors of the Partner's customers.

4 Results

The experimental design and the preliminary classification results related to the global process presented in this paper (Sect. 3) are described hereunder. This section is structured as follows: a brief description of the utilized data and the infrastructure is outlined in Sects. 4.1 and 4.2, next, the experimental setup is explained in Sect. 4.3 and, finally, the preliminary classification results are presented in Sect. 4.4.

4.1 Data

The analyzed database contains all data related to the Partner's customers of the Spanish region of Catalonia. These data have been retrieved in form of 251 csv files, likewise, this files compound 35 tables of 832 attributes in total. The size of the database is 1.48707619 TB (1487.076192 GB), it implies a real big data problem.

The most important characteristic of this database is that its content and relationships are divided in two: a scheme related to customer contracts and another with smart meters that collect consumption data. In each of the schemes, called *stars*, there is a central table that relates to the tables belonging to the same scheme. This table is also connected to the other central table. In this manner, both stars are linked by their centres and are joined to their corresponding tables (tips of the star). These stars are:

- Contract star: it is formed by the tables that contain all the information related to contracting, geolocation, invoicing, customers, files, consumption types, campaigns, and technicians' work in the field.
- Device star: it is made up of tables with information regarding devices and consumption load curves, as well as different tables containing different events, validation records, etc.

4.2   Infrastructure

Due to the storage capacity and computational power required, as well as to be able to work optimally distributed, we have used the following hardware and software facilities:

- A cluster composed by 72 processing cores, 64 of them Intel (R) Xeon (R) Xeon (R) E7- 4820 CPUs @ 2.00 GHz plus 8 Intel (R) Core (TM) i7-7700K CPUs @ 4.20 GHz.
- 3 GeForce GTX 1080 GPUs with 2560 cores, Nvidia CUDA and 8 GB GDDR5X memory each.
- 128 GB RAM: 64 GB DD3 and 64 GB DDR4.
- A total storage capacity of 8 TB.
- Nodes interconnected through a Gigabit Ethernet network with a bandwidth of 1 Gbit/s.
- AWS cloud computing services.
- Hadoop HDFS 2.8.0.
- Apache Spark framework 2.2.0.

4.3   Experimental Setup

For this preliminary experimental study, the larger customers are been taking into account, thus, in order to accomplish the task of classifying this kind of customers in fraudulent or normal behavior, we have obtained a consistent and significant set of customer's daily consumption curves with both fraud and without it. The two options were considered to build the datasets that will train and test the classification algorithm are the following.

Setup #A: Dataset with Subsequent Measurement. All the daily measurement that have occurred between a year before and three months after the start date of a sanction proceeding in the case of fraud customers have been selected.

For both the dataset to be balanced and to find a possible relationship between the consumption of the two types of customers, customers without fraud that have the same number of measurements and at the same time, have the same type of business activity and the same postcode, have been selected.

With the aim of produce a model based on load curves of a customer for the period, a process to change quarter-hourly measurements to hourly measurement was necessary to create. In addition, it was counted the total number of estimated measurements of the hourly measurement.

Therefore, the final dataset is composed of the following fields:

- **customer code**
- **business activity code**
- **postcode**
- **number of estimated measurements**
- **dx** (being x the day, between 1 and 454. The field value is the addition of the measurements of the day)
- **label** (label with value 1 for fraud customers or 0 for those who are not fraud)

Setup #B: Dataset Without Subsequent Measurement. On the contrary to the previous dataset, in this case, only measurements occurred until a year before the start date of a sanction proceeding have been selected. Again, for both the dataset to be balanced and to find a possible relationship between the consumption of the two types of customers, customers without fraud that have the same number of measurement and at the same time, have the same type of business activity and the same postcode, have been selected. To build the dataset, the same process as for the creation of the previous dataset has been followed. Therefore, in the final dataset we count with the following fields:

- **customer code**
- **business activity code**
- **postcode**
- **number of estimated measurements**
- **dx** (being x the day, between 1 and 364). The field value is the addition of the measurements of the day)
- **label** (label with value 1 for fraud customers or 0 for those who are not fraud).

4.4   Classification Results

The classification algorithm *Xgboost* has been chosen to be applied once the two datasets were created.

The *Xgboost* [4] algorithm trains, in an iterative way, decision trees to minimize a loss function. The specific method to tag again the records was defined by a loss function. The *Xgboost* algorithm decreases such loss function with the training data in every iteration.

For the purpose of establish the training, validation and testing procedure for the classification algorithm, the input datasets have been split into two parts: training-validation part (70%) and test part (30%). Next, the training-validation part has been split again into two parts: training part (70%) and validation part (30%). The training part has been used to the algorithm training, the validation part has been used to cross-validation procedure [10] and the test part has been used as an external test for the resulting model.

For each experimental setup, we have built two classification models: one with all the attributes of the dataset and another excluding the number of estimated measurements. The aim of this decision was to prove the importance of the number of estimated measurements in the model generation, considering that we had previously detected that this value is critical by means of clustering analysis of the data. For each generated model and for each proof method (Cross Validation or Test) we the accuracy (ACC), the true positive ratio (TPR) and the true negative ratio (TNR) have been shown.

Setup #A Results. We can find these in Table 1. The complete model is the best one in terms of ACC and TPR, on the contrary, the model that exclude the number of estimated measurements has better values of TPR. In general terms, the complete model offers better results, reinforcing the importance of the number of estimated measurements as a key factor to classify fraudulent and normal customers.

Table 1. Setup A results

Model	Proof method	ACC	TPR	TNR
Complete	Cross validation	91.01%	87.67%	7.33%
Complete	Test	92.36%	86.16%	4.54%
Excluding #em	Cross validation	72.26%	32.32%	8.96%
Excluding #em	Test	73.57%	39.51%	9.7%

Setup #B Results. They are presented in Table 2. In it, we can observe how the complete model is the best one in terms of ACC and TPR and how the model that exclude the number of estimated measurements has better values of TPR once again. The complete model offers once again better results in general terms, reinforcing the importance of the number of estimated measurements as a key factor to classify fraudulent and normal customers.

Table 2. Setup B results

Model	Proof method	ACC	TPR	TNR
Complete	Cross validation	89.16%	78.11%	5.45%
Complete	Test	87.55%	76.50%	6.58%
Excluding #em	Cross validation	72.48%	52.16%	17.99%
Excluding #em	Test	73.45%	52.28%	16.29%

5   Conclusions

A real case for fraud detection in electricity consumption has been addressed in this paper. In particular, data from Spanish users have been processed, making use of big data technologies. Up to 1.5 TB of raw data were initially retrieved from different sources. After intensive preprocessing, data were cleaning and transformed into useful information, thanks to experts guidance. Later, machine learning algorithms have been applied in order to discover fraud consumption patterns in users. Two different scenarios have been considered, both of them reaching accuracies above 80%. The entire process has been carried out in a HDFS-Spark architecture, making the most of current big data technologies. Future works are directed towards the generation of models for different types of users and geographic areas.

Acknowledgments. The authors would like to thank the Spanish Ministry of Economy and Competitiveness for the support under projects TIN2014-55894-C2-R and TIN2017-88209-C2-R.

References

1. Apache Parquet: A columnar storage format. https://parquet.apache.org
2. Armbrust, M., Xin, R.S., Lian, C., Huai, Y., Liu, D., Bradley, J.K., Meng, X., Kaftan, T., Franklin, M.J., Ghodsi, A., Zaharia, M.: Spark SQL: relational data processing in spark. In: Proceedings of the 2015 ACM SIGMOD International Conference on Management of Data, SIGMOD 2015, pp. 1383–1394. ACM, New York (2015)
3. Cabral, J.E., Pinto, J.O.P., Martins, E.M., Pinto, A.M.A.C.: Fraud detection in high voltage electricity consumers using data mining. In: Proceedings of the IEEE Transmission and Distribution Conference and Exposition, pp. 1–5 (2008)
4. Chen, T., Guestrin, C.: Xgboost: a scalable tree boosting system. In: Proceedings of the 22nd ACM SIGKDD International Conference on Knowledge Discovery and Data Mining, KDD 2016, pp. 785–794. ACM, New York (2016)
5. Cody, C., Ford, V., Siraj, A.: Decision tree learning for fraud detection in consumer energy consumption. In: Proceedings of the IEEE International Conference on Machine Learning and Applications, pp. 1175–1179 (2015)
6. Coma-Puig, B., Carmona, J., Gavald, R., Alcoverro, S., Martin, V.: Fraud detection in energy consumption: a supervised approach. In: Proceedings of the IEEE International Conference on Data Science and Advanced Analytics, pp. 120–129 (2016)
7. Costa, B.C., Alberto, B.L.A., Portela, A.M., Maduro, W., Eler, E.O.: Fraud detection in electric power distribution networks using an ANN-based knowledge discovery process. Int. J. Artif. Intell. Appl. 4(6), 17–23 (2013)
8. Fayyad, U., Piatetsky-Shapiro, G., Smyth, P.: The KDD process for extracting useful knowledge from volumes of data. Commun. ACM **39**(11), 27–34 (1996)
9. Ford, V., Siraj, A., Eberle, W.: Smart grid energy fraud detection using artificial neural networks. In: Proceedings of the IEEE Symposium on Computational Intelligence Applications in Smart Grid, pp. 1–6 (2014)

10. Golub, G.H., Heath, M., Wahba, G.: Generalized cross-validation as a method for choosing a good ridge parameter. Technometrics **21**(2), 215–223 (1979)
11. Google: Snappy: A fast compressor/decompressor. https://google.github.io/snappy/
12. Lawi, A., Wungo, S.L., Manjang, S.: Identifying irregularity electricity usage of customer behaviors using logistic regression and linear discriminant analysis. In: Proceedings of the International Conference on Science in Information Technology, pp. 552–557 (2017)
13. Meng, X., Bradley, J., Yavuz, B., Sparks, E., Venkataraman, S., Liu, D., Freeman, J., Tsai, D., Amde, M., Owen, S., et al.: MLlib: machine learning in Apache Spark. J. Mach. Learn. Res. **17**(1), 1235–1241 (2016)
14. Monedero, I., Biscarri, F., Len, C., Guerrero, J.I., Biscarri, J., Milln, R.: Detection of frauds and other non-technical losses in a power utility using Pearson coefficient, Bayesian networks and decision trees. Int. J. Electric. Power Energy Syst. **34**, 90–98 (2012)
15. Shvachko, K., Kuang, H., Radia, S., Chansler, R.: The Hadoop distributed file system. In: 2010 IEEE 26th Symposium on Mass Storage Systems and Technologies (MSST), pp. 1–10, May 2010
16. Spiric, J.V., Docic, M.B., Stankovic, S.S.: Fraud detection in registered electricity time series. Int. J. Electr. Power Energy Syst. **71**, 42–50 (2016)

Multi-class Imbalanced Data Oversampling for Vertebral Column Pathologies Classification

José A. Sáez[1][(✉)], Héctor Quintián[2], Bartosz Krawczyk[3], Michał Woźniak[4], and Emilio Corchado[1]

[1] Department of Computer Science and Automatics, University of Salamanca, Plaza de los Caídos s/n, 37008 Salamanca, Spain
{joseasaezm,escorchado}@usal.es
[2] Department of Industrial Engineering, University of A Coruña, Avda. 19 de Febrero s/n, 15405 Ferrol-Coruña, Spain
hector.quintian@udc.es
[3] Department of Computer Science, School of Engineering, Virginia Commonwealth University, 401 West Main Street, Richmond, VA 23284, USA
bkrawczyk@vcu.edu
[4] Department of Systems and Computer Networks, Wrocław University of Science and Technology, Wybrzeże Wyspiańskiego 27, 50-370 Wrocław, Poland
michal.wozniak@pwr.edu.pl

Abstract. Medical data mining problems are usually characterized by examples of some of the classes appearing more frequently. Such a learning difficulty is known as imbalanced classification problems. This contribution analyzes the application of algorithms for tackling multi-class imbalanced classification in the field of vertebral column diseases classification. Particularly, we study the effectiveness of applying a recent approach, known as *Selective Oversampling for Multi-class Imbalanced Datasets* (SOMCID), which is based on analyzing the structure of the classes to detect those examples in minority classes that are more interesting to oversample. Even though SOMCID has been previously applied to data belonging to different domains, its suitability in the difficult vertebral column medical data has not been analyzed until now. The results obtained show that the application of SOMCID for the detection of pathologies in the vertebral column may lead to a significant improvement over state-of-the-art approaches that do not consider the importance of the types of examples.

1 Introduction

Medical data mining is a highly challenging task, as it combines the importance of impacting human health, complex data, and need to work closely with a physician. Decisions derived from these analyses are of great importance because

© Springer International Publishing AG, part of Springer Nature 2018
F. J. de Cos Juez et al. (Eds.): HAIS 2018, LNAI 10870, pp. 131–142, 2018.
https://doi.org/10.1007/978-3-319-92639-1_12

they directly influence human beings on both individual and population levels. Machine learning methods are used to process complex medical data in a fast and effective way, as they are capable of analyzing massive amounts of data in a short-time and can offer a valuable decision support capabilities [8,11]. Even though machine learning techniques are widely used in medical classification, the general application of these techniques in *traumatic orthopedics*, in which this contribution focuses, is rather sparse in the specialized literature [15].

Most of medical classification data, included that on vertebral column pathologies classification used in this research, are problems in which some of the classes (which commonly corresponds to non-healthy patients) are likely to contain much less examples than the other classes [12]. Since standard classification methods assume an approximately balanced class distribution, they usually wrongly classify the minority class examples [13]. However, minority classes are many times the most interesting ones, particularly in the case of medical data. In order to overcome the negative impacts of class imbalance various techniques have been proposed, particularly for binary problems [7,13]. The main research lines include *inbuilt mechanisms* [13], which change the classification strategies to ease the recognition of the minority class and *data preprocessing methods* [2,7], which modify the data distribution to change the balance between classes.

However, many imbalanced medical datasets are focused on distinguishing, apart from healthy and non-healthy patients, different subtypes within a given pathology [15]. In these cases, we are dealing with imbalanced classification in multi-class problems, a significantly more complex scenario. Usually, the reported solutions for binary cases are not directly applicable to it [18]. In this context, there are some proposals, such as Static-SMOTE [5], which tries to increase the importance of minority classes within the dataset by resampling their instances, and AdaBoost.NC [18], an ensemble learning algorithm which combines AdaBoost with negative correlation learning.

This contribution analyzes the application of some of these techniques in the field of vertebral column pathologies classification and proposes the usage of a recent methodology, SOMCID, designed by Sáez et al. [16], which is particularly oriented to work with multi-class imbalanced datasets, and may be the most suitable to work with this kind of data providing competitive performance results. SOMCID has shown to provide competitive results in a wide variety of different domains, but it has not been applied yet to difficult medical classification data, such as those of vertebral column pathologies. This methodology, which is described in Sect. 3, is based on analyzing the structure of the classes in the dataset in order to detect subsets of most difficult examples that should be subject to the oversampling procedure in each of the minority classes.

The rest of this research is organized as follows. Next section presents an overview on imbalanced classification. Section 3 describes the methodology recommended to deal with vertebral column pathologies classification. Section 4 describes the details of the experimental framework. Section 5 presents the analysis of the results obtained and, finally, Sect. 6 presents the concluding remarks.

2   Imbalanced Datasets in Classification

Many standard machine learning methods are guided by the predictive accuracy and thus they tend to fail when the data is strongly imbalanced. This leads to a poor recognition of the minority classes. In the imbalance learning two scenarios can be considered depending on the data complexity: binary and multi-class.

Binary problems establish a clear relationship between classes, the predominant one being the majority class and other being the minority class. Basically, techniques for binary imbalanced datasets are divided into two groups [9,17]:

1. **Inbuilt mechanisms.** They adapt existing classifiers to the problem of imbalanced datasets and bias them towards favoring the minority class. This task is usually difficult, because it requires a depth knowledge about the classifiers. Solutions in this field include the design of one-class classifiers, which can learn the minority class model and treating majority objects as outliers [9] and cost-sensitive methods, which are based on a loss function which informs about the misclassification cost [10].
2. **Data preprocessing.** These try to equalize the number of the examples of the classes. They may generate new objects from the minority class (oversampling) or remove examples from the majority class (undersampling). Among oversampling strategies, the best known method is SMOTE [2]. On the other hand, among undersampling strategies, ENN (*Edited Nearest Neighbour*) [3], which removes a majority example it if there are no more majority class instances among its nearest neighbors is one of the most well-known.

Multi-class problems may produce a lose of performance on one class while trying to gain it on other [18]. Binary classification metrics cannot be directly applied to multi-class imbalanced datasets. Therefore, new specific metrics are needed in this context, such as multi-*Area Under an ROC Curve* (multi-AUC) and average accuracy [6].

Multi-class imbalanced problems are much more challenging and there is only but a handful of methods dedicated to solving them. One of them is Static-SMOTE [5], which applies the resampling procedure in several iterations, duplicating the examples of those underrepresented classes in the dataset. Another is AdaBoost.NC [18], which is designed as an ensemble learning algorithm which uses negative correlation learning for multi-majority and multi-minority cases. Other methods, such as cost-sensitive neural networks based on undersampling, oversampling, and moving threshold have also been adapted to the multi-class imbalanced task [20]. A recent approach which attracts our attention to be applied to the multi-class imbalanced problem of vertebral column pathologies dataset was proposed in [16]. This method, which will be further denoted as *Selective Oversampling for Multi-class Imbalanced Datasets* (SOMCID), is based on paying attention to the different types of examples that may be present in each class. The SOMCID method clearly establishes a different influence of difficult examples (such as those lying on the class borders) when there are more than two classes present in an imbalanced dataset. This is the main foundation of SOMCID, which is described in the next section.

3 Selective Oversampling for Multi-class Imbalanced Datasets

The SOMCID technique addresses an oversampling task for multi-class imbalanced problems. In [16], where SOMCID was proposed, it was tested over 21 datasets belonging to different fields and having different complexities regarding the number of examples, classes and class imbalanced ratios. SOMCID was compared to the two most representative and well-known methods for multi-class imbalance preprocessing: Static-SMOTE [5] and AdaBoost.NC [18]. SOMCID obtained a performance of 72.56% compared to Static-SMOTE (68.69%) and AdaBoost.NC (67.98%) – the reader may consult [16] to check the full results. These results show the potential and the generalization capabilities of SOMCID working with data of diverse domains. On the other hand, this research focuses on the study of SOMCID for the classification of vertebral column pathologies, which has not been previously studied and has shown to be a difficult to solve problem in previous works [1,15].

This section describes the SOMCID method. It is based on 4 main steps, each one presented in a different section:

1. *Determining the type of each example in the dataset* (Sect. 3.1). The first tasks consist of tagging each one of the examples in the dataset with a different type (safe, borderline, rare and outlier) depending to their relative position to other examples of different classes.
2. *Choosing which classes and examples can be oversampled* (Sect. 3.2). The second task involves an analysis of the internal structure of the dataset looking for which classes and types of the examples previously identified are chosen to perform the oversampling.
3. *Oversampling the data using all its valid configurations* (Sect. 3.3). The next steps oversamples, for each one of the configurations identified in the above step, the original data using an approach such as SMOTE.
4. *Selection of the best schemes of oversampling* (Sect. 3.4). Finally, once the oversampling schemes have been applied, the best of those are selected based on the performance evaluation carried by classification algorithms.

3.1 Determining the Type of Each Example in the Dataset

The first task performed by SOMCID consists of distinguishing different types of examples in each one of the classes of the dataset.

In order to determine the type of an example e, its class label is compared to that of the other examples of its neighborhood, which is computed using the k-nearest neighbors. We will consider the *Heterogeneous Value Difference Metric* (HVDM) [19] to compute the distance between examples (since this measure is valid for both nominal and numerical attributes) and the value $k = 5$ to analyze the neighborhood of each example. Even though the size of the neighborhood can be adapted to the characteristics of each dataset, this value of k has been traditionally used in the imbalanced data literature and, in our case, it shows a good potential distinguishing the type of each example and a low computational time is required. The types of examples and how they are identified are described below:

- **Safe examples.** An example is a safe one if it is placed in an homogeneous area with respect to other examples of its class. In this research, an example is identified as safe if 4 or 5 of its nearest neighbors share its class label.
- **Borderline examples.** These are located close to the boundaries of the classes, where different classes may overlap. An example is considered as a borderline one if 2 or 3 of its 5 nearest neighbors share its class label.
- **Rare examples.** These are few examples of a class placed in areas belonging to a different class. They are characterized by having only 1 neighbor from its class. Additionally, this only neighbor may have as much as another neighbor of its same class.
- **Outliers.** These are isolated examples of a concrete class surrounded by examples of a different class. Minority classes are sometimes exclusively represented or characterized in a high degree of this type of examples.

3.2   Choosing Which Classes and Examples Can Be Oversampled

The second step consists of determining which classes and types of examples can be oversampled. In order to do this, each example within each class is labeled as *safe, borderline, rare* or *outlier* following the aforementioned procedure.

Let be a *configuration* to be oversampled a pair of { *class, types of examples* } with the following characteristics:

- *class*: it is a concrete class of the dataset (excluding the majority class).
- *types of examples*: they are a combination, considering the preprocessing or not, of each one of the 4 types of examples (*safe, borderline, rare* and *outlier*) for the chosen class. It will be noted as *true* if a concrete type of examples is preprocessed and *false* if not. For example, the combination of types of examples { *safe = false, borderline = true, rare = false, outlier = true* } means that only borderline examples and outliers are chosen examples to generate new examples in the dataset.
- A configuration will be valid if the chosen class has examples of the types chosen to preprocess.

3.3   Oversampling the Data Using All Its Valid Configurations

The dataset is preprocessed, considering all the valid configurations found in the previous step. The preprocessing will consist of the application of an oversampling procedure, following an scheme to generate the new synthetic examples similar to that used by SMOTE [2] in binary imbalanced problems.

The oversampling procedure considers the class and the types of examples chosen to generate new synthetic examples. This method creates new synthetic examples of the given class until reaching the size of the majority class.

In order to create a new synthetic example, a random example x of the class and types of examples of interest is iteratively chosen. Then, a random example y among its $k = 5$ nearest neighbors is chosen. With these two examples x and y, the new synthetic example is created by interpolation as SMOTE does.

3.4 Selection of the Best Schemes of Oversampling

Finally, the last step of SOMCID consists of the comparison of the preprocessed datasets of the previous step considering the performance obtained by some classification algorithms. This research is focused on classifiers of decision trees built by C4.5 [14].

In this last step, the performance of the C4.5 algorithm over the dataset considering the preprocessing of a concrete class and combination of types of examples and not preprocessing at all will be compared. This will show which are the most interesting classes and types of examples to oversample to improve the performance of the classifiers.

4 Experimental Framework

This section shows the details of the experimental framework in order to check the performance of SOMCID with the vertebral column pathologies classification dataset. Section 4.1 describes the real-world data used in the study. Section 4.2 presents the multi-class imbalanced methods considered for the experimental comparisons, along with their parameter setup. Finally, Sect. 4.3 describes the methodology followed to analyze the results.

4.1 Vertebral Column Pathologies Dataset

As it was previously mentioned, the experimentation carried out in this research uses a dataset consisting of vertebral column pathologies. The vertebral column is a system composed by a group of vertebras, nerves, muscles, medulla and joints.

The real-world dataset used was collected by the medical center called *Centre Médico-Chirurgical de Réadaptation des Massues* (Lyon, France) [1]. It contains information of 310 patients obtained from sagittal panoramic radiographies of the spine. Each patient, that is, each example in the dataset, is described using 6 continuous biomechanical attributes of the spino-pelvic system, which have shown to have some relationship to vertebral column pathologies [1]. These are the angle of pelvic incidence, the angle of pelvic tilt, the lordosis angle, the sacral slope, the pelvic radius and the grade of slipping.

Thus, pathologies in the vertebral column are conditioned to the characteristics of the pelvis-spine system. The real-world dataset used in this contribution focuses on two of the most common pathologies in this context: *disc hernia* and *spondylolisthesis*. On the one hand, disc hernia appears when the core of the intervertebral disc migrates from its place, implying the compression of the nervous and, consequently, the patient's backache. On the other hand, spondylolisthesis occurs when one of the vertebras slips in relation to the others, causing pain or irritation of the nervous roots. This dataset classifies each one of the patients as either not having any pathology in their spines (100 examples), suffering from disc hernia (60 examples) or suffering from spondylolisthesis (150 examples). In the rest of this contribution, each of these classes have been denoted with the following indices: disc hernia = 1, normal patient = 2 and spondylolisthesis = 3.

4.2   Multi-class Imbalanced Classification Techniques

In order to check the suitability of the SOMCID approach for vertebral column pathologies classification, the following methods to deal with multi-class imbalanced classification datasets have been chosen as comparison techniques because they are some of the most well-known and representative methods in the field according to recent publications:

1. *No sampling or algorithm-level modifications* (None). This does not consider the application of any preprocessing or specifically designed method to deal with multi-class imbalanced data. Thus, the original dataset is used by None.
2. *Multi-class Random Oversampling* (MCROS) [16]. This method is based on the algorithm originally proposed for binary imbalanced data. It oversamples each one of the minority classes by replicating random examples from these classes until reach the size of the majority class in the dataset.
3. *Static-Synthetic Minority Over-sampling Technique* (Static-SMOTE) [5]. This applies a resampling procedure in M steps, where M is the number of classes in the dataset. In each iteration, it selects the minimum size class, and duplicates the number of instances of the class in the original dataset.
4. *AdaBoost with negative correlation* (AdaBoost.NC) [18]. This uses an ensemble learning algorithm for multi-majority and multi-minority cases, in which the well-known AdaBoost method is combined with negative correlation.

The configuration of the parameters for these algorithms has been set following the recommendations of their corresponding authors. For MCROS, Static-SMOTE and SOMCID we have considered 5 neighbors to generate new synthetic examples. For AdaBoost.NC, we have set up the penalty strength $\lambda = 2$ and the number of classifiers composing the ensemble to 51.

4.3   Methodology of Analysis

The performance estimation of each of the aforementioned methods in the vertebral column pathologies dataset is obtained by means of 4 runs of a 5-fold stratified cross-validation, averaging its test performance results. Since AdaBoost.NC is designed to work with ensembles of decision trees, the performance is computed considering the C4.5 decision trees generator [14] for all the other methods. The average accuracy [6], which is commonly employed in imbalanced domains for multi-class problems, is used as the measure to estimate the performance:

$AvgAcc = \frac{\sum_{i=1}^{CL} TPR_i}{CL}$, where CL is the number of classes of the dataset and TPR_i is the True Positive Rate of the i-th class (noted in percentage).

Furthermore, statistical comparisons of the performance results of the algorithms are performed using Wilcoxon's test [4]. In order to do that, the performance results of each run of the cross-validation for each algorithm are compared using Wilcoxon's test and the sum of ranks in favor of each algorithm (R^+ and R^-) and the associated p-value are computed.

5   Analisis of Results

This section presents the analysis of the results after carrying the experimental study described in Sect. 4. The analysis is divided into two main parts. First, Sect. 5.1 focuses on the results obtained from the oversampling of the different valid configurations using the SOMCID method versus not applying any preprocessing (None). Then, Sect. 5.2 shows the analysis of the comparison of results of the best configuration obtained by SOMCID versus the other multi-class imbalanced methods considered (MCROS, Static-SMOTE and AdaBoost.NC).

5.1   Results of SOMCID versus Not Preprocessing

Figure 1 shows the average accuracy results after oversampling each one of the valid configurations with SOMCID (vertical bars) against the results of not applying any preprocessing (horizontal line). This graphic shows all the data related to the configuration oversampled: the preprocessing of the safe, borderline, rare and outlier examples, and the class (1 or 2) associated to the oversampling. This figure is completed with the numerical results shown in Table 1.

Figure 1 shows a general view of the great differences and the importance of choosing the correct examples and classes when applying oversampling in our vertebral column pathologies dataset. As it can be appreciated from these results, the preprocessing of some classes and types of examples can improve or deteriorate the results of not considering any preprocessing at all. Because of this, the analysis of the types of examples and classes to be oversampled seems of interest to classify column vertebral diseases.

As it was mentioned in Sect. 4.1, the classes preprocessed with SOMCID were class 1 (patients suffering from a disc hernia) and class 2 (healthy patients). According to the preprocessing of the two aforementioned classes for each configuration of examples oversampled, the results from Table 1 show that, in general, it is more interesting oversampling those examples corresponding to patients with a disc hernia (class 1) than those not suffering from any disease in the column (class 2) – thus, the performance obtained in 13 out of the 16 configurations is better when preprocessing those examples of patients with a disc hernia. This result seems to be in concordance with the balance among classes in the data, because we only have 60 examples of patients with disc hernia compared to 250 examples belonging to other classes (100 without any disease and 150 with spondylolisthesis). This fact is a common situation in many medical datasets, in which the examples of patients with concrete diseases usually represent a

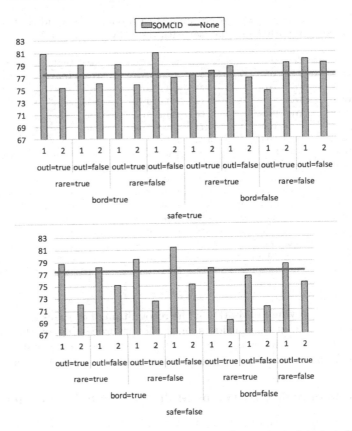

Fig. 1. Oversampling of classes and types of examples compared to not *None* – preprocessing (upper figure) and not (lower figure) the safe examples.

much smaller number than that of the other classes, even though the former can be sometimes more important from the medical point of view. As derived from the above analysis, the examples of patients with disc hernia seems the most interesting ones to preprocess.

Regarding the types of examples that should be oversampled, Table 1 shows that the worse performance is obtained for the preprocessing of safe and outlier examples. Safe examples are, by definition, placed in homogeneous areas of the class and do not pose difficulties for the classifier building. Outliers are usually very few isolated examples and they can even represent errors. Therefore, the preprocessing of these types of examples may be irrelevant in some cases (in the case of the safe examples, since these areas may not need to be reinforced) or even counterproductive (if the outliers represent errors in the data).

The best configuration of examples to preprocess corresponds to that focused only on borderline examples, ignoring the oversampling of the other types examples. This results seems to be logical, since the strengthening of the boundaries of the classes can contribute to create the classifier with more confidence.

Table 1. Numerical results after oversampling each one of the valid configurations of the vertebral column pathologies dataset with the SOMCID method.

Safe	Borderline	Rare	Outlier	Class	AvgAcc	Safe	Borderline	Rare	Outlier	Class	AvgAcc
True	True	True	True	1	80.89	False	True	True	True	1	78.78
True	True	True	True	2	75.33	False	True	True	True	2	72.11
True	True	True	False	1	79.11	False	True	True	False	1	78.11
True	True	True	False	2	76.00	False	True	True	False	2	75.11
True	True	False	True	1	79.11	False	True	False	True	1	79.44
True	True	False	True	2	75.78	False	True	False	True	2	72.56
True	True	False	False	1	81.00	**False**	**True**	**False**	**False**	**1**	**81.33**
True	True	False	False	2	76.89	False	True	False	False	2	75.22
True	False	True	True	1	77.56	False	False	True	True	1	77.89
True	False	True	True	2	78.00	False	False	True	True	2	69.33
True	False	True	False	1	78.67	False	False	True	False	1	76.56
True	False	True	False	2	76.78	False	False	True	False	2	71.56
True	False	False	True	1	74.78	False	False	False	True	1	78.56
True	False	False	True	2	79.22	False	False	False	True	2	75.44
True	False	False	False	1	79.89	False	False	False	False	1	77.44
True	False	False	False	2	79.22	False	False	False	False	2	77.44

Thus, when applying SOMCID for the recognition of diseases in the vertebral column, the best solution considers only the preprocessing of the borderline examples in the minority class, that is, that corresponding to patients with a disc hernia. This configuration is selected for further analysis in next section.

5.2   Results of SOMCID versus Multi-class Imbalanced Methods

Table 2 shows the performance comparing SOMCID to MCROS, Static-SMOTE and AdaBoost.NC. The results obtained by the Wilcoxon's test (sum of ranks R^+ in favor of the method of row, sum of ranks R^- in favor of the method of column and the p-value associated with the comparison) are also shown when comparing each pair of algorithms. The best results are remarked in bold-face.

As it is shown in Table 2, the SOMCID method is that providing the best performance when detecting the different diseases of the vertebral column (81.33%). The next best results corresponds to MCROS, with a performance of 78.56%. Since MCROS performs a random oversampling of each class (except the majority one), it can preprocess, by chance, some of the most interesting examples for our dataset, which are those corresponding to the borderline examples of the minority class, as indicated by the analysis made in Sect. 5.1. The Static-SMOTE and AdaBoost.NC methods are those obtaining the worse results (with a difference of 7.11% and 9% with respect to SOMCID, respectively).

The results of the Wilcoxon's test show similar conclusions to those of performance: SOMCID is better than the rest of the methods with statistical significant differences, while MCROS is better than Static-SMOTE and AdaBoost.NC, among which no statistical significant differences are detected. These results show the great importance of analyzing and determining the best examples to preprocess in the dataset about diseases of the vertebral column considered.

Table 2. Average accuracy and results of the Wilcoxon's test.

Method		MCROS	Static-SMOTE	AdaBoost.NC	SOMCID
AvgAcc		78.56	74.22	72.33	**81.33**
R^+/R^-	MCROS	-	**205/5**	**200/10**	10/200
	Static-SMOTE	5/205	-	110/100	0/210
	AdaBoost.NC	10/200	100/110	-	0/210
	SOMCID	**200/10**	**210/0**	**210/0**	-
p-value	MCROS	-	**1.91E–05**	**8.20E–05**	**8.20E–05**
	Static-SMOTE	**1.91E–05**	-	7.89E–01	**1.91E–06**
	AdaBoost.NC	**8.20E–05**	7.89E-01	-	**1.91E–06**
	SOMCID	**8.20E–05**	**1.91E–06**	**1.91E–06**	-

6   Concluding Remarks

In this research we have analyzed the effect of some multi-class imbalanced methods applied on the difficult problem of recognizing vertebral column diseases. We paid special effort to conducting a guided oversampling that considers the types of minority class instances. The SOMCID method allowed for creating a fine pre-processed and balanced training set that could be used by the classifiers.

Results obtained from the experimental study have shown the importance of the analysis of the structure of the classes in this problem. By focusing on specific instances, instead of conducting a blind oversampling, we were able to significantly alleviate the issue of skewed class distributions. The preprocessing of the borderline examples of those patients suffering from a disc hernia (the minority class), that is, those placed in the boundaries of the class, has shown to be the best solution for the data considered. The comparison of SOMCID to the rest of the methods (MCROS, Static-SMOTE and AdaBoost.NC) shows the benefits of SOMCID for the detection of pathologies in the vertebral column according to the average accuracy results and the results of the Wilcoxon's test.

The results in this contribution are proposed for a certain type of multi-class imbalance situation coming from the difficult problem of vertebral diseases, in which 3 classes are considered and one of them represents less than a 20% of the examples in the dataset. These results show the relevance of considering different classes and example types when handling this particular problem. They are highly satisfactory and prove that SOMCID can be a valuable part of a medical decision support system applied in hospitals.

Acknowledgment. José A. Sáez holds a *Juan de la Cierva-formación* fellowship (*Ref. FJCI-2015-25547*) from the Spanish Ministry of Economy, Industry and Competitiveness. Bartosz Krawczyk and Michał Woźniak are partially supported by the Polish National Science Center under the grant no. UMO-2015/19/B/ST6/01597.

References

1. Berthonnaud, E., Dimnet, J., Roussouly, P., Labelle, H.: Analysis of the sagittal balance of the spine and pelvis using shape and orientation parameters. J. Spinal Disord. Tech. **18**(1), 40–47 (2005)
2. Chawla, N.V., Bowyer, K.W., Hall, L.O., Kegelmeyer, W.P.: SMOTE: synthetic minority over-sampling technique. J. Artif. Intell. Res. **16**, 321–357 (2002)
3. Davies, E.: Training sets and a priori probabilities with the nearest neighbour method of pattern recognition. Pattern Recognit. Lett. **8**(1), 11–13 (1988)
4. Demšar, J.: Statistical comparisons of classifiers over multiple data sets. J. Mach. Learn. Res. **7**, 1–30 (2006)
5. Fernández-Navarro, F., Hervás-Martínez, C., Gutiérrez, P.A.: A dynamic over-sampling procedure based on sensitivity for multi-class problems. Pattern Recognit. **44**(8), 1821–1833 (2011)
6. Ferri, C., Hernández, J., Modroiu, R.: An experimental comparison of performance measures for classification. Pattern Recognit. Lett. **30**(1), 27–38 (2009)
7. Kang, Q., Chen, X., Li, S., Zhou, M.: A noise-filtered under-sampling scheme for imbalanced classification. IEEE Trans. Cybern. **47**(12), 4263–4274 (2017)
8. Krawczyk, B., Schaefer, G.: A hybrid classifier committee for analysing asymmetry features in breast thermograms. Appl. Soft Comput. **20**, 112–118 (2014)
9. Krawczyk, B., Woźniak, M., Cyganek, B.: Clustering-based ensembles for one-class classification. Inf. Sci. **264**, 182–195 (2014)
10. Krawczyk, B., Woźniak, M., Schaefer, G.: Cost-sensitive decision tree ensembles for effective imbalanced classification. Appl. Soft Comput. **14**, 554–562 (2014)
11. Krawczyk, B., Filipczuk, P.: Cytological image analysis with firefly nuclei detection and hybrid one-class classification decomposition. Eng. Appl. Artif. Intell. **31**, 126–135 (2014)
12. Li, J., Fong, S., Wong, R.K., Chu, V.W.: Adaptive multi-objective swarm fusion for imbalanced data classification. Inf. Fusion **39**, 1–24 (2018)
13. Menardi, G., Torelli, N.: Training and assessing classification rules with imbalanced data. Data Min. Knowl. Discov. **28**(1), 92–122 (2014)
14. Quinlan, J.R.: C4.5: Programs for Machine Learning. Morgan Kaufmann Publishers, San Francisco (1993)
15. da Rocha Neto, A.R., Sousa, R., de A. Barreto, G., Cardoso, J.S.: Diagnostic of pathology on the vertebral column with embedded reject option. In: Vitriá, J., Sanches, J.M., Hernández, M. (eds.) IbPRIA 2011. LNCS, vol. 6669, pp. 588–595. Springer, Heidelberg (2011). https://doi.org/10.1007/978-3-642-21257-4_73
16. Sáez, J.A., Krawczyk, B., Wozniak, M.: Analyzing the oversampling of different classes and types of examples in multi-class imbalanced datasets. Pattern Recognit. **57**, 164–178 (2016)
17. Sardari, S., Eftekhari, M., Afsari, F.: Hesitant fuzzy decision tree approach for highly imbalanced data classification. Appl. Soft Comput. **61**, 727–741 (2017)
18. Wang, S., Yao, X.: Multiclass imbalance problems: analysis and potential solutions. IEEE Trans. Syst. Man Cybern. Part B Cybern. **42**(4), 1119–1130 (2012)
19. Wilson, D.R., Martinez, T.R.: Improved heterogeneous distance functions. J. Artif. Intell. Res. **6**(1), 1–34 (1997)
20. Zhou, Z., Liu, X.: Training cost-sensitive neural networks with methods addressing the class imbalance problem. IEEE Trans. Knowl. Data Eng. **18**(1), 63–77 (2006)

Bio-inspired Models and Evolutionary Computation

A Hybrid Genetic-Bootstrapping Approach to Link Resources in the Web of Data

Andrea Cimmino$^{(\boxtimes)}$ and Rafael Corchuelo$^{(\boxtimes)}$

University of Seville, ETSI Informática, Avda. Reina Mercedes, s/n, Sevilla, Spain
{cimmino,corchu}@us.es

Abstract. In the Web of Data, real-world entities are represented by means of resources, for instance the southern Spanish city "Seville" that is represented by means of the resource that is available at http://es.dbpedia.org/page/Sevilla in the DBpedia dataset. Link rules are intended to link resources that are different, but represent the same real-world entities; for instance the resource that is available at https://www.wikidata.org/wiki/Q8717 represents exactly the same real-world entity as the resource aforementioned. A link rule may establish that two resources that represent cities should be linked as long as the GPS coordinates are the same. Such rules are then paramount to integrating web data, because otherwise programs would deal with every resource independently from the other. Knowing that the previous resources represent the same real-world entity allows them to merge the information that they provide independently (which is commonly known as integrating link data). State-of-the-art link rules are learnt by genetic programming systems and build on comparing the values of the attributes of the resources. Unfortunately, this approach falls short in cases in which resources have similar values for their attributes, but represent different real-world entities. In this paper, we present a proposal that hybridises a genetic programming system that learns link rules and an ad-hoc filtering technique that bootstraps them to decide whether the links that they produce must be selected or not. Our analysis of the literature reveals that our approach is novel and our experimental analysis confirms that it helps improve the F_1 score, which is defined in the literature as the harmonic mean of precision and recall, by increasing precision without a significant penalty on recall.

1 Introduction

The Web of Data has made it possible for programs to have access to a variety of data about real-world entities. Furthermore, the Linked-Data principles [4]

Supported by the Spanish R&D programme (grants TIN2013-40848-R and TIN2013-40848-R). The computing facilities were provided by the Andalusian Scientific Computing Centre (CICA). We are grateful to Dr. Carlos R. Rivero and Dr. David Ruiz for earlier ideas that led to the results in this paper. We also thank Dr. Francisco Herrera for his hints on statistical analyses and sharing his software with us.

F. J. de Cos Juez et al. (Eds.): HAIS 2018, LNAI 10870, pp. 145–157, 2018.
https://doi.org/10.1007/978-3-319-92639-1_13

support the idea that resources that are different but represent the same real-world entities must be linked so as to facilitate data integration. Link rules are intended to help link resources automatically.

The literature provides several proposals to machine learn link rules by means of genetic programming systems [2,11,12,19,20]. Such rules build on transformation and similarity functions that are applied to the values of the attributes of two resources to check if they can be considered similar enough (by attributes we mean their datatype properties); if they are, then the input resources are assumed to represent the same real-world entity and are then linked; if they are not, then the input resources are kept apart. Our experience confirms that such link rules fall short because some resources that represent different real-world entities have attributes with similar values. For instance, think of the many different authors who have the same name or the many different films that have similar titles.

In this paper, we present a hybrid approach to the problem: first, we use a state-of-the-art genetic programming system to learn a set of link rules; we then select a base link rule and apply it in order to obtain a collection of candidate links; the remaining rules are then bootstrapped to analyse the neighbours of the resources involved in each candidate link (the neighbours are the resources that can be reached by means of their object properties); finally, we analyse how similar they are in order decide which of the candidate links must be selected as true positives and which must be discarded as false positives. Our analysis of the related work unveils that this is a novel approach since current state-of-the-art link rules do no take the neighbours into account. Our experimental analysis confirms that precision can be improved by 68% in average, with an average -10% impact on recall; overall, the average improvement regarding the F_1 score is 47%. We also conducted the Iman-Davenport test to check that these differences are statistically significant regarding precision and the F_1 score, but not regarding recall. Our conclusion is that ours is a very good approach to help programs integrate the data that they fetch from the Web of Data.

The rest of the article is organised as follows: Sect. 2 reports on the related work; Sect. 3 provides the details of our proposal; Sect. 4 presents our evaluation results; finally, Sect. 5 summarises our conclusions.

2   Related Work

The earliest techniques to learn link rules were devised in the field of traditional databases, namely: de-duplication [7,17], collective matching [1,3,6,14,21], and entity matching [15]. They set a foundation for the researchers who addressed the problem in the context of the Web of Data, where data models are much richer and looser than in traditional databases.

Some of the proposals that are specifically-tailored to web data work on a single dataset [10,16], which hinders their general applicability; there are a few that attempt to find links between different datasets [5,8,9,13], but they do not take the neighbours of the resources being linked into account, only the values of the attributes; that is, they cannot make resources with similar values for their attributes apart in cases in which they represent different real-world entities. An additional problem is that they all assume that data are modelled by means of OWL ontologies. Unfortunately, many common datasets in the Web of Data do not rely on OWL ontologies, but on simple RDF vocabularies that consists of classes and properties whose relationships and constraints are not made explicit.

The previous problems motivated several authors to work on techniques that are specifically tailored to work with RDF datasets. Most such proposals rely on genetic programming algorithms [11,12,19,20] in which chromosomes encode the link rules as trees, which facilitates performing cross-overs and mutations. They differ regarding the expressivity of the language used to encode the link rules and the heuristics used to implement the selection, replacement, cross-over, and mutation operators, as well as the performance measure on which the fitness function relies. Isele and Bizer [11,12] contributed with a supervised proposal called Genlink. It is available with the Silk framework [24], which is gaining impetus thanks to many real-world projects [23]. It uses a tournament selection operator, a generational replacement operator, custom cross-over and mutation operators, and its fitness function relies on the Matthews correlation coefficient. It can use a variety of custom string transformation functions and the Levenshtein, Jaccard, Numeric, Geographic, and Date string similarity measures. An interesting feature is that the size of the link rules must not be pre-established at design time, but it is dynamically adjusted during the learning process. Ngomo and Lyko [19] contributed with a supervised proposal called Eagle, which is available with the LIMES framework [18]. It uses a tournament selection operator, a $\mu + \beta$ replacement operator, tree cross-over and mutation operators, and its fitness function relies on the F_1 score. It does not use transformation functions, but the Levenshtein, Jaccard, Cosine, Q-Grams, Overlap, and Trigrams string similarity functions. The maximum size of the link rules must be pre-established at design time. Nikolov et al. [20] contributed with an unsupervised proposal. It uses a roulette-wheel selection operator, an elitist replacement operator, a tree cross-over operator, a custom mutation operator, and a pseudo F_1 fitness function. Transformations are not taken into account, but the library of similarity functions includes Jaro, Levenshtein, and I-Sub. The maximum size of the link rules is also set at design time. There is a diverging proposal by Soru and Ngomo [22]. It supports the idea of using common machine-learning techniques on a training set that consists in a vectorisation of the Cartesian product of the resources in terms of the similarity of their attributes. Transformation functions are not taken into account and the only string similarity functions considered are Q-Grams, Cosine, and Levenshtein. Whether the size of the rules must be pre-set or not depends on the underlying machine learning technique. Unfortunately, none of the proposals that work on RDF datasets take the neighbours of the resources into account.

The previous analysis, makes it clear that the state of the art does not account for a proposal to link resources in RDF datasets that takes their neighbours into account. Our proposal is specifically-tailored to work with such datasets and it is novel in that it is not intended to generate link rules, but leverages the rules that are learnt with other proposals and bootstraps them in order to analyse the neighbours, which our experimental analysis confirms that has a positive impact on precision without degrading recall.

3   Our Proposal

Our proposal consists in two components, namely: the first one learns link rules and the second one filters out the links that they produce.

The link rule learner is based on Genlink [12], which is a state-of-the-art genetic programming system that has proven to be able to learn good link rules for many common datasets. It is well-documented in the literature, so we focus on describing the second one, which constitutes our original contribution. The filter is an ad-hoc component that works as follows: it takes a link rule and executes it to produce a set of candidate links; then, it analyses the neighbours of the resources involved in each candidate link by bootstrapping the remaining rules; links in which the corresponding neighbours are similar enough are preserved as true positive links while the others are discarded as false positive links.

Below, we present the details of the filter, plus an ancillary method that helps measure how similar the neighbours of two resources are.

Example 1. Figure 1 presents two sample datasets that are based on the DBLP and the NSF datasets. The resources are depicted in greyed boxes whose shapes encode their classes (i.e., the value of property *rdf:type*), the properties are represented as labelled arrows, and the literals are encoded as strings. The genetic programming component learns the following link rules in this scenario, which we represent using a Prolog-like notation for the sake of readability:

r_1: $link(A, R)$ if rdf: $type(A) = dblp{:}Author, rdf{:}type(R) = nsf{:}Researcher,$
$$N_A = dblp{:}name(A), N_R = nsf{:}name(R),$$
$$levenstein(lfname(N_A), lfname(N_R)) > 0.80.$$

r_2: $link(A, P)$ if rdf: $type(A) = dblp{:}Article, rdf{:}type(P) = nsf{:}Paper,$
$$T_A = dblp{:}title(A), T_P = nsf{:}title(P),$$
$$jaccard(lowercase(T_A), lowercase(T_P)) > 0.65.$$

where *levenstein* and *jaccard* denote the well-known string similarity functions (normalised to interval $[0.00, 1.00]$), *lfname* is a function that normalises people's names as "last name, first name", and *lowercase* is a function that changes a string into lowercase.

Intuitively, link rule r_1 is applied to a resource A of type *dblp:Author* and a resource R of type *nsf:Researcher*; it computes the normalised Levenshtein similarity between the normalised names of the author and the researcher; if it is greater than 0.80, then the corresponding resources are linked. Link rule r_2

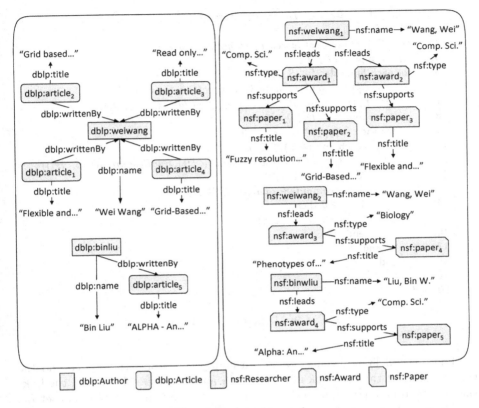

Fig. 1. Running example.

should now be easy to interpret: it is applied to a resource A of type *dblp:Article* and a resource P of type *nsf:Paper* and links them if the normalised Jaccard similarity amongst the lowercase version of the title of article A and the title of paper P is greater than 0.65.

It is not difficult to realise that link rule r_1 links resources *dblp:weiwang* and *nsf:weiwang₁* or *dblp:binliu* and *nsf:binwliu*, which are true positive links, but also *dblp:weiwang* and *nsf:weiwang₂*, which is a false positive link. In cases like this, the only way to make a difference between such resources is to analyse their neighbours, be them direct (e.g., *dblp:weiwang* and *dblp:article₂*) or transitive (e.g., *nsf:weiwang₁* and *nsf:paper₂*). ■

3.1 Filtering Links

Figure 2 presents the method to filter links. It works on a base link rule r, a set of supporting link rules S, a source dataset D_1, a companion dataset D_2, and a threshold θ that we explain later. It returns K, which is the subset of links produced by base link rule r that seem to be true positive links.

```
 1: method filterLinks(r, S, D_1, D_2, θ) returns K
 2:     K := ∅
 3:     (C_1, C_2) := (sourceClasses(r), targetClasses(r))
 4:     L_1 := apply(r, D_1, D_2)
 5:     for each link rule r' ∈ S do
 6:         (C'_1, C'_2) := (sourceClasses(r'), targetClasses(r'))
 7:         (P_1, P_2) := (findPath(C_1, C'_1, D_1), findPath(C_2, C'_2, D_2))
 8:         L_2 = apply(r', D_1, D_2)
 9:         for each (p_1, p_2) ∈ P_1 × P_2 do
10:             for each link (a, b) ∈ L_1 do
11:                 (A, B) := (findResources(a, p_1, D_1), findResources(b, p_2, D_2))
12:                 E := L_2 ∩ (A × B)
13:                 w := computeSimilarity(A, B, E)
14:                 if w ≥ θ then
15:                     K := K ∪ {(a, b)}
16:                 end
17:             end
18:         end
19:     end
20: end
```

Fig. 2. Method to filter links.

The method first initialises K to an empty set, stores the source and the target classes of the base link rule in sets C_1 and C_2, respectively, and the links that result from applying it to the source and the companion datasets in set L_1.

The main loop then iterates over the set of supporting link rules using variable r'. In each iteration, it first computes the sets of source and target classes involved in link rule r', which are stored in variables C'_1 and C'_2, respectively; next, it finds the set of paths P_1 that connect the source classes in C_1 with the source classes in C'_1 in dataset D_1; similarly, it finds the set of paths P_2 that connect the target classes in C_2 with the target classes in C'_2 in dataset D_2. By path between two sets of classes, we mean a sequence of object properties that connect resources with the first set of classes to resources with the second set of classes, irrespective of their direction. Simply put: the idea is to find the way to connect the resources linked by the base link rule with the resources linked by the supporting link rule, which is done by the intermediate and the inner loops.

The intermediate loop iterates over the set of pairs of paths (p_1, p_2) from the Cartesian product of P_1 and P_2. If there is at least a pair of such paths, it then means that the resources involved in the links returned by base link rule r might have some neighbours that might be linked by supporting link rule r'.

The inner loop iterates over the collection of links (a, b) in set L_1. It first finds the set of resources A that are reachable from resource a using path p_1 in source dataset D_1 and the set of resources B that are reachable from resource b using path p_2 in the companion dataset D_2. Next, the method applies supporting link rule r' to the source and the companion dataset and intersects the resulting links with $A \times B$ so as to keep resources that are not reachable from a or b apart; the result is stored in set E. It then computes the similarity of sets A and B; intuitively, the higher the similarity, the more likely that resources a and b refer to the same real-world entity. If the similarity is equal or greater than threshold θ, then link (a, b) is added to set K; otherwise, it is filtered out. When the main loop finishes, set K contains the collection of links that involve neighbours that are similar enough according to the supporting rules.

We do not provide any additional details regarding the algorithms to find paths or resources since they can be implemented using Dijkstra's algorithm to find the shortest paths in a graph. Computing the similarity coefficient is a bit more involved, so we devote a subsection to this ancillary method below.

Example 2. In our running example, link rule r_1 is the base link rule, i.e., we are interested in linking authors and researchers, and we use link rule r_2 as the support link rule, i.e., we take their articles and papers into account. Their source classes are $C_1 = \{dblp{:}Author\}$ and $C_1' = \{dblp{:}Article\}$, respectively, and their target classes are $C_2 = \{nsf{:}Researcher\}$ and $C_2' = \{nsf{:}Paper\}$, respectively. Link rule r_1 returns the following links: $L_1 = \{(dblp{:}weiwang, nsf{:}weiwang_1),$ $(dblp{:}weiwang, nsf{:}weiwang_2), (dblp{:}binliu, nsf{:}binwliu)\}$; note that the first and the third links are true positive links, but the second one is a false positive link. Link rule r_2 returns the following links: $L_2 = \{(dblp{:}article_1, nsf{:}paper_3),$ $(dblp{:}article_2, nsf{:}paper_2), (dblp{:}article_4, nsf{:}paper_2), (dblp{:}article_5,$ $nsf{:}paper_5)\}$, which are true positive links.

The sets of paths between the source and target classes of r_1 and r_2 are $P_1 = \{\langle dblp{:}writtenBy\rangle\}$ and $P_2 = \{\langle nsf{:}leads, nsf{:}supports\rangle\}$. Now, the links in L_1 are scanned and the resources that can be reached from the resources involved in each link using the previous paths are fetched.

Link $l_1 = (dblp{:}weiwang, nsf{:}weiwang_1)$ is analysed first. The method finds $A = \{dblp{:}article_1, dblp{:}article_2, dblp{:}article_3, dblp{:}article_4\}$ by following resource $dblp{:}weiwang$ through path $\langle dblp{:}writtenBy\rangle$; similarly, it finds $B = \{nsf{:}paper_1, nsf{:}paper_2, nsf{:}paper_3\}$ by following resource $nsf{:}weiwang_1$ through path $\langle nsf{:}leads, nsf{:}supports\rangle$. Now supporting link rule r_2 is applied and the results are intersected with $A \times B$ so as to keep links that are related to l_1 only; the result is $E = \{(dblp{:}article_1, nsf{:}paper_3), (dblp{:}article_2,$ $nsf{:}paper_2), (dblp{:}article_4, nsf{:}paper_2)\}$. Then, the similarity of A and B in the context of E is computed, which returns 0.67; intuitively, there are chances that l_1 is a true positive link.

Link $l_2 = (dblp{:}weiwang, nsf{:}weiwang_2)$ is analysed next. The method finds $A = \{dblp{:}article_1, dblp{:}article_2, dblp{:}article_3, dblp{:}article_4\}$ by following resource $dblp{:}weiwang$ through path $\langle dblp{:}writtenBy\rangle$; next, it finds $B = \{nsf{:}paper_4\}$ by following resource $nsf{:}weiwang_2$ through path $\langle nsf{:}leads,$ $nsf{:}supports\rangle$. Now supporting link rule r_2 is applied and the result is intersected with $A \times B$, which results in $E = \emptyset$. In such a case the similarity is zero, which intuitively indicates that it is very likely that l_2 is a false positive link.

Link $l_3 = (dblp{:}binliu, nsf{:}binwliu)$ is analysed next. The method finds $A = \{dblp{:}article_5\}$ by following resource $dblp{:}binliu$ through path $\langle dblp{:}writtenBy\rangle$;

```
1: method computeSimilarity(A, B, E) returns d
2:    A' := reduce(A, E)
3:    B' := reduce(B, E)
4:    W := intersect(A', B', E)
5:    d := |W|/ min{|A'|, |B'|})
6: end
```

Fig. 3. Method to compute similarity.

next, it finds $B = \{nsf\text{:}paper_5\}$ by following resource $nsf\text{:}binwliu$ through path $\langle nsf\text{:}leads, nsf\text{:}supports \rangle$. Now supporting link rule r_2 is applied and the result is intersected with $A \times B$, which results in $E = \{(dblp\text{:}article_5, nsf\text{:}paper_5)\}$. The similarity is now 1.00, i.e., it is very likely that link l_3 is a true positive link.

Assuming that θ is set to, e.g., 0.50, the $filterLinks$ method would return $K = \{(dblp\text{:}weiwang, nsf\text{:}weiwang_1), (dblp\text{:}binliu, nsf\text{:}binwliu)\}$. Note that the previous value of θ is intended for illustration purposes only because the running example must necessarily have very little data.

3.2 Computing Similarity

Figure 3 shows our method to compute similarities. Its input consists of sets A and B, which are two sets of resources, and E, which is a set of links between them. It returns the Szymkiewicz-Simpson overlapping coefficient, namely:

$$overlap(A, B) = \frac{|A \cap B|}{\min\{|A|, |B|\}}$$

The previous formula assumes that there is an implicit equality relation to compute $A \cap B$, $|A|$, or $|B|$. In our context, this relation must be inferred from the set of links E by means of Warshall's algorithm to compute the reflexive, commutative, transitive closure of relation E, which we denote as $E^\star$.

The method to compute similarities relies on two ancillary functions, namely: $reduce$, which given a set of resources X and a set of links E returns a set whose elements are subsets of X that are equal according to $E^\star$, and $intersect$, which given two reduced sets of resources X and Y and a set of links E returns the intersection of X and Y according to $E^\star$. Their definitions are as follows:

$$reduce(X, E) = \{W \mid W \propto W \subseteq X \wedge W \times W \subseteq E^\star)\}$$
$$intersect(X, Y, E) = \{W \mid W \propto W \subseteq X \wedge \exists W' : W' \subseteq Y \wedge W \times W' \in E^\star\}$$

where $X \propto \phi$ denotes the maximal set X that fulfils predicate ϕ, that is:

$$X \propto \phi \iff \phi(X) \wedge (\not\exists X' : X \subseteq X' \wedge \phi(X'))$$

The method to compute similarities then works as follows: it first reduces the input sets of resources A and B according to the set of links E; it then computes the intersection of both reduced sets; finally, it computes the similarity using Szymkiewicz-Simpson's formula on the reduced sets.

Example 3. Analysing link $l_1 = (dblp{:}weiwang, nsf{:}weiwang_1)$ results in sets $A = \{dblp{:}article_1, dblp{:}article_2, dblp{:}article_3, dblp{:}article_4\}$, $B = \{nsf{:}paper_1,$ $nsf{:}paper_2, nsf{:}paper_3\}$, and $E = \{(dblp{:}article_1, nsf{:}paper_3), (dblp{:}article_2,$ $nsf{:}paper_2), (dblp{:}article_4, nsf{:}paper_2)\}$. If E is interpreted as an equality relation by computing its reflexive, symmetric, transitive closure, then it is not difficult to realise that $dblp{:}article_2$ and $dblp{:}article_4$ can be considered equal, because $dblp{:}article_2$ is equal to $nsf{:}paper_2$ and $nsf{:}paper_2$ is equal to $dblp{:}article_4$. Thus, set A is reduced to $A' = \{\{dblp{:}article_1\}, \{dblp{:}article_2,$ $dblp{:}article_4\}, \{dblp{:}article_3\}\}$ and set B is reduced to $B' = \{\{nsf{:}paper_1\},$ $\{nsf{:}paper_2\}, \{nsf{:}paper_3\}\}$. As a conclusion, $|A' \cap B'| = |\{\{dblp{:}article_1,$ $nsf{:}paper_3\}, \{dblp{:}article_2, dblp{:}article_4, nsf{:}paper_2\}\}| = 2$, $|A'| = 3$, and $|B'| = 3$; so the similarity is 0.67.

When link $l_2 = (dblp{:}weiwang, nsf{:}weiwang_2)$ is analysed, $A = \{dblp{:}article_1, dblp{:}article_2, dblp{:}article_3, dblp{:}article_4\}$, $B = \{nsf{:}paper_4\}$, and $E = \emptyset$. Since the equality relation E^* is then empty, the similarity is zero because the intersection between the reductions of sets A and B is empty.

In the case of link $l_3 = (dblp{:}binliu, nsf{:}binwliu)$, $A = \{dblp{:}article_5\}$, $B = \{nsf{:}paper_5\}$, and $E = \{(dblp{:}article_5, nsf{:}paper_5)\}$. As a conclusion, $|A' \cap B'| = |\{\{dblp{:}article_5, nsf{:}paper_5\}\}| = 1$, $|A'| = 1$, and $|B'| = 1$, where A' and B' denote, respectively, the reductions of sets A and B; so the similarity is 1.00. ∎

4   Experimental Analysis

In this section, we first describe our experimental environment and then comment on our results.

Computing facility: We run our experiments on a virtual computer that was equipped with four Intel Xeon E5-2690 cores at 2.60 GHz and 4 GiB of RAM. The operating system was CentOS Linux 7.3.

Prototype: We implemented our proposal with Java 1.8 and the following components: the Genlink implementation from the Silk Framework 2.6.0 to generate link rules, Jena TDB 3.2.0 to work with RDF data, ARQ 3.2.0 to work with SPARQL queries, and Simmetrics 1.6.2, SecondString 2013-05-02, and JavaStringSimilarity 1.0.1 to compute string similarities.

Evaluation datasets:[1] We used the following datasets: DBLP, NSF, BBC, DBpedia, IMDb, RAE, Newcastle, and Rest. We set up the following scenarios: (1) DBLP–NSF, which focuses on the top 100 DBLP authors and 130 principal NSF researchers with the same names; (2) DBLP–DBLP, which focuses on the 9 076 DBLP authors with the same names who are known to be different people; (3) BBC–DBpedia, which focuses on 691 BBC movies and 445 DBpedia films that have similar titles; (4) DBpedia–IMDb, which focuses on 96 DBpedia movies and 101 IMDb films that have similar titles; (5) RAE–Newcastle, which focuses

---

[1] The datasets are available at https://goo.gl/asvKQV.

on 108 RAE publications and 98 Newcastle papers that are similar; and (6) Rest–Rest, which focuses on 113 and 752 restaurants published by OAEI.

Baseline: Our baseline was the Genlink implementation from the Silk Framework 2.6.0, which is a state-of-the-art genetic programming system to learn link rules.

Measures: We measured the number of links returned by each proposal ($Links$), precision (P), recall (R), and the F_1 score (F_1) using 2-fold cross validation. We also computed the normalised differences in precision (ΔP), recall (ΔR), and F_1 score (ΔF_1), which measure the ratio from the difference found between the baseline and our proposal and the maximum possible difference for each performance measure. We also applied Iman-Davenport's test and computed the corresponding p-values to check if the differences found are statistically significant or not at the standard confidence level ($\alpha = 0.05$).

Parameters: We set $\theta = 0.01$. We explored a large portion of the parameter space and our conclusion was that setting θ to this small value helps our proposal perform the best. Note that it is very small, which means that it generally suffices to find a single link amongst the neighbours of the resources involved in another link so that it can be considered a true positive link.

Scenario	Genlink				Our proposal							
	Links	P	R	F_1	Links	P	R	F_1	ΔLinks	ΔP	ΔR	ΔF_1
DBLP - NSF	127	0.25	0.97	0.40	52	0.62	0.97	0.75	-75	0.49	0.00	0.41
DBLP - DBLP	78,348	0.12	1.00	0.21	9210	0.98	1.00	0.99	-69,138	0.98	0.00	0.01
BBC - DBpedia	525	0.85	1.00	0.92	461	0.96	1.00	0.98	-64	0.74	0.00	0.24
DBpedia - IMDb	118	0.27	0.55	0.36	42	0.67	0.48	0.56	-76	0.54	-0.13	0.69
RAE-Newcastle	404	0.22	0.82	0.35	68	0.72	0.45	0.56	-336	0.64	-0.45	0.68
Rest - Rest	103	0.90	0.83	0.87	96	0.97	0.83	0.89	-7	0.68	0.00	0.78
								Average Δ		0.68	-0.10	0.47
								Iman-Davenport's test		0.00	0.25	0.00

Fig. 4. Experimental results.

Results: The results are presented in Fig. 4. We analyse them regarding precision, recall, and the F_1 score below.

The results regarding precision clearly show that our technique improves the precision of the rules learnt by GenLink in every scenario. In average, the difference in precision is 68%. The worst improvement is 49% in the DBLP–NSF scenario since these datasets are clearly unbalanced: the top authors in DBLP have about 500 papers in average, but NSF records an average of 7 papers in the projects in which they are involved; this obviously makes it difficult for our proposal to find enough context to make a decision. The best improvement is 98% in the DBLP–DBLP scenario since there are 9 076 authors with very similar names, which makes it almost impossible for GenLink to generate rules with good precision building solely on the attributes of the resources. Note that the p-value computed by Iman-Davenport's test is 0.00; since it is clearly smaller than the standard confidence level, we can interpret it as a strong indication that there is enough evidence in our experimental data to confirm the hypothesis that our proposal works better than the baseline regarding precision.

The normalised difference of recall ΔR shows that our proposal generally retains the recall of the link rules learnt by GenLink, except in the DBpedia–IMDb and the Rae–Newcastle scenarios. The problem with the previous scenarios was that there are many incomplete resources, that is many resources without neighbours. For instance, there are 43 papers in the Newcastle dataset that are not related to any authors. The incompleteness of data has also a negative impact on the recall of the base link rules. In our prototype, we are planning on implementing a simple check to identify incomplete resources so that the links in which they are involved are not discarded as false positives, but identified as cases on which our proposal cannot make a sensible decision. Note that Iman-Davenport's test returns 0.25 as the corresponding p-value; since it is larger than the standard confidence level, it may be interpreted as a strong indication that the differences in recall found in our experiments are not statistically significant. In other words, the cases in which data are that incomplete do not seem to be common-enough for them to have an overall impact on our proposal.

We also studied ΔF_1, which denotes the normalised difference in F_1 score. Note that it is 47% in average and that the corresponding Iman-davenport's p-value is 0.00, which can be interpreted as a strong indication that the difference is significant from a statistical point of view. Overall, this result confirms that our proposal helps improve precision without degrading recall.

5   Conclusions

Programs need to link the resources that they find on the Web of Data so that they can enrich the data about a real-world entity that is found in a source dataset with data that comes from companion datasets. Current link rules take the values of the attributes of the resources into account, but not their neighbours, which sometimes results in false positives that have a negative impact on their precision. We have presented a hybrid proposal[2] that learns a set of link rules using a genetic programming approach and then bootstraps them. Our proposal may be fed with rules generated by any genetic programming approach from the literature, the afterwards bootstrap that performs has proven to improve the overall F_1 score.

References

1. Ananthakrishna, R., Chaudhuri, S., Ganti, V.: Eliminating fuzzy duplicates in data warehouses. In: VLDB, pp. 586–597 (2002)

---

[2] The prototype is available at https://github.com/AndreaCimminoArriaga/Teide.

2. Back, T.: Evolutionary algorithms in theory and practice: evolution strategies, evolutionary programming, genetic algorithms. Oxford University Press, New York (1996)
3. Bhattacharya, I., Getoor, L.: Collective entity resolution in relational data. TKDD **1**(1), 1–36 (2007)
4. Bizer, C., Heath, T., Berners-Lee, T.: Linked Data: principles and state of the art. In: WWW (Invited talks) (2008). https://www.w3.org/2008/Talks/WWW2008-W3CTrack-LOD.pdf
5. Cruz, I.F., Antonelli, F.P., Stroe, C.: AgreementMaker: efficient matching for large real-world schemas and ontologies. PVLDB **2**(2), 1586–1589 (2009)
6. Dong, X., Halevy, A.Y., Madhavan, J.: Reference reconciliation in complex information spaces. In: SIGMOD, pp. 85–96 (2005)
7. Hernández, M.A., Stolfo, S.J.: The merge/purge problem for large databases. In: SIGMOD Conference, pp. 127–138 (1995)
8. Holub, M., Proksa, O., Bieliková, M.: Detecting identical entities in the semantic web data. In: Italiano, G.F., Margaria-Steffen, T., Pokorný, J., Quisquater, J.-J., Wattenhofer, R. (eds.) SOFSEM 2015. LNCS, vol. 8939, pp. 519–530. Springer, Heidelberg (2015). https://doi.org/10.1007/978-3-662-46078-8_43
9. Hu, W., Qu, Y.: Falcon-AO: a practical ontology matching system. J. Web Semant. **6**(3), 237–239 (2008)
10. Huber, J., Sztyler, T., Nößner, J., Meilicke, C.: CODI: Combinatorial optimization for data integration. In: OM, pp. 134–141 (2011)
11. Isele, R., Bizer, C.: Learning expressive linkage rules using genetic programming. PVLDB **5**(11), 1638–1649 (2012)
12. Isele, R., Bizer, C.: Active learning of expressive linkage rules using genetic programming. J. Web Semant. **23**, 2–15 (2013)
13. Jiménez-Ruiz, E., Grau, B.C.: LogMap: logic-based and scalable ontology matching. In: Aroyo, L., Welty, C., Alani, H., Taylor, J., Bernstein, A., Kagal, L., Noy, N., Blomqvist, E. (eds.) ISWC 2011. LNCS, vol. 7031, pp. 273–288. Springer, Heidelberg (2011). https://doi.org/10.1007/978-3-642-25073-6_18
14. Kalashnikov, D.V., Mehrotra, S., Chen, Z.: Exploiting relationships for domain-independent data cleaning. In: SDM, pp. 262–273 (2005)
15. Köpcke, H., Rahm, E.: Frameworks for entity matching: a comparison. Data Knowl. Eng. **69**(2), 197–210 (2010)
16. Lacoste-Julien, S., Palla, K., Davies, A., Kasneci, G., Graepel, T., Ghahramani, Z.: SIGMa: simple greedy matching for aligning large knowledge bases. In: KDD, pp. 572–580 (2013)
17. Monge, A.E., Elkan, C.: The field matching problem: algorithms and applications. In: KDD, pp. 267–270 (1996)
18. Ngomo, A.C.N., Auer, S.: LIMES: A time-efficient approach for large-scale link discovery on the web of data. In: IJCAI, pp. 2312–2317 (2011)
19. Ngomo, A.-C.N., Lyko, K.: EAGLE: efficient active learning of link specifications using genetic programming. In: Simperl, E., Cimiano, P., Polleres, A., Corcho, O., Presutti, V. (eds.) ESWC 2012. LNCS, vol. 7295, pp. 149–163. Springer, Heidelberg (2012). https://doi.org/10.1007/978-3-642-30284-8_17
20. Nikolov, A., d'Aquin, M., Motta, E.: Unsupervised learning of link discovery configuration. In: Simperl, E., Cimiano, P., Polleres, A., Corcho, O., Presutti, V. (eds.) ESWC 2012. LNCS, vol. 7295, pp. 119–133. Springer, Heidelberg (2012). https://doi.org/10.1007/978-3-642-30284-8_15
21. Rastogi, V., Dalvi, N.N., Garofalakis, M.N.: Large-scale collective entity matching. PVLDB **4**(4), 208–218 (2011)

22. Soru, T., Ngomo, A.C.N.: A comparison of supervised learning classifiers for link discovery. In: SEMANTICS, pp. 41–44 (2014)
23. Szekely, P., et al.: Building and using a knowledge graph to combat human trafficking. In: Arenas, M., Corcho, O., Simperl, E., Strohmaier, M., d'Aquin, M., Srinivas, K., Groth, P., Dumontier, M., Heflin, J., Thirunarayan, K., Staab, S. (eds.) ISWC 2015. LNCS, vol. 9367, pp. 205–221. Springer, Cham (2015). https://doi.org/10.1007/978-3-319-25010-6_12
24. Volz, J., Bizer, C., Gaedke, M., Kobilarov, G.: Silk: a link discovery framework for the web of data. In: LDOW (2009). http://ceur-ws.org/Vol-538/ldow2009_paper13.pdf

Modelling and Forecasting of the ^{222}Rn Radiation Level Time Series at the Canfranc Underground Laboratory

Iván Méndez-Jiménez and Miguel Cárdenas-Montes[✉]

Centro de Investigaciones Energéticas Medioambientales y Tecnológicas,
Madrid, Spain
{ivan.mendez,miguel.cardenas}@ciemat.es

Abstract. The ^{222}Rn level at underground laboratories, where Physics experiments of low-background are installed, is the largest source of background; and it is the main distortion for obtaining high accuracy results. At Spain, the Canfranc Underground Laboratory hosts ground-breaking experiments, such as Argon Dark Matter-1t aimed at the dark matter direct searches. For the collaborations exploiting these experiments, the modelling and forecasting of the ^{222}Rn level are very relevant tasks for efficient planning activities of installation and maintenance. In this paper, four years of values of ^{222}Rn level from the Canfranc Underground Laboratory are analysed using methods such as Holt-Winters, AutoRegressive Integrated Moving Averages, Seasonal and Trend Decomposition using Loess, Feed-Forward Neural Networks, and Convolutional Neural Networks. In order to evaluate the performance of these methods, both the Mean Squared Error and the Mean Absolute Error are used. Both metrics determine that the Seasonal and Trend Decomposition using Loess no periodic, and the Convolutional Neural Networks, are the techniques which obtain the best predictive results. This is the first time that the mentioned data are investigated, and it constitutes an excellent example of scientific time series with relevant implications for the quality of the scientific results of the experiments.

Keywords: Time series analysis · Convolutional neural network
^{222}Rn Measurements · Canfranc Underground Laboratory · Forecasting

1 Introduction

Time Series analysis, as part of the Big Data ecosystem, is becoming more and more relevant. The increasing data volume from industry, e-commerce, social media and science demands continuous efforts for understanding the new data sets, for extracting information, and for timely processing. In scientific disciplines, detectors and facilities are being populated of sensors, with increasing rate of data taking, providing larger and more complex data sets. Many of data sets emerging from those scenarios have chronological order, and therefore, can

© Springer International Publishing AG, part of Springer Nature 2018
F. J. de Cos Juez et al. (Eds.): HAIS 2018, LNAI 10870, pp. 158–170, 2018.
https://doi.org/10.1007/978-3-319-92639-1_14

be analysed with time series techniques, including classic ones and those arisen from deep learning area.

In this work, the ^{222}Rn level at the Canfranc Underground Laboratory (LSC), for a period of four years —from July 2013 to June 2017—, is analysed. This includes the efforts in preprocessing and cleaning the data. And later, the application of time series techniques for modelling and forecasting the ^{222}Rn values. This is the first time that these data are investigated, and it constitutes an excellent example of scientific time series with relevant implications for the quality of the scientific results of the experiments.

The modelling and forecasting of ^{222}Rn in underground laboratories are very relevant tasks. ^{222}Rn is a radionuclide produced by the ^{238}U and ^{232}Th decay chains. Being gas at room temperature, it can be emanated by the rocks and concrete of the underground laboratory, diffusing in the experimental hall. This contamination in the air is a potential source of background, both directly and through the long life radioactive daughters produced in the decay chain, which can stick to the experimental surfaces. The ^{222}Rn contamination in air can be reduced by orders of magnitude only in limited closed areas, flushing pure N_2 or "Rn-free" air produced by dedicated structures. In the deep underground laboratories the average activity depends on the local conditions and must be constantly monitored. It typically ranges from tens to hundreds of Bq/m^3, with periodic and non-periodic variations. Strong seasonal dependence has been observed in some cases [1]. A detail understanding of the ^{222}Rn periodicity can be fundamental for a precise understanding of the background of rare-event search experiments. This is particularly true in case of the dark matter direct searches, whose distinctive feature is the annual modulation of the signal foreseen by the hypothesis of a weakly interactive massive particle (WIMP) halo model. At the same time, the prediction of the evolution of the ^{222}Rn concentration in the laboratory is relevant in order to correctly organize the operations foreseeing the exposure of the detector materials to the air, minimizing, in such a way, the deposition of the radionuclide on the surfaces.

The rest of the paper is organized as follows: Sect. 2 summarizes the Related Work and previous efforts done. A brief description of the techniques and algorithms used for analysing the data is presented in Sect. 3. The most relevant points investigated and the results obtained are presented and analysed in Sect. 4. Finally, Sect. 5 contains the conclusions of this work.

2   Related Work

As part of the previous effort to study the ^{222}Rn level at LSC, a study over a year is presented in [2]. This study provides enough information to know the radioactivity state in the different halls of the LSC, but previous efforts have not been done to forecast the future behaviour of this gas.

Secondly, an example of how to deal with an environmental variable in scientific facilities is shown in [3]. In that work, an analysis of temperature taken at two telescopes located at the Observatorio del Roque de Los Muchachos in the

Canary Islands[1] is carried out. Its final aim is to better understand the influence of wide scale parameters on local meteorological data.

Many researchers use these forecasting techniques when they work with time series and there are missed observations or future values that are interesting to know. For instance, in studies [4–8], traditional and advanced forecasting methods are used to predict future values for different time series, such as: the Thai Stock Price Index Trend [4], the Feed Grain Demand [5], the Beijing Haze Episodes [6], the Tourism Economy in a country [7] or for Financial Forecasting in Microscopic High-Speed Trading Environment [8]. All these studies try to establish a comparison between traditional forecasting models and methods, most of them based on Neural Networks, which try to improve the results of the forecasts.

In this work, the following R packages [9] have been used when implementing seasonal ARIMA, Holt-Winters Exponential Smoothing, Seasonal and Trend decomposition using Loess, and Artificial Neural Networks: forecast [10,11], fpp [12], astsa [13], and stR [14]; and Keras [15] when implementing Convolutional Neural Networks.

3   Methodology

3.1   Data Preprocessing

The LSC (Spain) is composed of diverse halls for hosting scientific experiments with requirements of very low-background. The two main halls, Hall A and Hall B —which are contiguous—, have instruments for measuring the level of ^{222}Rn, particularly there is an Alphaguard P30 in each hall recording the radioactivity level[2] every 10 min, as well as the temperature, the humidity and the air pressure. Measurements are available from July 2013. In this work, four years are analysed. As an example, in Fig. 1 the values of ^{222}Rn level corresponding to 2015 are represented.

With regard to the measurements, few missing values are in the data set, as well as a gap of several days in some years. The large gaps appear in July 2014 with 913 missing values, in June 2015 with 1053, and in January 2016 with 585. In the worst case, the gap spans over a week (7.3 days). However, the missing values are not representative in comparison with the total number of observations (more than 200,000), nor the number of observations per month ($\approx$ 4,000).

In Fig. 2(a), the monthly boxplots of the ^{222}Rn level at Hall A of the LSC are represented. Visual inspection of this figure shows a certain seasonality in the medians. This seasonal behaviour is better appreciated when gathering the values of the months independently of the year (Fig. 2(b)). Rawly ^{222}Rn level tends to be higher in summer, from June to August, than in winter. This seasonal

---

[1] Telescopio Nazionale Galileo (TNG) and Carlsberg Automatic Meridian Circle Telescope (CAMC).

[2] Alphaguard P30 device takes values between 2 and $2 \cdot 10^6 Bq/m^3$.

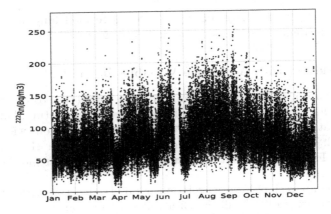

Fig. 1. ^{222}Rn concentration for 2015 at Hall A of the LSC.

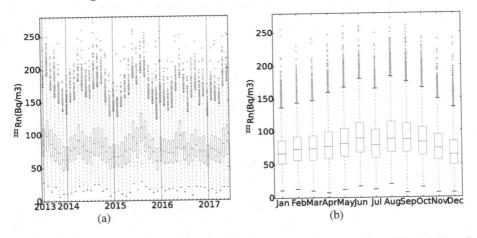

Fig. 2. Monthly box-plots of ^{222}Rn level at Hall A of the LSC, by year (Fig. 2(a)) and gathering the months independently of the year (Fig. 2(b)). Data taking corresponds to the period from July 2013 to June 2017. Hereafter, the monthly medians are the values used for creating the time series, and therefore, for further analyses.

behaviour is of high interest for the experiments hosted at LSC, and it is the main motivation for the current work.

The choice of the medians as indicator of ^{222}Rn level for the months carries out some benefits. Among others, the choice of the medians is more robust than the means in presence of outlier values. Furthermore, in presence of gaps, the median of the observed values is representative for the month.

Previous measurements of the ^{222}Rn level by ANAIS experiment, also at LSC, show a period of $T = 379 \pm 20$ days (May 2011 – Nov. 2012) [16], and $T = 385 \pm 1$ days (Jan. 2012 - Jan. 2016) [17]. This reinforces the intuition of the presence of a seasonal pattern on monthly medians.

From this point, the target of the research is to modelling and forecasting the monthly medians of ^{222}Rn level at Hall A of the LSC.

3.2 Correlation with Ambiance Variables in the Hall

In order to improve the modelling capacity, the correlation of the ^{222}Rn level with the ambiance variables in the Hall is analysed. Temperature, humidity and air pressure at Halls A and B are controlled, so that correlations with ^{222}Rn are not expected. In Fig. 3, the monthly mean of the temperature (Fig. 3(a)), of the air pressure (Fig. 3(b)), and of the humidity (Fig. 3(c)) and the monthly medians of ^{222}Rn are shown.

As can be appreciated, no correlation is observed between the ^{222}Rn level and the temperature of the hall (Fig. 3(a)), nor with the air pressure (Fig. 3(b)). Only the humidity (Fig. 3(c)) exhibits a certain correlation with the ^{222}Rn level. The Peason coefficient confirms that only this last case shows some correlation (Table 1).

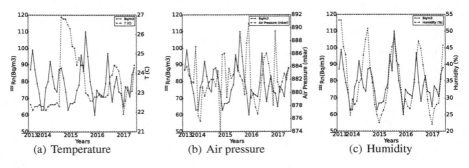

(a) Temperature   (b) Air pressure   (c) Humidity

Fig. 3. Correlation between ^{222}Rn monthly medians and the monthly means of temperature (Fig. 3(a)), air pressure (Fig. 3(b)) and humidity (Fig. 3(c)) in the Hall A. In all cases solid line corresponds to ^{222}Rn, and the dashed line to the ambiance variables.

Table 1. Pearson coefficient for the correlations between the monthly medians of ^{222}Rn values and the monthly mean values of humidity, pressure and temperature at Hall A.

	Pearson Coefficient
Temperature	−0.11
Air pressure	−0.05
Humidity	0.70

This analysis points forwards further research in the correlation between the ^{222}Rn level and the meteorological variables at surface responsible of humidity at LSC. Even the incorporation of this information for improving the forecasting capacity is proposed as future work.

3.3   Correlation Hall a and Hall B

A second unit for measuring the ^{222}Rn level is installed at Hall B of the LSC. This second unit is identical to one presented at Hall A. If the measurements of both units are highly correlated, then the gaps in the values from one unit can be covered by the other one. In Fig. 4(a), the monthly medians of both units are represented. As can be visually appreciated, medians almost overlap. In Fig. 4(b), it can be observed how the measurements of both units behave equally, being both measurements highly correlated. The value of the Pearson coefficient achieves 0.98.

This last point indicates that the influence factors on the modulation of ^{222}Rn level are not local to the halls, but they should come from the environment of the laboratory. Therefore, the exploration of the influence of meteorological variables at surface in the nearby of the laboratory could be valuable for improving the modelling and forecasting capacity.

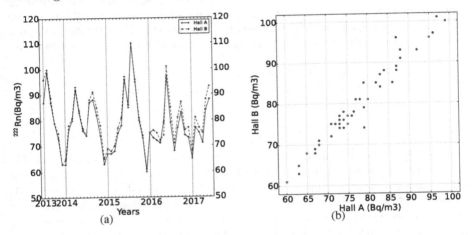

Fig. 4. Correlation between values observed of ^{222}Rn at Hall A and Hall B. The value of Peason correlation coefficient is 0.98.

3.4   Seasonal ARIMA

ARIMA (Auto-Regressive Integrated Moving Average) methods are popular and general models for modelling and forecasting time series, independently of its complexity or the presence or not of seasonality [18]. When seasonality is presented, ARIMA models are labelled as ARIMA(p,d,q)(P,D,Q), where p is the order of the autoregressive part, d is the degree of first differencing involved, q is the order of the moving average part, all of them for the non-seasonal part of the time series; and their capital corresponding are for seasonal part.

One of hardest point to achieve in the ARIMA models is the election of the suitable combination of parameters (p,d,q)(P,D,Q). This is usually done by minimizing the *Akaike's Information Criterion* and the *Maximum Likelihood*

Estimation. For an in-depth discussion about this type of models, readers are referred to [18].

When applying ARIMA to the first three years of data set, by using the R function `auto.arima`, the most suitable model is ARIMA(1,0,0)(0,1,0)(12).

3.5   Holt-Winters Exponential Smoothing

In contrast to moving average techniques, where all the past observations in a window frame are equally weighted, in exponential techniques the previous observations are less relevant as far as they are from actual one. The Holt-Winters method (Eq. 1) extends the Holt's method by allowing to capture the seasonality of the series [19,20]. It includes the forecasting method and three smoothing equations. These are for the level l_t, for the trend b_t, and for the seasonal component s_t, and their smoothing parameters α, β^*, and γ. This method should be viewed as part of the efforts of the smoothing methods for capturing time series with increasingly complexity: *simple exponential smoothing, double exponential smoothing* and *triple exponential smoothing* or Holt-Winters method. Holt-Winters method is the appropriate one when seasonality is present in data.

$$
\begin{aligned}
\hat{y}_{t+h|t} &= l_t + h \cdot b_t + s_{t-m+h_m^+} \\
l_t &= \alpha(y_t - s_{t-m}) + (1-\alpha)(l_{t-1} + b_{t-1}) \\
b_t &= \beta^*(l_t - l_{t-1}) + (1-\beta^*)b_{t-1} \\
s_t &= \gamma(y_t - l_{t-1} - b_{t-1}) + (1-\gamma)s_{t-m}
\end{aligned}
\tag{1}
$$

where $+h_m^+ = \lfloor (h-1) \mod m \rfloor + 1$, m is the period of the seasonality, and it assures that the seasonal indexes come from the correct period ; and where the initial values are: $l_0 = \frac{1}{m}(y_1 + \cdots + y_m)$, $b_0 = \frac{1}{m}(\frac{y_{m+1}-y_1}{m} + \cdots + \frac{y_{m+m}-y_m}{m})$, and $s_0 = \frac{y_m}{l_0}, s_{-1} = \frac{y_{m-1}}{l_0}, \ldots, s_{-m+1} = \frac{y_1}{l_0}$. The application of this method to the first three years of data set arises the following values for the smoothing parameters: $\alpha = 0.09$, $\beta^* = 0$, and $\gamma = 1$. The value $\beta^* = 0$ indicates that there are not changes expected for the trend $b_t = b_{t-1}$, as far as the time increasing. The value of $\gamma = 1$ leads to $s_t = (y_t - l_{t-1} - b_{t-1})$.

3.6   STL Decomposition

The intuition behind the time series decomposition is that the time series is the composition of three more elementary series. On the one hand, a trend (T_t), which is responsible of long-term increase or decrease of data. It does not have to be linear. On the other hand, a seasonal pattern is the second component (S_t). It is influenced by seasonal factors, such as: the month, the day of the week, or the quarter of the year. Finally, the third component is the remainder or random component (R_t). If the decomposition is additive, then the values of the time series (Y_t) can be modelled as $Y_t = T_t + S_t + R_t$.

Diverse techniques for time series decomposition have been proposed. STL, *Seasonal and Trend decomposition using Loess* [21], was proposed taking into account the limitations of previous classical decomposition methods, for example X-12-ARIMA. STL can handle any type of seasonality, not only monthly or quarterly. The seasonal component can change over time, being the amount of change controlled by a parameter of the algorithm. Besides, the smoothness of the trend component can be also controlled by the algorithm.

In Fig. 5, the STL decomposition no periodic for the four years of data set is shown. It is appreciated how the contribution of the trend is the most relevant. Seasonal and random components have an appreciable contribution in the order of the 20%. When forecasting using STL method in the next section, only the first three years will be used for modelling the series with the STL method, whereas the last year is used for evaluating the prediction. When evaluating the capacity for forecasting of STL, configurations allowing smooth changes in the trend and in the seasonal components, and other ones without changes allowed are tested.

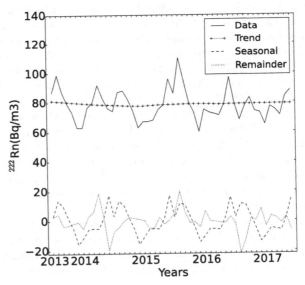

Fig. 5. STL decomposition no periodic for the four years of data, allowing smooth changes in trend and seasonal components.

3.7   Artificial Neural Network

Artificial Neural Networks (ANN) are biological-inspired composition of neurons, as fundamental elements, grouped in layers. Each layer is a non-linear combination of non-linear functions from the previous layer. Layers are ordered from the initial one, which receives the input data, to the last one, which produces the output, by passing from some intermediary ones or hidden, which map both data. The input of any neuron can be expressed as $f(w, \mathbf{x})$, where $\mathbf{x}$ is the input vector to the neuron, and w is the matrix of weights which is optimized trough the training process. By increasing the number of the hidden layers, more complicated relationships can be established.

Inspired by previous works [22,23], in this work **nnetar** function from **forecast** has been used for testing the prediction capacity of ANN for forecasting the ^{222}Rn values. This function implements a feed-forward neural networks with a single hidden layer with a number of neurons equal to the half of input plus one. In the current work, the network is trained with 100 epochs.

3.8   Convolutional Neural Networks

Convolutional Neural Networks (CNN) are specialized Neural Networks with special emphasis in image processing [24], although nowadays they are also employed in time series analysis [25,26]. The CNN consists of a sequence of convolutional layers, the output of which is connected only to local regions in the input. These layers alternate convolutional, non-linear and pooling-based layers which allow extracting the relevant features of the class of objects, independently of their placement in the data example. The CNN allows the model to learn filters that are able to recognize specific patterns in the time series, and therefore it can capture richer information from the series. It also embodies three features which provide advantages over ANN: sparse interactions, parameter sharing and equivariance to translation [24].

Although Convolutional Neural Networks are frequently associated to image classification —2D grid examples—, it can also be applied to time series analysis —1D grid examples—. When processing time series, instead of a set of images, the series has to be divided in overlapping contiguous time windows. These windows constitute the examples, where the CNN aims at finding patterns. At the same time, the application to time series modelling requires the application of 1D convolutional operators.

In this work, Keras [15] has been used for implementing the CNN for modelling the ^{222}Rn time series. The CNN employed is composed of two convolutional layers of 32 and 64 filters with **relu** as activation function, **MaxPooling1D**, 12 observations for the time window, and trained with 10 epochs.

4   Experimental Results and Models Comparison

As mentioned, the collected data are divided into two sets: the training data set —including the first three years, from July 2013 to June 2016— and the testing data set, which includes the fourth year, from July 2016 to June 2017.

In Fig. 6, the predicted values for the test set for the forecasting methods analysed in the current work are shown. As can be appreciated, all the methods can reproduce the period from October to June. However, most of the methods fail to predict the fall down in August 2016.

The month of August 2016 behaves differently that in the previous years. As can be appreciated in Fig. 2(a), in the previous years the level of ^{222}Rn increases in this month, but in 2016 it clearly falls down. Most of the methods are unable

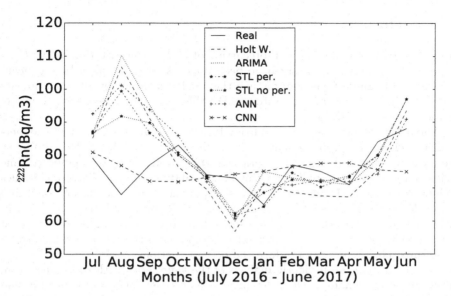

Fig. 6. Real values and forecasting for the test set —the fourth year, from July 2016 to June 2017— for the methods used in this study.

to capture information from previous values of the ^{222}Rn time series to correctly predict this change. Only the CNN is able to predict the different behaviour of August 2016, proposing a small reduction of ^{222}Rn level in comparison with July.

For evaluating the overall performance of the forecasting methods, the Mean Squared Error (MSE) and the Mean Absolute Error (MAE) are employed as figures of merit. In Table 2, the values of the MSE and the MAE after 15 independent executions are presented. Among the methods used for forecasting, the STL decomposition without periodic seasonal component —allowing smooth changes in the trend and the seasonal components—, and specially the CNN achieve the lowest MSEs and MAEs.

Table 2. Mean and standard deviation of Mean Squared Error and Mean Absolute Error for 15 independent runs for the test data set (fourth year).

Method	MSE	MAE
Holt-Winters (0.093,0,1)	195.1 ± 0	10.5 ± 0
ARIMA(1,0,0)(0,1,0)[12]	215.3 ± 0	9.7 ± 0
STL Decomposition periodic	117.5 ± 0	7.3 ± 0
STL Decomposition no periodic	87.7 ± 0	7.0 ± 0
Feed-Forward Neural Networks	175.4 ± 10.2	9.6 ± 0.3
CNN	59.4 ± 4.2	6.6 ± 0.6

5   Conclusions

In this study, the modelling and forecasting of ^{222}Rn concentration at the Canfranc Underground Laboratory have been done. As mentioned, a high concentration of this kind of gas, can disturb the results obtained by the experiments hosted in the underground laboratories. For this reason, it is necessary to model and forecast the ^{222}Rn levels.

The collected data set expands from July 2013 to June 2017. The tasks performed include the preprocessing and cleaning, and later the understanding of the most relevant features of the series. In this work, classic techniques of time series analysis are also applied for this understanding process. These techniques include seasonal ARIMA, Holt-Winters Exponential Smoothing, Seasonal and Trend decomposition using Loess, Feed-Forward Neural Networks, and Convolutional Neural Networks. If the data set is divided in a training set —composed of the first three years— and a test set —composed of the last twelve months— these three last techniques produce the best predictions.

As future work, an in-depth analysis of the time series with techniques from Deep Learning domain, such as Recurrent Neural Networks and Convolutional Neural Networks —configuration improvement— is proposed. It is expected that these techniques capture more information from the time series, and therefore, improve the overall performance. Besides, the extension of this work will evaluate and, when appropriately, incorporate additional information from exogenous variables, for example the meteorological information in the nearby of the laboratory placement. Obviously the incorporation of additional values of ^{222}Rn level, corresponding to the fifth year, is also proposed. This additional research requires an in-depth analysis to ascertain if the improvements are significant or not. Therefore, the use of non-parametric statistical inference for ascertain the significance of the efficiency of the proposed improvements is also proposed as future work.

Acknowledgment. The research leading to these results has received funding by the Spanish Ministry of Economy and Competitiveness (MINECO) for funding support through the grant FPA2016-80994-C2-1-R, and "Unidad de Excelencia María de Maeztu": CIEMAT - FÍSICA DE PARTÍCULAS through the grant MDM-2015-0509.

IMJ is co-funded in a 91.89 percent by the European Social Fund within the Youth Employment Operating Program, for the programming period 2014–2020, as well as Youth Employment Initiative (IEJ). IMJ is also co-funded through the Grants for the Promotion of Youth Employment and Implantation of Youth Guarantee in Research and Development and Innovation (I+D+i) from the MINECO.

The authors would like to thank Roberto Santorelli, Pablo García Abia and Vicente Pesudo for useful comments regarding the Physics related aspects of this work, and the Underground Laboratory of Canfranc by providing valuable feedback.

References

1. Bettini, A.: New underground laboratories: Europe, Asia and the Americas. Phys. Dark Universe **4(Supplement C)**, 36–40 (2014). DARK TAUP2013
2. Bandac, I., Bettini, A., Borjabad, S., Núñez-Lagos, R., Pérez, C., Rodríguez, S., Sánchez, P., Villar, J.: Radón y radiación ambiental en el Laboratorio Subterráneo de Canfranc (LSC). Revista de la sociedad española de protección radiológica 21 (2014)
3. Lombardi, G., Zitelli, V., Ortolani, S., Pedani, M.: El Roque de Los Muchachos site characteristics. 1. temperature analysis. Publ. Astron. Soc. Pac. **118**, 1198 (2006)
4. Inthachot, M., Boonjing, V., Intakosum, S.: Artificial neural network and genetic algorithm hybrid intelligence for predicting thai stock price index trend. Comput. Intell. Neurosci. **2016**, 8 (2016)
5. Yang, T., Yang, N., Zhu, C.: A forecasting model for feed grain demand based on combined dynamic model. Comput. Intell. Neurosci. **2016**, 6 (2016)
6. Yang, X., Zhang, Z., Zhang, Z., Sun, L., Xu, C., Yu, L.: A long-term prediction model of Beijing haze episodes using time series analysis. Comput. Intell. Neurosci. **2016**, 7 (2016)
7. Yu, Y., Wang, Y., Gao, S., Tang, Z.: Statistical modeling and prediction for tourism economy using dendritic neural network. Comput. Intell. Neurosci. **2016**, 9 (2016)
8. Huang, C., Li, H.: An evolutionary method for financial forecasting in microscopic high-speed trading environment. Comput. Intell. Neurosci. **2016**, 18 (2016)
9. R Core Team: R: A language and environment for statistical computing. R Foundation for Statistical Computing, Vienna, Austria (2014)
10. Hyndman, R., Khandakar, Y.: Automatic time series forecasting: the forecast package for R. J. Stat. Softw. Art. **27**(3), 1–22 (2008)
11. Hyndman, R.J.: forecast: Forecasting functions for time series and linear models. R package version 8.0 (2017)
12. Hyndman, R.J.: FPP: Data for "Forecasting: principles and practice". R package version 0.5 (2013)
13. Stoffer, D.: ASTSA: Applied Statistical Time Series Analysis. R package version 1.7 (2016)
14. Dokumentov, A., Hyndman, R.J.: stR: STR Decomposition. R package version 0.3 (2017)
15. Chollet, F., et al.: Keras (2015). https://github.com/fchollet/keras
16. Cuesta Soria, C.: ANAIS-0: Feasibility study for a 250 kg NaI(Tl) dark matter search experiment at the Canfranc Underground Laboratory. PhD thesis, Universidad de Zaragoza (2013)
17. Olivan Monge, M.A.: Design, scale-up and characterization of the data acquisition system for the ANAIS dark matter experiment. PhD thesis, Universidad de Zaragoza (2015)
18. Hyndman, R.J., Athanasopoulos, G.: Forecasting: principles and practice. OTexts (2014)
19. Brown, R.G.: Exponential smoothing for predicting demand (1956)
20. Winters, P.R.: Forecasting sales by exponentially weighted moving averages. Manag. Sci. **6**(3), 324–342 (1960)
21. Cleveland, R.B., Cleveland, W.S., McRae, J., Terpenning, I.: STL: A seasonal-trend decomposition procedure based on loess. J. Off. Stat. **6**(1), 3–73 (1990)
22. Tang, Z., Fishwick, P.A.: Feedforward neural nets as models for time series forecasting. ORSA J. Comput. **5**(4), 374–385 (1993)

23. Frank, R.J., Davey, N., Hunt, S.P.: Time series prediction and neural networks. J. Intell. Robot. Syst. **31**(1–3), 91–103 (2001)
24. Goodfellow, I., Bengio, Y., Courville, A.: Deep Learning. MIT Press, Massachusetts (2016). http://www.deeplearningbook.org
25. Gamboa, J.C.B.: Deep learning for time-series analysis (2017). CoRR abs/ 1701.01887
26. Wang, Z., Yan, W., Oates, T.: Time series classification from scratch with deep neural networks: A Strong Baseline (2016). CoRR abs/1611.06455

Sensor Fault Detection and Recovery Methodology for a Geothermal Heat Exchanger

Héctor Alaiz-Moretón[1]([⊠]), José Luis Casteleiro-Roca[2],
Laura Fernández Robles[3], Esteban Jove[2], Manuel Castejón-Limas[3],
and José Luis Calvo-Rolle[2]

[1] Departamento de Ingeniería Eléctrica y de Sistemas y Automática,
Edificio Tecnológico, Universidad de León,
Campus de Vegazana s/n, 24071 León, Spain
hector.moreton@unileon.es
[2] Departamento de Ingeniería Industrial, Universidade da Coruña,
Avda. 19 de febrero s/n, 15495 Ferrol, A Coruña, Spain
[3] Departamento de Ingenierías Mecánica, Informática y Aeroespacial, Edificio
Tecnológico, Universidad de León, Campus de Vegazana s/n, 24071 León, Spain

Abstract. This research addresses a sensor fault detection and recovery methodology oriented to a real system as can be a geothermal heat exchanger installed as part of the heat pump installation at a bioclimatic house. The main aim is to stablish the procedure to detect the anomaly over a sensor and recover the value when it occurs. Therefore, some experiments applying a Multi-layer Perceptron (MLP) regressor, as modelling technique, have been made with satisfactory results in general terms. The correct election of the input variables is critical to get a robust model, specially, those features based on the sensor values on the previous state.

Keywords: MLP · Fault detection · Recovery · Heat exchanger
Heat pump · Geothermal exchanger

1 Introduction

Regardless of whether a process is automatic or manual [1,22,23], the sensors failures detection turns out to be very important [10]. Also, in general terms, with the aim to make the systems more robust and fault tolerant, it is necessary to recover the wrong read with a right value [9,11,26]. The main objective of the present research was to implement a fault detection [7,9] and recovery methodology of sensors installed at a geothermal heat exchanger. This exchanger is part of a heat pump installation placed on a real bioclimatic house. To accomplish this goal a hybrid intelligent approach based on sensor fault detection was developed.

Both for detection and recovery tasks [18], an essential step consists on accomplishing a representative model of the system [17]. The main idea is to

© Springer International Publishing AG, part of Springer Nature 2018
F. J. de Cos Juez et al. (Eds.): HAIS 2018, LNAI 10870, pp. 171–184, 2018.
https://doi.org/10.1007/978-3-319-92639-1_15

predict the sensor measurement [14,19]. If the deviation between the real value and the predicted one is significant, the measurement is replaced by the prediction. Different methods were taken into account for the model implementation.

Multiple Regression Analysis (MRA) techniques are used at most of the accepted regression methods [6,8,15]. Despite the limitations and the irregular performance of these techniques, they are still being used in many different applications [12,27]. Intelligent techniques are employed with the aim of improving the model performance. The initial dataset used to train the model can be divided in different groups. Then, a local model of each group is obtained. Hence, simple or hybrid proposals are used in order to improve the MRA techniques [2,5,20,21].

Usually, the great effort is focused on the different techniques used to achieve a good performance during the modeling stage. The present research tries to make special emphasis on the different strategies when the available variables and their previous values are chosen.

After the present introduction, the case of study explains the geothermal heat exchanger of a bioclimatic house. Then, the fault detection and recovery approach is presented. The approach is implemented over the real case and another section is created to describe the experiments and the results. Finally, the conclusions are presented.

2   Case of Study

The approach is used to make fault detection over a geothermal heat exchanger sensor matrix. This is one of the systems installed at a bioclimatic house. The system is described in the following subsections.

2.1   Sotavento Bioclimatic House

Sotavento bioclimatic house is a bioclimatic house research project of Sotavento Galicia Foundation. The building is located at the *Sotavento Wind Farm*, which is a dissemination center of alternative energy and energy saving. It is located at the Xermade council (Lugo), in the regional area of Galicia (Spain). It is at coordinates 43° 21' North, 7° 52' West, at an elevation of 640 m over sea and at 30 Km from the coast. The thermal installation, based on renewable energy systems, has 3 different sources (geothermal, biomass and solar) that serve the domestic hot water (DHW) and the heating systems. For the electrical installation two are the renewable sources: wind and photovoltaic. Also, the house has one connection to the power grid. The electricity is used to supply the lighting and power systems of the house.

The specific case of the thermal installation could be classified into 3 groups [4]:

- Generation: Solar thermal, biomass boiler and geothermal.
- Accumulation: Preheating, and solar and inertial accumulator.
- Consumption: DHW and Underfloor heating.

2.2   The Geothermal System

This section gives the description of the geothermal operation system and its parts.

Equipment. Figure 1 shows the Heat Pump and the horizontal heat exchanger. The Heat Pump has two different circuits; the first one gives the energy from the ground (the geothermal exchanger), and the other one connect the unit and the inertial accumulator. The equipment includes two sensor to measure the energy taken from the ground to transfer to the inertial accumulator.

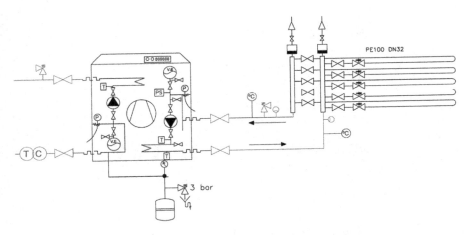

Fig. 1. Heat Pump and horizontal exchanger layout

Geothermal exchanger. The horizontal exchanger has five different circuits. The installation has several temperature sensors located along the heat exchanger, distributed into four different loops. The main aim of these sensors is to measure the ground temperature where the heat exchanger is laid. The geothermal exchanger sensors layout is shown in Fig. 2. In this figure it is possible to appreciate the sensors in the geothermal exchanger, and also the reference temperature of the ground, S401, placed separated from the exchanger. Moreover the sensors S28 and S29 to measure the energy taken from the ground as previously explained.

2.3   The Dataset

The employed dataset is a set of a year of the above explained temperature sensors, acquired with a sample rate of 10 min. During the acquisition phase, the sensors did not have any problem; all the data was considered as valid data for the fault detection system. However, despite of the data are valid, some samples were saved with a mark that indicate a wrong sample (bad sample time, bad range, open wire...).

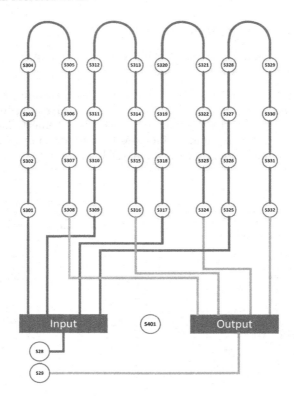

Fig. 2. Geothermal exchanger sensors layout

With the aim to avoid these wrong samples, the dataset was filtered to discard the erroneous data. Consequently, the samples were reduced from 52,705 to 52,699. The dataset consist on the temperatures measured by the sensors located at the geothermal heat exchanger and the other ones at the connection with the heat pump (Figs. 1 and 2). This last two sensors, as they are installed inside the house, the temperature need a filtering to take into account only the data when the heat pump is on. This two sensors measure the temperature in the input and the output of the geothermal exchanger, but, as they are installed inside the house, when the heat pump is off, the sensors measure the temperature inside the house.

3   Fault Detection and Recovery (FDR) Approach

The scheme defined for fault detection and recovery approach is shown in Fig. 3. It is possible to divide the figure into two parts: the model and the fault detection and recovery block. The first one gives the prediction of each sensor based on the measurements made by the rest of the sensors. The second one compares the prediction with the real measurement, and analyze the deviation based on a

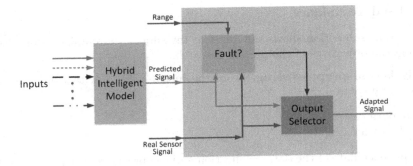

Fig. 3. Fault detection and recovery approach

defined range. If there is a significant deviation the valid signal is the prediction, otherwise the real measurement is set at the output.

3.1 FDR Steps

In this subsection are explained the necessary steps to accomplish the FDR developed approach.

Sensor fault detection. Initially, it is used a simple methodology for accomplishing sensor fault detection technique. The method followed is shown in Fig. 4 under a graphical point of view. In this case, it is allowed a specific configurable range deviation. The continuous line represents the sensor reading and the dotted lines the establish the limit range. If the measured sample is out of the range deviation, then a fault is labeled. The deviation percentage is relativized taking into account the operating temperature range.

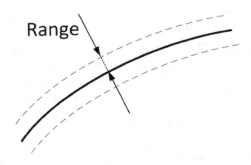

Fig. 4. Range deviation

Recovery. If a fault is detected, then it is necessary to recover the wrong sample with a value prediction. This prediction could be based on the other sensors readings, their previous values, and so on. To accomplish the recovery, a model must be implemented with the aim to predict the most accurate value.

3.2   Used Techniques

The present subsection shows the different techniques contemplated for accomplishing the objectives of the present research.

Analysis and preprocessing. From the initial raw data, two different subclasses were created:

1. Day data cases.
2. Night data cases.

Knowing each date of the data recollection and the precise location of the installation under study, the sunrise and sunset times can be obtained. This is the criteria used to split the raw data in the two subclasses.

With the aim of obtaining a representative model, some variables of the raw dataset have been selected. Also, the previous state of some signals is included as an artificial input, for developing each experiment shown in Sect. 4.

The fact of including this kind of extra information, can be more beneficial than obtaining the model with original data features only. The election of these artificial features is always based on expert knowledge about the system behavior [25].

Based on a data description of the new dataset generated from the raw data, a common pre-processing procedure have been developed, including those experiments with previous values of different sensor like artificial variables.

The criterion for data normalizing is shown in Eq. (1):

$$\frac{X_i - mean(x)}{stdev(x)} \tag{1}$$

The Standard Scaler data input pre-processing has been implemented with Python library *sklearn.preprocessing.StandardScaler* [13]. The main goal of the Normalization step is to avoid the very soon convergence in the first iterations, when the training process of a particular regression method begins [16].

Regression technique. The recovery methodology has included the best way construction of a regression model. The experiments oriented to create a regression model for a sensor in a malfunction state, have been based on the Multilayer Perceptron (MLP). This is a supervised learning algorithm that learns a function $f(\cdot) : R^m \rightarrow R^0$. In this research, the Class MLPRegressor by Python Scikit-Learn implements a Multi-Layer Perceptron (MLP) which is trained using backpropagation with linear activation in the output layer [3].

Due to this, the Normalization data preprocessing in the Eq. 1 is purposed like a good practice when the Multi-Layer Perceptron is the method chosen to get a regression model. Due to its robustness and its simple structure, MLP has been one of the most used ANN. The good performance of this kind of ANN has been proven in similar works such as [8,11,19].

With the aim of generating a sequence of transformations in the input data, Pipeline tool by Python Scikit-Learn is utilized. This technique makes easier to define a set of steps for pre-processing the raw dataset and for making the train and validation process later. The Pipeline tool assembles several steps that can be handled together in the cross-validated process. Cross validation ensures training several MLP with different parameters, choosing the best ones to construct the final model regression.

4 Experiments and Results

This section describes how the solution has been implemented. A total of twelve experiments have been designed and implemented to obtain two regression models per experiment, one global and another one hybrid.

4.1 Accomplished Experiments

Each model has been obtained by two ways:

1. Global model, using all day cases.
2. Hybrid model, composed by two sources, as separation of daily data between day and night.

Due to this, a total of three sub-models per experiment are obtained. Being the experiments implemented:

- Experiment A: prediction of sensor S-315 using S-309 to S-316 signals.
- Experiment B: prediction of sensor S-315 using S-309 to S-316 signals, adding the previous state from S-309 to S-316 as artificial variable.
- Experiment C: prediction of sensor S-315 using S-309 to S-316 signals, adding the previous state of S-315 as artificial variable.
- Experiment D: prediction of sensor S-315 from previous state of S-309 to S-316 and also, the previous state of S-315 as artificial variables.

In relation with the fault detection 20 fictitious faults have been created to check the model. The criteria established for the faults was give values out of the configured range.

4.2 Experiments Setup

For each experiment a split in two slices has been done. So the 30% of pre-processing data are oriented to the final testing of the model and the 70% for training purposes ir order to know the error measures and get the best model per experiment.

Cross validation has been developed applying *Ten fold* with the aim to obtain the best parameters for the MLP predictive model. The *Scoring Measured* chosen has been the *Negative Mean Squared Error* (*neg_MSE*). Best neg_MSE measured of the *Ten fold* procedure is captured with the goal to compare with a new *Ten fold* out-put, one per each combination of parameters. In this way, the best combinations of MLP parameters can be known for each experiments. Then, the MLP model will be configured with the best parameters to make a prediction with the data split, chosen for validation purposes.

Parameters combination is extracted from a parameter grid, each possible value of these parameters are fixed for each iteration of the grid search cross validation. These dynamic parameters of grid are composed by the following values:

- Early Stopping = True, False.
- Nesterovs momentum = True, False.
- Solver = lbfgs, sgd.
- Hidden layer sizes = 3 to 12.

The parameter named *Early Stopping* ensures that the possibility of an early best result can be captured if this fact occurs before the last training epochs.

The *Nesterov Mometum* [24] parameter indicates if the gradient descent with momentum is handled implementing two steeps. First step implements a significant jump in the same direction of the previous accumulated gradient. Then a measure the gradient is made where you ended up before for making the pertinent correction.

Solver parameter for weight optimization is configured for two possible values, the first one, *lbgs*, is an optimizer in the family of *quasi-Newton* methods while the second one *sgd* refers to stochastic gradient descent.

Finally, the number of possible neurons in the hidden layer goes from 3 to 12. So, the cross validation tests and chose the optimal number of number in the only one hidden layer of the MLP.

4.3   Results

The results presented in Tables 1, 2, 3 and 4, show that the preprocessing procedure based on incorporating artificial variables thanks to expert knowledge, improves quality of models. Also the error measures demonstrates that the hybrid model has better performance than the global one. A good prediction and recovery information is achieved when the obtained model is tested using the over the validation data group.

Graphical representation of real output (red color) versus predicted output (blue color) per each experiment are presented in Figs. 5, 6, 7 and 8. The "y" axis represents the temperature in Celsius degrees while the "x" axis represents each data sample of the final test data split.

They have been checked the 20 faults for the created model. In all cases the proposal detect the fictional faults.

The best way for recovering data missing in malfunction state sensor is utilizing like data input previous state of S-309 to S-316 and also, the previous state of S-315 as artificial variables (Experiment D), two models for each experiments have been obtained. The first one oriented to data collect for all day and the second one, a hybrid model composed of two submodels as it can be observed in

Table 1. Experiment A

	Night	Day	All day
LMLS	4.063507e−07	0.001086	7.355953e−05
MAE	0.000549	0.011771	0.005142
MAPE	6.659421e−05	0.001422	0.001565
MSE	8.127033e−07	0.002242	0.000147
NMSE	0.115293	0.082099	0.004510
SMAPE	6.659902e−05	0.001407	0.000622

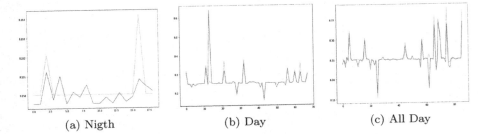

(a) Nigth   (b) Day   (c) All Day

Fig. 5. Experiment A (Color figure online)

Table 2. Experiment B

	Night	Day	All day
LMLS	8.426745−05	0.001560	0.001560
MAE	0.007760	0.012936	0.012936
MAPE	0.000940	0.001565	0.001565
MSE	0.000168	0.003280	0.003280
NMSE	0.504032	0.148473	0.148473
SMAPE	0.000941	0.001589	0.001589

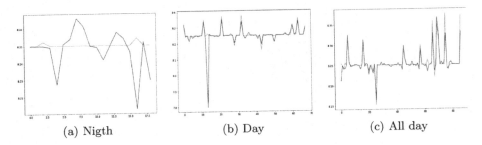

(a) Nigth   (b) Day   (c) All day

Fig. 6. Experiment B (Color figure online)

Table 3. Experiment C

	Night	Day	All day
LMLS	5.843426e−06	0.026457	5.832837e-05
MAE	0.002314	0.055992	0.005156
MAPE	0.000280	0.006770	0.000622
MSE	1.168699e−05	0.145421	0.000116
NMSE	0.233426	0.271015	0.003434
SMAPE	0.000280	0.008083	0.000623

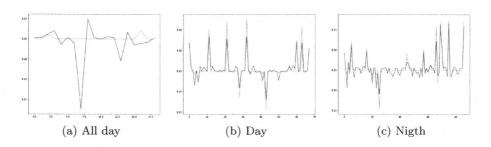

(a) Nigth   (b) Day   (c) All Day

Fig. 7. Experiment C (Color figure online)

Table 4. Experiment D

	Night	Day	All day
LMLS	4.0175248e−05	7.281459e−05	7.159417e−05
MAE	0.004411	0.006977	0.008986
MAPE	0.000534	0.000843	0.001088
MSE	8.037019e−05	0.000145	0.000143
NMSE	0.527251	0.004306	0.012549
SMAPE	0.000535	0.000843	0.001088

(a) All day   (b) Day   (c) Nigth

Fig. 8. Experiment D (Color figure online)

the error measures of the Table 4, the best predictive model is the hybrid one, composed by two sub-models, one per dataset of the day-time and another one for the dataset collected at night. The parameters of MLP regressor trained for getting the hybrid model are:

- One hidden layer.
- 10000 training epochs.
- Activation function of hidden layer the hyperbolic tangent function.
- Nesterov and Early Stopping options activated.
- Solver: lbfgs optimizer in the family of quasi-Newton methods.
- Number of neurons in the hidden layer of MLP: 6 neurons for the day sub-model and 5 for the night submodel.

While the most values for each parameter are the same for the day and night submodels of the hybrid model, different values have been obtained as best values in the grid search cross validation procedure for the parameter called *Number of neurons in the hidden layer of MLP*, 6 neurons for the day submodel and 5 for the night submodel. Therefore, when the MLP is configured with this set of values, the *negative mean square error* measure is optimized. The *negative mean square error* is a standard optimizing measure in the grid search cross validation algorithm.

5   Conclusions and Future Works

A methodology for recovering data missing in malfunction state sensor and the fault sensor detection have been addressed in this research successfully.

Sensor fault detection procedure is relaying on tagging of data as fault, when a measured sample is out of the range derivation. Moreover, the procedure for recovering data missing is based on the implementation of several experiments with the aim to get the best way to define a model when it is trying to get measurements of a sensor with problems. Input data features election is relevant when a robust regression model wants to be created to predict missing data in process where the temperature is involved. More concretely the election of new features and how these are estimated or calculated. In this research new artificial features based on the sensor values on the previous state are added to achieve and compare a global versus hybrid model, for recovering data missing of a sensor.

Results prove that hybrid model implemented with a one hidden layer MLP regressor, composed by day and night submodels including previous state values as artificial features, is the best way for recovering data missing.

Future works will address the improving the sensor fault detection procedure with one class SVM algorithm. From the point of view of recovering data missing, new experiments based on time series oriented to prevent the use of previous state information will be implemented. Some new and complex models will be used also in the next research phase.

Acknowledgments. We would like to thank the 'Instituto Enerxético de Galicia' (INEGA) and 'Parque Eólico Experimental de Sotavento' (Sotavento Foundation) for their technical support on this work.

References

1. Alaiz Moretón, H., Calvo Rolle, J., García, I., Alonso Alvarez, A.: Formalization and practical implementation of a conceptual model for PID controller tuning. Asian J. Control **13**(6), 773–784 (2011)
2. Basden, A.G., Atkinson, D., Bharmal, N.A., Bitenc, U., Brangier, M., Buey, T., Butterley, T., Cano, D., Chemla, F., Clark, P., Cohen, M., Conan, J.M., de Cos, F.J., Dickson, C., Dipper, N.A., Dunlop, C.N., Feautrier, P., Fusco, T., Gach, J.L., Gendron, E., Geng, D., Goodsell, S.J., Gratadour, D., Greenaway, A.H., Guesalaga, A., Guzman, C.D., Henry, D., Holck, D., Hubert, Z., Huet, J.M., Kellerer, A., Kulcsar, C., Laporte, P., Le Roux, B., Looker, N., Longmore, A.J., Marteaud, M., Martin, O., Meimon, S., Morel, C., Morris, T.J., Myers, R.M., Osborn, J., Perret, D., Petit, C., Raynaud, H., Reeves, A.P., Rousset, G., Sanchez Lasheras, F., Sanchez Rodriguez, M., Santos, J.D., Sevin, A., Sivo, G., Stadler, E., Stobie, B., Talbot, G., Todd, S., Vidal, F., Younger, E.J.: Experience with wavefront sensor and deformable mirror interfaces for wide-field adaptive optics systems. Mon. Not. R. Astron. Soc. **459**(2), 1350–1359 (2016). https://doi.org/10.1093/mnras/stw730
3. Buitinck, L., Louppe, G., Blondel, M., Pedregosa, F., Mueller, A., Grisel, O., Niculae, V., Prettenhofer, P., Gramfort, A., Grobler, J., Layton, R., VanderPlas, J., Joly, A., Holt, B., Varoquaux, G.: API design for machine learning software: experiences from the scikit-learn project. In: ECML PKDD Workshop: Languages for Data Mining and Machine Learning, pp. 108–122 (2013)
4. Cabrerizo, J.A.R., Santos, M.: ParaTrough: modelica-based simulation library for solar thermal plants. Rev. Iberoam. Autom. Inform. Ind. RIAI **14**(4), 412–423 (2017). http://www.sciencedirect.com/science/article/pii/S1697791217300481
5. Calvo-Rolle, J.L., Casteleiro-Roca, J.L., Quintián, H., del Carmen Meizoso-Lopez, M.: A hybrid intelligent system for PID controller using in a steel rolling process. Expert Syst. Appl. **40**(13), 5188–5196 (2013). http://www.sciencedirect.com/science/article/pii/S0957417413001632
6. Calvo-Rolle, J.L., Fontenla-Romero, O., Pérez-Sánchez, B., Guijarro-Berdinas, B.: Adaptive inverse control using an online learning algorithm for neural networks. Informatica **25**(3), 401–414 (2014). http://www.mii.lt/informatica/htm/INFO1028.htm
7. Calvo-Rolle, J.L., Quintian-Pardo, H., Corchado, E., del Carmen Meizoso-López, M., García, R.F.: Simplified method based on an intelligent model to obtain the extinction angle of the current for a single-phase half wave controlled rectifier with resistive and inductive load. J. Appl. Logic **13**(1), 37–47 (2015)
8. Casteleiro-Roca, J., Calvo-Rolle, J., Meizoso-Lpez, M., Pin-Pazos, A., Rodrguez-Gmez, B.: Bio-inspired model of ground temperature behavior on the horizontal geothermal exchanger of an installation based on a heat pump. Neurocomputing **150, Part A**, 90–98 (2015). http://www.sciencedirect.com/science/article/pii/S0925231214012417
9. Casteleiro-Roca, J.L., Calvo-Rolle, J.L., Méndez Pérez, J.A., Roqueñí Gutiérrez, N., de Cos Juez, F.J.: Hybrid intelligent system to perform fault detection on bis sensor during surgeries. Sensors **17**(1), 179 (2017)

10. Casteleiro-Roca, J.L., Jove, E., Sánchez-Lasheras, F., Méndez-Pérez, J.A., Calvo-Rolle, J.L., de Cos Juez, F.J.: Power cell SOC modelling for intelligent virtual sensor implementation. J. Sens. **2017**, 10 (2017)
11. Casteleiro-Roca, J.L., Quintián, H., Calvo-Rolle, J.L., Corchado, E., del Carmen Meizoso-López, M., Piñón-Pazos, A.: An intelligent fault detection system for a heat pump installation based on a geothermal heat exchanger. J. Appl. Logic **17**, 36–47 (2016)
12. Crespo-Ramos, M.J., Machn-Gonzlez, I., Lpez-Garca, H., Calvo-Rolle, J.L.: Detection of locally relevant variables using SOMNG algorithm. Eng. Appl. Artif. Intell. **26**(8), 1992–2000 (2013). http://www.sciencedirect.com/science/article/pii/S095219761300078X
13. Scikit-learn Developers: scikit-learn v0.19.1 (2017). http://scikit-learn.org/stable/modules/generated/sklearn.preprocessing.StandardScaler.html
14. Fernández-Serantes, L.A., Vázquez, R.E., Casteleiro-Roca, J.L., Calvo-Rolle, J.L., Corchado, E.: Hybrid intelligent model to predict the SOC of a LFP power cell type. In: Polycarpou, M., de Carvalho, A.C.P.L.F., Pan, J.-S., Woźniak, M., Quintian, H., Corchado, E. (eds.) HAIS 2014. LNCS (LNAI), vol. 8480, pp. 561–572. Springer, Cham (2014). https://doi.org/10.1007/978-3-319-07617-1_49
15. Fernndez, J.A., Muiz, C.D., Nieto, P.G., de Cos Juez, F., Lasheras, F.S., Roque, M.: Forecasting the cyanotoxins presence in fresh waters: a new model based on genetic algorithms combined with the MARS technique. Ecol. Eng. **53**, 68–78 (2013). http://www.sciencedirect.com/science/article/pii/S0925857412003692
16. Géron, A.: Hands-On Machine Learning with Scikit-Learn and TensorFlow: Concepts, Tools, and Techniques for Building Intelligent Systems. O'Reilly Media (2017). https://books.google.es/books?id=I6qkDAEACAAJ
17. Gonzalez-Cava, J.M., et al.: A machine learning based system for analgesic drug delivery. In: Pérez García, H., Alfonso-Cendón, J., Sánchez González, L., Quintián, H., Corchado, E. (eds.) SOCO/CISIS/ICEUTE -2017. AISC, vol. 649, pp. 461–470. Springer, Cham (2018). https://doi.org/10.1007/978-3-319-67180-2_45
18. Jove, E., Blanco-Rodríguez, P., Casteleiro-Roca, J.L., Moreno-Arboleda, J., López-Vázquez, J.A., de Cos Juez, F.J., Calvo-Rolle, J.L.: Attempts prediction by missing data imputation in engineering degree. In: Pérez García, H., Alfonso-Cendón, J., Sánchez González, L., Quintián, H., Corchado, E. (eds.) SOCO/CISIS/ICEUTE -2017. AISC, vol. 649, pp. 167–176. Springer, Cham (2018). https://doi.org/10.1007/978-3-319-67180-2_16
19. Jove, E., Gonzalez-Cava, J.M., Casteleiro-Roca, J.L., Pérez, J.A.M., Calvo-Rolle, J.L., de Cos Juez, F.J.: An intelligent model to predict ANI in patients undergoing general anesthesia. In: Pérez García, H., Alfonso-Cendón, J., Sánchez González, L., Quintián, H., Corchado, E. (eds.) SOCO/CISIS/ICEUTE -2017. AISC, vol. 649, pp. 492–501. Springer, Cham (2018). https://doi.org/10.1007/978-3-319-67180-2_48
20. Juez, F.J., Lasheras, F.S., Roque, N., Osborn, J.: An ANN-based smart tomographic reconstructor in a dynamic environment. Sensors **12**(7), 8895–8911 (2012). http://www.mdpi.com/1424-8220/12/7/8895
21. Quintián, H., Calvo-Rolle, J.L., Corchado, E.: A hybrid regression system based on local models for solar energy prediction. Informatica **25**(2), 265–282 (2014). http://www.mii.lt/informatica/htm/INFO1024.htm
22. Quintian Pardo, H., Calvo Rolle, J.L., Fontenla Romero, O.: Application of a low cost commercial robot in tasks of tracking of objects. Dyna **79**(175), 24–33 (2012)
23. Rolle, J., Gonzalez, I., Garcia, H.: Neuro-robust controller for non-linear systems. Dyna **86**(3), 308–317 (2011)

24. Ruder, S.: An overview of gradient descent optimization algorithms, pp. 1–14 (2016). http://arxiv.org/abs/1609.04747

25. Tuv, E.: Feature selection with ensembles, artificial variables, and redundancy elimination. J. Mach. Learn. Res. **10**, 1341–1366 (2009). http://www.jmlr.org/papers/volume10/tuv09a/tuv09a.pdf

26. Vilar-Martinez, X.M., Montero-Sousa, J.A., Calvo-Rolle, J.L., Casteleiro-Roca, J.L.: Expert system development to assist on the verification of "TACAN" system performance. Dyna **89**(1), 112–121 (2014)

27. Zotes, F.A., Peas, M.S.: Heuristic optimization of interplanetary trajectories in aerospace missions. Rev. Iberoam. Autom. Inform. Ind. RIAI **14**(1), 1–15 (2017). http://www.sciencedirect.com/science/article/pii/S1697791216300486

Distinctive Features of Asymmetric Neural Networks with Gabor Filters

Naohiro Ishii[1]([⊠]), Toshinori Deguchi[2], Masashi Kawaguchi[3],
and Hiroshi Sasaki[4]

[1] Aichi Institute of Technology, Toyota, Japan
ishii@aitech.ac.jp
[2] Gifu National College of Technology, Gifu, Japan
deguchi@gifu-nct.ac.jp
[3] Suzuka National College of Technology, Mie, Japan
masashi@elec.suzukact.ac.jp
[4] Fukui University of Technology, Fukui, Japan
hsasaki@fukui-ut.ac.jp

Abstract. To make clear the mechanism of the visual motion detection is important in the visual system, which is useful to robotic systems. The prominent features are the nonlinear characteristics as the squaring and rectification functions, which are observed in the retinal and visual cortex networks. Conventional models for motion processing, are to use symmetric quadrature functions with Gabor filters. This paper proposes a new motion processing model of the asymmetric networks. To analyze the behavior of the asymmetric nonlinear network, white noise analysis and Wiener kernels are applied. It is shown that the biological asymmetric network with nonlinearities is effective for generating the directional movement from the network computations. Further, responses to complex stimulus and the frequency characteristics are computed in the asymmetric networks, which are not derived for the conventional energy model.

Keywords: Asymmetric neural network · Gabor filter · Wiener analysis
Linear and nonlinear pathways

1 Introduction

In the biological neural networks, the sensory information is processed effectively. Reichard [1] evaluated the sensory information by the auto-correlations in the neural networks. The nonlinear characteristics as the squaring function and rectification function, which are observed in the retina and visual cortex networks [2–6, 8]. Conventional models for cortical motion sensors are to use symmetric, quadrature functions, which are called energy model. Recent study by Hess and Bair [3] discusses quadrature is not necessary nor sufficient under stimulus condition. Then, minimal models for sensory processing are expected. This paper proposes a new motion processing model based on the biological asymmetric networks. The nonlinear function exists in the asymmetrical neural networks. To investigate cells function in the

F. J. de Cos Juez et al. (Eds.): HAIS 2018, LNAI 10870, pp. 185–196, 2018.
https://doi.org/10.1007/978-3-319-92639-1_16

biological networks, white noise stimulus [9–13] are often used in physiological experiments. In this paper, to analyze the behavior of the asymmetric network with nonlinearity, white noise analysis and Wiener kernels are applied. It is shown that the asymmetric network with nonlinearities is effective for the movement detection from the network computations. We analyze the asymmetric network based on the retinal circuit of the catfish [13, 14, 16, 17], from the point of the optimization of the network model. It is shown that the asymmetric network with nonlinearity has the ability for the directional movement of the stimulus, which can be written in directional equations by Wiener kernels. Then, it is shown that the directional equations obtained are selective for the preferred and null direction stimulus in the asymmetric network. Further, the directive equation for the complex stimulus is developed and the frequency characteristics are computed in the asymmetric network, which are applicable to V1 and MT cortex neural networks.

2 Biological Neural Networks

The asymmetric neural network is extracted from the catfish retinal network [14]. The asymmetric structure network with a quadratic nonlinearity is shown in Fig. 1, which composes of the pathway from the bipolar cell B to the amacrine cell N and that from the bipolar cell B, via the amacrine cell N with squaring function to the N cell [14].

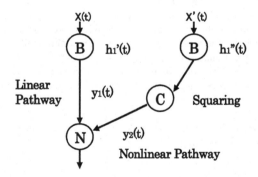

Fig. 1. Asymmetric network with linear and squaring nonlinear pathways

Figure 1 shows a network which plays an important role in the movement perception as the fundamental network. It is shown that N cell response is realized by a linear filter, which is composed of a differentiation filter followed by a low-pass filter. Thus, the asymmetric network in Fig. 1 is composed of a linear pathway and a nonlinear pathway. Here, the stimulus with Gaussian distribution is assumed to move from the left side to the right side in front of the network in Fig. 1, as shown in Fig. 2. $x''(t)$ is mixed with $x(t)$. Then, we indicate the right stimulus by $x'(t)$. By introducing a mixed ratio, α, the input function of the right stimulus, is described in the following

B1 cell   **B2 cell**

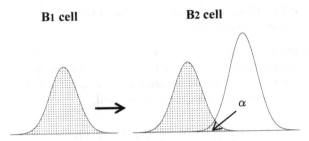

Fig. 2. Stimulus movement from the left to the right side

equation, where $0 \leq \alpha \leq 1$ and $\beta = 1 - \alpha$ hold. Then, Fig. 2 shows that the moving stimulus is described in the following equation,

$$x'(t) = \alpha x(t) + \beta x''(t) \tag{1}$$

Let the power spectrums of $x(t)$ and $x''(t)$, be p and p', respectively an equation $p = kp''$ holds for the coefficient k. Figure 2 shows that the slashed light is moving from the receptive field of B_1 cell to the field of the B_2 cell with mixed ratio, α. The stimulus on both cells in Fig. 2 is shown in the schematic diagram in Fig. 3.

B1 cell   B2 cell α

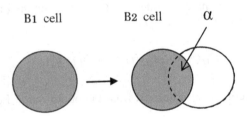

Fig. 3. Schematic diagram of the preferred stimulus direction

The output $y(t)$ of the cell N in Fig. 1 is shown in the following equation.

$$y(t) = \int h_1'''(\tau)(y1(t - \tau) + y2(t - \tau))d\tau + \varepsilon \tag{2}$$

where $y_1(t)$ shows the linear information on the linear pathway $y_2(t)$ shows the non-linear information on the nonlinear pathway and ε shows error value.

$$y_1(t) = \int_0^\infty h_1'(\tau)x(t - \tau)d\tau \tag{3}$$

$$y_2(t) = \int_0^\infty \int_0^\infty h1''(\tau_1)h1''(\tau_2)x'(t - \tau_1)x'(t - \tau_2)d\tau_1 d\tau_2 \tag{4}$$

2.1   Directional Equations from Optimized Conditions in the Asymmetric Networks

Under the assumption that the impulse response functions, $h_1'(t)$ of the cell B_1, $h_1''(t)$ of the cell B_2 and moving stimulus ratio α in the right to be unknown, the optimization of the network is carried out. By the minimization of the mean squared value ξ of ε in Eq. (2), the following necessary equations are derived,

$$\frac{\partial \xi}{\partial h_1'(t)} = 0, \frac{\partial \xi}{\partial h_2''(t)} = 0, \frac{\partial \xi}{\partial \alpha} = 0 \tag{5}$$

Then, the following three equations are derived for the optimization of Eq. (5).

$$
\begin{aligned}
E[y(t)x'(t - \lambda)] &= \alpha p h_1'(\lambda) \\
E[(y(t) - C_0)x'(t - \lambda_1)x'(t - \lambda_2)] &= 2\{(\alpha^2 + k\beta^2)p^2 h_1''(\lambda_1)h_1''(\lambda_2)\} \\
E[(y(t) - C_0)x(t - \lambda_1)x(t - \lambda_2)] &= 2\alpha^2 p^2 h_1''(\lambda_1)h_1''(\lambda_2) \\
E[(y(t) - C_0)x''(t - \lambda_1)x''(t - \lambda_2)] &= 2\beta^2 (kp)^2 h_1''(\lambda_1)h_1''(\lambda_2)
\end{aligned}
\tag{6}
$$

where C_0 is the mean value of, $y(t)$. Here, the Eqs. (6) can be rewritten by applying Wiener kernels, which are related with input and output correlations method developed by Lee and Schetzen [15]. First, we can compute the 0-th order Wiener kernel $C0$, the 1-st order one and $C_{11}(\lambda)$, the 2-nd order one $C_{21}(\lambda_1, \lambda_2)$ on the linear pathway by the cross-correlations between $x(t)$ and. $y(t)$.

$$C_{11}(\lambda) = \frac{1}{p}E[y(t)x(t - \lambda)] = h_1'(\lambda) \tag{7}$$

The 2-nd order kernel is also derived from the optimization Eq. (8) as follows,

$$
\begin{aligned}
C_{21}(\lambda_1, \lambda_2) &= \frac{1}{2p^2}E[(y(t) - C_0)x(t - \lambda_1)x(t - \lambda_2)] \\
&= \alpha^2 h_1''(\lambda_1)h_1''(\lambda_2)
\end{aligned}
\tag{8}
$$

Second, we can compute the 0-th order kernel, C_0 the 1-st order kernel $C_{12}(\lambda)$ and the 2-nd order kernel.

$$
\begin{aligned}
C_{12}(\lambda_1, \lambda_2) &= \frac{1}{p(\alpha^2 + k\beta^2)}E[y(t)x'(t - \lambda)] \\
&= \frac{\alpha}{\alpha^2 + k(1 - \alpha)^2}h_1'(\lambda)
\end{aligned}
\tag{9}
$$

and

$$C_{22}(\lambda_1, \lambda_2) = h_1''(\lambda_1)h_1''(\lambda_2) \tag{10}$$

The motion problem is how to detect the direction of the stimulus in the increase of the ratio α in Fig. 3. The second order kernels C_{21} and C_{22} are abbreviated in the representation of Eqs. (8) and (10).

$$(C_{21}/C_{22}) = \alpha^2 \tag{11}$$

holds. Then, from the Eq. (13) the ratio α is shown as follows

$$\alpha = \sqrt{\frac{C_{21}}{C_{22}}} \tag{12}$$

The Eq. (12) is called here α - equation, which implies the directional stimulus on the network From the first order kernels C_{11} and C_{12}, and the second order kernels in the above derivations, the directional equation from the left to the right, holds as shown in the following,

$$\frac{C_{12}}{C_{11}} = \frac{\sqrt{\frac{C_{21}}{C_{22}}}}{\frac{C_{21}}{C_{22}} + k(1 - \sqrt{\frac{C_{21}}{C_{22}}})^2} \tag{13}$$

3 Conventional Quadrature Energy Model

Motion detection of the conventional quadrature models under the same conditions in this paper is analyzed. The quadrature model in Fig. 4 is well known as the energy model for motion detection [2, 3], which is a symmetric network model. Only the second order kernels are computed in Fig. 4. On the left pathway in Fig. 4, the second order kernel $C_{21}(\lambda_1, \lambda_2)$ is computed as follows,

$$
\begin{aligned}
C_{21}(\lambda_1, \lambda_2) &= \frac{1}{2p^2} \iint h_1(\tau)h_1(\tau')E[x(t-\lambda)x(t-\lambda')x(t-\lambda_1)x(t-\lambda_2)]d\tau d\tau' \\
&+ \frac{1}{2p^2} \iint h_1'(\tau)h_1'(\tau')E[x'(t-\lambda)x'(t-\lambda')x(t-\lambda_1)x(t-\lambda_2)]d\tau d\tau' \\
&= h_1(\lambda_1)h_1(\lambda_2) + \alpha^2 h_1'(\lambda_1)h_1'(\lambda_2)
\end{aligned}
\tag{14}
$$

On the right pathway in Fig. 5, the 2nd order kernel $C_{22}(\lambda_1, \lambda_2)$ is computed similarly,

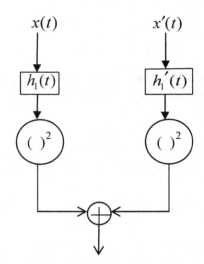

Fig. 4. Quadrature energy model with Gabor filters

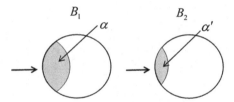

Fig. 5. Schematic diagram of the two stimulus for both cells

$$C_{22}(\lambda_1, \lambda_2) = \frac{\alpha^2}{(\alpha^2 + k\beta^2)^2} h_1(\lambda_1)h_1(\lambda_2) + h_1'(\lambda_1)h_1'(\lambda_2) \tag{15}$$

In the conventional energy model of motion [2, 3], the Gabor functions are given as

$$h_1(t) = \frac{1}{\sqrt{2\pi}\sigma} \exp(-\frac{t^2}{2\sigma^2}) \sin(2\pi\omega t) \quad h_1'(t) = \frac{1}{\sqrt{2\pi}\sigma} \exp(-\frac{t^2}{2\sigma^2}) cos(2\pi\omega t) \tag{16}$$

When Gabor functions in the quadrature model are given as the Eq. (16), the motion parameter α and the motion equation are computed as follows,

$$\alpha = \sqrt{\frac{C_{21}(\lambda_1, \lambda_2) - h_1(\lambda_1)h_1(\lambda_2)}{h_1'(\lambda_1)h_1'(\lambda_2)}} \tag{17}$$

In the Eq. (17), if impulse functions $h_1(t)$ and $h_1'(t)$ are given in the quadrature model, α is computed. Since these impulse functions are given as Gabor functions, α is computed. Gabor functions in the Eq. (16) are independent, but, they are not orthogonal.

$$C_{22}(\lambda_1, \lambda_2) = \frac{\alpha^2}{(\alpha^2 + k\beta^2)^2} h_1(\lambda_1) h_1(\lambda_2) + h_1'(\lambda_1) h_1'(\lambda_2) \qquad (18)$$

The Eq. (20) is satisfied if the Gabor functions are given. Note that the conventional quadrature model generates the motion Eq. (18) under the condition of the given Gabor functions (16), while the asymmetric network in Fig. 1 generate the motion Eq. (13) without the condition of the Gabor functions. Thus, the asymmetric network has a general ability of the motion compared to the conventional quadrature model.

4 Relations for Complex Stimulus Changes

We assume the stimulus in Fig. 3 is changed to that in Fig. 5, in which different shadowed stimulus on the B_1 and B_2 cells are moved to the arrowed direction. The stimulus on the B_1 cell is shown in the Eq. (19), while that on the B_2 cell is shown in the Eq. (20).

$$x'(t) = \alpha x(t) + \beta x''(t) \qquad (19)$$

$$x'''(t) = \alpha' x(t) + \beta' x''(t) \qquad (20)$$

This stimulus is shown in Fig. 5, in which the stimulus $x'(t)$ is inputted on the B_1 cell of the linear pathway in Fig. 1 and also $x'''(t)$ is inputted on the B_2 cell of the nonlinear pathway.

The final output of the asymmetric network by the two stimulus, become

$$y(t) = y_1(t) + y_2(t)$$
$$= \int h_1'(\tau) x'(t - \tau) d\tau + \iint h_1''(\tau_1) h_1''(\tau_2) x'''(t - \tau_1) x'''(t - \tau_2) d\tau_1 d\tau_2 \qquad (21)$$

where $y_1(t)$ is the output of the linear pathway, while $y_2(t)$ is that of the nonlinear pathway. The mean square of the error for the optimization of the network is given by

$$\xi = \int \{y(t) - y_1(t) - y_2(t)\}^2 dt \qquad (22)$$

The conditions of the minimization of the Eq. (22) is given in the following equations.

$$\frac{\partial \xi}{\partial h_1'(t)} = 0, \quad \frac{\partial \xi}{\partial h_1''(t)} = 0, \quad \frac{\partial \xi}{\partial \alpha} = 0, \quad \frac{\partial \xi}{\partial \alpha'} = 0 \qquad (23)$$

From the first equation in (23), the Eq. (24) is derived.

$$E[y(t)x'(t - \tau)] = h'_1(\tau)(\alpha^2 + k\beta^2)p \tag{24}$$

From the second equation in (23), the Eq. (25) is derived.

$$E[(y(t) - C_0)x'''(t - \tau_1)x'''(t - \tau_2)] = 2h''_1(\tau_1)h''_1(\tau_2)(\alpha'^2 + k\beta'^2)p^2 \tag{25}$$

From the third equation in (25), the Eq. (28) is derived.

$$E[y(t)x(t - \tau)] = h'_1(\tau)\alpha p$$

and

$$E[y(t)x''(t - \tau)] = h'_1(\tau)(\alpha\alpha' + k\beta\beta')p \tag{26}$$

The optimization Eqs. (23), are derived to the following Eq. (27) using Wiener kernels.

$$\frac{C_{12}}{C_{11}} = \sqrt{\frac{C_{21}}{C_{22}} \frac{\{\alpha^2 + k(1 - \alpha)^2\}}{\{\alpha'^2 + k(1 - \alpha')^2\}}} \tag{27}$$

Similarly, the Eq. (28) is derived.

$$(E[y(t)x(t - \tau))^2/E[(y(t) - C_0)x(t - \tau_1)x(t - \tau_2)] = (G_s(\tau))^2\alpha^2/2G_c(\tau_1)G_c(\tau_2)\alpha'^2 \tag{28}$$

From the Eqs. (27) and (28), the parameters α and α' are derived in the case of the Gabor filters $G_s(t)$ and $G_c(t)$ to be given. In the case of the conventional energy model with Gabor filters, the parameters α and α' are not derived.

5 Extraction of Frequency Characteristics in the Asymmetric Networks

Prof. Heeger derived the output of the conventional energy model with Gabor filters [7], in the frequency domain which shows only the Gabor energy without the phase-information as shown in the following. To the given sine-wave stimulus with the angular frequency $\bar{\omega}$ and the phase ϕ, the squared-output of the sine-phase Gabor filter convolved with the sine-wave input is derived [7] as follows,

Input stimulus: $\sin(2\pi\bar{\omega}x + \phi)$

Output of the sine-phase Gabor filter:

$$\int_{-\infty}^{\infty} |G_s(\omega_0, \sigma) * \sin(2\pi\bar{\omega}x + \phi)|^2 dx = \int_{-\infty}^{\infty} |F\{G_s(\omega_0, \sigma)\}F\{\sin(2\pi\bar{\omega}x + \phi)\}|^2 d\omega$$

$$= (1/8)[exp[-2\pi^2\sigma^2(\bar{\omega} - \omega_0)^2 - exp[-2\pi^2\sigma^2(\bar{\omega} + \omega_0)^2]^2$$

$$\tag{29}$$

while the out put of the cosine-Gabor filter becomes

$$\int_{-\infty}^{\infty} |G_c(\omega_0, \sigma) * \sin(2\pi\bar{\omega}x + \phi)|^2 dx$$
$$= (1/8)[exp[-2\pi^2\sigma^2(\bar{\omega} - \omega_0)^2 + exp[-2\pi^2\sigma^2(\bar{\omega} + \omega_0)^2]^2 \tag{30}$$

Combining Eqs. (37) and (38) becomes the output of the energy model in the frequency domain which gives the phase-independent energy as follows [7],

$$(1/4)\{exp[-4\pi^2\sigma^2(\bar{\omega} - \omega_0)^2] + exp[-4\pi^2\sigma^2(\bar{\omega} + \omega_0)^2]\} \tag{31}$$

In the asymmetric network with Gabor filters, similar derivation as the Eq. (29) is carried out. Then, the output of the sine-phase Gabor filter becomes

$$\sin \phi \cdot \{exp[-2\pi^2\sigma^2(\bar{\omega} - \omega_0)^2 - exp[-2\pi^2\sigma^2(\bar{\omega} + \omega_0)^2\} \tag{32}$$

,while the output of the cosine-Gabor filter is the same as the Eq. (30). Thus, the output $R(\omega)$ of Eqs. (29) and (32) in the asymmetric networks becomes

$$R(\omega) = (\sin \phi + (1/8)) \cdot \{exp[-2\pi^2\sigma^2(\bar{\omega} - \omega_0)^2 - exp[-2\pi^2\sigma^2(\bar{\omega} + \omega_0)^2\} \tag{33}$$

The Eq. (33) shows the two peak values are shown at the angular frequencies, $(\bar{\omega} - \omega_0)$ and $(\bar{\omega} + \omega_0)$. From the given frequency ω_0 of the Gabor filters, the peak value at $(\bar{\omega} - \omega_0)$, the angular frequency $\bar{\omega}$ of the input sinusoidal stimulus is obtained from $R(\omega)$. From the first peak value of $R(\omega)$, the phase information of the input sinusoidal stimulus, $\sin\phi$ is obtained from the Eq. (32). Compared with the conventional energy model, the Eq. (31) only derives the frequency information, $\bar{\omega}$, not its phase one, ϕ in the stimulus.

6 Application of Asymmetric Networks to Biological Neural Networks

Here, we present an example of layered neural network in Fig. 6(a), which is developed from the neural network in the brain cortex [6]. Figure 6 is a connected network model of V1 followed by MT, where V1 is the front part of the total network, while MT is the rear part of it. Figure 6(a) is transformed to the approximated one as follows.

$$f(x) = \frac{1}{1 + e^{-\eta(x-\theta)}} \tag{34}$$

By Taylor expansion of the Eq. (34) at $x = \theta$, the Eq. (31) is derived as

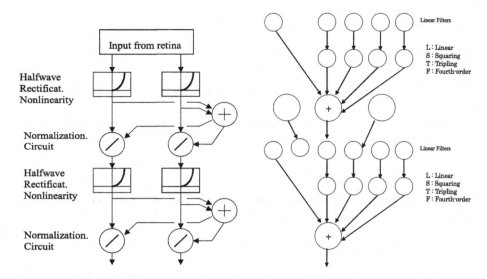

Fig. 6. (a) Neural model of V1 and MT [6] (b) Approximated model shown in (a)

$$f(x)_{x=0} = f(\theta) + f'(\theta)(x - \theta) + \frac{1}{2!}f''(\theta)(x = \theta)^2 + \dots$$
$$= \frac{1}{2} + \frac{\eta}{4}(x - \theta) + \frac{1}{2!}(-\frac{\eta^2}{4} + \frac{\eta^2 e^{-\eta\theta}}{2})(x - \theta)^2 + \dots \tag{35}$$

In the Eq. (34), the sigmoid function is approximated as the half-squaring non-linearity. The asymmetric network consists of the 1-st order nonlinearity(odd order nonlinearity) on the left pathway and the 4-th order(even order nonlinearity) on the right pathway. On the linear and nonlinear pathways(4-th order nonlinearity), the following Eqs. (36) and (37) are derived.

$$C_{21}(\lambda_1, \lambda_2) = \alpha^2 C_{22}(\lambda_1, \lambda_2) \tag{36}$$

$$C_{12}(\lambda) = \frac{\alpha}{(\alpha^2 + k\beta^2)} C_{11}(\lambda) \tag{37}$$

From the Eq. (36), the α Eq. (38) is derived.

$$\alpha = \sqrt{\frac{C_{21}}{C_{22}}} \tag{38}$$

From the Eqs. (37) and (38), the directional Eq. (39) is derived.

$$\therefore \quad \frac{C_{12}}{C_{11}} = \frac{\sqrt{\frac{C_{21}}{C_{22}}}}{\frac{C_{21}}{C_{22}} + k(1 - \sqrt{\frac{C_{21}}{C_{22}}})^2} \tag{39}$$

The α - Eq. (38) and the directional Eq. (39), are same to those of (12) and (13), respectively, in the asymmetric network with 1-st and 2-nd orders nonlinearities. The 4-th order kernels, become

$$C_{41}(\lambda_1, \lambda_2, \lambda_3, \lambda_4) = \alpha^4 h_1''(\lambda_1) h_1''(\lambda_2) h_1''(\lambda_3) h_1''(\lambda_4)$$
$$C_{42}(\lambda_1, \lambda_2, \lambda_3, \lambda_4) = h_1''(\lambda_1) h_1''(\lambda_2) h_1''(\lambda_3) h_1''(\lambda_4) \tag{40}$$

From the Eq. (40), the α - equation is derived as follows, which is equivalent to the Eq. (12).

$$\alpha = \left(\frac{C_{41}}{C_{42}} \right)^{\frac{1}{4}} = \left(\sqrt{\frac{C_{21}}{C_{22}}} \right) \tag{41}$$

Thus, the asymmetric network with 1-st and 4-th orders nonlinearities, are equivalent to that with 1-st and 2-nd orders nonlinearities. Similarly, it is shown the asymmetric networks with the odd order nonlinearity on the one pathway and the even order nonlinearity on the other pathway, has both α- equation and the directional movement equation.

7 Conclusion

The neural networks are analyzed to make clear functions of the biological asymmetric neural networks with nonlinearity. This kind of networks exits in the biological network as retina and brain cortex of V1 and MT areas. In this paper, the behavior of the asymmetrical network with nonlinearity, is analyzed to detect the directional stimulus from the point of the neural computation. For the motion detection, the asymmetrical network proposed here is compared with the conventional quadrature energy model with Gabor filters. It was shown that the quadrature model works with Gabor filters, while the asymmetrical network does not need their conditions that the impulse functions are Gabor functions. The complex stimulus responses and the frequency characteristics are computed in the asymmetrical network.

References

1. Reichard, W.: Autocorrelation, A principle for the evaluation of sensory information by the central nervous system. Rosenblith Edition. Wiley, New York (1961)
2. Adelson, E.H., Bergen, J.R.: Spatiotemporal energy models for the perception of motion. J. Opt. Soc. Am. A **2**(2), 284–298 (1985)

3. Heess, N., Bair, W.: Direction Opponency, Not Quadrature, Is Key to the 1/4 Cycle Preference for Apparent Motion in the Motion Energy Model. J. Neurosci. **30**(34), 11300–11304 (2010)
4. Chubb, C., Sperling, G.: Drift-balanced random stimuli, a general basis for studying non-Fourier motion. J. Opt. Soc. Am. A **5**(11), 1986–2006 (1988)
5. Taub, E., Victor, J.D., Conte, M.: Nonlinear preprocessing in short-range motion. Vision. Res. **37**, 1459–1477 (1997)
6. Simonceli, E.P., Heeger, D.J.: A Model of Neuronal Responses in Visual Area MT. Vision. Res. **38**, 743–761 (1996)
7. Heeger, D.J.: Models of Motion Perception, University of Pennsylvania, Department of Computer and Information Science, Technical Report No.MS-CIS-87-91, Sept1987
8. Heeger, D.J.: Normalization of cell responses in cat striate cortex. Vis. Neurosci. **9**, 181–197 (1992)
9. Marmarelis, P.Z., Marmarelis, V.Z.: Analysis of Physiological Systems – The White Noise Approach. Plenum Press, New York (1978)
10. Marmarelis, V.Z.: Nonlinear Dynamic Modeling of Physiological Systems. Wiley, New Jersey (2004)
11. Marmarelis, V.Z.: Modeling Methodology for Nonlinear Physiological Systems. Ann. Biomed. Eng. **25**, 239–251 (1997)
12. Wiener, N.: Nonlinear Problems in Random Theory. The MIT press, Cambridge (1966)
13. Sakuranaga, M., Naka, K.-I.: Signal Transmission in the Catfish Retina. III. Transmission to Type-C Cell. J. Neurophysiol. **53**(2), 411–428 (1985)
14. Naka, K.-I., Sakai, H.M., Ishii, N.: Generation of transformation of second order nonlinearity in catfish retina. Ann. Biomed. Eng. **16**, 53–64 (1988)
15. Lee, Y.W., Schetzen, M.: Measurements of the Wiener kernels of a nonlinear by cross-correlation. Int. J. of Control **2**, 237–254 (1965)
16. Ishii, N., Deguchi, T., Kawaguchi, M.: Neural computations by asymmetric networks with nonlinearities. In: Beliczynski, B., Dzielinski, A., Iwanowski, M., Ribeiro, B. (eds.) ICANNGA 2007. LNCS, vol. 4432, pp. 37–45. Springer, Heidelberg (2007). https://doi.org/10.1007/978-3-540-71629-7_5
17. Ishii, N., Deguchi, T., Kawaguchi, M., Sasaki, H.: Application of asymmetric networks to movement detection and generating independent subspaces. In: Boracchi, G., Iliadis, L., Jayne, C., Likas, A. (eds.) EANN 2017. CCIS, vol. 744, pp. 267–278. Springer, Cham (2017). https://doi.org/10.1007/978-3-319-65172-9_23

Tuning CNN Input Layout for IDS
with Genetic Algorithms

Roberto Blanco[2], Juan J. Cilla[2], Pedro Malagón[1,2(✉)], Ignacio Penas[2],
and José M. Moya[1,2]

[1] LSI-Universidad Politecnica de Madrid, Madrid, Spain
[2] CCS-Center for Computational Simulation, Madrid, Spain
{r.bandres,juanje,malagon,nacho,josem}@die.upm.es

Abstract. Intrusion Detection Systems (IDS) are implemented by service providers and network operators to monitor and detect attacks. Many machine learning algorithms, stand-alone or combined, have been proposed, including different types of Artificial Neural Networks (ANN). This work evaluates a Convolutional Neural Network (CNN), created for image classification, as an IDS that can be deployed in a router, which has not been evaluated previously. The layout of the features in the input matrix of the CNN is relevant. A Genetic Algorithm (GA) is used to find a high-quality solution by rearranging the layout of the input features, reducing the features if required. The GA improves the capacity of intrusion detection from 0.71 to 0.77 for normalized input featuress, similar to existing algorithms. For scenarios where data normalization is not possible, many input layouts are useless. The GA finds a solution with an intrusion detection capacity of 0.73.

Keywords: CNN · Genetic Algorithm · UNSW · Cybersecurity · IDS

1 Introduction

Security concerns of service providers include attacks to their infrastructures, which can affect their service availability, client or industrial privacy, integrity or reliability of their solutions. Moreover, the appearance of the Internet of Things has lead to an exponential growth of the number of devices connected to the Internet. The challenges related to protect our services, networks and devices are increasing in complexity drastically. Every year new services appear and new attacks or ways of implementing existing attacks appear as well. In the last decades signature-based and anomaly-based machine learning algorithms have proven to be more effective in an early automatic detection of attacks, known or new ones, than rule-based protection mechanisms such as firewalls. These algorithms are used to implement the Intrusion Detection System (IDS), which monitor and detect attacks, anomalies and misuse sniffing packets and collecting data from all over the network. The IDS classifies the data into categories using

© Springer International Publishing AG, part of Springer Nature 2018
F. J. de Cos Juez et al. (Eds.): HAIS 2018, LNAI 10870, pp. 197–209, 2018.
https://doi.org/10.1007/978-3-319-92639-1_17

different methods. It distinguishes normal from abnormal data. The abnormal data of the system can be classified among different threats.

In order to train and test the performance, efficiency and accuracy of the classifiers, there are public datasets available with real and simulated network labeled samples including multiple attacks with many features. Once a model has been generated to detect attacks, an online IDS needs to obtain the required features from real traffic packets passing through a firewall to feed the detection algorithm with data.

Many machine learning algorithms have been proposed for implementing the classifier of IDS, including Artificial Neural Networks (ANN). An ANN is an interconnected assembly of neurons (processing elements) that detect certain patterns from complex data as the human brain would process information.

Currently, deep learning algorithms have appeared that achieve high level abstractions in data by using a complex architecture or composition of non-linear transformations. Using deep learning we can acquire a high detection rate.

In this paper we consider using a deep learning algorithm in an IDS: the convolutional neural network (CNN). CNNs were created for image classification or object detection. We adapt the features available in a network monitoring system to be input of a CNN (an image) to exploit correlation between features. We consider a binary IDS (normal traffic or attack). The optimal layout of the input features in the image is a combinatorial problem with more than 20 input features. We consider using a heuristic evolutionary algorithm to select a local optimum solution to the feature layout problem.

In the remainder of this paper: the related work on IDS is introduced in Sect. 2. Our proposal and the algorithms used in our study are explained in Sect. 3. The experimental setup is depicted in Sect. 4. We discuss on the results in Sect. 5. Some conclusions are drawn in Sect. 6.

2   Related Work

The most widely used dataset for IDS training is the KDD99 [1]. It was created by processing nine weeks of raw tcpdump of the 1998 DARPA Intrusion Detection System (IDS) evaluation dataset. It has five million connection records, containing 24 attack types that fall into 4 major categories. Each record includes 41 features and is labeled either as normal or as an attack, with exactly one specific attack type. The NSL-KDD dataset was published [2] as a subset of the KDD99 dataset which does not include redundant records. Evaluating network IDS using KDD99 and NSL-KDD presents two major issues: their lack of modern records, including both attack and normal traffic scenarios, and a different distribution of training and testing sets.

To address these issues, the UNSW-NB15 dataset was published [3]. This data set has nine types of the modern attacks fashions and new patterns of normal traffic. It contains 49 attributes or features that comprise the flow based between hosts and the network packets inspection to discriminate between the observations, either normal or abnormal. It has been recently used to evaluate different proposals with simple or combined machine learning algorithms including ANN [4,5].

For the last decades different ANN implementations have been proposed to as classifiers in IDS. For supervised training, the Multi-layer Perceptron (MLP) is the most popular proposal [6,7]. The MLP has been evaluated with KDD99, considering new attacks [8], with NSL-KDD [9] and with UNSW-NB15 [4].

Modern proposals include the combination of machine learning algorithms with MLP, extracting knowledge from the features with decision trees [10], or in cascade [5].

Recently, more complex ANN for supervised training have been evaluated, including GRNN, PNN, RBNN [11]. Deep networks have been evaluated using KDD and NSL-KDD in [12], with only six features. Other deep networks with discriminative architectures, Recurrent Neural Networks (RNN), have been proposed using KDD based datasets in [13].

A two level detection approaches have been done on IDS. In [14] authors use PCA for feature selection and SVM for multi-class detection. Reducing the input features improves the detection accuracy for some classes, but reduces the overall performance. In [15] authors consider Deep Belief Network for feature selection, with 92.84% of accuracy, using NSL-KDD. In [16] authors use Self-taught learning, which has 2 stages, with 88.39% accuracy for binary detection and 79.10% for multi-class classification.

Genetic Algorithms have been recently used for feature selection in [17], for a SVM based detection algorithm, and in [18] for an ANN. According to [19], and to the best of our knowledge, CNN have not been evaluated for IDS, neither stand-alone nor in conjunction with genetic algorithms.

3 Proposal

Our proposal is to implement an intrusion detection system for TCP connections using CNN for classification. We consider that the order of the features in the layout of the CNN input can be critical for its correct training and behavior. A genetic algorithm is used when training the classifier for feature selection and placement in the CNN algorithm. The following sections describe the CNN, our approach for IDS and the hybrid approach using a genetic algorithm with CNN.

3.1 CNN

Artificial Neural Network (ANN) is a processing unit for information which was inspired by the functionality of human brains [20]. Typically neural networks are organized in layers which are made up of a number of interconnected nodes which contain an activation function. Selected features are presented to the ANN through an input layer, which has as many neurons as input features. The input layer neurons are connected to one or more hidden layer neurons via a system of weighted links. The hidden layers do the actual processing and then link to an output layer for producing the detection result as output, with as many neurons as desired outputs. The amount of hidden layers and hidden layer neurons is a parameter that can be tuned for a better performance. The most basic ANN is the Multi-layer Perceptron (MLP), also knwon as a fully connected network. Each neuron (node) in one layer has directed links to the neurons of the subsequent layer. The neurons apply an activation function (traditionally a sigmoid function). Each neuron of the output layer corresponds to one of the classification groups, having one normal and one abnormal output neuron.

In the training stage, the output values are compared with the known correct answer to compute the magnitude of the error made in the prediction, the gradient. The error is then fed back through the network, modifying the weights of the links backwards according to the gradient, sample by sample (stochastic gradient descent). This process might be slow, as we do it offline.

In the classification stage, the output neuron with the highest value indicates the class of the input trace. As there is no backward propagation, the evaluation consists on a set of multiplications that can be processed online.

A Convolutional Neural Network (CNN) is an ANN inspired by the visual cortex neurons. The CNN are used for emulating the human image processing, where each layer extracts concrete information, being more accurate and fine-grained as they are closed to the output. CNN are used in deep learning algorithms to extract features from raw information, specially in image processing [21] because they are more effective in artificial vision tasks. This kind of ANN includes a previous convolutional step to generate features that are inputs of a MLP. The convolution stride is used to preserve the spatial relationship between pixels by learning image features using small squares of input data.

Figure 1a depicts the architecture of a CNN divided in two stages. First, a succession of convolutional filters meant to extract and process the information in the input. The second stage is a MLP. The number on neurons of the final perceptron defines the different values or classes the complete CNN could classify. This input must be an image or an image-type bidimensional matrix with a global depth or the same number of values in each pixel. The applied filters are convolutions of fixed weights that are shifted all over the image, as shown in the example of Fig. 1b. This method usually generates a new matrix with reduced

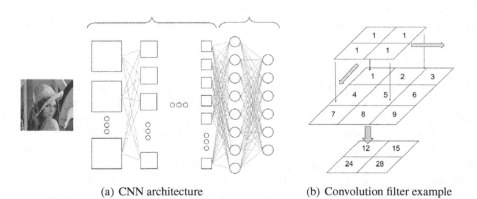

(a) CNN architecture   (b) Convolution filter example

Fig. 1. Convolutional neural networks

size and increased depth from the previous one. At the end of each convolution step it is common a pooling process. It consists of a new filter which takes the maximum or the average value of the input matrix in its window range instead of a convolution with its values. This filter is shifted all over the input just as the convolutional one. Due to the nature of convolutions, the CNN extracts information not only about the data itself but also about its location. This is the reason why CNNs are mainly applied to images in computer vision. Nevertheless, CNNs analyzes any kind of data ordered in a matrix configuration.

3.2   CNN on IDS

We propose to use a CNN to implement a binary classifier that distinguishes connections that are performing an attack to an element of the network (abnormal) from connections with regular traffic (normal). We start considering the features available in the UNSW dataset. However, we detect some restrictions on the system that imposes a feature reduction. First, the CNN exploits the distance between values of the features. As the numerical distance between values of categorical features, such as the protocol, are not showing any distance information, we decide to focus on a training a model for a concrete protocol (TCP). We select TCP because 58.86% of the UNSW traffic is TCP. The process we evaluate would be similar for other protocols, such as UDP or ICMP. Using the protocol information, the correct model would be selected. Focusing on TCP traffic, there are features for concrete application protocols, such as FTP or HTTP. We remove this features to treat all the connection with the same input features. Other categorical variables are IP addresses and ports. We use them as an identifier to obtain information on connections (timing, packets per connection), but not as input features to the CNN.

Second, the final goal is to implement an online attack detector that generates the features from the traffic passing through a firewall. We implement a sniffer and feature generation prototype to check which features to consider. There are features which we weren't able to obtain with our prototype, such as jitter. In other features the values we are obtaining are not similar to the ones published in UNSW, as it happens with TCP window, packets loss or load.

Therefore, we consider 23 connection oriented traces, a subset of the features available in the UNSW dataset. They are basic features (duration, number of bytes, ttl, packet count), TCP features (sequence number, mean packet size) and the connection features. The connection features are intended to sort accordingly with the last time feature to capture similar characteristics of the connection records for each 100 connections sequentially ordered.

As a CNN needs a bidimensional matrix as input, we arrange each 23 feature vector in a matrix of 5 by 5. We arrange the features following the UNSW order, adding two zeros at the end of the matrix. The CNN we propose consists of a filter convolution with 3×3 dimension and a depth of 4. This filter takes the 5×5 input and outputs a new one of 3×3 with 4 values per pixel. A second filter convolution filter has 2×2 dimension and a depth of 8. It takes the $3 \times 3x4$

output and generates a new $2 \times 2 \times 8$ matrix. Then, the CNN includes a max-pooling layer that keeps the depth and reduces the dimensionality of each 2×2 matrix taking the maximum value of its range. Therefore, it generates a $1 \times 1 \times 8$ matrix. The maxpooling is followed by a dropout layer, which is a regularization method that makes a more robust network. We tune a rate value of 0.25 that sets the fraction of the inputs to drop. At this point we flatten the dropout output and get a 1×8 vector that will be the input of the MLP. For the MLP, we use one hidden layer of 64 neurons followed by a new dropout layer of 0.5, ended by a softmax layer which give us the predicted class.

All neurons and filter convolutions include activation functions. We consider two functions: the Rectified Linear Unit (RELU) and the hyperbolic tangent (TANH). The RELU function requires less resources when processing and it typically offers better results than TANH. However, with non-normalized input features, the training of a RELU-based CNN might have problems of lack of convergence to a stable solution, because its output is unconstrained. On the other hand, the TANH function is constrained, granting the convergence.

As we propose a binary classifier, the kind of connection predicted can be positive, for connections classified as abnormal, or negative, for connection considered to be normal. There are only four possible scenarios when classifying the packets as attacks or not: True Positive (TP), which is a detected attack; False Positive (FP), which is a regular packet classified as an attack. True Negative (TN), regular traffic considered regular by the classifier; and False Negative (FN), which is a non-detected attack. Based on this four combinations there are many metrics available to evaluate an IDS. We use the CAP metric [22], which shows the behavior of the model considering the dataset distribution, related to B (fraction of attacks in dataset), PPV (Positive Predictive Value) and NPV (Negative Predictive Value).

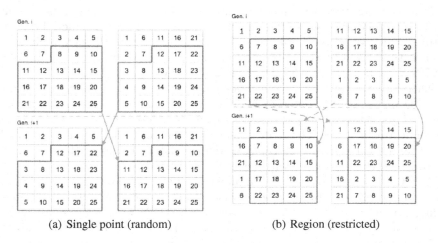

(a) Single point (random)   (b) Region (restricted)

Fig. 2. Selected crossover algorithms

3.3   Genetic Algorithm on CNN

We tackle the problem of the layout of the features in the input of the CNN using a Genetic Algorithm (GA). We propose using a GA to generate different input features layouts, train CNN models for each of the generated layouts and select the input layout which produces the best CNN model for IDS. The best CNN model will be the one used for online classification of traffic.

GAs are heuristic methods to solve complex optimizations problems based on modifications of existing solutions and evaluation of their fitness to select the best solutions (survival of the fittest): the solutions with lower value in the fitness function.

A GA starts with a set of individual solutions, called population. Each generation, the individuals evolve, by mutating and mixing with each other, to create new individuals. Each individual is evaluated, and those that better fit the objective will be selected for the next generation with higher probability.

In order to select the input configurations with better classification results, we select a fitness function related to the accuracy metric, which top value is 1. The accuracy is appropriate if the selected dataset is balanced, and it is faster than the CAP. When there is no classification done, assigning every sample to the same class, the value provided by the accuracy is 0.5. Therefore, the selected fitness function of the GA is:

$$Fitness \equiv 1 - accuracy \qquad (1)$$

We consider two different Genetic Algorithms according to the mutation and mixing operations: restricted and random. The restricted version considers only solutions including the 23 input features, while the random version considers solutions with reduce set of features, which in implementation would save resources and time.

In the random version, the mutation consists on a random selection of one feature for a random coordinate of the input layout. The mixing consists on the random selection of a single point in the two parent solutions, exchanging the trailing set of features, as depicted in Fig. 2a. The random version might reduce the number of different features selected for the CNN, allowing multiple repetitions of features in different coordinates. This algorithm can perform a feature reduction.

The mixing operation in the restricted version randomly selects a region of the size of the CNN filter. That region is kept untouched from each of the parents for the two new individuals. The remaining region is mixed, replacing the features that are already in the kept region by features that are not present, as depicted in Fig. 2b. The mutation operation swaps the features of two random selected coordinates of an individual. Therefore, in the restricted version the 23 features are always present, and the resulting layout keeps more than the random version the structure related to the convolutional stage of the CNN and the filter.

4   Experimental Setup

4.1   Dataset

We use the UNSW dataset with only 23 of the available 49 input features, as introduced in Sect. 3.2. Using only TCP traces the full dataset comprises 1495074 traces. In order to maximize the training fitness, the subset is balanced, considering every attack trace and the same amount of regular traces. This dataset includes 116368 traces. We divide the dataset in 10 subsets to implement a K-fold cross validation to estimate, with K being 10, with 11636 traces each.

4.2   Setup

We conduct a set of experiments using genetic algorithms to optimize the input layout of the CNN. Every experiment uses the CNN described in Sect. 3.2. First, as a reference for our optimization, we consider the layout of the features following the appearance in the UWS dataset, adding two extra zero to fill the 5×5 input matrix. We consider two different mutation algorithms introduced in Sect. 3.3: random, with repetitions and feature reduction allowed, and restricted, where every feature is present. Moreover, we consider two sets of inputs: normalized and non-normalized dataset. We consider a reduced set of experiments using TANH activation function as a reference, relevant for non-normalized input datasets.

Each RELU-based experiment has been executed using a 4 island approach with random populations. The resulting populations are then joined together in a last GA execution, which we call the hunger games. The TANH-based experiments have been executed only in one-island.

The Genetic Algorithm uses an accuracy-based fitness function (1). The algorithm evaluates 100 generations with a population of 500 solutions. We use a subset of the balanced dataset (one of the subsets generated for 10-fold cross validation). It increases the speed of the execution by reducing the quality of the fitness function. The original solution, the solution after the island stage and the solution after the hunger games have been 10-fold cross-validated and tested with the whole non-balanced TCP dataset.

The experiments presented in this paper have been tested in a single computer based on an Intel® Core™ i7-6700K CPU @ 4.00GHz×8, 16 GB, running Ubuntu 16.04 LTS. The process is divided in two different parts, client and server, connected using a TCP socket. The server side implements the CNN, both training and testing. It has been developed using Python3.5 with the following libraries: Keras version 2.1.2, TensorFlow version 1.1.0, Numpy version 1.13.0, Scipy version 0.19.0 and Pandas version 0.20.2. The code used is available at github[1] The client side implements the genetic algorithms that generates a population of different input layout and sends requests to the CNN server to evaluate them using the accuracy. The genetic algorithm has been developed using the Hero[2] library based in Java. The evaluation of a signle model trained

---

[1] https://github.com/greenlsi/HAIS18_code.

[2] https://github.com/jlrisco/hero.

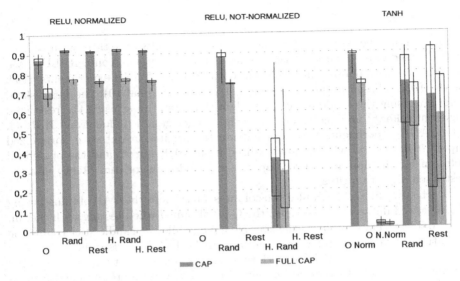

Fig. 3. CAP and deviation using 10-fold cross validation

with a subset of the 10-fold cross validation takes between 3 and 6 seconds, as it includes training and testing a CNN model. Considering the full cross validation, the evaluation time is over a minute. Using the full non-balanced dataset for testing increases the evaluation time of each classifier up to 40–50 s (10 classifiers when doing cross validation). Once the best CNN model is found, using it for classification purposes is similar to other ANN approaches, and it is suitable for online IDS.

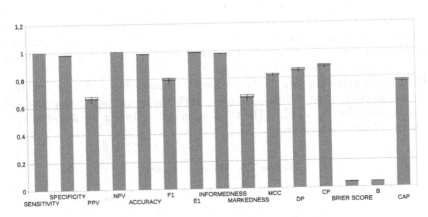

Fig. 4. Performance of the best solution

4.3   Results

Figure 3 shows the results of the experiments. The results have three groups: first, RELU-based experiments with normalized input dataset; second, RELU-based experiments with non-normalized input dataset; third, TANH experiments. Each of the RELU-based groups have five entries: the original (O), the best island solutions using the random mutation algorithm (Rand) and using the restricted algorithm (Rest) and the results after the hunger games (H.Rand and H.Rest). The TANH experiments have four entries: the original layout with normalized input (O.Norm) and with non-normalized input (O.N.Norm), and the optimized solution for the non-normalized input using random algorithm (Rand) and the restricted algorithm (Rest).

Each entry in Fig. 3 includes two columns: on the left, the CAP obtained using the balanced dataset; on the right, the CAP obtained when testing with the full TCP dataset. The bar shows the average CAP, while the candle sticks show the deviation between the different executions in the 10-fold cross validation.

Figure 4 shows the detailed performance of the best solution, found using a normalized input with the random algorithm. The results are obtained using a 10-fold cross validation including in the testing the full TCP UNSW dataset. Each entry corresponds to one of the metrics described in Sect. 3.2.

5   Discussion

Using a CNN for IDS shows interesting results. The best solution is found for a normalized input dataset, as expected. Figure 3 shows a CAP 0.87 (0.71 with the whole dataset) for the original layout. Using our optimization GA, we improve the performance of the CNN by optimizing the layout of the features. Both mutation algorithms show good and similar results.

The best solution is found using the random mutation algorithm, with a CAP of 0.92 (0.77 with the whole dataset), with a reduced deviation of the performance depending on the input dataset. Adding training phases to the CNN and increasing the amount of neurons in the hidden layer we obtain a CAP of 0.93 (0.79 with the whole dataset).

Table 1 compares the capability to detect attacks of our approach with the results obtained for the reduced dataset, with the same input features, using algorithms available in WEKA. The CNN shows very similar results after tuning it with the Genetic Algorithm.

Table 1. CAP comparative

Algorithm	J48	Rand. forest	MLP	Bayes net	Logistic function	CNN with GA
CAP	0.9356	0.9376	0.9344	0.929	0.9292	0.9309

Figure 4 shows the values of the different metrics. The True Negatives ratio is perfect. However, the False Positives ratios is around 70%. As the Full dataset includes approximately 3.85% of normal traces, the impact in the accuracy is reduced. The CAP considers multiple metrics to show a summary of the capacity to detect. Adding a lot of normal traces that are not included in the 10-fold cross validation introduces many False Positives, as some of them are classified as abnormal traces.

Using non-normalized inputs we obtain worse results. The original layout with the RELU activation function doesn't converge. The net tends to assume that all traces are regular connections, detecting no attack. Using the restricted mutation algorithm, that includes every feature, on the RELU-based CNN, shows the same behavior. The different ranges of values of the input features without bounding the limits of the output makes it difficult to extract information with the filter convolution. However, using the random mutation algorithm we find a solution with a very good performance: a CAP of 0.88 (0.73 using the whole dataset) with a reduced deviation, in spite of the magnitude variation between features. The genetic algorithm finds an optimal layout discarding features with incompatible ranges. This solution is interesting for online real-time anomaly detection, as the input requires less preprocessing. Using the TANH activation function the solution converges but it is useless, and after optimizing we obtain worse results than with RELU.

Table 2 compares the features that appear more frequently in the best solutions with the top 5 features evaluated with attribute evaluation algorithms available in WEKA [23] that show a feature relevance ranking. The numbers shown correspond to the UNSW feature notation.

Table 2. Feature relevance

Algorithm	SURF	OneR	InfoGain	GainRatio	Correlation	GA
Top 5	11,10,34,35,2	11,10,35,34,2	11,35,10,34,2	11,10,34,35,29	11,10,35,34,42	11,24,34,36,10

All the algorithms have top features related to TTL, mean size and connection setup time, while only the GA emphasizes if the port is the same.

6   Conclusions

We propose using Convolutional Neural Networks (CNN) to implement Intrusion Detection Systems (IDS) with a subset of the UNSW features. Our proposal is to arrange these features to create a 5×5 pixel image. The layout order is important for the CNN performance. Using normalized input features on a naïve layout we obtain a RELU-based CNN classifier with a CAP of 0.71 using the TCP connections of the UNSW dataset. Using non-normalized data, a RELU-based CNN classifier is not possible and a TANH-based classifier is useless.

We optimize the CNN classifiers using a Genetic Algorithm (GA) to find a better layout of the input features. Using a random mutation algorithm we obtain the best results for normalized input data with a CAP of 0.77 (8% of improvement). The results are similar to using other algorithms after applying the GA. For scenarios where data can't be normalized, due to resource constraints, the GA finds a solution with a CAP of 0.73, better than the original normalized solution.

Acknowledgements. This work was supported by the Spanish Ministry of Economy and Competitiveness under contracts TIN-2015-65277-R, AYA2015-65973-C3-3-R and RTC-2016-5434-8.

References

1. KDD cup (1999). http://kdd.ics.uci.edu/databases/kddcup99/kddcup99.html
2. NSL-KDD data set for network-based intrusion detection systems. http://nsl.cs.unb.ca/NSL-KDD/
3. Moustafa, N., Slay, J.: UNSW-NB15: a comprehensive data set for network intrusion detection systems (UNSW-NB15 network data set). In: Military Communications and Information Systems Conference (MilCIS), 2015, pp. 1–6. IEEE Stream (2015)
4. Moustafa, N., Slay, J.: The evaluation of network anomaly detection systems: statistical analysis of the UNSW-NB15 data set and the comparison with the KDD99 data set. Inf. Secur. J. Global Perspect. **25**(1–3), 18–31 (2016)
5. Baig, M.M., Awais, M.M., El-Alfy, E.M.: A multiclass cascade of artificial neural network for network intrusion detection. J. Intell. Fuzzy Syst. **32**(4), 2875–2883 (2017)
6. Ryan, J., Lin, M.J., Miikkulainen, R.: Intrusion detection with neural networks. In: Proceedings of the 1997 Conference on Advances in Neural Information Processing Systems 10. NIPS 1997, pp. 943–949. MIT Press, Cambridge (1998)
7. Moradi, M., Zulkernine, M.: A neural network based system for intrusion detection and classification of attacks. In: IEEE International Conference on Advances in Intelligent Systems - Theory and Applications (2004)
8. Kukielka, P., Kotulski, Z.: Analysis of neural networks usage for detection of a new attack in IDS. In: Annales UMCS, Information, vol. 10, no. 1, pp. 51–59, January 2010
9. Revathi, S., Malathi, A.: A detailed analysis on NSL-KDD dataset using various machine learning techniques for intrusion detection. Int. J. Eng. Res. Technol. (2013). ESRSA Publications
10. Guevara, C., Santos, M., López, V.: Intrusion detection with neural networks based on knowledge extraction by decision tree. In: International Joint Conference SOCO 2016-CISIS 2016-ICEUTE 2016, San Sebastián, Spain, pp. 508–517, October 2016
11. Devaraju, S., Ramakrishnan, S.: Performance comparison for intrusion detection system using neural network with KDD dataset. ICTACT J. Soft Comput. **4**(3) (2014)
12. Tang, T.A., Mhamdi, L., McLernon, D., Zaidi, S.A.R., Ghogho, M.: Deep learning approach for network intrusion detection in software defined networking. In: 2016 International Conference on Wireless Networks and Mobile Communications (WINCOM), pp. 258–263, October 2016
13. Kim, J., Kim, J., Thu, H.L.T., Kim, H.: Long short term memory recurrent neural network classifier for intrusion detection. In: 2016 International Conference on Platform Technology and Service (PlatCon), pp. 1–5. IEEE (2016)

14. Heba, F.E., Darwish, A., Hassanien, A.E., Abraham, A.: Principle components analysis and support vector machine based intrusion detection system. In: 2010 10th International Conference on Intelligent Systems Design and Applications. pp. 363–367, November 2010

15. Salama, M.A., Eid, H.F., Ramadan, R.A., Darwish, A., Hassanien, A.E.: Hybrid intelligent intrusion detection scheme. In: Gaspar-Cunha, A., Takahashi, R., Schaefer, G., Costa, L. (eds.) Soft Computing in Industrial Applications. AINSC, vol. 96, pp. 293–303. Springer, Heidelberg (2011). https://doi.org/10.1007/978-3-642-20505-7_26

16. Javaid, A., Niyaz, Q., Sun, W., Alam, M.: A deep learning approach for network intrusion detection system. In: Proceedings of the 9th EAI International Conference on Bio-inspired Information and Communications Technologies (Formerly BIONETICS), BICT 2015, ICST, Brussels, Belgium, Belgium, pp. 21–26. ICST (Institute for Computer Sciences, Social-Informatics and Telecommunications Engineering) (2016)

17. Gharaee, H., Hosseinvand, H.: A new feature selection ids based on genetic algorithm and SVM. In: 2016 8th International Symposium on Telecommunications (IST), pp. 139–144, September 2016

18. Guha, S., Yau, S.S., Buduru, A.B.: Attack detection in cloud infrastructures using artificial neural network with genetic feature selection. In: 2016 IEEE 14th International Conference on Dependable, Autonomic and Secure Computing, 14th International Conference on Pervasive Intelligence and Computing, 2nd International Conference on Big Data Intelligence and Computing and Cyber Science and Technology Congress (DASC/PiCom/DataCom/CyberSciTech), pp. 414–419, August 2016

19. Hodo, E., Bellekens, X.J.A., Hamilton, A., Tachtatzis, C., Atkinson, R.C.: Shallow and deep networks intrusion detection system: A taxonomy and survey. CoRR abs/1701.02145 (2017)

20. Haykin, S.: Neural Networks: A Comprehensive Foundation, 2nd edn. Prentice Hall PTR, Upper Saddle River (1998)

21. Simard, P.Y., Steinkraus, D., Platt, J.C.: Best practices for convolutional neural networks applied to visual document analysis. In. International Conference on Document Analysis and Recognition, pp. 958–963 (2003)

22. Gu, G., Fogla, P., Dagon, D., Lee, W., Skorić, B.: Measuring intrusion detection capability: an information-theoretic approach. In: Proceedings of the 2006 ACM Symposium on Information, Computer and Communications Security, pp. 90–101. ACM (2006)

23. Hall, M., Frank, E., et al.: The weka data mining software: an update. ACM SIGKDD 11(1), 10–18 (2009)

Improving the Accuracy of Prediction Applications by Efficient Tuning of Gradient Descent Using Genetic Algorithms

Arturo Duran-Dominguez[ID], Juan A. Gomez-Pulido$^{(\boxtimes)}$[ID],
and David Rodriguez-Lozano[ID]

University of Extremadura, 10003 Caceres, Spain
{arduran,jangomez,drlozano}@unex.es

Abstract. Gradient Descent is an algorithm very used by Machine Learning methods, as Recommender Systems in Collaborative Filtering. It tries to find the optimal values of some parameters in order to minimize a particular cost function. In our research case, we consider Matrix Factorization as application of Gradient Descent, where the optimal values of two matrices must be calculated for minimizing the Root Mean Squared Error criterion, given a particular training dataset. However, there are two important parameters in Gradient Descent, both constant real numbers, whose values are set without any strict rule and have a certain influence on the algorithm accuracy: the learning rate and regularization factor. In this work we apply a evolutionary metaheuristic for finding the optimal values of these two parameters. To that end, we consider as experimental framework the Prediction Student Performance problem, a problem tackled as Recommender System with training and test datasets extracted for real cases. After performing a direct search of the optimal values, we apply a Genetic Algorithm obtaining best results of the Gradient Descent accuracy with less computational effort.

Keywords: Gradient descent · Recommender systems · Prediction
Learning rate · Regularization factor · Genetic algorithm
Matrix factorization

1 Introduction

Machine Learning (ML) is a concept that involves a set of tools and methods designed to detect automatically patterns in data [1] in order to predict future of uncovered data, among other possibilities, by optimizing a performance criterion according to test data or past experience [2].

Nowadays, we can store and process high amount of data, under the concept of Big Data (BD). This information is collected from tasks where many users are involved. Thus, ML can process the data in order to extract useful knowledge.

© Springer International Publishing AG, part of Springer Nature 2018
F. J. de Cos Juez et al. (Eds.): HAIS 2018, LNAI 10870, pp. 210–221, 2018.
https://doi.org/10.1007/978-3-319-92639-1_18

There are three types of ML techniques: Supervised Learning (SL), Non-Supervise Learning (NSL) and Reinforcement Learning (RL). SL learns from the mapping of a set of X inputs to Y outputs, NSL considers not-labelled input data, and RL solves decision problems, where the learning is feed with positive or negative scores according to the decision taken.

Collaborative Filtering (CL) belongs to NSL. This method predicts the future behaviour of a user according to the past information of his activity in several tasks, and the activity of the other users in the same task. Among CL techniques, Recommender Systems (RS) [3] arises as a popular method that elaborates personalized recommendations to large database users, mainly according to the behavior when they request and handle information.

Recommender Systems focuses on prediction purposes; hence, they are applied to other systems where the knowledge of users behavior is important, not only for recommendation purposes, but for predicting analysis. We use as case of study for our research the Predicting Student Performance (PSP) problem, where the student performance is predicted for some tasks in the academic process [4] as a ranking prediction problem in RS.

The goal of the prediction method is to find the best model able to calculate accurate unknown performances. To this end, we consider the Matrix Factorization (MF) [5] technique for the relationship "student – performs – task", a very useful method for prediction in RS. In order to obtain the prediction model by MF, we consider the Gradient Descent (GD) method [6].

The prediction model based on MF considers the number of latent factors K implicit in the relationship mentioned before. The model is implicitly able to encode latent factors of students and tasks. The intuition behind using MF is that there should be some latent features that determine how a student performs a task, but it is difficult to establish the proper number of such latent factors.

There are two important parameters involved in the GD code: the learning rate β and the regularization factor λ. The first one is needed to determine the gradient, whereas the second one prevent the over-fitting. Both are constant real numbers, although their values are set without any strict rule and have a certain influence on the prediction accuracy. Therefore, in this work we propose a method based on an evolutionary metaheuristic for finding good values of β and λ with low computational effort.

2 Related Work

Matrix Factorization allows to build a prediction model in terms of the product of two matrices [7]. These matrices are calculated by GD iteratively. We choose GD for that purpose because it is very efficient dealing with large data sets [8].

When predicting student performance, many factors affect the prediction accuracy. For example, the parameters involved in the stability and sensitivity of prediction models based on ML are studied in [9], and selecting the adequate values for certain parameters allows a more accurate prediction for the slip and guess probabilities of students, as [10] shows.

In our case, two parameters were identified to have a high influence in the prediction accuracy. The selection of a good learning rate, β, is a key matter that affects the convergence of GD. We can adjust automatically β between iterations of GD if it does not converge (i.e. the cost function increases), or for accelerating the convergence (i.e. changing β results in a lower value of the cost function). This method can be easily programmed in the GD code for particular cases. Nevertheless, there are several sophisticated techniques to set that value adaptively in some cases where other algorithms different than GD are considered and require β. For example, a statistical method generates adaptive learning rates for modeling methods that use exponentially weighted moving average algorithm [11]. Also, an adaptive learning rate for stochastic variational inference is proposed in [12] without the need of previous tuning.

There is not a general rule to set the right learning rate. The usual way is to perform previous experimentations with different values of β in order to choose the best one, for a particular case. Nevertheless, many works chose β in the range from 0 to 1, typically 0.01.

3   Gradient Descent

The main terms used in the prediction technique are: performance (or score) p of user s for task i, number of users S, number of tasks I, performance matrix P of size $S \times I$, known performances D^{knw}, unknown performances D^{unk}, observed or training performances D^{train}, test performances D^{test}, and performance predictions $\hat{P}$. D^{train} is a subset of D^{knw} used to train the model in order to predict the unknown performance. The more training values (observed data) we use, the better the model we would get. On the other hand, D^{test} are known values chosen to validate the mathematical model using a particular criterion, as Root Mean Squared Error criterion (RMSE) (1). D^{test} is usually much smaller than D^{train}, and it is used for the performance predictions $\hat{p}$ over D^{test}.

$$RMSE = \sqrt{\frac{\sum_{s,i \in D^{test}} (p_{s,i} - \hat{p}_{s,i})^2}{|D^{test}|}} \tag{1}$$

Matrix factorization approaches P as the product of two smaller matrices W_1 and W_2 ($P \approx W_1 W_2^T$) of sizes $S \times K$ and $I \times K$, respectively. In this way, the performance $p_{s,i}$ of user s for task i is predicted by $\hat{p}_{s,i}$ (2).

$$\hat{p}_{s,i} = \sum_{k=1}^{K} (w_{s,k}^{(1)} w_{i,k}^{(2)}) = (W_1 W_2^T)_{s,i} \tag{2}$$

Gradient Descent updates W_1 and W_2 repeatedly in the learning phase minimizing the difference between real and predicted values of D^{train}. Once the model is finally obtained, we can predict the unknown performance of P. In addition, we can calculate RMSE for the predicted values of D^{test} in order to obtain the model accuracy.

The algorithm starts initializing W_1 and W_2 with random values, usually positive real numbers generated by a normal distribution $N(0, \sigma^2))$ with standard deviation $\sigma^2 = 0.01$. Next, we calculate the difference $e_{s,i}$ (3) between real and predicted values in order to obtain the error (4), which is minimised updating $W1$ and $W2$ iteratively. The algorithm knows, for each data, in which direction to update $w1_{s,k}$ and $w2_{i,k}$ calculating the gradient of $e_{s,i}^2$ applied to $w1_{s,k}$ and $w2_{i,k}$ (5) and (6). After obtaining the gradient, we update $w1_{s,k}$ and $w2_{i,k}$ to $w1'_{s,k}$ and $w2'_{i,k}$ in the opposite direction to the gradient, according to equations (7) and (8) respectively, where β is the learning rate.

$$e_{s,i} = p_{s,i} - \hat{p}_{s,i} = p_{s,i} - \sum_{k=1}^{K}(w1_{s,k}w2_{i,k}) \tag{3}$$

$$err = \sum_{(s,i) \in D^{train}} e_{s,i}^2 \tag{4}$$

$$\frac{\partial}{\partial w1_{s,k}}e_{s,i}^2 = -2e_{s,i}w2_{i,k} = -2(p_{s,i} - \hat{p}_{s,i})w2_{i,k} \tag{5}$$

$$\frac{\partial}{\partial w2_{i,k}}e_{s,i}^2 = -2e_{s,i}w1_{s,k} = -2(p_{s,i} - \hat{p}_{s,i})w1_{s,k} \tag{6}$$

$$w1'_{s,k} = w1_{s,k} - \beta\frac{\partial}{\partial w1_{s,k}}e_{s,i}^2 = w1_{s,k} + 2\beta e_{s,i}w2_{i,k} \tag{7}$$

$$w2'_{i,k} = w2_{i,k} - \beta\frac{\partial}{\partial w2_{i,k}}e_{s,i}^2 = w2_{i,k} + 2\beta e_{s,i}w1_{s,k} \tag{8}$$

The process ends when a predefined number of iterations is reached. We also can consider as stop criterion when the error is greater than the previous one, which means its minimum value has been reached; this criterion provides higher accuracy, although it could imply more computing effort in some cases.

In order to prevent over-fitting, we add a term (9) defined by the regularisation factor λ to $e_{s,i}^2$. Therefore, the newer gradients (10) and (11) update the values in $W1$ and $W2$ according to (12) and (13) respectively.

$$e_{s,i}^2 = (p_{s,i} - \hat{p}_{s,i})^2 + \lambda(||W1||^2 + ||W2||^2) \tag{9}$$

$$\frac{\partial}{\partial w1_{s,k}}e_{s,i}^2 = -2e_{s,i}w2_{i,k} + \lambda w1_{s,k} \tag{10}$$

$$\frac{\partial}{\partial w2_{i,k}}e_{s,i}^2 = -2e_{s,i}w1_{s,k} + \lambda w2_{i,k} \tag{11}$$

$$w1'_{s,k} = w1_{s,k} - \beta\frac{\partial}{\partial w1_{s,k}}e_{s,i}^2 = w1_{s,k} + \beta(2e_{s,i}w2_{i,k} - \lambda w1_{s,k}) \tag{12}$$

$$w2'_{i,k} = w2_{i,k} - \beta\frac{\partial}{\partial w2_{i,k}}e_{s,i}^2 = w2_{i,k} + \beta(2e_{s,i}w1_{s,k} - \lambda w2_{i,k}) \tag{13}$$

Once W_1 and W_2 are available, the users' performance for the task i is predicted by (2).

4   Methods for Adjusting Learning Rate and Regularization Factor

The learning rate and regularization factor have a strong influence on the prediction accuracy. In this section, we expose two methods to find optimal values for both parameters (it means minimum RMSE). The first method is a simple direct search involving a computational effort directly related to the number of pairs (β, λ) evaluated. The second method is our proposal based on GA in order to surpass the accuracy obtained by the direct search with less computation effort.

4.1   Direct Search

The Direct Search (DS) is a very simple method to obtain an optimal pair (β, λ) among a set of particular pairs. We calculate RMSE for all these pairs, where each value for β and λ is incremented by the intervals h_β and h_λ respectively, from the minimum values β_{min} and λ_{min} to the maximum values β_{max} and λ_{max} respectively. This procedure is coded by two nested loops; therefore, all the pairs generated by DS are processed to calculate the corresponding RMSE.

This method has the disadvantage of exploring all the space of generated solutions (β, λ), wasting so a high computation effort in calculating RMSE for those areas where the possible optimal is far away. However, DS allows to find a good solution among the generated ones, which can be in the nearness of the possible local or global optimal solutions.

4.2   Genetic Algorithms

Metaheuristics [13] are approximate algorithms usually applied to solve optimization problems, since they perform heuristic search to explore the space of solutions efficiently, by focusing the search in the nearness of a promising solution. The metaheuristics are classified as trajectory or population algorithms. Among population-based metaheuristics, Evolutionary Algorithms (EAs) [14] search the optimal solutions by evolving individuals according to nature-inspired rules. One of the most known EA are Genetic Algorithms [15], widely applied with success to solve many optimization problems.

GA performs a stochastic search considering as individual a solution of the optimization problem. The individual X is composed of several decision variables x_i. In our case, we have two decision variables: $x_1 = \beta$ and $x_2 = \lambda$. Therefore, a solution of the optimization problem is a pair (β, λ). In this context, the phenotype P of individual X is defined by their decision variables, coded into the genotype G according to a determined alphabet. Since we consider real numbers, both phenotype and genotype are the same.

A population is a set of individuals that evolves along generations. It is characterized by the chromosome G, composed of the genotypes of all their individuals. The population evolves following a strategy of minimizing an objective function $f(X)$. An additional fitness function transforms the objective function into a measure of relative fitness, although we consider as the fitness of an individual its objective value with regard to the entire population. In our problem, we work with RMSE as fitness function.

The GA starts generating an initial population, whose individuals are evaluated. From here on, the generations start to evolve, following several phases in each one. The first phase assigns a fitness value to each individual. Next, in the selection phase the best individuals (parents) are chosen for crossover according to their fitness values. The third phase (recombination) crosses the parents to generate new individuals (offspring), updating the fitness values. After that, particular mutations can be applied to some individuals of the offspring (updating the fitness accordingly). The offspring is evaluated in the evaluation phase and included in the population in the reinsertion phase, going back to the assignment phase again in order to start the next generation. The GA finishes when a stop criterion is reached (number of generations or computation time, for example).

5   Case of Study

We consider as experimental framework a real case, specifically the on-line campus of the University of Extremadura, Spain (CVUEx). It is composed of several software tools and services implemented in the Moodle platform [16], a learning environment widely used in the academic community.

Among the 3,500 existing virtual classrooms in CVUEx where 67,208 students and 4,329 teachers interact, we have chosen a dataset extracted from the subject "Introduction to Computers", coming from the first course of the Degree in Computer Engineering. This course had enrolled 207 students along the academic year 2016/2017.

The dataset was carefully filtered in order to remove those students and tasks with limited academic activity; otherwise, these data could introduce noise in the prediction. Thus, the results obtained after applying certain filters guarantee a good representation of the available data for the experiments.

The performance matrix P built after applying the filters represents an academic environment where there is not any student with less than 88% of activity in all the evaluation tasks, and the task with the minimum students' activity has 80% of participation. Our experimental instance IC-1617 consists of:

- $S=107$ students.
- $I=8$ evaluation tasks.
- $D^{knw}=800$ known scores.
- $D^{unk}=56$ unknown scores.
- $D^{train}=800$ training scores (we consider $D^{train}=D^{knw}$).
- $D^{test}=102$ scores (we chose as test scores one for each student following consecutive columns (tasks) and rows (students) in P; in case of coincidence with unknown score, we skip to the next row (student).

Figure 1a shows the first 18 students and the corresponding scores: light-gray cells show unknown scores and dark-gray cells contain test scores.

The cells show evaluations from 0 to 10, although GD normalizes them from 0 to 1. We can see in Fig. 1b an example of prediction (unknown scores calculated), where we have applied MF and GD considering $K=64$ latent factors, $\beta=0.8$, and $\lambda=0.06$, obtaining RMSE=0.37.

If we compare both sides of Fig. 1, we can analyse the expected behaviour of each student for each evaluation task considering the unknown performances. This information is very useful for students and teachers in order to detect the strengths and weakness of the learning process in the corresponding subject.

tasks

(a)

students	Teo1 1	Teo5 2	Lab1 3	Lab2 4	Lab3 5	Lab4 6	Lab5 7	Lab6 8
1	9.50	10.00	8.21	7.90	8.89	8.75	10.00	10.00
2	6.00	5.69	6.61	5.81	10.00	6.38	7.50	9.50
3	7.85	8.03	8.33	8.21	9.38	7.25	10.00	10.00
4	5.38	7.06	5.00	4.53	6.35	6.69		2.00
5	8.30		8.33	7.91	9.26	2.88	9.30	5.00
6	9.50	8.92	7.56	6.49	8.63	9.25	2.00	10.00
7	6.95		6.79	6.72	8.17	8.75	9.00	10.00
8	8.83	7.67	7.98	7.93	9.26	4.75	6.50	3.00
9	8.70	7.78	4.52	1.41	5.00	4.25	7.00	6.00
10		3.64	5.71	1.69	8.26	4.63	3.00	3.00
11	7.85	7.14	3.10	1.30	7.50	8.13	8.00	10.00
12	7.80	9.42	5.12	5.76	8.82	7.56	5.00	9.00
13	8.68		6.96	5.78	9.44	6.25	8.00	2.00
14	8.29	7.69	4.58	3.16	5.00	7.13		7.50
15	6.55	5.83	6.01	5.74	10.00	5.88	8.50	3.00
16	9.20	8.94	7.14	8.84	8.89	9.13	7.00	3.00
17	6.75	8.33	7.98	7.04	6.46	4.63	2.00	
18	9.05		6.73	5.49	2.94	9.25	9.00	8.00

(b)

Teo1 1	Teo5 2	Lab1 3	Lab2 4	Lab3 5	Lab4 6	Lab5 7	Lab6 8
9.50	10.00	8.21	7.90	8.89	8.75	10.00	10.00
6.00	5.69	6.61	5.81	10.00	6.38	7.50	9.50
7.85	8.03	8.33	8.21	9.38	7.25	10.00	10.00
5.38	7.06	5.00	4.53	6.35	6.69	3.43	2.00
8.30	3.37	8.33	7.91	9.26	2.88	9.30	5.00
9.50	8.92	7.56	6.49	8.63	9.25	2.00	10.00
6.95	3.45	6.79	6.72	8.17	8.75	9.00	10.00
8.83	7.67	7.98	7.93	9.26	4.75	6.50	3.00
8.70	7.78	4.52	1.41	5.00	4.25	7.00	6.00
4.03	3.64	5.71	1.69	8.26	4.63	3.00	3.00
7.85	7.14	3.10	1.30	7.50	8.13	8.00	10.00
7.80	9.42	5.12	5.76	8.82	7.56	5.00	9.00
8.68	3.33	6.96	5.78	9.44	6.25	8.00	2.00
8.29	7.69	4.58	3.16	5.00	7.13	4.66	7.50
6.55	5.83	6.01	5.74	10.00	5.88	8.50	3.00
9.20	8.94	7.14	8.84	8.89	9.13	7.00	3.00
6.75	8.33	7.98	7.04	6.46	4.63	2.00	5.06
9.05	3.61	6.73	5.49	2.94	9.25	9.00	8.00

Fig. 1. Performance matrices before (a) and after (b) prediction

6   Experimental Results

We consider that each experiment with DS or GA is composed of several runs of the same configuration, since the initialization phase of GD contains random values from a normal distribution. Hence, for a particular configuration, we can choose the model from the best run. The goal of the experiments with DS and GA is to obtain the best pair (β, λ), which corresponds with the minimum RMSE found. We will compare both methods in terms of accuracy and computing effort.

6.1   Direct Search

First, we establish four cases (A, B, C and D) where different search areas are selected, setting the corresponding bottom and upper limits for β and λ. Direct Search generates pairs (β, λ) iteratively by incrementing them using particular intervals h_β and h_λ. These steps determine the numbers N_β and N_λ of values for β and λ respectively; therefore, the number of predictions evaluated is $N_\beta \times N_\lambda$.

The limits of the search areas and the number of predictions calculated in each case are shown in Fig. 2 and Table 1, as well as the minimum RMSE found and their corresponding optimal pair (β, λ) and the time elapsed in performing the DS experiment. Case A corresponds with the wider search area, which is successively reduced in cases B, C and D, as Fig. 2 shows.

Obviously, the more predictions calculated, the more computing effort done. The successively search areas allow us to perform a more efficient optimal search. On the other hand, we note that the plots in Fig. 2 display the GRMSE metric, that is RMSE after applying a correcting function that highlights the maximum and minimum peaks, maintaining the proportion among all the points, in order to make easier sensing the existence of a minimums.

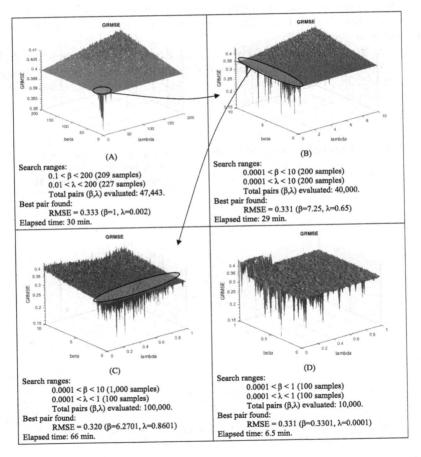

Fig. 2. Direct search of the minimum RSME corresponding with the best pair (β, λ) in four cases, reducing the search zone from case (A) to (D)

We think that the high number of minimum peaks in the plots suggests that the prediction accuracy is very sensitive to the pair setting, so we would need perform deeper searchs. Nevertheless, DS wastes a high computational effort performing predictions in wide areas where minimum values of RMSE are not present, moving us to consider more efficient search alternatives, as GA.

6.2 Genetic Algorithms

We have performed many experiments for tuning the more important GA parameters, according to the characteristics of the optimization problem (mainly, the convergence speed) and computers used: population size (100) and maximum number of generations as stop criterion (60).

Figure 3 shows an example of convergence to an optimal solution of a simple GA run. The best RMSE found in each generation is displayed as a dot, and the trend line is drawn as well.

We have studied the behaviour of many GA runs for the four cases, concluding that the convergence to an optimal solution is usually faster in cases A and B than C and D. The reason is simple: cases A and B consider larger search areas, so GA evolves its population faster to more reduced areas.

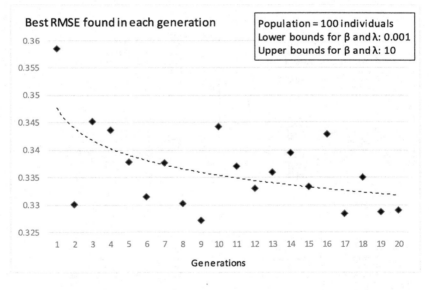

Fig. 3. Example of convergence to an optimal solution of a GA run

6.3 Comparison

Table 1 shows a comparison between the best solutions found by DS and GA. Each GA experiment consists of several runs, due to its non-deterministic nature,

Table 1. Comparison between Direct Search and Genetic Algorithm, for the four cases considered of bottom and upper limits of β and λ. In general, the best solution found in each case was obtained with less computation effort (number of prediction calculus) by GA, while the accuracy (RMSE) is slightly better with GA too

	CASE:	A	B	C	D
Search limits	β lower bound	0.0001	0.0001	0.0001	0.0001
	β upper bound	200	10	10	1
	λ lower bound	0.0001	0.0001	0.0001	0.0001
	λ upper bound	200	10	1	1
Direct search	RMSE (opt.)	0.333	0.331	0.320	0.331
	β (opt.)	1	7.250	6.270	0.330
	λ (opt.)	0.002	0.650	0.860	0.0001
	Predictions	47,443	40,000	100,000	10,000
Genetic algorithm	RMSE (opt.)	0.323	0.321	0.319	0.324
	mean	0.325	0.325	0.323	0.327
	st.deviation	8.1e-6	8.2e-5	5.3e-5	4.3e-5
	β (opt.)	1.439	0.232	2.928	0.789
	λ (opt.)	0.219	0.066	0.658	0.375
	Predictions	6,100	6,100	6,100	6,100

so the corresponding basic statistical results for RMSE are shown. Analyzing the table, we check how GA provides minimum RMSE in all the cases, although these values are very similar to the solutions obtained by DS. In any case, the number of predictions performed by GA is much lesser than DS. In this sense, GA deserves to be applied to those cases where large search areas are considered and DS spends much time.

We can conclude that, in general, GA improves the accuracy of the prediction when we apply the values of β and λ corresponding with the minimum RMSE foung by it. On the other hand, these values were obtained with lesser computational effort with regard to a simple direct search.

7   Conclusions

We have studied the advantages of considering a metaheuristic as Genetic Algorithm to obtain a good tuning for Gradient Descent instead of a simple direct search, where this tuning involves to determine the best values for two important parameters: learning rate and regularization factor. As Gradient Descent is very useful in many Machine Learning applications with prediction purposes, we think our proposal can contribute to improve the accuracy of the prediction in some problems, as Predicting Student Performance. We have checked experimentally the advantages of considering Genetic Algorithm instead of direct search: we obtain better results of the prediction error, which allows setting optimal values for both parameters. In addition, the metaheuristic is able to provide the optimal values with less computing effort in terms of computing time.

As future research works, we would like to consider mores experimental instances of different sizes and characteristics, in order to check if the advantages of a genetic algorithm with regard to a simple direct search are the same in all the cases. In addition, other metaheuristics different than GA can be considered.

Acknowledgments. This work was partially funded by the Government of Extremadura under the project IB16002, and by the AEI (State Research Agency, Spain) and the ERDF (European Regional Development Fund, EU) under the contract TIN2016-76259-P.

References

1. Murphy, K.P.: Machine Learning. A Probabilistic Perspective. The Massachusetts Institute of Technology Press, Cambridge (2012)
2. Alpaydin, E.: Introduction to Machine Learning. The Massachusetts Institute of Technology Press, Cambridge (2010)
3. Jannach, D., Zanker, M., Felfernig, A., Friedrich, G.: Recommender Systems. An Introduction. Cambridge University Press, New York (2011)
4. Thai-Nghe, N., Drumond, L., Horvath, T., Krohn-Grimberghe, A., Nanopoulos, A., Schmidt-Thieme, L.: Factorization techniques for predicting student performance. In: Educational Recommender Systems and Technologies: Practices and Challenges, pp. 129–153. IGI-Global (2012)
5. Koren, Y., Bell, R., Volinsky, C.: Matrix factorization techniques for recommender systems. Computer **42**, 30–37 (2009)
6. Koren, Y.: Factor in the neighbors: scalable and accurate collaborative filtering. ACM Trans. Knowl. Discov. from Data **4**, 1–24 (2010)
7. Rendle, S., Schmidt-Thieme, L.: Online-updating regularized kernel matrix factorization models for large-scale recommender systems. In: Proceedings of the 2008 ACM Conference on Recommender Systems, pp. 251–258 (2008)
8. Bottou, L.: Large-scale machine learning with stochastic gradient descent. In: Lechevallier, Y., Saporta, G. (eds.) Proceedings of COMPSTAT 2010, pp. 177–186. Springer, Heidelberg (2010). https://doi.org/10.1007/978-3-7908-2604-3_16
9. Tempelaar, D.T., et al.: Stability and sensitivity of Learning Analytics based prediction models. In: Helfert, M., Restivo, M.T., Zvacek, S., Uhomoibhi, J. (eds.) 7th International Conference on Computer Supported Education Lisbon, Portugal, pp. 156–166 (2015)
10. Baker, R.S.J., Corbett, A.T., Aleven, V.: More accurate student modeling through contextual estimation of slip and guess probabilities in bayesian knowledge tracing. In: Woolf, B.P., Aïmeur, E., Nkambou, R., Lajoie, S. (eds.) ITS 2008. LNCS, vol. 5091, pp. 406–415. Springer, Heidelberg (2008). https://doi.org/10.1007/978-3-540-69132-7_44
11. Zhang, R., Gong, W., Grzeda, V., Yaworski, A., Greenspan, M.: An adaptive learning rate method for improving adaptability of background models. IEEE Signal Proc. Lett. **20**, 1266–1269 (2013)

12. Ranganath, R., Wang, C., Blei, D.M., Xing, E.P.: An adaptive learning rate for stochastic variational inference. In: Proceedings of 30th International Conference on Machine Learning, pp. 1–9 (2013)
13. Gendreau, M., Potvin, J. (eds.): Handbook of Metaheuristics. Springer, New York (2010). https://doi.org/10.1007/978-1-4419-1665-5
14. Fogel, G., Corne, D.: Evolutionary Computation in Bioinformatics. Morgan Kaufmann Publishers, San Francisco (2003)
15. Reeves, C., Rowe, J.: Genetic Algorithms. Principles and perspectives. A guide to GA Theory. Kluwer Academic Publisher, Boston (2003)
16. Wild, I.: Moodle 3.x Developer's Guide. Packt Publishing, Birmingham (2017)

Mono-modal Medical Image Registration with Coral Reef Optimization

E. Bermejo[1]([✉]), M. Chica[2], S. Damas[3], S. Salcedo-Sanz[4], and O. Cordón[1,5]

[1] Department of Computer Science and Artificial Intelligence,
University of Granada, 18071 Granada, Spain
{ebermejo,ocordon}@decsai.ugr.es
[2] School of Electrical Engineering and Computing, The University of Newcastle,
Callaghan 2380, Australia
manuel.chicaserrano@newcastle.edu.au
[3] Department of Software Engineering, University of Granada, 18071 Granada, Spain
sdamas@ugr.es
[4] Department of Signal Theory and Communications,
University of Alcalá, 28805 Alcalá de Henares, Spain
sancho.salcedo@uah.es
[5] Research Center on Information and Communication Technologies,
University of Granada, 18071 Granada, Spain

Abstract. Image registration (IR) involves the transformation of different sets of image data having a shared content into a common coordinate system. To achieve this goal, the search for the optimal correspondence is usually treated as an optimization problem. The limitations of traditional IR methods have boomed the application of metaheuristic-based approaches to solve the problem while improving the performance. In this contribution, we consider a recent bio-inspired method: the Coral Reef Optimization Algorithm (CRO). This novel algorithm simulates the natural phenomena underlying a coral reef. We adapt the algorithm following two different approaches: feature-based and intensity-based designs and perform a thorough experimental study in a medical IR problem considering similarity transformations. The results show that CRO overcome the state-of-the-art results in terms of robustness, accuracy, and efficiency considering both approaches.

1 Introduction

In medical imaging there is a special interest in relating information from different images, which are frequently used for diagnosis, disease monitoring, or assisted surgery, among many other applications. These applications commonly require the integration or fusion of visual information acquired from different devices or conditions. Therefore, Image Registration (IR) [16] is an essential preprocessing task in medical imaging, as it allows the alignment of multiple images having a shared content.

© Springer International Publishing AG, part of Springer Nature 2018
F. J. de Cos Juez et al. (Eds.): HAIS 2018, LNAI 10870, pp. 222–234, 2018.
https://doi.org/10.1007/978-3-319-92639-1_19

Usually, IR methods consider a spatial transformation to align images by mapping their overlapping areas. The problem is treated as an iterative optimization procedure that explores the space of possible transformations. The quality of a solution is defined by the degree of resemblance between images after the transformation, which is measured by a similarity metric [5].

The procedure either makes use of the entire image (intensity-based approaches) or it is based on salient or distinctive parts of the images (features). Feature-based approaches [16] aim to expedite the optimization process, reducing the complexity of the problem by only using a portion of the images. However, they are highly dependent on the feature extraction process, a stage that may lead to inevitable errors if the features are not able to provide representative information. On the other hand, intensity-based approaches can deal with a larger amount of data at the expense of increasing the computational requirements [5]. The alignment is guided by the distribution of intensity values (gray levels) of the images, which makes these approaches more precise but sensitive to intensity changes when illumination changes and noise are present.

Traditional IR methods, such as the Iterative Closest Point (ICP) algorithm [2] and gradient-based methods are prone to be trapped in local minima by the influence of noise, discretization, and misalignment, among other factors. Methods based on evolutionary algorithms and other metaheuristics (MHs) are able to overcome some of these drawbacks. Thus, MH approaches are often considered to tackle medical IR problems due to their robust performance.

Several broad studies have been proposed to compare different MHs applied to the IR problem [4, 12]. These comparisons tackled the 3D registration of different human skulls models and 3D medical images. In both studies, a scatter search (SS) memetic approach proposed in [12] stands out for its results in the comparison with the state-of-the-art. This proposal was later adapted in [15] to an intensity-based scheme and compared both SS-based approaches against relevant IR methods: ICP, adaptive stochastic gradient descent (ASGD), and genetic algorithms (GAs). The excellent results delivered by SS consolidated its dominance as state-of-the-art evolutionary over conventional IR methods.

Recently, a novel bio-inspired evolutionary-type MH called coral reef optimization algorithm (CRO) was proposed in [11]. CRO is an evolutionary algorithm based on the artificial simulation of the natural phenomena taking place in the formation and reproduction of coral reefs. During their life, corals undergo different phases, such as reproduction, larval settlement, or fight for a space in the reef where they can survive. The proposed CRO approaches emulate these processes favoring a powerful trade-off between diversity and specificity, which makes it suitable for tackling complex optimization problems.

Since CRO has been successfully applied to different real-world problems by demonstrating a robust performance, we consider that its scheme can also be applied to complex IR problems and behave adequately. Thus, in the present work, we design a novel IR method based on CRO in which both the objective function and the coding scheme were adapted to a specific medical IR problem.

In order to validate our proposal, we develop an exhaustive experimental setup comparing CRO against some state-of-the-art evolutionary IR methods in different feature- and intensity-based medical IR scenarios. This comparison is an extension of our original proposal adapting CRO to intensity-based IR [1]. Each problem instance is generated using pairs of mono-modal images extracted from the well-known BrainWeb database at McGill University [3].

2   Image Registration: Problem Statement

In medical IR problems, we are usually provided with two images, a reference image, also called *model* (I_M), and the image that is transformed to reach the model geometry, known as *scene* (I_S). The objective of the registration process is to find a geometric transformation T making the model I_M and the transformed scene $T(I_S)$ be as similar as possible. The degree of resemblance between the considered images is measured using a similarity metric. Hence, the IR problem can be formulated as a maximization problem over the space of transformations:

$$\underset{T \in \text{ Transformations}}{\text{argmax}} \quad \text{Similarity}\,(I_M, T(I_S)).$$

Any IR method involves an iterative procedure of three main components: the transformation model, the similarity metric, and the optimization procedure. First, the optimizer estimates a candidate geometrical transformation to align the images, which is determined by the *transformation model*. Figure 1 illustrates the effects of applying different transformations to a brain MRI scan. The appropriate model depends on both the specific application and the nature of the images involved. In certain contexts a simple model (e.g., a translation transformation) can be an adequate choice, while other applications involving motion estimation require deformable models. Anyhow, the specific kind of transformation has to be carefully chosen, not only in terms of its number of parameters determining the computational effort to estimate them, but also considering the resulting transformation effects.

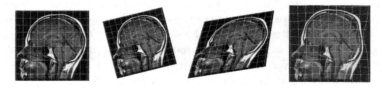

Fig. 1. From left to right: Images obtained from an initial scene by applying different transformations: similarity, affine, and B-spline.

Then, the *similarity metric* measures the quality of the alignment between the transformed scene and the model. The choice of this metric is a crucial step in the design of any IR method and it depends on the considered registration app-roach. Feature-based methods usually consider the mean square error (MSE) as similarity metric to measure the distance between corresponding features. Intensity-based methods consider the relationship between intensity distribu-tions to evaluate the alignment. There are multiple metrics available (normalized correlation (NC), mutual information (MI), among others) and their suitability is determined by the acquisition procedure and the relationship between images.

Last, the *optimizer* refines the solution until a particular stopping criteria is reached (e.g., when a suitable solution has been found or the algorithm per-forms a determined number of iterations). Hence, the optimization of the spatial transformation involves an iterative process to explore the search space of the geometrical transformation. According to the characteristics of the search space, we can find two different strategies. When the search is performed in the space of the transformation parameters, the registration is handled as a continuous optimization problem by parameter-based approaches. Alternatively, the search can also be performed on the space of correspondences by matching features (feature-based) or areas of the image (intensity-based) to align the images.

2.1 Relevant Evolutionary Algorithms for Medical IR

Intensity-Based IR Approaches

Valsecchi et al.'s r-GA*. r-GA* is based on a *genetic algorithm* [14]. The opti-mizer follows a real coded design in which a solution stores the transformation parameters in a real vector. The genetic operators are also real coded: blend crossover (BLX-α) [13] and random mutation. r-GA* is able to handle different similarity metrics and transformation models. In addition, it integrates the use of multiple resolutions combined with a restart and a search space adaptation mechanism. These mechanisms aim to speed up the optimization process and mitigate a premature convergence of the algorithm.

Valsecchi et al.'s SS*. SS* [15] is a variant of the original *scatter search* design, where the reference set is divided in two tiers containing the most high quality and diverse solutions. The methods integrated in the template were specifically designed for IR [12]. Thus, SS* integrates a diversification generation method based on a frequency memory, the solution combination method is the BLX-α crossover, and the improvement method is based on the PMX-α [8] operator. The reference set update method is responsible for classifying the solutions according to their quality and their diversity, and for maintaining the best ones of each category in the reference set. Last, the duplication control method prevents the appearance of identical copies of a solution in the reference set.

Feature-Based Evolutionary IR Approaches

He and Narayana's HE-GA. This method was proposed in [6] and consid-ers a real coding genetic scheme combined with dividing rectangle, a global optimization method based on branch and bound. The GA follows an elitist generational model with arithmetic crossover and uniform mutation operators, and it is applied to estimate a preliminary solution. Then, the dividing rectangle method refines this solution by means of a local search procedure. A restart mechanism is also used in case of premature convergence.

Santamaría et al.'s SS. This SS-based IR method was proposed in [12] and it was originally designed to tackle range IR problems. The authors adopted a similarity metric based on the median square error (MedSE), robust to low-overlapping images. It also integrates a data structure to speed up the computation of the fitness function, and a crossover-based local search (XLS) as improvement method. This SS variant also considers a restart mechanism to avoid local minima.

3   Coral Reefs Optimization Algorithm

CRO [10,11] is an evolutionary-type algorithm based on the behavior of the processes occurring in a coral reef. Let $\mathcal{R}$ be the reef represented by an $R_1 \times R_2$ grid, where each position (i, j) of $\mathcal{R}$ is able to allocate a coral or a colony of corals, $\mathcal{C}_{i,j}$, standing for solutions to the current optimization problem at hand. The CRO algorithm first initializes some random positions of $\mathcal{R}$ with random corals, and leaves some other positions empty. These holes in the reef are available to host new corals that will be able to freely settle and grow in later phases of the algorithm. The rate between free/occupied positions in $\mathcal{R}$ at the beginning of the algorithm is a parameter of the CRO algorithm, denoted as ρ_0 with $0 < \rho_0 < 1$.

The second phase simulates the processes of reproduction and reef formation. The different reproduction mechanisms available in nature are recreated by sequentially applying different operators:

1. **External sexual reproduction or Broadcast Spawning.** Broadcast spawning consists of the following steps at each iteration k of the algorithm:
   1.a. A random fraction of the existing corals is selected uniformly, turning them into broadcast spawners. The fraction of spawners with respect to the overall amount of existing corals in the reef will be denoted as F_b.
   1.b. Several coral larvae are formed. Two broadcast spawners are selected and a crossover operator is applied. Once two corals have been selected, they are not chosen anymore at iteration k for reproduction purposes. Corals' selection can be done randomly, uniformly, or using any fitness proportionate selection approach (e.g. roulette wheel).
2. **Internal sexual reproduction or Brooding.** Hermaphrodite corals mainly reproduce by brooding. It is modelled by any kind of mutation mechanism on a fraction of corals of $1 - F_b$. A percentage P_i of the coral is mutated.
3. **Larvae setting.** Once the larvae are formed, they will try to set in a random location and grow in the reef. Each larva will set in a position (i, j) if the location is free. Otherwise, the new larva will set only if its health function (fitness) is better than that of the existing coral. The algorithm defines a parameter η to determine the maximum number of tries a larva can attempt to occupy a position at each iteration k.

4. **Asexual reproduction.** Corals reproduce asexually by budding or fragmentation. CRO models this mechanism in the following way: the whole set of corals in the reef are sorted according to their level of health value (given by $f(\mathcal{C}_{ij})$). Then, a small fraction (denoted as F_a) of the available corals are duplicated and mutated (with probability P_a) to provide variability, and try to settle in a different part of the reef as in Step 3.

5. **Depredation.** Corals may die during the reef's formation. Therefore, at the end of each reproduction iteration k, a small number of corals in the reef can be depredated, thus liberating space in the reef for next coral generation (iteration $k + 1$). The depredation operator is applied with a very small probability (P_d) to a fraction (F_d) of the corals in the reef with worse health.

4   Experimental Study

We design an experimental study by considering the algorithms described in Sect. 2.1. These methods outperformed conventional IR methods [1,4,14,15], so our work is focused around the best performing evolutionary algorithms for medical IR. We aim to provide an exhaustive analysis, considering optimization capabilities of the methods tackling a synthetic medical IR problem.

4.1   Image Dataset and Problem Scenarios

The mono-modal images used in this experiment were obtained from the well-known BrainWeb public repository at McGill University [3]. This repository consists of a simulated brain database providing synthetic MRIs computationally generated. In order to consider different degrees of complexity, some images include noise (up to 5% relative to the brightest tissue) and multiple sclerosis lesions. All the images have the same size ($60 \times 181 \times 217$ voxels).

Table 1. Parameters of the transformations: rotation angle (λ), rotation axis (a_x, a_y, a_z), translation vector (t_x, t_y, t_z), and uniform scaling factor s.

	λ	a_x	a_y	a_z	t_x	t_y	t_z	s
T_1	115	−0.863	0.259	0.431	−26	15.5	−4.6	1
T_2	168	0.676	−0.290	0.676	6	5.5	−4.6	0.8
T_3	235	−0.303	−0.808	0.505	16	−5.5	−4.6	1
T_4	276.9	−0.872	0.436	−0.218	−12	5.5	−24.6	1.2

For each of the original images we extracted a set of points containing relevant curvature information. These set of points were obtained using a 3D crest-line edge detector [9]. In feature-based approaches, the heuristic values of the crest-line points will guide the optimizer to estimate the registration transformation.

In order to create different IR problem scenarios, each image was transformed using one of the four similarity transformations (involving rotation, translation, and uniform scaling) shown in Table 1. Therefore, the experimental study features a total number of sixteen IR problem instances that were created by pairing images with different transformations. The scenarios are labelled as: I_1 versus $T_i(I_2)$, I_1 versus $T_i(I_3)$, I_1 versus $T_i(I_4)$, and I_2 versus $T_i(I_4)$, for $i = 1, 2, 3, 4$.

4.2 Experimental Design and Parameter Configuration

The considered algorithms differ in their optimization procedure: feature-based approaches are typically guided by MSE or MedSE and consider only feature points, while intensity-based approaches are guided by either MI or NC over the intensity values of the images. The resulting values of different metrics are not directly comparable. Therefore, a standardized framework is required in order to objectively evaluate the registration results. We considered a common measure to evaluate the quality of all solutions. Once the algorithm reaches a solution, the MSE distance between the transformed scene's features and the model's features is computed over a set of anatomical landmarks (crest-line points). This value will be used as final quality comparison for all methods without interfering with the particularities of each IR approach.

A similarity transformation relates the images, considering $[-30, 30]$ as the parameters' range for the translation, and $[0.75, 1.25]$ for the scaling factor. No restriction to the rotation axis was applied so the actual range for each component of the rotation versor is $[-1, 1]$, with a rotation angle in the $[0, 360]$ range. Thereby, the transformation is encoded using seven real-coded parameter solution if the approach is intensity-based: rotation versor $(\theta_x, \theta_y, \theta_z)$, translation vector (t_x, t_y, t_z), and uniform scaling s. In case of feature-based approaches, instead of using a versor, the rotation was encoded using three parameters for the rotation axis in the interval $[-1, 1]$, and a rotation angle in the $[0, 360]$ range. Thus, the solutions use eight-dimensional real coded transformations.

In order to design our experimental setup as comprehensively as possible, we adapted CRO and the state-of-the-art IR methods (see Sect. 2.1) to both feature- and intensity-based approaches. Hence, we will compare six different IR methods: (i) *Intensity-based methods*[1]: r-GA*, SS*, and CRO*. (ii) *Feature-based methods*: HE-GA, SS, and CRO.

Due to the differences in computational requirements of both approaches, the optimal parameter configuration of the algorithms varies for each approach (see Table 2). These parameters were manually adjusted considering a different set of problem instances to avoid experimentation bias while following the general guidelines provided by the authors [11], or from previous studies [4]. Next, we detail the distinct characteristics of the considered approaches.

---

[1] For the sake of clarity we noted every intensity-based method including an asterisk (*) after its name. Note that the differences between methods are only relative to the specific implementation and its components, not the main scheme of the algorithm.

Table 2. Parameter configuration for considered IR methods.

r-GA*		SS*		CRO*		HE-GA		SS		CRO	
Individuals	100	Psize	12	Reef size	80	Individuals	60	Psize	30	Reef size	70
Generations	75	gen.	18	gen.	65	Restarts	5	Restarts	5	Restarts	15
Restarts	5	Restarts	3	Restarts	8	α	0.3	XLS iters.	100	XLS	100
tourn. size	3	PMX iters.	12	PMX	15	cross. prob.	0.7	α	0.3	ρ_0	0.6
Blend factor α	0.3	α	0.3	ρ_0	0.6	mut. prob.	0.2	refe. set s	8	F_{broad}	0.9
cross. prob.	0.5	ref. set	4	F_{broad}	0.8					P_d	0.05
mut. prob.	0.1			P_d	0.15					k	3

Intensity-based methods design: Every intensity-based method was implemented in Elastix [7], an open-source and widely-used toolbox specifically designed for intensity-based medical IR. The CRO method was proposed in [1] following this design with promising preliminary results. The optimizer of intensity-based approaches is guided by the normalized mutual information (NMI) metric:

$$\mathrm{NMI}(I_A, I_B) = \frac{\sum_{a \in I_A} \sum_{b \in I_B} p_{AB}(a,b) \log(p_A(a)p_B(b))}{\sum_{a \in I_A} \sum_{b \in I_B} p_{AB}(a,b) \log p_{AB}(a,b)},$$

where p_{AB} is the joint probability and p_A, p_B are the marginal discrete probabilities of the intensity values of the images.

In addition, the design of the intensity-based approaches integrates two specific components: (i) A **multi-resolution strategy** to reduce the computational cost of the process, applying both down-sampling and Gaussian smoothing to create two image representations (pyramids). In the first resolution, the optimizer selects a low-detail pyramid to obtain an approximation of the desired transformation. The second resolution acts as refinement phase, improving the quality of the previous transformation. (ii) A **restart mechanism** to ensure the process can recover from stagnation and find a good final solution. During the first resolution, the optimization procedure is performed a fixed number of times, restarting the population at the end of each run. This solution is carried out to the second resolution, where the refinement is performed. Due to the substantial amount of data that intensity-based approaches consider, the running time of each algorithm was limited at 180 s, including both resolutions.

Feature-Based Methods Design: Feature-based methods were implemented in C++ and compiled with the GNU/g++ tool following a similar structure to the intensity-based. Since MSE was chosen as quality metric for the final comparison, we considered a different optimization measure to avoid favoring feature-based methods. Hence, we adopted MedSE, which was specifically designed as similarity metric for feature-based medical IR approaches:

$$F(f, I_s, I_m) = w_1 \cdot (1/(1 + \sum_{i=1}^{N} ||(sR\boldsymbol{p}_i + \boldsymbol{t}_i) - \boldsymbol{p}_j^{'}||)) + w_2 \cdot (1/(1 + |p_c^s - p^m|)),$$

where I_s and I_m are the scene and model images; f is the solution encoding the transformation parameters; p_i is the i-th 3D point from the scene and p_j its corresponding closest point in the model obtained with a grid closest point data structure; w_1 and w_2 ($w_1 + w_2 = 1$) weigh the importance of each function term; p_c^s is the radius of the sphere wrapping up the transformed scene image; and p^m the radius of the sphere wrapping up the model image. Note that the first term of F corresponds to the MedSE between neighbor features.

Regarding specific components integrated within the original scheme of the feature-based algorithms, a restart mechanism is applied when the optimizer detects stagnation in the population, i.e., after 15 iterations without improvement. As feature-based approaches deal with a small set of feature points, no multi-resolution strategy is required. Every proposal in this experimentation was implemented using eight real-coded parameter solutions. The stopping criteria was limited to 20 s, due to the reduced sizes of the feature sets.

4.3 Analysis of Experimental Results

Table 3 reports the MSE results calculated between sets of feature points extracted from the images, and the partial ranking of each IR method for the sixteen scenarios. As mentioned in Sect. 4.2, the minimum achievable MSE error of each registration scenario accounts for the level of noise and lesion between the compared images. To clarify the interpretation of the results, the optimal MSE value per scenario is highlighted in brackets in Table 3. A statistical analysis is also performed over the mean MSE results in order to find significant differences between the algorithm with the best rank and the rest. Ranking, Bonferroni-Dunn's test, and Holm's test results are included in Table 4.

In general, feature-based approaches outperform their intensity-based counterparts in most of the occasions. The main reason for this behavior is a particular optimization design, which allow these methods to achieve a better accuracy in less time. Hence, feature-based approaches usually obtain the best minimum results, at the expense of increasing the amount of variance. On the other hand, the results provided by the intensity-based methods (except r-GA*) are more robust, showing lower standard deviation values than feature-based approaches.

Both genetic approaches performed with an unstable behavior but were able to deliver good minimum results. r-GA* ranked in last place (5.63), scoring the largest MSE values in the comparison. Despite its mediocre mean performance, HE-GA was able to achieve the best overall minimum result in three of the sixteen scenarios considering all the methods, with a ranking score of 4.94.

SS* and SS performed considerably better than the genetic methods, behaving consistently except in four different scenarios due to outlier results. Thus, SS-based methods reached medium ranking scores (3.0 and 3.56, respectively). Even though both methods perform similarly in terms of averaged MSE and failed to converge in some occasions, SS ranked above SS* by delivering better minimum outcomes despite a high standard deviation.

Regarding the CRO approaches, CRO* was able to obtain better mean MSE results that both SS approaches, but when considering minimum MSE values, it

Table 3. Minimum (m), mean (μ), and standard deviation (sd) results for the MSE values (in mm) and mean Ranking (R) for each IR instance. Colored rows correspond to the intensity-based approach of the algorithms.

Algorithm	I_1 vs $T_1(I_2)$ [31]				I_1 vs $T_2(I_2)$ [31]				I_1 vs $T_3(I_2)$ [31]				I_1 vs $T_4(I_2)$ [31]			
	m	μ	sd	R	m	μ	sd	R	m	μ	sd	R	m	μ	sd	R
r-GA*	37.32	134	>99	6	39.60	86.43	44	5	43.29	5506	>99	6	32.19	84.36	91	6
HE-GA	42.69	101	47	5	**31.85**	44.27	17	4	**31.96**	42.19	9	4	32.88	59.23	28	5
SS*	36.01	38.14	5.9	4	36.53	38.96	2.9	3	36.78	1755	>99	5	32.65	32.96	0.7	4
SS	**32.06**	32.84	2.2	2	32.11	119	>99	6	32.17	41.08	47	2	**32.00**	32.25	0.1	2
CRO*	36.35	36.79	0.3	3	36.26	36.78	0.3	2	40.72	41.41	0.5	3	32.67	32.79	0.1	3
CRO	32.11	**32.25**	0.1	1	32.10	**33.96**	5.3	1	32.14	**32.27**	0.1	1	32.14	**32.24**	0.0	1

Algorithm	I_1 vs $T_1(I_3)$ [42]				I_1 vs $T_2(I_3)$ [42]				I_1 vs $T_3(I_3)$ [42]				I_1 vs $T_4(I_3)$ [42]			
	m	μ	sd	R	m	μ	sd	R	m	μ	sd	R	m	μ	sd	R
r-GA*	50.61	142	>99	6	45.61	97.75	81	6	66.00	8470	>99	6	42.93	86.11	53	5
HE-GA	61.65	122	48	5	**42.91**	61.01	24	4	56.52	74.44	18	5	58.74	114	42	6
SS*	50.20	61.34	50	4	43.65	46.36	3.2	3	55.22	57.85	3.2	4	44.68	45.18	0.3	4
SS	**42.96**	58.69	59	3	43.08	83.82	>99	5	43.06	50.95	41	2	42.95	43.44	0.6	2
CRO*	51.23	51.77	0.3	2	43.54	**43.73**	0.1	1	55.78	56.53	0.4	3	44.61	45.17	0.3	3
CRO	43.01	**43.29**	0.2	1	42.94	44.06	2.9	2	**42.97**	**43.45**	0.2	1	**42.87**	**43.39**	0.3	1

Algorithm	I_1 vs $T_1(I_4)$ [46]				I_1 vs $T_2(I_4)$ [46]				I_1 vs $T_3(I_4)$ [46]				I_1 vs $T_4(I_4)$ [46]			
	m	μ	sd	R	m	μ	sd	R	m	μ	sd	R	m	μ	sd	R
r-GA*	51.06	151	>99	6	58.92	124	98	4	60.76	9428.7	>99	6	**46.44**	82.74	45	5
HE-GA	60.07	119	51	5	843	1011	>99	6	51.18	78.92	25	5	51.35	92.56	30	6
SS*	52.53	53.67	1.3	3	46.48	48.12	1.5	2	54.32	60.10	5.2	3	46.65	47.58	0.5	3
SS	46.79	61.77	77	4	46.52	154	>99	5	46.87	64.26	68	4	46.67	**47.33**	0.5	1
CRO*	52.78	53.24	0.3	2	**46.02**	**46.37**	0.1	1	57.46	58.02	0.3	2	47.03	47.70	0.3	4
CRO	**46.66**	**48.85**	6.0	1	46.77	48.80	3.4	3	**46.80**	**47.57**	0.6	1	46.64	47.37	0.6	2

Algorithm	I_2 vs $T_1(I_4)$ [28]				I_2 vs $T_2(I_4)$ [28]				I_2 vs $T_3(I_4)$ [28]				I_2 vs $T_4(I_4)$ [28]			
	m	μ	sd	R	m	μ	sd	R	m	μ	sd	R	m	μ	sd	R
r-GA*	32.15	3293	>99	6	32.20	68.26	33	6	33.38	3656	>99	6	29.53	57.89	32	5
HE-GA	33.59	94.40	35	5	29.94	46.64	41	5	29.88	37.67	14	3	31.34	78.25	33	6
SS*	34.80	35.66	0.5	3	30.36	33.07	3.7	3	38.85	43.70	9.6	5	**28.44**	29.69	0.5	4
SS	28.54	44.21	79	4	**28.61**	35.27	35	4	**28.46**	**28.80**	0.1	1	28.57	**28.75**	0.1	1
CRO*	35.06	35.54	0.2	2	30.25	30.47	0.1	2	39.11	40.66	0.5	4	29.36	29.64	0.1	3
CRO	**28.32**	**28.79**	0.3	1	**28.61**	**29.01**	0.6	1	28.57	28.82	0.1	2	28.59	28.76	0.1	2

is not able to reach their accuracy level. On the other hand, the feature-based approach CRO outperformed its intensity-based counterpart and the current state-of-the art evolutionary IR method in this comparison (SS). In terms of mean MSE values, CRO was able to outperform SS in twelve scenarios. Therefore, CRO* ranked in second place with 2.5 while CRO was first with a rank of 1.38, displaying an excellent exploration/exploitation trade-off. Both CRO-based proposals were able to outperform the current state-of-the-art evolutionary IR methods with the feature-based approaches obtaining the best performance.

According to the statistical analysis, the result of applying Friedman's test is $\chi^2_F = 92.0$, and the corresponding p-value is $<10^{-16}$. Given that the p-value is lower than the considered level of significance ($\alpha = 0.01$), the test concludes that there are significant differences among the results. We complemented the analysis with Holm's test (Table 4) comparing CRO (control) with the rest of the methods. The p-value results for both tests reveal that CRO present significant differences with both SS-based and genetic-based methods.

Finally, we include a comparison of different registration solutions obtained by the IR methods in one of the most complex scenarios to provide a visual assessment of the quality of the results. Figure 2 presents a visualization of the overlapping between the images regarding the best solution obtained by each

Table 4. Friedman's test ranking and statistical p-values with CRO as control method for Bonferroni's test, and Holm's test according to the mean MSE value.

	Mean Rank	Bonferroni-Dunn p	Holm p
CRO	1.38	-	-
CRO*	2.50	0.44	0.09
SS	3.00	0.07	< 0.05
SS*	3.56	< 0.01	< 0.01
HE-GA	4.94	< 0.01	< 0.01
r-GA*	5.63	$< 10^{-10}$	$< 10^{-10}$

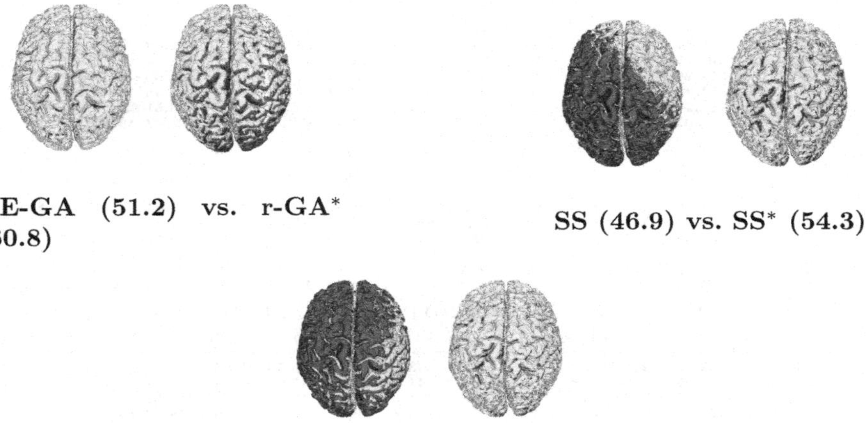

HE-GA (51.2) vs. r-GA* (60.8)

SS (46.9) vs. SS* (54.3)

CRO (46.8) vs. CRO* (57.5)

Fig. 2. Visual results of the overlapping between model (blue) and the scene (yellow). Figures provide a comparison of the results obtained by the feature-based (left) and the intensity-based (right) methods regarding the $11th$ scenario (I_1 vs. $T_3(I_4)$). The minimum MSE value is included in parenthesis.

algorithm. In general, the intensity-based version of the methods achieved a better overlapping than its feature-based counterpart[2], even though it obtained a higher MSE, outlining the good overlapping between images achieved by CRO-based methods. It is appreciable how CRO greatly improves the accuracy of the alignment provided by SS.

5   Conclusion

In this work, we described the design and implementation of a novel nature-inspired technique, known as CRO algorithm, to solve the problem of 3D medical

---

[2] If an optimal transformation is achieved, the visualization shows a brain of interlaced colors, where yellow (light gray) represents the transformed image and blue (dark gray) depicts the reference image.

IR in an efficient and robust way. In particular, we tackled this problem considering two different approaches, using the intensity distribution of the images and considering a reduced set of feature points, each one presenting different levels of complexity. CRO integrates a powerful and balanced exploration strategy in its design, which has allowed it to obtain excellent results in different complex real-world problems. Hence, we adapted and applied CRO to the IR problem following both intensity- and feature-based approaches to analyze its behavior.

In order to evaluate the performance of our proposal, we compared both implementations of CRO with some of the best performing evolutionary IR methods in the literature. We tackled an exhaustive experimental study involving sixteen mono-modal IR scenarios obtained by pairing four different synthetic brain MRI images, with different levels of complexity regarding noise and sclerosis lesions. These images were extracted from the BrainWeb database at McGill University [3]. CRO provided outstanding results considering both registration approaches, outperforming classical and well-consolidated methods and demonstrating the efficiency, robustness and suitability of the proposed algorithm when tackling complex optimization problems such as medical IR.

Acknowledgments. This work has been partially supported by the projects TIN2015-67661-P, including European Regional Development Funds (ERDF), from the Spanish Ministery of Economy, and TIN2014-54583-C2-2-R from the Spanish Ministerial Commission of Science and Technology (MICYT).

References

1. Bermejo, E., Chica, M., Salcedo-Sanz, S., Cordon, O.: Coral reef optimization for intensity-based medical image registration. In: IEEE Congress on Evolutionary Computation, CEC 2017, Proceedings, pp. 533–540. IEEE (2017)
2. Besl, P.J., McKay, N.D.: A method for registration of 3-D shapes. IEEE Trans. Pattern Anal. Mach. Intell. **14**(2), 239–256 (1992)
3. Collins, D.L., Zijdenbos, A.P., Kollkian, V., Sled, J.G., Kabani, N.J., Holmes, C.J., Evans, A.C.: Design and construction of a realistic digital brain phantom. IEEE Trans. Med. Imaging **17**, 463–468 (1998)
4. Damas, S., Cordón, O., Santamaría, J.: Medical image registration using evolutionary computation: an experimental survey. IEEE Comput. Intell. Mag. **6**(4), 26–42 (2011)
5. Goshtasby, A.A.: 2-D and 3-D Image Registration. Wiley Interscience, Hoboken (2005)
6. He, R., Narayana, P.A.: Global optimization of mutual information: application to three-dimensional retrospective registration of magnetic resonance images. Comput. Med. Imaging Graph. **26**(4), 277–292 (2002)
7. Klein, S., Staring, M., Murphy, K., Viergever, M.A., Pluim, J.P.W.: Elastix: a toolbox for intensity-based medical image registration. IEEE Trans. Med. Imaging **29**(1), 196–205 (2010)
8. Lozano, M., Herrera, F., Krasnogor, N., Molina, D.: Real-coded memetic algorithms with crossover hill-climbing. Evolut. Comput. **12**(3), 273–302 (2004)

9. Monga, O., Benayoun, S., Faugeras, O.: From partial derivatives of 3-D density images to ridge lines. In: Proceedings 1992 IEEE Computer Society Conference on Computer Vision and Pattern Recognition, vol. 1808, pp. 354–359. IEEE, Champaign (1992)

10. Salcedo-Sanz, S.: A review on the coral reefs optimization algorithm: new development lines and current applications. Prog. Artif. Intell. **6**(1), 1–15 (2017)

11. Salcedo-Sanz, S., Del Ser, J., Landa-Torres, I., Gil-López, S., Portilla-Figueras, J.A.: The coral reefs optimization algorithm: a novel metaheuristic for efficiently solving optimization problems. Sci. World J. **2014**, 1–15 (2014)

12. Santamaría, J., Cordón, O., Damas, S., García-Torres, J., Quirin, A.: Performance evaluation of memetic approaches in 3D reconstruction of forensic objects. Soft Comput. **13**(8–9), 883–904 (2009)

13. Takahashi, M., Kita, H.: A crossover operator using independent component analysis for real-coded genetic algorithms. In: Proceedings of the 2001 Congress on Evolutionary Computation (IEEE Cat. No.01TH8546), vol. 1, pp. 643–649 (2001)

14. Valsecchi, A., Damas, S., Santamaria, J., Marrakchi-Kacem, L.: Genetic algorithms for voxel-based medical image registration. In: 2013 Fourth International Workshop on Computational Intelligence in Medical Imaging (CIMI), pp. 22–29, Aprril 2013

15. Valsecchi, A., Damas, S., Santamaría, J., Marrakchi-Kacem, L.: Intensity-based image registration using scatter search. Artif. Intell. Med. **60**(3), 151–163 (2014)

16. Zitová, B., Flusser, J.: Image registration methods: a survey. Image Vis. Comput. **21**(11), 977–1000 (2003)

Evaluating Feature Selection Robustness on High-Dimensional Data

Barbara Pes[(✉)]

Dipartimento di Matematica e Informatica, Università degli Studi di Cagliari,
Via Ospedale 72, 09124 Cagliari, Italy
pes@unica.it

Abstract. With the explosive growth of high-dimensional data, feature selection has become a crucial step of machine learning tasks. Though most of the available works focus on devising selection strategies that are effective in identifying small subsets of predictive features, recent research has also highlighted the importance of investigating the robustness of the selection process with respect to sample variation. In presence of a high number of features, indeed, the selection outcome can be very sensitive to any perturbations in the set of training records, which limits the interpretability of the results and their subsequent exploitation in real-world applications. This study aims to provide more insight about this critical issue by analysing the robustness of some state-of-the-art selection methods, for different levels of data perturbation and different cardinalities of the selected feature subsets. Furthermore, we explore the extent to which the adoption of an ensemble selection strategy can make these algorithms more robust, without compromising their predictive performance. The results on five high-dimensional datasets, which are representatives of different domains, are presented and discussed.

Keywords: Feature selection robustness · Ensemble techniques
High-dimensional data

1 Introduction

In the context of high-dimensional data analysis, feature selection aims at reducing the number of attributes (features) of the problem at hand, by removing irrelevant and redundant information as well as noisy factors, and thus facilitating the extraction of valuable knowledge about the domain of interest. The beneficial impact of feature selection on the performance of learning algorithms is widely discussed in the literature [1] and has been experimentally proven in several application areas such as bio-informatics [2], text categorization [3], intrusion detection [4] or image analysis [5].

There exists currently a large body of feature selection methods, based on distinct heuristics and search strategies, and several works have investigated their strengths and weaknesses on both real [6] and artificial data [7]. Most of the existing studies, however, concentrate on the effectiveness of the available algorithms in selecting small subsets of predictive features, without taking into account other relevant aspects that only recently have gained attention, such as the scalability [8], the costs associated to

© Springer International Publishing AG, part of Springer Nature 2018
F. J. de Cos Juez et al. (Eds.): HAIS 2018, LNAI 10870, pp. 235–247, 2018.
https://doi.org/10.1007/978-3-319-92639-1_20

the features [9] or the robustness (stability) of the selection process with respect to changes in the input data [10]. This last issue has been recognized to be especially important when the high-dimensionality of data is coupled with a comparatively small number of instances: in this setting, actually, even small perturbations in the set of training records may lead to strong differences in the selected feature subsets.

Though the literature on feature selection robustness is still limited, an increasing number of studies recognize that a robust selection outcome is often equally important as good model performance [11, 12]. Indeed, if the outcome of the selection process is too sensitive to variations in the set of training instances, the interpretation (and the subsequent exploitation) of the results can be very difficult, with limited confidence of domain experts and final users. Moreover, as observed in [13], the robustness of feature selection may have practical implications for distributed applications where the algorithm should produce stable results across multiple data sources.

Further research, from both a theoretical and empirical point of view, should be devoted to better characterizing the degree of robustness of state-of-art selection algorithms in multiple settings, in order to achieve a better understanding of their applicability/utility in knowledge discovery tasks. On the other hand, the definition of feature selection protocols which can ensure a better trade-off between robustness and predictive performance is still an open issue, though a number of studies [11, 14] seem to suggest that the adoption of an ensemble selection strategy can be useful in this regard.

To give a contribution to the field, this work presents a case study which aims to provide more insight about the robustness of six popular selection methods across high-dimensional classification tasks from different domains. Specifically, for each method, we evaluate the extent to which the selected feature subsets are sensitive to some amount of perturbation in the training data, for different levels of perturbation and for different cardinalities of the selected subsets.

In addition, for each selection algorithm, we implement an "ensemble version" whose output is built by a *bagging* procedure similar to that adopted in the context of multi-classifier systems [15], i.e. (i) different versions of the training set are created through a re-sampling technique, (ii) the feature selection process is carried out separately on each of these versions and (iii) the resulting outcomes are combined through a suitable aggregation function. The studies so far available on the robustness of this ensemble approach are limited to a single application domain [11, 16], to a single selection method [14] or to a given number of selected features [17], so it is worth providing the interested reader with a more comprehensive evaluation which encompasses different kinds of data, different selection heuristics (both univariate and multivariate) and different subset sizes.

The results of our experiments clearly show that, when comparing the overall performance of the considered selection methods, the differences in robustness can be significant, while the corresponding differences in accuracy (or other metrics, such as the AUC) are often null or negligible. In the choice of the best selector for a given task, hence, the degree of robustness of the selection outcome can be a discriminative criterion. At the same time, our study shows that the least stable methods can benefit, at least to some extent, from the adoption of an ensemble selection strategy.

The rest of this paper is organized as follows. Section 2 summarizes background concepts and related works. Section 3 describes all the materials and methods relevant to our study, i.e. the methodology used for the robustness analysis, the ensemble strategy and the selection algorithms here considered, and the datasets used as benchmarks. The experimental results are presented and discussed in Sect. 4. Finally, Sect. 5 gives the concluding remarks.

2 Background and Related Work

As discussed in [10], the robustness (or stability) of a given selection method is a measure of its sensitivity to changes in the input data: a robust algorithm is capable of providing (almost) the same outcome when the original set of records is perturbed to some extent, e.g. by adding or removing a given fraction of instances.

Recent literature has investigated the potential causes of selection instability [18] and has also focused on suitable methodologies [19] for evaluating the degree of robustness of feature selection algorithms. This evaluation basically involves two aspects: (a) a suitable protocol to generate a number of datasets, different to each other, which overlap to a great ("soft" perturbation) or small ("hard" perturbation) extent with the original set of records; (b) a proper consistency index to measure the degree of similarity among the outputs that are produced (in the form of feature weightings, feature rankings or feature subsets) when a given algorithm is applied to the above datasets. The higher the similarity, the more robust the selection method.

As regards the data perturbation protocols, simple re-sampling procedures are adopted in most cases, though some studies have investigated how to effectively measure and control the variance of the generated sample sets [13]. The influence of the amount of overlap between these sets is discussed by Wang et al. [20], who propose a method for generating two datasets of the same size with a specified degree of overlap.

As regards the similarity measure used to compare the selection outcomes, various approaches have been proposed [10, 21, 22], each expressing a slightly different view of the problem. For example, the *Pearson*'s correlation coefficient can be used if the output is given as a weighting of the features, the *Spearman*'s rank correlation coefficient if the output is a ranking of the features, the *Tanimoto* distance or the *Kuncheva* index if the output is a feature subset. A good review of stability measures can be found in [18].

From an experimental point of view, a number of studies have compared the robustness of different selection methods on high-dimensional datasets [23–25]. This work extends and complements the available studies by encompassing different application domains; besides, stability patterns are derived for feature subsets of different cardinalities and for different levels of data perturbation. As a further contribution, we investigate the impact, in terms of selection robustness, of using an ensemble selection strategy; though presented as a promising approach to achieve more stable results, indeed, it has been so far evaluated in a limited number of settings, particularly with biomedical/genomic data [11, 14, 16, 17].

3 Materials and Methods

In our study we focus on selection techniques that provide, as output, a feature ranking, i.e. a list (usually referred as *ranked list*) where the available features appear in descending order of relevance. In turn, the ranked list can be cut at a proper threshold point to obtain a subset of highly predictive features. In the context of high-dimensional problems, indeed, this ranking-based approach is a de facto standard to reduce the dimensionality of the feature space; then, the filtered space can be either refined through more sophisticated (and computationally expensive) techniques or directly used for predictive and knowledge discovery purposes.

The robustness of six popular ranking techniques is here evaluated in a two-fold setting (simple and ensemble ranking), according to the methodology presented in Subsect. 3.1; next, Subsect. 3.2 provides some details on the chosen techniques and describes the datasets used as benchmarks and the specific settings of the experiments.

3.1 Methodology for Robustness Evaluation: Simple vs Ensemble Ranking

Leveraging on best practices from the literature, we evaluate the robustness of the selection process in conjunction with the predictive performance of the selected subsets. Both the aspects, indeed, must be taken into account when assessing the suitability of a given selection approach (actually, stable but not accurate solutions would be not meaningful; on the other hand, accurate but not stable results could have limited utility for domain experts and final users).

In more detail, given the input dataset, we repeatedly perform random sampling (without replacement) to create m different training sets, each containing a fraction f of the original records. For each training set, a test set is also formed using the remaining fraction $(1 - f)$ of the instances. The feature selection process is then carried out in a two-fold way:

– *Simple ranking.* A given ranking method is applied separately on each training set to obtain m distinct ranked lists which in turn produce, when cut at a proper threshold (t), m different feature subsets (here referred as simple subsets).
– *Ensemble ranking.* An ensemble version of the same ranking method is implemented using a bagging-based approach, i.e. each training set is in turn sampled (with replacement) to construct b samples of the same size (*bootstraps*). The considered ranking method is then applied to each bootstrap, which results in b distinct ranked lists that are finally combined (through a mean-based aggregation function [26]) into a single ensemble list. In turn, this list is cut at a proper threshold (t) to obtain an ensemble subset of highly discriminative features. Overall, m ensemble subsets are selected, one for each training set.

For both the simple and the ensemble setting, the robustness of the selection process is measured by performing a similarity analysis on the resulting m subsets.

Specifically, for each pair of subsets S_i and S_j ($i, j = 1, 2, ..., m$), we use a proper consistency index [21] to quantify their degree of similarity:

$$sim_{ij} = \left(\left|S_i \cap S_j\right| - t^2/n\right)/\left(t - t^2/n\right) \qquad (1)$$

where t is the size of the subsets (corresponding to the cut-off threshold) and n the overall number of features. Basically, the similarity sim_{ij} expresses the degree of overlapping between the subsets, i.e. the fraction of features which are common to them ($\left|S_i \cap S_j\right|/t$), with a correction term reflecting the probability that a feature is included in both subsets simply by chance. The need for this correction, which increases as the subset size approaches the total number of features, is experimentally demonstrated for example in [27]. The resulting similarity values are then averaged over all pair-wise comparisons, in order to evaluate the overall degree of similarity among the m subsets and, hence, the robustness of the selection process.

At the same time, in both simple and ensemble settings, a classification model is built on each training set using the selected feature subset, and the model performance is measured (through suitable metrics such as accuracy and AUC) on the corresponding test set. By averaging the accuracy/AUC of the resulting m models, we can obtain an estimate of the effectiveness of the applied selection approach (simple or ensemble) in identifying the most discriminative features. This way, the trade-off between robustness and predictive performance can be evaluated for different values of the cut-off threshold.

3.2 Ranking Techniques, Datasets and Settings

The above methodology can be applied in conjunction with any ranking method. To obtain useful insight on the robustness of different selection approaches, as well as on the extent to which the ensemble implementation affects their outcome, we included in our study six algorithms that are representatives of quite different heuristics. In particular, we considered three univariate methods (*Symmetrical Uncertainty, Gain Ratio* and *OneR*), which evaluate each feature independently from the others, and three multivariate methods (*ReliefF, SVM-AW* and *SVM-RFE*) which take into account the inter-dependencies among the features. More details on these techniques and their pattern of agreement can be found in [28]. In brief:

- *Symmetrical Uncertainty* (SU) and *Gain Ratio* (GR) both leverage the concept of information gain, that is a measure of the extent to which the class entropy decreases when the value of a given feature is known. The SU and GR definitions differ for the way they try to compensate for the information gain's bias toward features with more values.
- *OneR* (OR) ranks the features based on the accuracy of a rule-based classifier that constructs a simple classification rule for each feature.
- *ReliefF* (RF) evaluates the features according to their ability to differentiate between data points that are near to each other in the attribute space.
- *SVM_AW* exploits a linear *Support Vector Machine* (SVM) classifier, which has an embedded capability of assigning a weight to each feature (based on the

contribution the feature gives to the decision function induced by the classifier); the absolute value of this weight (AW) is used to rank the features.

- *SVM_RFE*, in turn, relies on a linear SVM classifier, but adopts a recursive feature elimination (RFE) strategy that iteratively removes the features with the lowest weights and repeats the overall weighting process on the remaining features (the percentage of features removed at each iteration is 50% in our implementation).

Each of the above methods has been applied, in its simple and ensemble version, on five high-dimensional datasets, chosen to be representatives of different domains. In particular:

- The *Gastrointestinal Lesions* dataset [29] contains 1396 features extracted from a database of colonoscopy videos; there are 76 instances of lesions, distinguished in 'hyperplasic', 'adenoma' and 'serrated adenoma'.
- The *Voice Rehabilitation* dataset [30] contains 310 features resulting from the application of speech processing algorithms to the voices of 126 Parkinson's disease subjects, who followed a rehabilitative program with 'acceptable' or 'unacceptable' results.
- The *DLBCL Tumour* dataset [31] contains 77 samples, including 'follicular lymphoma' and 'diffuse large b-cell lymphoma' samples, each described by the expression level of 7129 genes.
- The *Ovarian Cancer* dataset [32] contains 15154 features describing proteomic spectra generated by mass spectrometry; the instances are 253, divided in 'normal' and 'cancerous'.
- The *Arcene* dataset, in turn, is a binary classification problem where the task is to distinguish 'cancerous' versus 'normal' patterns from mass spectrometric data. Unlike the *Ovarian Cancer* dataset, it results from the combination of different data sources; a number of noisy features, having no predictive power, were also added in order to provide a challenging benchmark for the NIPS 2003 feature selection challenge [33]. The overall dimensionality is 10000, while the number of instances is 200.

Note that all the above datasets are characterized by a large number of features and a comparatively small number of records, which makes it difficult to achieve a good trade-off between predictive performance and robustness.

According to the methodology described in Subsect. 3.1, different training/test sets have been built for each dataset; specifically, we set $m = 20$. As regards the amount of data perturbation, i.e. the fraction of the original instances randomly included in each training set, we explored the values $f = 0.70$, $f = 0.80$ and $f = 0.90$. For the number of bootstraps involved in the construction of the ensemble subsets, we also explored different values, i.e. $b = 20$, $b = 50$ and $b = 80$. Further, for both the simple and the ensemble subsets, different values of the cut-off threshold (i.e. different subset sizes) have been considered, ranging from 0.5% to 7% of the original number of features.

4   Experimental Study: Results and Discussion

In this section, we summarize the main results of our robustness analysis. First, it is interesting to consider the effect of varying the amount of perturbation introduced in the input data, i.e., in our setting, the effect of including in the training sets only a fraction f of the original records. Limited to the simple ranking, Fig. 1 shows the robustness of the six selection methods here considered (SU, GR, OR, RF, SVM-AW, SVM-RFE) on the *Gastrointestinal Lesions* dataset, for different values of f and different subset sizes.

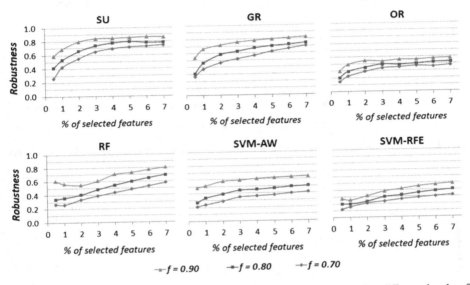

Fig. 1. *Gastrointestinal Lesions* dataset: robustness of simple ranking, for different levels of data perturbation ($f = 0.90, f = 0.80, f = 0.70$)

As we can see, even a small amount of perturbation ($f = 0.90$) affects the stability of the selection outcome in a significant way, since the average similarity among the 20 feature subsets (selected from the $m = 20$ training sets built from the original dataset) is far lower than the maximum value of 1. As the amount of perturbation increases, the degree of robustness dramatically falls off, for all the selection methods, though some of them exhibit a somewhat better behaviour. Similar considerations can be made for the other datasets here considered (whose detailed results are omitted for the sake of space), thus confirming that the instability of the selection outcome is a very critical concern when dealing with high-dimensional problems.

A further point to be discussed is the extent to which the adoption of an ensemble strategy improves the robustness of the selection process. Figs. 2, 3, 4, 5 and 6 show, for the five datasets included in our study, the stability of both the simple and the ensemble subsets, with a data perturbation level of $f = 0.80$. In particular, for each selection method, three ensembles have been implemented with different numbers of bootstraps ($b = 20, b = 50, b = 80$), but only the results for $b = 20$ (*20b-ensemble*) and $b = 50$ (*50b-ensemble*) have been reported here, since a higher value of b does not further improve the robustness in an appreciable way.

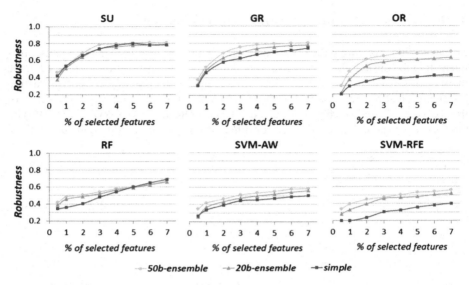

Fig. 2. *Gastrointestinal Lesions* dataset: robustness of simple and ensemble ranking ($f = 0.80$)

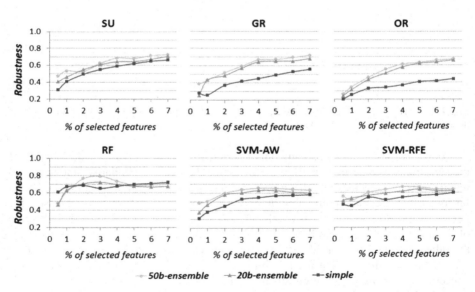

Fig. 3. *Voice Rehabilitation* dataset: robustness of simple and ensemble ranking ($f = 0.80$)

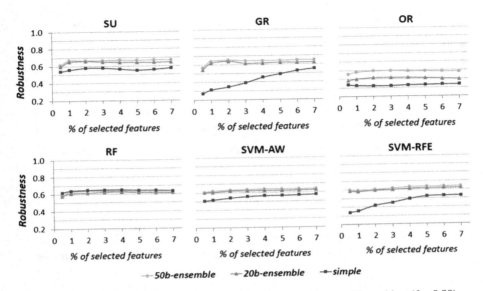

Fig. 4. *DLBCL Tumour* dataset: robustness of simple and ensemble ranking ($f = 0.80$)

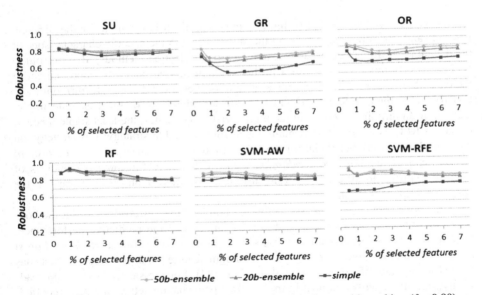

Fig. 5. *Ovarian Cancer* dataset: robustness of simple and ensemble ranking ($f = 0.80$)

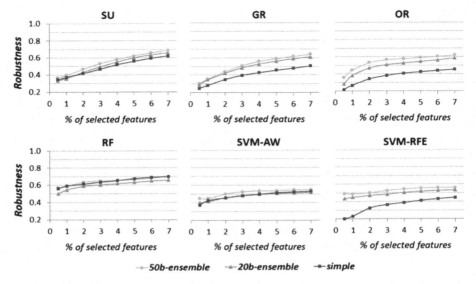

Fig. 6. *Arcene* dataset: robustness of simple and ensemble ranking ($f = 0.80$)

As we can see, the impact of the ensemble approach is different for the different methods and varies in dependence on the subset size and the specific characteristics of the data at hand. In particular, among the univariate selection methods, SU turns out to be intrinsically more robust, with a further (though limited) stability improvement in the ensemble version. The other univariate approaches, i.e. GR and OR, turn out to be less robust in their simple form and take greater advantage of the ensemble implementation. In turn, in the group of the multivariate approaches, the least stable method, i.e. SVM-RFE, is the one that benefits most from ensemble strategy; this strategy, on the other hand, is not beneficial for the RF method, except that in the *Gastrointestinal Lesions* and in the *Voice Rehabilitation* datasets, but only for some percentages of selected features. In all cases, it is not useful to use more than 50 bootstraps in the ensemble implementation.

The above robustness analysis has been complemented, according to the methodology presented in Subsect. 3.1, with a joint analysis of the predictive performance. Specifically, the selected feature subsets have been used to train a *Random Forest* classifier (parameterized with $\log_2(t) + 1$ random features and 100 trees), which has proved to be very effective in several domains [34]. For the sake of space and readability, only the results obtained in the $f = 0.80$ perturbation setting are here reported; specifically, Table 1 summarizes the AUC performance (averaged over the $m = 20$ training/test sets) achieved with both the simple and the 50b-ensemble subsets, limited to a threshold $t = 5\%$ of the original number of features (but the AUC results obtained with feature subsets of different cardinalities confirm what observed in Table 1).

Table 1. AUC analysis ($f = 0.80$, $b = 50$, $t = 5\%$ of n)

		SU	GR	OR	RF	SVM-AW	SVM-RFE
Gastrointestinal Lesions dataset	Simple	0.797	0.806	0.790	0.800	0.785	0.773
	Ensemble	0.778	0.790	0.795	0.814	0.784	0.781
Voice Rehabilitation dataset	Simple	0.870	0.856	0.857	0.912	0.884	0.904
	Ensemble	0.880	0.868	0.860	0.908	0.905	0.911
DLBCL Tumour dataset	Simple	0.960	0.956	0.955	0.981	0.987	0.982
	Ensemble	0.957	0.956	0.960	0.988	0.985	0.982
Ovarian Cancer dataset	Simple	1.000	1.000	1.000	1.000	1.000	1.000
	Ensemble	1.000	1.000	1.000	1.000	1.000	1.000
Arcene dataset	Simple	0.820	0.817	0.858	0.845	0.745	0.831
	Ensemble	0.819	0.809	0.859	0.831	0.803	0.814

When comparing the overall performance of the six selection methods, in their simple form, it is clear that the differences in AUC are much smaller (and often negligible) than the corresponding differences in robustness. In cases like these, where the AUC/accuracy is not a discriminative factor, the outcome stability can then be assumed as a decisive criterion for the choice of the best selector.

A further important observation is that no significant difference exists between the AUC performance of the simple and the ensemble version of the considered selection methods. Indeed, irrespective of the application domain, each selection algorithm achieves almost the same AUC outcome in both the implementations. When looking at the trade-off between the predictive performance and the robustness of the selection process, we can then conclude that the adoption of an ensemble strategy can lead to more stable feature subsets without compromising at all the predictive power of these subsets.

5 Conclusions

This work emphasized the importance of evaluating the robustness of the selection process, besides the final predictive performance, when dealing with feature selection from high-dimensional data. The stability of the selection outcome, indeed, is important for practical applications and can be a useful (and objective) criterion to guide the choice of the proper selection method for a given task. Further, the proposed study contributed to demonstrate that the adoption of an ensemble selection strategy can produce better results even in those domains where the selection of robust subsets is intrinsically harder, due to a very low instances-to-features ratio. The beneficial impact of the ensemble approach is more significant for the selection methods that turn out to be less stable in their simple form (e.g., the univariate Gain Ratio and the multivariate SVM-RFE). Actually, the stability gap between the different methods tend to become much smaller (or sometimes null) when they are used in the ensemble version. This is noteworthy for practitioners and final users that, in the ensemble setting, could exploit different, but equally robust, selection methods.

Acknowledgments. This research was supported by Sardinia Regional Government, within the projects "DomuSafe" (L.R. 7/2007, annualità 2015, CRP 69) and "EmILIE" (L.R. 7/2007, annualità 2016, CUP F72F16003030002).

References

1. Guyon, I., Elisseeff, A.: An introduction to variable and feature selection. J. Mach. Learn. Res. **3**, 1157–1182 (2003)
2. Saeys, Y., Inza, I., Larranaga, P.: A review of feature selection techniques in bioinformatics. Bioinformatics **23**(19), 2507–2517 (2007)
3. Forman, G.: An extensive empirical study of feature selection metrics for text classification. J. Mach. Learn. Res. **3**, 1289–1305 (2003)
4. Bolón-Canedo, V., Sánchez-Maroño, N., Alonso-Betanzos, A.: Feature selection and classification in multiple class datasets: an application to kdd cup 99 dataset. Expert Syst. Appl. **38**(5), 5947–5957 (2011)
5. Staroszczyk, T., Osowski, S., Markiewicz, T.: Comparative analysis of feature selection methods for blood cell recognition in leukemia. In: Proceedings of the 8th International Conference on Machine Learning and Data Mining in Pattern Recognition, pp. 467–481 (2012)
6. Tang, J., Alelyani, S., Liu, H.: Feature selection for classification: a review. In: Aggarwal, C. C. (ed.) Data Classification: Algorithms and Applications, pp. 37–64. CRC Press, Boca Raton (2014)
7. Bolón-Canedo, V., Sánchez-Maroño, N., Alonso-Betanzos, A.: A review of feature selection methods on synthetic data. Knowl. Inf. Syst. **34**(3), 483–519 (2013)
8. Bolón-Canedo, V., Rego-Fernández, D., Peteiro-Barral, D., Alonso-Betanzos, A., Guijarro-Berdiñas, B., Sánchez-Maroño, N.: On the scalability of feature selection methods on high-dimensional data. Knowl. Inf. Syst. 1–48 (2018). https://link.springer.com/article/10.1007/s10115-017-1140-3
9. Maldonado, S., Pérez, J., Bravo, C.: Cost-based feature selection for support vector machines: an application in credit scoring. Eur. J. Oper. Res. **261**(2), 656–665 (2017)
10. Kalousis, A., Prados, J., Hilario, M.: Stability of feature selection algorithms: a study on high-dimensional spaces. Knowl. Inf. Syst. **12**(1), 95–116 (2007)
11. Saeys, Y., Abeel, T., Van de Peer, Y.: Robust feature selection using ensemble feature selection techniques. In: Daelemans, W., Goethals, B., Morik, K. (eds.) ECML PKDD 2008, Part II. LNCS (LNAI), vol. 5212, pp. 313–325. Springer, Heidelberg (2008). https://doi.org/10.1007/978-3-540-87481-2_21
12. Pes, B.: Feature selection for high-dimensional data: the issue of stability. In: 26th IEEE International Conference on Enabling Technologies: Infrastructure for Collaborative Enterprises, WETICE 2017, pp. 170–175 (2017)
13. Alelyani, S., Zhao, Z., Liu, H.: A dilemma in assessing stability of feature selection algorithms. In: IEEE 13th International Conference on High Performance Computing and Communications, pp. 701–707 (2011)
14. Abeel, T., Helleputte, T., Van de Peer, Y., Dupont, P., Saeys, Y.: Robust biomarker identification for cancer diagnosis with ensemble feature selection methods. Bioinformatics **26**(3), 392–398 (2010)
15. Dietterich, T.: Ensemble methods in machine learning. In: Proceedings of the 1st International Workshop on Multiple Classifier Systems, pp. 1–15 (2000)

16. Kuncheva, L.I., Smith, C.J., Syed, Y., Phillips, C.O., Lewis, K.E.: Evaluation of feature ranking ensembles for high-dimensional biomedical data: a case study. In: IEEE 12th International Conference on Data Mining Workshops, pp. 49–56. IEEE (2012)

17. Haury, A.C., Gestraud, P., Vert, J.P.: The influence of feature selection methods on accuracy, stability and interpretability of molecular signatures. PLoS ONE 6(12), e28210 (2011)

18. Zengyou, H., Weichuan, Y.: Stable feature selection for biomarker discovery. Comput. Biol. Chem. 34, 215–225 (2010)

19. Awada, W., Khoshgoftaar, T.M., Dittman, D., Wald, R., Napolitano, A.: A review of the stability of feature selection techniques for bioinformatics data. In: IEEE 13th International Conference on Information Reuse and Integration, pp. 356–363. IEEE (2012)

20. Wang, H., Khoshgoftaar, T.M., Wald, R., Napolitano, A.: A novel dataset-similarity-aware approach for evaluating stability of software metric selection techniques. In: Proceedings of the IEEE International Conference on Information Reuse and Integration, pp. 1–8 (2012)

21. Kuncheva, L.I.: A stability index for feature selection. In: 25th IASTED International Multi-Conference: Artificial Intelligence and Applications, pp. 390–395. ACTA Press Anaheim (2007)

22. Somol, P., Novovicova, J.: Evaluating stability and comparing output of feature selectors that optimize feature subset cardinality. IEEE Trans. Pattern Anal. Mach. Intell. 32(11), 1921–1939 (2010)

23. Dessì, N., Pascariello, E., Pes, B.: A comparative analysis of biomarker selection techniques. BioMed. Res. Int. 2013, Article ID 387673 (2013)

24. Drotár, P., Gazda, J., Smékal, Z.: An experimental comparison of feature selection methods on two-class biomedical datasets. Comput. Biol. Med. 66, 1–10 (2015)

25. Wang, H., Khoshgoftaar, T.M., Seliya, N.: On the stability of feature selection methods in software quality prediction: an empirical investigation. Int. J. Soft. Eng. Knowl. Eng. 25, 1467–1490 (2015)

26. Wald, R., Khoshgoftaar, T.M., Dittman, D.: Mean aggregation versus robust rank aggregation for ensemble gene selection. In: 11th International Conference on Machine Learning and Applications, pp. 63–69 (2012)

27. Cannas, L.M., Dessì, N., Pes, B.: Assessing similarity of feature selection techniques in high-dimensional domains. Pattern Recogn. Lett. 34(12), 1446–1453 (2013)

28. Dessì, N., Pes, B.: Similarity of feature selection methods: an empirical study across data intensive classification tasks. Expert Syst. Appl. 42(10), 4632–4642 (2015)

29. Mesejo, P., Pizarro, D., Abergel, A., Rouquette, O., et al.: Computer-aided classification of gastrointestinal lesions in regular colonoscopy. IEEE Trans. Med. Imaging 35(9), 2051–2063 (2016)

30. Tsanas, A., Little, M.A., Fox, C., Ramig, L.O.: Objective automatic assessment of rehabilitative speech treatment in Parkinson's disease. IEEE Trans. Neural Syst. Rehabil. Eng. 22, 181–190 (2014)

31. Shipp, M.A., Ross, K.N., Tamayo, P., Weng, A.P., et al.: Diffuse large B-cell lymphoma outcome prediction by gene-expression profiling and supervised machine learning. Nat. Med. 8(1), 68–74 (2002)

32. Petricoin, E.F., Ardekani, A.M., Hitt, B.A., Levine, P.J., et al.: Use of proteomic patterns in serum to identify ovarian cancer. Lancet 359, 572–577 (2002)

33. Guyon, I., Gunn, S.R., Ben-Hur, A., Dror, G.: Result analysis of the NIPS 2003 feature selection challenge. In: Advances in Neural Information Processing Systems, vol. 17, pp. 545–552. MIT Press (2004)

34. Rokach, L.: Decision forest: twenty years of research. Inf. Fusion 27, 111–125 (2016)

Learning Algorithms

Generalized Probability Distribution Mixture Model for Clustering

David Crespo-Roces[1], Iván Méndez-Jiménez[1], Sancho Salcedo-Sanz[2], and Miguel Cárdenas-Montes[1(✉)]

[1] Department of Fundamental Research, Centro de Investigaciones Energéticas Medioambientales y Tecnológicas, Madrid, Spain
david.crespo.roces@gmail.com,
{ivan.mendez,miguel.cardenas}@ciemat.es
[2] University of Alcalá, Alcalá de Henares, Madrid, Spain
sancho.salcedo@uah.es

Abstract. Gaussian Mixture Model is a popular clustering method based on modelling the data through a set of Gaussian probability distributions. This method is able to correctly identify non-spheroidal overlapping or not overlapping clusters with different sizes and densities. Due to its performance, it has achieved a high popularity among the practitioners. In this work, the first efforts for extending Gaussian Mixture Models toward the mixture of other probability distributions are presented. At this point, it includes the use of diverse probability distributions, such as Gaussian, Exponential, Weibull, and t-Student Probability Distributions. Instead of the Expectation-Maximization algorithm, for optimizing the parameters of the Gaussian mixture, the parameters of the mixture of diverse probability distributions are optimized using the Coral Reef Optimizer meta-heuristic.

Keywords: Coral Reef Optimizer · Clustering · Optimization
Probability distribution · Gaussian Mixture Model

1 Introduction

In [1], the authors define the *model-based clustering methods* as "Model-based clustering methods attempt to optimize the fit between the given data and some mathematical model." In *Gaussian Mixture Model* (GMM), the data are modelled based on a set of Gaussian probability distributions with different parameters: mean and covariance matrix. Objects modelled through GMM have a non-null probability to be in more than a cluster. The model assumes that the objects are generated by a mixture of probability distributions. Each individual distribution of the mixture is termed as *component distribution*. Although some flexibility is inherent to the model, it is restricted to the variation of parameters of a Gaussian probability distributions.

© Springer International Publishing AG, part of Springer Nature 2018
F. J. de Cos Juez et al. (Eds.): HAIS 2018, LNAI 10870, pp. 251–263, 2018.
https://doi.org/10.1007/978-3-319-92639-1_21

Clustering methods can be employed in two main tasks. On the one hand for inferring the similarity degree among a set of examples [1]; and on the other hand for finding outliers [2]. GMM can be applied for both.

In the current work, a *Generalized Probability Distribution Mixture Model* (GPDMM), as extension of GMM is proposed. This extension aims at widening the probability distributions handled by the clustering method. This modification provides an increasing potential for producing a higher-quality fitting between the model emerging from the clustering method and the data.

Usually, GMM is linked to Expectation-Maximization (EM) algorithm to find the optimal values for the unknown parameters (see Sect. 3.2). Given a set of objects and a GMM configuration, the EM aims at estimating the GMM parameters which better fit the probability distributions mixture and the objects. In practice, the most commonly used fitness function is the maximization of the likelihood (Sect. 3.4). In the current approach, the EM algorithm is replaced by the Coral Reef Optimizer (CRO) for selecting both the most suitable probability distributions mixture and the optimal parameters of the distributions (Sect. 3.1). The CRO is a bio-inspired metaheuristic [3] based on the simulation of the processes in coral reefs, such as coral reproduction, growing, fighting for space in the reef or depredation. The CRO has the capacity to fit not only the parameters of a model for the data, but also to select among a set of models the most suitable combination of components.

To the authors' knowledge, a similar extension of GMM-EM algorithm for clustering has not been proposed in the past.

The rest of the paper is organized as follows: Sect. 2 summarizes the Related Work and previous efforts done. The methodology is described in Sect. 3. The Results and the Analysis are shown in Sect. 4. Finally, Conclusions and Future Work are presented in Sect. 5.

2   Related Work

Among the works aimed at modifying the couple GMM-EM, in [4] a genetic-based expectation-maximization algorithm for learning Gaussian mixture models from multivariate data is proposed. The genetic-based expectation-maximization algorithm allows escaping from local minima. The proposed implementation profits from the capacity of GA for exploring the search space, in such a way, it overcomes one of the flaw of the EM. In this work, the modified GMM incorporates the capacity for selecting the most suitable model among a set of optimized models. This is done through the minimum description length criterion.

In [5] a greedy method to optimize the Gaussian mixture mode is proposed. In the proposal, the model starts with an unique Gaussian probability distribution. Next it incorporates one by one the additional distributions until the convergence is reached. In [6], a review of the mixtures of probability distributions and their properties is presented. Also it introduces the difficulty associated to find suitable parameters in multivariate Gaussian mixture. Some efforts to this purpose can be found in [5, 7].

CRO algorithm is introduced in [3]. In [8], the CRO algorithm is applied to solve a Mobile Network Deployment Problem (MNDP), in which the control of the electromagnetic pollution plays an important role.

An in-depth description of the probability distributions used in this work can be found in [9].

3   Methodology

3.1   The Coral Reef Optimizer

The CRO is an evolutionary-type metaheuristic inspired by the reproduction of the corals and the coral reefs formation [10]. The essential of this algorithm is as follows:

- The coral reef (population of candidate solutions), Λ, is modelled as a grid of $N \times M$.
- Each position of the grid (i, j) can hold a coral, $\Omega_{i,j}$.
- Initially a set of randomly generated solutions partially populate the reef, whereas the remaining positions of the reef are kept empty. The ratio of positions initially populated and empty positions is a parameter of the algorithm, ρ, $0 < \rho < 1$.
- For each candidate solutions or corals, the fitness is calculated.

After the reef initialization, the reproductive cycle starts. Following the nature inspiration of the algorithm, two sexual reproduction mechanisms (broadcast spawning and brooding) and the asexual reproduction one (budding) of the corals are simulated. Additionally to the reproduction, the depredation of the corals is also simulated.

In broadcast spawning (external sexual reproduction), a fraction, F_b, of the existing corals, ρ_k, is randomly selected. For each couple of corals, sexual crossover is performed, and a larva is generated. A larva is a new candidate solution. Corals from the fraction F_b can not be selected more than one for sexual crossover. The generated larvae are released out to the water.

The fraction of the corals not selected for broadcast spawning, $1 - F_b$, is selected for brooding (internal sexual reproduction). Brooding consists of a random mutation of the coral for generating a new larva, which is also release out to the water.

Once the reproduction has taken place, the *larvae setting* phase starts. The larvae generated in the previous reproductive steps, try to randomly setting in a position of the reef. If the position of the grid is empty, then the larva is settled in the position independently of its fitness. Oppositely, if the position is already occupied, the new larva will set only if it is better suited for the problem, i.e. it has a better fitness than the coral in the position. A restriction for this last step is established. To avoid the dominance of a single coral type, a maximum number of the identical type is configured. The larvae have κ tries to be part of the reef. After these number of tries, they are depredated.

Next an asexual reproduction process is performed. The corals in the reef duplicate itself, at the same time that a mutation is undertaken, and try to establish in the reef following the same process described in the previous paragraph.

Finally, corals in the reef coral can be depredated. This process is performed over the less suitable corals. This allows liberating space for further generations.

Some constraints have been implemented in the application of CRO to GPDMM. For example, the minimum number of corals of a given type in the reef is two. A lower number of corals of a type impedes the reproductive function. The reef size used in this study is a bi-dimensional grid of 12×12 positions. The broadcast probability is $F_b = 0.9$, and the probability for asexual reproduction is $F_a = 1 - F_b = 0.1$. Number of opportunities for a new coral to settle in the reef is $\kappa = 3$. The depredation operator is applied with a probability $P_d = 0.01$ to a fraction $F_d = 0.1$ of the worse corals. The ratio of occupied corals to total corals at the beginning of CRO is 0.6. And the number of generations in the optimization process is 250.

3.2   Gaussian Mixture

Gaussian Mixture Models (GMM) are probabilistic models constructed with the mixture of a set of Gaussian probability distribution (Eq. 1). The mixture is a weighted sum of terms. Each term of the sum is composed of a weight w_i, and a Gaussian function $N(\boldsymbol{x}|\mu_i, \Sigma_i)$.

$$p(\boldsymbol{x}|w_i, \boldsymbol{\mu_i}, \Sigma_i) = \sum_{i=1}^{M} w_i \cdot N(\boldsymbol{x}|\boldsymbol{\mu_i}, \Sigma_i) \tag{1}$$

The weights, w_i, corresponds to the probability of the point i to belong to the distribution $N(\boldsymbol{x}|\mu_i, \Sigma_i)$. The accumulated probability for any point to be in the set of distribution should be the unit, $\sum_{i=1}^{M} w_i = 1$, where $w_i \geq 0$. Through w_i meaning, each object has a certain probability to be member of a given cluster.

Each component of the weighted sum is a Gaussian function with dimensionality of the problem's dimensionality (Eq. 2).

$$N(\boldsymbol{x}|\boldsymbol{\mu_i}, \Sigma_i) = \frac{1}{(2\pi)^{D/2}|\Sigma_i|^{1/2}} \cdot e^{-\frac{(x-\mu_i)^T \Sigma_i^{-1}(x-\mu_i)}{2}} \tag{2}$$

where D is the dimensionality of the problem, $\boldsymbol{\mu_i}$ is the vector of the mean of the distribution, and Σ_i is the covariance matrix.

The flexibility of GMM holds on the variety of covariances types: spherical, diagonal, full or tied. In the current work, the proposed algorithm aims at increasing the flexibility in the distribution choice, being the GMM a case of the algorithm GPDMM.

GMM has been frequently used for unsupervised learning and outlier detection. Their strengths include its high speed, and the low tend to be biased.

Oppositely, among its weaknesses it can cited the need to declare the number of clusters, which requires a priori knowledge of the data.

Usually GMM appears associated to Expectation-Maximization (EM) algorithm for fitting the values of their parameters, $\{w_i, \mu_i, \Sigma_i\}$. EM is an iterative algorithm which after the random initialization follows two steps: Expectation and Maximization [1]. During the Expectation step, each object x is assigned to a cluster C_k with a certain probability which depends on the object and the cluster's parameters, $\{\mu, \Sigma\}$, (Eq. 3). Next, in the Maximization step, the weight of the objects are recalculated (Eq. 4).

$$P(x_i \in C_k) = p(C_k | x_i) = \frac{p(C_k)p(x_i | C_k)}{p(x_i)} \qquad (3)$$

$$w_i = \frac{1}{n} \sum_{i=1}^{n} \frac{x_i P(x_i \in C_k)}{\sum_j P(x_i \in C_j)} \qquad (4)$$

EM algorithm has a fast convergence, however the convergence to the global minimum is not guaranteed, it can fall in local minimum. In order to avoid this flaw, usually the implementations make more than one start for selecting the most suitable minimum, and therefore, for approaching to the global minimum.

By ending the process, a model for describing the data set is produced. This model provides a probabilistic value for the association of each object to each cluster. This information can be used for stating the probability of an object belongs to a certain cluster, or for labelling as outlier the objects with low-probability of belonging to all the clusters as outliers.

Mixture models assume a specific probability distribution, and in practice, it is the Gaussian one. Note that this model is simple, and then it is easy and fast to optimize its parameters. However, it might lead to an low-quality fitting between the model and the data.

On the other hand, higher-complexity models, involving a larger number of parameters, can lead to harder-to-learn models. An example of this scenario can be found when using a GMM in a high-dimensional space if the covariance matrix, Σ, is defined as full covariance matrix or only the diagonal matrix. The scenario can be simpler and faster to learn, if the distribution is considered as a product of diverse one-dimensional Gaussian distributions. In this last scenario, the correlation among the different dimensions is ignored. In GPDMM, the richness of the model is maintained, at the same time that the use of the CRO allows obtaining high-quality fitting to the data within a reasonable processing time. Since the EM is an iterative model, it becomes slow when dealing with large data sets. The replacement of the EM by the CRO meta-heuristic also allows overcome this penalty. Note that there are some previous works to infer the parameters of the mixture by introducing evolutionary algorithms in the EM algorithm [11].

3.3   Structure of the Candidate Solutions

In order to model a set of objects based on a catalogue of probability distributions, it is necessary to keep information about the distribution as well as its parameters. A possible structure for the candidate solutions might be a set of identifiers for the probability distribution plus the values of the parameters of these distributions: $\{distribution\,identifiers|distribution\,parameters\}$. The total number of probability distributions identifiers correspond with the number of clusters.

As an example, for a problem in which the objects are modelled based on two clusters[1], a solution composed of two Exponential probability distributions holds a structure as $\{exp, exp|\lambda, \lambda\}$; whereas for a solution composed of an Exponential and a Weibull probability distribution $\{exp, wei|\lambda, shape, scale\}$.

In adopting this structure for the solutions, two different types of optimizations have to be performed: discrete and continuous one. The discrete optimization arises from the selection for the probability distribution, whereas the continuous one from the optimization of the parameters of the selected distribution. The election of the CRO as optimizer algorithm for this problem is based on its capacity for finding the most suitable combination of probability distributions and their parameters. Therefore, the CRO has the capacity to deal with the discrete and the continuous optimization processes.

The data structure for holding this information consists of a list of corals, each one stores the number of components for each type of distribution, their parameters, and the fitness of the coral.

3.4   Fitness Function

Similarly to GMM, for GPDMM, and based on the independence of the objects, the likelihood, $\mathcal{L}$, can be used as basement of the fitness function. The likelihood is also used as fitness function in GMM (Eq. 5).

$$p(\mathcal{X}|\boldsymbol{\mu}, \sigma) = \prod_{t=1}^{T} p(\boldsymbol{x}_t|\boldsymbol{\mu}, \boldsymbol{\sigma}) \tag{5}$$

For GPDMM the Eq. 5 can be modified, in such a way that the Gaussian distribution, $p(\boldsymbol{x}_t|\boldsymbol{\mu}, \boldsymbol{\sigma})$, is replaced by the probability distribution pointed by the identifier of the candidate solution. Finally, to get a fitness function able to be minimized, the negative logarithm of the likelihood is proposed (Eq. 6). In Eq. 6 the probability of the object for a particular probability distribution, $p(\boldsymbol{x}_t)$, is weighted with the probability of belonging of the object to the distribution, w_i.

$$\mathcal{F} = -log(\mathcal{L}) = -log\left(\prod_t p(\boldsymbol{x}_t)\right) = -\sum_t log\left(\sum_k w_k p(\boldsymbol{x}_t|dist_k)\right) \tag{6}$$

---

[1] In the current state of this work, the structure of candidate solutions is restricted to two identical distributions. Additional efforts are undertaken for eliminating this constraint, allowing individuals composed of two different probability distributions.

The methodology described in this section has been implemented by using Python 2.7.12, and the numerical experiments executed at a CPU i7-6700T at 2.80 GHz.

4   Results and Discussion

The initial tests conducted aim at evaluating the capacity of GPDMM to correctly recover the parameter of previously generated synthetic data sets. For this purpose, diverse synthetic data sets are generated and later evaluated. They are generated from two bi-dimensional probability distributions of a single type, for example two Gaussian distributions, each one containing 100 points.

The first data set is generated with two Gaussian probability distributions, $f(x; \eta, \sigma) = \frac{1}{\sigma\sqrt{2\pi}} e^{-\frac{1}{2}\left(\frac{x-\mu}{\sigma}\right)^2}$, with $\mu_x = \mu_y = 3.0$ and $\sigma_x^2 = \sigma_y^2 = 0.8$ for the first cluster, and $\mu_x = \mu_y = 8$ and $\sigma_x^2 = \sigma_y^2 = 0.5$ for the second cluster. The second data set is generated with two Exponential probability distributions, $f(x; \alpha) = \frac{1}{\alpha} e^{-x/\alpha}$, in the same positions $(3,3)$ and $(8,8)$ with $\alpha = 0.8$ for the first cluster and $\alpha = 2.0$ for the second one. The third data set is generated with two Weibull probability distributions, $f(x; \eta, \sigma) = \frac{\eta}{\sigma} \cdot \left(\frac{x}{\sigma}\right)^{\eta-1} \cdot e^{-\left(\frac{x}{\sigma}\right)^\eta}$, with parameters shape $\eta = 2.0$ and scale $\sigma = 0.8$ for the first cluster, and $\eta = 1.0$ and scale $\sigma = 0.5$ for the second one. Finally, the fourth data set is generated with two t-Student probability distributions, $f(x; n) = \frac{\Gamma(\frac{n+1}{2})}{\sqrt{n\pi}\Gamma(n/2)}\left(1 + \frac{x^2}{n}\right)^{-\frac{n+1}{2}}$, at positions $(3,3)$ and $(10,10)$ with 3 and 8 degrees of freedom respectively.

In Fig. 1 the fitness evolution and the number of corals in function of the number of generations are represented. Rawly, the fitness evolution (Figs. 1a, c, e and g) demonstrates the capacity of GPDMM to critically reduce the fitness at the initial generations, particularly the first fifth; whereas the remaining generations have a low effectiveness in the improvement of the fitness.

With regard to the most represented type of coral, the proposed GPDMM is able to correctly reproduce the underlying data structure. Independently of the probability distributions mixture used for generating the input, the most represented coral rapidly reaches the correct model: Fig. 1b for Gaussian distributions mixture, Fig. 1d for Exponential distributions mixture, Fig. 1f for Weibull distributions mixture, and Fig. 1h for t-Student distributions mixture. Although in the first generations other coral types are present, even in a larger number than the correct coral, their number rapidly decreases with the generations.

The success rate, when the synthetic input data set has been generated through a mixture of Gaussian distributions, is shown in Table 1. In all cases the underlying data model is perfectly captured.

In Table 2, the success rate, when the synthetic input data set has been generated through a mixture of two Exponential distributions, for diverse configurations for the number of generations and the size of the reef is shown. The success rate is measured as the percentage of executions that end with an Exponential best coral, and therefore, it can be considered that the proposed GPDMM correctly captures the underlying data model [12]. As can be appreciated, nor a

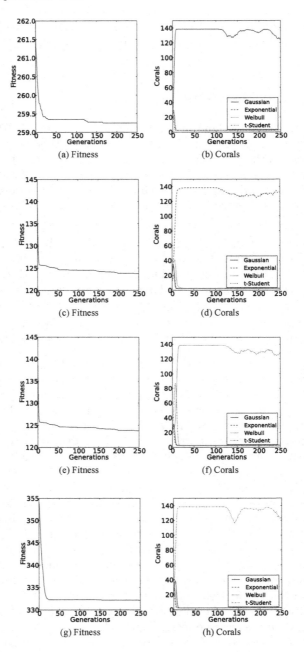

Fig. 1. Fitness evolution of best coral (Fig. 1a) and number of corals of each type of probability distribution (Fig. 1b) as function of the number of generations when the synthetic input data set has been generated through a mixture of two Gaussian probability distributions; and its equivalents for Exponential (Figs. 1c and d), Weibull (Figs. 1e and f), and t-Student (Figs. 1g and h). The mean values of 20 independent runs are shown.

Table 1. Percentage of executions, after 20 executions, that end with a Gaussian best coral when the synthetic input data set has been generated through a mixture of two Gaussian probability distributions.

Generations	Reef size							
	36	64	100	144	196	256	324	400
50	100	100	100	100	100	100	100	100
100	100	100	100	100	100	100	100	100
150	100	100	100	100	100	100	100	100
200	100	100	100	100	100	100	100	100
250	100	100	100	100	100	100	100	100
300	100	100	100	100	100	100	100	100
350	100	100	100	100	100	100	100	100
400	100	100	100	100	100	100	100	100

Table 2. Percentage of executions, after 20 executions, that end with an Exponential best coral when the synthetic input data set has been generated through a mixture of two Exponential probability distributions.

Generations	Reef size							
	36	64	100	144	196	256	324	400
50	55	65	90	70	85	65	70	80
100	75	90	80	75	80	80	55	75
150	70	80	85	70	90	65	80	65
200	55	65	90	90	75	75	65	65
250	75	40	80	70	75	100	65	90
300	70	95	85	65	80	80	60	65
350	70	85	90	85	80	80	95	65
400	70	80	70	80	90	90	85	90

larger number of generations, nor a larger reef seem increasing the success rate. This case arises as the most difficult to reproduce.

In Table 3 the success rate, when the synthetic input data set has been generated through a mixture of two Weibull distributions, for diverse configurations for the number of generations and the size of the reef is shown. In this case, an increment in the size of the reef and a larger number of generations lead to a clear improvement of the success rate; whereas smaller configurations lead to low performance.

Finally, for the case of two t-Student distributions (Table 4), most of the configurations perfectly reproduce the underlying data model; and there are not cases below 90% of success rate.

Table 3. Percentage of executions, after 20 executions, that end with a Weibull best coral when the synthetic input data set has been generated through a mixture of two Weibull probability distributions.

Generations	Reef size							
	36	64	100	144	196	256	324	400
50	55	95	100	100	100	100	100	100
100	65	90	95	100	100	100	100	100
150	70	75	100	95	100	100	100	100
200	60	95	100	100	100	100	100	100
250	90	85	90	100	100	100	100	100
300	85	80	95	100	100	100	100	100
350	75	90	100	100	100	100	100	100
400	55	100	95	100	100	100	100	100

Table 4. Percentage of executions, after 20 executions, that end with a t-Student best coral when the synthetic input data set has been generated through a mixture of two t-Student probability distributions.

Generations	Reef size							
	36	64	100	144	196	256	324	400
50	100	95	100	100	100	100	100	100
100	100	95	100	100	100	100	100	100
150	100	100	100	100	100	100	100	100
200	95	95	95	100	100	100	100	100
250	100	100	100	100	100	100	100	100
300	100	100	100	100	100	100	100	100
350	100	100	100	100	100	100	100	100
400	100	90	100	100	100	100	100	100

With regard to the processing time, when analysing data from two Gaussian distributions the process time takes the shortest (Table 5), with an increasing of computational charge for the cases of Exponential distributions, Weibull distributions; being the most adverse scenario when analysing data from two t-Student distributions (Table 6). For the sake of the brevity, only the two extreme cases are presented.

Table 5. Mean execution time (seconds) for the runs (20 runs) that end with a Gaussian best coral when the synthetic input data set has been generated through a mixture of two Gaussian probability distributions.

Generations	Reef size							
	36	64	100	144	196	256	324	400
50	3.80	6.19	9.61	14.53	20.42	27.68	35.49	46.55
100	6.88	11.83	18.57	29.11	41.20	54.17	70.86	93.02
150	10.21	17.58	28.01	43.88	63.60	87.16	113.67	143.93
200	12.94	23.03	36.87	59.30	82.70	110.97	149.04	201.54
250	17.17	28.63	46.17	73.95	103.75	144.23	186.33	253.67
300	19.34	36.41	57.29	88.80	128.32	169.40	230.55	305.40
350	22.54	40.39	64.64	102.01	153.01	206.01	266.83	359.13
400	26.29	45.58	72.70	114.33	165.54	234.04	318.52	409.50

Table 6. Mean execution time (seconds) for the runs (20 runs) that end with a t-Student best coral when the synthetic input data set has been generated through a mixture of two t-Student probability distributions.

Generations	Reef size							
	36	64	100	144	196	256	324	400
50	4.10	7.63	15.15	30.28	59.47	118.28	205.76	352.26
100	7.78	15.87	34.52	74.54	158.35	343.26	603.13	1108.30
150	11.41	24.58	54.16	121.88	262.07	537.35	1017.29	1916.92
200	14.97	33.26	72.56	166.36	357.37	753.76	1441.24	2639.11
250	18.83	41.16	92.09	215.71	483.02	941.07	1852.94	3383.60
300	22.71	48.95	112.55	267.61	595.84	1166.61	2263.51	4266.90
350	25.66	57.35	130.20	314.87	675.48	1375.87	2684.97	4937.79
400	28.99	66.49	151.59	356.79	786.22	1613.65	3428.42	5756.57

5   Conclusions

In this paper a generalization of GMM for clustering is proposed. GMM is constrained to Gaussian probability distributions for the mixing. In this work the initial effort for relaxing this constraint is presented. This initial effort includes the incorporation in the mixture of other probability distribution, and particularly Exponential, Weibull, and t-Student probability distributions have been added.

As a result of this wider portfolio of probability distributions, the mechanism for finding the most suitable parameters combination of the probability distributions mixture is changed from the traditional Expectation-Maximization to a meta-heuristic based approach: the Coral Reef Optimizer. By using the CRO the proposal benefits of its capacity to fit the parameters of a model for the data, and at the same time to select among a set of models the most suitable combination of components. From the experimental tests performed, it can be stated that this optimizer is able to reach an excellent agreement with the underlying data model.

As previously mentioned, this work encompasses the initial efforts for relaxing the constraints in GMM. Only a reduced part of the total effort has been presented in this publication. For instance, parameters optimization of the CRO has been skipped. Future work lines include to allow a structure for the candidate solutions composed of different probability distributions. Furthermore, the evaluation of the proposed methodology for real problems is also being undertaken. These two last tasks could require the modification of the CRO operators. Finally, the application of GPDMM to outlier detection is also proposed as Future Work.

Acknowledgment. IMJ is co-funded in a 91.89% by the European Social Fund within the Youth Employment Operating Program, for the programming period 2014–2020, as well as Youth Employment Initiative (IEJ). IMJ is also co-funded through the Grants for the Promotion of Youth Employment and Implantation of Youth Guarantee in Research and Development and Innovation (I+D+i) from the MINECO.

SSS has been partially supported by the project TIN2014-54583-C2-2-R of the Spanish Ministerial Commission of Science and Technology (MICYT).

MCM has received funding by the Spanish Ministry of Economy and Competitiveness (MINECO) for funding support through the grants FPA2016-80994-C2-1-R, and "Unidad de Excelencia María de Maeztu": CIEMAT - FÍSICA DE PARTÍCULAS through the grant MDM-2015-0509.

References

1. Han, J., Kamber, M.: Data Mining: Concepts and Techniques. Morgan Kaufmann, San Francisco, CA, USA (2000). http://dblp.uni-trier.de/rec/bib/books/mk/HanK2000. ISBN 1-55860-489-8
2. Aggarwal, C.C.: Outlier Analysis. Springer, Heidelberg (2013)
3. Salcedo-Sanz, S., Ser, J.D., Landa-Torres, I., Gil-López, S., Portilla-Figueras, J.: The coral reefs optimization algorithm: a novel metaheuristic for efficiently solving optimization problems. Sci. World J. **2014**, 15 (2014)
4. Pernkopf, F., Bouchaffra, D.: Genetic-based EM algorithm for learning gaussian mixture models. IEEE Trans. Pattern Anal. Mach. Intell. **27**(8), 1344–1348 (2005)
5. Verbeek, J.J., Vlassis, N.A., Kröse, B.J.A.: Efficient greedy learning of Gaussian mixture models. Neural Computat. **15**(2), 469–485 (2003)
6. Everitt, B., Hand, D.J.: Finite Mixture Distributions. Chapman and Hall, London (1981)
7. Kalai, A.T., Moitra, A., Valiant, G.: Efficiently learning mixtures of two Gaussians. In: Proceedings of the 42nd ACM Symposium on Theory of Computing, STOC 2010, Cambridge, Massachusetts, USA, pp. 553–562, 5–8 June 2010
8. Salcedo-Sanz, S., García-Díaz, P., Portilla-Figueras, J., Ser, J.D., Gil-López, S.: A coral reefs optimization algorithm for optimal mobile network deployment with electromagnetic pollution control criterion. Appl. Soft Comput. **24**, 239–248 (2014)

9. Walck, C.: Hand-book on statistical distributions for experimentalists. Technical report, Particle Physics Group, Fysikum, University of Stockholm (1996)
10. Salcedo-Sanz, S.: A review on the coral reefs optimization algorithm: new development lines and current applications. Prog. AI **6**(1), 1–15 (2017)
11. Martínez, A.M., Vitrià, J.: Learning mixture models using a genetic version of the EM algorithm. Pattern Recognit. Lett. **21**(8), 759–769 (2000)
12. Eiben, A., Smith, J.: Introduction to Evolutionary Computing, 2nd edn. Springer, Heidelberg (2015)

A Hybrid Approach to Mining Conditions

Fernando O. Gallego[✉] and Rafael Corchuelo[✉]

ETSI Informática, University of Seville,
Avda. Reina Mercedes, s/n, Sevilla, Spain
{fogallego,corchu}@us.es

Abstract. Text mining pursues producing valuable information from natural language text. Conditions cannot be neglected because it may easily lead to misinterpretations. There are naive proposals to mine conditions that rely on user-defined patterns, which falls short; there is only one machine-learning proposal, but it requires to provide specific-purpose dictionaries, taxonomies, and heuristics, it works on opinion sentences only, and it was evaluated very shallowly. We present a novel hybrid approach that relies on computational linguistics and deep learning; our experiments prove that it is more effective than current proposals in terms of F_1 score and does not have their drawbacks.

1 Introduction

Text mining pursues processing natural language text to produce useful information. Unfortunately, current state-of-the-art text miners do not take conditions into account, which may easily result in misinterpretations. For instance, given sentence "Let it happen and John will leave Acme", current entity-relation extractors [6,12] return fact ("John","will leave","Acme"); similarly, current opinion miners [17,19] return a negative score since "will leave" typically conveys a negative opinion. Neglecting the conditions clearly results in misinterpretations.

The simplest approach to mine conditions consists in searching for user-defined patterns [3,10], which falls short regarding recall because there are many common conditions do not fit common patterns. There is only one machine-learning approach [13], but it must be customised with several specific-purpose dictionaries, taxonomies, and heuristics, and it mines conditions regarding opinions only, not to mention that it was evaluated very shallowly.

In this paper, we present a proposal to mine conditions that hybridises computational linguistics and deep learning, without any of the previous problems. We have performed a comprehensive experimental analysis on a dataset with 3 779 000 sentences on 15 common topics in English and Spanish; our results prove that our approach is comparable to others in terms of precision [5], but improves recall enough to beat them in terms of F_1 score [22].

Supported by Opileak.com and the Spanish R&D programme (grants TIN2013-40848-R and TIN2013-40848-R). The computing facilities were provided by the Andalusian Scientific Computing Centre (CICA). We also thank Dr. Francisco Herrera for his hints on statistical analyses and sharing his software with us.

F. J. de Cos Juez et al. (Eds.): HAIS 2018, LNAI 10870, pp. 264–276, 2018.
https://doi.org/10.1007/978-3-319-92639-1_22

The rest of the paper is organised as follows: Sect. 2 provides an insight into the related work; Sect. 3 describes our proposal; Sect. 4 reports on our experimental analysis; finally, Sect. 5 presents our conclusions.

2 Related Work

Narayanan et al. [14] range amongst the first authors who realised the problem with conditions in the field of opinion mining. However, they did not report on a proposal to mine them.

The simplest approaches to mine conditions build on searching for user-defined patterns. Mausam et al. [10] studied the problem in the field of entity-relation extraction and suggested that conditions might be identified by locating adverbial clauses whose first word is one of the sixteen one-word condition connectives in English; unfortunately, they did not report on the effectiveness of their approach to mine conditions, only on the overall effectiveness of their proposal for entity-relation extraction. Chikersal et al. [3] proposed a similar, but simpler approach: they searched for sequences of words in between connectives "if", "unless", "until", and "in case" and the first occurrence of "then" or a comma. Unfortunately, the previous proposals are not generally appealing because hand-crafting such patterns is not trivial and the results typically fall short regarding recall, as our experimental analysis confirms.

The only existing machine-learning approach was introduced by Nakayama and Fujii [13], who worked in the field of opinion mining. They devised a model that is based on features that are computed by means of a syntactic parser and a semantic analyser. The former identifies so-called "bunsetus", which are Japanese syntactic units that consists of one independent word and one or more ancillary words, as well as their inter-dependencies; the latter identifies opinion expressions, which requires to provide several specific-purpose dictionaries, taxonomies, and heuristics. They used Conditional Random Fields and Support Vector Machines to learn classifiers that make "bunsetus" that can be considered conditions apart from the others. Unfortunately, their approach was only evaluated on a small dataset with 3 155 Japanese sentences regarding hotels and the best F_1 score attained was 0.5830. As a conclusion, this proposal is not generally applicable and its effectiveness is poor for practical purposes.

Our conclusion is that mining conditions is a problem to which researchers are paying attention recently because it is a must for software agents to mine text properly so as to avoid misinterpretations. Unfortunately, the few existing techniques have many drawbacks that hinder their general applicability. This motivated us to work on a new approach that overcomes their weaknesses and outperforms them by means of a hybrid approach that combines computational linguistics and deep learning; our proposal only requires a stemmer, a dependency parser, and a word embedder, which are readily-available components.

3   A Hybrid Approach to Mining Conditions

In this section, we first describe the main methods of our proposal, which work in co-operation to learn a regressor that assesses the candidate conditions in a sentence before the most promising ones are returned; then, we describe some ancillary methods to generate candidate conditions, to compute their scores, to set up a regressor using a deep neural network, and to remove overlapping candidate conditions. We use sentence "If you're someone who likes cakes, then try John's." as a running example where appropriate.

3.1   Description of the Main Methods

The main methods are sketched in Fig. 1, namely: method *train*, which is used to learn a regressor that assesses candidate conditions, and method *apply*, which selects the best candidate conditions in a sentence and returns them.

Method *train* takes a dataset ds as input and returns a regressor r. The input dataset is of the form $\{(s_i, L_i)\}_{i=1}^n$, where each s_i denotes a sentence and each L_i denotes a set of labels that identify the conditions in that sentence ($n \geq 0$). The output regressor is a function that given a candidate condition returns a score that assesses how likely it is an actual condition. The method first initialises training set T to the empty set and then loops over dataset ds; for each sentence s and set of labels L in ds, it first computes a set of candidate conditions; then, for each condition c, it computes a score σ and stores a tuple of the form (c, σ) in training set T. When the main loop finishes, it learns a regressor from T using a deep-learning approach.

Method *apply* takes a sentence s, a regressor r, and a threshold θ as input and returns a set R of tuples of the form $\{(c_i, \sigma_i)\}_{i=1}^m$, where each c_i denotes a condition and σ_i its corresponding score, which must be equal or greater than the threshold ($m \geq 0$). The method first generates the candidate conditions in s, stores them in set C, and initialises the result R to an empty set; it then iterates over set C; for each candidate condition c in set C, it first computes its score by applying regressor r to it; if it is equal or greater than threshold θ, then

```
 1: method train(ds) returns r              1: method apply(s, r, θ) returns R
 2:     T := ∅                               2:     C := generateCandidates(s)
 3:     for each (s, L) ∈ ds do              3:     R := ∅
 4:         C := generateCandidates(s)       4:     for each c ∈ C do
 5:         for each c ∈ C do                5:         σ := apply r to c
 6:             σ := computeScore(c, L)       6:         if σ > θ then
 7:                 T := T ∪ {(c, σ)}         7:             R := R ∪ {(c, σ)}
 8:         end                              8:     end
 9:     end                                  9:     end
10:     r := learnRegressor(T)              10:     R := removeOverlaps(R)
11: end                                     11: end
```

Fig. 1. Main methods of our proposal.

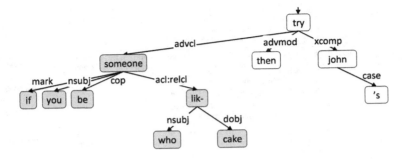

Fig. 2. Sample dependency tree.

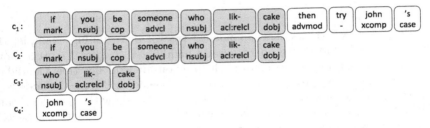

Fig. 3. Sample candidate conditions.

candidate condition c is added to the result set. When the main loop finishes, R provides a collection of candidates and scores; before returning it, we must remove the conditions that overlap others with a higher score.

3.2   Method to Generate Candidates

Our first ancillary method is *generateCandidates*, which takes a sentence as input and returns a set of candidate conditions. A naive approach would simply generate as many sub-strings as possible, but it would be very inefficient because a sentence with n words has $O(n^2)$ such sub-strings. In order to reduce the candidate space we use a dependency tree to generate them since conditions are clauses from a grammatical point of view and the non-leaf nodes of a dependency tree typically represent many such clauses.

Method *generateCandidates* first computes the dependency tree of the input sentence, then changes the words in its nodes to lowercase, and finally stems them. Now, for each non-leaf node in the dependency tree, we compute all of the sequences of tokens that originate from that node; a token is a tuple of the form (w, d), where w denotes a stem and d the dependency tag that links its corresponding node in the dependency tree to its parent, if any. Note that we do not select leaf nodes because we have not found a single example in which one word can be considered a condition. As a conclusion, a condition is modelled as a sequence of the form $\langle (w_i, d_i) \rangle_{i=1}^{n}$ where each w_i is a stem and d_i is its corresponding dependency tag $(n \geq 2)$.

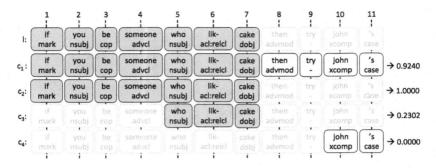

Fig. 4. Sample matchings.

Example 1. Figure 2 shows the dependency tree of our running example; the nodes that correspond to the condition are highlighted in grey. Figure 3 shows the candidates that are generated from the previous dependency tree. Candidate c_1 is generated from the root node, candidate c_2 is generated from the node with stem "someone", candidate c_3 is generated from the node with stem "lik-", and candidate c_4 is generated from the node with stem "john". Note that, as expected, the condition is confined to one of the nodes in the dependency tree.

3.3   Method to Compute Scores

Our second ancillary method is *computeScore*, which takes a candidate condition c and a set of labels L as input and returns its corresponding score. A naive approach would simply return 0.0000 if c does not exactly match any of the labels in L and 1.0000 otherwise, but that is too crisp. An approach in which a candidate gets a score in range $[0.0000, 1.0000]$ better captures the chances that it is an actual condition. We use an approach that is based on the well-known F_1 score in order to balance the precision and the recall of candidate conditions.

The F_1 score is computed as $\frac{2\,tp}{(tp+fp)+(tp+fn)}$, where tp, fp, and fn denote, respectively, the number of true positives, false positives, and false negatives. Given a candidate condition c and a label l, it makes sense to interpret the tokens that they have in common as true positive tokens, the tokens in c that are not in l as false positive tokens, and the tokens in l that are not in c as false negative tokens. We also realised that the first few tokens in a condition typically provide a landmark that characterises it. Thus, we decided to measure the degree of matching between a condition and a label as follows:

$$match(c, l) = \sum_{i=1}^{|l|} \left\{ \begin{array}{ll} 1/i & \text{if } l_i \in c \\ 0 & \text{otherwise} \end{array} \right\} \tag{1}$$

Simply put: let l_i denote the i-th token in the label $(i = 1..|l|)$; if l_i is in the candidate, we then add $1/i$ to the score and zero otherwise. This way, the first few tokens in the label contribute much more to the score than the remaining ones.

That is, given a candidate condition c and a label l, $match(c, l)$ is a measure of the number of true positive tokens in c.

Given the previous definition, the maximum degree of matching for a candidate condition or a label x is defined as follows:

$$match^*(x) = \sum_{i=1}^{|x|} {}^1\!/_i \qquad (2)$$

Realise that given a candidate condition c, $match^*(c)$ is a measure of the number of true positive tokens (the tokens that belong to both the candidate condition and the label) and the false positive tokens (the tokens that belong to the candidate, but not to the label); similarly, given a label l, $match^*(l)$ is a measure of the number of true positive tokens (the tokens that belong to both the label and the candidate condition) and the number of false negative tokens (the tokens that belong to the label, but not to the candidate condition).

Our proposal to compute the score of candidate condition c with respect to the set of labels L is then as follows:

$$score(c, L) = \max_{l \in L} \frac{2\ match(c, l)}{match^*(c) + match^*(l)} \qquad (3)$$

Note the similarity to the F_1 score since $match(c, l)$, $match^*(c)$, and $match^*(l)$ are measures of tp, $tp + fp$, and $tp + fn$, respectively; the difference is that we do not count the actual number of true positive, false positive, or false negative tokens, but a measure that puts an emphasis on the first few tokens and decays asymptotically.

Example 2. Figure 4 shows label l, which corresponds to the condition in our running example, and how the candidate conditions match it. Candidate condition c_1 represents the whole input sentence, which obviously contains the label, i.e., seven true positive tokens, but also four false positive tokens, which results in a score of 0.9240. Candidate c_2 matches the label perfectly, i.e., it matches seven true positive tokens and no false positive or false negative token, which results in a score of 1.0000. Candidate c_3 is a partial match with three true positive tokens and three false negative tokens, which results in a score of 0.2302. Finally, condition c_4 does not match any true positive token, but two false positive tokens and seven false negative tokens, which results in a score of 0.0000.

3.4   Model to Learn a Regressor

Our third ancillary method is *learnRegressor*, which takes a training set T as input and returns a regressor r. We implemented it using a deep-learning approach because of its natural ability to transform data into feature-based representations that help learn good regressors.

Prior to learning a regressor, the candidate conditions in the training set must be vectorised. Our labelled dataset, which is presented in Sect. 4, suggests that the length of common conditions ranges from a few tokens to a few dozens,

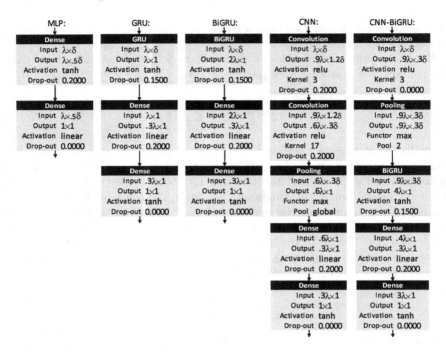

Fig. 5. Neural networks for learning regressors.

so it makes sense to represent them as large-enough fixed-size sequences. Given a candidate condition of the form $\langle (w_i, d_i) \rangle_{i=1}^{n}$, we transform it into a sequence of the form $\langle \overline{w_i} \oplus \overline{d_i} \rangle_{i=1}^{\lambda}$, where $\overline{w_i}$ denotes the vectorisation of stem w, $\overline{d}$ denotes the vectorisation of dependency tag d, $\overline{w} \oplus \overline{d}$ the vector that results from catenating the previous ones, and λ denotes the size of the longest possible condition; note that padding tokens need to be added if the original condition is shorter than λ.

Given a stem w, we compute its vectorisation $\overline{w}$ by using word embedding [11], which can unsupervisedly produce vectors that preserve some semantic relationships amongst the original stems; to reduce the stem space, we replaced numbers, email addresses, URLs, and stems whose frequency is equal or smaller than five by class words "NUMBER", "EMAIL", "URL", and "UNK", respectively. Given dependency tag d, we compute its vectorisation $\overline{d}$ by means of one-hot encoding [23], which vectorises a finite set of tags using binary features. Note that the vectorisation of a condition can then be interpreted as a matrix with λ rows and δ columns, where δ denotes the dimensionality of the word embedding vectorisation plus the dimensionality of the one-hot vectorisation.

Our baseline architecture was a multi-layer perceptron (MLP) with two dense layers. We devised a dozen more architectures, but the best ones were based on the following components: gated recurrent units (GRU), bi-directional gated recurrent units (BiGRU), convolutional neural networks (CNN), and a hybrid approach that combines convolutional neural networks and bi-directional gated recurrent units (CNN-BiGRU).

GRUs are a kind of recurrent neural network (RNN) [8] and BiGRUs are a kind of bi-directional recurrent neural networks (BiRNN) [20]. In both RNNs and BiRNNs the connections between units form a directed cycle, which allows to apply them to sequences of varying length; the difference is that RNNs cannot take future elements in a sequence into account whereas BiRNNs can. Unfortunately, both RNNs and BiRNNs suffer from the so-called exploding/vanishing gradient problems [16], which is overcome by using gated recurrent units (GRUs) [4] or bi-directional gated recurrent units (BiGRUs) [15] that help control the amount of data that is passed on to the next epoch or forgotten. The CNN network [18] includes two convolutional layers and a pooling layer. The former consists of several filtering units that take a small region of the input data as input and applies a non-linear function to it; the latter consists of pooling units that apply a merging method to the results of the previous layer. Our proposal is to use a convolutional layer with a large number of filters in order to create a wide range of first-level features, but a smaller number of filters in the second convolutional layer to obtain a more specific range of second-level features that combine de first ones. Finally, the pooling layer combines the previous deep features using a global maximum function as the global pooling strategy since our experiments prove that it performs better than others. The CNN-BiGRU network uses a convolutional layer with a number of filters similar to the input length, and then applies a local pooling that captures the most relevant features only. We then apply a BiGRU layer that takes the dependencies between tokens into account, from both the beginning to the end of the sentences and vice versa.

Figure 5 summarises the previous architectures. The boxes represent the layers and provide information about the corresponding parameters, namely: in all cases, the input and output dimensions in terms of λ and δ (rounding to the closest natural number is assumed); in all cases, but pooling layers, the activation function and the drop-out ratio; in the case of convolutions, the kernel size; and, in the case of pooling layers, the functor and the pool used. The parameters were computed using the Stochastic Gradient Descent method [9] with batch size equal to 32. In order to prevent over-fitting as much as possible, we used some drop-out regularisations [21] and early stopping [2] when the loss did not improve enough after 10 epochs. We used the Mean Squared Error as the loss function since it is very common in regression problems [1]. We did not apply a decay momentum because we observed that the loss always converges smoothly, even if it needs more epochs for some networks than for others.

3.5   Criteria to Remove Overlaps

The fourth ancillary method is *removeOverlaps*, which takes a set of tuples of the form (c, σ) as input, where c denotes a candidate and σ its corresponding score, and filters some of them out.

It is the simplest method in our proposal. Basically, it works as follows: it iterates over the set of input tuples and removes those whose conditions overlap a condition with a higher score. In other words, given the input set of tuples R, it computes the following subset:

$$\{(c, \sigma) \mid (c, \sigma) \in R \wedge \nexists(c', \sigma') : (c', \sigma') \in R \wedge c' \cap c \neq \langle\rangle \wedge \sigma' > \sigma\} \qquad (4)$$

Example 3. Assume that the threshold to select the best candidates is set to $\theta = 0.5000$. In our running example, method *apply* would return candidate conditions c_1 and c_2 since they are the only whose scores exceed the threshold, cf. Fig. 4. Note that both candidate conditions overlap, so the one with the lowest score is filtered out. In this case, method *apply* would then return condition c_2 only, which, indeed, represents the condition in our running example.

4   Experimental Analysis

Computing Facility: We run our experiments on a virtual computer that was equipped with one Intel Xeon E5-2690 core at 2.60 GHz, 2 GiB of RAM, and an Nvidia Tesla K10 GPU accelerator with 2 GK-104 GPUs at 745 MHz with 3.5 GiB of RAM each; the operating system was CentOS Linux 7.3.

Prototype Implementation:[1] We implemented our proposal with Python 3.5.4 and the following components: Snowball 1.2.1 to stemmise words, the Stanford NLP Core Library 3.8.0 to generate dependency trees, Gensim 2.3.0 to compute word embedders using a Word2Vec implementation, and Keras 2.0.8 with Theano 1.0.0 to learn the regressors.

Evaluation Dataset:[2] We used a dataset with 3 779 000 sentences in English and Spanish that were randomly gathered from the Web between April 2017 and May 2017. The sentences were classified into 15 topics according to their sources, namely: adults, baby care, beauty, books, cameras, computers, films, headsets, hotels, music, ovens, pets, phones, TV sets, and video games. None of the conditions that we found in this dataset was smaller than two tokens or longer than 50 tokens, so we set those limits to vectorise candidate conditions.

Baselines: We used the proposals by Mausam et al. [10] and Chikersal et al. [3] as baselines. The proposal by Nakayama and Fujii [13] was not considered because it is not clear if it can be customised to deal with languages other than Japanese and its best F_1 was 0.5830; neither could we find an implementation.

Performance Measures: We measured the standard performance measures, namely: precision, recall, and the F_1 score. Regarding the baselines, we computed the measures from our dataset since there is no machine-learning involved; regarding our proposals, we computed the measures using 5-fold cross-validation. We computed the measures independently for each of our approaches and set threshold θ to 0.2500, 0.5000, and 0.7500.

---

[1] Available at https://github.com/FernanOrtega/HAIS18.

[2] Available at https://www.kaggle.com/fogallego/reviews-with-conditions.

Lang	Proposal	θ = 0.2500			θ = 0.5000			θ = 0.7500		
		P	R	F₁	P	R	F₁	P	R	F₁
	MB	0.6270	0.6144	0.6206	0.6270	0.6144	0.6206	0.6270	0.6144	0.6206
	CB	0.7979	0.4642	0.5870	0.7979	0.4642	0.5870	0.7979	0.4642	0.5870
	Averages	0.7125	0.5393	0.6038	0.7125	0.5393	0.6038	0.7125	0.5393	0.6038
	MLP	0.4741	0.7799	0.5897	0.5612	0.5271	0.5436	0.5739	0.4582	0.5096
en	GRU	0.9999	0.4421	0.6131	0.9999	0.4421	0.6131	0.9999	0.4421	0.6131
	BiGRU	0.5448	0.5262	0.5353	0.8999	0.4421	0.5929	0.9999	0.4421	0.6131
	CNN	0.5908	0.7546	0.6628	0.6211	0.6278	0.6244	0.6571	0.5432	0.5948
	CNN-BiGRU	0.5586	0.8052	0.6596	0.6318	0.6529	0.6422	0.7327	0.4914	0.5883
	Averages	0.6336	0.6616	0.6121	0.7428	0.5384	0.6033	0.7927	0.4754	0.5838
	MB	0.6699	0.5285	0.5909	0.6699	0.5285	0.5909	0.6699	0.5285	0.5909
	CB	0.7953	0.4399	0.5665	0.7953	0.4399	0.5665	0.7953	0.4399	0.5665
	Averages	0.7326	0.4842	0.5787	0.7326	0.4842	0.5787	0.7326	0.4842	0.5787
	MLP	0.4232	0.8295	0.5604	0.5382	0.5678	0.5526	0.5771	0.4465	0.5034
es	GRU	0.5246	0.7483	0.6168	0.7089	0.4304	0.5356	0.9999	0.4153	0.5869
	BiGRU	0.5321	0.7451	0.6209	0.6335	0.4692	0.5391	0.9999	0.4153	0.5869
	CNN	0.5997	0.7519	0.6672	0.6606	0.6521	0.6563	0.7065	0.5467	0.6164
	CNN-BiGRU	0.5227	0.8221	0.6390	0.6195	0.6968	0.6559	0.6843	0.5369	0.6017
	Averages	0.5205	0.7794	0.6209	0.6321	0.5633	0.5879	0.7935	0.4721	0.5790

Fig. 6. Experimental results in comparison with baselines.

		θ = 0.2500		
Proposal	Ranking	Comparison	z	p-value
CNN	1.0000	CNN x CNN	-	-
CNN-BiGRU	2.0000	CNN x CNN-BiGRU	1.4142	0.1573
BiGRU	3.5000	CNN x BiGRU	3.5355	0.0008
MLP	4.1000	CNN x MLP	4.3841	0.0000
GRU	4.4000	CNN x GRU	4.8083	0.0000
		(a)		

		θ = 0.5000		
Proposal	Ranking	Comparison	z	p-value
CNN-BiGRU	1.4000	CNN-BiGRU x CNN-BiGRU	-	-
CNN	1.6000	CNN-BiGRU x CNN	0.2828	0.7773
MLP	3.1000	CNN-BiGRU x MLP	2.4042	0.0324
BiGRU	4.2000	CNN-BiGRU x BiGRU	3.9598	0.0002
GRU	4.7000	CNN-BiGRU x GRU	4.6669	0.0000
		(b)		

		θ = 0.7500		
Proposal	Ranking	Comparison	z	p-value
CNN	1.3000	CNN x CNN	-	-
CNN-BiGRU	1.7000	CNN x CNN-BiGRU	0.5657	0.5716
MLP	3.0000	CNN x MLP	2.4042	0.0324
GRU	4.5000	CNN x GRU	4.5255	0.0000
BiGRU	4.5000	CNN x BiGRU	4.5255	0.0000
		(c)		

Proposal	Ranking	Comparison	z	p-value
$CNN_{0.25}$	1.4000	$CNN_{0.25}$ x $CNN_{0.25}$	-	-
$CNN\text{-}BiGRU_{0.50}$	1.8000	$CNN_{0.25}$ x $CNN\text{-}BiGRU_{0.50}$	0.5657	0.5716
MB	3.4000	$CNN_{0.25}$ x MB	2.8284	0.0094
$CNN_{0.75}$	3.7000	$CNN_{0.25}$ x $CNN_{0.75}$	3.2527	0.0034
CB	4.7000	$CNN_{0.25}$ x CB	4.6669	0.0000
		(d)		

Fig. 7. Statistical analysis based on Hommel's test.

Experimental Results: The experimental results are presented in Fig. 6. MB and CB refer to Mausam et al.'s and Chikersal et al.'s baselines, respectively. The greyed cells highlight the approaches that beat the best baseline.

The precision of the baselines is relatively good taking into account that they are naive approaches to the problem that rely on handcrafted user-defined patterns; Mausam et al.'s proposal achieves a recall that is similar to its precision, but Chikersal et al.'s proposal falls short regarding recall. Our approaches do not generally beat the baselines regarding precision, but attain results that are almost similar. It is regarding recall that most of our approaches beat the baselines since they are able to learn patterns that are more involved; thanks to our deep learning approach, the input sentences are projected onto a rich feature space that can capture many patterns that an expert cannot easily spot. Note that the improvement regarding recall is enough for the F_1 score to improve the baselines. Regarding the value of threshold θ, note that increasing it increases the average precision of our approaches, but decreases their average recall.

To make a decision regarding which of the approaches performs the best, we used a stratified strategy that builds on Hommel's test [7]. In Figs. 7a, b and c, we report on the results of the statistical analysis regarding our proposal; our goal was to select the best ones for each of the values of threshold θ. The previous figures show the experimental rank of each approach, and then the comparisons between the best one and the others; for every comparison, we show the value of the z statistic and its corresponding adjusted p-value. Note that the experimental results do not provide any evidences that the best-ranked approach is different from the second one since the adjusted p-value is greater than the significance level; however, there is enough evidence to prove that it is different from the remaining ones since the adjusted p-value is smaller than the significance level. Our conclusion is that CNN is the best approach when $\theta = 0.2500$, CNN-BiGRU is the best approach when $\theta = 0.5000$, and CNN is again the best approach when $\theta = 0.7500$. In Fig. 7d, we present the results of comparing the previous best approaches and the baselines (we denote the corresponding value of θ using subindices). According to Hommel's test, CNN with $\theta = 0.2500$ is similar to CNN-BiGRU with $\theta = 0.5000$, but they are better than both baselines and CNN with $\theta = 0.7500$.

Our experimental analysis confirms that our best alternatives are CNN with $\theta = 0.2500$ or CNN-BiGRU with $\theta = 0.5000$, that they are similar to the other proposals in terms of precision, but improve recall enough to beat them in terms of F_1 score. In summary, it confirms that our approach is very promising.

5   Conclusions

We have presented a novel proposal to mine conditions. It relies on a hybrid approach that merges computational linguistics and deep learning as a means to overcome the problems that we have found in the literature, namely: it does not rely on user-defined patterns, it does not require any specific-purpose dictionaries, taxonomies, or heuristics, and it can mine conditions in both factual and opinion sentences. Furthermore it relies on a number of components that are readily available, namely: a stemmer, a dependency parser, and a word embedder. We have also performed a comprehensive experimental analysis on a dataset with 3 779 000 sentences on 15 common topics in English and Spanish. Our results confirm that our proposal can beat the state-of-the-art proposals in terms of recall and F_1 score.

References

1. Aravkin, A.Y., Burke, J.V., Chiuso, A., Pillonetto, G.: Convex vs non-convex estimators for regression and sparse estimation: the mean squared error properties of ARD and GLasso. J. Mach. Learn. Res. **15**(1), 217–252 (2014)
2. Caruana, R., Lawrence, S., Giles, C.L.: Overfitting in neural nets: backpropagation, conjugate gradient, and early stopping. In: NIPS, pp. 402–408 (2000)
3. Chikersal, P., Poria, S., Cambria, E., Gelbukh, A.F., Siong, C.E.: Modelling public sentiment in Twitter. In: CICLing, vol. 2, pp. 49–65 (2015)
4. Chung, J., Gülçehre, Ç., Cho, K., Bengio, Y.: Empirical evaluation of gated recurrent neural networks on sequence modeling. CoRR abs/1412.3555 (2014)
5. Cummins, R.: On the inference of average precision from score distributions. In: CIKM, pp. 2435–2438 (2012)
6. Etzioni, O., Fader, A., Christensen, J., Soderland, S., Mausam: Open information extraction: the second generation. In: IJCAI, pp. 3–10 (2011)
7. Garcia, S., Herrera, F.: An extension on "statistical comparisons of classifiers over multiple data sets" for all pairwise comparisons. J. Mach. Learn. Res. **9**, 2677–2694 (2008)
8. Han, H., Zhang, S., Qiao, J.: An adaptive growing and pruning algorithm for designing recurrent neural network. Neurocomputing **242**, 51–62 (2017)
9. Mandt, S., Hoffman, M.D., Blei, D.M.: Stochastic gradient descent as approximate Bayesian inference. J. Mach. Learn. Res. **18**, 134:1–134:35 (2017)
10. Mausam, Schmitz, M., Soderland, S., Bart, R., Etzioni, O.: Open language learning for information extraction. In: EMNLP-CoNLL, pp. 523–534 (2012)
11. Mikolov, T., Sutskever, I., Chen, K., Corrado, G.S., Dean, J.: Distributed representations of words and phrases and their compositionality. In: NIPS, pp. 3111–3119 (2013)
12. Mitchell, T.M., Cohen, W.W., Hruschka, E.R., Talukdar, P.P., Betteridge, J., Carlson, A., Mishra, B.D., Gardner, M., Kisiel, B., Krishnamurthy, J., Lao, N., Mazaitis, K., Mohamed, T., Nakashole, N., Platanios, E.A., Ritter, A., Samadi, M., Settles, B., Wang, R.C., Wijaya, D.T., Gupta, A., Chen, X., Saparov, A., Greaves, M., Welling, J.: Never-ending learning. In: AAAI, pp. 2302–2310 (2015)
13. Nakayama, Y., Fujii, A.: Extracting condition-opinion relations toward fine-grained opinion mining. In: EMNLP, pp. 622–631 (2015)
14. Narayanan, R., Liu, B., Choudhary, A.N.: Sentiment analysis of conditional sentences. In: EMNLP, pp. 180–189 (2009)
15. Nußbaum-Thom, M., Cui, J., Ramabhadran, B., Goel, V.: Acoustic modeling using bidirectional gated recurrent convolutional units. In: Interspeech 2016, pp. 390–394 (2016)
16. Pascanu, R., Mikolov, T., Bengio, Y.: On the difficulty of training recurrent neural networks. In: ICML, vol. 3, pp. 1310–1318 (2013)
17. Ravi, K., Ravi, V.: A survey on opinion mining and sentiment analysis: tasks, approaches and applications. Knowl. Based Syst. **89**, 14–46 (2015)
18. dos Santos, C.N., Xiang, B., Zhou, B.: Classifying relations by ranking with convolutional neural networks. In: ACL, vol. 1, pp. 626–634 (2015)
19. Schouten, K., Frasincar, F.: Survey on aspect-level sentiment analysis. IEEE Trans. Knowl. Data Eng. **28**(3), 813–830 (2016)
20. Schuster, M., Paliwal, K.K.: Bidirectional recurrent neural networks. IEEE Trans. Signal Process. **45**(11), 2673–2681 (1997)

21. Srivastava, N., Hinton, G.E., Krizhevsky, A., Sutskever, I., Salakhutdinov, R.: Drop-out: a simple way to prevent neural networks from overfitting. J. Mach. Learn. Res. **15**(1), 1929–1958 (2014)
22. Zhang, D., Wang, J., Zhao, X.: Estimating the uncertainty of average F_1 scores. In: ICTIR, pp. 317–320 (2015)
23. Zhang, X., Zhao, J.J., LeCun, Y.: Character-level convolutional networks for text classification. In: NIPS, pp. 649–657 (2015)

A First Attempt on Monotonic Training Set Selection

J.-R. Cano[1(✉)] and S. García[2]

[1] Department of Computer Science, University of Jaén, EPS of Linares,
Avenida de la Universidad S/N, 23700 Linares, Jaén, Spain
jrcano@ujaen.es
[2] Department of Computer Science and Artificial Intelligence,
University of Granada, 18071 Granada, Spain
salvagl@decsai.ugr.es

Abstract. Monotonicity constraints frequently appear in real-life problems. Many of the monotonic classifiers used in these cases require that the input data satisfy the monotonicity restrictions. This contribution proposes the use of training set selection to choose the most representative instances which improves the monotonic classifiers performance, fulfilling the monotonic constraints. We have developed an experiment on 30 data sets in order to demonstrate the benefits of our proposal.

Keywords: Monotonic classification · Ordinal classification
Training set selection · Data preprocessing · Machine learning

1 Introduction

Monotonic classification is an ordinal classification problem where monotonic constraints are present in the sense that a higher value of a feature in an instance, fixing the other values, should not decrease its class assignment [1,2]. These relationships between the inputs features and the response are termed as monotonic.

In the specialized literature we can find multiple monotonic classifiers proposed, like neural networks and support vector machines [3], classification trees and rule induction [4–6] and instance-based learning [7–9]. As a restriction, some of them require the training set to be purely monotone to work properly. Other classifiers can handle non-monotonic data sets, but they do not guarantee monotone predictions.

In addition, real-life data sets are likely to have noise, which obscures the relationship between features and the class [10,11]. This fact affects the prediction capabilities of the learning algorithms which learn models from those data sets.

In order to address these shortcomings and to test the prediction competences of the monotonic classifiers, the usual trend is to generate data sets which completely satisfy the monotonicity conditions [12]. The intuitive idea behind this

© Springer International Publishing AG, part of Springer Nature 2018
F. J. de Cos Juez et al. (Eds.): HAIS 2018, LNAI 10870, pp. 277–288, 2018.
https://doi.org/10.1007/978-3-319-92639-1_23

is that the models trained on monotonic data sets should offer better predictive performance than the models trained on the original data.

Training Set Selection (TSS) is known as an application of instance selection methods [13–16] over the training set used to build any predictive model. Thus, TSS can be employed as a way to improve the behavior of predictive models, precision and interpretability [17,18].

In this contribution we propose a TSS algorithm to manage monotonic classification problems, called Monotonic Training Set Selection (MonTSS). It is a data preprocessing technique which, by means of a suitable TSS process for monotonic domains, offers an alternative without modifying the class labels of the data set, it instead removes harmful instances. MonTSS incorporates proper measurements to identify and select the most suitable instances in the training set to enhance both the accuracy and the monotonic nature of the models produced by different classifiers. We have compared the results offered by well-known classical monotonic classifiers over 30 data sets with and without the use of MonTSS as a data preprocessing stage. The results show that MonTSS is able to select the most representative instances, which leads monotonic classifiers to always offer equal or better results than without preprocessing.

The contribution is organized as follows. Section 2 describes the proposed MonTSS algorithm. Section 3 describes the experimental framework, with respect to data sets, parameters and quality metrics considered. In Sect. 4 the results and analysis are included. Finally, Sect. 5 concludes the contribution.

2   Proposal for Monotonic Training Set Selection

In TSS the aim is to reduce the size of the training data set by selecting the most representative instances. The effects produced by such data preprocessing, according to [19–21], are: reduction in space complexity, decrease in computational cost and the selection of the most representative instances by discarding noisy ones.

In monotonic classification, besides the previously enumerated advantages, the most representative instances must also keep or improve the monotonicity constraints of the training set. In this manner, the learned models from the data sets resulting from the TSS process are expected to improve their monotonic condition with respect to those obtained from the original training data set.

2.1   Probabilistic Collision Removal

Two instances produce a collision if they do not satisfy the monotonicity constraints. Before carrying out the TSS, we apply an ordered probabilistic removal based on the collisions produced by these instances. The process is stochastic to avoid falling in local optima. For this reason, two parameters are introduced: *Candidates*, which points out the rate of the best candidates to be selected, and *CollisionsAllowed*, which represents the minimal rate of collisions permitted to stop the removal process. The probabilistic algorithm is given in Algorithm 1.

Algorithm 1. Probabilistic Collision Removal algorithm.

1: **function** PROBCOLREMOVAL(T - training data, $Candidates$ - candidate rate, $CollisionsAllowed$ - minimal rate of collisions allowed)
2:     **initialize:** $totalCollisions=0$, $S=T$
3:     $[Col[],ColMatrix[][],totalCollisions]$=Calculate_Collisions($S$)
4:     $maxCandidates = \#S \cdot Candidates$
5:     $minCol = totalCollisions \cdot CollisionsAllowed$
6:     $newCol = totalCollisions$
7:     **repeat**
8:         S=sort(S,$Col[]$)
9:         $candSelected=$ Rand($1,maxCandidates$)
10:        $S = S \setminus \{x_{candSelected}\}$
11:        $newCol = newCol - Col[candSelected]$
12:        $Col[candSelected] = 0$
13:        **for all** $x_j \in T$ **do**
14:            **if** ($ColMatrix[j][candSelected] = true$) **then**
15:                $ColMatrix[j][candSelected] = false$
16:                $Col[j] = Col[j] - 1$
17:            **end if**
18:        **end for**
19:    **until** $newCol < minCol$
20:    **return** S
21: **end function**
22: **function** CALCULATE_COLLISIONS(T - training data)
23:     **for all** $x_i \in T$ **do**
24:         $Col[i]$=0
25:         **for all** $x_j \in T$ **do**
26:             $ColMatrix[i][j]$=0
27:         **end for**
28:     **end for**
29:     **for all** $x_i \in T$ **do**
30:         **for all** $x_j \in T$ **do**
31:             **if** ($i \neq j$) **then**
32:                 **if** ($x_i \preceq x_j$ and $Y(x_i) > Y(x_j)$) or ($x_i = x_j$ and $Y(x_i) \neq Y(x_j)$) **then**
33:                     $Col[i]$=$Col[i] + 1$
34:                     $Col[j]$=$Col[j] + 1$
35:                     $ColMatrix[i][j] = true$
36:                     $ColMatrix[j][i] = true$
37:                 **end if**
38:             **end if**
39:         **end for**
40:     **end for**
41:     $totalCollisions$=0
42:     **for all** $x_i \in T$ **do**
43:         $totalCollisions$=$totalCollisions+Col[i]$
44:     **end for**
45:     **return** $[Col[], ColMatrix[][] , totalCollisions]$
46: **end function**

The efficiency of the algorithm is $O(n^2)$, where n is the number of instances in the training set T.

The more collisions an instance has, the higher is its probability to be removed. After this process, the resulting data is formed by instances with the allowed number of collisions. Although taking this process to the extreme could obtain collisions-free data, it is desirable to provide greater freedom of action to the quality metrics described next.

2.2   Quality Metrics Selection

We introduce two new metrics: *Delimitation* (Del) and *Influence* (Infl), to assess how representational the instances are. Both are calculated for each instance resulting from the previous stage and are used to determine which instances must be included in the final TSS.

The instances located at the boundary of decision classes are of great importance in monotonic classification. They should be conserved to guarantee that the relationship of monotonicity between the inputs and the response remains constant. Hence, both metrics have been proposed to identify the subset of most important instances which belong to the decision boundaries. They are defined for an instance $x_i \in D$ as follows:

- *Delimitation* (Del):

$$\mathrm{Del}(x_i) = \frac{|\mathrm{Dom}(x_i) - \mathrm{NoDom}(x_i)|}{\mathrm{Dom}(x_i) + \mathrm{NoDom}(x_i)}, \tag{1}$$

$$\mathrm{Dom}(x_i) = \#X', x' \in X' \Leftrightarrow x_i \prec x' \wedge Y(x_i) = Y(x'), \tag{2}$$

$$\mathrm{NoDom}(x_i) = \#Z', x' \in Z' \Leftrightarrow x_i \succeq x' \wedge Y(x_i) = Y(x'), \tag{3}$$

where $\mathrm{Dom}(x_i)$ represents the number of instances of the same class as x_i that dominate x_i, and $\mathrm{NoDom}(x_i)$ is the number of instances of the same class as x_i dominated by x_i. The range of $\mathrm{Del}(x_i)$ is $[0, 1)$. A value closer to 0 means that the instance is located near the central region of the class and the value near to 1 means that it is closer to the boundaries.

- *Influence* (Infl). The *Influence* of each instance x_i is computed by using the number of neighbors of any different class label and their distance to x_i. First, the k nearest neighbors are obtained and their corresponding distance to x_i is computed. Each neighbor has an associated weight (nWeight), which depends on its distance to the instance x_i, normalized by the sum of the distances of all the neighbors (see Eq. 5). nWeight is greater as closer the neighbor is to the instance x_i. The value of nWeight for each neighbor is normalized using the Eq. 6 so the *Influence* metric is situated in range $[0,1]$. The *Influence* value for the instance x_i is calculated by considering only the neighbors of different class than x_i among its k nearest neighbors (see Eq. 7):

$$kNN_{x_i} = k \text{ nearest neighbors of } x_i \tag{4}$$

$$\mathrm{nWeight}(x_j) = \frac{\sum_{l=1}^{k} \mathrm{Distance}(x_i, x_l) - \mathrm{Distance}(x_i, x_j)}{\sum_{l=1}^{k} \mathrm{Distance}(x_i, x_l)}, \forall x_j \in kNN_{x_i}, \tag{5}$$

$$\mathrm{influenceWeight}(x_j) = \frac{\mathrm{nWeight}(x_j)}{\sum_{l=1}^{k} \mathrm{nWeight}(x_l)}, \forall x_j \in kNN_{x_i}, \tag{6}$$

$$\mathrm{Infl}(x_i) = \sum_{j=1}^{k} \mathrm{influenceWeight}(x_j),$$

$$\text{where } Y(x_i) \neq Y(x_j) \wedge x_j \in kNN_{x_i}, \tag{7}$$

The value 0 reflects that the instance is surrounded by instances of the same class, showing a high degree of redundancy. The value 1 represents that it is surrounded by neighbor instances of other classes, which means that it is an instance that could introduce separation between classes.

2.3 The MonTSS Algorithm

The whole process is composed of three stages:

1. The MonTSS process starts with a preprocessing step where MonTSS analyzes the original data set by quantifying the relationship between each input feature and the output class. This relation is estimated with a metric called Rank Mutual Information (RMI) defined in [22]. With it, we know the features which have a real direct or inverse monotonic relation with the class or no relation as well (including unordered categorical features). The RMI value is evaluated in the training data set to decide which features are used in the computation of collisions between instances.

   In essence, rank mutual information can be considered as the degree of monotonicity between features $A_1,...,A_f$ and the feature class Y. Given any feature A_j and feature class Y, the value of RMI for the feature A_j is calculated as follows:

$$\text{RMI}(A_j, Y) = -\frac{1}{n} \sum_{i=1}^{n} log \frac{\#[x_i]_{A_j}^{\leq} \cdot \#[x_i]_{Y}^{\leq}}{n \cdot \#([x_i]_{A_j}^{\leq} \cap [x_i]_{Y}^{\leq})} \tag{8}$$

   where n is the number of instances in data set D, $[x_i]_{A_j}^{\leq}$ is the set formed by all the instances of the set D whose feature A_j is less or equal than feature A_j of instance x_i, and $[x_i]_{Y}^{\leq}$ is the set composed of the instances of the set D whose feature class Y is less or equal than feature class Y of instance x_i.

2. In the second stage, the probabilistic collision removal mechanism is applied (see Sect. 2.1), which eliminates most of the instances which produce collisions. The remaining instances are used as input in the last stage.

3. Here, the quality metrics are computed and based on them, the selection procedure is developed considering the following rule:

$$\text{Select} x_i = \begin{cases} \text{true} & \textbf{if } \text{Del}(x_i) < \text{Infl}(x_i) \\ & \text{or } \text{Del}(x_i) \geq 0.9 \\ \text{false} & \text{otherwise.} \end{cases} \tag{9}$$

The rationale behind this rule is to retain the instances which are closer to the class boundaries, using a straightforward threshold of 0.9 which is independent from the diversity of their neighborhood. Furthermore, a relationship between $\text{Del}(x_i)$ and $\text{Infl}(x_i)$ can be easily established as they represent a measurement in the same range of the relative rate of the situation and the neighborhood variety of every instance. In this respect, the rule is built as a function of both measures. As a result, for instance, the rule preserves the instances belonging to central areas if there are instances of other classes around.

3   Experimental Framework

In this section we introduce the data sets used to test our proposed algorithm, the parameters fixed in the algorithms and the performance metrics considered to evaluate the results.

3.1   Data Sets

The data sets have been collected from the KEEL-data set[1] [23] and UCI Repository [24]. They are split into two blocks. In the first one, we enumerate 20 standard classification data sets and the second block is composed of 10 regression data sets whose class feature is discretized into 4 categorical values, keeping the class distribution balanced. The algorithms are evaluated using a 10-fold cross validation schema (10-fcv). Table 2 shows the name of the data sets.

3.2   Parameters

The classical monotonic classification algorithms considered in this study are: Ordinal Learning Model (OLM [25]), Ordinal Stochastic Dominance Learner (OSDL [26,27]), Monotone Induction of Decision trees (MID [4]) and Monotonic k-Nearest Neighbor Classifier (MkNN [7]).

The parameters of the algorithms appear in Table 1 and are fixed for all the data sets. The values for those parameters have been selected according to the recommendations of the authors of each method. The parameter values of our proposal have been empirically determined to achieve a proper balance between reduction and prediction capabilities.

Both for MkNN an MonTSS algorithms, the euclidean distance is adopted to measure the similarity in numerical and categorical ordinal features.

3.3   Performance Metrics

To analyze the results offered by the monotonic classification algorithms, with and without data preprocessing, we have considered the following Mean Absolute Error and data related metrics:

- Mean Absolute Error (MAE): This measure intends to evaluate the prediction capabilities of the algorithms. It is defined as the sum of the absolute values of the errors, divided by the number of classifications. The literature in various studies concludes that MAE is one of the best performance metrics in ordered classification [28].
- Data related metrics: In this case, the aim is to assess the monotonicity of the training data sets offered by the preprocessing proposal.

---

[1]   http://www.keel.es/datasets.php.

Table 1. Parameters for the algorithms considered.

Algorithm	Parameters
MonTSS	Candidates = 0.01, CollisionsAllowed = 0.01,
	k for Influence estimation = 5, distance = euclidean
MkNN	distance = euclidean, k = 3
OLM	modeClassification = conservative
	modeResolution = conservative
MID	C4.5 as base classifier
	2 items per leaf, confidence = 0.25, R = 1
OSDL	balanced = No, classificationType = media
	tuneInterpolationParameter = No, weighted = No,
	upperBound = 1, lowerBound = 0
	interpolationParameter = 0.5,
	interpolationStepSize = 10

- Non-Monotonicity Index (NMI [29]), calculated as the number of non-monotone instances divided by the total number of instances:

$$\text{NMI} = \frac{1}{n} \sum_{x \in D} \text{Collision}(x) \qquad (10)$$

where Collision$(x) = 0$ if x does not collide with any instances in D, and 1 otherwise.
- Non-Comparable, defined as the number of pairs of non comparable instances in the data set. Two instances x and x' are non-comparable if they do no satisfy $x \preceq x' \wedge x \neq x'$. The reason for considering it as a metric is based on the fact that it is harder to build accurate models as the number of non-comparable pairs increases.
- Size of the training set selected using the preprocessing algorithm proposed. It is included to analyze the reduction capabilities of the method.

4   Results and Analysis

The average results of the 10-fcv evaluations are shown in Table 2. In the first column of this table, we present the name of the data set. The second column contains the results corresponding to the combination of Relabeling+MkNN. In the third column we present the results associated with the evaluation of the original data set directly by the monotonic classifiers considered. The fourth column is devoted to the combination of MonTSS and monotonic classifiers. Fifth and sixth columns summarize the monotonicity indexes obtained using the original data sets and after the use of MonTSS. These metrics are NMI, Non-Comparable

Table 2. Predictive MAE and Monotonic Metrics results for the 30 data sets.

Data sets	Relabeling	Original Data Set				MonTSS				Original Data Set			MonTSS		
	+ MkNN	MkNN	OLM	OSDL	MID	+ MkNN	+ OLM	+ OSDL	+ MID	NMI	Non comparable	Size	NMI	Non-comparable	Size
appendicitis	0.735	0.359	0.198	0.226	0.176	0.124	0.105	0.206	0.159	0.980	710.3	95.4	0.000	194.1	48.2
australian	0.271	0.230	0.306	0.264	0.174	0.265	0.349	0.203	0.175	0.152	92,424.6	621.0	0.000	85,242.1	594.8
balance	0.995	0.950	0.990	0.923	0.245	0.239	0.354	0.656	0.235	0.992	79,435.5	562.5	0.000	18,327.7	303.1
breast	0.354	0.326	0.318	0.455	0.325	0.286	0.286	0.336	0.300	0.477	12,342.5	249.3	0.006	6,243.6	180.3
cleveland	0.787	0.771	0.848	0.647	0.710	0.626	0.764	0.664	0.721	0.156	22,814.7	267.3	0.000	16,765.9	235.0
contraceptive	0.856	0.826	0.847	0.763	0.720	0.705	0.747	0.684	0.724	0.864	534,957.4	1325.7	0.104	256,392.9	920.0
era	2.135	2.481	2.150	1.285	1.363	1.746	2.169	1.422	1.830	1.000	291,577.0	900.0	0.269	183.5	24.1
esl	0.608	1.739	0.473	0.361	0.342	0.361	0.486	0.344	0.455	0.798	18,108.9	439.2	0.088	1,933.0	124.4
german	0.359	0.370	0.289	0.626	0.308	0.366	0.293	0.583	0.304	0.071	169,934.6	900.0	0.000	160,463.0	876.6
haberman	0.519	0.709	0.650	0.284	0.288	0.314	0.344	0.295	0.279	0.918	11,117.9	275.4	0.124	1,179.8	97.1
heart	0.285	0.293	0.333	0.374	0.244	0.285	0.293	0.363	0.226	0.034	14,373.2	243.0	0.000	11,663.1	224.9
lev	1.072	1.226	0.668	0.392	0.399	0.541	0.774	0.455	0.525	0.991	213,431.0	900.0	0.124	10,517.9	226.5
mammographic	0.386	0.466	0.193	0.371	0.201	0.173	0.183	0.164	0.170	0.851	54,129.3	747.0	0.213	5,548.7	188.3
pima	0.263	0.267	0.289	0.344	0.273	0.260	0.271	0.344	0.290	0.258	98,662.6	691.2	0.008	72,386.5	582.3
spectfheart	0.299	0.295	0.802	0.217	0.239	0.284	0.794	0.206	0.237	0.024	9,442.8	240.3	0.000	9,444.5	240.3
swd	0.447	1.084	0.763	0.437	0.468	0.519	0.648	0.457	0.487	0.975	234,339.2	900.0	0.060	35,797.1	398.3
wdbc	0.652	0.687	0.661	0.643	0.065	0.035	0.587	0.575	0.051	0.813	57,925.5	512.1	0.000	60,413.9	512.1
wine	0.202	0.247	0.954	0.949	0.084	0.045	0.893	0.618	0.095	0.220	8,417.3	160.2	0.000	8,191.4	160.2
winequality-red	0.726	0.560	1.674	2.529	0.454	0.377	1.410	1.537	0.449	0.313	663,745.0	1439.1	0.003	613,713.1	1392.6
wisconsin	0.036	0.035	0.113	0.041	0.054	0.047	0.159	0.050	0.051	0.023	15,597.8	614.7	0.000	7,091.4	305.4
auto-mpg4cl	1.252	0.964	1.082	0.696	0.248	0.252	0.462	0.365	0.245	0.795	46,346.5	352.8	0.000	10,636.4	208.5
baseball4cl	0.670	0.625	0.587	0.603	0.480	0.590	0.771	0.513	0.480	0.316	29,290.0	303.3	0.000	20,336.4	241.9
bostonhousing4cl	0.546	0.595	1.409	1.014	0.389	0.352	0.595	0.694	0.387	0.128	77,750.9	455.4	0.007	51,657.6	405.4
dee4cl	0.513	0.468	0.784	1.493	0.422	0.312	0.682	1.428	0.387	0.284	38,527.3	328.5	0.000	20,911.0	268.1
diabetes4cl	0.508	1.033	0.542	1.025	1.075	0.583	0.542	1.058	0.608	0.811	212.2	38.7	0.000	71.8	20.3
ele1.4cl	1.660	1.203	1.515	1.107	1.081	1.122	1.140	1.186	1.056	0.991	26,996.2	445.5	0.470	6,346.3	171.2
ele2.4cl	0.067	0.108	0.073	0.905	0.044	0.107	0.118	0.911	0.106	0.105	51,730.7	950.4	0.000	6,836.6	336.8
mortgage4cl	0.043	0.043	1.056	1.433	0.049	0.166	0.407	0.989	0.105	0.000	332,281.2	944.1	0.000	35,731.5	632.1
stock4cl	0.564	0.583	0.993	1.008	0.417	0.397	1.221	1.444	0.417	0.327	269,251.2	855.0	0.006	249,201.5	823.9
treasury4cl	0.043	0.043	1.085	1.443	0.073	0.184	0.446	0.984	0.114	0.003	332,476.0	944.1	0.000	28,968.8	596.3
Average	0.595	0.652	0.754	0.761	0.380	0.388	0.609	0.657	0.388	0.489	126,944.977	590.040	0.049	60,413.037	377.967

and Size. The last row in the table contains the average results over the 30 data sets evaluated. The best values for each row have been highlighted in bold.

To strengthen the analysis, we have included the statistical outcome based on non-parametric procedures to compare the performances offered by our proposal with respect to the original classifiers without preprocessing. For this, the Wilcoxon signed-rank test has been used to conduct pairwise comparisons between two algorithms [30]. The Wilcoxon test involves ranking all nonzero difference scores disregarding sign, reattaching the sign to the rank, and then evaluating the mean of the positive and the mean of the negative ranks. The Wilcoxon test was specially recommended by Demsar in [31] to make performance comparisons of pairs of algorithms on multiple data sets.

Observing Tables 2 and 3, we come up with the following analysis:

- In most of the cases, the prediction capabilities of the classifiers are improved when the TSS is applied. This conclusion is confirmed when statistical tests are applied, as Table 3 reflects. Using the Wilcoxon test we compare each classifier with its preprocessed version and the table indicates to us that MonTSS improves MkNN with and without relabeling, OLM and OSDL. MID is robust to noise and its predictive performances are the same with or without TSS, indicating that MonTSS does not change its behavior. The greatest improvement noted is the combination of MonTSS with MkNN, achieving similar performances to MID.
- Regarding data related metrics, in Table 2 it must be mentioned that the monotonic models extracted using our proposal are those which best satisfy the monotonicity constraints (with NMI near to 0). In addition, MonTSS significantly reduces the number of non-comparable pairs of instances and the size of the training data set. This occurs in all data sets evaluated.
- It is worth mentioning that relabeling clearly improves standard MkNN, a fact that was suggested in [7]. Nevertheless, MonTSS is able to improve even more the performance of MkNN, becoming a competitive algorithm.

Table 3. Wilcoxon test to analyze the MAE results on classifiers with and without our preprocessing proposal.

Algorithm	Ranking	Adjusted p-value	vs. Algorithm
MonTSS+MkNN	R^+ 382.0	**0.000**	Relabeling+MkNN
	R^- 53.0		
MonTSS+MkNN	R^+ 427.0	**0.000**	MkNN
	R^- 38.0		
MonTSS+OLM	R^+ 335.5	**0.010**	OLM
	R^- 100.0		
MonTSS+OSDL	R^+ 327.5	**0.016**	OSDL
	R^- 107.5		
MonTSS+MID	R^+ 231.0	1.000	MID
	R^- 234.0		

5   Concluding Remarks

This contribution aims to present a proposal of training set selection for monotonic classification called MonTSS. It selects the most representative instances focusing on the class distribution and monotonicity conditions of those instances. The experimental evaluation highlights some remarkable characteristics of the proposal.

MonTSS is able to select the most representative instances independently of the classifier to be applied later. This leads monotonic classifiers to always offer equal or better results than without preprocessing. Furthermore, data related metrics are notably improved, fully satisfying the monotonicity restrictions without affecting or modifying the nature of the original data. At the same time, it reduces the number of non-comparable pairs of instances and the size of the training data sets before the learning stage starts.

As future work we are interested in the use of the proposal in big data [32].

Acknowledgement. This work was supported by TIN2014-57251-P, by the Spanish "Ministerio de Economía y Competitividad" and by "Fondo Europeo de Desarrollo Regional" (FEDER) under Project TEC2015-69496-R and the Foundation BBVA project 75/2016 BigDaPTOOLS.

References

1. Kotłowski, W., Słowiński, R.: On nonparametric ordinal classification with monotonicity constraints. IEEE Trans. Knowl. Data Eng. **25**(11), 2576–2589 (2013)
2. Gutiérrez, P.A., García, S.: Current prospects on ordinal and monotonic classification. Prog. Artif. Intell. **5**(3), 171–179 (2016)
3. Chen, C.C., Li, S.T.: Credit rating with a monotonicity-constrained support vector machine model. Expert Syst. Appl. **41**(16), 7235–7247 (2014)
4. Ben-David, A.: Monotonicity maintenance in information theoretic machine learning algorithms. Mach. Learn. **19**, 29–43 (1995)
5. Potharst, R., Bioch, J.: Decision trees for ordinal classification. Intell. Data Anal. **4**, 97–111 (2000)
6. Alcalá-Fdez, J., Alcalá, R., González, S., Nojima, Y., García, S.: Evolutionary fuzzy rule-based methods for monotonic classification. IEEE Trans. Fuzzy Syst. **25**(6), 1376–1390 (2017)
7. Duivesteijn, W., Feelders, A.: Nearest neighbour classification with monotonicity constraints. In: Daelemans, W., Goethals, B., Morik, K. (eds.) ECML PKDD 2008, Part I. LNCS (LNAI), vol. 5211, pp. 301–316. Springer, Heidelberg (2008). https://doi.org/10.1007/978-3-540-87479-9_38
8. García, J., Albar, A., Aljohani, N., Cano, J.R., García, S.: Hyperrectangles selection for monotonic classification by using evolutionary algorithms. Int. J. Comput. Intell. Syst. **9**(1), 184–201 (2016)
9. García, J., Fardoun, H.M., Alghazzawi, D.M., Cano, J.R., García, S.: Mongel: monotonic nested generalized exemplar learning. Pattern Anal. Appl. **20**(2), 441–452 (2017)
10. Frénay, B., Verleysen, M.: Classification in the presence of label noise: a survey. IEEE Trans. Neural Netw. Learn. Syst. **25**(5), 845–869 (2014)

11. Triguero, I., González, S., Moyano, J.M., García, S., Alcalá-Fdez, J., Luengo, J., Fernández, A., del Jesús, M.J., Sánchez, L., Herrera, F.: Keel 3.0: an open source software for multi-stage analysis in data mining. Int. J. Comput. Intell. Syst. **10**(1), 1238–1249 (2017)
12. Feelders, A.: Monotone relabeling in ordinal classification. In: IEEE International Conference on Data Mining (ICDM), pp. 803–808 (2010)
13. García, S., Derrac, J., Cano, J.R., Herrera, F.: Prototype selection for nearest neighbor classification: taxonomy and empirical study. IEEE Trans. Pattern Anal. Mach. Intell. **34**(2), 417–435 (2012)
14. Silva, D.A., Souza, L.C., Motta, G.H.: An instance selection method for large datasets based on markov geometric diffusion. Data Knowl. Eng. **101**, 24–41 (2016)
15. García, S., Luengo, J., Herrera, F.: Tutorial on practical tips of the most influential data preprocessing algorithms in data mining. Knowl. Based Syst. **98**, 1–29 (2016)
16. Cano, J.R., Aljohani, N.R., Abbasi, R.A., Alowidbi, J.S., García, S.: Prototype selection to improve monotonic nearest neighbor. Eng. Appl. Artif. Intell. **60**, 128–135 (2017)
17. Cano, J.R., Herrera, F., Lozano, M.: Stratification for scaling up evolutionary prototype selection. Pattern Recogn. Lett. **26**(7), 953–963 (2005)
18. Cano, J.R., García, S., Herrera, F.: Subgroup discover in large size data sets preprocessed using stratified instance selection for increasing the presence of minority classes. Pattern Recogn. Lett. **29**(16), 2156–2164 (2008)
19. García, S., Luengo, J., Herrera, F.: Data Preprocessing in Data Mining. Springer, Heidelberg (2015). https://doi.org/10.1007/978-3-319-10247-4
20. Cano, J.R., Herrera, F., Lozano, M.: On the combination of evolutionary algorithms and stratified strategies for training set selection in data mining. Appl. Soft Comput. **6**(3), 323–332 (2006)
21. Nanni, L., Lumini, A., Brahnam, S.: Weighted reward-punishment editing. Pattern Recogn. Lett. **75**, 48–54 (2016)
22. Hu, Q., Che, X., Zhang, L., Zhang, D., Guo, M., Yu, D.: Rank entropy-based decision trees for monotonic classification. IEEE Trans. Knowl. Data Eng. **24**(11), 2052–2064 (2012)
23. Alcalá, J., Fernández, A., Luengo, J., Derrac, J., García, S., Sánchez, L., Herrera, F.: Keel data-mining software tool: Data set repository, integration of algorithms and experimental analysis framework. J. Mult. Valued Logic Soft Comput. **17**(255–287), 11 (2010)
24. Bache, K., Lichman, M.: UCI machine learning repository (2013)
25. Ben-David, A., Serling, L., Pao, Y.: Learning and classification of monotonic ordinal concepts. Comput. Intell. **5**, 45–49 (1989)
26. Lievens, S., De Baets, B., Cao-Van, K.: A probabilistic framework for the design of instance-based supervised ranking algorithms in an ordinal setting. Ann. Oper. Res. **163**, 115–142 (2008)
27. Lievens, S., De Baets, B.: Supervised ranking in the weka environment. Inf. Sci. **180**(24), 4763–4771 (2010)
28. Gaudette, L., Japkowicz, N.: Evaluation methods for ordinal classification. In: Gao, Y., Japkowicz, N. (eds.) AI 2009. LNCS (LNAI), vol. 5549, pp. 207–210. Springer, Heidelberg (2009). https://doi.org/10.1007/978-3-642-01818-3_25
29. Milstein, I., Ben-David, A., Potharst, R.: Generating noisy monotone ordinal datasets. Artif. Intell. Res. **3**(1), 30–37 (2014)
30. Gibbons, J.D., Chakraborti, S.: Nonparametric statistical inference. In: Lovric, M. (ed.) International Encyclopedia of Statistical Science. Springer, Heidelberg (2011). https://doi.org/10.1007/978-3-642-04898-2_420

31. Demšar, J.: Statistical comparisons of classifiers over multiple data sets. J. Mach. Learn. Res. **7**, 1–30 (2006)
32. Triguero, I., Peralta, D., Bacardit, J., García, S., Herrera, F.: Mrpr: a mapreduce solution for prototype reduction in big data classification. Neurocomputing **150**, 331–345 (2015)

Dealing with Missing Data and Uncertainty in the Context of Data Mining

Aliya Aleryani[1,2(✉)], Wenjia Wang[1], and Beatriz De La Iglesia[1]

[1] University of East Anglia, Norwich NR4 7TJ, UK
A.Aleryani@uea.ac.uk
[2] King Khalid University, Abha 61421, Saudi Arabia

Abstract. Missing data is an issue in many real-world datasets yet robust methods for dealing with missing data appropriately still need development. In this paper we conduct an investigation of how some methods for handling missing data perform when the uncertainty increases. Using benchmark datasets from the UCI Machine Learning repository we generate datasets for our experimentation with increasing amounts of data Missing Completely At Random (MCAR) both at the attribute level and at the record level. We then apply four classification algorithms: C4.5, Random Forest, Naïve Bayes and Support Vector Machines (SVMs). We measure the performance of each classifiers on the basis of complete case analysis, simple imputation and then we study the performance of the algorithms that can handle missing data. We find that complete case analysis has a detrimental effect because it renders many datasets infeasible when missing data increases, particularly for high dimensional data. We find that increasing missing data does have a negative effect on the performance of all the algorithms tested but the different algorithms tested either using preprocessing in the form of simple imputation or handling the missing data do not show a significant difference in performance.

Keywords: Missing data · Classification algorithms
Complete case analysis · Single imputation

1 Introduction

Many real-world datasets have missing or incomplete data [24]. Since the accuracy of most machine learning algorithms for classification, regression, and clustering is affected by the completeness of datasets, processing and dealing with missing data is a significant step in the Knowledge Discovery and Data Mining (KDD) process. Some strategies have been devised to handle incomplete data as explained in [5,8,14]. In particular, for regression, where missing data has been more widely studied (e.g. [9]), multiple imputation has shown advantage over other methods [22,23]. However, much work is still needed to solve this problem

© Springer International Publishing AG, part of Springer Nature 2018
F. J. de Cos Juez et al. (Eds.): HAIS 2018, LNAI 10870, pp. 289–301, 2018.
https://doi.org/10.1007/978-3-319-92639-1_24

in the context of data mining tasks and multiple imputation in particular needs some research to show if it is equally applicable to data mining.

Before we investigate multiple imputation and data mining, which is our long term aim, in this research we want to deliver a thorough understanding of how the different methods for handling missing data affect the accuracy of data mining algorithms when the uncertainty increases, i.e. the amount of missing data increases. We create an experimental environment using the university of California Irvine (UCI) Machine learning repository [13], by removing data from a number of UCI datasets completely at random (MCAR). We select increasing number of attributes at random to remove data from and we also increase the number of records at random from which we remove data in the attributes selected. Therefore, we produce a number of experimental datasets which contain increasing amounts of data MCAR.

Researchers have used a number of different methods to treat missing data in the data preprocessing phase. In this paper, we study the performance of classification algorithms in the context of increasing missing data under different pre-processing scenarios. In particular, we investigate how increasing the amount of missing data affects the performance for complete case analysis, and single imputation for a number of classification algorithms. We also compare that to the performance of algorithms with an internal mechanisms to handle the missing data, such as C4.5, and Random Forest.

The rest of this paper is organised as follows: Sect. 2 presents the problem of missing data and Sect. 3 presents the mechanisms used in Data Mining to address the problem. The methods used in our paper to set up our experimental environment are discussed in Sect. 4. Section 5 analyses the results. A discussion of the results is in Sect. 6. Finally, Sect. 7 presents our conclusions.

2   The Problem of Missing Data

Little and Rubin [14] have defined missing data based on the mechanism that generates the missing values into three main categories as follows: Missing Completely at Random (MCAR), Missing at Random (MAR), and Missing not at Random (MNAR). Missing Completely at Random (MCAR) occurs when the probability of an instance missing for a particular variable is independent from any other variable and independent from the missing data so missing is not related to any factor known or unknown in the study. Missing at Random (MAR) occurs when the probability of an instance having a missing value for an attribute may depend on the known values but not on the value of the missing data itself. Missing not at Random (MNAR) occurs when the probability of the instance having a missing value depends on unobserved values. This is also termed a *non-ignorable* process and is the most difficult scenario to deal with. In this paper we focus on generating missing data using the MCAR mechanism. Further work will investigate the other mechanisms.

Horton et al. [9] have further categorized the patterns of missing data into monotone and non-monotone. They state that the patterns are concerned with which values are missing, whereas, the mechanisms are concerned with why data is missing. We can state that we have monotone patterns of missing data if the same data points have missing values in one or more features. We focus in this study on non-monotone missing data.

3 Dealing with Missing Data

In practice, there are three popular approaches that are commonly used to deal with incomplete data:

1. **Complete Case Analysis:** This approach is the default in many statistical packages but should be only used when missing is under MCAR [14]. All incomplete data points are simply omitted from the dataset and only the complete records are used for model building [14]. The approach results in decreasing the size of data and the information available to the models and may also bias the results [20]. Tabachnick and Fidell [21] assumed that both the mechanisms and the patterns of missing values play a more significant role than the proportion of missing data when complete case analysis is used.

2. **Imputation:** Imputation means that missing values are replaced in some way prior to the analysis [14]. Mean or median imputation is commonly used with numerical instances and mode imputation with the nominal instances. Such simple imputation methods have been criticized widely [4,18], because they do not reflect the uncertainty in the data and may introduce bias in the analysis. On the other hand, multiple imputation [17], a more sophisticated method, replaces missing values with a number of plausible values which reflect the uncertainty although the technique may have higher computational complexity. A method for combining the results of the analysis on multiple datasets is also required. For regression analysis, Rubin [17] defined some rules to estimate parameters from multiple imputation analysis. For application to data mining, good methods for pooling the analysis may be required.

3. **Model Approach:** A number of algorithms have been constructed to cope with missing data, that is, they can develop models in the presence of incomplete data. The internal mechanisms for dealing with missing data are discussed in the context of the algorithms used in this study.

3.1 The Classification Algorithms and Missing Data

We focus on the following well known classification algorithms, some of which have been identified as top data mining algorithms [25]: Decision Trees (C4.5), Naïve Bayes (NB), Random Forest (RF) and Support Vector Machines (SVMs). Further, we will explain how different algorithms and their implementations in *Weka*, our platform of choice, can treat missing values at both the building and the application phase.

C4.5 is one of the most influential decision trees algorithm. The algorithm was modified by Quinlan [15,16] to treat missing data using *fractional* method in which the proportion of missing values of an attribute are used to modify the *Information gain* and *Split ratio* of the attribute's *Gain ratio*. After making the decision for splitting on an attribute with the highest gain ratio, any instance with missing values of that attribute is split into several fractional instances which may travel down different branches of the tree. When classifying an instance with missing data, the instance is split into several fractional instances and the final classification decision is a combination of the fractional cases [6]. We use the *Weka* implementation, J48, which uses the fractional method [7].

Naïve Bayes algorithm is based on the Bayes theorem of probabilities using the simplification that the features are independent of one another. Naïve Bayes ignores features with missing values thus only the complete features are used for classification [2,11]. Therefore, it uses complete case analysis instead of handling missing data internally.

Random Forest is an ensemble algorithm which produces multiple decision trees and can be used for classification and regression. It is considered as a robust algorithm and produces high classification accuracies. This is because random forest splits training samples to a number of subsets then builds a tree for each subset, rather than building one tree [1] and combines their decision. Random Forest, uses the *fractional* method [1,10] for missing data in a similar manner to C4.5. The implementation of the algorithm in Weka also uses the fractional method as in C4.5 algorithm.

SVMs are used for binary classification and can be extended to higher dimensional datasets using the Kernel function [19]. SVMs maximize the margin between the separating hyperplane and the classes. The decision function is determined by a subset of training samples which are the support vectors. We use a Weka implementation called SMO (Sequential Minimal Optimization), a modification of the algorithm that solves the problem of Quadratic Programming (QP) when training SVMs in higher dimensions without extra storage or optimization calculations. Although SVMs do not deal with missing values [12], the SMO implementation performs simple imputation by globally replacing the missing values with the mode if the attribute is nominal or with the mean if the attribute is continuous [7].

4   Methods

For our study, a collection of 17 benchmark datasets are collected from UCI machine learning repository [13]. The datasets have different sizes and feature types (numerical continuous, numerical integer, categorical and mixed) as shown in Table 1. None of the datasets have missing values in their original form so this enables us to study how missing data affects the accuracy and performance of classification algorithms.

Data values are then removed completely at random as follows to generate increasing amounts of missing data. First, 10% (then 20%, 50%) of the attributes are randomly selected then missing values are artificially generated by removing values randomly in 5%, 30% and 50% of the records, respectively. As a result,

Table 1. The details of the datasets collected for the experiments.

No.	Dataset	#Features	#Instances	#Classes	Feature types
1	Post-Operative Patient	8	90	4	Integer, Categorical
2	Ecoli	8	336	8	Real
3	Tic-tac-toe	9	958	2	Categorical
4	Breast Tissue	10	106	6	Real
5	Statlog	20	1000	2	Integer, Categorical
6	Flags	30	194	8	Integer, Categorical
7	Breast Cancer Wisconsin	32	569	2	Real
8	Chess	36	3196	2	Categorical
9	Connectionist Bench	60	208	2	Real
10	Spect	69	287	2	Categorical
11	Hill Valley	101	606	2	Real
12	Urban Land Cover	148	168	9	Integer, Real
13	Epileptic Seizure Recognition	179	11500	5	Integer, Real
14	Semeion	256	1593	2	Integer
15	LSVT Voice Rehabilitation	309	126	2	Real
16	HAR Using Smartphones	561	10299	6	Real
17	Isolet	617	7797	26	Real

Table 2. Experimental scenarios with missing data artificially created.

Scenario	%Features	%Missing
Scenario 1	10	5
Scenario 2		30
Scenario 3		50
Scenario 4	20	5
Scenario 5		30
Scenario 6		50
Scenario 7	50	5
Scenario 8		30
Scenario 9		50

nine artificial datasets are produced for each of the original datasets with multiple levels of missing data. In total, we have 153 datasets. Table 2 summarises the experimental scenarios artificially created.

For testing the models, 10-fold cross-validation was used and performed 10 times. All results reported represent the average of the 10 experiments with 10-fold cross-validation.

In the complete case analysis, all the incomplete records are omitted. This often results in datasets that are too sparse to be used for classification. The datasets that are left with enough records for classification are considered feasible.

To test simple imputation, the numerical attributes are replaced with their mean and the categorical attributes with their mode. Then the produced datasets after imputation are used for classification model building.

We use the classifiers: J48, Naïve Bayes, RandomForest and SMO implemented in *Weka* with their default options for classifying the data. We use the classification accuracy as a metric for our experiments. To further compare performance of the classifiers, we compute the average of the percentage difference in accuracy between a classifier obtained with the original (complete) datasets and the datasets with increasing missing data as follows:

$$\%\text{Diff} = (((Acc\_Sce_i - Acc\_Org_j)/Acc\_Org_j) * 100) \tag{1}$$

where Acc_Sce_i represents the classifier accuracy for a specific scenario, in our experiment we have 9 scenarios, and Acc_Org_j represents the classifier accuracy of the corresponding original dataset.

We perform two different statistical tests when evaluating the performance of classifiers over the datasets as follows:

1. When comparing differences in accuracy for each scenario we first use Wilcoxon Signed Rank test with a significance level at $\alpha = 0.05$.
2. We then compare multiple classifiers over multiple datasets using the method described by Demšar [3], including the Friedman test and the post hoc Nemenyi test which is presented as a Critical Difference diagram, with a significance level of $\alpha = 0.05$.

5   Results

Figure 1 shows the average accuracy of classifiers and standard deviation (as error bars) for each of the original complete datasets along with the baseline *majority class model* accuracy. Models perform better than the baseline in most of the datasets except Post-Operative Patient, Breast Tissue, Spect, and LSVT Voice Rehabilitation, where default accuracy is similar or slightly better than that obtained by the models. We use the Friedman test for statistical differences. The resulting p-value <0.05, so we proceed with Nemenyi test. The Critical Difference diagram for the Nemenyi test is shown in Fig. 2. The Figure illustrates that SMO and RandomForest behave better than J48 and Naïve Bayes although there is no statistical differences within each group.

5.1   Complete Case Analysis

The datasets that are not feasible for classification after removing missing records are marked with ✗ whereas the feasible are marked with ✓ as shown in Table 3. Datasets are ordered by increasing number of attributes (dimensionality) and then number of records. Only two low dimensional datasets are feasible for classification in all scenarios: Ecoli and Tic-tac-toe. In contrast, datasets with increasing dimensionality are not feasible for classification when increasing the

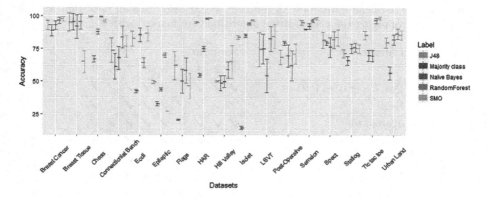

Fig. 1. The average accuracy of classifiers and standard deviation (as error bars) for each of the original (complete) datasets along with majority class.

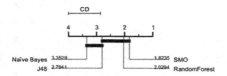

Fig. 2. Critical Difference diagram shows the statistical difference between the classifiers. The bold line connecting classifiers means that they are not statistically different.

amount of missing data due to widespread sparsity. For example, Hill Valley, UrbanLandCover, Epileptic Seizure Recognition, Semeion, LSVT Voice Rehabilitation, HAR Using Smartphones and Isolet all become mostly infeasible.

Figure 3 illustrates the average accuracy of the classifiers and standard deviation for the datasets that are feasible for classification. In scenario 1, the average and the standard deviation are nearly equal to those on the original data. However, with a decreasing number of feasible datasets, the standard deviation increases and the classifiers' performance deteriorate as we increase missing data.

Table 4 shows the average %Diff in accuracy between classifiers obtained with the original (complete) data and the datasets with increasing missing data for the different data handling approaches and algorithms. For complete case analysis, the deterioration in accuracy reached more than 18% for J48, RF, and SMO in different scenarios. However, Naïve Bayes behaved better gaining 2% in some scenarios. We do not produce statistical analysis due to the small number of datasets that produce a feasible classification with complete analysis.

5.2   Simple Imputation

Table 4 also shows the average of all the percentage differences in accuracy (%Diff) between a classifier obtained with the original (complete) datasets and the imputed data for each scenario and each algorithm. %Diff increases when

Table 3. The artificial datasets with different scenarios of missing data that are not feasible when applying the classification algorithms are marked with ✗.

Dataset	Scenario								
	1	2	3	4	5	6	7	8	9
Post-Operative Patient	✓	✓	✓	✓	✓	✓	✓	✓	✗
Ecoli	✓	✓	✓	✓	✓	✓	✓	✓	✓
Tic-tac-toe	✓	✓	✓	✓	✓	✓	✓	✓	✓
Breast Tissue	✓	✓	✓	✓	✓	✓	✓	✓	✗
Statlog	✓	✓	✓	✓	✓	✓	✓	✓	✗
Flags	✓	✓	✓	✓	✓	✗	✓	✗	✗
Breast Cancer Wisconsin	✓	✓	✓	✓	✗	✗	✓	✗	✗
Chess	✓	✓	✓	✓	✓	✓	✓	✗	✗
Connectionist Bench	✓	✓	✗	✓	✗	✗	✓	✗	✗
Spect	✓	✓	✓	✓	✓	✗	✓	✗	✗
Hill Valley	✓	✓	✗	✓	✗	✗	✓	✗	✗
UrbanLandCover	✓	✗	✗	✓	✗	✗	✗	✗	✗
Epileptic Seizure Recognition	✓	✓	✗	✓	✗	✗	✓	✗	✗
Semeion	✓	✗	✗	✓	✗	✗	✗	✗	✗
LSVT Voice Rehabilitation	✓	✗	✗	✗	✗	✗	✗	✗	✗
HAR Using Smartphones	✓	✗	✗	✓	✗	✗	✗	✗	✗
Isolet	✗	✗	✗	✗	✗	✗	✗	✗	✗

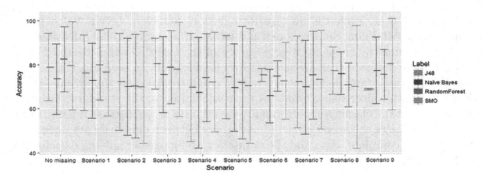

Fig. 3. The average accuracy of classifiers and standard deviation (as error bars) for all artificial datasets in all scenarios of missing data including the original (complete) datasets when applying complete case analysis.

missing data increases in all classifiers, however simple imputation performs much better than complete case analysis. Accuracy decreased in a small range between $[-0.54, -5.59]$ for J48 and by -6.94% for RandomForest in the worst case. For Naïve Bayes, the differences with the original data where smaller with

Table 4. Average % diff in accuracy with respect to complete data. Wilcoxon Signed Rank is used to test statistical significance with significant results marked by *.

Scenario #	Complete case				Simple imputation				Algorithms only			
	J48	NB	RF	SMO	J48	NB	RF	SMO	J48	NB	RF	SMO
Scenario 1	−3.27	0.26	−2.28	−2.07	−0.54*	0.01	−0.10	−0.57*	−0.19	0.05	−0.41*	−0.59*
Scenario 2	−6.88	−3.92	−12.08	−8.00	−0.57	0.11	−0.82	−1.34	−0.38	0.22	−0.72	−1.38
Scenario 3	−1.82	−3.81	2.82	−4.04	−0.96*	−0.30	−1.10*	−2.03*	−0.64	−0.48	−1.33*	−2.04*
Scenario 4	−11.50	−9.43	−14.29	−6.75	−0.83*	−0.17	−0.56	−1.05*	−0.53	0.00	−0.66*	−1.08*
Scenario 5	−8.99	−10.01	−11.26	−12.97	−1.24*	−0.14	−1.62*	−2.26*	−0.56	0.00	−1.50*	−2.31*
Scenario 6	−8.35	−16.03	−6.58	−12.03	−1.62*	−0.84	−2.31*	−3.07	−0.99	−0.78	−1.85*	−2.95
Scenario 7	−6.80	−4.11	−4.27	−1.14	−1.27*	0.22	−2.03*	−1.39	−1.02*	0.10	−1.94*	−1.25
Scenario 8	−3.64	−2.30	−8.75	−14.04	−3.86*	−1.41	−5.11*	−5.11*	−2.56*	−1.15*	−4.78*	−5.04*
Scenario 9	−18.17	−2.10	−4.31	−13.83	−5.59*	−2.67*	−6.94*	−5.71*	−3.95*	−1.42*	−5.85*	−5.82*

a maximum deterioration of −2.67%. SMO sees deteriorations of up to 5.71% in the scenarios of most missing data. We applied the Wilcoxon Signed Rank test to check statistical significance over the differences. Significant values are marked with * and tend to be those for the higher scenarios, except for SMO where the differences are more often statistically significant. From this we can conclude that simple imputation may work well for low amounts of missing data, and is beneficial over complete case analysis, but performance deteriorates significantly when the amount of missing data increases.

We also applied the Friedman test described by Demšar [3] and found statistically significant differences over multiple datasets in all scenarios except scenario 9 so we proceeded with the Nemenyi Test. We perform the post test between the classifiers over the imputed datasets for each scenario separately. The resulting Critical Difference diagrams in most scenarios in Fig. 4 show that RandomForest and SMO outperform J48 and Naïve Bayes. Random Forest seems to outperform SMO as the amount of missing data increases but not significantly. There is no statistical difference between RandomForest, SMO, and J48 in most scenarios. Overall, RandomForest was the most accurate classifier when the uncertainty increases and Naïve Bayes was the worst.

5.3 Building Models with Missing Data

In Sect. 3.1 we discussed that some of this algorithm have their own ways of dealing with missing data. We therefore pass all the data including missing data to the algorithms without preprocessing. We again compare (%Diff) in accuracy between a classifier obtained with the original (complete) datasets and the models built with missing data and show results in Table 4 with statistically significant differences marked by *. %Diff increases when missing data increases in all classifiers. However, for J48 in most scenarios the deterioration is within a small range [−0.19%, −3.95%] and similarly for RandomForest [−0.41%, −5.85%]. Naïve Bayes only ignores the missing values when computing the probability and the differences ranged between [+0.22%, −1.42]. SMO uses (mean/mode) imputation so behaves similarly to the imputed data performance in Table 4.

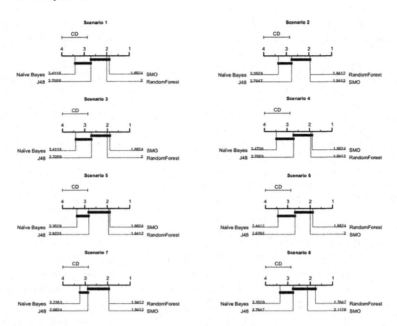

Fig. 4. Critical Difference diagrams show the statistical significant differences between classifiers using simple imputation. We exclude scenario 9 where all classifiers are not statistically different with the Friedman test.

In scenarios 8 and 9, the accuracy of all classifiers are statistically different from the classifiers' accuracy for the original datasets. Thus, the capabilities of classifiers dealing with missing data seem to deteriorate when the ratio of missing data increases.

As before we apply the Friedman test and Nemenyi post test. The resulting Critical Difference diagrams in most scenarios show that RandomForest and SMO outperform J48 and Naïve Bayes. However, there is no statistical difference between all classifiers in scenario 9 whereas no statistical significant between RandomForest, SMO and J48 and between Naïve Bayes and J48. SMO was the most accurate classifier in the first six scenarios, however, when increasing missing data RandomForest outperforms other classifiers and Bayes was the worst in all scenarios. Figure 5 represents the Critical Difference diagrams of all scenarios.

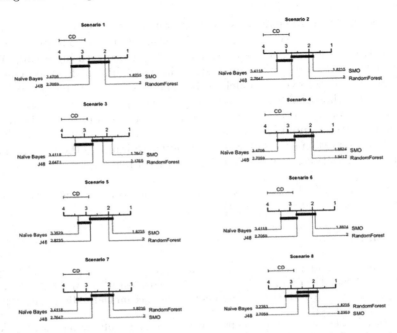

Fig. 5. Critical difference diagrams show the statistical difference between classifiers with no preprocessing of missing data, excluding scenario 9 where all classifiers are not statistically different.

6 Discussion

With complete data, Naïve Bayes and J48 perform worse than SMO and Random Forest. Complete case analysis results in many datasets becoming infeasible for analysis due to sparsity of the data for the algorithms we tested, thus it is not recommended if missing values are spread among records in high dimensional data. Simple imputation works well for low amounts of missing data but not when the amount of missing data increases substantially (scenarios 8,9), as the performance of all classifiers becomes statistically significantly worse than classifying with complete data. RandomForest and SMO behave better than J48 and Naïve Bayes in all scenarios (including when complete data is available). The capability to cope with missing data for RandomForest by using fractional method when uncertainty increases seems to outperform the SMO handling of missing data using mean/mode but not significantly.

7 Conclusion

Accuracy deteriorates for most classifiers when increasing percentages of missing data are encountered. Complete case analysis is not recommended if missing values are spread among (Features/Records) in high dimensional data. Simple imputation may help when a dataset has low ratio of missing values but not with increasing uncertainty. When applying the algorithms without preprocessing, again the trend is for some deterioration in performance with increasing missing data with those differences becoming statistically significant for the higher scenarios. So overall, we expect models to become worse as the amount of missing data increases though different algorithms do not perform significantly differently under those scenarios. As future work, we will expand on our imputation to include multiple imputation that combines models generated from multiple imputed datasets with data ensemble techniques to improve the performance of data mining classification algorithms for data with missing values.

Acknowledgments. This work was supported by the Economic and Social Research Council (grant number ES/L011859/1).

References

1. Breiman, L.: Random forests. Mach. Learn. **45**(1), 5–32 (2001)
2. Chai, X., Deng, L., Yang, Q., Ling, C.X.: Test-cost sensitive naive bayes classification. In: 2004 Fourth IEEE International Conference on Data Mining, ICDM 2004, pp. 51–58. IEEE (2004)
3. Demšar, J.: Statistical comparisons of classifiers over multiple data sets. J. Mach. Learn. Res. **7**(Jan), 1–30 (2006)
4. Fichman, A., Cummings, J.N.: Multiple imputation for missing data: Making the most of what you know. Organ. Res. Meth. **6**(3), 282–308 (2003)
5. García-Laencina, P.J., Sancho-Gómez, J.-L., Figueiras-Vidal, A.R.: Pattern classification with missing data: a review. Neural Comput. Appl. **19**(2), 263–282 (2010)
6. Gavankar, S., Sawarkar, S.: Decision tree: Review of techniques for missing values at training, testing and compatibility. In: 2015 3rd International Conference on Artificial Intelligence, Modelling and Simulation (AIMS), pp. 122–126. IEEE (2015)
7. George-Nektarios, T.: Weka classifiers summary. Athens University of Economics and Bussiness Intracom-Telecom, Athens (2013)
8. Grzymala-Busse, J.W., Hu, M.: A comparison of several approaches to missing attribute values in data mining. In: Ziarko, W., Yao, Y. (eds.) RSCTC 2000. LNCS (LNAI), vol. 2005, pp. 378–385. Springer, Heidelberg (2001). https://doi.org/10.1007/3-540-45554-X_46
9. Horton, N., Kleinman, K.P.: Much ado about nothing: a comparison of missing data methods and software to fit incomplete data regression models. Am. Stat. **61**, 79–90 (2007)
10. Khalilia, M., Chakraborty, S., Popescu, M.: Predicting disease risks from highly imbalanced data using random forest. BMC Med. Inf. Decis. Making **11**(1), 51 (2011)
11. Kohavi, R., Becker, B., Sommerfield, D.: Improving simple bayes. In: Proceedings of the European Conference on Machine Learning. Citeseer (1997)
12. Kotsiantis, S.B., Zaharakis, I., Pintelas, P.: Supervised machine learning: a review of classification techniques. Emerg. Artif. Intell. Appl. Comput. Eng. **160**, 3–24 (2007)
13. Lichman, M.: UCI machine learning repository (2013). http://archive.ics.uci.edu/ml

14. Little, R.J.A., Rubin, D.B.: Statistical Analysis With Missing Data. Wiley, Hoboken (2014)
15. Quinlan, J.R.: C4.5: Programs for Machine Learning. Elsevier, San Francisco (2014)
16. Quinlan, J.R., et al.: Bagging, boosting, and c4. 5. In: The Association for the Advancement of Artificial Intelligence (AAAI), vol. 1, pp. 725–730 (1996)
17. Donald, B.: Rubin. Multiple imputation after 18+ years. J. Am. Stat. Assoc. 91(434), 473–489 (1996)
18. Scheffer, J.: Dealing with missing data. Res. Lett. Inf. Math. Sci. 3(1), 153–160 (2002)
19. Schölkopf, B., Burges, C.J.C., Smola, A.J.: Advances in Kernel Methods: Support Vector Learning. MIT press, Cambridge (1999)
20. Soley-Bori, M.: Dealing with missing data: Key assumptions and methods for applied analysis. Boston University School of Public Health (2013)
21. Tabachnick, B.G., Fidell, L.S., Osterlind, S.J.: Using Multivariate Statistics. Allyn and Bacon, Boston (2001)
22. Tran, C.T., Zhang, M., Andreae, P., Xue, B., Bui, L.T.: Multiple imputation and ensemble learning for classification with incomplete data. In: Leu, G., Singh, H.K., Elsayed, S. (eds.) Intelligent and Evolutionary Systems. PALO, vol. 8, pp. 401–415. Springer, Cham (2017). https://doi.org/10.1007/978-3-319-49049-6_29
23. van der Heijden, G.J.M.G., Donders, A.R.T., Stijnen, T., Moons, K.G.M.: Imputation of missing values is superior to complete case analysis and the missing-indicator method in multivariable diagnostic research: a clinical example. J. Clin. Epidemiol. 59(10), 1102–1109 (2006)
24. Witten, I.H., Frank, E., Hall, M.A., Pal, C.J.: Data Mining: Practical Machine Learning Tools and Techniques. Morgan Kaufmann, Massachusetts (2016)
25. Wu, X., Kumar, V., Quinlan, J.R., Ghosh, J., Yang, Q., Motoda, H., McLachlan, G.J., Angus, N., Liu, B., Philip, S.Y., et al.: Top 10 algorithms in data mining. Knowl. Inf. Syst. 14(1), 1–37 (2008)

A Preliminary Study of Diversity
in Extreme Learning Machines Ensembles

Carlos Perales-González$^{(\boxtimes)}$, Mariano Carbonero-Ruz, David Becerra-Alonso,
and Francisco Fernández-Navarro

Department of Quantitative Methods, Universidad Loyola Andalucia, Sevilla, Spain
cperales@uloyola.es

Abstract. In this paper, the neural network version of Extreme Learning Machine (ELM) is used as a base learner for an ensemble meta-algorithm which promotes diversity explicitly in the ELM loss function. The cost function proposed encourages orthogonality (scalar product) in the parameter space. Other ensemble-based meta-algorithms from AdaBoost family are used for comparison purposes. Both accuracy and diversity presented in our proposal are competitive, thus reinforcing the idea of introducing diversity explicitly.

Keywords: Extreme learning machine · Diversity · Machine learning
Ensemble · AdaBoost

1 Introduction

In the area of *supervised learning*, the goal of machine learning models is to develop a mapping function from the input variables to the output targets [1]. Two types of problems could be found within this field: classification and regression problems [2]. The main difference lies in the nature of the target vector. If it is a continuous variable, then the problem is a regression problem, whereas if the output is represented by a categorical variable, the task is called classification.

In classification problems, ensemble methods are learning algorithms that estimate the parameters of a set of models, also known as base learners, and then classify new patterns by taking a (weighted) vote of their estimations [3]. The research done by the machine learning community within the field of ensemble learning has been specially intense during the last years [4,5]. The main finding is that ensembles are, in some circumstances, more accurate than the individual models that make them up [6].

The two main approaches to combine several classifiers into one predictive model are Bagging and Boosting. Bagging, which stands for bootstrap aggregating, is a learning method for generating several versions of a base learner by selecting some subsets from the training set and using these as new learning sets [7]. Thus, each training subset is used to train a different classifier. Individual models in the ensemble are combined by majority voting their estimations.

© Springer International Publishing AG, part of Springer Nature 2018
F. J. de Cos Juez et al. (Eds.): HAIS 2018, LNAI 10870, pp. 302–314, 2018.
https://doi.org/10.1007/978-3-319-92639-1_25

Boosting is a family of machine learning meta-algorithms which focus on combining base learners over several iterations and generate a weighted majority hypothesis [8]. Unlike Bagging, in the classical Boosting the training subsets are not randomly created but depend upon the performance of the previous models. Hence, a certain pattern is more likely to be included in the new subset created if it was misclassified by previous base learners. Arguably the best known of all boosting-based algorithms is the AdaBoost (Adaptive Boosting) method [8].

In this research work, we focus on ensemble-systems which use Extreme Learning Machine (ELM) models [9] as base learners. Extreme learning machine (ELM) was proposed as a new learning algorithm to train single-hidden-layer feedforward neural networks (SLFNs). In ELM, the learning parameters are estimated in two stages: the input weights and the hidden bias are randomly chosen, whereas the weights that connect the hidden and the output layer are analytically determined [9]. Recently, it was also proposed a kernel-version of the ELM framework [10].

Bagging ELM was proposed and tested in the prediction of the electricity price in [11]. Also, AdaBoost Extreme Learning Machine have been studied and tested with ordinal data sets in [12]. However, to the best of our knowledge, there is no approach in the ELM community in which diversity is directly promoted in the error function to be optimized. Instead, it is promoted indirectly through different sampling strategies in ELM state-of-the-art papers.

Motivated by this fact, we propose a novel ensemble method in which diversity is explicitly incorporated in the cost function. This paper also defines a novel measurement of diversity, termed as orthogonality in its parameters. The new diversity metric is based on the scalar product of the weights connecting the hidden and output layer. With the designed orthogonality, we further propose an ensemble ELM classifier, namely Diverse Extreme Learning Machine (DELM), to jointly suppress the training error of ensemble and improve the diversity between base learners.

This paper is organized as follows. In the next section, Extreme Learning Machine and its different versions are briefly introduced. The detailed description of the methodology proposed is given in Sect. 3. The experimental framework adopted in this paper is described in Sect. 4 whereas the results and empirical comparisons with related approaches are presented in Sect. 5. A conclusion is drawn in Sect. 6.

2   Extreme Learning Machine

The parameters of the ELM models are estimated from a training set $\mathcal{D} = \{(\boldsymbol{x}_i, \boldsymbol{y}_i)\}_{i=1}^n$, where $\boldsymbol{x}_i \in \mathbb{R}^m$ is the vector of attributes of the ith pattern, m is the dimension of the input space (number of attributes in the problem), $\boldsymbol{y}_i \in \mathbb{R}^J$ is the class label assuming the "1-of-J" encoding ($y_{ij} = 1$ if $\boldsymbol{x}_i$ is a pattern of the j-th class, $y_{ij} = 0$ otherwise) and J is the number of classes. Let us denote

Y as $Y = (Y_j, j = 1, \ldots, J) = \begin{pmatrix} y_1' \\ \vdots \\ y_n' \end{pmatrix}$, where Y_j is the j-th column of the Y

matrix. Under this formulation, the output function of the classifier is defined as:

$$f(x) = h'(x)\beta, \tag{1}$$

where $\beta = (\beta_j, j = 1, \ldots, J) \in \mathbb{R}^{d \times J}$ is the output matrix of coefficients, $\beta_j \in \mathbb{R}^d$ the weights of the j-th output node, $h : \mathbb{R}^m \to \mathbb{R}^d$ is the mapping function and d is the number of hidden nodes (the dimension of the transformed space). Let us also denote H as $H = (h'(x_i), i = 1, \ldots, n) \in \mathbb{R}^{n \times d}$, the transformation of the training set from the the input space to the transformed one.

ELM minimizes the following optimization problem [10]:

$$\min_{\beta \in \mathbb{R}^{d \times J}} \left(\|\beta\|^2 + C\|H\beta - Y\|^2 \right), \tag{2}$$

where $C \in \mathbb{R}$ is a user-specified parameter that promotes generalization performance.

Although the ELM problem is typically presented in its matrix form, the final optimization problem is the sum of J separable vector problems, one for each class. In fact, if we disaggregate the objective function by columns, it can be verified that:

$$\|\beta\|^2 + C\|H\beta - Y\|^2 = \sum_{j=1}^{J} \left(\|\beta_j\|^2 + C\|H\beta_j - Y_j\|^2 \right). \tag{3}$$

The optimization problem for each class can be reformulated as:

$$\begin{aligned} \|\beta_j\|^2 + C\|H\beta_j - Y_j\|^2 &= \beta_j'\beta_j + C(\beta_j'H' - Y_j')(H\beta_j - Y_j) \\ &= \beta_j'\beta_j + C\beta_j'H'H\beta_j - 2CY_j'H\beta_j + Y_j'Y_j \end{aligned} \tag{4}$$

Hence, the optimization problem can be rewritten for each class as:

$$\min_{\beta_j \in \mathbb{R}^d} \left(\beta_j'(I + CH'H)\beta_j - 2CY_j'H\beta_j \right), \tag{5}$$

as $Y_j'Y_j$ is constant. The solution to that optimization problem is well known:

$$\beta_j = \left(\frac{I}{C} + H'H \right)^{-1} H'Y_j \tag{6}$$

The final solution is obtained after grouping the β_j elements by columns:

$$\beta = \left(\frac{I}{C} + H'H \right)^{-1} H'Y \tag{7}$$

This solution can be alternatively written as:

$$\beta = H' \left(\frac{I}{C} + H'H \right)^{-1} Y \tag{8}$$

The two solutions are equivalent as it is shown below.

Lemma 1. *It can be verified that:*

$$H' \left(\frac{I}{C} + HH' \right)^{-1} Y = \left(\frac{I}{C} + H'H \right)^{-1} H'Y \tag{9}$$

Proof. It will suffice to prove that the matrices to the left of Y coincide on both sides of the equation:

$$H' \left(\frac{I}{C} + HH' \right)^{-1} = \left(\frac{I}{C} + H'H \right)^{-1} H' \tag{10}$$

Multiplying both sides by the corresponding inverse matrices, the proposition will be proven if

$$\left(\frac{I}{C} + H'H \right) H' = H' \left(\frac{I}{C} + HH' \right) \tag{11}$$

which is trivial.

There are two possible implementations of the ELM framework: the neural network and the kernel version. The main difference between these two approaches lies in the way of computing the $h(x)$ function (for pattern x) and consequently the H matrix. In the neural implementation of the framework, $h(x)$ can be explicitly computed and it is defined as:

$$h(x) = (\phi(x_i; \mathbf{w}_j, b_j), j = 1, \ldots, d), \tag{12}$$

where $\phi(\cdot; \mathbf{w}_j, b_j) : \mathbb{R}^m \rightarrow \mathbb{R}$ is the activation function of the j-th hidden node, $\mathbf{w}_j \in \mathbb{R}^m$ is the input weight vector associated to the j-th hidden node and $b_j \in \mathbb{R}$ is the bias of the j-th hidden node. The activation function is typically the sigmoidal one, i.e.:

$$\phi(x_i; \mathbf{w}_j, b_j) = g(\mathbf{w}'_j x_i + b_j), \tag{13}$$

where $g(t) = (1 + \exp(-t))^{-1}$. In the neural version of the framework, the input weights of the hidden nodes are randomly chosen, and the output weight matrix, β, is analytically determined using Eq. 7 or 8.

In the kernel implementation of the framework, the $h(x)$ function is an unknown feature mapping. Fortunately, there are certain functions $k(x_i, x_j)$ that compute the dot product in another space, $k(x_i, x_j) = \langle h(x_i), h(x_j) \rangle$ (for all x_i and x_j in the input space). Thus, we will reformulate the solution to group the $h(x)$ terms in dot products and apply the kernel trick to them (substituting those elements by their kernel functions). Hence, the output function of the ELM classifier (after applying the kernel trick) for a test pattern, x, can be written compactly as (Eq. 8):

$$
\begin{aligned}
f(x) &= h(x)\beta \\
&= h(x)\mathbf{H}'\left(\frac{\mathbf{I}}{C} + \mathbf{HH}'\right)^{-1}\mathbf{Y} \\
&= \mathbf{K}(x)'\left(\frac{\mathbf{I}}{C} + \mathbf{\Omega}_{\mathrm{ELM}}\right)^{-1}\mathbf{Y},
\end{aligned}
\tag{14}
$$

where $\mathbf{K} : \mathbb{R}^m \to \mathbb{R}^n$ is the vectorial kernel function

$$
\mathbf{K}(x) = (k(x, x_i), i = 1, \ldots, n).
$$

The Gaussian kernel function implemented in this study is

$$
k(u, v) = \exp(-\sigma \|u - v\|^2),
\tag{15}
$$

where $\sigma \in \mathbb{R}$ is the kernel parameter. The kernel matrix $\mathbf{\Omega}_{\mathrm{ELM}} = (\Omega_{i,j})_{i,j=1,\ldots,n}$ is defined as:

$$
\Omega_{i,j} = k(x_i, x_j).
\tag{16}
$$

Finally, it is important to clarify that the predicted class label, $\hat{y}(x)$, for a test pattern x is computed in the ELM framework as follows:

$$
\hat{y}(x) = \arg \max_{j=1,\ldots,J} \left(h'(x)\beta\right)_j.
\tag{17}
$$

3   Methodology Proposed

The aim in this section is to build an ensemble made of s instances of ELM. We will use the same transformation of the input space for all individuals in the ensemble (i.e. $\mathbf{H}$ and $h(\mathbf{x})$ are common for all individuals in the ensemble) and will consider the neural implementation of the ELM framework. In this scenario, diversity could be promoted using the metric previously described as all the output vectors are using the same transformed space. Thus, the goal is to estimate s matrices associated to the output matrices of s diverse neural networks, $\{\beta^1, \ldots, \beta^s\}$.

3.1   Diversity as a Metric

Diversity is an implicit attribute of an ensemble. Thus, in order to measure the diversity of solutions, a metric for diversity is needed. Many proposals have been made in this sense [13]. The one proposed in this article is based on the angle between output vectors.

It is known that the angle between two vectors and its dot product are related. Given non-zero vectors u and v, provided they have the same dimension size, their angle is given by

$$\cos\left(\angle\left(u, v\right)\right) = \frac{\langle u, v \rangle}{\|u\| \|v\|}.$$

These vectors will be most different when, $|\angle(u, v)| = \pi/2$, and therefore $\langle u, v \rangle = 0$. When most similar, $\langle u, v \rangle = \pm 1$. Taking this into account, a metric of diversity among u and v can be described as[1]

$$d\left(u, v\right) = 1 - \frac{\langle u, v \rangle^2}{\|u\|^2 \|v\|^2}$$

Although there is no standard definition for the angle between two matrices of the same size, for the purpose of the present work it will be defined as the average angles of its homologous columns.

Definition 1. *Given matrices* $A, B \in \mathbb{R}^{d \times J}$ *the diversity between them is defined as:*

$$d\left(A, B\right) = \frac{1}{J} \sum_{j=1}^{J} d\left(A_j, B_j\right) \tag{18}$$

where A_j *is the j-th column of matrix* A *and the same applies to* B.

As in vectors, the diversity d of two matrices is bounded between 0 ($\beta_1 \parallel \beta_2$) and 1 ($\beta_1 \perp \beta_2$). The definition can be extended to any s number of matrices, simply averaging the diversities obtained from all possible pairs of matrices.

Definition 2. *Given matrices* $A_1, \ldots, A_s$ *of the same size, their diversity is given by*

$$d\left(A_1, \ldots, A_s\right) = \frac{1}{\binom{s}{2}} \sum_{k<l} d\left(A_k, A_l\right) \tag{19}$$

where $\binom{s}{2}$ term is introduced in order to normalize the diversity between 0 y 1.

3.2   Optimization Problem

The proposed ensemble algorithm, named Diverse Extreme Learning Machine (DELM), searches for diverse β sequentially, disaggregating it by columns. The first component of the ensemble, β^1, is estimated using the solution in Eq. 7 whereas β^2 is obtained from β^1; β^3 from β^1 and β^2 and so on, up to β^s. Thus, $\beta^l, l = \{2, \ldots, s\}$, is the solution to the minimization problem

$$\min_{\beta^l \in \mathbb{R}^{d \times J}} \frac{1}{2} \left(\|\beta^l\|^2 + C\|H\beta^l - Y\|^2 + D \sum_{j=1}^{J} \sum_{k=1}^{l-1} \left\langle \beta_j^l, u_j^k \right\rangle^2 \right) \tag{20}$$

---

[1] The dot product is squared aiming to focus solely on the direction of the vector.

where $\boldsymbol{u}^k \in \mathbb{R}^{d \times J}$ is the column-by-column normalized $\boldsymbol{\beta}^k$ from the iteration k of the ensemble, $\boldsymbol{u}_j^k \in \mathbb{R}^d$ is the normalized vector associated to $\boldsymbol{\beta}_j^k \in \mathbb{R}^d$ (the j-th output vector of the k-th individual in the ensemble), and $D > 0$ is a hyperparameter (a parameter that can be cross validated), like C, optimizing for diversity.

The diversity term, $\sum_{j=1}^{J} \sum_{k=1}^{l-1} \left\langle \boldsymbol{\beta}_j^l, \boldsymbol{u}_j^k \right\rangle^2$, is appreciably smaller than the error term, $\|\boldsymbol{H}\boldsymbol{\beta}^l - \boldsymbol{Y}\|^2$. In order to balance the importance in minimization, a coefficient was added to the diversity term. This coefficient is directly proportional to the number of errors (the same as the number of training instances, n) and inversely proportional to the total number of $\boldsymbol{u}^k$, which is s. Therefore, minimization problem becomes

$$\min_{\boldsymbol{\beta}^l \in \mathbb{R}^{d \times J}} \frac{1}{2} \left(\|\boldsymbol{\beta}^l\|^2 + C\|\boldsymbol{H}\boldsymbol{\beta}^l - \boldsymbol{Y}\|^2 + \left(D + \frac{n}{s}\right) \sum_{j=1}^{J} \sum_{k=1}^{l-1} \left\langle \boldsymbol{\beta}_j^l, \boldsymbol{u}_j^k \right\rangle^2 \right) \quad (21)$$

3.3   Matrix Expression

In order to find the solution for problem defined in Eq. (21) it is necessary to rewrite it in matrix form. The sum of the dot products $\sum_{k=1}^{l-1} \left\langle \boldsymbol{\beta}_j^l, \boldsymbol{u}_j^k \right\rangle^2$ can be rewritten as:

$$\sum_{k=1}^{l-1} \left\langle \boldsymbol{\beta}_j^l, \boldsymbol{u}_j^k \right\rangle^2 = \sum_{k=1}^{l-1} \left(\boldsymbol{\beta}_j^l\right)' \boldsymbol{u}_j^k \left(\boldsymbol{u}_j^k\right)' \boldsymbol{\beta}_j = \left(\boldsymbol{\beta}_j^l\right)' \boldsymbol{M}_j^l \boldsymbol{\beta}_j^l \quad (22)$$

where $\boldsymbol{M}_j^l$ is defined as

$$\boldsymbol{M}_j^l \equiv \sum_{k=1}^{l-1} \boldsymbol{u}_j^k \left(\boldsymbol{u}_j^k\right)' \quad (23)$$

Taking into account the matrix form of the diversity term, the optimization problem of the DELM method could be rewritten as:

$$\min_{\boldsymbol{\beta}^l \in \mathbb{R}^{d \times J}} \frac{1}{2} \left(\left(\boldsymbol{\beta}_j^l\right)' \left(\boldsymbol{I} + \left(D + \frac{n}{s}\right) \boldsymbol{M}_j^l\right) \boldsymbol{\beta}_j^l + C \left(\boldsymbol{H}\boldsymbol{\beta}_j^l - \boldsymbol{Y}_j\right)^T \left(\boldsymbol{H}\boldsymbol{\beta}_j^l - \boldsymbol{Y}_j\right) \right) \quad (24)$$

Taking derivatives and equating them to zero gives:

$$\left(\boldsymbol{I} + \left(D + \frac{n}{s}\right) \boldsymbol{M}_j^l\right) \boldsymbol{\beta}_j^l - C\boldsymbol{H}' \left(\boldsymbol{Y}_j - \boldsymbol{H}\boldsymbol{\beta}_j^l\right) = 0$$

Hence, $\boldsymbol{\beta}_j^l$ could be obtained analytically as:

$$\boldsymbol{\beta}_j^l = \left(\frac{\boldsymbol{I}}{C} + \boldsymbol{H}'\boldsymbol{H} + \frac{1}{C}\left(D + \frac{n}{s}\right) \boldsymbol{M}_j^l\right)^{-1} \boldsymbol{H}'\boldsymbol{Y}_j \quad j = 1, \dots, J \quad (25)$$

This of course provided the inverse of the matrix exists.

Lemma 2. *Matrix $\frac{I}{C} + H^T H + \frac{1}{C}(D + \frac{n}{s})M^l_j$ is invertible.*

Proof. Due to $\left(u^k_j u^{k}_j{}'\right)^T = u^k_j u^{k}_j{}'$, matrix M^l_j is symmetric. Also, is positive-definite, because

$$x' M^l_j x = \sum_{k=1}^{l-1} x' u^k_j u^{k}_j{}' x = \sum_{k=1}^{l-1} \langle x, u^k_j \rangle^2 \geq 0.$$

Therefore all its eigenvalues are non negative. From other studies [10] we know that matrix $\frac{I}{C} + H^T H$ is also positive-definite. Since $\frac{1}{C}\left(D + \frac{n}{s}\right) > 0$ the analyzed matrix is definite positive, and thus, its inverse can be obtained.

For simplicity, the way of β^l is ensembled is using the same weights as in AdaBoost [14], which takes into account the accuracy of each β^l separately. Thus, the predicted class label, $\hat{y}(x)$, for a test pattern x is computed as follows:

$$\hat{y}(x) = \arg \max_{j=1,\dots,J} \left(\sum_{l=1}^{s} \alpha^{(l)}(h'(x)\beta^l) \right)_j, \qquad (26)$$

where $\alpha^{(l)}$ is computed as suggested in [14].

4 Experimental Framework

In this section, the experimental study performed to validate the new algorithm is presented. In Sect. 4.1, the datasets used in the experimental framework are detailed. Section 4.2 describes the metrics employed to evaluate the performance of the algorithms. Finally, Sect. 4.3 gives a description of the algorithms used for comparison purposes along with the configuration of their parameters.

4.1 Description of the Datasets

The datasets were extracted from the UCI Machine Learning [15] and the mldata.org repositories. Table 1 summarizes the properties of the selected datasets. For each dataset, the number of instances (Size), the number of input features (#Attr.), the number of classes (#Classes) and the number of instances per class (Class distribution) are shown.

The experimental design was conducted using a 10-fold cross-validation, with 5 repetitions per each fold. All nominal variables were transformed to binary variables. Finally, it is important to mention that we have carried out a simple linear rescaling of the input variables in the interval $[-1, 1]$.

Table 1. Characteristics of the data sets, ordered by size and number of classes

Data sets				
Dataset	Size	#Attr.	#Classes	Class distribution
car	1728	21	4	(1210, 384, 69, 65)
winequality-red	1599	11	6	(10, 53, 681, 638, 199, 18)
ERA	1000	4	9	(92, 142, 181, 172, 158, 118, 88, 31, 18)
LEV	1000	4	5	(93, 280, 403, 197, 27)
SWD	1000	10	4	(32, 352, 399, 217)
newthyroid	215	5	3	(30, 150, 35)
automobile	205	71	6	(3, 22, 67, 54, 32, 27)
squash-stored	52	51	3	(23, 21, 8)
squash-unstored	52	52	3	(24, 24, 4)
pasture	36	25	3	(12, 12, 12)

4.2 Description of the Metrics

In this paper, two metrics are considered for comparison: the accuracy and the diversity.

Acc Accuracy rate: It is the number of successful hits (correct classifications) relative to the number of total classifications. It is a common metric for classifiers [1]. The mathematical expression is:

$$\text{Acc} = \frac{1}{n} \sum_{i=1}^{n} I\left(\tilde{y}\left(\boldsymbol{x}_i\right) = y_i\right) \tag{27}$$

where $I(x = x')$ is 1 if the arguments are equal and 0 if they are not and $\tilde{y}\left(\boldsymbol{x}_i\right)$ is transformation of the $\hat{y}\left(\boldsymbol{x}_i\right)$ element to the "1-of-J" encoding.

d Diversity: For an ELM ensemble with s individuals, this metric is obtained applying Eq. (19) to the $\boldsymbol{\beta}^1, \ldots, \boldsymbol{\beta}^s$ matrices:

$$\text{d} = d\left(\boldsymbol{\beta}^1, \ldots, \boldsymbol{\beta}^s\right)$$

where $d = 0$ means all the base learners are the same, and $d = 1$ means they are as different as they can be.

These metrics aim to evaluate different characteristics of the ensembles. Accuracy quantifies how good an algorithm is at predicting, using the relation among predicted labels and real ones for the test set. Diversity involves the base learners in the ensembles and measure how similar they are to one another.

4.3 Description of the Algorithms

Four ensembles meta-algorithms are tested, using as base-learner the Neural Extreme Learning Machine. Our proposal is Diverse Extreme Learning Machine (DELM), and the other algorithms come from the AdaBoost implementation.

AELM AdaBoost Extreme Learning Machine, defined in [12].
BRELM Boosting Ridge Extreme Learning Machine, explained in [16].
NCELM Negative Correlation Extreme Learning Machine, defined in [17].
DELM Diverse Extreme Learning Machine, previously detailed in Sect. 3.2.

AELM, BRELM and NCELM come from a boosting approach. The reasons to use algorithms from this approach, and not others as bagging, is because all of them share the same H, so the β can be compared.

The same size has been considered for all the methods, $s = 5$. Hyperparameters were chosen through a grid search with a 5-fold cross-validation considering just the training set. Regulation hyperparameter C and neurons in the hidden layer for all the ensembles, where $C \in \{10^2, \ldots, 10^{-2}\}$ and $d = \{10, 20, \ldots, 200\}$. For BRELM, also hyperparameter $\lambda \in \{0.25, 0.5, 1, 5, 10\}$ was considered. For DELM, we set the range of $D \in \{10^2, \ldots, 10^{-2}\}$.

5   Results

As explained in Sect. 4, 10 folds were executed, with a 5-fold cross-validation of the hyperparameters. Each metric was averaged over 5 executions, in order to avoid randomness as using Neural Extreme Learning Machine as base learner.

From a purely descriptive point of view, our proposal method (DELM) improves the diversity in 9 out of the 10 datasets considered, just as we expected. Not only diversity was improved, but also in 8 of the data sets accuracy is better than in any boosting ensemble. NCELM appears to be the second most appealing method, considering both diversity and accuracy. Thus, NCELM is the most competitive ensemble method in the AdaBoost family when faced with DELM. In fact, it has been the only one outperforming DELM in diversity in at least one dataset. In view of these results, we can affirm that DELM emerges as a potential competitive alternative in the design of neural network ensembles. In order to assert the results from a statistical point of view, the Wilcoxon signed rank test have been performed for the two metrics, comparing DELM against the other methods by pairs. We have chosen a confidence level of $\alpha = 0.05$ and $N = 10$ data sets, which means the critical value for the Wilcoxon test is $\omega_{crit} = 11$. It is worth mentioning that our method outperforms significantly the rest of the methods, both in accuracy and in diversity (Table 2).

As it has been detailed in the methodological section, the proposed method needs the prior configuration of three hyper-parameters. With the exception of the NCELM method, the remaining ones has only two hyper-parameters. Thus, the competitive results in accuracy could be explained by the greater degree of freedom of the DELM method. Additionally, we also hypothetise that the incorporation of the diversity metric in the error function could also justify the mean accuracy yielded. ELM algorithm is based on convex optimization. Relation between data features and target labels does not have to be perfectly represented by a convex function, though it can present local convexities. DELM explores the data in each iteration, looking for a set of s β^l which are the most

Table 2. Test Diverse ELM and AdaBoost family, including the average over all the different splits

	Accuracy (Acc)			
	DELM	AELM	BRELM	NCELM
car	**0.929711**	0.834618	0.901805	*0.905111*
winequality-red	**0.853687**	*0.840085*	0.839670	0.837363
ERA	**0.829479**	0.822201	0.828019	*0.828428*
LEV	**0.836345**	0.786404	0.792371	*0.798220*
SWD	**0.787940**	*0.764487*	0.759893	0.760442
newthyroid	**0.932035**	0.817172	0.812035	*0.819509*
automobile	**0.867376**	0.834618	0.841636	*0.846499*
squash-stored	0.694286	**0.751429**	0.694063	*0.711937*
squash-unstored	*0.814286*	**0.830952**	0.813810	0.812381
pasture	**0.833333**	0.766667	0.811111	*0.826667*
	Diversity (d)			
	DELM	AELM	BRELM	NCELM
car	**0.999206**	0.180213	0.176621	*0.181212*
winequality-red	**0.926890**	0.152451	0.124054	*0.185804*
ERA	**0.968917**	0.138991	0.143748	*0.156551*
LEV	**0.992860**	0.098013	0.089886	*0.133830*
SWD	**0.980953**	0.130116	*0.138222*	0.137884
newthyroid	**0.886023**	0.043061	0.040141	*0.057340*
automobile	**0.932272**	0.314529	0.311612	*0.317588*
squash-stored	**0.668662**	0.216839	0.181023	*0.217834*
squash-unstored	**0.568838**	0.130116	0.145780	*0.155101*
pasture	0.081884	*0.181297*	0.175300	**0.187400**

The best result for each dataset is in bold face and the second one in italics

different among themselves. However, its main competitor, AdaBoost, exploits the data by iterating over the misclassified instances in each learning iteration.

Regarding the computational complexity, it is important to stress that the methods selected for comparison purposes all belong to the AdaBoost family of methods. The computationally most demanding part of AdaBoost based on ELM is the inversion of a $n \times n$ matrix, which is performed in $O(n^2)$. Thus, the computational complexity of those methods is $O(s \times n^2)$, as the inversion of the matrix is computed s times. In the DELM method, the inversion of the matrix is computed $J \times s$ times, and therefore, it is less computationally efficient than the comparison methods. Furthermore, the DELM method does require the optimization of three additional hyper-parameters.

6   Conclusions

The presented paper explores the possibility of introducing diversity explicitly in the ensemble development. DELM usually keeps the same level of accuracy, while it obtains an improving difference in diversity against the rest of ensembles in all the data sets. This is the result of introducing explicitly the diversity in the ELM loss function. Thus introducing a diversity term in the loss function seems reasonable in order to improve diversity, it also shows that it is better for accuracy in most of the data sets studied.

The experimental setting will be extended significantly (by including more datasets, including unbalanced ones, and comparison methods) in future research works. The goal is two fold: (i) firstly, to study the performance of the method proposed in unbalanced problems and (ii) secondly, to understand the competitive performance of the DELM method in a stronger experimental setting. Besides, having two metrics to explore (diversity and accuracy), we will study multiobjective methods to achieve to betters pair of solutions.

References

1. Witten, I.H., Frank, E., Hall, M.A., Pal, C.J.: Data Mining: Practical Machine Learning Tools and Techniques. Morgan Kaufmann, Burlington (2016)
2. Mohri, M., Rostamizadeh, A., Talwalkar, A.: Foundations of Machine Learning. MIT press, Cambridge (2012)
3. Dietterich, T.G.: Ensemble methods in machine learning. In: Kittler, J., Roli, F. (eds.) MCS 2000. LNCS, vol. 1857, pp. 1–15. Springer, Heidelberg (2000). https://doi.org/10.1007/3-540-45014-9_1
4. Zhang, L., Shah, S., Kakadiaris, I.: Hierarchical multi-label classification using fully associative ensemble learning. Pattern Recogn. **70**, 89–103 (2017)
5. Krawczyk, B., Minku, L.L., Gama, J., Stefanowski, J., Woźniak, M.: Ensemble learning for data stream analysis: a survey. Inf. Fusion **37**, 132–156 (2017)
6. Polikar, R.: Ensemble learning. In: Zhang, C., Ma, Y. (eds.) Ensemble Machine Learning, pp. 1–34. Springer, Boston (2012). https://doi.org/10.1007/978-1-4419-9326-71
7. Bbeiman, L.: Bagging predictors. Mach, Learn. **24**, 123–140 (1996)
8. Freund, Y., Schapire, R.E.: A decision-theoretic generalization of on-line learning and an application to boosting. J. Comput. Syst. Sci. **55**, 119–139 (1997)
9. Huang, G.-B., Zhu, Q.-Y., Siew, C.-K.: Extreme learning machine: theory and applications. Neurocomputing **70**(1), 489–501 (2006)
10. Huang, G.-B., Zhou, H., Ding, X., Zhang, R.: Extreme learning machine for regression and multiclass classification. IEEE Trans. Syst. Man Cybern. Part B Cybern. **42**(2), 513–29 (2012)
11. Tian, H., Meng, B.: A new modeling method based on bagging elm for day-ahead electricity price prediction. In: 2010 IEEE Fifth International Conference on Bio-Inspired Computing: Theories and Applications (BIC-TA), pp. 1076–1079. IEEE (2010)
12. Riccardi, A., Fernández-Navarro, F., Carloni, S.: Cost-sensitive AdaBoost algorithm for ordinal regression based on extreme learning machine. IEEE Trans. Cybern. **44**(10), 1898–1909 (2014)

13. Kuncheva, L.I., Whitaker, C.J.: Measures of diversity in classifier ensembles and their relationship with the ensemble accuracy. Mach. Learn. **51**(2), 181–207 (2003)
14. Freund, Y., Schapire, R.E.: A short introduction to boosting. J. Jpn. Soc. Artif. Intell. **14**(5), 771–780 (1999)
15. Lichman, M.: UCI machine learning repository (2013)
16. Ran, Y., Sun, X., Sun, H., Sun, L., Wang, X., Ran, Y.W.X., Sun, X., Sun, H., Sun, L.: Boosting ridge extreme learning machine. In: Proceedings - 2012 IEEE Symposium on Robotics and Applications, ISRA 2012, pp. 881–884 (2012)
17. Wang, S., Chen, H., Yao, X.: Negative correlation learning for classification ensembles. In: Proceedings of the International Joint Conference on Neural Networks (2010)

Orthogonal Learning Firefly Algorithm

Kadavy Tomas[⊠], Pluhacek Michal, Viktorin Adam,
and Senkerik Roman

Tomas Bata University in Zlin, T.G. Masaryka 5555, 760 01 Zlin,
Czech Republic
{kadavy,pluhacek,aviktorin,senkerik}@utb.cz

Abstract. In this paper, a proven technique, orthogonal learning, is combined
with popular swarm metaheuristic Firefly Algorithm (FA). More precisely with
its hybrid modification Firefly Particle Swarm Optimization (FFPSO). The
performance of the developed algorithm is tested and compared with canonical
FA and above mentioned FFPSO. Comparisons have been conducted on
well-known CEC 2017 benchmark functions, and the results have been evaluated for statistical significance using Friedman rank test.

Keywords: Firefly algorithm · Particle swarm optimization
Orthogonal learning · Friedman rank

1 Introduction

In general, the swarm-based metaheuristic optimization algorithms are still quite
popular amongst researchers. Nowadays [1], instead of developing a new heuristic
optimization algorithm, the utilization and smart improvement of existing ones are
more favorable. Several modern techniques were designed to improve the overall
performance. For example, the ensemble method [2, 3], the hybridization [4, 5], or
some adaptive control [6, 7] were already adopted. Generally speaking, all the mentioned methods are trying to analyze and improve the inner dynamic of an algorithm or
combine the different strategies borrowed from different optimization algorithms. For a
hybrid version of two different algorithms, the basic idea is that the resulting hybrid
should combine the advantages of both and eliminate their disadvantages. However,
achieving this ideal state is a difficult task, and this is a clear motivation behind the
presented research.

There are several typical representatives (e.g., Particle Swarm Optimization
(PSO) [8]) among the metaheuristic swarm-based optimization algorithms with a long
history and with many modifications [9, 10]. One of the possible adjustment is based
on the orthogonal learning [11]. The basic idea behind orthogonal learning lies in the
identification of combination or instead parts of solutions in particular dimensions
offering the best results and using them in the next optimization steps. The principle of
the orthogonal learning will be more explained in the relevant section.

For several years, another swarm-based algorithm proves its usefulness and is
becoming quite popular among researchers. The Firefly Algorithm (FA) was introduced
in 2010 [12]. Since that year, and like the mentioned PSO, many extensions and

© Springer International Publishing AG, part of Springer Nature 2018
F. J. de Cos Juez et al. (Eds.): HAIS 2018, LNAI 10870, pp. 315–326, 2018.
https://doi.org/10.1007/978-3-319-92639-1_26

modifications were proposed for this new optimization algorithm [13–15]. The popularity and versatility of both algorithms, PSO and FA, lead to the creation of hybrid which is called Firefly Particle Swarm Optimization (FFPSO) [16].

The exciting method already used with PSO, the orthogonal learning [11], could improve the performance of FA. However, the enhancement of canonical FA may be a difficult task, due to its nature. Luckily, thanks to mentioned hybrid FFPSO, the advantages of orthogonal learning are available, and the situation is much more simplified regarding the application. The proposed optimization algorithm is tested and statistically evaluated on well-known benchmark function CEC 2017 [17]. The comparisons with original hybridized algorithm FFPSO and with the original version of the FA and also PSO and Orthogonal Learning Particle Swarm Optimization (OLPSO) are presented and statistically evaluated.

The rest of the paper is structured as follows. Brief descriptions of PSO, OLSPO, FA, and FFPSO are in Sects. 2 and 3. These sections also cover a description of the proposed algorithm based on the orthogonal learning. In Sect. 4, the CEC benchmark is defined, and the parameter settings of tested algorithms are shown as well. The results and conclusion sections follow afterward.

2 Particle Swarm Optimization

The PSO is one of the current leading representatives on a field of swarm intelligence based algorithms. It was first published by Eberhart and Kennedy in 1995 [8]. This algorithm mimics the social behavior of swarming animals in nature. Despite the fact that its quite long time from its first appearance, its still plenty used across many optimization problems.

2.1 Canonical Particle Swarm Optimization

Every particle has a position in n-dimensional solution space, and this position represents the input parameters of the optimized problem. This position of particles changes over the time due to two factors. One of them is the current position of a particle. The second one is the velocity of a particle, labeled as v. Each particle also remembers its best position (solution of the problem) obtained so far. This solution is tagged as the *pBest*, personal best solution. Also, each particle has access to the global best solution, *gBest*, which is selected from all *pBests*. These variables set the direction for every particle and a new position in the next iteration. PSO usually stops after a number of iterations or a number of FEs (objective function evaluations).

In every iteration of the algorithm, the new positions of particles are calculated based on the previous positions and velocities. The new position of a particle is checked if it still lies in the space of possible solutions.

The position of particle x is calculated according to the formula (1).

$$x'_i = x_i + v_i \tag{1}$$

Where x_i' is a new position of particle i, x_i is the previous old position of a particle and v is the velocity of a particle. The velocity of a particle v is calculated according to (2).

$$v_i' = w \cdot v_i + c_1 \cdot r_1 \cdot (pBest_i - x_i) + c_2 \cdot r_2 \cdot (gBest - x_i) \tag{2}$$

Where w is inertia weight [18], c_1 and c_2 are learning factors, and r_1 and r_2 are random numbers of unimodal distribution in the range $< 0,1 >$.

2.2   Orthogonal Learning Particle Swarm Optimization

The Orthogonal Experimental Design (OED) mechanism was introduced to PSO to improve its learning strategy. This improvement leads to the creation of Orthogonal Learning PSO (OLPSO) [11]. As the name suggests, the traditional learning mechanism of PSO was replaced by novel Orthogonal Learning, which should help construct an efficient and promising exemplar for a particle to learn from. Each corresponding dimension of an optimized problem is regarded as a factor. The factor means that the OLPSO combines information of the particle's *pBest* and the *pBest*s of its neighborhoods to construct a guidance vector. The velocity Eq. (2) is then changed as (3).

$$v_i' = w \cdot v_i + c \cdot r \cdot (gVector_i - x_i) \tag{3}$$

Where *gVector* is the mentioned guidance vector, the values of *gVector* are just points to which *pBest* should be used in particular dimension. The guidance vector (learning exemplar) is used until it cannot improve the particles *pBest* solution for a certain number of generations which is called reconstruction gap G.

The brief process of the construction of the *gVector* is as follows:

1. An orthogonal array (OA) $L_M(2^D)$, where $M = 2^{\log_2(D+1)}$ is created using the procedure which is clearly and more detailed described in [11].
2. According to the OA, the M trial solutions are created by selecting the corresponding value from *pBest*s. For the trial solution can be chosen own *pBest* or *pBest* of the different particle.
3. Each trial solution is evaluated and the best solution is recorded as X_b.
4. Calculate the effect of each level on each factor and determine the best level for each factor. Based on these levels, the predictive solution X_P is created and evaluated.
5. The solution (X_b or X_P) is selected as the guidance vector *gVector* based on the obtained objective function value.

3   Firefly Algorithm

This optimization nature-based algorithm was developed and introduced by Yang in 2008 [12]. The fundamental principle of this algorithm lies in simulating the mating behavior of fireflies at night when fireflies emit light to attract a suitable partner.

3.1   Canonical Firefly Algorithm

The main idea of FA is that the objective function value that is optimized is associated with the flashing light of these fireflies. The author for simplicity set a couple of rules to describe the algorithm itself:

- The brightness of each firefly is based on the objective function value.
- The attractiveness of a firefly is proportional to its brightness. This means that the less bright firefly is lured towards, the brighter firefly. The brightness depends on the environment or the medium in which fireflies are moving and decreases with the distance between each of them.
- All fireflies are sexless, and it means that each firefly can attract or be lured by any of the remaining ones.

The movement of one firefly towards another one is then defined by Eq. (4). Where x_i' is a new position of a firefly i, x_i is the current position of firefly i and x_j is a selected brighter firefly (with better objective function value). The α is a randomization parameter and $sign$ simply provides random direction -1 or 1.

$$x_i' = x_i + \beta \cdot (x_j - x_i) + \alpha \cdot sign \qquad (4)$$

The brightness I of a firefly is computed by the Eq. (5). This equation of brightness consists of three factors mentioned in the rules above. On the objective function value, the distance between two fireflies and the last factor is the absorption factor of a media in which fireflies are.

$$I = \frac{I_0}{1 + \gamma r^m} \qquad (5)$$

Where I_o is the objective function value, the γ stands for the light absorption parameter of a media in which fireflies are and the m is another user-defined coefficient and it should be set $m \geq 1$. The variable r is the Euclidian distance (6) between the two compared fireflies.

$$r_{ij} = \sqrt{\sum_{k=1}^{d} (x_{i,k} - x_{j,k})^2} \qquad (6)$$

Where r_{ij} is the Euclidian distance between fireflies x_i and x_j. The d is current dimension size of the optimized problem.

The attractiveness β (7) is proportional to brightness I as mentioned in rules above and so these equations are quite similar to each other. The β_0 is the initial attractiveness defined by the user, the γ is again the light absorption parameter and the r is once more the Euclidian distance. The m is also the same as in Eq. (5).

$$\beta = \frac{\beta_0}{1 + \gamma r^m} \qquad (7)$$

The pseudocode 1 below shows the fundamentals of FA operations.

Pseudocode 1. Firefly Algorithm

```
1.  FA initialization
2.  while(terminal condition not met)
3.    for i = 1 to all fireflies
4.      for j = 1 to all fireflies
5.        if(Ij < Ii) then
6.          move xi to xj
7.          evaluate xi
8.        end if
9.      end for j
10.   end for i
11.   record the best firefly
12. end while
```

3.2 Hybrid of Firefly and Particle Swarm Optimization Algorithms

Another more advanced versions of the FA could be represented by its modifications, or more likely hybridizations, with others successful metaheuristic algorithms. The basic idea behind such an approach is that the new hybrid strategy can share advantages from both algorithms and hopefully eliminate their disadvantages.

The typical example is a hybrid of the FA and PSO algorithms, the FFPSO [16] introduced in late 2015 by Padmavathi Kora and K. Sri Rama Krishna. The central principle remains the same as in the standard FA, but the equation for firefly motion (4) is slightly changed according to PSO movement and is newly computed as (8).

$$x_i' = wx_i + c_1 e^{-r_{px}^2}(pBest_i - x_i) + c_2 e^{-r_{gx}^2}(gBest - x_i) + \alpha \cdot sign \tag{8}$$

Where w, c_1, and c_2 are control parameters transferred from PSO and their values often depends on the user. Also, the *pBest* and *gBest* are variables formerly belonging to PSO algorithm. They both represent the memory of the best position where *pBest* is best position of each particle and *gBest* is globally achieved best position so far. The remaining variables r_{px} (9) and r_{gx} (10) are distances between particle x_i and both $pBest_i$ and *gBest*.

$$r_{px} = \sqrt{\sum_{k=1}^{d} \left(pBest_{i,k} - x_{i,k}\right)^2} \tag{9}$$

$$r_{gx} = \sqrt{\sum_{k=1}^{d} \left(gBest_k - x_{i,k}\right)^2} \tag{10}$$

3.3 Orthogonal Learning Firefly Algorithm

Our proposed algorithm, OLFA, is based on FFPSO mentioned above and it uses the orthogonal learning technique. Our application of orthogonal learning is similar as in the Orthogonal Learning Particle Swarm Optimization (OLPSO). The OLFA also generates the promising learning exemplar by adopting an orthogonal learning strategy for each particle to learn from. This means that the equation of firefly moves (8) is slightly changed and does not contain the *pBest* and *gBest* any longer. The new Eq. (11) includes only the trial exemplar *gVector*.

$$x_i' = wx_i + c \cdot e^{-r^2}(gVector_i - x_i) + \alpha \cdot sign \tag{11}$$

Where Euclidian distance r is computed as (12) between firefly x_i and trial *gVector*.

$$r = \sqrt{\sum_{k=1}^{d} (gVector_{i,k} - x_{i,k})^2} \tag{12}$$

The *gVector* is used as the guide for each particle until it cannot improve the solution quality for a certain number of generations, which is called refreshing gap G. When the number of non-improved generations reaches the refreshing gap limit, the learning *gVector* is reconstructed. The (re)construction of *gVector* is the same as described in Sect. 2.2.

4   Experimental Setup

The experiments were performed on a set of well-known benchmark functions CEC'17 which are detailly described in [17]. The tested dimensions were 10 and 30. The maximal number of function evaluation was set as $10\,000 \cdot dim$ (dimension size). The lower and upper boundary was as $b^l = -100$ and $b^u = 100$ according to the CEC'17 definition. The number of particles (or fireflies) was set to 40 for all dimension sizes. Every test function was repeated for 51 independent runs and the results were statistically evaluated. The benchmark itself includes 30 test functions in four categories: unimodal, multimodal, hybrid and composite types.

The control parameters settings for all tested and compared algorithms are given in Table 1. Parameters were set to optimal values according to literature. The proposed OLFA algorithm adopted the parameters from its predecessors.

Table 1. Parameters of tested algorithms

Name	Description	Parameters
PSO	Particle Swarm Optimization	$w = 0.729$, $c_1 = c_2 = 1.49445$
OLPSO	Orthogonal Learning PSO	$w = 0.9 - 0.4$, $c = 2$, $G = 5$
FA	Firefly Algorithm	$\alpha = 0.5$, $\beta_0 = 0.2$, $\gamma = 1$,
FFPSO	Hybrid of FA and PSO	$\alpha = 0.5$, $\beta_0 = 0.2$, $\gamma = 1$ $w = 0.729$, $c = 1.49445$
OLFA	Orthogonal Learning FA	$w = 0.9 - 0.4$, $c = 2$, $G = 5$ $\alpha = 0.5$, $\beta_0 = 0.2$, $\gamma = 1$

5   Results

The results of all performed experiments are reported herein details. The results overview and simple statistical comparisons are presented in Table 3 and 4, depicting *mean* and *std. dev.* final objective function values. Further, the Wilcoxon rank-sum test pairwise comparisons are shown in Table 2 with statistical significance 0.05. Also, the overall performance of all tested algorithms on dimension sizes is evaluated and compared using Friedman ranks with critical distance assessed according to the Nemenyi Critical Distance post-hoc test for multiple comparisons. The visual outputs of various comparisons with rankings are given in Fig. 1. All Friedman rank test hypothesis are relevant with a p-value lower than 0.05. The dashed line represents the critical distance from the best-performed algorithm (the lowest mean rank). The lower the rank is, the better is the overall performance of that algorithm on particular dimension size.

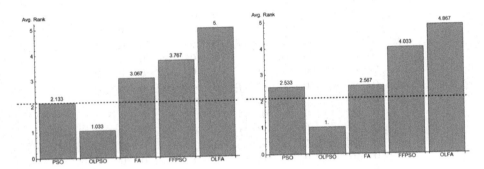

Fig. 1. Friedman rank tests for dimension size 10 (left) and 30 (right).

From the results, it is noticeable, that the OLPSO outperforms other algorithms on both dimension sizes. Also, the performance of OLFA seems to be worse compared to others firefly algorithms (FA and FFPSO). This may suggest that the orthogonal learning is not suitable (at least with current parameter setting) for FFPSO enhancement.

Table 2 contains test results for three tested algorithms (the compared algorithms in this order are FA, FFPSO, and OLFA) on all 30 benchmark functions and for both compared dimension sizes. Each cell of Table 2 contains a simple matrix of compared results. The number of columns and rows are equal to a number of compared algorithms. The "<" means that the algorithm outperformed the other compared algorithm from the test couple, the ">" means that the algorithm performs worse than the compared one. The "=" stand for the equivalent performance. The "0" symbol is present in Table 2 since the matrix is symmetrical by the main diagonal. For example, how to read the results in Table 2, for function f_1 in dimension 10, the first algorithm achieved better results than the other two and the second algorithm had a better result than the third algorithm. To conclude this example, the best performing algorithm for this test and dimension is the first algorithm FA, the second-best performing one is FFPSO, and the worst is the OLFA.

Table 2. Wilcoxon rank-sum test

F	Dimension						F	Dimension					
	10			30				10			30		
f_1	0	<	<	0	<	<	f_{16}	0	<	<	0	<	<
	0	0	<	0	0	<		0	0	<	0	0	<
	0	0	0	0	0	0		0	0	0	0	0	0
f_2	0	<	<	0	<	<	f_{17}	0	=	<	0	<	<
	0	0	<	0	0	<		0	0	<	0	0	<
	0	0	0	0	0	0		0	0	0	0	0	0
f_3	0	>	<	0	>	<	f_{18}	0	<	<	0	<	<
	0	0	<	0	0	<		0	0	<	0	0	<
	0	0	0	0	0	0		0	0	0	0	0	0
f_4	0	<	<	0	<	<	f_{19}	0	<	<	0	=	<
	0	0	<	0	0	<		0	0	<	0	0	<
	0	0	0	0	0	0		0	0	0	0	0	0
f_5	0	<	<	0	<	<	f_{20}	0	<	<	0	<	<
	0	0	<	0	0	<		0	0	<	0	0	<
	0	0	0	0	0	0		0	0	0	0	0	0
f_6	0	<	<	0	<	<	f_{21}	0	>	>	0	<	<
	0	0	<	0	0	<		0	0	<	0	0	<
	0	0	0	0	0	0		0	0	0	0	0	0
f_7	0	=	<	0	=	<	f_{22}	0	<	<	0	<	<
	0	0	<	0	0	<		0	0	<	0	0	<
	0	0	0	0	0	0		0	0	0	0	0	0
f_8	0	<	<	0	<	<	f_{23}	0	<	<	0	<	<
	0	0	<	0	0	<		0	0	=	0	0	>
	0	0	0	0	0	0		0	0	0	0	0	0
f_9	0	<	<	0	<	<	f_{24}	0	>	>	0	<	<
	0	0	<	0	0	<		0	0	<	0	0	>
	0	0	0	0	0	0		0	0	0	0	0	0
f_{10}	0	<	<	0	<	<	f_{25}	0	<	<	0	<	<
	0	0	<	0	0	<		0	0	<	0	0	<
	0	0	0	0	0	0		0	0	0	0	0	0
f_{11}	0	<	<	0	<	<	f_{26}	0	<	<	0	<	<
	0	0	<	0	0	<		0	0	<	0	0	<
	0	0	0	0	0	0		0	0	0	0	0	0
f_{12}	0	<	<	0	<	<	f_{27}	0	<	<	0	<	<
	0	0	<	0	0	<		0	0	>	0	0	>
	0	0	0	0	0	0		0	0	0	0	0	0
f_{13}	0	=	<	0	<	<	f_{28}	0	<	<	0	<	<
	0	0	<	0	0	<		0	0	<	0	0	<
	0	0	0	0	0	0		0	0	0	0	0	0
f_{14}	0	<	<	0	<	<	f_{29}	0	<	<	0	<	<
	0	0	<	0	0	<		0	0	<	0	0	<
	0	0	0	0	0	0		0	0	0	0	0	0
f_{15}	0	=	<	0	=	<	f_{30}	0	<	<	0	<	<
	0	0	<	0	0	<		0	0	<	0	0	<
	0	0	0	0	0	0		0	0	0	0	0	0

Table 3. Statistical results for dimension 10 (mean and std. dev.)

F	PSO		OLPSO		FA		FFPSO		OLFA	
f1	8.36E+08	3.38E+08	4.38E+07	1.99E+08	1.06E+09	3.01E+08	3.23E+09	7.66E+08	4.00E+09	1.59E+09
f2	9.22E+06	1.39E+07	8.32E+02	4.69E+03	1.26E+08	1.28E+08	4.93E+08	4.91E+08	4.16E+09	5.57E+09
f3	2.70E+03	9.22E+02	1.03E+02	1.26E+02	6.68E+03	1.69E+03	5.46E+03	1.27E+03	1.03E+04	3.47E+03
f4	4.51E+02	2.17E+01	5.30E+01	1.21E+02	4.68E+02	1.48E+01	5.82E+02	4.45E+01	6.58E+02	8.92E+01
f5	5.45E+02	1.15E+01	1.34E+02	1.81E+02	5.44E+02	5.34E+00	5.70E+02	8.10E+00	6.44E+02	4.25E+02
f6	6.19E+02	4.59E+00	1.05E+02	1.83E+02	6.25E+02	2.38E+00	6.36E+02	4.57E+00	6.18E+02	1.09E+02
f7	7.91E+02	1.39E+01	2.08E+02	2.94E+02	8.18E+02	1.18E+01	8.18E+02	1.12E+01	8.72E+02	4.44E+02
f8	8.42E+02	6.97E+00	2.22E+02	3.08E+02	8.48E+02	5.90E+00	8.50E+02	5.80E+00	9.12E+02	3.96E+02
f9	1.09E+03	6.32E+01	2.20E+02	3.29E+02	1.24E+03	7.60E+01	1.29E+03	9.81E+01	1.78E+03	3.73E+02
f10	2.27E+03	2.21E+02	2.80E+02	4.56E+02	2.34E+03	1.35E+02	2.43E+03	1.55E+02	2.44E+03	4.79E+02
f11	1.18E+03	2.77E+01	2.35E+02	3.85E+02	1.27E+03	3.99E+01	1.30E+03	6.56E+01	1.48E+03	5.18E+02
f12	1.21E+07	5.58E+06	5.19E+03	1.19E+04	2.24E+07	1.11E+07	3.06E+07	1.63E+07	7.65E+07	4.84E+07
f13	3.03E+04	2.05E+04	3.86E+03	7.21E+03	6.36E+04	4.73E+04	8.45E+04	7.77E+04	3.66E+05	5.32E+05
f14	1.50E+03	3.86E+01	7.13E+02	2.87E+02	1.61E+03	1.08E+02	1.80E+03	2.68E+02	2.24E+03	8.69E+02
f15	2.25E+03	6.20E+02	2.44E+03	7.67E+03	3.51E+03	1.12E+03	3.40E+03	9.27E+02	6.46E+03	3.17E+02
f16	1.73E+03	8.77E+01	4.15E+02	6.64E+02	1.74E+03	3.87E+01	1.87E+03	7.67E+01	1.90E+03	2.31E+02
f17	1.78E+03	1.51E+01	5.25E+02	7.19E+02	1.79E+03	1.62E+01	1.80E+03	1.68E+01	1.83E+03	2.66E+02
f18	5.41E+04	5.13E+04	2.01E+03	3.85E+03	9.92E+04	7.65E+04	8.13E+05	8.87E+05	1.85E+06	1.71E+06
f19	2.67E+03	8.63E+02	1.01E+03	2.54E+03	3.15E+03	9.05E+02	5.62E+03	2.95E+03	1.43E+04	1.29E+04
f20	2.08E+03	1.68E+01	6.30E+02	9.02E+02	2.10E+03	1.62E+01	2.13E+03	2.98E+01	2.14E+03	2.30E+02
f21	2.24E+03	5.58E+01	5.74E+02	7.70E+02	2.31E+03	3.12E+01	2.24E+03	1.11E+01	2.25E+03	2.19E+02
f22	2.38E+03	3.73E+01	4.38E+02	7.68E+02	2.43E+03	2.98E+01	2.50E+03	5.49E+01	2.47E+03	4.52E+02
f23	2.66E+03	5.26E+01	5.82E+02	8.66E+02	2.65E+03	4.65E+00	2.69E+03	6.07E+01	2.59E+03	4.25E+02
f24	2.67E+03	1.11E+02	7.15E+02	1.03E+03	2.77E+03	2.29E+01	2.65E+03	3.34E+01	2.68E+03	3.36E+02
f25	2.97E+03	2.55E+01	6.98E+02	1.12E+03	2.99E+03	1.25E+01	3.06E+03	3.53E+01	3.05E+03	3.53E+02
f26	3.10E+03	1.07E+02	1.00E+03	1.29E+03	3.10E+03	2.86E+01	3.36E+03	8.75E+01	3.24E+03	7.33E+02
f27	3.13E+03	2.47E+01	8.32E+02	1.21E+03	3.11E+03	1.95E+00	3.20E+03	1.66E+01	3.13E+03	3.05E+02
f28	3.25E+03	6.60E+01	9.32E+02	1.13E+03	3.31E+03	6.02E+01	3.44E+03	6.49E+01	3.41E+03	6.37E+02
f29	3.24E+03	3.69E+01	9.24E+02	1.21E+03	3.24E+03	1.93E+01	3.31E+03	3.78E+01	3.23E+03	6.07E+02
f30	1.12E+06	9.55E+05	6.03E+04	1.66E+05	1.47E+06	6.68E+05	3.44E+06	1.67E+06	5.69E+06	3.27E+06

Table 4. Statistical results for dimension 30 (mean and std. dev.)

F	PSO		OLPSO		FA		FFPSO		OLFA	
f1	2.30E + 10	4.69E + 09	1.11E + 09	2.65E + 09	2.22E + 10	2.05E + 09	3.71E + 10	3.51E + 09	5.47E + 10	1.02E + 10
f2	7.31E + 35	2.95E + 36	4.83E + 34	3.45E + 35	6.85E + 35	1.71E + 36	3.23E + 39	7.76E + 39	4.86E + 41	9.23E + 41
f3	5.29E + 04	1.09E + 04	1.21E + 04	2.25E + 04	9.14E + 04	1.11E + 04	7.10E + 04	9.37E + 03	1.15E + 05	2.98E + 04
f4	4.05E + 03	1.20E + 03	1.48E + 02	3.77E + 02	2.44E + 03	2.74E + 02	8.15E + 03	1.39E + 03	1.35E + 04	1.95E + 03
f5	8.12E + 02	3.74E + 01	1.35E + 02	2.43E + 02	7.94E + 02	1.28E + 02	8.74E + 02	1.90E + 01	9.32E + 02	6.82E + 01
f6	6.63E + 02	8.84E + 00	1.09E + 02	1.99E + 02	6.55E + 02	3.23E + 00	6.78E + 02	4.41E + 00	6.86E + 02	2.24E + 01
f7	1.39E + 03	9.63E + 01	2.14E + 02	3.42E + 02	1.59E + 03	5.47E + 01	1.58E + 03	5.98E + 01	2.05E + 03	2.90E + 02
f8	1.08E + 03	3.28E + 01	2.06E + 02	3.55E + 02	1.08E + 03	1.52E + 01	1.13E + 03	1.48E + 01	1.16E + 03	1.29E + 02
f9	7.54E + 03	1.39E + 03	8.99E + 02	1.83E + 03	7.33E + 03	7.76E + 02	1.03E + 04	1.16E + 03	1.51E + 04	4.13E + 03
f10	7.98E + 03	6.34E + 02	1.01E + 03	1.84E + 03	8.05E + 03	2.89E + 02	8.20E + 03	2.93E + 02	8.45E + 03	1.41E + 03
f11	2.90E + 03	4.67E + 02	3.64E + 02	9.90E + 02	4.53E + 03	3.91E + 01	4.69E + 03	5.25E + 02	7.77E + 03	2.53E + 03
f12	2.15E + 09	6.93E + 08	2.33E + 07	1.20E + 08	2.21E + 09	3.91E + 08	5.53E + 09	9.89E + 08	7.93E + 09	1.73E + 09
f13	5.07E + 08	2.31E + 08	3.82E + 06	1.45E + 07	8.66E + 08	2.65E + 08	2.85E + 09	7.51E + 08	4.14E + 09	1.40E + 09
f14	1.65E + 05	1.11E + 05	1.34E + 04	4.72E + 04	3.33E + 05	1.76E + 05	4.46E + 05	2.35E + 05	9.49E + 05	6.05E + 05
f15	2.48E + 07	1.43E + 07	3.93E + 03	8.35E + 03	9.23E + 07	3.88E + 07	8.70E + 07	3.93E + 07	3.13E + 08	1.52E + 08
f16	3.85E + 03	2.20E + 02	6.17E + 02	1.05E + 03	3.62E + 03	2.12E + 02	4.62E + 03	2.74E + 02	4.92E + 03	2.50E + 02
f17	2.52E + 03	1.68E + 02	5.02E + 02	8.60E + 02	2.59E + 03	1.22E + 02	3.02E + 03	1.68E + 02	3.34E + 03	2.72E + 02
f18	2.83E + 06	1.31E + 06	4.56E + 05	1.29E + 06	5.06E + 06	2.19E + 06	6.39E + 06	2.31E + 06	1.73E + 07	7.46E + 06
f19	4.95E + 07	2.84E + 07	1.71E + 05	9.88E + 05	1.36E + 08	5.77E + 07	1.46E + 08	6.87E + 07	4.96E + 08	2.03E + 08
f20	2.67E + 03	1.06E + 02	5.39E + 02	8.83E + 02	2.65E + 03	9.48E + 01	2.73E + 03	6.92E + 02	2.91E + 03	1.76E + 02
f21	2.60E + 03	3.64E + 02	3.86E + 02	7.65E + 02	2.57E + 03	1.24E + 01	2.65E + 03	4.24E + 02	2.62E + 03	4.39E + 02
f22	5.32E + 03	1.64E + 03	8.13E + 02	1.61E + 03	4.64E + 03	2.03E + 02	6.76E + 03	4.33E + 02	8.11E + 03	1.60E + 03
f23	3.16E + 03	6.74E + 01	5.91E + 02	1.02E + 03	2.97E + 03	1.31E + 01	3.43E + 03	4.22E + 01	3.38E + 03	5.65E + 01
f24	3.35E + 03	7.77E + 01	6.07E + 02	1.05E + 03	3.11E + 03	1.59E + 01	3.69E + 03	7.79E + 01	3.53E + 03	3.18E + 02
f25	3.93E + 03	2.64E + 02	8.27E + 02	1.22E + 03	4.11E + 03	1.72E + 02	4.61E + 03	2.33E + 02	6.40E + 03	1.34E + 03
f26	7.24E + 03	9.76E + 02	1.03E + 03	1.83E + 03	7.14E + 03	1.56E + 02	8.96E + 03	5.38E + 02	1.05E + 04	7.35E + 02
f27	3.71E + 03	1.18E + 02	6.50E + 02	1.20E + 03	3.41E + 03	2.47E + 01	4.15E + 03	1.13E + 02	3.91E + 03	3.37E + 02
f28	4.57E + 03	3.07E + 02	6.93E + 02	1.22E + 03	4.38E + 03	1.42E + 02	5.51E + 03	3.25E + 02	6.64E + 03	1.12E + 03
f29	4.84E + 03	2.23E + 02	5.36E + 02	1.23E + 03	5.04E + 03	1.57E + 02	5.62E + 03	2.05E + 02	5.92E + 03	5.10E + 02
f30	7.01E + 07	3.38E + 07	3.09E + 05	1.17E + 06	1.23E + 08	3.42E + 07	2.87E + 08	7.13E + 07	5.13E + 08	1.62E + 08

6 Conclusion

In this study, an unusual method, orthogonal learning, based on orthogonal design method is applied to the hybridized algorithm of PSO and FA. The final proposed algorithm is tested on advanced benchmark functions CEC 2017. The results are statistically evaluated and compared with other algorithms via the Friedman rank test and the Wilcoxon rank-sum test. The compared algorithms are canonical FA, FFPSO, PSO, and OLPSO. The algorithms PSO and OLPSO are for performance comparison of orthogonal learning.

The analyzed data suggest that the orthogonal learning performs on PSO algorithm better than on FFPSO. This may be affected by the nature of FA behavior, which seems to be less suitable than the mentioned PSO. However, the orthogonal learning could be improved by better parameter setting (refreshing gap).

All the presented results show that the original FFPSO hybrid algorithm offers a noticeable potential to many improvements. The proposed algorithm OLFA is only one of many. Nevertheless, more research including optimal parameter tuning is needed to increase the understandability of the algorithm behavior and its performance. Also, a design of the proposed algorithm offers several possible changes that could affect its performance. Majority of them are based on the parent algorithms FA and PSO, for example, Lévy flights, adaptive control parameters, comprehensive learning and more.

Acknowledgments. This work was supported by the Ministry of Education, Youth and Sports of the Czech Republic within the National Sustainability Programme Project no. LO1303 (MSMT-7778/2014), further by the European Regional Development Fund under the Project CEBIA-Tech no. CZ.1.05/2.1.00/03.0089 and by Internal Grant Agency of Tomas Bata University under the Projects no. IGA/CebiaTech/2018/003. This work is also based upon support by COST (European Cooperation in Science & Technology) under Action CA15140, Improving Applicability of Nature-Inspired Optimisation by Joining Theory and Practice (ImAppNIO), and Action IC1406, High-Performance Modelling and Simulation for Big Data Applications (cHiPSet). The work was further supported by resources of A.I. Lab at the Faculty of Applied Informatics, Tomas Bata University in Zlin (ailab.fai.utb.cz).

References

1. Fister, Jr., I., Mlakar, U., Brest, J., Fister, I.: A new population-based nature-inspired algorithm every month: is the current era coming to the end. In: Proceedings of the 3rd Student Computer Science Research Conference, pp. 33–37, University of Primorska Press (2016)
2. Du, W., Li, B.: Multi-strategy ensemble particle swarm optimization for dynamic optimization. Inf. Sci. **178**(15), 3096–3109 (2008)
3. Wang, H., Wu, Z., Rahnamayan, S., Sun, H., Liu, Y., Pan, J.: Multi-strategy ensemble artificial bee colony algorithm. Inf. Sci. **20**(279), 587–603 (2014)
4. Shelokar, P.S., Siarry, P., Jayaraman, V.K., Kulkarni, B.D.: Particle swarm and ant colony algorithms hybridized for improved continuous optimization. Appl. Math. Comput. **188**(1), 129–142 (2007)

5. Das, S., Abraham, A., Konar, A.: Particle swarm optimization and differential evolution algorithms: technical analysis, applications and hybridization perspectives. In: Advances of computational intelligence in industrial systems, pp. 1–38. Springer, Heidelberg (2008)

6. Clerc, M.: The swarm and the queen: towards a deterministic and adaptive particle swarm optimization. In: Proceedings of the 1999 Congress on Evolutionary Computation, 1999. CEC 99, vol. 3, pp. 1951–1957. IEEE (1999)

7. Shi, Y., Eberhart, R.C.: Fuzzy adaptive particle swarm optimization. In: Proceedings of the 2001 Congress on Evolutionary Computation, vol. 1, pp. 101–106, IEEE (2001)

8. Eberhart, R., Kennedy, J.A.: New optimizer using particle swarm theory (1995)

9. Lynn, N., Suganthan, P.N.: Heterogeneous comprehensive learning particle swarm optimization with enhanced exploration and exploitation. Swarm Evol. Comput. **24**, 11–24 (2015)

10. Nepomuceno, F.V., Engelbrecht, A.P.: A self-adaptive heterogeneous PSO for real-parameter optimization. IEEE (2013)

11. Zhan, Z.H., Zhang, J., Li, Y., Shi, Y.H.: Orthogonal learning particle swarm optimization. IEEE Trans. Evol. Comput. **15**(6), 832–847 (2011)

12. Yang, X.: Nature-inspired metaheuristic algorithms. Luniver press, UK (2010)

13. Gandomi, A.H., Yang, X.S., Talatahari, S., Alavi, A.H.: Firefly algorithm with chaos. Commun. Nonlinear Sci. Numer. Simul. **18**(1), 89–98 (2013)

14. Yang, X.S: Firefly Algorithm, Lévy Flights and Global Optimization. In: Bramer, M., Ellis, R., Petridis M. (eds.) Research and Development in Intelligent Systems XXVI, pp. 209–218. Springer, London (2010)

15. Farahani, S.M., Abshouri, A.A., Nasiri, B., Meybodi, M.R.: A Gaussian firefly algorithm. Int. J. Mach. Learn. Comput. **1**(5), 448 (2011)

16. Kora, P., Rama Krishna, K.S.: Hybrid firefly and Particle Swarm Optimization algorithm for the detection of Bundle Branch Block. Int. J. Cardiovasc. Acad. **2**(1), 44–48 (2016)

17. Awad, N.H., et al.: Problem Definitions and Evaluation Criteria for CEC 2017 Special Session and Competition on Single-Objective Real-Parameter Numerical Optimization (2016)

18. Kennedy, J.: The particle swarm: social adaptation of knowledge. In: Proceedings of the IEEE International Conference on Evolutionary Computation, pp. 303–308 (1997)

Multi-label Learning by Hyperparameters Calibration for Treating Class Imbalance

Andrés Felipe Giraldo-Forero[(⊠)], Andrés Felipe Cardona-Escobar,
and Andrés Eduardo Castro-Ospina

Grupo de Automática, Electrónica y Ciencias Computacionales,
Instituto Tecnológico Metropolitano, Medellín, Colombia
{felipegiraldo,andrescastro}@itm.edu.co,
andrescardona134713@correo.itm.edu.co

Abstract. Multi-label learning has been becoming an increasingly active area into the machine learning community due to a wide variety of real world problems. However, only over the past few years class balancing for these kind of problems became a topic of interest. In this paper, we present a novel method named *hyperparameter calibration* to treat class imbalance in a multi-label problem, to this aim we develop an extensive analysis over four real-world databases and two own synthetic databases exhibiting different ratios of imbalance. The empirical analysis shows that the proposed method is able to improve the classification performance when it is combined with three of the most widely used strategies for treating multi-label classification problems.

Keywords: Multilabel classification · Imbalanced learning
Support vector machine · Problem transformation

1 Introduction

Automatic classification is one of the most important tasks in data mining, where the aim is to associate a sample with the correct category from a set of features. The traditional classification task where each sample is assigned to only one category presents some limitations. First, one sample can not belong simultaneously to multiple categories. Second, the relationship between categories is not exploited. With the aim of overcoming these limitations, a new classification paradigm known as multi-label learning has emerged over the last years. Multi-label learning has attracted a significant attention given that a huge quantity of real world problems can have multi-label structures, from semantic annotation when one text can address several topics as politics and sciences, or images and video that can depict various types of environments simultaneously, as mountains and forests.

Some proposals have arisen in the state of art aiming to group multi-label methods, we selected the classic categorization proposed in [1], where methods are grouped in *problem transformation* methods [2,3] and *algorithm adaptation*

© Springer International Publishing AG, part of Springer Nature 2018
F. J. de Cos Juez et al. (Eds.): HAIS 2018, LNAI 10870, pp. 327–337, 2018.
https://doi.org/10.1007/978-3-319-92639-1_27

methods. The former group consists in decomposing multi-label problems as binary classification problems, while the latter extends and customizes an existing machine learning algorithm to solve multi-label problems. Several methods employ problem transformation strategies, inasmuch as are highly parallelizable and more versatile than methods based on algorithm adaptation. However, class imbalance problems are induced in the process [4]. Class imbalance has been successfully treated for several years in bi-class and multi-class classification scenarios [5], some of these techniques have been focused to operate on support vector machine (SVM) [6]. Nevertheless, it has only been in the last few years that relevance in multi-label classification problems has raise. A few works have addressed class imbalance problems in multi-label learning through re-sampling techniques [4,7,8], but generating new samples increases computational cost and removing samples can alter the majority class representation.

In this paper, we propose a new multi-label learning approach named *hyper-parameters calibration*, where information of multi-label samples is captured based on the assumption that in representation space multi-label samples should be in the transition region from one class to another. Our approach is applied over two own synthetic databases and four real-world databases using SVM and three of the main *problem transformation* methods, achieving promising results even with highly imbalanced classes.

The rest of the paper is organized as follows: Sect. 2 defines the tasks of multi-label classification and the methods used in the experimental evaluation. Experimental framework is presented in Sect. 3. In Sect. 4 are shown results and discussion. Finally, Sect. 5 presents the conclusions of this work.

2 Background and Related Work

Suppose a classification problem where each sample $x \in \mathcal{X}$ belongs to one or more classes. Now, consider a training set $\mathcal{T} = \{(\mathbf{x}_1, \mathbf{y}_1), \ldots, (\mathbf{x}_i, \mathbf{y}_i), \ldots, (\mathbf{x}_n, \mathbf{y}_n)\}$, with $\mathbf{x}_i \in \mathbb{R}^d$ and $\mathbf{y}_i \subseteq \mathcal{Y}$, being $\mathcal{Y}$ a label set. For representation reasons and without loss of generality, we denote the label set by matrix $\boldsymbol{Y} \in \mathbb{R}^{n \times q}$, with each row being a binary vector $\boldsymbol{y}_i = \{y_i^1, ..., y_i^j, ..., y_i^q\}$ where $y_i^j \in \{1, 0\}$ indicates whether the i-th sample must be associated to the j-th class or not. The goal of the multi-label classification is to get a function $\boldsymbol{h} : \mathcal{X} \rightarrow \mathcal{Y}$ that correctly assigns a subset of labels to new samples from the information given in $\mathcal{T}$. Several methods have emerged to decompose $\boldsymbol{h}$ into a set of binary classifiers h_k. Some of them are described below:

- **Binary Relevance (BR)**
  The binary relevance method is a problem transformation strategy that decomposes the multi-label learning problem into k independent binary classification problems, where each binary classification problem is responsible of predict the labels associated with each sample. The training process is performed with the whole set of features $\boldsymbol{x}_i \in \mathcal{T}$ while the set of labels for each sample is redefined as $\boldsymbol{Y}_k = \boldsymbol{y}^k$. At the end of the training stage, $h_k : \mathcal{X} \rightarrow \{1, 0\}$ functions will be obtained [1].

- **Label Powerset (LP)**
  This strategy decomposes the multi-label learning problem into $2^{|\mathcal{Y}|}$ subproblems corresponding to the powerset of $\mathcal{Y}$, where the operator $|\cdot|$ denotes cardinality of the label set. However, in practice, the number of new classifiers after the transformation is lower than the original. As an example, suppose a label set $\mathcal{Y} = \{a, b, c\}$ where prior to transformation there are three labels, but once the transformation is performed, four new labels are generated $\{a, b\}$ $\{a, c\}$ $\{b, c\}$ and $\{a, b, c\}$, where class $\{a, b\}$ is formed for $\{x_i : y_i^a = 1 \wedge y_i^b = 1 \wedge y_i^c = 0\}$ [1].
- **Calibrated Label Ranking (CLR)**
  Calibrated Label Ranking is a strategy for extending the common pairwise approach to multi-label learning. Two parts compose this strategy, first the multi-label learning problem is decomposed into $q(q-1)/2$ binary classifiers, which are generated by pair-wise comparison, one for each label pair $(y_j, y_k)(1 \leq j < k \leq q)$, in this stage multi-label samples are removed and a vote-based system is used as ranking. In the second part, a calibration label λ_0 is introduced to serve as an artificial splitting point between x_i's relevant and irrelevant labels, for calculate λ_0, q auxiliary binary classifiers are induced [2].

When a multi-label problem is transformed to traditional classification problem (binary), it can be solved by SVM classifiers. SVMs search for the hyperplane that maximizes the margin of the training data, minimizing the Eq. (1), known as primal form [9].

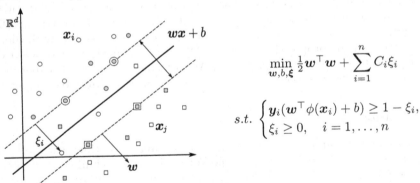

$$\min_{w,b,\xi} \tfrac{1}{2}w^\top w + \sum_{i=1}^{n} C_i \xi_i$$

$$s.t. \begin{cases} y_i(w^\top \phi(x_i) + b) \geq 1 - \xi_i, \\ \xi_i \geq 0, \quad i = 1, \ldots, n \end{cases}$$

Where C_i, ξ_i and $\phi(x_i)$ are the regularization parameter, slack variable and mapping to a high-dimensional space for each sample x_i, respectively. For accessibility reasons primal form is converted to dual form (2). Now, the goal is to minimize the vector $\boldsymbol{\alpha}$, corresponding to Lagrange multipliers in the Eq. (2). At the end of the process, only a few α_i takes values different from zero; the samples associated with those α are called support vectors SV, and the number of SV will depend on C.

$$\min_{\alpha} \tfrac{1}{2}\alpha^\top Q \alpha - e^\top \alpha \quad s.t. \begin{cases} y^\top \alpha = 0, \\ 0 \leq \alpha_i \leq C_i, i = 1, \ldots, n \end{cases} \tag{2}$$

Let e be a vector of all ones and Q an n by n matrix where each element is computed as $Q_{ij} = y_i y_j \phi(\boldsymbol{x}_i)^\top \phi(\boldsymbol{x}_j)$. In order to simplify the model, usually all C_i weights have the same value. Nevertheless, some works have treated class imbalance problems successfully, when C_i have different weight values per class in multiclass problems [9]. However, these strategies were not designed to capture the possible correlation between classes. We propose a novel method, named *hyperparameters calibration*. The method takes into account the relationship between classes, based on the assumption that in the representation space multi-label samples should be in the transition region from one class to another. We take advantage of the influence of C_i in the choice of SV by means of the next rules:

- Only the C_i values of samples related with target class are modified.
- Pure samples (belong ing exclusively to one class) are not modified, while multi-label samples associated with a major number of classes will have higher relevance.
- Three kinds of modifications of C_i are considered:
  - Linear decreasing weights (W$\searrow$), with $C_i = C/|\boldsymbol{y}_i|$
  - Linear increasing weights (W$\nearrow$), with $C_i = C \times |\boldsymbol{y}_i|$
  - Constant weights (W$\rightarrow$), with $C_i = C$

3   Experimental Setup

With the intent to assess the benefits of the proposed technique, an extensive experimental study was conducted. The experiments were designed to respond two essential questions: giving more weight to multi-label samples improves classification performance? and, how the performances are affected according to the degree of imbalance in the proposed approach?. All the experiments were developed in the R project for statistical computing. An implementation of our experiments are freely available at: https://afgiraldofo@bitbucket.org/afgiraldofo/tim.git

3.1   Databases

We created two different synthetic multi-label databases 30 times each following a hyper-spheres strategy [10]: Volvox and Wheel. In order to build each database, we generated a set of independent and identically distributed (i.i.d.) points, corresponding to the parameters radius (r) and angle (θ) in a polar coordinate system by a uniform probability distribution. The upper and lower boundaries for each uniform distribution are given according to Tables 1 and 2, where each table corresponds to a database, and each row to one class of power set of $\mathcal{Y}$. Then, parameters r and θ are mapped to the components x_1 and x_2 through the following equations $x_1 = \beta_1(r \cdot cos(\theta)) + c_1$ and $x_2 = \beta_2(r \cdot sin(\theta)) + c_2$, being (c_1, c_2) the center coordinates and $\beta = \beta_1/\beta_2$ a scale factor.

In the design of the databases we present two type of scenarios, one where there is a majority class and several minority classes (Volvox, see Fig. 1). The second scenario contains minority and majority classes (Wheel, see Fig. 2).

This was designed to take into account the two possible imbalance scenarios with different levels of immersion.

General information of databases is given in Tables 1 and 2, as samples number per class, total number of samples and the cardinality, i.e., an average of the number of labels associated to each sample. Given that the imbalance level in traditional classification tasks is measured taking into account only two classes unlike the multi-label classification that present several labels, with the aim of including all of them, we use max and mean imbalance ratio denoted as $MaxIR$ and $MeanIR$ respectively Eq (3) suggested in [11] as additional descriptors.

$$MaxIR = \frac{\max_j \sum_{i=1}^{n} y^j}{\min_j \sum_{i=1}^{n} y^j} \qquad MeanIR = \frac{1}{q} \sum_{j=1}^{q} \frac{\max_j \sum_{i=1}^{n} y^j}{\sum_{i=1}^{n} y^j} \qquad (3)$$

Table 1. Description and generation parameters for r and θ in the Volvox database.

Class	Color	Radius	Angles	Scale	Center	Samples
A	Red	$(0,5)$	$(0,2\pi)$	1	$(0,0)$	$100 - 1900$
B	Blue	$(0,3/2)$	$(0,2\pi)$	1	$(\frac{-5}{2\sqrt{2}}, \frac{5}{2\sqrt{2}})$	100
C	Green	$(0,3/2)$	$(0,2\pi)$	1	$(\frac{5}{2\sqrt{2}}, \frac{5}{2\sqrt{2}})$	100
D	Orange	$(0,3/2)$	$(0,2\pi)$	1	$(0, \frac{-5}{2})$	100
$A \cap B$	Gray	$(1,2)$	$(0,2\pi)$	1	$(\frac{-5}{2\sqrt{2}}, \frac{5}{2\sqrt{2}})$	100
$A \cap C$	Gray	$(1,2)$	$(0,2\pi)$	1	$(\frac{5}{2\sqrt{2}}, \frac{5}{2\sqrt{2}})$	100
$A \cap D$	Gray	$(1,2)$	$(0,2\pi)$	1	$(0, \frac{-5}{2})$	100
Cardinality			$1.43 - 1.13$		Total	$700 - 2500$

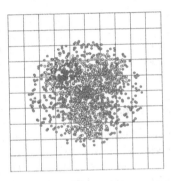

Fig. 1. Scatter plot for Volvox database.

Table 2. Description and generation parameters for r and θ in the Wheel database.

Class	Color	Radius	Angles	Scale	Center	Samples
A	Red	$(0,3/2)$	$(0,2\pi)$	1	$(0,0)$	100
B	Blue	$(0,3)$	$(-\pi/2, \pi/2)$	1	$(0,0)$	$100 - 700$
C	Green	$(0,3)$	$(\pi/2, 7\pi/6)$	1	$(0,0)$	$100 - 700$
D	Orange	$(0,3)$	$(7\pi/6, 11\pi/6)$	1	$(0,0)$	$100 - 700$
$A \cap B$	Gray	$(1,2)$	$(-\pi/2, \pi/2)$	1	$(0,0)$	100
$B \cap C$	Gray	$(1,2)$	$(\pi/2, 7\pi/6)$	1	$(0,0)$	100
$C \cap D$	Gray	$(1,2)$	$(7\pi/6, 11\pi/6)$	1	$(0,0)$	100
Cardinality			$1.43 - 1.05$		Total	$700 - 2500$

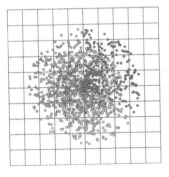

Fig. 2. Scatter plot for Wheel database.

In the majority of real-world databases, the ratio of multi-label samples is lower than pure samples, for this reason in our synthetic databases, the number of pure samples is higher than the number of multi-label samples.

In order to track the evolution of the proposed technique regarding the imbalance ratio, we generate the databases with seven different imbalance ratios, so that both databases have the same size, we increase the number of pure samples of the majority classes from 100 to 1900 with steps of 300 in Volvox and from 100 to 700 with steps of 100 in Wheel, different imbalance ratios were obtained due to the nature of each database, $MeanIR = \{1.75, 2.875, \ldots, 8.5\}$ for Volvox and $MeanIR = \{1.75, 1.25, 1, 1.1875, 1.375, 1.5625, 1.75\}$ for Wheel.

Additionally to synthetic generated databases, we evaluate our proposed algorithm against several widely known real-world databases, employing a 2×5 folds cross validation scheme[1]. A brief description of these is shown in Table 3.

Table 3. Description of real multi-label databases used in our experiments

Database	Cardinality	Samples	Features	Labels	MaxIR	MeanIR	Domain	Reference
Emotions	1.869	593	72	6	1.784	1.478	Music	[12]
Scene	1.074	2407	294	6	1.464	1.254	Image	[13]
Yeast	4.237	2417	103	14	53.412	7.197	Biology	[14]
Cal500	26.044	502	68	174	88.800	20.578	Music	[15]

3.2 Classification

In the classification stage we aim to compare three different approaches: equally weighted samples (W→), linear increasing weights (W↗) and linear decreasing weights (W↘), over three known multi-label methods, namely, BR, LP and CLR, described in Sect. 2. This comparison was performed in two own synthetic databases constructed under the criteria of Sect. 3.1 and in the real-world database emotions, scene, yeast and cal500 presented in Table 3.

As learning algorithm we use an SVM with Radial Basis function as kernel for running all the classification tests. This SVM classifier is trained with R language with an available modification of the *e1071* package [16][2], allowing the assignment of C value for each sample. The training process is performed in two steps, first each database is divided into train and test by a stratified 5-fold cross-validation, then each fold used as train is divided the same way to perform a hyperparameter tuning. The process for each train fold was performed via Grid Search, a commonly used method for this purpose, where kernel dispersion and trade-off penalization are tuned from the next parameter sequences: the Radial Basis function kernel band-width $\gamma = \{2^{-5}, 2^{-4}, \ldots, 2^3\}$ and the regularization parameter $C = \{10^{-1}, 10, \ldots, 10^3\}$ according to a suggestion of the literature [17]. This mechanism allows an estimation of the reliability of the

---

[1] Train and test sets for emotions, scene, yeast and cal500 databases were obtained from http://simidat.ujaen.es/~research/MLSMOTE/index.html#datasets.

[2] https://afgiraldofo@bitbucket.org/afgiraldofo/e1071.git.

model by computing the variability of the results through the five repetitions. The objective function in the tuning process was computed as the macro-average of the F_1 *measure*, inasmuch as it is a measure that presents a good compromise between specificity and sensitivity [18]. The calculation of macro-average was done by Eq. (4). Being $B(\cdot)$ the binary evaluation measure, tp, fp, tn, and fn are true positive, false positive, true negative, and false negative, respectively.

$$
\begin{aligned}
B_{micro} &= B\left(\sum_{j=1}^{q}|tp|_j, \sum_{j=1}^{q}|fp|_j, \sum_{j=1}^{q}|tn|_j, \sum_{j=1}^{q}|fn|_j\right) \\
B_{macro} &= \frac{1}{q}\sum_{j=1}^{q} B\left(|tp|_j, |fp|_j, |tn|_j, |fn|_j\right)
\end{aligned}
\tag{4}
$$

In order to compare (W→) against (W↗) and (W↘), we developed a paired t-test to two sided at 95% significance level. For doing so, each database was randomly generated 30 times, and results were computed for each trial.

4   Results and Discussion

In this section, we present the results from the experimental evaluation. For each synthetic database, method and approach, we present and discuss two type of experiments: First we compare three different ways to assign weights to multi-label samples (W↗), (W→) and (W↘). In this opportunity we employ the databases with the highest imbalance ratio 8.5 for Volvox database and 1.75 for Wheel database. For real-world databases only the results of the method BR over each approach is presented.

For the first experiment, we evaluated four label-based measures: *Recall*, *Precision*, the Matthews correlation coefficient (*Matthews*) and the harmonic mean between *Recall* and *Precision* (*F_1 measure*), using testing samples and hyperparameters obtained in the tuning process. Given the nature of the problem, the label-based measures were calculated by micro-average and macro-average [18] using Eq. (4). The result of the first experiment is shown in the Tables 4 and 5 for Wheel and Volvox databases, respectively. The best values of each approach are highlighted in bold. Overall, this first experimental

Table 4. The classification performance over Wheel database for an imbalance ratio of 1.75. $\oplus$ ($\ominus$) indicate which approch is significantly higer (lower) than (W→) while $\bigcirc$ without significant difference based on paired t-test at 95% significance level.

Criterion	Measures	BR			LP			CLR		
		W↗	W→	W↘	W↗	W→	W↘	W↗	W→	W↘
Macro	F_1 *score*	**0.801**$\oplus$	0.779	0.765$\ominus$	**0.786**$\oplus$	0.770	0.764$\ominus$	**0.798**$\oplus$	0.774	0.760$\ominus$
	Matthews	**0.752**$\oplus$	0.738	0.735$\bigcirc$	**0.732**$\oplus$	0.722	0.726$\bigcirc$	**0.748**$\oplus$	0.728	0.722$\ominus$
	Precision	**0.813**$\oplus$	0.771	0.753$\ominus$	**0.782**$\oplus$	0.754	0.745$\ominus$	**0.809**$\oplus$	0.762	0.742$\ominus$
	Recall	0.790$\bigcirc$	0.794	**0.806**$\oplus$	0.791$\bigcirc$	0.790	**0.799**$\oplus$	0.789$\bigcirc$	0.790	**0.797**$\oplus$
Micro	F_1 *score*	0.864$\ominus$	0.870	**0.877**$\oplus$	0.849$\ominus$	0.855	**0.865**$\oplus$	0.860$\bigcirc$	0.862	**0.864**$\bigcirc$
	Matthews	0.811$\ominus$	0.822	**0.833**$\oplus$	0.790$\ominus$	0.802	**0.818**$\oplus$	0.805$\ominus$	0.810	**0.817**$\oplus$
	Precision	**0.879**$\oplus$	0.851	0.841$\ominus$	**0.847**$\oplus$	0.831	0.828$\bigcirc$	**0.874**$\oplus$	0.840	0.826$\ominus$
	Recall	0.850$\ominus$	0.890	**0.916**$\oplus$	0.851$\ominus$	0.881	**0.906**$\oplus$	0.847$\ominus$	0.884	**0.906**$\oplus$

Table 5. The classification performance over Volvox database for an imbalance ratio of 8.5. ⊕ (⊙) indicate which approch is significantly higer (lower) than (W→) while ◯ without significant difference based on paired t-test at 95% significance level.

Criterion	Measures	BR			LP			CLR		
		W↗	W→	W↘	W↗	W→	W↘	W↗	W→	W↘
Macro	F_1 *score*	**0.573**⊕	0.548	0.520⊙	**0.547**⊕	0.531	0.521⊙	**0.574**⊕	0.543	0.524⊙
	Matthews	**0.389**⊕	0.364	0.338⊙	**0.361**⊕	0.346	0.341◯	**0.389**⊕	0.359	0.342⊙
	Precision	**0.600**⊕	0.537	0.488⊙	**0.551**⊕	0.512	0.491⊙	**0.602**⊕	0.532	0.492⊙
	Recall	0.549⊙	**0.562**	0.562◯	0.544◯	**0.553**	0.558◯	0.549⊙	0.556	**0.565**◯
Micro	F_1 *score*	**0.822**◯	0.821	0.815⊙	0.796◯	0.798	**0.800**◯	**0.822**⊕	0.817	0.814⊙
	Matthews	0.750◯	**0.752**	0.746⊙	0.718⊙	0.724	**0.728**◯	0.751◯	0.747	0.744◯
	Precision	**0.842**⊕	0.818	0.796⊙	**0.785**⊕	0.776	0.772⊙	**0.843**⊕	0.813	0.793⊙
	Recall	0.803⊙	0.824	**0.835**⊕	0.807⊙	0.822	**0.832**⊕	0.801⊙	0.822	**0.835**⊕

stage determines that linear increasing weights presents a superior behavior for *Matthews*, F_1 *measure* and *Precision* measures mainly for the macro-average since it is more sensitive to imbalance than the micro-average. Nonetheless, for *Recall* measure, the linear decreasing weights performed better.

On the other hand, the second experiment was aimed to assess how the imbalance degree affects the performance of the proposed approach. To this aim seven different imbalance ratios are analyzed using macro F_1 *measure*, the results are reported in Fig. 3 for Volvox database and Fig. 4 for Wheel database. The blue, orange and green bars depict (W↗), (W→) and (W↘), respectively. The black lines in the top of each bar represent the standard deviation. In Fig. 3 it can be seen how the linear increasing weights suffered less deterioration as the imbalance ratio increased, while for Fig. 4, the incremental approach experienced a slight superiority for low imbalance ratios. For higher imbalance ratios, the three problem transformation methods achieved a similar behavior.

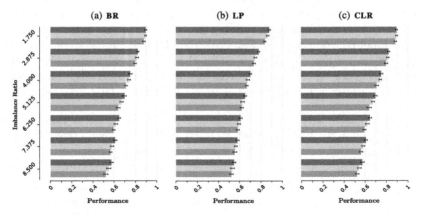

Fig. 3. Classification performances in terms of the macro F_1-*measure* over different imbalance ratios and problem transformation methods for the Volvox database. The blue, orange and green bars depict W↗, W→ and W↘, respectively. (Color figure online)

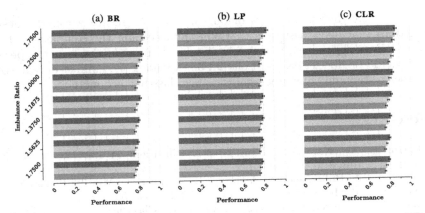

Fig. 4. Classification performances in terms of the macro F_1-*measure* over different imbalance ratios and problem transformation methods for the Wheel database. The blue, orange and green bars depict W↗, W→ and W↘ , respectively.

Table 6. The classification performance over real database

Criterion	Measures	Emotions			Scene			Yeast			Cal500		
		W↗	W→	W↘	W↗	W→	W↘	W↗	W→	W↘	W↗	W→	W↘
Macro	F_1 score	0.631	**0.634**	0.632	0.732	**0.740**	0.719	**0.362**	0.152	0.312	**0.155**	**0.155**	0.148
	Matthews	0.497	0.496	**0.498**	0.693	**0.699**	0.685	**0.2746**	0.0968	0.2319	0.0218	0.0217	**0.0222**
	Precision	0.579	**0.586**	0.580	0.639	**0.657**	0.612	**0.319**	0.159	0.270	**0.148**	0.147	0.136
	Recall	0.701	0.696	**0.702**	0.860	0.850	**0.875**	0.789	**0.958**	0.796	0.172	0.171	**0.177**
Micro	F_1 score	**0.649**	**0.649**	**0.649**	0.724	**0.731**	0.713	**0.639**	0.496	0.597	**0.353**	**0.353**	0.352
	Matthews	**0.515**	0.512	**0.515**	0.687	**0.693**	0.678	**0.528**	0.411	0.494	0.248	0.249	**0.256**
	Precision	0.596	**0.602**	0.597	0.632	**0.649**	0.606	**0.553**	0.369	0.497	**0.333**	0.332	0.319
	Recall	**0.714**	0.707	**0.714**	0.854	0.846	**0.866**	0.758	0.756	**0.761**	0.377	0.377	**0.400**

The results of real-world database are presented in Table 6 where the best values of each approach are highlighted in bold. Attained results show that the proposed method of incremental weights yields to better results for yeast and cal500 databases, presumably due to the fact that they present a bigger cardinality than emotions and scene databases.

5   Conclusions

A novel method named *hyperparameter calibration* was presented for treating class imbalance in a multi-label learning problem by modifications on parameters of the support vector machine classifier. Moreover, an experimental analysis over two own synthetic imbalanced databases and four real-world databases was performed showing that the proposed method is able to improve classification performance, quantified by means of four binary evaluation measures, when combined with *problem transformation* techniques and with different imbalance ratios. However, low levels of cardinality in data together with low imbalance ratio can affect the effectiveness of our method. As future work, the design of functions that allow assigning C_i taking into account the complexity of the data and inter-dependencies between classes to improve the learning performance should be done. Besides, it should be expanded the number of real-world databases tested and compared against other class-balance methods.

References

1. Tsoumakas, G., Katakis, I.: Multi-label classification: an overview. In. J. Data Warehous. Min. **3**(3), 1–3 (2006)
2. Fürnkranz, J., Hüllermeier, E., Mencía, E.L., Brinker, K.: Multilabel classification via calibrated label ranking. Mach. Learn. **73**(2), 133–153 (2008)
3. Read, J., Pfahringer, B., Holmes, G., Frank, E.: Classifier chains for multi-label classification. Mach. Learn. **85**(3), 333 (2011)
4. Giraldo-Forero, A., Jaramillo-Garzón, J., Castellanos-Dominguez, C.: A comparison of multi-label techniques based on problem transformation for protein functional prediction. In: Proceedings of the 35th Annual International Conference of the EMBS, pp. 2688–2691. IEEE (2013)
5. Koziarski, M., Woźniak, M.: CCR: A combined cleaning and resampling algorithm for imbalanced data classification. Int. J. Appl. Math. Comput. Sci. **27**(4), 727–736 (2017)
6. Batuwita, R., Palade, V.: Class Imbalance Learning Methods for Support Vector Machines (2013)
7. Charte, F., Rivera, A.J., del Jesus, M.J., Herrera, F.: MLSMOTE: Approaching imbalanced multilabel learning through synthetic instance generation. Knowl-Based Syst. **89**, 385–397 (2015)
8. Tahir, M.A., Kittler, J., Yan, F.: Inverse random under sampling for class imbalance problem and its application to multi-label classification. Pattern Recognit. **45**(10), 3738–3750 (2012)
9. Chang, C.C., Lin, C.J.: LIBSVM: a library for support vector machines. ACM Trans. Intell. Syst. Technol. (TIST) **2**(3), (2011). Article No. 27
10. Tomás, J.T., Spolaôr, N., Cherman, E.A., Monard, M.C.: A framework to generate synthetic multi-label datasets. Electron. Notes Theor. Comput. Sci. **302**, 155–176 (2014)
11. Charte, F., Rivera, A.J., del Jesus, M.J., Herrera, F.: Addressing imbalance in multilabel classification: measures and random resampling algorithms. Neurocomputing **163**, 3–16 (2015)
12. Wieczorkowska, A., Synak, P., Raś, Z.W.: Multi-label classification of emotions in music. In: Intelligent Information Processing and Web Mining, pp. 307–315. Springer, Heidelberg (2006)
13. Boutell, M.R., Luo, J., Shen, X., Brown, C.M.: Learning multi-label scene classification. Pattern Recognit. **37**(9), 1757–1771 (2004)
14. Elisseeff, A., Weston, J.: A kernel method for multi-labelled classification. In: Advances in Neural Information Processing Systems, pp. 681–687 (2002)
15. Turnbull, D., Barrington, L., Torres, D., Lanckriet, G.: Semantic annotation and retrieval of music and sound effects. IEEE Trans. Audio, Speech, Lang. Process. **16**(2), 467–476 (2008)
16. Meyer, D., Wien, F.T.: Support vector machines. R News **1**(3), 23–26 (2001)

17. Hsu, C.W., Chang, C.C., Lin, C.J., et al.: A Practical Guide to Support Vector Classification (2010)
18. Giraldo-Forero, A.F., Jaramillo-Garzón, J.A., Castellanos-Domínguez, C.G.: Evaluation of example-based measures for multi-label classification performance. In: Ortuño, F., Rojas, I. (eds.) IWBBIO 2015. LNCS, vol. 9043, pp. 557–564. Springer, Cham (2015). https://doi.org/10.1007/978-3-319-16483-0_54

Drifted Data Stream Clustering Based on *ClusTree* Algorithm

Jakub Zgraja[✉] and Michał Woźniak

Faculty of Electronics, Department of Systems and Computer Networks,
Wroclaw University of Science and Technology, Wybrzeże Wyspiańskiego 27,
50-370 Wrocław, Poland
jakub.zgraja@pwr.edu.pl

Abstract. Correct recognition of the possible changes in data streams, called *concept drifts* plays a crucial role in constructing the appropriate model learning strategy. This paper focuses on the unsupervised learning model for non-stationary data streams, where two significant modifications of the *ClusTree* algorithm are presented. They allow the clustering model to be adapted to the changes caused by a *concept drift*. An experimental study conducted on a set of benchmark data streams proves the usefulness of the proposed solutions.

Keywords: Concept drift · Data streams · *ClusTree*
On-line clustering

1 Introduction

Among several tasks studied in data streams [7], *clustering* is worth mentioning, especially because, on the one hand, information about class labels, which is one of the main limitations of *supervised learning*, does not appear and, on the other hand, this task has been employed in many practical applications such as sensor network analysis [8], predicting demand profiles for electrical supply [9], or analysis of medical data [13], to enumerate only a few.

When designing a data stream clustering system we have to take into consideration that several important issues should be resolved, including: (i) how to detect attributes which allow data to be classified into clusters; (ii) how to detect a *concept drift*, without knowledge of example labels; and (iii) how to adapt the model to data changes, i.e. how to implement forgetting mechanisms in the clustering algorithm.

We should also correctly address the problem of a limited amount of RAM, because a clustering model is usually stored in the computer's operating memory. Therefore in dealing with data streams, memory optimization is required. For instance, the above mentioned limited memory usage can be achieved by using sliding windows [12], where only the latest observations are memorized, while the older ones are removed. Data streams can also have various velocities – when data is generated at high velocity, there is a need to use sampling algorithms [6].

© Springer International Publishing AG, part of Springer Nature 2018
F. J. de Cos Juez et al. (Eds.): HAIS 2018, LNAI 10870, pp. 338–349, 2018.
https://doi.org/10.1007/978-3-319-92639-1_28

The main contributions of this paper are the propositions of two modifications of the *ClusTree* algorithm and their experimental evaluation on the basis of diverse benchmark data streams.

2   Method Description

The *ClusTree* algorithm uses the *R-tree* [11] structure and is designed to automatically adapt the model, independently of the data velocity, and with a small ratio of skipped samples. The model is represented by so-called *microclusters* [1], which contain a summary of similar objects and allow incremental updates. Every sample traverses the tree and, if applicable, is inserted to a similar *microcluster*. Otherwise, a new *microcluster* is created. The *ClusTree* algorithm employs a mechanism to phase out oldest the *microclusters* by using a half-life mechanism inspired by physics. Each *microcluster's* weight is controlled by the following function

$$\omega(\Delta t) = \beta^{-\lambda \Delta t} \tag{1}$$

where: λ stands for the decay rate (the higher λ, the faster the *microcluster* is phased out), t is the timestamp, β denotes base, set to 2 [10]. The older the *microclusters*, the lower their weight and the smaller their share is in the clustering process.

The proposed modifications are inspired by Gama's work [7], where the author describes the usage of sliding windows. Both modifications consider issues connected with analyzing data streams by using partial memory. One uses adaptation to update the model and other monitors whether the samples fit to the model built.

2.1   The *Hierarchical Clustering Drift Detection* Algorithm (HCDD)

Each sample traverses through the *ClusTree* internal tree structure and is put in a leaf represented by *microcluster*. The proposed modification has three parameters (the sliding window length, a threshold indicating samples as novelties, and a threshold indicating samples in a window as a *concept drift*). After each sample is added, it should be checked whether a given sample belongs to the nearest *microcluster*. Every *microcluster* contains information about its center of gravity and radius.

Checking if a sample belongs to a *microcluster* is verified if the vector representing the sample is within the *microcluster's* n-dimensional hypersphere.

$$\sum_{i=1}^{n}(x_i - s_i)^2 \leqslant r^2 \tag{2}$$

where: x_i is i-th element of the sample represented by the n-dimensional vector, s_i stands for i-th element of the *microcluster* gravity center, and r denotes its radius. Checking should be implemented on every tree level and it will return the signal if the sample is recognized as a novelty; if this is the case, then it would be indicated in the current window. Every *microcluster* which is created after a sliding window is started will be treated as novelty. If too many samples in a window are marked as novelties, then that situation is treated as a *concept drift*, and the algorithm resets the *R-tree* structure.

Algorithm 2.1. HCDD – learning algorithm

```
Input: collection of samples
Parameters: windowSize size, threshold of concept drift

function train(samples):
  if isLearningAfterDrift and size(conceptDriftWindow) == 0 then
    isLearningAfterDrift ← false
  end if
  conceptDriftWindow ← samples
  areThereNovelties ← insert(samples)

  if not isLearningAfterDrift
    if size(conceptDriftWindow) >= windowSize
      if size(areThereNovelties) / size(conceptDriftWindow) > threshold
        isLearningAfterDrift ← true
        reset ClusTree structure
        train(conceptDriftWindow)
        clear conceptDriftWindow
  end if
```

Algorithm 2.2. HCDD – tree update algorithm

```
Input: collection of samples
Output: list whether samples are novelties

function insert(samples):
  for sample in samples:
    insertionPoint ← ClusTreeInsert(samples)
    if sample is not fitting to sphere spanned by insertionPoint
      areThereNovelties ← sample
  return areThereNovelties
```

Algorithm 2.1 shows the learning procedure. When a *concept drift* occurs, there is a need to build a new model based on the samples from the buffer represented by the current sliding window. First, we check if it the buffer is empty, then learning after drift detection is completed. Incoming samples are added to a sliding window and the counter of samples marked as a novelty set to zero. The next step is to add samples to the *microcluster* tree, as described in Algorithm 2.2. The adding procedure is done by function ClusTreeInsert, which is presented in [10]. If during the addition a new *microcluster* is created, then we check if a sample is within the n-dimensional hypersphere determined by the *microcluster* of the parent node of that newly created *microcluster*. If the sample

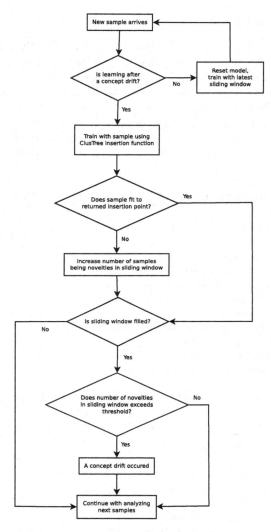

Fig. 1. Flowchart of the *HCDD* algorithm.

is not within the hypersphere, then its novelty counter is increased. After the inserting procedure, the novelty counter is divided by the number of the *micro-cluster* tree levels. If this number is greater than a given threshold, then the sample is marked as novelty. When the learning procedure does not occur after the *concept drift*, then the number of examples in the sliding window is checked. When the window is filled and the number of samples treated as novelties exceeds a given threshold, then a *concept drift* is detected. This causes the whole tree structure to be reset and a new one to be learnt on the basis of the samples from last sliding window.

2.2   The *HA* Algorithm

The second modification of the *ClusTree* works in a slightly different way and has two parameters (the sliding window length, a threshold of branch usage). After analysis of the sample, the usage counters on every level of the branch are updated, starting from the leaf where the sample is put. When the sliding window is filled, then every branch whose usage is lower than a given threshold is pruned. The usage value may vary, because it is the arithmetic mean of the following parameters: (i) minimum *microclusters* usage on the given level, (ii) maximum *microclusters* usage on the given level, (iii) the sliding window length.

Algorithm 2.3. HA – tree updating algorithm

```
Input: collection of samples
Parameters: number of maxInsertions (sliding window length)

function insert(samples)
  insertionPoint ← ClusTreeInsert(samples)
  numberOfInsertions ← numberOfInsertions + 1
  if numberOfInsertions % maxInsertions == 0
    numberOfInsertions ← 0
    entriesToUpdate ← getNodesToUpdateUsage()
    updateUsages(entriesToUpdate, false)
    removeUnusedEntries(root)
    updateUsages(entriesToUpdate, true)
  end if
```

Algorithm 2.4. HA – algorithm of building a node list to update the node usage

---

```
Output: list of entriesToUpdate

function getNodesToUpdateUsage()
  nodeQueue ← root
  while size(nodeQueue) != 0
    currentNode ← pop(nodeQueue)
    if getLevel(currentNode) == treeHeight
      entriesToUpdate ← getEntries(currentNode)
    else if not isLeaf(currentNode)
      nodeQueue ← all children of currentNode
    end if
  end while
  return entriesToUpdate
```

Algorithm 2.5. HA – node usage update algorithm

```
Input: list of entriesToUpdate, flag whether usages should be reset or
    ↪ not
Parameters: number of maxInsertions

function updateUsages(entriesToUpdate, reset)
  for entry in entriesToUpdate
    entry.usage = 0 if reset else entry.usage += 1
    updateUsages(parent(entry), reset)
  end for
```

Algorithm 2.6. Hierarchical Cluster Structure Adaptation – pruning nodes below the threshold algorithm

```
Input: starting node, from which pruning should start
Parameters: node usageThreshold, size of

function removeUnusedEntries(node)
  if not isLeaf(node)
    for entry in node
      if getChild(entry)
        removeUnusedEntries(getChild(entry))
      end if

      if entry.usage / ((getMinMaxUsagesOnLevel(getLevel(node) + size(
          ↪ slidingWindow)) / 3.0) < usageThreshold
        clear(entry.data)
      end if
    end for
  end if
```

Algorithm 2.7. HA – determining the highest usage on a given level

```
Input: tree level on which computation should be performed
Output: sum of minimal and maximal usage at given level

function getMinMaxUsagesOnLevel(level)
  nodeQueue ← root
  while size(nodeQueue) != 0
    currentNode = pop(nodeQueue)
    if currentNode is on level
      update min or max if entries in currentNode are minimum or maximum
    else if not isLeaf(currentNode)
      nodeQueue ← all children of currentNode
    end if
  end while
  return min + max
```

The *Hierarchical Cluster Structure Adaptation* algorithm modifies the sample insertion function in the tree structure of the *ClusTree* algorithm. It is shown in Algorithm 2.3. After inserting a sample by the original function ClusTreeInsert [10], it is checked whether the sliding window is filled. After filling the window, the list of leaf nodes for which the usage will be counted is created (Algorithm 2.4). The counting of node usage is presented in Algorithm 2.5 – which is run after a leaf node is analyzed. For every parent node, up to the root node the usage counters are increased. Then, the nodes whose usage does not exceed a given threshold are pruned from the tree. It is checked recursively, as shown in Algorithm 2.6. As stated earlier, to compute the threshold, the minimum and maximum usage count should be determined and it is done by Algorithm 2.7, where the tree is scanned recursively, up to a given level. The usage threshold is determined by the mean of the minimum and the maximum usage counts and the sliding window length. Pruning begins from a given node, up to leaf nodes.

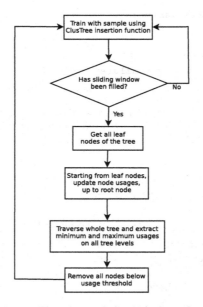

Fig. 2. Flowchart of the *HA* algorithm.

3   Experiments

The main objective of computer experiments is to compare the quality of the proposed modifications of the *ClusTree* algorithm with its original version. The comparison has been done using statistics generated by *MOA* framework [4] during the clustering procedure. MOA provides a few statistics for clustering, i.e. precision, recall, F1 statistics (F1-P and F1-R), purity. F1-R is the Sørensen-Dice coefficient. F1-P is a F1 statistic evaluated for every newly created cluster. In order to evaluate the clustering quality of the proposed modifications, a non-parametric statistical test has been chosen – the NxN Friedman test with *post-hoc* Shaffer test [2,3]. The two statistical significance levels have been analyzed: $\alpha = 0.05$ and $\alpha = 0.10$. Despite that, the clustering quality of the proposed modifications has also been evaluated by the paired *t*-test [5]. To compare the proposed modifications with the original version of the *ClusTree* algorithm, the following hypotheses have been formulated:

H$_0$ F1-R – for the proposed algorithm modifications there is no statistically significant difference for the F1-R statistic evaluation; **H$_1$ F1-R** – for the proposed algorithm modifications there is a statistically significant difference for the F1-R statistic evaluation.

H$_0$ Purity – for the proposed algorithm modifications there is no statistically significant difference for the cluster purity statistic evaluation; **H$_1$ Purity** – for the proposed algorithm modifications there is a statistically significant difference for the cluster purity statistic evaluation.

3.1 Setup of the Experiments

Experiments have been run on *Amazon Cloud Services*[1], with 16 cores and 64 GB of RAM, running *Linux Ubuntu* 16.04.

The proposed algorithms have been implemented in *MOA* framework [4]. The source code of the proposed algorithms is publicly available[2]. The modifications have been implemented in a copy of the *ClusTree* algorithm, under `MyClusTree` class name, to avoid collision of the clustering algorithms' names. Like the original algorithm, the one with modifications is derived from the `AbstractClusterer` class.

HCDD algorithm implementation in *MOA* framework has parameters marked as: (i) the sliding window length – `H`, (ii) a threshold indicating samples as novelties – `D` and (iii) a threshold indicating samples in a window as a *concept drift* – `C`.

HA algorithm implementation in MOA framework has parameters marked as: (i) the sliding window length – `I` and (ii) a threshold of branch usage – `U`.

As described above, algorithm modifications are parameterized. The parameters for both algorithms have been selected experimentally, i.e. as a part of the learning process. The plots with the comparison of the various parameter groups are also publicly available[3]. All experiments have been carried out on the set of the benchmark data streams described in Table 1.

---

[1] https://aws.amazon.com/.

[2] Source code of the proposed algorithms can be found at https://github.com/jagub2/mgr/tree/master/MyClusTree/src/moa.

[3] https://github.com/jagub2/mgr/tree/master/plots.

Table 1. Datasets used during experiments. Clus. stands for the number of clusters, Attr. for the number of attributes and Obj. for the number of objects respectively.

	Dataset	Clus.	Attr.	Obj.		Dataset	Clus.	Attr.	Obj.
1	ICDT	2	2	16 000	16	GEARS2C2D	2	2	200 000
2	1CHT	2	2	16 000	17	HyperPlane	5	10	100 000
3	1CSurr	2	2	55 283	18	Powersupply	24	2	29 928
4	2CDT	2	2	16 000	19	Electricity	2	6	45 312
5	2CHT	2	2	16 000	20	Airlines	18	13	539 383
6	4CE1CF	5	2	173 250	21	Circles[a]	4	4	100 000
7	4CR	4	2	144 400	22	CirclesS[a]	4	2	100 000
8	4CRE-V1	4	2	125 000	23	Sine[a]	2	4	100 000
9	4CRE-V2	4	2	183 000	24	SineS[a]	2	2	100 000
10	5CVT	5	2	40 000	25	STAGGER[a]	2	3	100 000
11	UG2C2D	2	2	100 000	26	LEDS[a]	10	7	100 000
12	MG2C2D	2	2	200 000	27	LED1DS[a]	10	7	100 000
13	FG2C2D	2	2	200 000	28	LED3DS[a]	10	7	100 000
14	UG2C3D	2	3	200 000	29	LED5DS[a]	10	7	100 000
15	UG2C5D	2	5	200 000	30	LED7DS[a]	10	7	100 000

[a]Datasets were generated by generators in MOA. Capital letters after dataset name means used switch, number before is value of switch – eg. LED1DS means LED generator, with parameters -d 1 -s.

Table 2. Best parameters selected experimentally – for HA and HCDD algorithms

	HA	HCDD
1	I100, U0.01	W1000, D0.01, C0.01
2	I200, U0.01	W300, D0.01, C0.01
3	I50, U0.01	W1000, D0.01, C0.03

3.2 Results

The plots presented in Fig. 3 show how the proposed modifications behave. For the *HA* algorithm strong influence of parameters on the quality of clustering is observed. Improperly selected parameters may increase fluctuations, which is not desirable behavior during clustering. The *HCDD* algorithm is much more stable, unlike the other algorithm. In this case less influence of parameters on the quality of clustering could be observed. Selecting extremely improper parameters can cause the noise to be detected as a *concept drift*, which is also undesirable during clustering.

According to statistical tests (Tables 3 and 4) hypotheses H_0 F1-R, H_0 Purity for all statistical significance levels for selected parameter groups for the *HA* algorithm cannot be rejected, because there is no statistical significance in com-

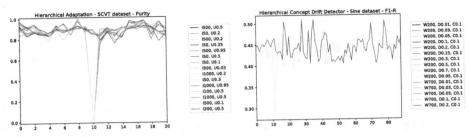

Fig. 3. Exemplary evaluations of both modifications.

Table 3. P-values for F1-R statistic

Hypothesis	p	$p_{Shaffer}$
ClusTree vs. W300, D0.01, C0.01	0.000000	0.000005
ClusTree vs. W1000, D0.01, C0.03	0.266238	24.493864
ClusTree vs. W1000, D0.01, C0.01	0.684196	62.261797
ClusTree vs. I100, U0.01	1.000000	78.000000
ClusTree vs. I200, U0.01	1.000000	78.000000
ClusTree vs. I50, U0.01	1.000000	78.000000

parison to the original algorithm. For the *HCDD* algorithm, several parameter setups which are able to outperform the original version of the *ClusTree* can be observed. However, post-hoc Shaffer method showed that none of the parameter setups for the HCDD algorithm has statistical significance for the cluster purity statistic. For the F1-R statistic one parameter setup (W300, D0.01, C0.01) which demonstrates statistical significance can be observed (Table 3).

On the basis of the results presented in e.g. Fig. 4 we may also draw the following conclusions. The proposed algorithms have a slightly worse purity statistic, because their structure is reset or firmly modified. The *HA* algorithm does not show any significant improvement over the original version of the *ClusTree* algorithm, but it is also not outperformed by the original *ClusTree*. For the majority of data streams used, the *HCDD* algorithm presents a better median and stan-

Table 4. P-values for cluster purity statistic

Hypothesis	p	$p_{Shaffer}$
ClusTree vs. W1000, D0.01, C0.03	0.009994	1.199237
ClusTree vs. W1000, D0.01, C0.01	0.021173	2.223160
ClusTree vs. W300, D0.01, C0.01	0.065196	5.998071
ClusTree vs. I100, U0.01	1.000000	78.000000
ClusTree vs. I200, U0.01	1.000000	78.000000
ClusTree vs. I50, U0.01	1.000000	78.000000

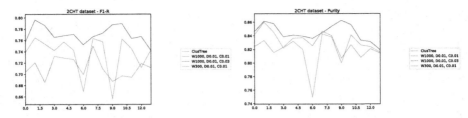

Fig. 4. Exemplary comparison of ClusTree and HCDD modification.

dard deviation of the F1 statistic than the original *ClusTree*. It is possible to train the *HCDD* algorithm, e.g. for the following parameter setups: W1000, D0.01, C0.01 or W1000, D0.01, C0.03, which for most of the data streams are more stable than the original algorithm, i.e. with fewer fluctuations than the original algorithm, which is a very desirable characteristic for clustering.

The proposed modifications of the *ClusTree* algorithm are strongly parametrized and while solving practical tasks their parameters should be set properly as a part of the learning process. The *Hierarchical Clustering Drift Detection* algorithm is able to outperform the original version of *ClusTree*. Even if the size of the sliding window is big, this modification is not outperformed by the original algorithm, offering a more stable model at the same time. Statistical analysis of the F1-R statistic showed that the *HCDD* algorithm behaves significantly better than the original *ClusTree* for simpler streams, which do not contain too much noise. Statistical analysis of the cluster purity showed that the *HCDD* algorithm is better than the original version of *ClusTree* for some datasets. It is also shown that for a narrower sliding window the cluster purity statistic increased. When *concept drifts* occur and they are detected, this statistic is lower, because the internal model representation would be reset, which is confirmed by that statistic evaluation analysis.

4   Final Remarks

Designing and implementing mechanisms for improving clustering quality with *concept drifts* is a very challenging, non-trivial issue which has a huge impact on the clustering quality. It is worth mentioning that this research is one of the first attempts at using the hierarchical data stream clustering algorithm. We have proposed two modifications of the *ClusTree* algorithm and especially the *HCDD* method seems to be very attractive, because it can increase the effectiveness of the *ClusTree* algorithm, especially when the *concept drift* phenomenon appears. The main issue of the proposed modification is proper algorithm parameter setting, which should be a part of the learning process. We may observe that the sliding window length cannot be too narrow, because it would lower the evaluation quality, especially because noise could be detected as a *concept drift*. However, a sliding window should also not be very wide, because then *concept drifts* may be skipped.

The proposed algorithms are open for modifications. They could be extended with dynamic sliding window length adaptation or by detecting recurring concept drifts.

Acknowledgments. This work was supported by Statutory Fund of the Department of Systems and—Computer Networks, Faculty of Electronics, Wroclaw University of Science and Technology.

References

1. Aggarwal, C.C., Han, J., Wang, J., Yu, P.S.: A framework for clustering evolving data streams. In: Proceedings of the 29th International Conference on Very Large Data Bases, VLDB 2003, vol. 29, pp. 81–92. VLDB Endowment (2003)
2. Alcalá-Fdez, J., Fernandez, A., Luengo, J., Derrac, J., García, S., Sánchez, L., Herrera, F.: KEEL data-mining software tool: data set repository, integration of algorithms and experimental analysis framework. J. Mult.-Valued Log. Soft Comput. **17**(2–3), 255–287 (2011)
3. Alcalá-Fdez, J., Sánchez, L., García, S., del Jesus, M.J., Ventura, S., Garrell, J.M., Otero, J., Romero, C., Bacardit, J., Rivas, V.M., Fernández, J.C., Herrera, F.: KEEL: a software tool to assess evolutionary algorithms to data mining problems. Soft. Comput. **13**(3), 307–318 (2009)
4. Bifet, A., Holmes, G., Kirkby, R., Pfahringer, B.: MOA: massive online analysis. J Mach. Learn. Res. **11**, 1601–1604 (2010)
5. Demšar, J.: Statistical comparisons of classifiers over multiple data sets. J. Mach. Learn. Res. **7**, 1–30 (2006)
6. Domingos, P., Hulten, G.: Mining high-speed data streams. In: Proceedings of the Sixth ACM SIGKDD International Conference on Knowledge Discovery and Data Mining, KDD 2000, pp. 71–80. ACM, New York (2000)
7. Gama, J.: Knowledge Discovery from Data Streams. CRC Press, Boca Raton (2010)
8. Gama, J., Gaber, M.: Learning from Data Streams: Processing Techniques Insensor Networks. Springer, Heidelberg (2007). https://doi.org/10.1007/3-540-73679-4
9. Gama, J., Rodrigues, P.P.: Stream-based electricity load forecast. In: Kok, J.N., Koronacki, J., Lopez de Mantaras, R., Matwin, S., Mladenič, D., Skowron, A. (eds.) PKDD 2007. LNCS (LNAI), vol. 4702, pp. 446–453. Springer, Heidelberg (2007). https://doi.org/10.1007/978-3-540-74976-9_45
10. Kranen, P., Assent, I., Baldauf, C., Seidl, T.: The ClusTree: indexing micro-clusters for anytime stream mining. Knowl. Inf. Syst. **29**(2), 249–272 (2011)
11. Manolopoulos, Y., Nanopoulos, A., Papadopoulos, A.N., Theodoridis, Y.: R-Trees: Theory and Applications. Springer, Heidelberg (2005). https://doi.org/10.1007/978-1-84628-293-5
12. Ren, J., Ma, R.: Density-based data streams clustering over sliding windows. In: 2009 Sixth International Conference on Fuzzy Systems and Knowledge Discovery, vol. 5, pp. 248–252, August 2009
13. Sun, J., Sow, D., Hu, J., Ebadollahi, S.: A system for mining temporal physiological data streams for advanced prognostic decision support. In: Proceedings of the 2010 IEEE International Conference on Data Mining, ICDM 2010, pp. 1061–1066, Washington, DC, USA. IEEE Computer Society (2010)

Featuring the Attributes in Supervised Machine Learning

Antonio J. Tallón-Ballesteros[1]([✉]), Luís Correia[2], and Bing Xue[3]

[1] Department of Languages and Computer Systems,
University of Seville, Seville, Spain
atallon@us.es

[2] BioISI - Faculdade de Ciências, Universidade de Lisboa, Lisboa, Portugal

[3] School of Engineering and Computer Science, Victoria University of Wellington,
Wellington, New Zealand

Abstract. This paper introduces an approach to feature subset selection which is able to characterise the attributes of a supervised machine learning problem into two categories: essential and important features. Additionally, the fusion of both kinds of features yields to an overcoming in the prediction task, where some measures such as accuracy and Receiver Operating Characteristic curve (ROC) have been reported. The test-bed is composed of eight binary and multi-class classification problems with up to five hundred of attributes. Several classification algorithms such as Ridor, PART, C4.5 and NBTree have been tested to assess the proposal.

1 Introduction

Supervised Machine Learning (SML) via classification requires that every object has a label associated [6]. Essentially, classification partitions the whole feature space (the space of all possible attribute value combinations) into different regions, one for each class. The properties involved in a classification procedure may not always be manageable, which is more prone to happen when their number is high. Removing some of them alleviates the load of the learning machine induction and might lead a more accurate classification model. Lesser useful attributes for classification are detected and discarded, which is the operation performed by an attribute selection procedure.

The objective of this paper is to propose a new approach to feature selection, splitting it into two sequential stages: selection of *essential* attributes and selection of *important* attributes from the set of non-essential ones. The merge of these two sets is used to train the classifier. To evaluate the approach we use it with four different classifiers, namely Ridor, PART, C4.5 and Naïve Bayes tree (NBTree). This allows to assess the influence of using trees, rules and/or probabilistic approaches for the attribute subset selection model proposed.

The remaining of this article is arranged as follows. Section 2 provides a brief overview of different concepts about feature selection. Section 3 details the proposal. Section 4 describes the experimentation by means of the approach setting,

© Springer International Publishing AG, part of Springer Nature 2018
F. J. de Cos Juez et al. (Eds.): HAIS 2018, LNAI 10870, pp. 350–362, 2018.
https://doi.org/10.1007/978-3-319-92639-1_29

problems and classifiers used. Then, Sect. 5 depicts the empirical results. Lastly, Sect. 6 draws some conclusions.

2   Feature Selection

Attribute selection is a specially important process for mining big data. Doing feature selection before a learning algorithm is applied has numerous benefits. By eliminating a significant amount of attributes it becomes easier to train learning machines. The computational time of the induction is reduced and the resulting model will usually be simpler and easier to interpret. It is also frequently the case that simpler models generalise better. Therefore, a model employing fewer features is likely to perform better. This is a process to determine from the instance set which attributes are more relevant to predict or explain the data, and conversely which attributes are redundant or provide little information [11]. Finally, the identification of the most relevant attributes can be useful in its own right providing valuable information about the problem in hand.

Generally speaking, three types of approaches might be used for attribute selection [9]: (a) *Filter methods*, which select the best individual attributes usually assuming they are independent given the class. In this case some statistical measure is used to assess the quality of the attributes; (b) *Wrapper methods* that use a machine learning algorithm to select a sub-set of the attributes. Usually this involves selection and evaluation of different sub-sets under some accuracy measure; (c) *Embedded methods* combine the model creation problem with the attribute selection. These methods include in the induction model some bias towards fewer attributes.

By its part, the filter approaches may be divided into feature ranking and feature subset selection methods depending on the output which may be an ordered list of the attributes or a subset of attributes. This article focuses on filter method to obtain feature subsets. The main contribution of the current work is the ability to characterise groups of attributes into two types of categories, namely essential and important feature subsets.

3   Proposal

This paper proposes a way to categorise the features in supervised machine learning problems. According to our approach, there are two kinds of features: (i) Essential features which represent the core properties to be collected from the new instances belonging to the problem; (ii) Important features which constitute additional information that may be interesting to be reported on unseen instances. The procedure is as follows: (a) first, the data set is divided into two sets: training and test sets, (b) the feature selection method is applied to the training set and as an outcome we have the essential features which are those that have been picked up by the data preparation method and, on other bag, we have the non-selected properties that may not be thought to be very relevant in terms of aid to the predictive data mining method, (c) feature selection is performed on the non-selected attributes to extract the best from the not very promising features, (d) the attributes from steps (b) and (c) are merged which will be the next characteristic space for classifier, (e) the list of attributes is projected into the test set as it was originally and (f) the usual assessment in data mining is conducted: we start training the classifier with the reduced training set and the evaluation takes place on the reduced test set. Figure 1 depicts the approach which has been named Characterisation of Features through Feature Subset Selection (ChaF2S2).

It is important to remark that some connections may be found with a previous work [21] due to the use of the merge operation. It is straightforward that in this paper no kind of overlapping [19] may occur, which may do the new approach more applicable and even more oriented to the goal that we have marked for the current contribution.

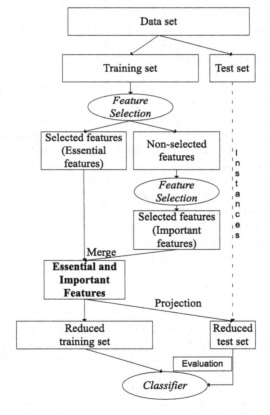

Fig. 1. Proposal: characterisation of features through feature subset selection (ChaF2S2).

4   Experimentation

4.1   Approach Setting

It is true that the amount of literature coping with many different measures is very extensive. We opt for correlation measures since the behaviour is very good and also has been one of the most commonly used by the data mining community. Moreover, our previous experience showed that the correlation is very convenient for supervised machine learning tasks [16]. Table 1 describes the methods to evaluate the current proposal which are founded on Correlation-based Feature Selection (CFS) [5] and Fast Correlation-Based Filter (FCBF) [25]. The reason of this chosen is motivated by the good performance of these feature subset selectors.

Specifically, we use the implementations provided by Weka tool [1] which are called CfsSubsetEval and FCBFSearch working with SymmetricalUncertAttributeSetEval, respectively for CFS and FCBF. CFS procedure has been used for FSS1 and FSS2 methods whereas FCBF has been utilised for FSS3 and FSS4 selectors. FSS1 and FSS3 capture the subset of essential features and FSS2 and FFS4 incorporate an extra subset of attributes, which we have called important features, to the solutions got by FSS1 and FSS3, respectively. FSS2 and FSS4 are the more complete options within their category and are the base of the current contribution. As an additional breakthrough, the distinction between essential and important features has been outlined. Table 2 reports on the parameters and properties to set up the method and also to ease the reproducibility of the experiments. It is important to remark that the experiments have been conducted with the default values parameters because the own authors have recommended them. Moreover, we also tested for CFS three deeper levels for the number of expanded nodes such as 6, 7 and 8; since there are not differences in the reached solutions we keep the number of expanded nodes to 5.

Table 1. Feature subset selectors for the experimentation

Abbreviation	Method	Essential features	Important features
FSS1	CFS	Yes	No
FSS2	CFS	Yes	Yes
FSS3	FCBF	Yes	No
FSS4	FCBF	Yes	Yes

4.2   Problems

A good range of problems have been tested to evaluate the performance of the new proposal. Table 3 summarises the test-bed. Their source is varied since some of them are available in the very well-known repository from the University of California (UC) at Irvine [22], MADELON has been proposed in NIPS 2003

Table 2. Parameter values and description of CFS and FCBF feature subset selectors

Method	Parameter/Property	Value
CFS	Attribute evaluation measure	Correlation
	Search method	Best First
	Consecutive expanded nodes without improving	5
	Search direction	Forward
FCBF	Attribute evaluation measure	Correlation
	Attribute evaluator	Symmetrical Uncertainty
	Search method	FCBFSearch

challenge [4] and STAD is a Bioinformatics problem [23] that stands for STom-ach ADenocarcinoma. There are five multi-class problems and the remaining are binary, throwing there an average close to 4. The dimensionality goes from around 10 to 500 with an average over 82 whereas the data size fluctuates from one hundred to nineteen hundred. Nowadays, the number of attributes that we may have in problems at hand is very high and it is not strange to have thousands of features [18].

The data partition in some problems such as Led24 and SPECTF [8] follows the original pre-arrangement [14] and in most of the cases has been obtained with a stratified hold-out keeping the original data distribution in both sets, namely training and testing sets. Regarding the data imputation, a single imputation method called mean or mode imputation [13] has been applied which imputes a missing value with the mean or the mode within the class. We have adopted this strategy since the amount of missing values is very small. The data preparation method at the feature level has been only conducted to the training set and hence to get the reduced test only the projection operator is applied.

Table 3. Supervised machine learning problems

Problem	Classes	Instances			Features				
		Total	Tra.	Tes.	Ori.	CFS		FCBF	
						Ess.	Imp.	Ess.	Imp.
B. tissue	6	106	81	25	9	6	2	4	2
CTG	3	2126	1594	532	22	7	11	8	3
Led24	10	3200	200	3000	24	6	1	6	1
MADELON	2	2000	1500	500	500	12	4	7	2
Magic	2	19020	14265	4755	10	4	2	2	1
SPECTF	2	267	80	187	44	12	8	6	5
STAD	3	100	75	25	14	4	1	4	1
Waveform	3	5000	3750	1250	40	14	4	5	2
Average	3.9	3977.4	2693.1	1284.3	82.9	8.1	4.1	5.3	2.1

Tra. = Training Tes. = Testing Ori. = Original Ess. = Essential Imp. = Important

4.3   Classification Algorithms

Different classifiers based on rules and trees have been used in this work, namely, Ridor and PART, from the former category, and C4.5 and NBTree, from the latter type, to assess how the new approach to feature subset selection performs in various conditions. We briefly review their characteristics. The proposal has been tested in two classic classifiers such as C4.5 and PART due to their good mixture with feature selection based on correlation as a previous contribution [20] to HAIS 2016 [10] reported. Ridor is an effective classification algorithm. Finally, NBTree is a very powerful classifier according to a very recent study in medium and high-dimensionality problems [15].

Ridor [3] is a ripple-down rule learner. It creates a default rule first and then the exceptions for the default rule with the least (weighted) error rate. Then it builds the best exceptions for each exception and iterates until pure. Thus it performs a tree-like expansion of exceptions. The exceptions are a set of rules that predict classes other than the default. PART [2] is a learning algorithm that generates a rule classifier using tree generation in the process. To generate a rule, a pruned decision tree is constructed for the current set of instances, then a rule is generated representing the leaf with the largest coverage, and the tree is discarded. The instances covered by the rule are removed and the process is repeated. The main advantage of PART is simplicity which allows it to scale with a high performance. C4.5 [12] builds a decision tree choosing for each node the attribute with the highest entropy. It can handle both discrete and continuous attributes. The implementation used also includes a pruning phase to reduce the tree structure and to improve the generalisation performance. The robustness and good interpretable results of this algorithm made it a popular choice in a variety of machine learning problems, and therefore we also chose it for this study. A NBTree classifier [7] generates a tree with Naïve Bayes (NB) classifiers at the leaves. To define each new node a univariate split is tested and the attribute with the highest utility is selected for that node. There is an exception when the utility is not significantly better than the utility of the current node, in which case a NB classifier is created for the current node. This model is as interpretable as trees and NB models while often showing better performance in large problems. It uses NB that is proven to be an optimal classifier under some circumstances [26] and it is usually taken as a reference classifier. This is the main reason for having chosen NBTree for testing our attribute selection model.

5   Results

This section compares every couple of related feature subset selection methods. Concretely, on the one hand, FSS1 and FSS2 are compared and, on the other hand, FSS3 faces FSS4. For the aforementioned classifiers, we report on the accuracy and Receiver Operating Characteristic curve (ROC) measure on the test set for each problem of the test-bed under all the scenarios described in 4.1.

5.1   Application of the Proposal on Correlation-Based Feature Selection (CFS)

Table 4 shows the test results of the proposal on CFS with Ridor classifier. The number of wins is higher than the losses. There are also some scenarios with ties. The proposal helps to enhance one or both assessment measures in most cases.

Table 5 depicts the performance of supervised machine learner PART. The scenario has completely changed from the previous classifier. Improvements have been reached in five out of the problems. Besides, the effect of the No-free lunch theorem is drawn around because only one measure is overcame in the half of the test-bed [24]; in particular, it happens improvement for B. tissue, CTG, Led24 and STAD in accuracy or ROC metric, exclusively.

Table 6 reports on the test results for classifier C4.5. In most of the data sets, it takes place improvement not only in accuracy but also in ROC. Moreover, there are two problems that may hint to be very difficult because it happen a worsening with both measures. STAD is a complex data set because: (i) there are only 25 instances in the test set which means that every error in the prediction scores a negative 4%, (ii) there are 3 classes and (iii) is a Bio-informatics problem whose data have been collected very recently and the number of available measures is very low which makes the study a very challenging task. The results for STAD suggest that C4.5 is not a good option for this data set probably due to the cut-off values to create a decision node. SPECTF is a particular case because the important features may be discarded safely with no difference in performance; in addition, if we test with the data set without any kind of pre-processing the results are a bit better what suggests that feature selection may not be a good approach to deal with this problem [17].

Table 7 represents the behaviour of NBTree approach. The accuracy is enhanced in most cases, more concretely if the single tie is excluded, in five out of seven problems there is an overcoming. On the other way round, the ROC measure is often decreased what in any sense suggests to explore new ways or even to think about the option of only incorporating some of the important features. The good news here is that in two out of the top-3 problems in terms of features such as MADELON and Waveform a very noticeable progress has taken place.

5.2   Application of the Proposal on Fast Correlation-Based Filter (FCBF)

Table 8 exhibits the performance via Ridor. Accuracy has been improved six times whereas ROC has been overcame four times. For those cases with negative outcomes the differences are very small which makes the approach very convenient and handy for the majority of the test-bed.

Table 9 shows the results with unseen data for the classification algorithm PART which is based on rules. There are many wins and only one or two losses for accuracy and ROC, respectively. In five out of the problems both measures are enhanced simultaneously which is very noticeable.

Table 4. Test results for the approach on CFS with Ridor

Problem	Accuracy			ROC		
	FSS1	FSS2	Diff.	FSS1	FSS2	Diff.
B. Tissue	60.00	56.00	−4.00	0.9000	0.9000	0.0000
CTG	78.20	80.64	2.44	0.7792	0.8560	0.0769
Led24	67.40	66.50	−0.90	0.8857	0.8275	−0.0582
MADELON	68.20	73.00	4.80	0.6820	0.7300	0.0480
Magic	79.89	81.30	1.41	0.7782	0.7450	−0.0332
SPECTF	63.64	65.24	1.60	0.6806	0.6893	0.0087
STAD	64.00	64.00	0.00	0.6654	0.6654	0.0000
Waveform	76.88	76.72	−0.16	0.7552	0.7728	0.0176
W/T/L	4/1/3			4/2/2		

Table 5. Test results for the approach on CFS with PART

Problem	Accuracy			ROC		
	FSS1	FSS2	Diff.	FSS1	FSS2	Diff.
B. tissue	56.00	48.00	−8.00	0.9250	0.9300	0.0050
CTG	81.20	81.95	0.75	0.9190	0.8674	−0.0516
Led24	68.50	68.53	0.03	0.9227	0.9094	−0.0132
MADELON	60.80	62.60	1.80	0.7104	0.7295	0.0191
Magic	81.91	83.32	1.41	0.8712	0.8797	0.0086
SPECTF	70.05	72.19	2.14	0.6459	0.7000	0.0541
STAD	52.00	36.00	−16.00	0.5331	0.6581	0.1250
Waveform	77.04	76.80	−0.24	0.8432	0.8426	−0.0007
W/T/L	5/0/3			5/0/3		

Table 6. Test results for the approach on CFS with C4.5

Problem	Accuracy			ROC		
	FSS1	FSS2	Diff.	FSS1	FSS2	Diff.
B. tissue	68.00	56.00	−12.00	0.9250	0.8350	−0.0900
CTG	78.38	83.65	5.26	0.8967	0.9145	0.0177
Led24	68.10	68.80	0.70	0.8905	0.9079	0.0174
MADELON	70.60	73.60	3.00	0.7414	0.7826	0.0412
Magic	82.42	83.79	1.37	0.8653	0.8646	−0.0007
SPECTF	66.84	66.84	0.00	0.5519	0.5519	0.0000
STAD	72.00	52.00	−20.00	0.7574	0.6838	−0.0735
Waveform	74.40	76.16	1.76	0.7884	0.7879	−0.0005
W/T/L	5/1/2			3/1/4		

Table 7. Test results for the approach on CFS with NBTree

Problem	Accuracy			ROC		
	FSS1	FSS2	Diff.	FSS1	FSS2	Diff.
B. tissue	52.00	64.00	12.00	0.9250	0.9000	−0.0250
CTG	76.50	76.69	0.19	0.8413	0.7505	−0.0908
Led24	70.73	70.73	0.00	0.9685	0.9685	0.0000
MADELON	71.20	75.80	4.60	0.7693	0.8106	0.0413
Magic	81.93	83.11	1.18	0.8647	0.8747	0.0101
SPECTF	72.19	67.91	−4.28	0.7649	0.7103	−0.0547
STAD	64.00	56.00	−8.00	0.7096	0.6912	−0.0184
Waveform	76.88	81.36	4.48	0.8696	0.8916	0.0220
W/T/L	5/1/2			3/1/4		

Table 10 displays the behaviour of classifier C4.5. There are from 4 up to 5 wins according to the concrete metric and there is one tie. The situation for STAD problem has not been changed compared to the approach based on CFS; it seems that STAD may not be combined with a split criterion founded on entropy as C4.5 has.

Table 11 reports the test results for NBTree which is a tree-based approach built via the Bayes theorem. The outcome is very similar to the previous scenario although the differences for negative cases are smaller which leads to think that a probabilistic model is more suitable than traditional C4.5 algorithm, especially for STAD problem.

Table 8. Test results for the approach on FCBF with Ridor

Problem	Accuracy			ROC		
	FSS3	FSS4	Diff.	FSS3	FSS4	Diff.
B. Tissue	60.00	56.00	−4.00	0.9000	0.9000	0.0000
CTG	78.38	79.32	0.94	0.7804	0.7895	0.0091
Led24	67.37	68.37	1.00	0.8857	0.8569	−0.0289
MADELON	55.20	57.40	2.20	0.5520	0.5740	0.0220
Magic	77.60	81.47	3.87	0.6970	0.7558	0.0589
SPECTF	59.89	68.45	8.56	0.6907	0.6764	−0.0143
STAD	64.00	64.00	0.00	0.6654	0.6654	0.0000
Waveform	74.16	75.44	1.28	0.7489	0.7515	0.0026
W/T/L	6/1/1			4/2/2		

Table 9. Test results for the approach on FCBF with PART

Problem	Accuracy			ROC		
	FSS3	FSS4	Diff.	FSS3	FSS4	Diff.
B. Tissue	48.00	48.00	0.00	0.9150	0.9300	0.0150
CTG	77.26	79.51	2.26	0.8909	0.9388	0.0479
Led24	68.50	68.90	0.40	0.9227	0.9373	0.0146
MADELON	60.40	60.40	0.00	0.6307	0.6227	−0.0080
Magic	79.26	82.54	3.28	0.8385	0.8648	0.0264
SPECTF	64.71	72.73	8.02	0.6983	0.6151	−0.0831
STAD	52.00	36.00	−16.00	0.5331	0.6581	0.1250
Waveform	74.00	74.24	0.24	0.8494	0.8561	0.0067
W/T/L	5/2/1			6/0/2		

Table 10. Test results for the approach on FCBF with C4.5

Problem	Accuracy			ROC		
	FSS3	FSS4	Diff.	FSS3	FSS4	Diff.
B. Tissue	48.00	56.00	8.00	0.8000	0.8350	0.0350
CTG	77.82	79.89	2.07	0.8546	0.9215	0.0668
Led24	68.10	68.10	0.00	0.8905	0.8905	0.0000
MADELON	58.60	59.20	0.60	0.6041	0.6097	0.0056
Magic	79.71	82.42	2.71	0.8174	0.8594	0.0420
SPECTF	67.91	66.84	−1.07	0.7116	0.6717	−0.0399
STAD	72.00	52.00	−20.00	0.7574	0.6838	−0.0735
Waveform	74.72	75.68	0.96	0.8636	0.8482	−0.0154
W/T/L	5/1/2			4/1/3		

Table 11. Test results for the approach on FCBF with NBTree

Problem	Accuracy			ROC		
	FSS3	FSS4	Diff.	FSS3	FSS4	Diff.
B. Tissue	48.00	52.00	4.00	0.9000	0.9350	0.0350
CTG	78.76	80.64	1.88	0.8384	0.8631	0.0247
Led24	70.73	70.73	0.00	0.9685	0.9685	0.0000
MADELON	61.20	61.00	−0.20	0.6393	0.6400	0.0007
Magic	80.13	82.73	2.61	0.8487	0.8731	0.0243
SPECTF	71.12	70.59	−0.53	0.7579	0.7083	−0.0496
STAD	64.00	56.00	−8.00	0.7096	0.6912	−0.0184
Waveform	75.60	79.20	3.60	0.8760	0.8912	0.0152
W/T/L	4/1/3			5/1/2		

Once the results under two different scenarios have been depicted for the proposal, we must remark that CFS and FCBF are very good candidates to be used in future works although the performance of FCBF is stronger than CFS what may make the new approach an interesting option for data sets with a huge number of features.

6   Conclusions

This paper introduced a new approach to feature subset selection that is able to distinguish between essential and important attributes. Moreover, the combination of both types of features on CFS and FCBF feature subset selectors yielded to an enhanced performance of classifiers such as PART, Ridor -in an outstanding way-, C4.5 and NBTree compared to the selection of only essential attributes. The main idea achieved by this research is that there are some attributes which are crucial to have a good generalisation capacity; at the same time those attributes that seems not to be very promising are handy to be lead through a feature subset selection procedure in order to keep the best of the not so good potential attributes that may be called important features, which is the second best kind of attribute according our new approach. The empirical study was conducted on eight binary and multi-class problems from different areas and sources. The results revealed that some progress took place in terms of performance at the price of increasing a bit the characteristic space. Lastly, it must be mentioned that FCBF takes a greater advantage than CFS with the proposal. Nonetheless, the approach is also very convenient for CFS.

Acknowledgments. This work has been partially subsidized by TIN2014-55894-C2-R project of the Spanish Inter-Ministerial Commission of Science and Technology (MICYT), FEDER funds, the P11-TIC-7528 project of the "Junta de Andalucía" (Spain) and by FCT, Portugal, under Grant UID/Multi/04046/2013.

References

1. Bouckaert, R.R., Frank, E., Hall, M.A., Holmes, G., Pfahringer, B., Reutemann, P., Witten, I.H.: Weka–experiences with a Java open-source project. J. Mach. Learn. Res. **11**, 2533–2541 (2010)
2. Frank, E., Witten, I.H.: Generating accurate rule sets without global optimization. In: Shavlik, J. (ed.) Fifteenth International Conference on Machine Learning, pp. 144–151. Morgan Kaufmann (1998)
3. Gaines, B.R., Compton, P.: Induction of ripple-down rules applied to modeling large databases. J. Intell. Inf. Syst. **5**(3), 211–228 (1995)
4. Guyon, I., Gunn, S., Ben-Hur, A., Dror, G.: Result analysis of the NIPS 2003 feature selection challenge. In: Advances in Neural Information Processing Systems, pp. 545–552 (2005)
5. Hall, M.A.: Correlation-based feature selection for machine learning. Ph.D. thesis, University of Waikato, Hamilton, New Zealand (1999)
6. Kacprzyk, J., Pedrycz, W.: Springer Handbook of Computational Intelligence. Springer, Heidelberg (2015). https://doi.org/10.1007/978-3-662-43505-2

7. Kohavi, R.: Scaling up the accuracy of Naive-Bayes classifiers: a decision-tree hybrid. In: KDD, pp. 202–207 (1996)
8. Koller, D., Sahami, M.: Toward optimal feature selection. Technical report, Stanford InfoLab (1996)
9. Liu, H., Motoda, H.: Feature Extraction, Construction and Selection: A Data Mining Perspective, vol. 453. Springer Science & Business Media, New York (1998). https://doi.org/10.1007/978-1-4615-5725-8
10. Martínez-Álvarez, F., Troncoso, A., Quintián, H., Corchado, E.: Hybrid Artificial Intelligent Systems: 11th International Conference, HAIS 2016, Seville, Spain, April 18-20, 2016, Proceedings, vol. 9648. Springer, Heidelberg (2016). https://doi.org/10.1007/978-3-319-32034-2
11. Olafsson, S., Li, X., Wu, S.: Operations research and data mining. Eur. J. Oper. Res. **187**(3), 1429–1448 (2008)
12. Quinlan, J.R.: C4.5: Programs for Machine Learning, vol. 1. Morgan Kaufmann, Stanford (1993)
13. Schafer, J.L.: Analysis of Incomplete Multivariate Data. CRC Press, Boca Raton (1997)
14. Somol, P., Grim, J., Pudil, P.: The problem of fragile feature subset preference in feature selection methods and a proposal of algorithmic work around. In: 2010 20th International Conference on Pattern Recognition (ICPR), pp. 4396–4399. IEEE (2010)
15. Tallon-Ballesteros, A.J., Correia, L.: Medium and high-dimensionality attribute selection in Bayes-type classifiers. In: 2017 International Work Conference on Bioinspired Intelligence (IWOBI), pp. 121–126. IEEE (2017)
16. Tallón-Ballesteros, A.J., Hervás-Martínez, C., Riquelme, J.C., Ruiz, R.: Improving the accuracy of a two-stage algorithm in evolutionary product unit neural networks for classification by means of feature selection. In: Ferrández, J.M., Álvarez Sánchez, J.R., de la Paz, F., Toledo, F.J. (eds.) IWINAC 2011. LNCS, vol. 6687, pp. 381–390. Springer, Heidelberg (2011). https://doi.org/10.1007/978-3-642-21326-7_41
17. Tallón-Ballesteros, A.J., Hervás-Martínez, C., Riquelme, J.C., Ruiz, R.: Feature selection to enhance a two-stage evolutionary algorithm in product unit neural networks for complex classification problems. Neurocomputing **114**, 107–117 (2013)
18. Tallón-Ballesteros, A.J., Ibiza-Granados, A.: Simplifying pattern recognition problems via a scatter search algorithm. Int. J. Computat. Methods Eng. Sci. Mech. **17**(5–6), 315–321 (2016)
19. Tallón-Ballesteros, A.J., Riquelme, J.C.: Low dimensionality or same subsets as a result of feature selection: an in-depth roadmap. In: Ferrández Vicente, J.M., Álvarez-Sánchez, J.R., de la Paz López, F., Toledo Moreo, J., Adeli, H. (eds.) IWINAC 2017. LNCS, vol. 10338, pp. 531–539. Springer, Cham (2017). https://doi.org/10.1007/978-3-319-59773-7_54
20. Tallón-Ballesteros, A.J., Riquelme, J.C., Ruiz, R.: Accuracy increase on evolving product unit neural networks via feature subset selection. In: Martínez-Álvarez, F., Troncoso, A., Quintián, H., Corchado, E. (eds.) HAIS 2016. LNCS (LNAI), vol. 9648, pp. 136–148. Springer, Cham (2016). https://doi.org/10.1007/978-3-319-32034-2_12
21. Tallón-Ballesteros, A.J., Riquelme, J.C., Ruiz, R.: Merging subsets of attributes to improve a hybrid consistency-based filter: a case of study in product unit neural networks. Connect. Sci. **28**(3), 242–257 (2016)
22. ML UCI: Repository, the UC Irvine machine learning repository (2017)

23. Wang, K., Yuen, S.T., Xu, J., Lee, S.P., Yan, H.H.N., Shi, S.T., Siu, H.C., Deng, S., Chu, K.M., Law, S., et al.: Whole-genome sequencing and comprehensive molecular profiling identify new driver mutations in gastric cancer. Nature Genet. **46**(6), 573 (2014)
24. Wolpert, D.H.: The supervised learning no-free-lunch theorems. In: Roy, R., Köppen, M., Ovaska, S., Furuhashi, T., Hoffmann, F. (eds.) Soft Computing and Industry, pp. 25–42. Springer, London (2002). https://doi.org/10.1007/978-1-4471-0123-9_3
25. Yu, L., Liu, H.: Feature selection for high-dimensional data: a fast correlation-based filter solution. In: Proceedings of the 20th International Conference on Machine Learning (ICML-03), pp. 856–863 (2003)
26. Zhang, H.: The optimality of Naive Bayes. In: Barr, V., Markov, Z. (eds.) Proceedings of the Seventeenth International Florida Artificial Intelligence Research Society Conference (FLAIRS 2004). AAAI Press (2004)

Applying VorEAl for IoT Intrusion Detection

Nayat Sanchez-Pi[1], Luis Martí[2], and José M. Molina[2,3(✉)]

[1] Institute of Mathematics and Statistics,
Rio de Janeiro State University, Rio De Janeiro, Brazil
nayat@ime.uerj.br
[2] Institute of Computing, Fluminense Federal University, Niterói, Brazil
lmarti@ic.uff.br
[3] Computer Science Department, Carlos III University of Madrid, Madrid, Spain
molina@ia.uc3m.es

Abstract. Smart connected devices create what has been denominated as the Internet of Things (IoT). The combined and cohesive use of these devices prompts the emergence of Ambient Intelligence (AmI). One of the current key issues in the IoT domain has to do with the detection and prevention of security breaches and intrusions. In this paper, we introduce the use of the Voronoi diagram-based Evolutionary Algorithm (VorEAl) in the context of IoT intrusion detection. In order to cope with the dimensions of the problem, we propose a modification of VorEAl that employs a proxy for the volume that approximates it using a heuristic surrogate. The proxy has linear complexity and, therefore, highly scalable. The experimental studies carried out as part of the paper show that our approach is able to outperform other approaches that have been previously used to address the problem of interest.

Keywords: IDS · IoT · Machine learning · Predictive analysis
Time series

1 Introduction

Connected smart devices create what has been denominated as the Internet of Things (IoT). The combined and cohesive use of these devices prompts the emergence of Ambient Intelligence (AmI). AmI envisions a future information society where users are proactively but sensibly provided with services that support their activities in everyday life. To achieve this goal, AmI systems embed a multitude of sensors in the environment that acquire and exploit data in order to generate an adequate response through actuators, using communication systems and computational processes.

It has been forecasted that the number of connected smart devices will grow up to 50 billion active devices by 2020 [23]. This massive number of online devices raises new security and privacy challenges that, combined with the current state of world affairs, call for special attention to these issues.

© Springer International Publishing AG, part of Springer Nature 2018
F. J. de Cos Juez et al. (Eds.): HAIS 2018, LNAI 10870, pp. 363–374, 2018.
https://doi.org/10.1007/978-3-319-92639-1_30

As evidence of this fact, a recent global-level cyber attack relied on IoT devices to launch the attacks. Security and privacy threats to citizens, industries, and governments derived from these potential weaknesses call for decisive actions to deal with it. Just to mention an example, in [20] IoT devices (in particular, smart lights) were used to exfiltrate data from a highly secure office building.

Because of this, it is obvious that preventive and corrective actions should be taken to address the issue of intrusion detection in the IoT context.

Anomaly-based intrusion detection systems (IDSs) rely on a given model for normal system operation that is able to tell apart anomalies. Anomaly-based IDSs have employed different statistics, machine learning and bio-inspired methods [7,21], including evolutionary computation, to detect anomalies and classify breaches. An example of intrusion detection and challenges also come from Denial of Service (DoS) and reduced quality of service (QoS), while maintaining high true positive rates for detection and avoid high false positive alarm rates [25].

This model of operation can be constructed either from (i) a fully supervised intrusion (anomalies) dataset, (ii) can be inferred from normal operation data in an unsupervised way, or (iii) can be learned in a semisupervised scheme combining both approaches.

The Voronoi diagram-based Evolutionary Algorithm (VorEAl) [17,18] was proposed as an anomaly detection algorithm with success. VorEAl evolves Voronoi diagrams that are used to classify data in an anomaly detection context [7]. VorEAl applies a multi-objective optimization principle that allows it to build models of operation that are a compact representation of normal operation while still taking into account known anomalies.

VorEAl is particularly suitable for the IoT domain because of the low computational footprint at exploitation time. This is because it represents the areas of the input space as Voronoi cells. These cells can be represented as a k-d tree [2]. Therefore, the computational complexity of computing a VorEAl prediction for a Voronoi diagram of m cells is, on the average case, of $O(\log m)$ and, in the worst case, $O(m)$.

However, the original formulation of VorEAl relies intensively on the computation of the volume of Voronoi cells. This hampers the applicability of the algorithm to relatively high-dimensional problems as the computation of the volume is a $\#P$-hard problem.

In this paper, we introduce the use of VorEAl in the context of IoT intrusion detection. In order to cope with the dimensions of the problem, we propose a modification of VorEAl that employs a proxy for the volume that approximates it using a heuristic. The proxy has liner complexity and, therefore, highly scalable.

The rest of this paper is organized as follows. In the next section, we introduce the foundation elements that are necessary for our discussion. After that, in Sect. 3 we briefly describe the VorEAl algorithm. Subsequently, in Sect. 4 we describe in detail the volume proxy that is used to handle high-dimensional problems. The experimental study is carried out in Sect. 5, where we apply VorEAl and other methods to the Australian Defence Force Academy Linux Dataset (ADFA-LD) [10]. Finally, in Sect. 6 we provide some conclusive comments and final remarks.

2   Foundations

As mentioned, Ambient Intelligence (AmI) envisions a future Information Society where users are proactively, but sensibly [1], provided with services that support their activities in everyday life. AmI scenarios described by the European Commission Information Society Technologies Advisory Group (ISTAG) depicts intelligent environments capable of recognizing and responding to the presence of different individuals in a seamless, unobtrusive and often invisible way [12]. AmI is strongly founded on the concept of Ubiquitous Computing (UC), introduced by Weiser in the early 90s, which presents a world where a multitude of computational objects communicate and interact in order to help humans in daily activities [26].

Each AmI project needs a set of sensors and actuators deployed to develop the "intelligence" of the environment. Sensors and actuators are specific for an AmI project, but the infrastructure should be developed using IoT platforms. An IoT platform involves sensing/actuating, communications, computational processes and, actually, these developed should be based on the cloud.

2.1   IoT Intrusion Detection

Cybersecurity systems try to minimize the attack surface of a given computer system over time. The attack surface refers to the portion of a system that has vulnerabilities. Attackers attempt to influence the system's nominal state and operation by varying their interactions with the attack surface in a non-compliant, hard-to-detect manner.

An *intrusion detection system* (IDS) [19] is a device or software application that monitors network and/or system activities for malicious activities or privacy/policy violations and raises alarms when such activities are detected. Intrusion can be launched from outside a network during the Penetration phase of a DoS or APT, or as an insider Reconnaissance activity after Penetration during Escalation, wherein the adversary has already penetrated and from within, using malware, is trying to find a suitable target on the network. Methods for detection of intrusion attacks [21] can be grouped in two main classes:

- *signature-based IDSs*, that look for *a priori* known patterns of attacks in system activities,
- *anomaly-based IDSs*, which model the normal behavior of the system/network under supervision and flag deviations from normal as anomalous, and hence, possible attacks.

Signature-based IDSs can detect known attacks for which patterns have been discerned. However, it is impossible for them to detect new or unknown attacks, as, by their very nature, they do not possess a known pattern for such attacks. This fact limits the applicability of this class of IDS in IoT scenarios, where low or little maintenance can be expected and the multiplicity of devices implies that many more patterns should be elaborated than are practically discovered and maintained.

Anomaly-based IDSs learn to model normal system behaviour and detect deviations from it, they are capable of detecting known and unknown attacks. This is a special case of semi-supervised learning [8].

Anomaly detection can be posed as a particular case of the classification problem in which data items must be tagged as either 'normal' or 'anomalous'. That is, relying on a dataset

$$\Psi = \left\{ \boldsymbol{x}^{(i)}, y^{(i)} \right\} \tag{1}$$

in which, without loss of generality, we have $\boldsymbol{x} \in \mathbb{R}^n$ and $y^{(i)} \in$ {normal, anomaly}, we describe a classifier that correctly detects instances that correspond to each of the two categories. Because of this fact, the existing metrics devised to assess the quality of a classification algorithm are also applicable in this context.

2.2 Evolutionary Multi-Objective Optimization

As stated, a MOP is an optimization problem where a set of objective functions $f_1(\boldsymbol{x}), \ldots, f_M(\boldsymbol{x})$ should be jointly optimized; formally,

$$\min \boldsymbol{F}(\boldsymbol{x}) = \langle f_1(\boldsymbol{x}), \ldots, f_M(\boldsymbol{x}) \rangle ; \, \boldsymbol{x} \in \mathcal{S} ; \tag{2}$$

where $\mathcal{S} \subseteq \mathcal{D}$ is known as the feasible set and could be expressed as a set of restrictions over the decision or search space $\mathcal{D}$. The image set $\mathcal{O} \subseteq \mathbb{R}^M$ of $\mathcal{S}$ produced by the vector-valued function $\boldsymbol{F}(\cdot)$ is called feasible objective set or criterion set.

The solution to this type of problem is a set of trade-off points. The optimality of a solution can be expressed in terms of the Pareto dominance relation. The solution of (2) is $\mathcal{S}^*$, the non-dominated subset of $\mathcal{S}$. $\mathcal{S}^*$ is known as the *efficient set* or *Pareto-optimal set* [6]. Its image in objective space is known as the *Pareto-optimal front*, $\mathcal{O}^*$.

As finding the explicit formulation of $\mathcal{S}^*$ is often impossible, generally, an algorithm solving (2) yields a discrete non-dominated set, $\mathcal{P}^*$, that approximates $\mathcal{S}^*$. The image of $\mathcal{P}^*$ in objective set, $\mathcal{PF}^*$, is known as the *non-dominated front*.

A recent generation of EMOAs exploits existing performance indicators for their selection processes. The $\mathcal{S}$-metric selection evolutionary multiobjective optimization algorithm (SMS-EMOA) [4] belongs to that group of approaches. SMS-EMOA is a steady-state algorithm. Which means that, in every iteration, only one individual is created and only one has to be deleted from the population in each generation. The hypervolume is not computed exactly. Instead, the k-greedy strategy is employed. These decisions were made in the hope of tackling the high computational demands of computing the hypervolume.

Another promising line comes from the reference-point-based many-objective version of the nondominated sorting genetic algorithm (NSGA-II), denominated NSGA-III [11]. Similar to NSGA-II, NSGA-III employs the Pareto non-dominated sorting to partition the population into a number of fronts. In the last front, instead of using the crowding distance to determine the selected solutions a novel niche-preservation operator is applied. This niche-preservation operator relies on reference points organized in a hyperplane to promote diversification of the population. Therefore, solutions associated with less crowded reference points are more likely to be selected. A sophisticated normalization is incorporated into NSGA-III to effectively handle objective functions of different scales.

3   The Voronoi Diagram Evolutionary Algorithm

As stated before, anomaly detection can be posed as a particular case of classification problems. Here data items must be tagged either as 'normal' or 'anomalous'. That is, relying on a dataset $\Psi = \{x^{(i)}, y^{(i)}\}$, where, without loss of generality, we can state that $x \in \mathbb{R}^n$ and $y^{(i)} \in \{normal; anomaly\}$ obtain a classifier, $M(x; \phi) \rightarrow \{normal; anomaly\}$, that correctly detects instances that correspond to each of the two categories. Because of this fact, the existing metrics devised to assess the quality of a classification algorithm are also applicable in this context. For this particular problem, the most relevant metrics are accuracy, recall and specificity, although many more could also be of use. These metrics rely on the number of true positives, t_p, false positives, f_p, false negatives, f_n, and true negatives t_n produced by a given model M. Accuracy measures the proportion of true results (both true positives and true negatives) regard to the total number of elements in the dataset by computing $a(M) = (t_p + t_n)/(t_p + f_p + f_n + t_n)$. On the other hand, recall gauges the ratio between the true positives and false negatives as $r(M) = t_p/(t_p + f_n)$.

Finally, specificity quantifies the proportion of negatives that are correctly identified $s(\mathfrak{M}) = tn/tn + fp$.

Voronoi diagrams are geometrical constructs that partition a given space and can be used for classification. Any set of points, known as *Voronoi sites*, in a given n-dimensional Euclidean space $\mathcal{E}$ defines a *Voronoi diagram*, i.e., a partition of that space into *Voronoi cells*: the cell corresponding to a given site S is the set of points whose closest site is S. The boundaries between Voronoi cells are the medians of the $[S_i S_j]$ segments, for neighbor Voronoi sites S_i and S_j. Though originally defined in two or three dimensions, there exist several algorithmic procedures to efficiently compute Voronoi diagrams in any dimension (see e.g., [5]).

Voronoi diagrams offer a compact classification representation by attaching to each Voronoi cell (or, equivalently, to the corresponding Voronoi site), a Boolean label. The resulting Voronoi diagram is a partition of the space into 2 subsets: the 'normal' cells are the shape/volume, and the 'anomalous' cells are the outside of the shape/volume.

This representation allows Voronoi diagrams to be evolved in order to use them for anomaly detection. In this case, the genotype is a (variable length) list of labeled Voronoi sites, and the phenotype is the corresponding partition in the space into two subsets. More generally, any piece-wise constant function on the underlying space can be represented by a similar representation by using real-valued labels.

Consequently, every individual represents a Voronoi diagram as a set of sites, $\mathcal{I} = \{S_i\}$, where each site has an associated label, $S.\ell \in \{\text{normal}, \text{anomaly}\}$. Relying on that, that individual can be used as a classifier as

$$\text{clfy}(\mathcal{I}, \boldsymbol{x}) = \boldsymbol{S}^*.\ell \text{ with } \boldsymbol{S}^* = \arg\min_{S_i \in \mathcal{I}} \|\boldsymbol{x} - \boldsymbol{S}_i\|. \tag{3}$$

It is possible to prompt the Voronoi diagrams (individuals) to represent the known data in a form as compact as possible by expressing that as the relation between the volumes of the Voronoi cell and the convex hull of the training data that it contains. Maximizing the empty regions will allow to cover the space as much as possible with one region, this region will detect possible anomalies in the future.

Let $\mathcal{I} = \{S_i, i = 1 \ldots n_{\mathcal{I}}\}$ be a Voronoi diagram, and, for each cell C_i, let $v_i \in \mathbb{R}^+$ be its volume and $\mathcal{D}_i$ the set of data points it contains, i.e., $\mathcal{D}_i = \{\boldsymbol{x} \in \Psi; d(x, S_i) \le d(x, S_j) \forall i \neq j\}$, d being the n-dimensional Euclidean distance. We can then define the individual compactness as the sum, for each cell, of the ratio of the volume of the convex hulls of $\mathcal{D}_i$ and the volume of the cell,

$$c(\mathcal{I}) = \begin{cases} \sum_i (|\mathcal{D}_i| - n) \frac{\text{volume}(\text{convex\_hull}(\mathcal{D}_i))}{v_i} & \text{if } |\mathcal{D}_i| > n, \\ 0 & \text{in other case.} \end{cases} \tag{4}$$

Maximizing compactness will produce cells that contain the data in a form as tight as possible. However, the compactness objective can be complemented by one that promotes the existence of empty cells that represent areas of the input domain that are now present in the training data. Such objective would take care of sites with small $\mathcal{D}_i$'s and promote that they become empty as the evolution takes place. A form of representing this is by computing the total volume of cells with an anomaly label of an individual and rate it by the number of elements it contains,

$$v(\mathcal{I}) = \sum_{i, S_i.\ell = \text{anomaly}} \frac{v_i}{1 + 2\ln(|\mathcal{D}_i| + 1)}. \tag{5}$$

Following the above presentation, the problem of finding Voronoi diagrams for anomaly detection can be formalized as the many-objective optimization problem

$$\max \boldsymbol{F} = \langle a(\mathcal{I}), r(\mathcal{I}), c(\mathcal{I}), v(\mathcal{I}) \rangle; \ \mathcal{I} \in \mathcal{D}, \tag{6}$$

where $\mathcal{D}$ is the set of all possible Voronoi diagrams in the problem domain.

Two variation operators have been put forward to operate on Voronoi diagrams. The mutation operator acts on two levels. At an individual level, a new Voronoi site can be added, at a randomly chosen position, with a random label; or a randomly chosen Voronoi site can be removed. At a site level, Voronoi sites can be moved around in the space (the well-known self-adaptive Gaussian mutation has been chosen here, inspired by evolution strategies) or the label of a Voronoi site can be changed.

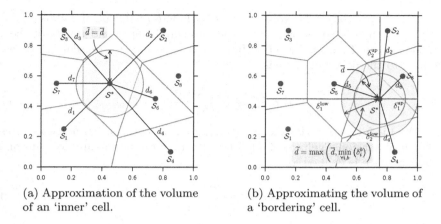

(a) Approximation of the volume of an 'inner' cell.

(b) Approximating the volume of a 'bordering' cell.

Fig. 1. Volume proxies inner and bordering Voronoi cells.

The crossover operator takes two Voronoi diagrams as argument and respects the locality of the representation. Voronoi sites that are close to each other should have more chance to stay together than Voronoi sites that are far apart. This is achieved by the geometric crossover that operates by creating a random cutting hyperplane, and exchanging the Voronoi sites from both sides of the hyperplane, ensuring that the resulting diagrams meet the minimum length bound, $n_{\min}$. Both operators are described in detail in [17,18].

4   Adapting VorEAl for High-Dimensional Problems

One evident drawback of the volume-based objective functions has to do with its computational complexity. The exact computation of the volume of a Voronoi cell implies (i) calculating the vertices of the cell and, then, (ii) computing the volume of the convex polytopes they define, as well as computing the volume of the convex hull of the data points that are classified by the cell.

Regretfully, both steps have a high computational complexity. For that reason, the proxy volume function $\tilde{V}(\cdot)$ was proposed. $\tilde{V}(\cdot)$ approximates the volume of a cell relying on the volume of an n-dimensional ball of radius equal to half of the mean distance of the corresponding site to its neighbor sites (defined as the cells that share a common ridge with the one of interest). Inner and bordering cells should be treated differently. For inner cell it is sufficient to just compute the mean of the distances, as explained in Fig. 1a. In the case of bordering cells (see Fig. 1b), the mean distance to the neighboring sites should be contrasted with how far is the site from the nearest endpoint of $\mathcal{D}$.

These ideas are consolidated in as,

$$\tilde{V}(S) = \begin{cases} B_n\left(\max\left(\bar{d}, \min_{\forall i,b}(\delta_i^b)\right)\right) & \text{if } S \in \mathcal{B} \\ B_n\left(\bar{d}\right) & \text{in other case} \end{cases}, \tag{7}$$

where $B_n(r)$ calculates the volume of an n-dimensional ball of radius r, δ_i^{low} and δ_i^{up} are the distances from S to the lower and upper end points of the i-th dimension of $\mathcal{D}$, $\mathcal{B}$ is the set of bordering sites and $\bar{d}$ represents the mean distances to the neighbor sites as,

$$\bar{d} = \frac{\sum_{S_i \in N(S)} d(S, S_i)}{2 |N(S)|}, \tag{8}$$

where $N(\cdot)$ is a function that, for a given site, returns the set of neighbor sites.

The n-dimensional volume of a Euclidean ball of radius r in an n-dimensional Euclidean space is defined as

$$B_n(r) = \frac{\pi^{\frac{n}{2}}}{\Gamma\left(\frac{n}{2} + 1\right)} r^n, \tag{9}$$

where Γ is Leonhard Euler's gamma function, that is extension of the factorial function to noninteger arguments.

Transforming (9) the volume can be expressed as the one-dimension recursion formula

$$\begin{aligned}
B_{2k}(r) &= r\pi \frac{(2k - 1)!!}{2^k k!} B_{2k-1}(r), \\
B_{2k+1}(r) &= 2r \frac{2^k k!}{(2k + 1)!!} B_{2k}(r).
\end{aligned} \tag{10}$$

$\tilde{V}(\cdot)$ can be implemented using a k-d tree [2]. Therefore, the computational complexity of computing it for a for Voronoi diagram of m cells is, on the average case, of $O(\log m)$ and, in the worst case, $O(m)$.

5   Experiments

In order to validate VorEAl as an intrusion detection method in the context of an IoT application, it is necessary to resort to an existing data set that can be used as a valid ground for comparison. The main challenge, in this case, is derived from the lack of community-accepted IoT intrusion detection data sets.

That is why, for this study, we used the Australian Defence Force Academy Linux Dataset (ADFA-LD) [10]. This data set captures a context similar to IoT devices as they mostly rely on Unix-based systems for their operation.

The data set consists of traces Linux of 230 system calls. There are 833 traces that represent normal system function data meant for training, 773 are validation normal system data meant for assessing the methods and 60 traces are intrusion traces that corresponds to different attacks (10 each), in particular, *password brute force (FTP)*, *password brute force (SSH)*, *add new superuser*, *Java-based meterpreter*, *Linux meterpreter payload* and *C100 Webshell*.

In order to capture the semi-supervised nature of the optimization process, the data set was configured in a way that preserved most of the intrusions for validation time. In particular, we included 18 randomly chosen attack traces (three for each attack) in the training set and preserved the rest for validation.

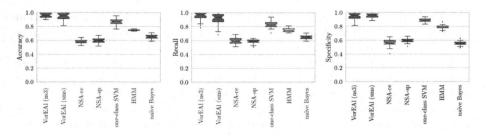

Fig. 2. Performance of the different algorithms involved in the experiment measured as accuracy, recall and specificity.

As stated above, the selection of SMS-EMOA and of NSGA-III are applied and compared as they have been shown to yield substantially better results when confronted with many-objective optimization problems. The rest of the configuration of VorEAl was set as population size of 500, mating probability of 0.3 and the mutation probabilities set to $p_s = 0.5$, $p_f = 0.2$, $p_t = 0.1$, $p_+ = 0.05$ and $p_- = 0.05$ with learning rate $\eta = 0.5$. These values are obtained after performing a cross-validation-based grid search optimization on the considered training dataset. The number of Voronoid cells depends on the size of the genotype (variable length).

Other methods were included in the experiments to provide grounds for comparison with related bio-inspired approaches as well as 'classical' or well-known ones. In particular, we included the negative selection algorithm (NSA) [15] using both variable-sized hyper-spheres and hyper-rectangles. For fair comparisons, we applied the NSA_{sp}^+ and NSA_{re}^+ in which non-self training samples are subsequently used to enrich the detector library generated by NSA.

Similarly, we have included in the experiments three well-known algorithms previously used for intrusion detection in other contexts: one-class vector machines (SVMs) [22], hidden Markov models (HMMs) [9] and the naive Bayes classifier [16]. SVMs and HMMs have are particularly capable of yielding adequate results as reported in some previous works, while the naive Bayes classifier would serve as a baseline.

As the traces are of variable length they should be pre-processed in order to be used as inputs of the algorithms. In these experiments, we employed a "bag of calls" approach where we create a vector of dimension equal to the number of system calls (230 in our case) where each position in the vector indicates the appearance of the system call in the trace.

For all cases, 100 experiment runs were performed. The results of VorEAl with SMS and NSGA-III-based selection and the other methods are reported as box plots in Fig. 2. Here it can be asserted how the two variants of VorEAl managed to outperform the other methods.

The stochastic nature of the algorithms being analyzed calls for the use of an experimental methodology that relies on statistical hypothesis tests. Using those tests, we are able to determine in a statistically sound way if one algorithm instance outperforms another. The topic of assessing stochastic classification algorithms is studied in depth in [14]. There, it is shown that the Bergmann–Hommel [3] procedure is the most suitable for our class of problem.

The results of the statistical hypothesis tests are reported in Table 1. Here it can be clearly asserted the improvement of VorEAl with regard to the other methods. This can be attributed to the fact that VorEAl exploits all the knowledge available. Similarly, these results are also valuable as they endorse the volume proxy used as a substitute for the exact one. Thanks to that we are able to reach higher dimensions at a linear computation cost.

Table 1. Results of the Bergmann–Hommel procedure statistical test. Cases where the algorithm in the row was significantly better than the column are marked with a plus (+) sign, if the one on the column was better than the row are marked with a minus (−) sign. Cases where no significance difference was established are marked with a similar (~) sign.

Accuracy

	VorEAl (ns3)	VorEAl (sms)	NSA-re	NSA-sp	one-class SVM	HMM	naive Bayes
VorEAl (ns3)		~	+	+	+	+	+
VorEAl (sms)	~		+	+	+	+	+
NSA-re	−	−		~	−	−	−
NSA-sp	−	−	~		−	−	−
one-class SVM	−	−	+	+		+	+
HMM	−	−	+	+	−		+
naive Bayes	−	−	+	+	−	−	

Recall

	VorEAl (ns3)	VorEAl (sms)	NSA-re	NSA-sp	one-class SVM	HMM	naive Bayes
VorEAl (ns3)		~	+	+	+	+	+
VorEAl (sms)	~		+	+	+	+	+
NSA-re	−	−		~	−	−	−
NSA-sp	−	−	~		−	−	−
one-class SVM	−	−	+	+		+	+
HMM	−	−	+	+	−		+
naive Bayes	−	−	+	+	−	−	

Specificity

	VorEAl (ns3)	VorEAl (sms)	NSA-re	NSA-sp	one-class SVM	HMM	naive Bayes
VorEAl (ns3)		~	+	+	+	+	+
VorEAl (sms)	~		+	+	+	+	+
NSA-re	−	−		~	−	−	+
NSA-sp	−	−	~		−	−	+
one-class SVM	−	−	+	+		+	+
HMM	−	−	+	+	−		+
naive Bayes	−	−	−	−	−	−	

6 Conclusions

In this paper, we have extended VorEAl, a multi-objective evolutionary algorithm that relies on Voronoi diagrams for classifying anomalies. The extension was conceived in order to make viable the application of VorEAl in the context of IoT intrusion detection systems.

This extension consisted of the substitution of the exact volume computation by a proxy that has linear complexity. The experimental study carried out shows the viability of the approach and validates the working hypothesis of the proposal.

In spite of the encouraging results obtained so far, there are many areas that should be further studied and explored. From an algorithmic point of view, we should explore other classification objectives (metrics). Perhaps devising new ones with this particular problem in mind, in such was as to get a better and more condensed representation of anomalous situations. In the context of anomaly detection, another possibility could have been to add some *specificity pressure*, similar to the parsimony pressure. On-going work will address this issue, in a way similar to [24].

Certainly, more experiments, with more data sets are necessary to fully assert VorEAl performance by confronting it with other problems and algorithms. It is critical the lack of intrusion detection data sets in the IoT context.

As a future objective, we are setting up a living lab with real-life IoT devices in order to gather data closer to the original context as well validate the viability of our for deployment as a part of the network. In this regard, our intention is to explore the implementation of an autonomous evolving device that could adapt to the IoT environment where it is embedded. We also envision the use of Field Programmable Gate Array (FPGAs) [13] for dynamic hardware support.

Acknowledgements. This work was supported in part by Project MINECO TEC2017-88048-C2-2-R, FAPERJ APQ1 Project 211.500/2015, FAPERJ APQ1 Project 211.451/2015, CNPq Universal 430082/2016-9, FAPERJ JCNE E-26/203.287/2017.

References

1. Augusto, J., Shapiro, D.: Advances in Ambient Intelligence, vol. 164. IOS Press Inc., Amsterdam (2007)
2. Bentley, J.L.: Multidimensional binary search trees used for associative searching. Commun. ACM **18**(9), 509–517 (1975). https://doi.org/10.1145/361002.361007
3. Bergmann, B., Hommel, G.: Improvements of general multiple test procedures for redundant systems of hypotheses. In: Bauer, P., Hommel, G., Sonnemann, E. (eds.) Multiple Hypothesenprüfung/Multiple Hypotheses Testing. MEDINFO, vol. 70, pp. 100–115. Springer, Heidelberg (1988). https://doi.org/10.1007/978-3-642-52307-6_8
4. Beume, N., Naujoks, B., Emmerich, M.: SMS-EMOA: multiobjective selection based on dominated hypervolume. Eur. J. Oper. Res. **181**(3), 1653–1669 (2007). http://ideas.repec.org/a/eee/ejores/v181y2007i3p1653-1669.html
5. Boissonnat, J.D., Yvinec, M.: Algorithmic Geometry. Cambridge University Press, New York (1998)
6. Branke, J., Miettinen, K., Deb, K., Słowiński, R. (eds.): Multiobjective Optimization. LNCS, vol. 5252. Springer, Heidelberg (2008). https://doi.org/10.1007/978-3-540-88908-3
7. Chandola, V., Banerjee, A., Kumar, V.: Anomaly detection: a survey. ACM Comput. Surv. (CSUR) **41**(3), 15 (2009)
8. Chapelle, O., Schlkopf, B., Zien, A.: Semi-Supervised Learning, 1st edn. The MIT Press, Cambridge (2010)
9. Cho, S.B., Park, H.J.: Effcient anomaly detection by modeling privilege flows using hidden Markov model. Comput. Secur. **22**(1), 45–55 (2003)
10. Creech, G.: Developing a high-accuracy cross platform Host-Based Intrusion Detection System capable of reliably detecting zero-day attacks. Ph.D. thesis (2014)
11. Deb, K., Jain, H.: An evolutionary many-objective optimization algorithm using reference-point-based nondominated sorting approach, part I: solving problems with box constraints. IEEE Trans. Evol. Comput. **18**(4), 577–601 (2014). http://ieeexplore.ieee.org/lpdocs/epic03/wrapper.htm?arnumber=6600851

12. Ducatel, K., Bogdanowicz, M., Scapolo, F., Leijten, J., Burgelman, J.: Scenarios for ambient intelligence 2010, ISTAG report, European commission. Institute for Prospective Technological Studies, Seville (2001). ftp://ftp.cordis.lu/pub/ist/docs/istagscenarios2010.pdf

13. Farooq, U., Marrakchi, Z., Mehrez, H.: FPGA architectures: an overview. In: Farooq, U., Marrakchi, Z., Mehrez, H. (eds.) Tree-Based Heterogeneous FPGA Architectures, pp. 7–48. Springer, Heidelberg (2012). https://doi.org/10.1007/978-1-4614-3594-5_2

14. García, S., Herrera, F.: An extension on "statistical comparisons of classifiers over multiple data sets" for all pairwise comparisons. J. Mach. Learn. Res. 9, 2677–2694 (2008)

15. Ji, Z., Dasgupta, D.: Real-valued negative selection algorithm with variable-sized detectors. In: Deb, K. (ed.) GECCO 2004. LNCS, vol. 3102, pp. 287–298. Springer, Heidelberg (2004). https://doi.org/10.1007/978-3-540-24854-5_30

16. Manning, C.D., Raghavan, P., Schütze, H., et al.: Introduction to Information Retrieval, vol. 1. Cambridge University Press, Cambridge (2008)

17. Martí, L., Fansi-Tchango, A., Navarro, L., Schoenauer, M.: Anomaly detection with the voronoi diagram evolutionary algorithm. In: Handl, J., Hart, E., Lewis, P.R., López-Ibáñez, M., Ochoa, G., Paechter, B. (eds.) PPSN 2016. LNCS, vol. 9921, pp. 697–706. Springer, Cham (2016). https://doi.org/10.1007/978-3-319-45823-6_65

18. Martí, L., Fansi-Tchango, A., Navarro, L., Schoenauer, M.: VorAIS: a multi-objective Voronoi diagram-based artificial immune system. In: Proceedings of the 2016 Annual Conference on Genetic and Evolutionary Computation (GECCO 2016), pp. 11–12. ACM Press, New York (2016)

19. Northcutt, S., Novak, J.: Network Intrusion Detection. Sams Publishing, Indianapolis (2002)

20. Ronen, E., Shamir, A.: Extended functionality attacks on IoT devices: the case of smart lights. In: 2016 IEEE European Symposium on Security and Privacy (EuroS&P) (2016)

21. Shafi, K., Abbass, H.A.: Biologically-inspired complex adaptive systems approaches to network intrusion detection. Inf. Secur. Tech. Rep. 12(4), 209–217 (2007)

22. Tax, D.M.J., Duin, R.P.W.: Support vector data description. Mach. Learn. 54(1), 45–66 (2004)

23. Vestberg, H.: CEO to shareholders: 50 billion connections by 2020, March 2010. https://www.ericsson.com/thecompany/press/releases/2010/04/1403231

24. Wagner, M., Neumann, F.: Parsimony pressure versus multi-objective optimization for variable length representations. In: Coello, C.A.C., Cutello, V., Deb, K., Forrest, S., Nicosia, G., Pavone, M. (eds.) PPSN 2012. LNCS, vol. 7491, pp. 133–142. Springer, Heidelberg (2012). https://doi.org/10.1007/978-3-642-32937-1_14

25. Wechsler, H.: Cyberspace security using adversarial learning and conformal prediction. Intell. Inf. Manag. 7(04), 195 (2015)

26. Weiser, M.: The computer for the 21st century. Sci. Am. 265(3), 94–104 (1991)

Visual Analysis and Advanced Data Processing Techniques

Evaluation of a Wrist-Based Wearable Fall Detection Method

Samad Barri Khojasteh[1,2], José R. Villar[2(✉)], Enrique de la Cal[2],
Víctor M. González[3], Javier Sedano[4], and Harun Reşit Yazğan[1]

[1] Department of Industrial Engineering, Sakarya University, Sakarya, Turkey
samad.khojasteh@ogr.sakarya.edu.tr, yazgan@sakarya.edu.tr
[2] Computer Science Department, University of Oviedo, EIMEM, Oviedo, Spain
{villarjose,delacal}@uniovi.es
[3] Control and Automatica Department, University of Oviedo, EPI, Gijón, Spain
vmsuarez@uniovi.es
[4] Instituto Tecnológico de Castilla y León,
Pol. Ind. Villalonquejar, 09001 Burgos, Spain
javier.sedano@itcl.es

Abstract. Fall detection represents an important issue when dealing with Ambient Assisted Living for the elder. The vast majority of fall detection approaches have been developed for healthy and relatively young people. Moreover, plenty of these approaches make use of sensors placed on the hip. Considering the focused population of elderly people, there are clear differences and constraints. On the one hand, the patterns and times in the normal activities -and also the falls- are different from younger people: elders move slowly. On the second hand, solutions using uncomfortable sensory systems would be rejected by many candidates. In this research, one of the proposed solutions in the literature has been adapted to use a smartwatch on a wrist, solving some problems and modifying part of the algorithm. The experimentation includes a publicly available dataset. Results point to several enhancements in order to be adapted to the focused population.

Keywords: Fall detection · Wearable devices
Ambient Assisted Living

1 Introduction

Fall Detection (FD) is a very active research area, with many applications to healthcare, work safety, etc. Even though there are plenty of commercial products, the best rated products only reach a 80% of success. There are basically two types of FD systems: context-aware systems and wearable devices [11]. FD has been widely studied using context-aware systems, i.e. video systems [25]; nevertheless, the use of wearable devices is crucial because the high percentage of elderly people and their desire to live autonomously in their own house [14].

© Springer International Publishing AG, part of Springer Nature 2018
F. J. de Cos Juez et al. (Eds.): HAIS 2018, LNAI 10870, pp. 377–386, 2018.
https://doi.org/10.1007/978-3-319-92639-1_31

Wearables-based solutions include, mainly, tri-axial accelerometers (3DACC) either alone or combined with other sensors. Several solutions incorporate more than one sensory element; for instance, Sorvala et al. [18] proposed two sets of a 3DACC and a gyroscope, one on the wrist and another on the ankle, detecting the fall events with two defined thresholds. The use of 3DACC and a barometer in a necklace was also reported in [3]; similar approaches have been developed in several commercial products.

Several solutions using wearable devices combining 3DACC have been reported, i.e., identifying the fall events using Support Vector Machines [26]. In [10] several classifiers are compared using the 3DACC and the inertial sensor within a smartphone to sample the data. A similar solution is proposed in [24], using some different transformations of the 3DACC signal. The main characteristic in all these solutions is that the wearable devices are placed on the wrist. The reason for this location is that it is much easier to detect a fall using the sensory system in this placement. Nevertheless, this type of devices lacks in usability and the people tend to dismiss them in the bedside table. Thus, this research limits itself to use a single sensor -a marketed smartwatch- placed on the wrist in order to promote its usability.

Interestingly, the previous studies do not focus on the specific dynamics of a falling event: although some of the proposals report good performances, they are just machine learning applied to the focused problem. There are studies concerned with the dynamics in a fall event [1,6], establishing the taxonomy and the time periods for each sequence. Additionally, Abbate et al. proposed the use of these dynamics as the basis of the FD algorithm [1]. A very interesting point of this approach is that the computational constraints are kept moderate, although this solution includes a high number of thresholds to tune. Nevertheless, we consider this solution as valid, representing the starting point of this research.

2   Adapting Fall Detection to a Wrist-Based Solution

Abbate et al. [1] proposed the following scheme to detect a candidate event as a fall event (refer to Fig. 1). A time t corresponds to a **peak time** (point 1) if the magnitude of the acceleration a is higher than $th_1 = 3 \times g, g = 9.8\ m/s$. After a peak time there must be a period of 2500 ms with relatively calm (no other a value higher than th_1). The **impact end** (point 2) denotes the end of the fall event; it is the last time for which the a value is higher than $th_2 = 1.5 \times g$. Finally, the **impact start** (point 3) denotes the starting time of the fall event, computed as the time of the first sequence of an $a <= th_3$ ($th_3 = 0.8 \times g$) followed by a value of $a >= th_2$. The impact start must belong to the interval $[impact\ end - 1200\ ms, peak\ time]$. If no impact end is found, then it is fixed to peak time plus 1000 ms. If no impact start is found, it is fixed to peak time.

Whenever a peak time is found, the following transformations should be computed:

- Average Absolute Acceleration Magnitude Variation, $AAMV = \sum_{t=is}^{ie} \frac{|a_{t+1}-a_t|}{N}$, with is being the impact start, ie the impact end, and N the number of samples in the interval.
- Impact Duration Index, $IDI = impact\ end - impact\ start$. Alternatively, it could be computed as the number of samples.
- Maximum Peak Index, $MPI = max_{t \in [is,ie]}(a_t)$.
- Minimum Valley Index, $MVI = min_{t \in [is-500,ie]}(a_t)$.
- Peak Duration Index, $PDI = peak\ end - peak\ start$, with peak start defined as the time of the last magnitude sample below $th_{PDI} = 1.8 \times g$ occurred before peak time, and peak end is defined as the time of the first magnitude sample below $th_{PDI} = 1.8 \times g$ occurred after peak time.
- Activity Ratio Index, ARI, measuring the activity level in an interval of 700 ms centered at the middle time between impact start and impact end. The activity level is calculated as the ratio between the number of samples not in $[th_{ARIlow}0.85 \times g, th_{ARIIhigh} = 1.3 \times g]$ and the total number of samples in the 700 ms interval.
- Free Fall Index, FFI, computed as follows. Firstly, search for an acceleration sample below $th_{FFI} = 0.8 \times g$ occurring up to 200 ms before peak time; if found, the sample time represents the end of the interval, otherwise the end of the interval is set 200 ms before peak time. Secondly, the start of the interval is simply set to 200 ms before its end. FFI is defined as the average acceleration magnitude evaluated within the interval.
- Step Count Index, SCI, measured as the number of peaks in the interval $[peak\ time - 2200, peak\ time]$. SCI is the step count evaluated 2200 ms before peak time. The number of valleys are counted, defining a valley as a region with acceleration magnitude below $th_{SCIlow} = 1 \times g$ for at least 80 ms, followed by a magnitude higher than $th_{SCIhigh}1.6 \times g$ during the next 200 ms. Some ideas on computing the time between peaks [23] were used when implementing this feature.

Evaluating this approach was proposed as follows. The time series of acceleration magnitude values are analyzed searching for peaks that marks where a fall event candidate appears. When it happens to occur, the *impact end* and the *impact start* are determined, and thus the remaining features. As long as this fall events are detected when walking or running, for instance, a Neural Network (NN) model is obtained to classify the set of features extracted.

In order to train the NN, the authors made use of an Activities of Daily Living (ADL) and FD dataset, where each file contains a Time Series of 3DACC values corresponding to an activity or to a fall event. Therefore, each dataset including a fall event or a similar activity -for instance, running can perform similarly to falling- will generate a set of transformation values. Thus, for a dataset file we will detect something similar to a falling, producing a row of the transformations computed for each of the detected events within the file. If nothing is detected within the file, no row is produced. With this strategy, the Abbate et al. obtained the training and testing dataset to learn the NN.

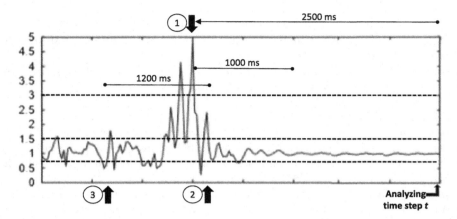

Fig. 1. Evolution of the magnitude of the acceleration -y-axis, extracted from [1]. Analyzing at time stamp *t*, the three conditions described in the text must be found in order to detect a fall. Graph elaborated from [1].

2.1 The Modifications on the Algorithm

As stated in [8,21], the solutions to this type of problems must be ergonomic: the users must feel comfortable using them. We considered that placing a device on the waist is not comfortable, for instance, it is not valid for women using dresses. When working with elderly people, this issue is of main relevance. Therefore, in this study, we placed the wearable device on the wrist. This is not a simple change: the vast majority of the literature reports solutions for FD using waist based solutions. Moreover, according to [7,20] the calculations should be performed on the smartwatches to extend the battery life by reducing the communications. Therefore, these calculations should be kept as simple as possible.

A second modification is focused on the training of the NN. The original strategy for the generation of the training and testing dataset produced a highly imbalanced dataset: up to 81% of the obtained samples belong to the class FD, while the remaining belong to the different ADL similar to a fall event.

To solve this problem a normalization stage is applied to the generated imbalanced dataset, followed by a SMOTE balancing stage [5]. This balancing stage will produce a 60%(FALL)–40%(no FALL) dataset, which would allow avoiding the over-fitting of the NN models. As usual, there is a compromise between the balancing of the dataset and the synthetic data samples introduced in the dataset.

3   Experiments and Results

An ADL and FD dataset is needed to evaluate the adaptation, so it contains time series sample from ADL and for falls. This research made use of the UMA-FALL dataset [4] among the publicly available datasets. This dataset includes data for several participants carrying on with different activities and performing forward, backward and lateral falls. Actually, these falls are not real falls -demonstrative videos have been also published-, but they can represent the initial step for evaluating the adapted solution problem. Interestingly, this dataset includes multiple sensors; therefore, the researcher can evaluate the approach using sensors placed on different parts of the body.

The thresholds used in this study are exactly the same as those mentioned in the original paper. All the code was implemented in R [16] and caret. The parameters for SMOTE were perc.over set to 300 and perc.under set to 200 -that is, 3 minority class samples are generated per original sample while keeping 2 samples from the majority class-. These parameters produce a balanced dataset that moves from a distribution of 47 samples from the minority class and 200 from the majority class to a 188 minority class versus 282 majority class (40%/60% of balance). To obtain the parameters for the NN a grid search was performed [15,17,19]; the final values were p_size set to 11, p_decay set to 10^{-6} and maximum number of iterations 1000.

Both 5×2 cross validation (cv) and 10-fold cv were performed to analyze the robustness of the solution. The latter cv would allow us to compare with existing solutions, while the former shows the performance of the system with an increase in the number of unseen samples. The results are shown in Table 1 and 2 for 10-fold cv and 5×2 cv, respectively. The boxplots for the statistical measurements Accuracy, Kappa factor, Sensitivity and Specificity are shown in Fig. 2.

Table 1. 10 fold cv results. From left to right, the columns stand for the fold number, the classification error, the accuracy, the Kappa factor, the sensitivity, the specificity, and the True Positive, False Positive, False Negative and True Negative results.

Fold	Error %	Acc	Ka	Sens	Spec	TP	FP	FN	TN
1	0.0426	0.9574	0.9117	0.9474	0.9643	18	1	1	27
2	0.0625	0.9375	0.8681	0.9444	0.9333	17	2	1	28
3	0.0213	0.9787	0.9562	0.9500	1.0000	19	0	1	27
4	0.0417	0.9583	0.9144	0.9048	1.0000	19	0	2	27
5	0.0851	0.9149	0.8203	0.9412	0.9000	16	3	1	27
6	0.0217	0.9783	0.9539	1.0000	0.9655	17	1	0	28
7	0.0638	0.9362	0.8664	0.9444	0.9310	17	2	1	27
8	0.0426	0.9574	0.9131	0.9048	1.0000	19	0	2	26
9	0.0870	0.9130	0.8175	0.8889	0.9286	16	2	2	26
10	0.1489	0.8511	0.6934	0.8000	0.8889	16	3	4	24
Mean	0.0617	0.9383	0.8715	0.9226	0.9512				
Median	0.0525	0.9475	0.8899	0.9428	0.9488				
Std	0.0382	0.0382	0.0792	0.0532	0.0412				

Table 2. 5×2 cv results. From left to right, the columns stand for the fold number, the classification error, the accuracy, the Kappa factor, the sensitivity, the specificity, and the True Positive, False Positive, False Negative and True Negative results.

Fold	Error %	Acc	Ka	Sens	Spec	TP	FP	FN	TN
1	0.0723	0.9277	0.8511	0.8812	0.9627	89	5	12	129
2	0.0723	0.9277	0.8511	0.8812	0.9627	89	5	12	129
3	0.0766	0.9234	0.8421	0.8800	0.9556	88	6	12	129
4	0.0766	0.9234	0.8421	0.8800	0.9556	88	6	12	129
5	0.0851	0.9149	0.8252	0.8627	0.9549	88	6	14	127
6	0.0426	0.9574	0.9113	0.9468	0.9645	89	5	5	136
7	0.0638	0.9362	0.8673	0.9158	0.9500	87	7	8	133
8	0.0809	0.9191	0.8348	0.8571	0.9692	90	4	15	126
9	0.07234	0.9277	0.8496	0.9053	0.9429	86	8	9	132
10	0.0340	0.9660	0.9288	0.9674	0.9650	89	5	3	138
Mean	0.0677	0.9323	0.8603	0.89778	0.9583				
Median	0.0723	0.9277	0.8503	0.8812	0.9591				
Std	0.0166	0.0166	0.0336	0.0360	0.0080				

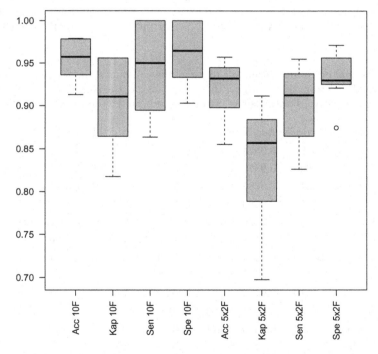

Fig. 2. Boxplot for the different measurements -Accuracy, Kappa, Sensitivity and Specificity-, both for 10 fold cv (four boxplots to the left) and 5×2 cv (four boxplots to the right).

3.1   Discussion on the Results

As seen from the previous figures, the classification results after performing the SMOTE are really impressive, with very reduced miss classified samples. As expected, the 10-fold cv results are a bit better than those depicted for the 5×2 cv; nevertheless, the robustness of the method seems pretty good.

After these results, can the problem of FD be considered solved after several minor changes? Answering this question needs discussing some other topics: (i) the nature of the dataset, (ii) the basis of the method, and (iii) the deployment issues and the computational requirements.

The UMA-Fall dataset used in this study [4] was generated with young participants using a very deterministic protocol of activities. The falls were performed with the participants standing still and letting them fall in the forward/backward/lateral direction. Consequently, there could be severe differences between the movements and the data gathered from a real unexpected fall. For instance, after the fall, the participant kept quiet: this is not normal when a person falls unless he/she faints. There are more publicly available datasets, the majority of these datasets have been gathered with healthy volunteers [12,13]. However, a real-world fall and activity of daily living dataset is published in [2], where a comparison of the different methods published so far is also included. Therefore, the method described in this study needs to be validated with more datasets, more specifically, with data from real fall events.

Concerning with the basis of the Abbate et al. method, the number of predefined thresholds represents its main drawback. These thresholds have been set for young participants: they might not be valid for a different population. Even if the thresholds are valid, perhaps the classification models must be specific for groups of people according to their movement characteristics [9]. Certainly, there are clear differences in the walking between a young participant and an elder person, even between two elder persons. For instance, using crutches means totally different dynamics. All of these issues must be analyzed in order to validate the approach. Determining the thresholds sets needs further research, trying to find general thresholds -if any-; alternatively, clustering the population according to a given criteria and then finding specific thresholds -and classification models- for each group.

Besides, the eHealth and wearable applications deployment issues have been study in the literature [20]. According to the published results, there is a trade off between the mobile computation and the communication acts to extend the battery charge as long as possible. Consequently, it has been found that moving all the preprocessing and modelling issues to the mobile part could be advantageous provided the computational complexity of the solution is kept low. The consequences of these findings shall be reflected in the transformations and in the models, reducing complex floating point operations as much as possible [22].

4   Conclusions

In this study a fall detection method using a wearable device placed on the wrist is described; the classification method was originally proposed to use a waist located accelerometer. This original approach has been adapted with minor enhancements in the computation of some features; the idea underneath is to detect peaks; for each detected peak a set of 8 features are computed, generating a sample. This sample is labelled according to what has happened -either a fall or the corresponding activity-. With the available samples, the classifying feed-forward Neural Network model can be learned. Nevertheless, this solution clearly generates imbalanced data, which was not considered. In this study, SMOTE was used to balance the training/testing dataset.

The good performance of the method on the UMA-Fall dataset shows this method can represent the basis for a good FD method using wrist-based wearables. However, there are still work to be done, basically, coping with datasets including real falls time series. Also, new solutions need to be proposed to tackle with the thresholds tuning. Furthermore, different solutions for the classification models are needed in order to reduce the floating point operations, so the battery charging cycles could be elongated.

Acknowledgments. This research has been funded by the Spanish Ministry of Science and Innovation, under project MINECO-TIN2014-56967-R and MINECO-TIN2017-84804-R, and FC-15-GRUPIN14-073 (Regional Ministry of Principality of Asturias).

References

1. Abbate, S., Avvenuti, M., Bonatesta, F., Cola, G., Corsini, P.: AlessioVecchio: a smartphone-based fall detection system. Pervasive Mob. Comput. **8**(6), 883–899 (2012)
2. Bagala, F., Becker, C., Cappello, A., Chiari, L., Aminian, K., Hausdorff, J.M., Zijlstra, W., Klenk, J.: Evaluation of accelerometer-based fall detection algorithms on real-world falls. PLoS One **7**(5), e37062 (2012)
3. Bianchi, F., Redmond, S.J., Narayanan, M.R., Cerutti, S., Lovell, N.H.: Barometric pressure and triaxial accelerometry-based falls event detection. IEEE Trans. Neural Syst. Rehabil. Eng. **18**(6), 619–627 (2010)
4. Casilari, E., Santoyo-Ramón, J.A., Cano-García, J.M.: UMAFALL: a multisensor dataset for the research on automatic fall detection. Procedia Computer Science **110(Supplement C)**, 32–39 (2017). http://www.sciencedirect.com/science/article/pii/S1877050917312899. In: 14th International Conference on Mobile Systems and Pervasive Computing (MobiSPC 2017)/12th International Conference on Future Networks and Communications (FNC 2017)/Affiliated Workshops
5. Chawla, N.V., Bowyer, K.W., Hall, L.O., Kegelmeyer, W.P.: Smote: synthetic minority over-sampling technique. J. Artif. Intell. Res. **16**, 321–357 (2002)
6. Delahoz, Y.S., Labrador, M.A.: Survey on fall detection and fall prevention using wearable and external sensors. Sensors **14**(10), 19806–19842 (2014). http://www.mdpi.com/1424-8220/14/10/19806/htm

7. Gil-Pita, R., Ayllón, D., Ranilla, J., Llerena-Aguilar, C., Díaz, I.: A computationally efficient sound environment classifier for hearing aids. IEEE Trans. Biomed. Eng. **62**(10), 2358–2368 (2015). https://doi.org/10.1109/TBME.2015.2427452
8. González, S., Sedano, J., Villar, J.R., Corchado, E., Herrero, Á., Baruque, B.: Features and models for human activity recognition. Neurocomputing **167**, 52–60 (2015)
9. González, S., Villar, J.R., Sedano, J., Terán, J., Alonso-Álvarez, M.L., González, J.: Heuristics for apnea episodes recognition. In: Proceedings of the International Conference on Soft Computing Models in Industrial and Environmental Applications. Springer, Cham (2015) (accepted)
10. Hakim, A., Huq, M.S., Shanta, S., Ibrahim, B.: Smartphone based data mining for fall detection: analysis and design. Procedia Comput. Sci. **105**, 46–51 (2017). http://www.sciencedirect.com/science/article/pii/S1877050917302065
11. Igual, R., Medrano, C., Plaza, I.: Challenges, issues and trends in fall detection systems. Biomed. Eng. Online **12**, 66 (2013). http://www.biomedical-engineering-online.com/content/12/1/66
12. Igual, R., Medrano, C., Plaza, I.: A comparison of public datasets for acceleration-based fall detection. Med. Eng. Phys. **37**(9), 870–878 (2015). http://www.sciencedirect.com/science/article/pii/S1350453315001575
13. Khan, S.S., Hoey, J.: Review of fall detection techniques: a data availability perspective. Med. Eng. Phys. **39**, 12–22 (2017). http://www.sciencedirect.com/science/article/pii/S1350453316302600
14. Kumari, P., Mathew, L., Syal, P.: Increasing trend of wearables and multimodal interface for human activity monitoring: a review. Biosens. Bioelectron. **90**(15), 298–307 (2017)
15. Montañés, E., Quevedo, J.R., Díaz, I., Ranilla, J.: Collaborative tag recommendation system based on logistic regression. In: Proceedings of ECML PKDD (The European Conference on Machine Learning and Principles and Practice of Knowledge Discovery in Databases) Discovery Challenge 2009, Bled, Slovenia, 7 September 2009. http://ceur-ws.org/Vol-497/paper_20.pdf
16. R Development Core Team: R: A Language and Environment for Statistical Computing. R Foundation for Statistical Computing, Vienna, Austria (2008). http://www.R-project.org, ISBN 3-900051-07-0
17. Sanchez-Lasheras, F., de Andres, J., Lorca, P., et al.: A hybrid device for the solution of sampling bias problems in the forecasting of firms' bankruptcy. Expert Syst. Appl. **39**, 7512–7523 (2012)
18. Sorvala, A., Alasaarela, E., Sorvoja, H., Myllyla, R.: A two-threshold fall detection algorithm for reducing false alarms. In: Proceedings of 2012 6th International Symposium on Medical Information and Communication Technology (ISMICT) (2012)
19. Turrado, C.C., López, M.D.C.M., Lasheras, F.S., Gómez, B.A.R., Rollé, J.L.C., Juez, F.J.D.C.: Missing data imputation of solar radiation data under different atmospheric conditions. Sensors **14**, 20382–20399 (2014)
20. Vergara, P.M., de la Cal, E., Villar, J.R., González, V.M., Sedano, J.: An IoT platform for epilepsy monitoring and supervising. J. Sens. **2017**, 18 (2017)
21. Villar, J.R., González, S., Sedano, J., Chira, C., Trejo, J.M.: Human activity recognition and feature selection for stroke early diagnosis. In: Pan, J.-S., Polycarpou, M.M., Woźniak, M., de Carvalho, A.C.P.L.F., Quintián, H., Corchado, E. (eds.) HAIS 2013. LNCS (LNAI), vol. 8073, pp. 659–668. Springer, Heidelberg (2013). https://doi.org/10.1007/978-3-642-40846-5_66

22. Villar, J.R., Vergara, P., Menéndez, M., de la Cal, E., González, V.M., Sedano, J.: Generalized models for the classification of abnormal movements in daily life and its applicability to epilepsy convulsion recognition. Int. J. Neural Syst. **26**(6), 1650037 (2016)

23. Villar, J.R., González, S., Sedano, J., Chira, C., Trejo-Gabriel-Galán, J.M.: Improving human activity recognition and its application in early stroke diagnosis. Int. J. Neural Syst. **25**(4), 1450036–1450055 (2015)

24. Wu, F., Zhao, H., Zhao, Y., Zhong, H.: Development of a wearable-sensor-based fall detection system. Int. J. Telemed. Appl. **2015**, 11 (2015). https://www.hindawi.com/journals/ijta/2015/576364/

25. Zhang, S., Wei, Z., Nie, J., Huang, L., Wang, S., Li, Z.: A review on human activity recognition using vision-based method. J. Healthc. Eng. **2017**, 31 (2017)

26. Zhang, T., Wang, J., Xu, L., Liu, P.: Fall detection by wearable sensor and one-class SVM algorithm. In: Huang, D.S., Li, K., Irwin, G.W. (eds.) Intelligent Computing in Signal Processing and Pattern Recognition. LNCIS, vol. 345, pp. 858–863. Springer, Heidelberg (2006). https://doi.org/10.1007/978-3-540-37258-5_104

EnerVMAS: Virtual Agent Organizations to Optimize Energy Consumption Using Intelligent Temperature Calibration

Alfonso González-Briones[1(✉)], Javier Prieto[1], Juan M. Corchado[1,2,3], and Yves Demazeau[4]

[1] BISITE Digital Innovation Hub, University of Salamanca, Salamanca, Spain
{alfonsogb, javierp, corchado}@usal.es
[2] Osaka Institute of Technology, Osaka, Japan
[3] Universiti Malaysia Kelantan, Kelantan, Malaysia
[4] University of Grenoble-Alps, CNRS-LIG, Grenoble, France
yves.demazeau@imag.fr

Abstract. One of the problems we encounter when dealing with the optimization of household energy consumption is how to reduce the consumption of air conditioning systems without reducing the comfort level of the residents. The systems that have been proposed so far do not succeed at optimizing the electricity consumed by heating and air conditioning systems because they do not monitor all the variables involved in this process, often leaving users' comfort aside. It is therefore necessary to develop a solution that monitors the factors which contribute to greater energy consumption. Such a solution must have a self-adaptive architecture with the capacity of self-organization which will allow it to adapt to changes in user temperature preferences. The methodology that is the most suitable for the development of such solution are virtual agent organizations, they allow for the management of wireless sensor networks (WSN) and the use of Case-Based Reasoning (CBR) for predicting the presence of people at home. This work presents an energy optimization system based on virtual agent organizations (VO-MAS) that obtains the characteristics of the environment through sensors and user behavior pattern using a CBR system. A case study was carried out in order to evaluate the performance of the proposed system, the results show that 22.8% energy savings were achieved.

Keywords: Energy savings · Virtual organization · CBR system
Sensor-based monitoring · Ambient computing

1 Introduction

At present, climate change is one of the most worrying problems facing our planet. Sadly, not everyone agrees with this scientifically proven fact; there are many skeptics around the world who deny the very existence of global warming or the idea of humans being able to positively or negatively influence these changes. However, it has been proven that this undoubtedly complex problem can be resolved to a large degree if we

© Springer International Publishing AG, part of Springer Nature 2018
F. J. de Cos Juez et al. (Eds.): HAIS 2018, LNAI 10870, pp. 387–398, 2018.
https://doi.org/10.1007/978-3-319-92639-1_32

individually take measures in the correct direction. Simple solutions, such as saving and using our energy in a more efficient way at home will have a positive impact on reducing the adverse effects of climate change on our planet. Some proposals were aimed at using energy in buildings more efficiently. This solution makes it possible to reduce electricity bills, accounting for around 70% of the annual bill payment. On the basis of these factors temperature can be adjusted to reduce unnecessary energy consumption while maintaining the residents' comfort.

This article proposes a new approach to the analysis of parameters that influence energy consumption in heating and air conditioning systems. The system will make decisions that optimize energy consumption based on user habits and preferences by coordinating the agents that are part of virtual organizations (VO). The main contributions of this work are the following: the non-intrusive acquisition of resident information, thanks to the use of sensors; the conjunction of the current indoor and outdoor temperature with the future temperature in order to prevent the heating, ventilation and air conditioning (HVAC) system from making drastic temperature changes, since this causes a high increase in the consumption. This system has been developed using PANGEA [29] for the technical implementation of the architecture.

This article is organized as follows: Sect. 2 describes the state of the art on energy optimization, Sect. 3 describes the proposal, and Sect. 4 presents the results and conclusions.

2 Energy Background and Virtual Organizations

This section details the need for comprehensive home temperature control. With the recent introduction of energy performance certificates for homes our need for such a system has become even greater [8]. The following subsections review the current state of the art on this type of systems.

2.1 Factors Involved in Household Energy Consumption

Previous works have failed to make an in-depth study of the behaviour of residents in homes. These simplistic studies are owing to the higher cost of automating a house in the past [19, 24]. They looked at lifestyle and comfort preferences of the residents which would allow to program and accommodate systems according to user needs. The times and frequency with which users opened and closed the buildings was also evaluated, simulating occupancy in the building. It is evident that the behaviour of users has a major impact on the energy balance of a building. Some studies have carried out simulations of the user's behaviour in order to optimise the design phase of the building. In the study by Hoes et al. [20] the parameters that influenced the opening of windows such as season, outdoor and indoor temperature, time of day, presence was studied, however, the results cannot be applied to all buildings since the type of building, location, climate or culture have influence over these variables. For efficient energy management, Bayesian networks were used to predict user behaviour from a multimodal sensor [18]. Therefore, an automated system must be developed in order to independently obtain the variables that influence energy consumption, taking into account the presence of people in the building.

2.2 Sensor Data Acquisition

As shown above, the behaviour of users and changes indoor and outdoor temperatures influence energy consumption [1], we must also take into account the comfort preferences of the house's residents. It is necessary to deploy a WSN which allows to obtain parameters such as temperature, presence of people in the home, etc. Since the number of people in the house is unknown, the proposed system must be able to negotiate the preferences of each user of the house in order to establish the correct conditions in a shared space. This will provide the system with the knowledge it needs to control and manage the temperature of the air conditioning equipment by increasing or reducing the temperature to optimize the consumption always within the user's comfort parameters.

It is convenient to complement the system with new data such as the temperature forecast for the next few days. Information on the presence or absence of people at home needs to be included in order to minimize energy consumption when users are not at home. In this regard, a network of distributed sensors is required since it allows to obtain and analyse house data. This network must consist of a motion sensor, an acoustic sensor and presence sensors [12, 13].

2.3 Data Communication Protocol

Existing commercial home automation systems are aimed at achieving savings in the objectives they propose, however, they do not provide a global vision that would allow for the addition of new objectives in the system [21]. To be able to connect these devices and use them for a common purpose: to communicate, send information and create universal systems, a suitable communication protocol is required. In this regard, there are three standard open protocols – ModBus, LonWorks and BACnet – which allow to include new devices. These protocols have a set of characteristics which make them suitable for different uses and applications [27]. According to a survey conducted on the Building Operating Management website in 2011 [12, 14] 30% of respondents in the USA had at least one application using Modbus, 40% for Lon-Works and 62% for BACNet.

2.4 Virtual Organizations of Agents

In order to be able to manage all the factors involved in energy consumption, data acquisition through sensors and communication between the sensorisation and the platform and data analysis, it is necessary to use a methodology that allows for these activities to be carried out in a simple, transparent and modular way. For this reason, the use of virtual agent organizations is the option adopted for system design.

The proposed system made it possible to manage energy without relinquishing the objective of achieving energy efficiency and creating intelligent buildings. Cai *et al.* employed an agent system for building management in centralized air-conditioning systems [7]. In this work, the optimization problem was reformulated into several sub-problems, each of which was solved by an individual agent. However, the success rate of this system would have been higher if virtual agent organizations were used. This is because agents would have provided a range of solutions for a single problem, allowing for the choice of the solution that renders the best results. The proposal of Al-Daraiseh *et al.*, optimized energy by predicting the times at which HVAC systems switched on/off in educational institutions. In that work, the authors considered the influence of certain factors, such as external climatic conditions and the presence of people [2]. However, the proposed system lacked automation and values had to be entered manually, this limitation made it impossible to achieve large energy savings.

Given the shortcomings of previous works, it is necessary to develop a system based on VOs which will provide numerous solutions for configuring the temperature inside a home and will allow to maintain the user's comfort level. VOs allow all kinds of multi agent systems (MASs) to be modelled according to the structural model of human organizations. In this way, it is possible to limit the unpredictability of the system within a group of agents that is subjected to a series of institutional rules. VOs are intended to limit the autonomy of the agents themselves. Clausen *et al.* [9, 10] wrote an article in which VOs were used to obtain energy savings. The authors made an agent-based integration of Distributed Energy Resources in Virtual Power Plants. The approach models Distributed Energy Resources and Virtual Power Plants as agents with multi-objective, multi-issue reasoning. This allowed for the modeling of the VPPs constituting complex and heterogeneous Distributed Energy Resources with multiple, local objectives and decision points. These authors' approach was similar to ours, however our aim is to develop a system with the least possible complexity, making it possible for the user to manage the system efficiently. To this end, in our approach different agent organizations are created, they will be in charge of obtaining the user's preference values, a building's indoor and outdoor temperature or the presence of people.

3   EnerVMAS Architecture Overview

This section details the technical of the architecture of Energy Virtual Multi-Agent Systems (EnerVMAS). It describes aspects related to the collection of data by sensors, data transmission and communication, the conversion of data into useful information, as well as the use of this information in decision making for the achievement of efficient energy use.

3.1   Sensor Deployment and Data Acquisition

The sensors implement by the system allow to detect the presence of people in the home and collect outdoor and indoor temperature values.

- **PIR sensors.** A display has been made in the PIR house [23] so that could be detected the presence of people in each room, and in the house in general. These sensors are called passive sensors because, instead of emitting radiation, they receive it. They capture presence in the home by detecting the difference between the heat emitted by users and the space they are in. Once deployed, these sensors need a period of adaptation, during which these processes "get used" to the infrared radiation of the environment. This simple system consists of the PIR sensor and a led, it has a 100 Ω resistance and is powered by a 3.3 V signal. To avoid false detections by sunlight or other light sources, the sensor incorporates a small plastic lens that acts as a special light filter to eliminate this possibility. When no presence is detected, a signal is sent to the PIR sensor in the data acquisition organization, so that the system knows there are no people at home. Thus, if no presence is detected in the house, the system changes its behaviour when adjusting the temperature.

- **Temperature Sensors.** It is also necessary to deploy temperature sensors that allow us to obtain the indoor and outdoor temperature of each room and provide this information to the temperature adjustment algorithm. For this purpose, DHT22 sensors [4] were used to obtain temperature and humidity values. This sensor can measure temperatures between −40 °C and 125 °C with an accuracy of 0.5 °C, humidity between 0% and 100% with an accuracy of 2–5% and a sampling frequency of 0.5 Hz. In cases where rooms are exposed to the outside, the system also needs to collect the outside temperature. It is important to obtain external temperature data since this allows us to find out how the heat escapes or how cold air enters from the outside, in this way the system is able to regulate the temperature better.

The collection of data on the presence of people in the home, indoor and outdoor temperature, and weather forecasting, is necessary for the correct adjustment of temperature and for correct decisions to the made by the proposed system. However, prior knowledge on weather forecasts and presence in the household is equally important. Knowledge on whether there will be people in the building is key for the system since it allows it to lower the temperature or turn off radiators and in this way, reduce energy consumption. However, the system will not switch off all the heating since then more energy would be required to heat the house before the residents return; this means that the amount of energy required to establish optimal temperature in the building again would be greater than the energy saved.

3.2 CBR System for Learning Timetables

This section presents the proposed CBR framework for the prediction of the residents' work schedules [11, 30]. This CBR is better detail in [31]. CBR allows the system to automatically learn schedules from the cases collected in the baseline period and thus predict when users leave the home and when they come back.

The proposed system predicts the schedule of each resident in a two-stage process. In the first stage, data are obtained from the deployed presence sensors. The second stage is more sophisticated and it consists in predicting a schedule for each of the residents, generating an entry and exit pattern. Once several cases have been recorded for each inhabitant of the house (hours of entry and exit from the house), the database that stores the CBR cases is ready to be used. A distance-based algorithm was used in the retrieve stage. This is done through a weighted variant of the k-Nearest Neighbors (kNN) algorithm, which grants to solid probabilistically-interpretable out-put. The system is able to learn from revised cases, using the revision-retaining procedure. The CBR agent has been developed using the jColibri2 framework [26] in order to provide it with the ability to predict the daily schedules of the residents.

3.3   Energy Optimization Algorithm

The developed savings algorithm analyses the data collected for the variables that have an influence over temperature in a home. The algorithm has double functionality: one for conditioning (adjusting the temperature) and another for making decisions (turning on/off the air conditioning or individual radiators). Thermostats keep the temperature steady, they are programmed according to sensor and weather forecast data. They optimize consumption since they prevent sudden drops or increases in temperature (consumption increases considerably if temperature varies by a few degrees in a short time period). The majority of programmable thermostats found in the market provide four programmable parameters for creating setback periods: comfort temperature, energy saving temperature, energy saving period, temperature setback. This algorithm makes use of the knowledge provided by the CBR in order to know when the residents leave home every day, $time_{leave}$, and the time that elapses until they return $time_{back}$, also the time that elapses inside the home $time_{backPeriod}$. The temperature parameters are obtained by the sensors, this temperature variables are indoor temperature tmp_{indoor}, outdoor temperature $tmp_{outdoor}$, desired temperature $tmp_{desired}$ and forecast temperature $tmp_{forecast}$. The system must also consider whether air conditioning units and radiators are switched on or off, $device_{status}$. The temperature adjustment period of the HVAC system will increase or decrease the temperature according to the inhabitants' preferences.

3.4   VO-MAS Architecture

The VOs of agents use their communication, coordination and cooperation capacities to obtain data from sensors, other devices even act as "crawlers" to obtain information from external data sources. These data are sent from agents of some organizations to others to perform the analysis and decision-making tasks in scenarios as distinct from energy optimization as Bioinformatics [6] or even drug detection [5]. These characteristics allow to program gradual changes in temperature and to prevent sudden changes which cause high energy consumption. For this reason, the EnerVMAS aims to optimally manage the temperature of a home, including the monitoring and control of radiators and air conditioning in real time. It also establishes temperature patterns which comply with the energy efficiency frameworks and the comfort of residents. PANGEA has been chosen for the development of virtual agent organizations [28]. PANGEA incorporates agents that manage security at the system level, unlike other types of systems in which it is necessary to develop this type of measures so that the information collected is really the one that is transmitted and analyzed [16]. The system has four VO, as observed in Fig. 1.

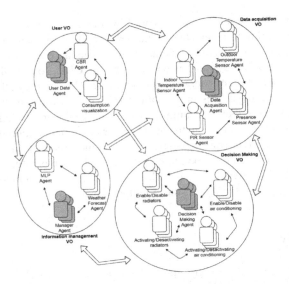

Fig. 1. EnerVMAS architecture based on virtual agent organizations.

- **Data acquisition VO.** The agents in this organization are in charge of communicating and obtaining data from the sensors, these data include indoor temperature, outdoor temperature and presence. This data collects temperature every 60 s. These agents connect to the distributed sensors through the middleware. Information on the presence of users at home is obtained through the PIR sensors.
- **Information management VO.** This organization is responsible for all the data analysis processes, decisions are taken by the system on the basis of these processes. Specifically, the difference in the interior temperature of rooms that are exposed to the outside of the building (the walls of these rooms are part of the building's façade). The weather forecasting agent requests meteorological information from OpenWeatherMap, this information is used to adjust temperature for the next few days. In this way, the system can gradually increase or decrease the temperature of a home.
- **Decision Making VO.** This organization makes decisions on the basis of information received from the other organizations that make up the system. There is an agent for activating/deactivating the air conditioning and another agent for turning the radiators on and off. There is also an agent whose task is to increase/decrease the temperature of radiators/air conditioning agent in each of the rooms and in the common area. These actions can be carried out thanks to communication between agents and the intelligent thermostat.
- **User VO.** This organization contains the CBR. The tasks performed by the CBR consist of four different steps, this workflow allows the CBR to adapt to the conditions of the presented problem. First, the CBR agent system retrieves the relevant

cases to solve the problem using workflow memory. Once a relevant case is recovered the workflow is adapted to the new problem; to changes in temperature, presence of the user(s), events in the academic calendar, etc. before reusing it. The system first simulates the solutions provided by the CBR, so only the most relevant actions for the reduction of energy consumption are chosen, they allow to obtain the greatest energy savings. This organization has User Data agent, the user interacts with it in order to see their consumption data, and the data collected by each sensor in real time. In this organization, the agents enter into a negotiation process in which they find out of the temperature preferences of all the residents. In case two or more inhabitants have the same temperature preferences that temperature is chosen, if not the average of all temperatures is established. And choose the temperature preferred by the majority or an average of these temperatures.

4   Case Study

A case study was designed in order to prove the feasibility of EnerVMAS, to this end it was implemented in a home [17].

4.1   Experimental Set-Up

The case study was divided into two phases, each lasted one month: an energy consumption control phase during which the system only monitored variables but did not make any decisions. And an evaluation phase during which the proposed system was implemented in the home and made decisions. The environment in which the case study was carried out consisted in a 69.98 m^2 flat constructed by BHS, (http://www.barcelonahousingsystems.com/es/) and whose orientation is North. The dependent variables are the objective variables of the system; the system acts on them to obtain energy savings. Independent variables are factors that can have an impact on dependent variables. For the two phases described above, these variables were collected in order to analyse how the system's actions influenced over the value of these variables.

The experiment was divided into 2 phases (baseline, evaluation), each phase lasted one month, allowing us to properly analyse the efficacy of the system (from October 23th to December 23th). The date of the experiment was chosen because in this period the number of holidays per month is similar and the maximum and minimum temperatures barely vary from week to week. In this way, the conditions of the experiment were very similar in the two phases. In Fig. 2 we can see how the sensors used for data collection were distributed in the case study's home. An outdoor temperature sensor was placed at each window of the house (The values collected by all sensors are averaged.), two sensors in each room, except the two bathrooms where only one was placed, a presence sensor in each room except the living room where two sensors were placed.

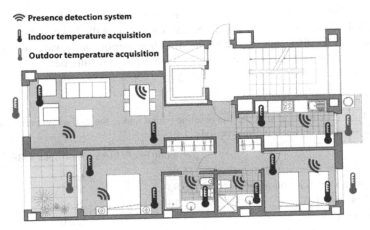

Fig. 2. An example of how sensors were deployed

4.2 Results

The following data were collected by the system every 15 min: indoor (tmp_{indoor}), outdoor ($tmp_{outdoor}$) temperature and presence of people (presence) in the home. This information is managed by the Data acquisition VO. This information is linked to the weather forecast or period type (forecast, time period) and the other variables used by the algorithm, obtained by the Multi-Layer Perceptron (MLP), such as $time_{back}$, $time_{backPeriod}$, $time_{leave}$, $time_{leavePeriod}$. The rest of the information that the algorithm needs is obtained by the Information Management VO. The variables used by the optimization algorithm ($room_{type}$, $thermostat_{status}$, $thermostat_{tmp}$, $time_{now}$, $time_{periodDay}$, tmp_{now}, $tmp_{desired}$) are obtained by the Decision-Making VO thanks to the communication it maintains with User VO agents.

In the baseline period, which lasted one month, the system had only collected household data without taking any action. In the evaluation period, the system not only collected the values of all the variables but also made appropriate decisions on the basis of these values, as can be seen in Table 1.

Table 1. Data collected by the system in each phase

		Case study home
Baseline period	Outdoor avg. temp (°C)	10.54
	Indoor avg. temp (°C)	23.68
	Energy consumption (kWh)	159.64
Evaluation period	Outdoor avg. temp (°C)	6.38
	Indoor avg. temp (°C)	24.71
	Energy consumption (kWh)	192.79

Table 2 shows electrical consumption for each stage (Baseline and evaluation period). Table 3 shows the results of the Student's test t and the Levene test for equality of variances. The difference between the Baseline period and evaluation period is significantly lower with a p-value close to 0.000. These results demonstrate that the algorithm achieves energy savings efficiently.

Table 2. Total consumption in Wh in the baseline period and in the evaluation period, as well as the savings achieved between the two periods.

	Without EnerVMAS	With EnerVMAS
Baseline period consumption (Wh)	15.06	15.04
Evaluation period consumption (Wh)	13.93	11.61
Difference (Wh)	1.13	3.43
Savings (%)	7.5	22.80

Table 3. Result of the Student's t-test and Levene's test performed to assess the difference of means and variances between the baseline data and the evaluation period.

Baseline		Evaluation period		t	p-Value (2-Tailed)	F	p-Value
Mean	Std.	Mean	Std.				
6.2192	0.19155	5.9369	0.23755	0.928	0.257	6.056	0.017

5 Conclusions

This work presented an innovative approach based on VO of agents for the optimization of energy consumption in homes. To reduce the overall amount of energy consumed by a home, our proposal focused on making air conditioning systems more energy efficient. The use of VOs was fundamental in achieving these objectives, since it provided a simple method for monitoring the home air conditioning system. The system was fed back with weather forecast and home occupancy data, a few days ahead in order to optimally configure the temperature of radiators and air-conditioning.

In the conducted case study, it was demonstrated how the EnerVMAS correctly adjusted the temperature of radiators and the air conditioning system, this was done by turning them on/off. It prevented any drastic changes in the temperature of the house, since this would increase electrical consumption significantly. The case study was carried out over two months and it demonstrated that the proposed solution was capable of achieving significant energy savings, with the implementation of the system electrical consumption lowered by 22.8%. These savings could be achieved thanks to the algorithm which had been specifically developed for making temperature optimisation decisions in the air conditioning system.

Acknowledgements. This research has been partially supported by the European Regional Development Fund (FEDER) within the framework of the Interreg program V-A Spain-Portugal 2014-2020 (PocTep) under the IOTEC project grant 0123_IOTEC_3_E and by the Spanish Ministry of Economy, Industry and Competitiveness and the European Social Fund under the ECOCASA project grant RTC-2016-5250-6. The research of Alfonso González-Briones has been co-financed by the European Social Fund (Operational Programme 2014-2020 for Castilla y León, EDU/310/2015 BOCYL).

References

1. Afram, A., Janabi-Sharifi, F.: Theory and applications of HVAC control systems–a review of model predictive control (MPC). Build. Environ. **72**, 343–355 (2014)
2. Al-Daraiseh, A., El-Qawasmeh, E., Shah, N.: Multi-agent system for energy consumption optimisation in higher education institutions. J. Comput. Syst. Sci. **81**(6), 958–965 (2015)
3. Annunziato, M., Pierucci, P.: The emergence of social learning in artificial societies. Appl. Evol. Comput. 293–294 (2003)
4. Bogdan, M.: How to use the DHT22 sensor for measuring temperature and humidity with the arduino board. ACTA Universitatis Cibiniensis **68**(1), 22–25 (2016)
5. Briones, A.G., González, J.R., de Paz Santana, J.F.: A drug identification system for intoxicated drivers based on a systematic review. ADCAIJ Adv. Distrib. Comput. Artif. Intell. J. **4**(4), 83–101 (2015)
6. Briones, A.G., González, J.R., de Paz Santana, J.F., Rodríguez, J.M.C.: Obtaining relevant genes by analysis of expression arrays with a multi-agent system. ADCAIJ Adv. Distrib. Comput. Artif. Intell. J. **3**(3), 35–42 (2014)
7. Cai, J., Kim, D., Putta, V.K., Braun, J.E., Hu, J.: Multi-agent control for centralized air conditioning systems serving multi-zone buildings. In: American Control Conference (ACC), pp. 986–993. IEEE (2015)
8. Casals, X.G.: Analysis of building energy regulation and certification in Europe: their role, limitations and differences. Energy Build. **38**(5), 381–392 (2006)
9. Clausen, A., Demazeau, Y., Jørgensen, B.N.: Load management through agent based coordination of flexible electricity consumers. In: Demazeau, Y., Decker, K.S., Bajo Pérez, J., de la Prieta, F. (eds.) PAAMS 2015. LNCS (LNAI), vol. 9086, pp. 27–39. Springer, Cham (2015). https://doi.org/10.1007/978-3-319-18944-4_3
10. Clausen, A., Umair, A., Demazeau, Y., Jørgensen, B.N.: Agent-based integration of complex and heterogeneous distributed energy resources in virtual power plants. In: Demazeau, Y., Davidsson, P., Bajo, J., Vale, Z. (eds.) PAAMS 2017. LNCS (LNAI), vol. 10349, pp. 43–55. Springer, Cham (2017). https://doi.org/10.1007/978-3-319-59930-4_4
11. Corchado, J.M., Laza, R.: Constructing deliberative agents with case-based reasoning technology. Int. J. Intell. Syst. **18**(12), 1227–1241 (2003)
12. Dodier, R.H., Henze, G.P., Tiller, D.K., Guo, X.: Building occupancy detection through sensor belief networks. Energy Build. **38**(9), 1033–1043 (2006)
13. Dong, B., Lam, K.P.: Building energy and comfort management through occupant behaviour pattern detection based on a large-scale environmental sensor network. J. Build. Perform. Simul. **4**(4), 359–369 (2011)
14. FacilitiesNet: Building Owners & Facility Executives - Building Operating Management - Facilities Management Magazine (2011). http://www.facilitiesnet.com/bom/. Accessed 12 Jan 2018

15. Faia, R., Pinto, T., Abrishambaf, O., Fernandes, F., Vale, Z., Corchado, J.M.: Case based reasoning with expert system and swarm intelligence to determine energy reduction in buildings energy management. Energy Build. **155**, 269–281 (2017)
16. González Briones, A., Chamoso, P., Barriuso, A.L.: Review of the main security problems with multi-agent systems used in E-commerce applications (2016)
17. Guerrisi, A., Martino, M., Tartaglia, M.: Energy saving in social housing: an innovative ICT service to improve the occupant behaviour. In: International Conference on Renewable Energy Research and Applications (ICRERA), pp. 1–6. IEEE (2012)
18. Harris, C., Cahill, V.: Exploiting user behaviour for context-aware power management. In: IEEE International Conference on Wireless and Mobile Computing, Networking and Communications (WiMob 2005), vol. 4, pp. 122–130. IEEE (2005)
19. Herkel, S., Knapp, U., Pfafferott, J.: A preliminary model of user behaviour regarding the manual control of windows in office buildings. In: Proceedings of the 9th International IBPSA Conference BS 2005, Montréal, pp. 403–410 (2005)
20. Hoes, P., Hensen, J.L.M., Loomans, M.G.L.C., De Vries, B., Bourgeois, D.: User behavior in whole building simulation. Energy Build. **41**(3), 295–302 (2009)
21. Li, Y.C., Hong, S.H.: BACnet–EnOcean smart grid gateway and its application to demand response in buildings. Energy Build. **78**, 183–191 (2014)
22. McArthur, S.D.J., Davidson, E.M., Catterson, V.M., Dimeas, A.L., Hatziargyriou, N.D., Ponci, F., Funabashi, T.: Multi-agent systems. IEEE Trans. Power Syst. **22**(4), 1753–1759 (2007)
23. Moghavvemi, M., Seng, L.C.: Pyroelectric infrared sensor for intruder detection. In: TENCON 2004. 2004 IEEE Region 10 Conference, vol. 500, pp. 656–659. IEEE (2004)
24. Mozer, M.C.: The neural network house: an environment hat adapts to its inhabitants. In: Proceedings of the AAAI Spring Symposium on Intelligent Environments, vol. 58 (1998)
25. Ramos, J., Castellanos-Garzón, J.A., González-Briones, A., de Paz, J.F., Corchado, J.M.: An agent-based clustering approach for gene selection in gene expression microarray. Interdiscip. Sci. Comput. Life Sci. **9**(1), 1–13 (2017)
26. Recio-García, J.A., González-Calero, P.A., Díaz-Agudo, B.: jcolibri2: a framework for building case-based reasoning systems. Sci. Comput. Program. **79**, 126–145 (2014)
27. Samad, T., Frank, B.: Leveraging the web: a universal framework for building automation. In: American Control Conference, ACC 2007, pp. 4382–4387. IEEE (2007)
28. Wang, Z., Yang, R., Wang, L.: Multi-agent control system with intelligent optimization for smart and energy-efficient buildings. In: IECON 2010-36th Annual Conference on IEEE Industrial Electronics Society, pp. 1144–1149. IEEE (2010)
29. Zato, C., Villarrubia, G., Sánchez, A., Bajo, J., Corchado, J.M.: PANGEA: a new platform for developing virtual organizations of agents. Int. J. Artif. Intell. **11**(A13), 93–102 (2013)
30. González-Briones, A., Prieto, J., De La Prieta, F., Herrera-Viedma, E., Corchado, J.M.: Energy optimization using a case-based reasoning strategy. Sensors **18**(3), 865 (2018)
31. González-Briones, A., Chamoso, P., Yoe, H., Corchado, J.M.: GreenVMAS: virtual organization based platform for heating greenhouses using waste energy from power plants. Sensors **18**(3), 861 (2018)

Tool Wear Estimation and Visualization Using Image Sensors in Micro Milling Manufacturing

Laura Fernández-Robles[✉], Noelia Charro, Lidia Sánchez-González,
Hilde Pérez, Manuel Castejón-Limas, and Javier Alfonso-Cendón

Departamento de Ingenierías Mecánica, Informática y Aeroespacial,
Universidad de León, León, Spain
{l.fernandez,lidia.sanchez,hilde.perez,manuel.castejon,
javier.alfonso}@unileon.es, ncharg00@estudiantes.unileon.es

Abstract. This paper presents a reliable machine vision system to automatically estimate and visualize tool wear in micro milling manufacturing. The estimation of tool wear is very important for tool monitoring systems and image sensors configure a cheap and reliable solution. This system provides information to decide whether a tool should be replaced so the quality of the machined piece is ensured and the tool does not collapse. In the method that we propose, we first delimit the area of interest of the micro milling tool and then we delimit the worn area. The worn area is visualized and estimated while errors are computed against the ground truth proposed by experts. The method is mainly based on morphological operations and k-means algorithm. Other approaches based on pure morphological operations and on Otsu multi threshold algorithms were also tested. The obtained result (a harmonic mean of precision and recall 90.24 (±2.78)%) shows that the machine vision system that we present is effective and suitable for the estimation and visualization of tool wear in micro milling machines and ready to be installed in an on-line system.

Keywords: Tool wear · Micro milling · Wear estimation
Wear visualization

1 Introduction

In recent years, manufacturing systems have evolved dramatically from a system based on the operator to the automation of the entire process. That industry 4.0 with smart factories is the desire of the existing companies in order to succeed in the market. To deal with that automation, different efforts have been made: new production procedures, resource optimization, monitoring of the whole process and use of cyber-physical systems, among others.

© Springer International Publishing AG, part of Springer Nature 2018
F. J. de Cos Juez et al. (Eds.): HAIS 2018, LNAI 10870, pp. 399–410, 2018.
https://doi.org/10.1007/978-3-319-92639-1_33

In high precision machining, i.e. micro milling, these cyber-physical systems need to include smart cutting tools and smart machining to cope with machining dynamics, process variations and complexity [4]. Tool wear influences significantly on the cutting forces and on the machining outcomes.

Force, vibration and machine sound sensors [1] as well as acoustic emission signals [18] are used for tool wear monitoring. Moreover, image sensors to acquire digital images from the tool to analyze them to predict tool wear are also employed [13]. Those images need to be processed to analyze them automatically using active contours [16], linear regression learning [11], neural networks [6], Hough transform to detect edges [7] or COSFIRE filters [8].

In this paper we propose a method to detect wear in micro milling tools by using digital image processing techniques. This aspect is really important to optimize resources in manufacturing systems since affects the outcomes and is related to the cutting forces [4]. Micro milling also requires high precision techniques to satisfy the demand at this micrometer level [3]. To deal with that, there are authors who estimate the wear area by using a threshold processing based on Morphological Component Analysis and by computing a rotation invariant feature of the wear area [19]. Other works [17] are based on the extraction of a set of wear features such as the width of the crater wear, the depth of the crater or the distance between the center of the crater and the cutting edge among others, and using a fuzzy statistical method to analyze the feature vector.

We present a method that evaluates tool wear at the resting position of a micro milling machine using image sensors and processing techniques. It automatically determines the worn area of the tool and provides an easy visualization of it. We present three different approaches to solve this issue and evaluate and compare them. We apply morphological operations, k-means [14] and Otsu multi level thresholding [15] algorithms, among others. Besides, the system does not need image references to compare to, so from one single image is able to determine the worn area of the tool.

The paper is organized as follows. Section 2 describes the method proposed in this paper. Experimental details are presented in Sect. 3 and discussed in Sect. 4. Finally, conclusions are gathered in Sect. 5.

2   Method

In this section, we describe the supervised machine vision method for estimating and visualizing tool wear of a milling tool. The method is two-fold and a schema is shown in Fig. 1. First the area of interest is isolated. We consider the area of interest as the detachment surface and, therefore, it can be damaged. Then, the worn area is delimited. The estimation of the worn area is done using three different approaches: (i) purely based on morphological operations; (ii) using k-means and morphological operations; and (iii) using Otsu multi level thresholding and morphological operations. Even if the steps that comprise the pipeline are well-known, the pipeline itself is an intelligent-based method that can learn, by means of a supervised training, specifically for the application at hand.

2.1   Delimitation of the Area of Interest

We define the area of interest as shown in Fig. 2. This area is delimited by the main line, which is the cutting edge of the milling tool, the secondary line, which is due to the shape of the milling tool, and the contour of the tool. Generally, a milling tool gets worn in this area earlier than in others and that is why it is so important to control the wear in such region.

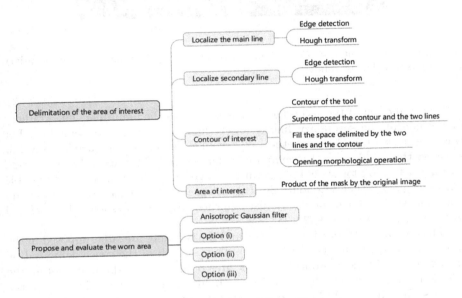

Fig. 1. Illustration of the method.

In order to localize the main line, we basically detect edges in the image and apply Hough transform [10]. Specifically, we use Sobel filter [5] for edge detection. Figure 3(a) shows an image sample of a milling tool after manufacturing 4 meters. Figure 3(b) shows the edges detected using Sobel filter for the same image. We also use Sobel edge map to define the contour of the milling tool. Then we apply Hough transform to the binary edge map in order to look for straight lines. Since the position of the milling tool and acquisition system do not vary, we can delimit the possible degrees of the line. Experimentally in the training set, we define the range of possible angles $[a_{1,1}, a_{1,2}]$, considering 0 degrees in the vertical and measuring in clockwise direction. We take the highest peak of the Hough transform as the main line, which is defined by the parameters ρ and θ in the parametric representation of a line, $\rho = x \cos(\theta) + y \sin(\theta)$. ρ is the distance from the origin to the line along a vector perpendicular to the line, and θ is the angle in degrees between the x-axis and this vector. The automatically detected main line for the image sample for angles [23, 29] appears overlapped to the edge map in Fig. 3(c).

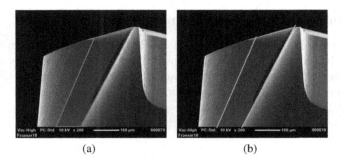

(a)   (b)

Fig. 2. (a) Area of interest. (b) The main line is marked in orange and the secondary line in green (color fig online).

The detection of the secondary line is similar to the main line. The gradient magnitude of the secondary line is lower than the one in the main line. For this reason, we use Canny method [2] for edge detection with a relatively low sensitive threshold, t_1, that is computed when training the system. We use Hough transform to look for straight lines in the edge map with a new set of possible angles $[a_{2,1}, a_{2,2}]$. The secondary lines always present higher slopes than the main lines and therefore can be easily distinguished. Figure 3(d) shows the detected secondary lines for angles [22, 23].

The contour of the tool using Sobel and the two lines are superimposed in the same binary image, as seen in Fig. 3(e). Then, we fill in the space delimited by the two lines and the contour, this is the area of interest. To do so, we obtain the middle point of the main line as reference point to indicate that the area on the left of such point should be filled in. The red cross in Fig. 3(e) indicates the left pixel neighbor of the middle point in the main line. Figure 3(f) shows the result of filling in the area of interest. We utilize an opening morphological operation to get rid of the pixels outside the area of interest. It consists of the dilation of the erosion of the image by a disk-shaped structuring element with radius 10. The result of this operation is the extraction of the mask that defines the area of interest. Figure 3(g) shows such mask for the example considered. The Hadamard product of the mask by the original image produces the an image in which the area of interest is visible and the background is black, see Fig. 3(h).

2.2 Delimitation of the Worn Area

The detection of the worn area is proposed and evaluated using three different approaches. Before, we pre-process the image to suppress horizontal features in the images. We use an Anisotropic Gaussian filter [9] with higher standard deviation along the x axis. Figure 3(i) shows the result of applying the Anisotropic Gaussian filter to the image example. Once the image is pre-processed, we define the three different approaches as follows:

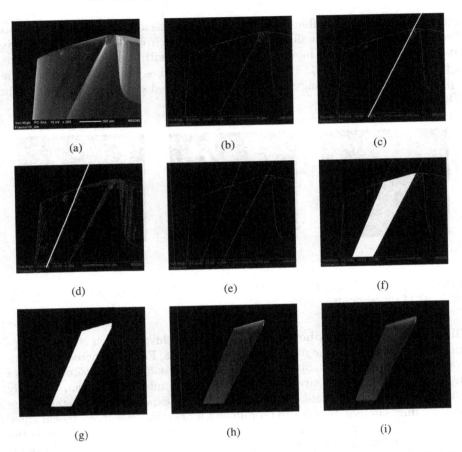

Fig. 3. (a) Micro milling tool after machining 1 meter. (b) Binary edge map with Sobel operator. (c) Blue line indicates the main line. (d) Blue line indicates the secondary line. (e) Main and secondary lines superimposed to the contour of the tool. The red cross indicates the pixel of reference to fill in the area of interest. (f) Filled in of the area of interest. (g) Mask of the area of interest. (h) Area of interest. (i) Area of interest after Anisotropic Gaussian filtering (color fig online).

(i) Morphological Operations. This approach is based on the use of morphological operations. First, we obtain the binary edge map of the image using Canny algorithm. Then we fill small gaps in the image using filledgegaps method [12]. We use it in an iterative way, increasing the edge gap size to fill in each iteration in order to not produce great deformations on the shape of the original edge map. In these experiments, edge gap sizes of 21, 31 and 41 were selected –only odd numbers are valid for the method–. Figure 4(a) shows the image sample after filling gaps. After, we fill in the worn area which is placed at the left of the middle point of the main line, see Fig. 4(b). Then we extract the largest connected component (object) from the binary image with a connectivity of 4 pixels. In order to get a uniform mask, we fill holes, considering a hole as a set of background pixels that cannot be reached by filling in the background from the edge of the image, and perform an opening morphological operation with a disk-shaped structuring element with radius 3. The resulting mask defines the worn area using the approach (i). The resulting masks on the considered examples can be seen in Fig. 4(c).

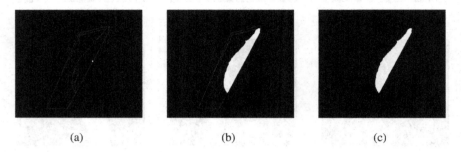

(a)   (b)   (c)

Fig. 4. Approach (i): (a) Image sample after filling gaps. (b) Filled in worn area. (c) Mask of the worn area.

(ii) k-means and morphological operations This approach is based on k-means algorithm and some morphological operations. First, we increase the contrast of the images by mapping the intensity values such that 1% of the data in the original images is saturated at low and high intensities. Then, we apply k-means with $k = 4$ clusters. We chose $k = 4$ at an experimentation stage using only training samples. In order to reproduce a more stable result, we repeat three times the clustering using new initial cluster centroid positions and keep the solution with the minimum within-cluster sums of point-to-centroid distances. Figure 5(a) shows the k-means labels obtained for the sample images. We select the pixels of the k-means algorithm map that share the same label as the label of the left neighbor pixel to the middle pixel in the main line, see Fig. 5(b). Then, we perform an opening morphological operation with a disk-shaped structuring element with radius 3. And we extract the largest connected component from the binary image with a connectivity of 4 pixels. Finally, we fill holes of the binary image and obtain the resulting mask that defines the worn area using the approach (ii). The masks of the worn area using approach (ii) are shown in Fig. 5(c).

(iii) Otsu Multi Level Thresholding and Morphological Operations. This approach is based on Otsu multi level thresholding algorithm and some morphological operations. First, we increase the contrast of the images as in approach (ii). Then, we apply Otsu multi level thresholding algorithm using 3 threshold values, therefore, obtaining 4 discrete levels in the image. Similarly to k-means, we obtain an image map with pixels being clustered in 4 groups. Figure 6(a) shows the Otsu multi level labels obtained for the sample images.

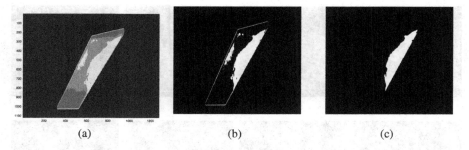

(a)   (b)   (c)

Fig. 5. Approach (ii): (a) k-means map. (b) Pixels belonging to the same label as the worn area. (c) Mask of the worn area.

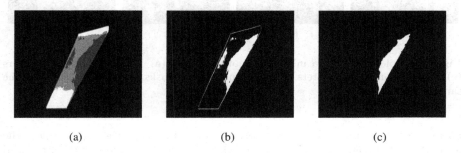

(a)   (b)   (c)

Fig. 6. Approach (iii): (a) Otsu multi level thresholding map. (b) Pixels belonging to the same label as the worn area. (c) Mask of the worn area.

We select the pixels of the Otsu multi level map that share the same label as the label of the left neighbor pixel to the middle pixel in the main line, see Fig. 6(b). The rest of the approach consists of the same steps as in the approach (ii): opening morphological operation, extraction of the largest connected component and fill holes. The masks of the worn area using approach (iii) are shown in Fig. 6(c).

3   Experiments

3.1   Dataset

Images of a micro milling tool are used in this work. The tool is symmetric and contains two cutting edges. The milling tool was in its vertical position and images were taken while the tool was in the resting position. The monochrome camera is placed in the same horizontal plane as the tool, so images similar to the ones shown in Fig. 7 are acquired. The milling tool is metallic and red LED lights are used to avoid shines. The images were taken after 1, 2, 3 and 4 meters of milling with different micro milling tools. The micro milling machine does not use lubricants, oils or other sources that may cause a filthy tool. Therefore, marks in the tool are caused by wear and images are clean. Together with the

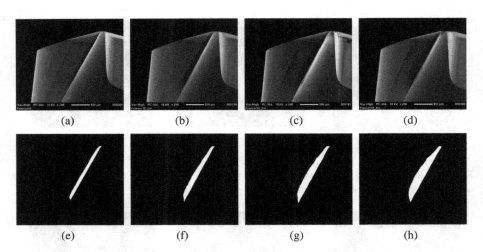

Fig. 7. Sample images of a milling tool after milling (a) 1 meter and (b) 2 meters (c) 3 meters and (d) 4 meters. (e–h) Ground truth masks of the worn areas for the sample images in (a–d).

dataset, experts manually created ground truth masks of the worn areas. The ground truth images are only used to check the performance of the proposed method but not for training the system. Sample images for 1, 2, 3 and 4 meters and their corresponding ground truth images are shown in Fig. 7.

3.2   Experimental Setup

We perform a hold-out validation in which half of the dataset was used for training and the rest for testing. In the training phase, the values of the parameters were set up. The proposed method with the parameters obtained in the training phase is applied to the test images and results are computed. We computed four error metrics:

Area error is defined as the rate of difference area between the ground truth and the automatically defined areas with respect to the ground truth area, as defined in Eq. 1.

$$\text{Area error} = \frac{\text{actual area} - \text{predicted area}}{\text{actual area}} \tag{1}$$

We refer to worn pixels as pixels that are identified by the method to belong to the worn area and we consider them as the positive class. Therefore, we define a true positive (TP) as a pixel that belongs to the worn area and is identified as worn by the method; a false positive (FP) as a pixel that does not belong to the worn area and is identified as worn; and a false negative (FN) as a pixel that belongs to the worn area and it is not identified as worn.

We define the precision as the sum of pixels in the intersection area between the ground truth and automatically detected worn areas, and it is defined in Eq. 2.

$$\text{Precision} = \frac{\text{TP}}{\text{TP} + \text{FP}} \qquad (2)$$

We define the recall as the sum of pixels in the intersection area between the inverse ground truth and automatically detected worn areas, and it is defined in Eq. 3.

$$\text{Recall} = \frac{\text{TP}}{\text{TP} + \text{FN}} \qquad (3)$$

We define the F-score as the harmonic mean of the precision and recall, Eq. 4.

$$\text{F-score}(\%) = \frac{2 \cdot \text{Precision} \cdot \text{Recall}}{\text{Precision} + \text{Recall}} \qquad (4)$$

4 Results

We present the results of the three approaches of our method for the training set in Fig. 8(a) and Table 1 and for the test set in Fig. 8(b) and Table 2. Figure 9 shows the visual results for two sample images.

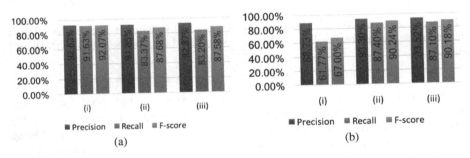

Fig. 8. Results for the (a) training set and (b) test set.

Table 1. Average results and standard deviations, shown as percentage, for the training images. Best results are shown in bold.

Approach	Area error	Recall	Precision	F-score
(i)	**6.29 ± 6.48**	**92.83 ± 2.95**	**91.63 ± 11.45**	**92.07 ± 7.45**
(ii)	10.50 ± 9.56	91.85 ± 3.86	83.37 ± 11.84	87.68 ± 8.49
(iii)	10.69 ± 9.60	92.87 ± 3.79	83.20 ± 11.89	87.58 ± 8.53

Table 2. Average results and standard deviations, shown as percentage, for the test images. Best results are shown in bold.

Approach	Area error	Recall	Precision	F-score
(i)	34.47 ± 41.36	88.73 ± 10.11	61.77 ± 40.60	67.00 ± 39.09
(ii)	**6.36 ± 2.47**	93.30 ± 1.64	**87.40 ± 3.76**	**90.24 ± 2.78**
(iii)	6.89 ± 2.26	**93.52 ± 1.57**	87.10 ± 3.54	90.18 ± 2.64

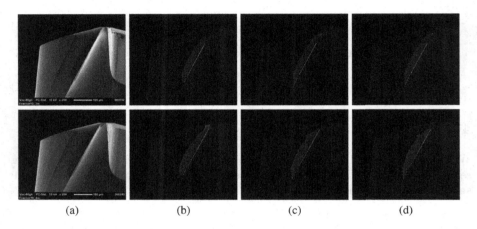

    (a)   (b)   (c)   (d)

Fig. 9. Visual results for two images of the dataset. (a) Original image. In the original image, areas out of the resulting masks are attenuated for approaches (i) (b), (ii) (c) and (iii) (d).

Approach (i) obtains very good results for training but poor for testing. Approaches (ii) and (iii) show highly similar results both in training and testing sets, which leads to methods that generalize well for a wide range of machining conditions. We achieved an area error of 6.36 (±2.47)% and a harmonic mean of precision and recall equals to 90.24 (±2.78)%, with precision (87.40 ± 3.76%) and recall (93.30 ± 1.64%). This is a very satisfactory result since experts find very difficult to determine the exact ground truth areas to pixel precision.

Approach (i) is hand crafted and lacks generalization abilities, it is quite sensitive to machining conditions and it is hard to find parameters that suit for different kind of micro milling processes. On the contrary, approach (ii) and (iii) are based on learning which provides better generalization abilities. Approach (iii) is computationally faster than approach (ii).

5   Conclusions

In this paper we presented a method for the estimation and visualization of tool wear in micro milling manufacturing. This method relies on machine vision techniques, in particular, we proposed three approaches based on: (i) morphological operations; (ii) k-means algorithm and morphological operations; and (iii) Otsu multi level thresholding algorithm and morphological operations. Approaches (ii) and (iii) yielded satisfactory results for the estimation and visualization of tool wear. The presented system can be set up on-line and it can be applied while the milling head tool is in a resting position. This system helps to model and visualize tool wear in order to automatically decide the replacement of the tool, which reduces the risk of tool collapse and assures a good quality of the machined pieces. In future, we will evaluate other methods, such as DBSCAN and Gaussian Mixture Expectation-Maximization. Morevoer, more views of the tool will be analyzed using several vision sensors such as endoscopy systems.

Acknowledgements. We gratefully acknowledge the financial support of Spanish Ministry of Economy, Industry and Competitiveness, through grant DPI2016-79960-C3-2-P.

References

1. Aliustaoglu, C., Ertunc, H.M., Ocak, H.: Tool wear condition monitoring using a sensor fusion model based on fuzzy inference system. Mech. Syst. Signal Process. **23**(2), 539–546 (2009). http://www.sciencedirect.com/science/article/pii/S0888327008000642
2. Canny, A.: A computational approach to edge detection. IEEE Trans. Pattern Anal. Mach. Intell. PAMI **8**(6), 679–698 (1986)
3. Cheng, K., Huo, D.H.: Micro Cutting: Fundamentals and Applications. Wiley, Chichester (2013)
4. Cheng, K., Niu, Z.C., Wang, R.C., Rakowski, R., Bateman, R.: Smart cutting tools and smart machining: development approaches, and their implementation and application perspectives. Chin. J. Mech. Eng. **30**(5), 1162–1176 (2017). https://link.springer.com/article/10.1007/s10033-017-0183-4
5. Danielsson, P.E., Seger, O.: Generalized and separable sobel operators. In: Freeman, H. (ed.) Machine Vision for Three-Dimensional Scenes, pp. 347–379. Academic Press (1990). https://www.sciencedirect.com/science/article/pii/B9780122667220500166
6. D'Addona, D., Teti, R.: Image data processing via neural networks for tool wear prediction. Procedia CIRP **12**, 252–257 (2013). http://www.sciencedirect.com/science/article/pii/S2212827113006859. Eighth CIRP Conference on Intelligent Computation in Manufacturing Engineering
7. Fernández-Robles, L., Azzopardi, G., Alegre, E., Petkov, N.: Machine-vision-based identification of broken inserts in edge profile milling heads. Rob. Comput. Integr. Manuf. **44**, 276–283 (2017). http://www.sciencedirect.com/science/article/pii/S0736584515300806
8. Fernández-Robles, L., Azzopardi, G., Alegre, E., Petkov, N., Castejón-Limas, M.: Identification of milling inserts in situ based on a versatile machine vision system. J. Manuf. Syst. **45**, 48–57 (2017). http://www.sciencedirect.com/science/article/pii/S0278612517301231
9. Geusebroek, J.M., Smeulders, A.W.M., van de Weijer, J.: Fast anisotropic gauss filtering. IEEE Trans. Image Process. **12**(8), 938–943 (2003)
10. Hough, P.: Method and Means for Recognizing Complex Patterns. U.S. Patent 3.069.654, December 1962

11. Karuppusamy, N.S., Pal Pandian, P., Lee, H.S., Kang, B.Y.: Tool wear and tool life estimation based on linear regression learning. In: 2015 IEEE International Conference on Mechatronics and Automation (ICMA), pp. 17–21, August 2015
12. Kovesi, P.: Fast almost-gaussian filtering. In: 2010 International Conference on Digital Image Computing: Techniques and Applications, pp. 121–125, December 2010
13. Leem, C.S., Dornfeld, D.A.: Design and implementation of sensor-based tool-wear monitoring systems. Mech. Syst. Signal Process. **10**(4), 439–458 (1996). http://www.sciencedirect.com/science/article/pii/S088832709690031X
14. Lloyd, S.: Least squares quantization in PCM. IEEE Trans. Inf. Theory **28**(2), 129–137 (1982)
15. Otsu, N.: A threshold selection method from gray-level histograms. IEEE Trans. Syst. Man Cybern. **9**(1), 62–66 (1979)
16. Schmitt, R., Cai, Y., Pavim, A.: Machine vision system for inspecting flank wear on cutting tools. ACEEE Int. J. Control Syst. Instrum. **3**, 27–31 (2012)
17. Yang, Z., Liu, L., Peng, K., Li, S., Zhang, J., Gai, L.: Monitoring method of high-speed tool wear level based on machine vision. Int. J. Signal Process. Image Process. Pattern Recogn. **10**(6), 23–38 (2017)
18. Yen, C.L., Lu, M.C., Chen, J.L.: Applying the self-organization feature map (SOM) algorithm to AE-based tool wear monitoring in micro-cutting. Mech. Syst. Signal Process. **34**(1), 353–366 (2013). http://www.sciencedirect.com/science/article/pii/S0888327012001859
19. Zhu, K., Yu, X.: The monitoring of micro milling tool wear conditions by wear area estimation. Mech. Syst. Signal Process. **93**, 80–91 (2017)

Compensating Atmospheric Turbulence with Convolutional Neural Networks for Defocused Pupil Image Wave-Front Sensors

Sergio Luis Suárez Gómez[1(✉)], Carlos González-Gutiérrez[2],
Enrique Díez Alonso[2], Jesús Daniel Santos Rodríguez[1],
Laura Bonavera[1], Juan José Fernández Valdivia[3],
José Manuel Rodríguez Ramos[3],
and Luis Fernando Rodríguez Ramos[4]

[1] Department of Physics, University of Oviedo, Oviedo, Spain
suarezsergio@uniovi.es
[2] Prospecting and Exploitation of Mines Department, University of Oviedo,
Oviedo, Spain
[3] Wooptix S.L., San Cristóbal de La Laguna, Spain
[4] Instituto de Astrofísica de Canarias, San Cristóbal de La Laguna, Spain

Abstract. Adaptive optics are techniques used for processing the spatial resolution of astronomical images taken from large ground-based telescopes. In this work are presented computational results from a modified curvature sensor, the Tomographic Pupil Image Wave-front Sensor (TPI-WFS), which measures the turbulence of the atmosphere, expressed in terms of an expansion over Zernike polynomials.

Convolutional Neural Networks (CNN) are presented as an alternative to the TPI-WFS reconstruction. This technique is a machine learning model of the family of artificial neural networks, which are widely known for its performance as modeling and prediction technique in complex systems. Results obtained from the reconstruction of the networks are compared with the TPI-WFS reconstruction by estimating errors and optical measurements (root mean square error, mean structural similarity and Strehl ratio).

Two different scenarios are set, attending to different resolutions for the reconstruction. The reconstructed wave-fronts from both techniques are compared for wave-fronts of 25 Zernike modes and 153 Zernike modes. In general, CNN trained as reconstructor showed better performance than the reconstruction in TPI-WFS for most of the turbulent profiles, but the most significant improvements were found for higher turbulent profiles that have the lowest r0 values.

Keywords: Adaptive optics · TPI-WFS · Convolutional Neural Networks

© Springer International Publishing AG, part of Springer Nature 2018
F. J. de Cos Juez et al. (Eds.): HAIS 2018, LNAI 10870, pp. 411–421, 2018.
https://doi.org/10.1007/978-3-319-92639-1_34

1 Introduction

In the field of grounded astronomical observation, there are two well-known techniques that lead to diffraction-limited imaging. Lucky Imaging (LI) [1, 2] allows reaching the limits of spatial resolution in the visible band in small ground telescopes [3]. One of the essential disadvantages of this method is the resolution limitations, as well as that most of the images are discarded, and consequently only relatively bright targets can be observed.

Adaptive Optics (AO) is the other main technique to improve the spatial resolution of large ground-based telescopes [4]. It has demonstrated excellent results on offering diffraction limited images in the near infrared, due to the minor effects of turbulence in this range.

In the optical bands AO systems on duty today are able to achieve high success in correcting the incoming wave-front errors at high degrees by using Shack-Hartmann Wave-front Sensor (SH-WFS) sensors, but they require very bright, and hence scarce, reference stars or a laser non-natural star [5].

As an alternative, a previous work has proposed a modified curvature sensor, the Tomographic Pupil Image Wave-front Sensor (TPI-WFS), to be used instead of a classical SH-WFS [6].

Artificial Neural Networks (ANN) are a type of machine learning technique, which are known for its capacity of modeling complex systems [7, 8]. In the field of AO, techniques as ANNs have been proven to be very useful with the use of SH-WFS as, for example, the Complex Atmospheric Reconstructor based on Machine lEarNing (CARMEN) [9]. This algorithm use information from the SH-WFS to estimate the atmospheric turbulence [10]. The improvements and characteristics of this artificial intelligence approach have been widely studied AO reconstruction [11, 12].

Convolutional Neural Networks (CNNs) are a machine learning technique of the family of ANNs, which are widely used in image recognition, language processing, etc., achieving great success [13, 14]. One of the best advantages that CNN provides is that it allows full images as inputs.

This work presents a comparison between a real time TPI-WFS restoration and a CNN reconstruction. The comparison includes different scenarios, regarding the resolution and capability of reconstruction of the deformable mirrors included in the AO systems.

The paper will enclose the following contents. In Sect. 2, an explanation about the techniques is presented. These techniques include an explanation about conventional AO, the TPI-WFS sensor and CNNs. In Sect. 3, the results of the performance of both methods of reconstruction are presented. These results are later compared in Sect. 4, along with the discussion and explanation of the behavior of the models. Finally, in Sect. 5 the conclusions of the work are presented, followed by some insights on the possible future lines that this work leads to.

2   Methods

2.1   Adaptive Optics

The technique of AO is essential for astronomical observation performed with grounded telescopes for visible wavelengths, since all the possible images are distorted to a certain degree due to the aberrations produced by the atmosphere in the light that passes through it. The aim of AO is to estimate the turbulences, compute the corrections that are needed in the image and calculate the position that a deformable mirror has to adopt for compensating the aberrations in the wave-front as fast as possible due to the extremely changing nature of the atmosphere [15].

The SH is used in astronomy to characterize an incoming wave-front. It consists of an array of lenses with the same focal length (called lenslets) each focused on a photon sensor. With this data, the aberration induced in the wave-front by atmospheric turbulence can be approximated in terms of Zernike Polynomials [16]. Reconstructors based on artificial intelligence techniques for Multi-object adaptive optics (MOAO) [17] showed promising results in AO field [9], both in simulation [18] and on-sky [11].

2.2   TPI-WFS

In contrast to SH-WFS sensors, the TPI-WFS sensor was developed [6] as an alternative to SH-WFS, where the photons from the incoming light are distributed amongst all the illuminated lenslets. This is a disadvantage since it sets a limitation for the reference stars magnitude and the correspondent wave-front reconstructions.

The main function of the TPI-WFS optics is to obtain two defocused images near the pupil plane, which are later acquired by the WFS camera and finally processed by the Wave-Front Reconstruction (WFR) software [6, 19]. The measurements obtained by the sensor are two defocused pupil images taken at two different planes, with the aim of obtaining values for the posterior calculations as the intensity of these images.

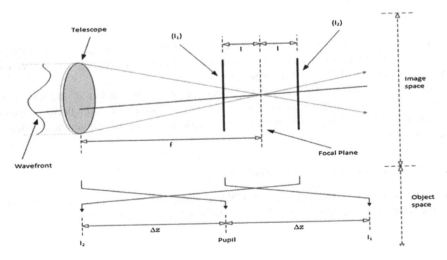

Fig. 1. Conceptual diagram of a curvature sensor. Design from [5].

Up to this day, this method presented reliable results with the implementation of computer simulations when measuring wave-fronts, obtaining reconstructions with down to 100 photons falling within each pupil image [19]. The WFR software is founded on the algorithm proposed by [20]. It is expected to perform a considerable improvement in the sensitivity when compared to the SH-WFS.

2.3   Convolutional Neural Networks

Artificial Neural Networks are a type of mathematical models that took as inspiration the biological archetype of neural networks as base. CNN models provide versatility to the systems of study and better performance in most scenarios such as document recognition [21], image classification [22] or speech recognition [23].

After the convolution, an activation function is applied, the most usual type is the Rectified Linear Unit, known as ReLU [24]. Frequently, the outputs of these layers are post-processed with a pooling layer [14], which reduces the size of the resultant images, by extracting the maximum or mean value from a certain region of pixels (Fig. 2).

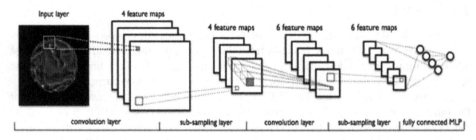

Fig. 2. Example of the topology and implementation of a convolutional neural network. After the sequences of convolution and sub-sampling layers, the output feature maps are connected to a multi-layer perceptron.

Training process for CNNs involves the adjustment of the weights in the connections between the neurons of adjacent layers of the Multi-Layer Perceptron, as well as the weights of the filters from the convolutional layers.

In this work we used 2 networks, depending on the resolution of the case considered, with 6 convolutional layers for the higher resolution case and a network with 4 convolutional layers for the lower resolution. These networks are explained in detail in the following section.

2.4   Simulations and Network Training

In this work are presented two different CNN models, and consequently the simulations for the training are adapted for improving the learning in each case.

In the first case, as the original TPI-WFS was designed for an AO system that had deformable mirrors with resolution to correct up to the first 25 Zernike modes, the simulations were performed to represent a telescope with a deformable mirror that could perform the correction of wave-fronts with the first 25 Zernike modes. Then, for the training of the first network, phases of 25 Zernike modes, with turbulence simulations with r0 ranging from 5 cm to 20 cm, and wavelength of 590 nm.

Also, the original simulated phase, or reference phase, was obtained to perform a later comparison with the phases obtained by both TPI-WFS and CNN reconstructors. Due to the resolution of the deformable mirror as limitation, in this case, the comparison considers the recovered phases using 25 Zernike modes with both reconstructions and the reference phase which contains 153 Zernike modes.

Specifically, the chosen network used for training 1500000 images of 56 pixels of side as inputs. These images had 2 channels, corresponding with the I1 and I2 presented in Fig. 1. The network had 4 convolutional layers, each of them with four filters of size 3×3, shifting 1 pixel horizontally in each step, and then, 1 vertically. In order to do this along all the image, 1 zero was padded outside of the edges of each image. After each convolutional layer, ReLU is applied. Also, after the second convolutional layer, Max-Pooling, a sample-based discretization process is applied in sections of 4×4, and after the fourth in sections of 7×7.

The number of images increased in 4 times for each convolutional layer (due to the 4 filters in each layer), leading to 512 images of size 2×2 at the end of the convolutions. This represents 2048 values which are the input values of the neurons that conformed the fully connected layers. Each of these input neurons were paired all the 2048 neurons set to conform the hidden layer. At last, these were connected with the 25 output neurons, which corresponds with the 25 desired estimations of the Zernike coefficients. Also, another set of 100000 different images was used for testing in the training process, and 80000 different images as validation set.

The second case, is simulated for a situation where the deformable mirror has higher resolution and it is able to correct wave-fronts modelled with Zernike representation up to 153 coefficients. For the second network, the simulations include values of r0 ranging from 5 cm to 20 cm for wave-fronts up to 153 Zernike modes. The other specifications of the simulations are the same as the previous case.

The chosen network for the higher resolution case had 6 convolutional layers with 5×5 kernels, followed by the application of ReLU. Max-Pooling of 2×2 is computed each 2 convolutional layers, resulting on 128 images of 7 pixels of size. These are the inputs of the fully connected layers, with 6272 neurons in the input layer, and a hidden layer of 3136 neurons. The training was performed with sets of the same size as in the 25 modes case.

3 Results and Discussion

Results obtained from the reconstruction of the CNN can be compared with the TPI-WFS reconstruction by means of optical measurements, such as Root Mean Square (RMS) error, Mean Structural Similarity (MSSIM) and Strehl ratio.

The RMS error measures the differences in the raw values of the pixels in the image; it is calculated with each of the recovered phases compared with a reference phase. MSSIM measures the image quality of the reconstructions obtained from both methods; its values ranging from 0 (totally dissimilar) to 1 (totally similar). Finally, the Strehl measures the image quality comparing peak intensities of the images, with values ranging from 0 to 1, with 1 corresponding to an unaberrated image. In the further subsections, there are disclosed the different cases considered.

3.1   Comparison and Results for the 25 Zernike Modes Case

As it can be seen in Fig. 3, the RMS error decreases in the less turbulent scenarios for both methods, as the value of r0 increases. In these terms, for the most turbulent scenarios, the CNN improves significantly the reconstruction.

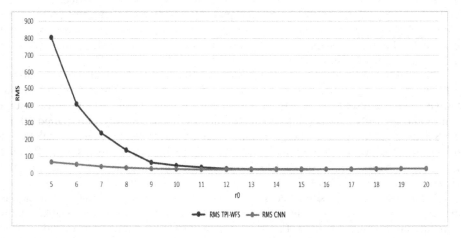

Fig. 3. RMS error of the difference between a recovered phase image and the reference one with the CNN reconstruction and the TPI-WFS reconstruction for 25 Zernike modes.

MSSIM is shown in Fig. 4, being the best results obtained by the CNN with values from 5 to 12 cm, as from this value, results of both technique are very similar. As happened with the RMS, the results of the CNN in the most turbulent scenarios improves the ones of the reconstructor in TPI-WFS. Overall, both methods work well in terms of image quality.

Results of the Strehl ratio are shown in Fig. 5. Both methods showed improvements as the value of r0 raises and consequently in less turbulent profiles. The CNN shows better performance than the TPI-WFS reconstruction in all the cases. Low values of Strehl are caused by the low number of Zernike modes considered.

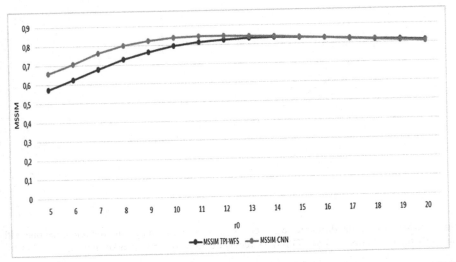

Fig. 4. MSSIM between a recovered phase image and the reference one with the CNN reconstruction and the TPI-WFS reconstruction for 25 Zernike modes.

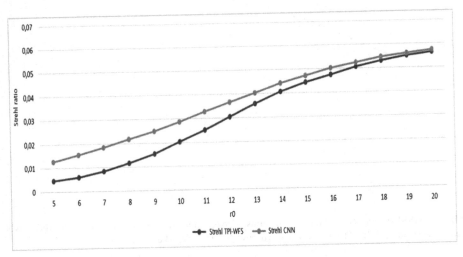

Fig. 5. Strehl ratio of the difference between a recovered phase image and the reference one, with the CNN reconstruction and the TPI-WFS reconstruction for 25 Zernike modes.

3.2 Comparison and Results for the 153 Zernike Modes Case

In this subsection, the mentioned measures are applied for the more complex scenario of considering 153 Zernike modes. In Fig. 6 is shown the RMS error from both reconstructors. The CNN reaches lower values of error, especially for the most turbulent scenarios the CNN shows better results for the reconstruction.

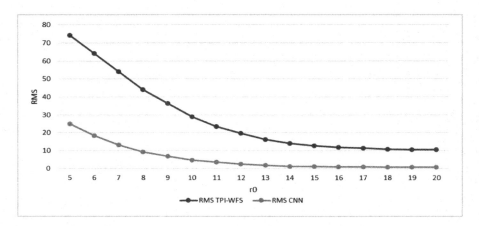

Fig. 6. RMS error of the difference between a recovered phase image and the reference one with the CNN reconstruction and the TPI-WFS reconstruction for 153 Zernike modes.

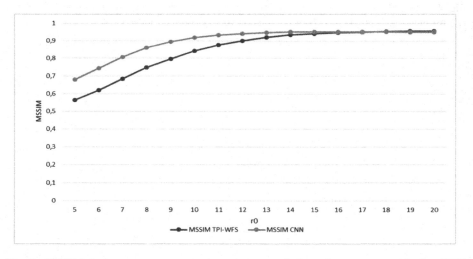

Fig. 7. MSSIM between a recovered phase image and the reference one with the CNN reconstruction and the TPI-WFS reconstruction for 153 Zernike modes.

Best results with the MSSIM similarity index are obtained by the CNN, as it is shown in Fig. 7. For the higher values of r0, results of both technique were very similar. As happened with the 25 Zernike modes case, the results of the CNN in the most turbulent scenarios improves the ones of the reconstructor in TPI-WFS.

The Strehl ratio is shown in Fig. 8. Both methods showed improvements as the value of r0 raises and consequently in less turbulent profiles. In this case, the number of Zernike modes allows both methods to reach better Strehl values, with the CNN showing better performance than the TPI-WFS reconstruction for all the r0 values.

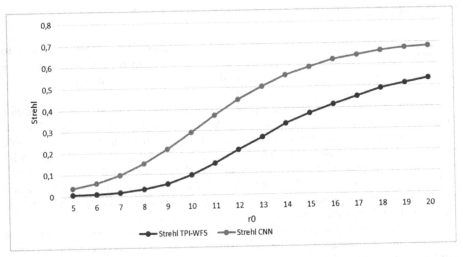

Fig. 8. Strehl ratio of the difference between a recovered phase image and the reference one, with the CNN reconstruction and the TPI-WFS reconstruction for 153 Zernike modes.

4   Discussion

For both the scenarios considered, the CNN reconstruction performed better than TPI-WFS reconstruction in general.

In the first case, where 25 Zernike modes were considered for reconstruction, the CNN showed better performance in RMS error and MSSIM index for r0 up to 12 cm; above that r0 value, both reconstructors showed satisfactory results, for example, achieving around 80% of image similarity, for the majority of turbulent profiles. Regarding Strehl, CNN improves slightly better in comparison.

Regarding the second case, where up to 153 Zernike modes were considered, the differences of both methods are more noticeable. This is a more complex scenario where the high number of Zernike modes with a small pupil size supposes an additional difficulty. For MSSIM index, better results provided by the CNN are clearer for more turbulent profiles, after that both methods have excellent results, achieving 95% of similarity; better than the reachable results in the less complex case of 25 Zernike modes. For RMS error results, the CNN performs clearly better in all the scenarios as also does for the Strehl ratio that now reaches values up to 0.7, which are much more apt values than in the case of 25 Zernike modes, due to the higher number of modes considered.

The main reason for the improvements achieved by the CNN relies in the training procedure. Since computational simulations were available, we could generate data with enough variability to cover all the turbulent profiles satisfactorily. In particular, the most turbulent scenarios, where the r0 reach its lowest values, has the same importance in the training as the rest of profiles. Consequently, the notable results in these cases are reasonable, concretely when considering direct measurements of the error, such as in the RMS case, where the differences are more noticeable.

5 Conclusions

The artificial network approach to the AO image reconstruction correspondent to TPI-WFS data was adequate. The usage of CNN as reconstruction technique gave excellent results thanks to the advantage of allowing the usage of images as inputs and the capability of extracting its main features.

Performance of both methods for phase reconstruction has been compared in this study with computational simulations, considering two main scenarios regarding different number of Zernike modes in the wave-fronts. Both methods showed notable result in most of turbulent profiles, however, the CNN improved the TPI-WFS reconstruction in general, giving better results for the strongest turbulence cases.

Comparing the cases of 25 modes and 153 modes, both methods reached better results for the more complex case, as a result of the increment in available information, especially the CNN, which reached the higher performances.

Usage of other techniques of artificial intelligence suppose the possibility of increasing the performance of the reconstruction. Methods as recurrent real-time learning, suggest that more information from the data can be learned in the training process on a neural network.

References

1. Fried, D.L.: Probability of getting a lucky short-exposure image through turbulence. JOSA **68**(12), 1651–1658 (1978)
2. Brandner, W., Hormuth, F.: Lucky imaging in astronomy. In: Boffin, Henri M.J., Hussain, G., Berger, J.-P., Schmidtobreick, L. (eds.) Astronomy at High Angular Resolution. ASSL, vol. 439, pp. 1–16. Springer, Cham (2016). https://doi.org/10.1007/978-3-319-39739-9_1
3. Oscoz, A., Rebolo, R., López, R., Pérez-Garrido, A., Pérez, J.A., Hildebrandt, S., Rodríguez, L.F., Piqueras, J.J., Villó, I., González, J.M., et al.: FastCam: a new lucky imaging instrument for medium-sized telescopes. In: Ground-based and Airborne Instrumentation for Astronomy II, vol. 7014, p. 701447 (2008)
4. Roddier, F.: Adaptive optics in astronomy. Cambridge University Press, Cambridge (1999)
5. Roddier, C., Roddier, F.: Wave-front reconstruction from defocused images and the testing of ground-based optical telescopes. JOSA A **10**(11), 2277–2287 (1993)
6. Colodro-Conde, C., Velasco, S., Fernández-Valdivia, J.J., López, R., Oscoz, A., Rebolo, R., Femenia, B., King, D.L., Labadie, L., Mackay, C., et al.: Laboratory and telescope demonstration of the TP3-WFS for the adaptive optics segment of AOLI. Mon. Not. R. Astron. Soc. **467**(3), 2855–2868 (2017)
7. Villar, J.R., Chira, C., Sedano, J., González, S., Trejo, J.M.: A hybrid intelligent recognition system for the early detection of strokes. Integr. Comput. Aided Eng. **22**(3), 215–227 (2015)
8. Villar, J.R., Menéndez, M., Sedano, J., de la Cal, E., González, V.M.: Analyzing accelerometer data for epilepsy episode recognition. In: Herrero, Á., Sedano, J., Baruque, B., Quintián, H., Corchado, E. (eds.) 10th International Conference on Soft Computing Models in Industrial and Environmental Applications. Advances in Intelligent Systems and Computing, vol. 368, pp. 39–48. Springer, Cham (2015). https://doi.org/10.1007/978-3-319-19719-7_4
9. Osborn, J., De Cos Juez, F.J., Guzman, D., Butterley, T., Myers, R., Guesalaga, A., Laine, J.: Using artificial neural networks for open-loop tomography. Opt. Express **20**(3), 2420 (2012)

10. de Cos Juez, F.J., Lasheras, F.S., Roqueñí, N., Osborn, J.: An ANN-based smart tomographic reconstructor in a dynamic environment. Sensors 12(7), 8895–8911 (2012)

11. Osborn, J., Guzman, D., Juez, F.J.D.C., Basden, A.G., Morris, T.J., Gendron, E., Butterley, T., Myers, R.M., Guesalaga, A., Lasheras, F.S., Victoria, M.G., Rodríguez, M.L.S., Gratadour, D., Rousset, G.: Open-loop tomography with artificial neural networks on CANARY: on-sky results. Mon. Not. R. Astron. Soc. 441(3), 2508–2514 (2014)

12. Suárez Gómez, S.L., Santos Rodríguez, J.D., Iglesias Rodríguez, F.J., de Cos Juez, F.J.: Analysis of the temporal structure evolution of physical systems with the self-organising tree algorithm (SOTA): application for validating neural network systems on adaptive optics data before on-sky implementation. Entropy 19(3), 103 (2017)

13. Mirowski, P.W., LeCun, Y., Madhavan, D., Kuzniecky, R.: Comparing SVM and convolutional networks for epileptic seizure prediction from intracranial EEG. In: IEEE Workshop on Machine Learning for Signal Processing, 2008. MLSP 2008, pp. 244–249 (2008)

14. Nagi, J., Ducatelle, F., Di Caro, G.A., Cireşan, D., Meier, U., Giusti, A., Nagi, F., Schmidhuber, J., Gambardella, L.M.: Max-pooling convolutional neural networks for vision-based hand gesture recognition. In: 2011 IEEE International Conference on Signal and Image Processing Applications (ICSIPA), pp. 342–347 (2011)

15. Guzmán, D., de Cos Juez, F.J., Myers, R., Guesalaga, A., Lasheras, F.S.: Modeling a MEMS deformable mirror using non-parametric estimation techniques. Opt. Express 18(20), 21356–21369 (2010)

16. Noll, R.J.: Zernike polynomials and atmospheric turbulence. JOsA 66(3), 207–211 (1976)

17. Vidal, F., Gendron, E., Rousset, G.: Tomography approach for multi-object adaptive optics. JOSA A 27(11), A253–A264 (2010)

18. Gómez, S.L.S., Gutiérrez, C.G., Rodríguez, J.D.S., Rodríguez, M.L.S., Lasheras, F.S., de Cos Juez, F.J.: Analysing the performance of a tomographic reconstructor with different neural networks frameworks. In: Madureira, A.M., Abraham, A., Gamboa, D., Novais, P. (eds.) ISDA 2016. AISC, vol. 557, pp. 1051–1060. Springer, Cham (2017). https://doi.org/10.1007/978-3-319-53480-0_103

19. van Dam, M.A., Lane, R.G.: Extended analysis of curvature sensing. JOSA A 19(7), 1390–1397 (2002)

20. van Dam, M.A., Lane, R.G.: Wave-front sensing from defocused images by use of wave-front slopes. Appl. Opt. 41(26), 5497–5502 (2002)

21. Lecun, Y., Bottou, L., Bengio, Y., Haffner, P.: Gradient-based learning applied to document recognition. Proc. IEEE 86(11), 2278–2323 (1998)

22. Krizhevsky, A., Sutskever, I., Hinton, G.E.: Imagenet classification with deep convolutional neural networks. In: Advances in Neural Information Processing Systems, pp. 1097–1105 (2012)

23. Graves, A., Mohamed, A., Hinton, G.: Speech recognition with deep recurrent neural networks. Icassp 3, 6645–6649 (2013)

24. Nair, V., Hinton, G.E.: Rectified linear units improve restricted boltzmann machines. In: Proceedings of the 27th International Conference on Machine Learning (ICML-10), pp. 807–814 (2010)

Using Nonlinear Quantile Regression
for the Estimation of Software Cost

J. De Andrés$^{(\boxtimes)}$, M. Landajo , and P. Lorca

University of Oviedo, Oviedo, Spain
{jdandres, landajo, plorca}@uniovi.es

Abstract. Estimation of effort costs is an important task for the management of software development projects. Researchers have followed two approaches – namely, statistical/machine-learning and theory-based– which explicitly rely on mean/median regression lines in order to model the relationship between software size and effort. Those approaches share a common drawback deriving from their inability to properly incorporate risk attitudes in the presence of heteroskedasticity. We propose a more flexible quantile regression approach that enables risk aversion to be incorporated in a systematic way, with the higher order conditional quantiles of the relationship between project size and effort being used to represent more risk adverse decision makers. A cubic quantile regression model allows consideration of economies/diseconomies of scale. The method is illustrated with an empirical application to a database of real projects. Results suggest that the shapes of higher order regression quantiles may sharply differ from that of the conditional median, revealing that the naive expedient of translating or multiplying some average norm (adding a safety margin to median estimates or including a multiplicative correction factor) is a potentially biased way to consider risk aversion. The proposed approach enables a more realistic analysis, adapted to the specificities of software development databases.

Keywords: Software cost estimation · Project size · Effort
Nonlinear quantile regression · Risk aversion

1 Introduction

One of the most important tasks when managing software projects is cost estimation. However, on many occasions important deviations take place. For example, Hu et al. [1] point out that 200 to 300 percent cost overrun and 100 percent schedule slippage would not be unusual in large software system development projects. A poor resource prediction may cause project failure [2], whereas underestimating software project effort causes schedule delays and cost overruns that in the end may also result in project failure. Conversely, overestimating software project effort can also be detrimental in effectively utilizing software development resources [3].

The main components of software project costs are hardware costs, training, travel, and effort costs, with the latter being the cost of paying software engineers. Effort costs are harder to assess in the earlier stages of a project [4]. A number of models have been proposed to estimate such costs.

© Springer International Publishing AG, part of Springer Nature 2018
F. J. de Cos Juez et al. (Eds.): HAIS 2018, LNAI 10870, pp. 422–432, 2018.
https://doi.org/10.1007/978-3-319-92639-1_35

However, these models share a common drawback, namely, that the relationship between software size and effort is modeled by using classical mean/median regression lines. Once the project size is estimated, conventional models deliver an estimate for the expected effort required. Then, that cost estimate is plugged in the procedures for the pricing of the project. Nevertheless, it is well known that decision makers should also include risk considerations when estimating project effort, as an error in that estimation may have important financial implications in the event that the project price does not cover the actual costs. Consideration of risk is a subjective issue, as different decision makers have various levels of risk aversion/appetite. Unfortunately, extant approaches –either theory or heuristic-based– do not allow the inclusion of risk aversion in the modelling process, so some corrections for risk aversion must be incorporated after the model has returned an estimate for project effort. This is usually carried out by adding a quantity or by multiplying the estimated effort by a suitable correction factor. As to be detailed below, under certain circumstances this kind of corrections may seriously mislead the evaluation process.

In this paper we propose an alternative way to tackle the above issue. We rely on a quantile regression approach that allows risk aversion of the decision maker to be explicitly incorporated into the process of estimating the relationship between software size and effort. This is readily carried out by fitting a selected battery of upper regression quantiles with orders exceeding 50%. The more risk adverse the decision maker is, the higher the quantile in the relationship between size and effort to be estimated. There are sound reasons for using this approach, as some prior research [5] has detected strong heteroskedasticity in the size/effort relationship and the use of additive/multiplicative factors may not be the best option for risk modeling under heteroskedasticity.

We tested the validity of our proposal on the ISBSG (International Software Benchmarking Standards Group) database, which comprises real projects. More specifically, we conducted exhaustive heteroskedasticity tests and estimated a selected battery of conditional upper quantiles that allowed us to consider several levels of risk aversion. We rely for the analysis on a parametric nonlinear structure, namely, a cubic quantile regression model that allows both scale economies (for smaller projects) and diseconomies (for those ones above a certain size limit) to be simultaneously taken into account.

The remainder of the paper is organized as follows: Sect. 2 includes a review of the relevant literature and explains the pertinence of our proposal. The main details on the database and the methodology for the empirical analysis are outlined in Sect. 3. Section 4 reports the main results of the analysis, and Sect. 5 includes a summary of research conclusions and some proposed research avenues.

2 Literature Review

The research on estimation of software effort costs has a long history. According to Pendharkar [6] researchers have followed two basic approaches: theory-based techniques and statistical/machine-learning models.

Theory-based methods see software development as an economic production process where inputs are converted into outputs. Thus, the problem of software cost estimation reduces to understanding the relationship between input and output variables [7]. If we follow that approach two key issues arise: (1) if the relationship among the variables is additive or multiplicative, and (2) whether economies/diseconomies of scale exist.

The first relevant theoretical model is the constructive cost model (COCOMO), developed by Boehm [8]. COCOMO estimates effort by using the following exponential specification:

$$Y = a(KLOC)^q \tag{1}$$

where Y is the effort (in person-months), $KLOC$ is software size (measured in thousands of lines of code), and a and q are constants determined by the environment and the complexity of the application to be developed. COCOMO has three levels (basic, intermediate and advanced) with increasing refinement in the estimation of a and q. The exponential approach has later been used in other models such as COCOMO II [9] and Constructive Systems Engineering Cost Model (COSYSMO) [10], among others.

Parameter q in the above exponential models has particular relevance as it determines the existence of economies/diseconomies of scale. The models generally assume $q > 1$, which means that diseconomies of scale are present no matter the software size we consider. Two serious drawbacks of COCOMO and related models are the subjectivity of the cost drivers used to determine parameters a and q [11] and the high correlations among such cost drivers [12].

Other models assume multiplicative relationships among variables. Among them we can highlight software lifecycle analysis (SLIM) [13] and the SELECT cost estimator [14].

Another theoretical model used for software effort estimation is the Cobb-Douglas production function, under the following general form:

$$Output = c_0(Input1)^{C_1}(Input2)^{C_2} \tag{2}$$

where $Input\ 1$ and $Input\ 2$ usually denote labor and capital, respectively, and c_i, $i = 0$, 1, 2 are unknown parameters. Note that the values of c_1 and c_2 determine the existence of increasing/decreasing/constant returns to scale.

In the specific case of software development, $Output$ is usually considered as an independent variable rather than the dependent variable of the model, as the size of the software product to be delivered depends on requirements assessment and client needs. Then the production function is transformed into the following cost function:

$$Y = A(x)^b(z)^c \tag{3}$$

where Y is the software effort, x is the software development capital and z is the output (namely, a suitable measure of the size of the software product). Parameters b and c define the shape of the function and the type of returns to scale. If both b and c are less than one, then software effort function exhibits increasing returns to scale. If b and c are greater than one then returns to scale are decreasing. This specification has been estimated by Hu [15], and Pendharkar et al. [7], among others.

We must highlight that the various theory-based models do not agree on the basic behavior of software effort costs. Some studies detect significant evidence of returns to scale [7]. On the contrary, other models (i.e., COCOMO and related) have found decreasing returns to scale. This inconsistency may be caused by the fact that most models do not consider the possibility that increasing returns to scale may prevail for smaller projects, while above a certain size threshold they are overweighed by factors causing decreasing returns to scale. Therefore, a nonlinear, s-shaped form can be expected for the relation between software size and effort.

Other researchers have proposed more flexible models, mainly based on statistical/machine learning methods. Those models involve fewer constraints on the functional form of the relationship among software size and effort, so more accurate measurement of the effect of the factors causing increasing/decreasing returns to scale across the various size levels should be enabled.

However, the above models do not take into consideration a fundamental aspect, namely, that the various kinds of errors in the estimation process do not have the same cost. This is because the estimation of software effort is a tool for resource allocation but also a means to gather information when bidding in competition with other software vendors. Thus, when the error conveys an overoptimistic estimation of software effort, and therefore underestimation of the development costs, this may lead to aggressive bids and therefore to the so-called winner's curse, meaning that bidders win only when they bid overoptimistically [16]. Development of a software project for which the costs have been underestimated may lead to substantial losses that eventually compromise the financial stability of the company. On the contrary an error implying overestimation of costs implies, according to the winner's curse, losing the bid. In that event the company still has resources that can be devoted to other projects.

Furthermore, as effort estimation models are part of a system to plan the development of software projects, they can be considered as tools for aiding in decision making when only partial information is available. Most decision makers are risk averse in that environment. This fact may be taken into account explicitly in order to help decision making, by the expedient of constructing a suitable model that delivers, for each size level, not only the most likely amount of effort required but also an adjusted estimation taking into account the risk appetite of the decision maker.

Mainstream models rely on estimating conditional means/medians by exploiting data from historical projects. The model fitting algorithms assign the same weight to both underperforming and overperforming projects. So, those models are unable to deal with asymmetry of costs and risk considerations. In order to take these considerations into account, a naïve approach could be to adjust the estimations from statistical/machine learning model by using an additive or multiplicative factor that reflects cost asymmetry and/or risk aversion.

The above additive/multiplicative corrections respectively assume that the shape of the relationship between size and effort is unique (up to translation/dilation) for both average-performing and underperforming projects. However, prior research [5, 17] suggests that heteroskedasticity is prevalent in databases containing information from software projects. Heteroskedasticity implies that dispersion varies with the various levels of the independent variables in the model (project size in our case). Therefore, the regression line defining the relationship between size and effort for underperforming projects may not follow the same pattern as its counterpart for mean/median performance projects. An illustration of this phenomenon can be seen in Fig. 1 (the meaning of the units in the X and Y axis will be further explained in Sect. 3.1).

Effort (hours)

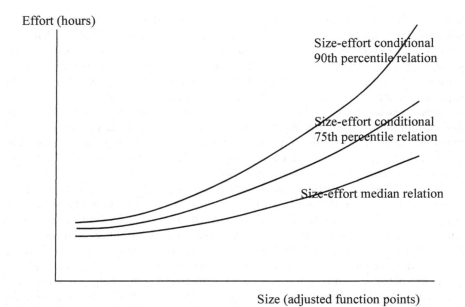

Size (adjusted function points)

Fig. 1. Heteroskedasticity in the relationship between size and effort (Source: own elaboration)

Therefore, it is reasonable to think that –in the presence of heteroskedasticity– naïve approaches based on additive/multiplicative corrections on the average performance line tend to produce incorrect, strongly biased results, so specific estimation of the line passing through the underperforming projects for each project size is needed. As clear from Fig. 1 above, such line may sharply differ from that obtained by naïve translation/scaling of the mean/median regression line. Quantile regression (QR; [18]) provides a systematic way to estimate separate regression lines for the underperforming projects.

As also suggested by Fig. 1 above, the effects of heterokedasticity may be further complicated by the possibly nonlinear nature of the map of conditional quantiles. In our case, nonlinearities may have different intensity depending on the specific quantiles we consider. This would occur because the same factors that render a software project inefficient may have effects of different magnitude depending on the size of the project. For instance, if the organization of the project is not adequately arranged but it has small size, many problems may be bridged through informal communication between the team members. On the contrary, that solution tends to be less feasible when the project is bigger, so the impact on effort costs tends to be larger.

3   Data Set and Methodology

3.1   The Data Set

For the empirical analysis we shall use the ISBSG Development & Enhancement Repository (Release 11). This is a huge database comprising 5,052 software projects from 24 countries. The ISBSG Repository is a consolidated dataset in the field of software metrics (further details can be found at http://www.isbsg.org/). It has been used in many papers dealing with effort estimation [5, 19]. Even though the use of this dataset may involve some limitations and risks concerning the process followed and the quality of the recorded data, it keeps a high potential in advancing software engineering research, as pointed out by Fernández-Diego and González-Ladrón-de-Guevara [20].

We shall measure the size of software projects in terms of so-called adjusted function points. A numerical score is assigned to each one of the functions performed by the software product and the overall scoring is then adjusted by using a value adjustment factor usually depending on the general system characteristics. In order to achieve homogeneity in the data we only considered projects measured using the IFPUG (International Function Point Users Group) counting approach. In order to ensure data quality we discarded the projects for which the available information is unreliable. In this regard, it is noticeable that the ISBSG Repository provides information about the reliability of the data of each project.

As for project effort, its measurement can be considered a challenging endeavor. The Project Management Institute [21] defines project effort metric as "the number of labor units required to complete an activity or other project element. This is usually expressed as staff hours, staff days, or staff weeks and should not be confused with duration". Thus, we measured project effort by using the 'Summary Work Effort' field that appears in the ISBSG Repository. This variable reports the total effort in hours recorded against the project. A limitation of that criterion -shared with the majority of prior research efforts- is that our models do not take into account that the experience of the developers has an effect on the costs of the project, provided that the wages of more experienced developers are generally higher.

The biggest 1% of projects was eliminated from the original sample, so 2,850 projects were left for the analysis. This was motivated mainly by the need to eliminate outliers that may adversely affect the estimates for the most extreme conditional quantiles.

3.2   Methodology

As pointed out in previous sections, our interest specifically concentrates on some 'one-sided' characteristics of the relationship between project cost (denoted by Y) and project effort (X hereafter), namely, the conditional median and the upper conditional quantiles of Y given X representing increasing levels of risk aversion. As pointed out in Sect. 2, the relationship between Y and X can be expected to follow an s-shaped curve. This conveys the fact that increasing returns to scale would be prevalent among smaller projects, whereas for projects exceeding a certain threshold the scale advantages would tend to be overrun by factors inducing the emergence of decreasing returns to scale. This led us to postulate a cubic statistical relationship between software effort/size, so we rely on the following cubic quantile regression (CQR) specification:

$$Q^q(x) = \beta_{0,q} + \beta_{1,q}x + \beta_{2,q}x^2 + \beta_{3,q}x^3 \tag{4}$$

where $Q^q(x)$ denotes the qth order ($0 < q < 1$) conditional quantile of Y given X and the various β_{jq} are the parameters to be estimated. We focused on a battery of 9 cases, namely, $q = 0.5$(the median), $0.55, 0.60, 0.65, 0.7, 0.75, 0.8, 0.85, 0.9$. The 50th conditional percentile of Y given X was used as a benchmark and was compared with conditional percentiles -ranging from 55% to 90%- that correspond to increasing levels of underperformance in project development, and therefore to several levels of risk aversion.

We were interested in testing a number of hypotheses related to heteroskedasticity of the CQR model. More specifically, we focused primarily on analyzing "upper" heteroskedasticity, i.e., heteroskedasticity for the upper quantiles in model (4) above. The classical heteroskedasticity test proposed by Buchinsky [22] may be readily applied to the selected battery of upper quantiles.

In addition, we were also interested in testing whether, at a certain point (x_0), the slopes of a given pair of conditional percentiles, namely those of orders $q1$ and $q2$, coincide. We focus on comparisons with median performance, so we kept $q1 = 0.5$ fixed and set $q2 = 0.55, 0.6, \ldots, 0.9$ sequentially in our analysis. If the slope of the 50% conditional percentile of Y given X significantly differs from that of any of the upper percentiles then our use of the proposed quantile regression approach as a means to include risk aversion in the model would be fully backed by empirical evidence.

The above test is readily implemented. Let $DQ^{qj}(x_0) = \beta_{1,qj} + 2\beta_{2,qj}x_0 + 3\beta_{3,qj}x_0^2$ be the first derivative of the qjth order conditional quantile of Y given X, $Q^{qj}(x); j = 1, 2$, evaluated at point x_0. We wish to test $H_0 : \delta = 0$ against $H_1 : \delta \neq 0$, where $\delta = DQ^{q2}(x_0) - DQ^{q1}(x_0)$. The raw test statistic is the sample analogue of δ, namely, $\hat{\delta} = D\hat{Q}^{q2}(x_0) - D\hat{Q}^{q1}(x_0)$, with $D\hat{Q}^{qj}(x_0) = \hat{\beta}_{1,qj} + 2\hat{\beta}_{2,qj}x_0 + 3\hat{\beta}_{3,qj}x_0^2$ being the sample analogue of $DQ^{qj}(x_0)$.

Classical results (e.g., Buchinsky [22]) ensure that (under the null) the standardized test statistic $z = \sqrt{(n)}\hat{\delta}/\hat{\sigma}_\delta$ -with n being sample size and $\hat{\sigma}_\delta$ being a suitable estimator for the asymptotic standard error of $\hat{\delta}$- has (under the null) a standard normal limiting distribution and delivers a consistent test.

As for the point x_0, in order to detect possible differences originated in project size, in Sect. 4 below we considered the following two cases:

(a) x_0 is set at the first quartile of the sample distribution of X. Thus, we focus on the bottom 25% of projects in terms of size. In our case this corresponds to projects having size $x_0 = 85$ adjusted function points.

(b) x_0 is set at the third sample quartile of X. Thus, we concentrate on 25% of biggest projects, with $x_0 = 421$ adjusted function points.

Following Buchinsky's advice, we use design matrix bootstrap -with $B = 500$ resamples- to compute heteroskedasticity-robust standard errors for all the CQR estimates and related statistical tests. All the calculations were implemented in MatlabR2017b.

4   Empirical Analysis

Table 1 below reports the main results of CQR estimation, including significance tests and 95% confidence limits for the parameters of each conditional quantile model. It is also included the result of Buchinsky's heteroskedasticity test, which detects very strong evidence of heteroskedasticity, this being congruent with the estimates for β_{1q}, β_{2q} and β_{3q}, clearly differing among quantiles. Therefore, the use of the quantile regression approach would be justified. The inherently nonlinear, s-shaped nature of the modeled relationship would also be confirmed by the QR estimates for β_{2q} and β_{3q}, which are statistically significant in most of the cases considered.

Table 1. Results of CQR estimation and heteroskedasticity testing.

Cond. Percentile	50	55	60	65	70	75	80	85	90
β_{0q} estimate	148.4848	148.244	188.2216	213.8079	221.4205	265.1933	335.4637	548.3785	926.8514
p-value	0.0081	0.0118	0.0033	0.0019	0.0018	0.0044	0.0030	0.0007	0.0012
Lower limits (95%)	27.4738	19.9435	52.3991	69.0353	72.2434	67.0496	95.7354	212.1803	329.6384
Upper limits (95%)	269.4958	276.5444	324.044	358.5804	370.5977	463.3371	575.192	884.5767	1524.0645
β_{1q} estimate	10.2178	11.3849	12.6405	14.6867	16.826	19.5587	22429	24.7935	27.8594
p-value	0.0000	0.0000	0.0000	0.0000	0.0000	0.0000	0.0000	0.0000	0.0000
Lower limits (95%)	8.2375	9.2186	10.549	12.1791	14.4087	16.611	18.7921	20.7417	20.8889
Upper limits (95%)	12.198	13.5513	14.732	17.1943	19.2434	22.5063	26.0659	28.8454	34.8299
β_{2q} estimate	-0.0055	-0.0052	-0.0057	-0.0070	-0.0084	-0.0093	-0.0101	-0.0097	-0.0083
p-value	0.0016	0.0068	0.0021	0.0006	0.0001	0.0002	0.0013	0.0072	0.0731
Lower limits (95%)	-0.0092	-0.0094	-0.0096	-0.0113	-0.0128	-0.0145	-0.0167	-0.0175	-0.0196
Upper limits (95%)	-0.0018	-0.0011	-0.0018	-0.0028	-0.0040	-0.0041	-0.0035	-0.0019	0.0029
β_{3q} estimate	1.57E-006	1.41E-006	1.56E-006	0,001869	2.29E-006	2.30E-006	2.34E-006	2.08E-006	1.91E-006
p-value	0.0086	0.0254	0.0097	0.0047	0.0015	0.0047	0.0147	0.0524	0.1222
Lower limits (95%)	2.79E-007	-4.31E-009	2.52E-007	4.59E-007	7.79E-007	5.65E-007	2.33E-007	-4.34E-007	-1.31E-006
Upper limits (95%)	2.86E-006	2.83E-006	2.88E-006	3.28E-006	0,003805	4.03E-006	4.45E-006	0,004597	5.13E-006
Buchinsky's heterosk. test	**Chisq. Stat.** = 445.5774				**p-value** = 0.000				

Table 2 below displays the results of the derivative-based test. In all cases the z-statistics have very large positive values that indicate statistically significant deviations from the null of equality of the derivatives of the conditional quantile lines. Not surprisingly, the effect is clearer for the biggest projects. Higher values of the test statistics are obtained when x_0 is set at the third sample quartile of X as compared with the first quartile. It is also observed in Table 2 that the z-statistics tend to grow monotonically with q2 until a certain threshold (about the 85th–90th conditional percentile) is reached, and thereafter they decrease sharply, although they remain strongly significant. This might reflect the fact that a limit exists to the level of underperformance that a project may exhibit: projects with extreme levels of underperformance are cancelled before completion and therefore they are not considered as suitable for inclusion in any database.

Table 2. Results of the derivative tests (base line $q1 = 0.5$).

Cond. Percentile order ($q2 \times 100$)	$x_0 = 85$ (1st quartile of X)		$x_0 = 421$ (3rd quartile of X)	
	Obs. z-stat.	p-value	Obs. z-stat.	p-value
55	3.5109	0.0004	5.0399	0.0000
60	5.1926	0.0000	7.5088	0.0000
65	6.2587	0.0000	8.2321	0.0000
70	8.5680	0.0000	9.2299	0.0000
75	9.6557	0.0000	10.7163	0.0000
80	9.6101	0.0000	12.6523	0.0000
85	9.9832	0.0000	11.4237	0.0000
90	6.3912	0.0000	9.9656	0.0000

5 Summary and Future Research

In this paper we have addressed effort estimation in software project development. Software effort is one of the main components of the cost in software projects and its accurate estimation previously to the development of the project is a key issue both for the management of the resources to be devoted to the project and for correct pricing.

Several methods have been proposed to assess effort. Apart from expert judgment and simpler methods such as analogies and percentages, a line of research has proposed more sophisticated approaches that rely on exploiting databases of cases comprising past projects as a means to estimate suitable mathematical model. Some of those methods impose theoretical assumptions (such as those from economic theory), while others (including a number of artificial intelligence models) rely on heuristic procedures.

Although those prior models surely provide mean/median estimates of the amount of effort needed for the completion of a project given an estimation of its size, we have seen that this may not be the most sensible procedure since in most cases the decision makers using those effort prediction models are risk averse. In addition, the costs caused by underestimation of the effort needed for a given project size are higher than those of overestimation, so a differentiated treatment of both kinds of errors – unavailable in mean/median regressions– is required. A model incorporating risk/cost asymmetry considerations -instead of merely providing central tendency estimations- could better satisfy the needs of managers and other users. A way to do this is by merely shifting the function obtained from traditional procedures, by applying a suitable additive/multiplicative correction factor. Nonetheless, we have postulated that this expedient would be largely inappropriate since prior studies strongly suggest the presence of heteroskedasticity in the software project data. Instead we have proposed an approach that relies on estimating a battery of quantile regression functions corresponding to selected conditional quantiles above the conditional median. This can be regarded as a "multinorm" approach, in the sense that instead of a single norm/regression line we rely on a multiplicity of them (each corresponding to a different conditional quantile of the relevant distribution), chosen in accordance with such issues as the level of risk aversion of the decision maker and the magnitude of cost asymmetry, which may eventually depend on specific features of the organization developing the software project. We have seen that –given the peculiarities of software development– economies of scale would be present only until a given project size limit is reached, whereas diseconomies of scale would be more likely beyond that point. This led us to postulate an s-shaped relationship, specified through a cubic quantile regression model.

The proposed methodology was applied to a benchmark data set, obtained from the ISBSG Repository. Results endorse the validity of the proposed approach, confirming the strongly heteroskedastic nature of the statistical relationship between project size and effort in the studied population and the need of relying on more flexible approaches. We have shown that the use of margins or percentages is inappropriate in presence of heteroskedasticity, with the approach proposed in this paper being able to incorporate the decision-maker's risk aversion into the budgeting process in a systematic, more accurate fashion.

References

1. Quing, H., Plant, R.T., Hertz, D.B.: Software cost estimation using economic production models. J. Manag. Inf. Syst. 15(1), 143–163 (1998)
2. Jørgensen, M., Halkjelsvik, T.: The effects of request formats on judgment/based effort estimation. J. Syst. Softw. 83(1), 29–36 (2010). https://doi.org/10.1016/j.jss.2009.03.076
3. Seo, Y.S., Bae, D.-H., Jeffrey, R.: AREION: Software effort estimation based on multiple regressions with adaptive recursive data partitioning. Inf. Softw. Technol. 55(10), 1710–1725 (2013). https://doi.org/10.1016/j.infsof.2013.03.007
4. Henrich, A.: Repository based software cost estimation. In: Hameurlain, A. (ed.) DEXA 1997. LNCS, vol. 1308, pp. 653–662. Springer, Heidelberg (1997). https://doi.org/10.1007/BFb0022073
5. Cuadrado-Gallego, J.J., Rodríguez, D., Sicilia, M.A., Garre Rubio, M.A., García Crespo, A.: Software project effort estimation based on multiple parametric models generated through data clustering. J. Comput. Sci. Technol. 22(3), 371–378 (2007). https://doi.org/10.1007/s11390-007-9043-5

6. Pendharkar, P.C.: Probabilistic estimation of software size and effort. Expert Syst. Appl. **37** (6), 4435–4440 (2010). https://doi.org/10.1016/j.eswa.2009.11.085

7. Pendharkar, P.C., Rodger, J.A., Subramanian, G.H.: An empirical study of the Cobb-Douglas production function properties of software development effort. Inf. Softw. Technol. **50**(12), 1181–1188 (2008). https://doi.org/10.1016/j.infsof.2007.10.019

8. Boehm, B.W.: Software Engineering Economics. Prentice-Hall, Englewood Cliffs (1981)

9. Boehm, B.W., Clar, B., Horowitz, B.C., Westland, C., Madachy, R., Selby, R.: Cost models for future software life cycle processes: COCOMO 2.0.0. Ann. Softw. Eng. **10**(1), 1–30 (1995)

10. Valerdi, R.: The Constructive Systems Engineering Cost Model (COSYSMO): Quantifying the Costs of Systems Engineering Effort in Complex Systems. VDM Verlag, Saarbrücken (2008)

11. Fenton, N., Pfleeger, S.: Software Metrics: A Rigorous & Practical Approach. PWS Publishing, Boston (1997)

12. Chulani, S., Boehm, B.W., Steece, B.: Bayesian analysis of empirical software engineering cost models. IEEE Trans. Softw. Eng. **25**(4), 573–580 (1999). https://doi.org/10.1109/32.799958

13. Putnam, L.H.: The real economics of software development. In: Goldberg, R. (ed.) The Economics of Information Processing. Wiley, New York (1982)

14. Boehm, B.W., Abts, C., Chulani, S.: Software development cost estimation approaches: a survey. Ann. Softw. Eng. **10**(1), 177–205 (2000). https://doi.org/10.1023/A:1018991717352

15. Hu, Q.: Evaluating alternative software production functions. IEEE Trans. Softw. Eng. **23** (6), 379–387 (1997). https://doi.org/10.1109/32.601078

16. Jørgensen, M.: How to avoid selecting bids based on overoptimistic cost estimates. IEEE Softw. **26**(3), 79–84 (2009). https://doi.org/10.1109/MS.2009.71

17. Stensrud, E., Foss, T., Kitchenham, B., Myrveit, I.: An empirical validation of the relationship between the magnitude of relative error and project size. In: Proceedings of the Eight Symposium on Software Metrics, Ottawa, Canada, pp. 3–12 (2002)

18. Koenker, R., Basset, G.W.: Regression quantiles. Econometrica **46**, 33–50 (1978)

19. Rodríguez, D., Sicilia, M.A., García, E., Harrison, R.: Empirical findings on team size and productivity in software development. J. Syst. Softw. **85**(3), 562–570 (2012). https://doi.org/10.1016/j.jss.2011.09.009

20. Fernández-Diego, M., González-Ladrón-de-Guevara, F.: Potential and limitations of the ISBSG dataset in enhancing software engineering research: a mapping review. Inf. Softw. Technol. **56**(6), 527–544 (2014). https://doi.org/10.1016/j.infsof.2014.01.003

21. Project Management Institute: A Guide to the Project Management Body of Knowledge (PMBOK Guide), 4th Edn. Project Management Institute, Pennsylvania (2008)

22. Buchinski, M.: Recent advances in quantile regression models: a practical guideline for empirical research. J. Hum. Resour. **23**(1), 88–126 (1998). https://doi.org/10.2307/146316

A Distributed Drone-Oriented Architecture for In-Flight Object Detection

Diego Vaquero-Melchor[✉], Iván Campaña, Ana M. Bernardos, Luca Bergesio, and Juan A. Besada

Information Processing and Telecommunications Center,
Universidad Politécnica de Madrid, ETSI Telecomunicación,
Av. Complutense 30, 28040 Madrid, Spain
{diego.vaquero,icampana,abernardos,luca.bergesio,besada}@grpss.ssr.upm.es

Abstract. Drones are increasingly being used to provide support to inspection tasks in many industrial sectors and civil applications. The procedure is usually completed off-line by the final user, once the flight mission terminated and the video streaming and conjoint data gathered by the drone were examined. The procedure can be improved with real-time operation and automated object detection features. With this purpose, this paper describes a cloud-based architecture which enables real-time video streaming and bundled object detection in a remote control center, taking advantage of the availability of high-speed cellular networks for communications. The architecture, which is ready to handle different types of drones, is instantiated for a specific use case, the inspection of a telecommunication tower. For this use case, the specific object detection strategy is detailed. Results show that the approach is viable and enables to redesign the traditional inspection procedures with drones, in a step forward between manual operation and full automation.

1 Introduction

The use of Unmanned Aerial Vehicles (UAV, drones) for scenario diagnosis and aerial imagery retrieval is currently in full swing, due to UAV increasing in-flight autonomy and sensing capabilities, and the cost reduction that may be achieved over traditional procedures. Surveillance [10] and reconnaissance [4, 13] are among the most common drone-based operations. In such applications, object search and detection in images or video streaming, and their posterior analysis and classification, are key tasks which, in many cases, are still manually performed. An obvious benefit for the final user can be achieved if automatic image recognition is integrated in the drone-enabled inspection workflow.

Automatic object detection and recognition may pose specific requirements over the overall system architecture. In particular, some operational scenarios ideally require that detection is performed almost real-time. The straightest approach to solve this is to have the drone's video retrieval module and the object

© Springer International Publishing AG, part of Springer Nature 2018
F. J. de Cos Juez et al. (Eds.): HAIS 2018, LNAI 10870, pp. 433–445, 2018.
https://doi.org/10.1007/978-3-319-92639-1_36

detection module onboard. The problem is that techniques currently used for object detection in images usually entail a high computing cost - e.g. some state-of-the-art procedures rely on heavy convolutional neural networks (CNN). A way to speed-up the computing tasks goes through the use of dedicated GPUs. Most onboard computers and microcomputer do not have those components integrated, which make them unsuitable for a reliable object detection. The rising ubiquity of high-speed mobile networks (4G-5G) allows bypassing the computing problem by transferring the image processing task from the drone itself to dedicated calculation centers. By doing this, the drone will only need the capacity to stream video, being unnecessary to carry the weight of a more powerful computer, thus saving energy (the battery keeps on being a bottleneck element in the performance of UAV).

In this paper, we propose a distributed cloud-based architecture for object detection, to be integrated in an unmanned aerial system to enable real-time search and detection operations. Some works as [11] have previously explored this possibility; we make a step forward to provide a distributed, hardware-independent architecture, general enough to serve as common ground for different operation scenarios with different object detection and analysis needs. The architecture proposal is evaluated on a specific use-case: the inspection and detection of anomalies in antennas in telecommunication towers.

The rest of this paper is organized as follows. Section 2 gives an overview of related work, reviewing UAV application scenarios in which object recognition is needed. In Sect. 3 the proposed architecture is presented, with its different modules. Section 4 details the particularities of the core Processing Center module, which contains the Object Detection capability. Algorithm design decisions are explained, together with the strategy used to expand the model training dataset through augmentation techniques (i.e. automated variations over the available training image set to increase its diversity). Section 5 focuses on the description of the testing scenario - telecommunication tower inspection -, gathers some performance results and analyzes the system limitations. Finally, Sect. 6 summarizes conclusions and presents how the system is being evolved.

2   Background

Object detection and recognition is key for some drone-enabled applications. It is the case of people search, infrastructure inspection or traffic monitoring. For example, drone-enabled people detection can be performed by either using images from a standard camera or a thermal one. In [4] it is described a system for automatic UAV pedestrian tracking in which the position of the target is determined by the combination of a pedestrian detector and a color-based particle filter. The pedestrian detector is employed to help the particle filter remain on the target. In [13] it is described a system that detects and tracks pedestrians using thermal infrared images captured by a UAV, since humans leave an important thermal signature. This type of detection and tracking is complicated: the images have a low resolution, there is instability in the image due to drone movement and the objects have a relatively small size.

With respect to infrastructure inspection to detect deterioration and damage, drones may simplify, accelerate and reduce the cost of traditional manual inspection procedures. In [2], power lines are detected on UAV imagery by using a combination of spectral and spatial methods. Instead, [14] describes the dynamics of a UAV for monitoring of structures and maintenance of bridges, presenting a novel control law based on computer vision for quasi-stationary flights above a planar target. UAVs also are a viable and less time-consuming alternative for monitoring traffic in real time [8]. In [9] it is described a classification of road situations using images obtained after a traffic accident through the set of features, facts and attributes that can occur in a situation.

Regarding image classification object recognition techniques, nowadays Convolutional Neural Networks (CNN) are intensively used [17]. The algorithms used in these neural networks are divided into two main groups. The first one is composed of those algorithms that incorporate a region proposal pipeline (e.g. R-CNN [5], Faster R-CNN [16] and R-FCN [3]). The second one groups those algorithms that do not incorporate that pipeline (e.g. YOLO [15] and SSD [12]). In [7], authors explored the trade-off between accuracy and speed comparing the previously mentioned algorithms. In order to compare the algorithms, they recreated training pipelines for the SSD, Faster-RCNN and R-FCN approaches. Results show that SSD and R-FCN models are faster on average, but that Faster R-CNN provides more accurate models.

Additionally, SSD models using Inception v2 [18] and Mobilenet [6] feature extractors are the most accurate of the fastest models. In addition, whether post-processing is ignored, Mobilenet is twice as fast as Inception v2 while, providing a slightly lower accuracy.

Finally, it is important to underline the computation cost problem that can arise in drone-based architectures using heavy image recognition techniques. A viable solution is to transfer the calculations from the drone to a more powerful device. A combined processing is proposed in [11], in which computation is transferred to an off-board cloud while the low-level object detection and navigation is completed onboard. Authors use Faster Regions with CNNs (R-CNNs) to perform object detection in almost real-time, taking into account the communication delay, which inserts a high and unpredictable lag.

In the following section, we present our own architecture to deal with remote object detection and recognition for real-time performance.

3   Architecture

This section details the architecture of the system, which main functional objective is to perform real-time detection of predefined objects in a video stream provided by a drone. In practice, the detected object will be bounded in a rectangular shape (bounding box or BB) and classified depending on its category. The proposed architecture enables to transfer the calculations needed to perform the object detection from the onboard device to a remote, more powerful computer. To do so, we propose a distributed cloud-based approach that may be

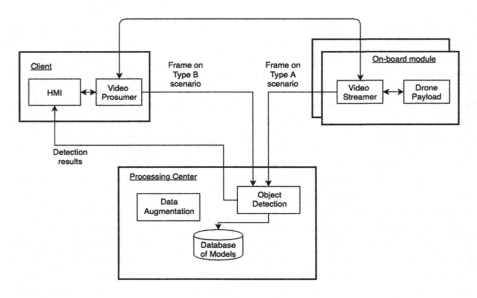

Fig. 1. System architecture

applied for a wide range of use cases. This architecture intends to be hardware-independent, so it may be implemented for different types of drones.

In brief, the requirements to consider for the final system are:

1. Real-time services: The system must provide real time video stream, object detection and telemetry.
2. Object detection algorithm-agnostic solution: The architecture should be designed in such way that at any moment it is possible to change the object detection model. This is a requirement because the operation scenario may vary, for example, from an urban field (pedestrians, cars and so on) to a rural one (crops, roads, rivers). In addition, it is required to be able to change the model's own parameters, and to be able to detect multiple types of objects of different types in the same image.
3. Web-based user interface: The architecture will be designed taking into account the philosophy of the cloud paradigm. Thus, the main access to the system must be done using web (some particularizations are explained at the end of this section).
4. UAV-platform independency: The architecture should be generalizable for as many kinds of drones and payloads as possible.

Taking these requirements in consideration, the architecture has been built on three main modules. The first module *On-board Module* represents the drone and its payload, which serves the video stream. The second module *Processing Center* executes the object detection. Finally, the third module is the *Client*, which is used by the user to interact with the system. Figure 1 shows the basic diagram of the designed architecture.

A. *On-board Module*: In Fig. 1 the drone payload is composed of two sub-modules (which may be merged into a single one). The sub-module labelled as *Drone Payload* represents an onboard computer. The *Video streamer* sub-module is a component that reads a video stream from the *Drone Payload*, outputting a single frame. The reason for this separation is that the architecture has been developed taking into account two different types of scenarios. The first one related to drones with customized payload (Type A) and the second one related to drones with a standard payload (Type B).

  - Type A scenario: The installed custom payload simplifies the video streaming process. The drone has an onboard computer, so it is possible to install a simple component that connects to the camera raw stream, extracts individual frames and send them to the Processing center.
  - Type B scenario: In this case the drone builds an ad-hoc Wi-Fi network for a computer or a mobile phone to connect to. The Video Prosumer component on the Client is responsible for opening the video stream from the drone and extracting individual frames, which are sent to the Processing Center Module.

B. *Processing Center Module*: This module is divided into three main sub-modules as follows:

  - The main sub-module is the one labelled as *Object Detection*. Taking as input an image (or a video frame), it feeds the object detection algorithm. Once it gets the results from the object detection algorithm, it facilitates their retrieval from the HMI (details are explained in Sect. 4.1). For the implementation of this submodule we used Tensorflow [1].
  - The *Database of Models* represents a collection of trained models for different applications. The Object Detection module can choose one or more models for the detection process. The models and their weights are saved as part of the architecture, so the Object Detection sub-module may load and query them.
  - The *Data Augmentation* sub-module does not participate in the real-time execution process. The purpose of this module is to take an image dataset and augment it, in order to help the training process of the object detection model (details are provided in Sect. 4.2).

c. *Client*: This module is divided into two sub-modules. The first one is the *HMI* (Human Machine Interface) that provides the detection results to the user. The second one is the sub-module labelled as *Video Prosumer*, which produces frames extracted either from a local file or a video stream from the drone and it consumes the video stream from the drone. From the HMI, the user is able to view and select the video source and may choose the classification model to apply. Once the detections are done, the HMI displays the resulted bounding boxes.

The architecture offers two possibilities when returning the detection results. The first one is to return a video stream that joins the input video and the drawn BB. The second possibility is to return only the BB information (position and size). Using the first alternative, the output video can be embedded in any compatible video player, but with a higher consumption of bandwidth. Instead, by returning only the BB graphic information the required bandwidth is much lower, but it is necessary to combine the input video and the detections result.

Although the full architecture has been thought to be cloud-based - thus requiring an continuous high-speed internet connection -, there are specific scenarios in which 3G-4G coverage cannot be taken for granted, thus being impossible to send the video stream to the Processing Center Module. Thanks to the modular architecture, the Object Detection Module and the Client can be installed in any laptop with enough computing power. This laptop can then locally connect to the drone Wi-Fi connection and the user can open a local version of the web HMI. The Video Consumer module will handle the video stream from the drone, and query the local Processing Center Module, which will have previously downloaded the models' dataset. If there is a mobile connection to which the drone could connect, and it has the proper hardware the architecture offers the possibility to directly stream video from the drone to the Processing Center module, with no need of using the Client module.

4   Processing Center Module

The Processing Center Module is based on an object detection architecture built using CNN. When using CNNs, the workflow is divided into three main stages: model design, model training and model querying. In this Section, we provide details about the first two stages, that are carried out in the Processing Center Module.

4.1   Object Detection Module

This sub-module receives an image as input, reporting the labelled image as result. This labelled image is composed of the original image data, plus the *bounding boxes* of the detected objects, and their respective labels. The core of this sub-module is the object detection algorithm. The criteria used for its selection are below exposed.

Using the API in [7], we trained some models with our own image dataset (detailed in Sect. 4.2), focusing on the SSD algorithms (said to be faster in [7]); both Mobilenet and Inception feature extractors were applied. Besides, we evaluated a supposedly slower, more accurate algorithm as Faster-RCNN. Execution times with these different options are shown in Table 1.

As expected, the Faster-RCNN algorithm requires a longer execution time. It is discarded because the sum of communication and computing delays may be too high to guarantee real-time. SSD provides the best computing time among the available options. Regarding the feature extraction strategy, both Mobilenet and Inception v2 delivers similar execution times. According to [7] and our own results, Mobilenet is faster than Inception v2 with a slight loss of accuracy, so it has been our final choice.

Table 1. Comparison of times

Algorithm	Feature extractor	Mean (ms)	Variance (ms)
SSD	Mobilenet	39.85	1.15
SSD	Inception v2	39.23	1.43
RFCN	Resnet 101	372.63	4.80
Faster-RCNN	Resnet 50	604.29	78.15

For testing purposes, images datasets of two different applications have been used to train two different models. The first model is used to detect antennas in telecommunication towers and the second one, to detect insulators in high-voltage towers (the case of the telecommunication antennas has served as case study and it is detailed in Sect. 5). The trained models are stored in the Processing Center Module and they can be queried simultaneously. This gives the possibility to change the model for comparison at any time, or even to combine the output from several models, each one of them detecting different types of objects. An example of this combination could be the use of one model trained to detect vehicles and another model trained to detect persons.

Following, Sect. 4.2 explains how the training of the model has been performed and details the Data Augmentation submodule.

4.2 Model Training and Data Augmentation

To train the model, it is necessary to have a set of images properly labelled. The larger and more diverse the image dataset is, the better the model fits. The goal is to get a good recognition rate even if the drone is not retrieving excellent video as it happens if the video stream is too bright (e.g. the drone's camera is facing the sun). In practice this means that the classification algorithm must be trained with images with abnormal conditions (a lot of brightness, out of focus, darkness...) that are frequent during flights. In the case of drone-taken imagery, these images may be difficult to obtain or may be not diverse enough to show all the possible situations. In order to make a more robust system against these usual situations, a possible solution is to manipulate the available images to extend the dataset. This is known as *data augmentation*. In practice, nowadays several libraries offer the possibility of manipulating images. OpenCV has gained the recognition of being one of the most powerful, and it contains the libraries that have been used to implement the subsequent transformations.

Depending on the object to detect and locate, the space of features to describe the object pattern varies, but standard characteristics that allow the detection and classification are usually pattern's shape, length, width, color and brightness. Image preprocessing for feature extraction is needed both in the training and the real-time object recognition. The data augmentation transformations that have been used in the training phase enable to enhance contrast (through brightness management) and to increase the blurring of the camera (through

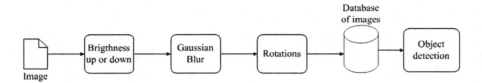

Fig. 2. Data augmentation pipeline

gaussian blur). Along with these transformations, rotations have included to further increase the database of images as it can be seen in Fig. 2. Analyzing the data augmentation architecture, the first transformation is the brightness variation, which can take the values −30%, −15%, 0%, 15% and 30%, so image can be modified by increasing or reducing the brightness level or remain the same. The next possible transformation is the Gaussian Blur, which make that image remains the same if it takes 0 as value, or image increases its blur if takes 5 or 10 as value. Finally, there are the rotations, which originate a large number of different images depending on the value of the rotation that can be −25°, −10°, 0°, 10° and 25°. Thanks to these transformations, situations that occur in real life outdoors are simulated, so it is possible to use the system outdoors with different climatic conditions. The structure is easily scalable and it allows to add other transformations that make changes in the contrast or resolution of the image, that add noise to the image or that simulate the presence of fog in the image, between many others. In Fig. 3 the effect of the different transformation options is shown.

In the case that is going to be explained in Sect. 5, the original database of images was composed of 115 pictures. After going through the transformation pipeline and removing the useless images (completely black or white), about 4000 images were obtained. These images were used to train the neural network.

5   Evaluation for the Inspection of Telecommunication Towers

The following case study is used to analyze the feasibility of our architecture. The operation to fulfill is an inspection of a telecommunications tower, in which the key task is to detect antennas and groups of them. This study case is thought to be used by the responsible of the inspection as a helping tool for routine maintenance. This responsible is located in a remote location, while the flight is performed by the pilot. The system provides real time information to the operator, who can perform RT checking of the received video stream. This allows the operator to ask the pilot for modifications of the flight plan near real time.

A flight is performed with this purpose, using our system for real-time object detection.

As the mission is to be completed within an area with low 3G coverage, an ad-hoc network between the drone and the monitoring computer is built. The

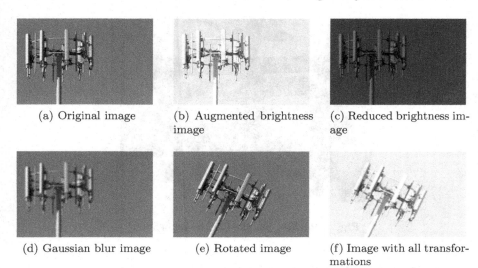

(a) Original image   (b) Augmented brightness   (c) Reduced brightness im-
                           image   age

(d) Gaussian blur image   (e) Rotated image   (f) Image with all transfor-
                                                      mations

Fig. 3. Example of images after the transformations of the data augmentation

computer is equipped with an NVIDIA Quadro K1000M graphic Card and an
Intel i7 processor. On top of it, the Processing Center Module has been installed,
together with the HMI Module. The ad-hoc network is enabled by Apple AirPort
Extreme (50 m of full performance; beyond this range, the video stream has very
low quality, dropping video frames and causing coding noise). Figure 4 shows the
custom payload built on the octocopter used for the experiment. It integrates a
Sony DSC-WX500 photo camera, a Flir A35 thermal camera and a Brick IMU
2.0 from Thinkerforge. The Sony camera has been selected due to it provides a
30x optical zoom and a 60x digital zoom. All those components were mounted
in a RONIN-MX gimbal from DJI. In addition, an Intel NUC was mounted on
the drone, which provide access to specific sensors.

In order to evaluate the trained model, we fed the algorithm with a set of
1600 images manually labelled. This set is completely independent of the training set, which consisted of 4000 images. The SSD algorithm is a threshold-based
one. We repeated the test for several thresholds; threshold values represent the
minimum percentage of accuracy that a labelled box must achieve to be considered as valid. To compare the results of the model and the manual labels, we
used (i) Correct recognition rate: percentage of images labelled with the correct
number of objects and (ii) Average of Intersect Over Union (also known as Jaccard similarity index) for all the images. The IoU or Jaccard similarity index
compares the similarity and diversity of sample tests by analyzing the percentage of intersection surface between ground-truth and automatically identified
bounding boxes, with respect to the union surface of both types of boxes. In
order to calculate it, we associated the manually identified ground-truth boxes
with the predicted ones, and calculated the intersect over union of them. Once

Fig. 4. Payload installed in the octocopter

this is done, all the indexes were averaged. The obtained results showed a similar behavior for different thresholds. From a minimum threshold of 70% to 90% the correct recognition rate was nearly the same. Given that situation, we chosen for the flight a threshold of 90%, which achieves a correct recognition rate of 93.1%.

Regarding false alarms, we used a set of 2492 images obtained using the same techniques as before, with no overlap with the training and validation sets. In these images didn't appear neither antennas nor groups of the same. Table 2 shows the results obtained. In the case of a minimum threshold of 90%, the percentage of images with false alarms is 0.69%.

Table 2. False alarms results

Minimum threshold (%)	5	10	30	50	90
Images with false alarms (%)	27.29	14.93	5.70	3.01	0.69

In order to measure the transmission delay, we carried out the following procedure. First, we focused the drone camera to a display. On that display they were shown at the same time a chronometer and the real time video stream from the drone. Making a screenshot we were able to calculate the difference of times between the one displayed at the chronometer and the one from the stream. These measurements have been taken in our laboratory in a controlled environment with a good Wi-Fi and 4G signal. No packet loss was recorded during the tests. The current implementation of the architecture using Wi-Fi has an average delay time of 366 ms, with a standard deviation of 65.01 ms, which is admissible to operate in real time. Regarding the use of 4G technology the system had an average delay of 434 ms and a standard deviation of 5 ms. When using 3G the average delay raised to 484.2 ms with a standard deviation of 76.14 ms. Aside from that delay, the camera is sending frames at 25 FPS.

(a) Blurred and high brightness   (b) Blurred and low brightness

Fig. 5. Detections results (Color figure online)

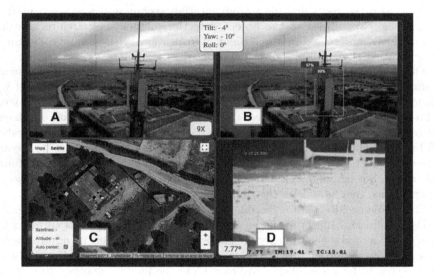

Fig. 6. HMI screenshot (Color figure online)

Thanks to the data augmentation techniques explained in Sect. 4.2, the trained model has a high noise tolerance. Figure 5 shows two images obtained by the UAV, and the detection results. The red boxes wrap individual antennas, and the green ones cover the group of them. Each one of these images was taken in a different time of the day. In the first one, the white balance of the camera was set on manual mode in a more light tolerant way. The second one was taken during sunset, with white balance in auto mode, so the light is lower. Due to stability problems, both images appear out of focus. Even with these issues, the system has been able to properly detect the antennas and the group. The system gives an estimated probability of 99% of being correctly labelled.

In the Fig. 6 the web HMI is shown. The interface is divided as follows. Tile A holds the video stream of the drone, obtained in real time. Tile B shows the results of the object detection; in this example two different kind of object are being detected: the red box shows that the system has detected an antenna with an estimated probability of 99%, and the green box is showing a group of antennas with an estimated probability of 67%. Tile C shows a Google Maps view, in which the current position of the drone is marked. Finally, tile D shows the thermal video, obtained in real time from a thermal camera mounted on the drone.

6 Conclusions and Future Work

In this paper we have introduced a distributed drone-oriented architecture for in-flight real-time object detection, together with the object detection strategy. We have also treated some data augmentation techniques, which we used to increase the capabilities of the object detection model. The architecture has been deployed and tested in a real flight of inspection, with specific constraints, demonstrating the feasibility of the approach.

Although the results of the evaluation have been promising, some improvements can be made. One of the main is that the object detection model will benefit from an increased number of images for training. This could be achieved by integrating more filters on the data augmentation stage. The more different and realistic filters, the more capacity to adapt to adverse situations the system will have. Further, the image encoding and compression issue must be treated more deeply. If the model was trained using, for example PNG images, we found that it does not respond well when the input image format was JPG. Although at first glance they may look the same, the content is not.

The whole architecture is flexible enough to be distributed using for example Docker. In this line, progresses are being made. Our intend is to make the deployment as easy and quickly as possible. With this idea in mind, trained models can be shared, giving the option to build a more complex shared library. Future steps to be accomplished are to explore the idea to incorporate a smartphone to the drone payload and to adapt this architecture to real time search and tracking use cases.

Acknowledgements. This work was supported in part by Universidad Politécnica de Madrid Project RP1509550C02, and by the Spanish Ministry of Economy and Competitiveness under Grants TEC2014-57022-C2-1-R and TEC2014-55146-R.

References

1. Abadi, M., Agarwal, A., Barham, P., Brevdo, E., Chen, Z., Citro, C., Corrado, G.S., Davis, A., Dean, J., Devin, M., et al.: TensorFlow: Large-scale machine learning on heterogeneous distributed systems. arXiv preprint arXiv:1603.04467 (2016)
2. Bhola, R., Krishna, N.H., Ramesh, K., Senthilnath, J., Anand, G.: Detection of the power lines in UAV remote sensed images using spectral-spatial methods. J. Environ. Manag. **206**, 1233–1242 (2018)

3. Dai, J., Li, Y., He, K., Sun, J.: R-FCN: object detection via region-based fully convolutional networks. CoRR abs/1605.06409 (2016)
4. De Smedt, F., Hulens, D., Goedeme, T.: On-board real-time tracking of pedestrians on a UAV. In: The IEEE Conference on Computer Vision and Pattern Recognition (CVPR) Workshops, June 2015
5. Girshick, R.B., Donahue, J., Darrell, T., Malik, J.: Rich feature hierarchies for accurate object detection and semantic segmentation. CoRR abs/1311.2524 (2013)
6. Howard, A.G., Zhu, M., Chen, B., Kalenichenko, D., Wang, W., Weyand, T., Andreetto, M., Adam, H.: MobileNets: efficient convolutional neural networks for mobile vision applications. CoRR abs/1704.04861 (2017)
7. Huang, J., Rathod, V., Sun, C., Zhu, M., Korattikara, A., Fathi, A., Fischer, I., Wojna, Z., Song, Y., Guadarrama, S., Murphy, K.: Speed/accuracy trade-offs for modern convolutional object detectors. CoRR abs/1611.10012 (2016)
8. Kanistras, K., Martins, G., Rutherford, M.J., Valavanis, K.P.: Survey of Unmanned Aerial Vehicles (UAVs) for traffic monitoring. In: Valavanis, K.P., Vachtsevanos, G.J. (eds.) Handbook of Unmanned Aerial Vehicles, pp. 2643–2666. Springer, Dordrecht (2015). https://doi.org/10.1007/978-90-481-9707-1_122
9. Kim, N., Bodunkov, N.: Automated decision making in road traffic monitoring by on-board unmanned aerial vehicle system. In: Favorskaya, M.N., Jain, L.C. (eds.) Computer Vision in Control Systems-3. ISRL, vol. 135, pp. 149–175. Springer, Cham (2018). https://doi.org/10.1007/978-3-319-67516-9_6
10. Kumar, R., Sawhney, H., Samarasekera, S., Hsu, S., Tao, H., Guo, Y., Hanna, K., Pope, A., Wildes, R., Hirvonen, D., et al.: Aerial video surveillance and exploitation. Proc. IEEE 89(10), 1518–1539 (2001)
11. Lee, J., Wang, J., Crandall, D., Šabanović, S., Fox, G.: Real-time, cloud-based object detection for unmanned aerial vehicles. In: IEEE International Conference on Robotic Computing (IRC), pp. 36–43. IEEE (2017)
12. Liu, W., Anguelov, D., Erhan, D., Szegedy, C., Reed, S.E., Fu, C., Berg, A.C.: SSD: single shot multibox detector. CoRR abs/1512.02325 (2015)
13. Ma, Y., Wu, X., Yu, G., Xu, Y., Wang, Y.: Pedestrian detection and tracking from low-resolution unmanned aerial vehicle thermal imagery. Sensors 16(4), 446 (2016)
14. Metni, N., Hamel, T.: A UAV for bridge inspection: visual servoing control law with orientation limits. Autom. Constr. 17(1), 3–10 (2007)
15. Redmon, J., Divvala, S.K., Girshick, R.B., Farhadi, A.: You only look once: unified, real-time object detection. CoRR abs/1506.02640 (2015)
16. Ren, S., He, K., Girshick, R.B., Sun, J.: Faster R-CNN: towards real-time object detection with region proposal networks. CoRR abs/1506.01497 (2015)
17. Szegedy, C., Liu, W., Jia, Y., Sermanet, P., Reed, S., Anguelov, D., Erhan, D., Vanhoucke, V., Rabinovich, A.: Going deeper with convolutions. In: IEEE Conference on Computer Vision and Pattern Recognition (CVPR), pp. 1–9, June 2015
18. Szegedy, C., Vanhoucke, V., Ioffe, S., Shlens, J., Wojna, Z.: Rethinking the inception architecture for computer vision. CoRR abs/1512.00567 (2015)

3D Gabor Filters for Chest Segmentation in DCE-MRI

I. A. Illan[1,2(✉)], J. Perez Matos[3], J. Ramirez[2], J. M. Gorriz[2], S. Foo[3], and A. Meyer-Baese[1]

[1] Scientific Computing Department, Florida State University, Tallahassee, FL 32306, USA
illan@ugr.es
[2] Departament of Signal Theory, Networking and Communications, Universidad de Granada, Granada, Spain
[3] Electrical and Computer Engineering, FAMU-FSU, Tallahassee, FL 32310, USA

Abstract. Computer aided applications in Dynamic contrast enhanced magnetic resonance imaging (DCE-MRI) are increasingly gaining attention as important tools to asses the risk of breast cancer. Chest wall detection and whole breast segmentation require effective solutions to increase the potential benefits of computer aided tools for tumor detection. Here we propose a 3D extension of Gabor filtering for detection of wall-like regions in medical imaging, and prove its effectiveness in chest-wall detection.

Keywords: Medical image processing · 3D gabor filter
Breast segmentation · DCE-MRI · Computer aided image processing

1 Introduction

Medical image processing frequently requires the detection and segmentation of regions, such as liver, abdominal wall or AC-PC line on the brain [5,13,18], separated by wall-like organic structures. In DCE-MRI for breast cancer diagnosis, internal organs produce strong noisy signals that may obstruct the detection of cancer-related signals originated in the breast. Automated segmentation of the whole-breast from the other parts imaged is a crucial preprocessing step for computer-aided diagnosis (CAD) applications. To this aim, the most challenging region to detect is the chest wall line, due to its overlap with other nearby regions and its mixed tissue composition. Automatic segmentation of whole breast has important applications, not only limited to CAD systems, but also extended to breast density estimation [2] or fibroglandular tissue extraction [1]. Recently, several approaches have been proposed in the literature to adress the problem of automatic whole-breast segmentation [7,10,17], dominantly by atlas-based approaches [9,11]. Automatic methods often rely on manually-defined benchmarks, either through specific anatomical points or through atlas [3]. Manual interventions are prone to error, and breast morphological variability hinder

© Springer International Publishing AG, part of Springer Nature 2018
F. J. de Cos Juez et al. (Eds.): HAIS 2018, LNAI 10870, pp. 446–454, 2018.
https://doi.org/10.1007/978-3-319-92639-1_37

benchmark definitions. Here, an automatic algorithm that do not require manually predefined landmarks is presented, based on a chest-wall enhancement processing by 3D Gabor filters.

Gabor filtering is a prolific texture-analysis technique, first proposed as a one dimensional tool for signal processing, and soon extended to the two dimensional case and widely extended for image processing. Lately, several successful attempts have been made to extend Gabor filters to the 3D case, as in the problem of motion detection [15] or face recognition [16]. In the context of medical imaging, the concept of 3D Gabor filter bank as a resource for texture analysis has been also exploited [14,18]. Here, the use of 3D Gabor filters is focused on wall-like structure enhancement rather than texture analysis, therefore requiring a different implementation from filter bank processing. Concretely, a 3D Gabor filter is applied to spatial windows in the image with an orientation previously obtained from the mentioned image. The main effect is that the information of surrounding structures is used for continuity in regions of overlapping.

2   Methods

2.1   3D Gabor Filter

Three dimensional complex Gabor functions in space domain are defined by:

$$h(\mathbf{x}) = h(x, y, z) = g(x, y, z) \cdot s(x, y, z) \tag{1}$$

where $s(x, y, z)$ is a complex sinusoid, known as the carrier and $g(x, y, z)$ is a Gaussian kernel, known as the envelope. The carrier, in its most general form can be expressed as:

$$s(\mathbf{x}) = \exp j(2\pi \mathbf{v}^T \mathbf{x} + P) \tag{2}$$

where $\mathbf{v}^T = (v_0, u_0, w_0)$ and P define the spatial frequency vector and the phase respectively. Without loss of generality, it is always possible to find a new coordinate system $\mathbf{x}'$ in which $\mathbf{v}'$ is aligned with the x' axes, so that:

$$s(\mathbf{x}') = \exp j(2\pi f x' + P) \tag{3}$$

where $f = \sqrt{v_0^2 + u_0^2 + w_0^2} = \|\mathbf{v}\|$. Therefore, the frequency is more precisely defined by the length of the vector $\mathbf{v}$.

The Gaussian kernel is usually defined as [4,15]:

$$g(x', y', z') = \frac{1}{(2\pi)^{2/3}\sigma_x\sigma_y\sigma_z} \exp\left[-\frac{x'^2}{2\sigma_x^2} - \frac{y'^2}{2\sigma_y^2} - \frac{z'^2}{2\sigma_z^2}\right] \tag{4}$$

with σ_x, σ_y and σ_z being the scale of the Gaussian in each respective direction. The exponent can be written as $-\frac{1}{2}\mathbf{x}'^T\mathbf{D}^{-1}\mathbf{x}'$, with $\mathbf{D} = \text{diag}(\sigma_x^2, \sigma_y^2, \sigma_z^2)$. However, in its most general form, the Gaussian kernel does not need to be aligned with the x, y, and z axes, but in general:

$$g(\mathbf{x}) = \frac{1}{(2\pi)^{2/3}\sqrt{\det(\mathbf{K})}} \exp\left[-\frac{1}{2}\mathbf{x}^T\mathbf{K}^{-1}\mathbf{x}\right] \tag{5}$$

where $\mathbf{K}$ is a symmetric matrix, related to $\mathbf{D}$ by the congruency transformation $\mathbf{D} = \mathbf{A}^T \mathbf{K} \mathbf{A}$. Being $\mathbf{D}$ positive definite, $\mathbf{A}$ is an element of the group $\mathbf{SO}(3)$, representing a special orthogonal transformation of the coordinate system $\mathbf{x} = \mathbf{A}\mathbf{x}'$. Note that, in general, this is not the coordinate transformation that simplifies the envelope function in Eq. (3).

The most general 3D Gabor filter, with the center of the Gaussian at the origin, will be expressed through Eq. (1), by combining the envelope function defined in Eq. (2) and the Gaussian kernel in Eq. (5).

The main interest of this problem is in spherically symmetric Gaussian kernels (SGK), obtained by assuming that $\sigma_x = \sigma_y = \sigma_z = \sigma$. In the SGK case, the frequency vector $\mathbf{v}$ in Eq. (2) breaks the spherical symmetry in an unique direction of the space, preserving rotational invariance on the orthogonal subspace, defining an *oriented filter*. The Fourier transform of the SGK filter is:

$$\hat{h}(\mathbf{u}) = \int_{\mathbb{R}^3} h(\mathbf{x}) \exp(-2\pi j \mathbf{u}^T \mathbf{x}) d\mathbf{x} \tag{6}$$

$$= \exp(-\frac{\sigma^2}{2}(\mathbf{u} - \mathbf{v})^T (\mathbf{u} - \mathbf{v})) \exp(jP) \tag{7}$$

It is easy to see (by taking the derivative respect $\mathbf{u}$) that the peak response of the filter occurs when $\mathbf{u} = \mathbf{v}$. In other words, the energy or magnitude of the Fourier transform $\|\hat{h}(\mathbf{u})\|$ attains its maximum in the frequency space for frequencies of magnitude $\|\mathbf{v}\|$, and direction given by the vector $\mathbf{v}$. Then, the maximum half magnitude bandwidth can be obtained from:

$$\exp(-\frac{\sigma^2}{2}(\mathbf{u} - \mathbf{v})^T (\mathbf{u} - \mathbf{v})) = 0.5 \tag{8}$$

when $\mathbf{v}$ and $\mathbf{u}$ are parallel. In that case:

$$(\|\mathbf{u}\| - f) = \frac{1}{\sigma}\sqrt{2\log(2)} \approx \frac{1.17}{\sigma} \tag{9}$$

The parameter σ scales the bandwidth of the filter, determining the half bandwidth as $2\frac{1.17}{\sigma}$. It is important to recall that if $f << 2\frac{1.17}{\sigma}$ the filter behaves effectively as a 3D Gaussian filter, and the envelope function has no effect.

Summarizing, three parameters are necessary to characterize the SGK 3D Gabor filter: the scale of the filter σ, the frequency $\|\mathbf{v}\|$ and the phase P, while two angles θ and φ will be sufficient to determine the spatial orientation of the frequency vector $\mathbf{v}$. Therefore, the convolution of the filter with a 3D image will acquire its maximum response when the filter is aligned with the spatial direction of frequency modulation and the magnitude of the frequency is comparable to the spatial frequency of the image. If these parameters are estimated from some properties of the 3D image, then this result can be exploited to enhance 3D structures within the image. Concretely, σ, $\|\mathbf{v}\|$ and P, will be estimated globally, to fit the wall-like properties of the image to enhance. On the other hand, the orientation of the frequency vector $\mathbf{v}$ will be estimated locally, by averaging the orientation field in the neighborhood of each point.

2.2 Orientation Field

The aim of using the oriented filter of Eq. (1) is to enhance the wall-like structure of the chest wall. In contrast with usual applications of 3D Gabor filtering [18], it is convenient to orient the filter parallel to the wall instead of using a filter bank. In texture analysis, several methods have been proposed to calculate the orientation of patterns [8].

We will follow here the method proposed by Hong, Wan and Jain [6] to obtain the orientation field. The angle in radians for each point (i, j) will be given by:

$$\theta(i,j) = \frac{\pi}{2} + \frac{1}{2} \tan^{-1} \frac{\partial_x(i,j) \cdot \partial_y(i,j)}{\partial_x(i,j)^2 - \partial_y(i,j)^2} \tag{10}$$

where ∂_x is the gradient in the direction x, and ∂_y is the gradient in the direction y. The gradients must be smoothed before calculating the orientation map. As in spherical coordinates, two angles θ and φ will be necessary to fix the filter orientation in space. While θ will provide the orientation in the **xy** plane, φ will determine the angle with respect to the vertical line as:

$$\varphi(i,j) = \frac{\pi}{2} + \frac{1}{2} \tan^{-1} \frac{H_{xy}(i,j) \cdot \partial_z(i,j)}{H_{xy}(i,j)^2 - \partial_z(i,j)^2} \tag{11}$$

where $H_{xy}(i,j) = \sqrt{\partial_x(i,j)^2 + \partial_y(i,j)^2}$. Figure 1 shows an example of orientation fields obtained from Eqs, (10) and (11) from a DCE-MRI image.

3 Experiments

The Medical Center at the University of Maastricht provided 32 breast DCE-MRI cases, belonging to different BI-RADS categories of lesions and densities. The MRI was performed with a 1.5 T system (Magnetom Vision, Siemens, Erlangen, Germany) equipped with a dedicated surface coil to enable simultaneous imaging of both breasts. The patients were placed in a prone position and transversal images were acquired with a short TI inversion recovery (STIR) sequence (TR = 5600 ms, TE = 60 ms, FA = 90°, IT = 150 ms, matrix size 228 × 182 pixels, slice thickness 3 mm). Then a dynamic T1-weighted gradient echo sequence (3D fast low angle shot sequence) was performed (TR = 4.9 ms, TE = 1.83 ms, FA = 12°) in transversal slice orientation with a matrix size of 352 × 352 pixels and an effective slice thickness of 1 mm. The dynamic study consisted of 5 measurements with an interval of 1.4 min. The first frame was acquired before injection of paramagnetic CA (gadopentatate dimeglumine, 0.1 mmol/kg body weight, Magnevist^{TM}, Schering, Berlin, Germany) immediately followed by the four other measurements. The first frame was used for algorithm evaluation, preprocessed with a median filter, in order to avoid unnecessary masking from enhancing tissues.

The validation metric used in this analysis is the dice similarity coefficient (DSC), commonly used [2,3] to asses the efficiency of segmentation algorithms (A) versus manually-generated (M) segmentations. It is defined as:

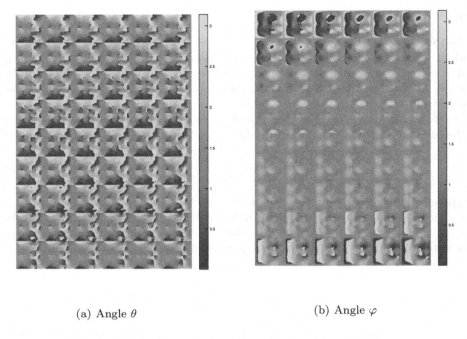

(a) Angle θ (b) Angle φ

Fig. 1. Orientation maps in radians displayed in axial slices.

$$DSC = 2 * \frac{A \bigcap M}{A \bigcup M} \qquad (12)$$

and measures the amount of overlap between segmentations with respect to the size of the segmented region.

For the execution of the algorithm, it is necessary to fix three parameters: the size of the window w, the frequency f and the phase P.

3.1   Results

The adjustable parameters of the filter f and w are problem dependent. Once they are set for a specific image modality, they don't need further adjustments. The frequency f optimal value (f=1.4) is dependent on the thickness of the wall-like structure to enhance. Frequency values above the optimal value will produce several hills and valleys for each wall to detect, while suboptimal values will not produce any enhancement of wall-like structures. The scale of the Gaussian σ is defined to be dependent on the size of the window, so that $3\sigma \simeq w$, corresponding to full width at half maximum (FWHM). The filter size is adjusted to the curvature of the wall structure to enhance, being higher values suitable for coarser structure enhancement, and lower values for fine ones. In this case, the value is set to fill a volume of approximately 15% of the total volume of the

Algorithm 1. Segmentation algorithm

1: **for** each $\theta \in [0, 2\pi]$ and $\varphi \in [0, \pi]$ **do**
2:    Create a filter bank $h(\theta, \varphi)$
3: **end for**
4: **for** each x_i $i = 1, ..,$ number of voxels **do**
5:    Calculate the values of θ_i and φ_i
6:    On windows of size $w \times w \times w$ centered at x_i convolve the filter $h(\theta_i, \varphi_i)$ with the image.
7: **end for**
8: Quantize the edge image into 3 different intensity levels (hills, valleys and background).
9: Binarize the quantized image and remove small objects.
10: Morphological open, fill and close the binary image to get a segmentation mask.

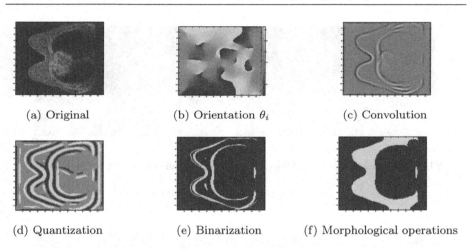

(a) Original   (b) Orientation θ_i (c) Convolution

(d) Quantization   (e) Binarization   (f) Morphological operations

Fig. 2. Steps followed in the segmentation process.

image. The phase P is set to 0 in order to have symmetric effects around the chest-wall and avoid orientation ambiguities at 0 and π radians.

Figures 2 and algorithm 1 summarize the main steps necessary to perform the SGK segmentation:

The Otsu' method [12] is used in the quantization step to set 2 global thresholds in the image. This method is simple and fast but might lead to edge discontinuities. This could be critical in some extreme cases, such as patients that underwent surgery.

The total DSC value of the database is 0.91 with a mean value of 0.90 and a standard deviation of 0.04. Higher DSC values have been reported in the literature [2,10]. However, the database and manual segmentation differences make direct comparisons unfeasible. In general, a DSC value over 0.9 is indicative of a high overlapping between manual and automatic segmentations [3]. Moreover, the proposed approach generalizes the 2D segmentation of Milenkovic et. al. [10] to the 3D case, more suitable for the nature of the problem.

Figures 3 and 4 compare the segmentation results of the proposed SGK method and the 2D approach of Milenkovic et. al., applied in a challenging case, with medium fibroglandular density and overlapping of chest wall and internal organs. The results in the Milenkovic et. al. case shows that the chest wall line does not follow the segmentation in every slice, but some other internal organs and bones are included in the segmented region. The inclusion of 3D information leads to the SGK segmentation of Fig. 4, that follows the chest wall line more accurately. Moreover, the proposed approach does not exclude the lymph nodes, that may contain relevant information for diagnosis, since it does not rely on landmarks for segmentation.

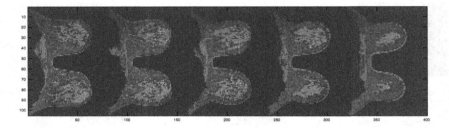

Fig. 3. Representative axial slices for the Milenkovic et.al. [10] segmentation

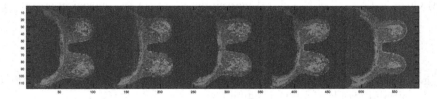

Fig. 4. Representative axial slices for the proposed SGK segmentation

4   Conclusions and Future Work

This contribution proposes a novel method for detecting challenging wall-like structures in medical imaging by 3D Gabor filtering, and is successfully applied to breast segmentation in DCE-MRI, having a potential impact addressing the risk of the most common cancer in women. The proposed method performs with DSC of 0.91 in a set of 32 images. Still, a margin of improvement is left in the

quantization step, where adaptive methods could be employed to avoid discontinuities in challenging cases. The proposed approach overcomes the limitations of
some other automatic methods, as the dependency on manually defined benchmarks or atlas, with enough versatility to generalize to other databases. It also
improves previously published methods that do not take into account 3D properties of the DCE-MRI images [10]. Moreover, its versatility also allows it to be
applied to other wall-like detection problems in medical imaging, as the detection
of mid-sagittal plane in the brain.

Acknowledgment. We'd like to thank Marc Lobbes for the provision of the DCE-
MRI database. This work is supported by Marie Sklodowska-Curie actions (MSCA-IF-
GF-656886).

References

1. Dalmış, M.U., Litjens, G., Holland, K., Setio, A., Mann, R., Karssemeijer, N.,
   Gubern-Mérida, A.: Using deep learning to segment breast and fibroglandular tissue in MRI volumes. Med. Phys. **44**(2), 533–546 (2017)
2. Gubern-Mérida, A., Kallenberg, M., Mann, R.M., Martí, R., Karssemeijer, N.:
   Breast Segmentation and density estimation in breast MRI: a fully automatic
   framework. IEEE J. Biomed. Health Inform. **19**(1), 349–357 (2015)
3. Gubern-Mérida, A., Wang, L., Kallenberg, M., Martí, R., Hahn, H.K.,
   Karssemeijer, N.: Breast segmentation in MRI: quantitative evaluation of three
   methods. vol. 8669, p. 86693G. International Society for Optics and Photonics,
   March 2013
4. Haq, I.U., Nagoaka, R., Makino, T., Tabata, T., Saijo, Y.: 3D gabor wavelet based
   vessel filtering of photoacoustic images. In: 2016 38th Annual International Conference of the IEEE Engineering in Medicine and Biology Society (EMBC), pp.
   3883–3886, August 2016
5. Heimann, T., Ginneken, B.V., Styner, M.A., Arzhaeva, Y., Aurich, V., Bauer, C.,
   Beck, A., Becker, C., Beichel, R., Bekes, G., Bello, F., Binnig, G., Bischof, H.,
   Bornik, A., Cashman, P.M.M., Chi, Y., Cordova, A., Dawant, B.M., Fidrich, M.,
   Furst, J.D., Furukawa, D., Grenacher, L., Hornegger, J., KainmÜller, D., Kitney,
   R.I., Kobatake, H., Lamecker, H., Lange, T., Lee, J., Lennon, B., Li, R., Li, S.,
   Meinzer, H.P., Nemeth, G., Raicu, D.S., Rau, A.M., Rikxoort, E.M.V., Rousson,
   M., Rusko, L., Saddi, K.A., Schmidt, G., Seghers, D., Shimizu, A., Slagmolen,
   P., Sorantin, E., Soza, G., Susomboon, R., Waite, J.M., Wimmer, A., Wolf, I.:
   Comparison and evaluation of methods for liver segmentation from CT datasets.
   IEEE Trans. Med. Imaging **28**(8), 1251–1265 (2009)
6. Hong, L., Wan, Y., Jain, A.: Fingerprint image enhancement: algorithm and performance evaluation. IEEE Trans. Pattern Anal. Mach. Intell. **20**(8), 777–789 (1998)
7. Jiang, L., Hu, X., Xiao, Q., Gu, Y., Li, Q.: Fully automated segmentation of whole
   breast using dynamic programming in dynamic contrast enhanced MR images.
   Med. Phys. **44**(6), 2400–2414 (2017)
8. Kass, M., Witkin, A.: Analyzing oriented patterns. Comput. Vis. Graph. Image
   Process. **37**(3), 362–385 (1987)
9. Lin, M., Chen, J.H., Wang, X., Chan, S., Chen, S., Su, M.Y.: Template-based
   automatic breast segmentation on MRI by excluding the chest region. Med. Phys.
   40(12), 122301 (2013)

10. Milenković, J., Chambers, O., Marolt Mušič, M., Tasič, J.F.: Automated breast-region segmentation in the axial breast MR images. Comput. Biol. Med. **62**, 55–64 (2015)
11. Ortiz, C.G., Martel, A.L.: Automatic atlas-based segmentation of the breast in MRI for 3D breast volume computation. Med. Phys. **39**(10), 5835–5848 (2012)
12. Otsu, N.: A threshold selection method from gray-level histograms. IEEE Trans. Syst. Man Cybern. **9**(1), 62–66 (1979)
13. Prima, S., Ourselin, S., Ayache, N.: Computation of the mid-sagittal plane in 3-D brain images. IEEE Trans. Med. Imaging **21**(2), 122–138 (2002)
14. Qian, Z., Metaxas, D.N., Axel, L.: Extraction and tracking of MRI tagging sheets using a 3D gabor filter bank. In: 2006 International Conference of the IEEE Engineering in Medicine and Biology Society, pp. 711–714, August 2006
15. Reed, T.R.: Motion analysis using the 3-D gabor transform. In: Conference Record of The Thirtieth Asilomar Conference on Signals, Systems and Computers. vol. 1, pp. 506–509, November 1996
16. Wang, Y., Chua, C.S.: Face recognition from 2D and 3D images using 3D Gabor filters. Image Vis. Comput. **23**(11), 1018–1028 (2005)
17. Wu, S., Weinstein, S.P., Conant, E.F., Schnall, M.D., Kontos, D.: Automated chest wall line detection for whole-breast segmentation in sagittal breast MR images. Med. Phys. **40**(4), 042301 (2013)
18. Xu, Z., Allen, W.M., Baucom, R.B., Poulose, B.K., Landman, B.A.: Texture analysis improves level set segmentation of the anterior abdominal wall. Med. Phys. **40**(12), 121901 (2013)

Fingertips Segmentation of Thermal Images and Its Potential Use in Hand Thermoregulation Analysis

A. E. Castro-Ospina[1(✉)], A. M. Correa-Mira[1], I. D. Herrera-Granda[2],
D. H. Peluffo-Ordóñez[3], and H. A. Fandiño-Toro[1]

[1] Instituto Tecnológico Metropolitano - Medellín,
Grupo de Investigación Automática, Electrónica y Ciencias Computacionales,
Carrera 31 #54-22, Medellín, Colombia
{andrescastro,hermesfandino}@itm.edu.co
[2] Universidad Técnica del Norte, Ibarra, Ecuador
[3] Yachay Tech, Urcuquí, Ecuador

Abstract. Thermoregulation refers to the physiological processes that maintain stable the body temperatures. Infrared thermography is a noninvasive technique useful for visualizing these temperatures. Previous works suggest it is important to analyze thermoregulation in peripheral regions, such as the fingertips, because some disabling pathologies affect particularly the thermoregulation of these regions. This work proposes an algorithm for fingertip segmentation in thermal images of the hand. By using a supervised index, the results are compared against segmentations provided by humans. The results are outstanding even when the analyzed images are highly resized.

Keywords: Thermorregulation · Thermal hand images
Fingertip segmentation · NPR measurement

1 Introduction

Thermoregulation refers to the physiological processes that the human body releases as a response to a thermal stress. An example where thermoregulation intervenes is the body rewarming in response to low temperatures exposure. Any thermal recovery produces temperature patterns that can be visualized with infrared thermography images. Some pathologies as cold intolerance [1,2], Raynaud's Phenomenon (RP) [3,4] and carpal tunnel syndrome [5] produce neurological disorders which affect the thermoregulation of peripheral regions, particularly at fingertips.

This work is carried out under grants provided by "Análisis de la termorregulación de la mano, en pacientes sanos mediante imágenes de termografía infrarroja" (Programa Nacional de Jóvenes Investigadores e Innovadores – COLCIENCIAS – Convocatoria 706 de 2015).

However in practice, there are few methods dedicated to specifically segment these regions, much less to validate the effectiveness of such segmentations. In this work, we present a simple yet effective and precise strategy for fingertip segmentation from hand infrared thermal images. A performance index is used to compare the proposed strategy against segmentations made by humans, leading to excellent results, highlighting the similarities of segmentations achieved.

2   Related Work

A common strategy for visualizing the hand thermoregulation implies a thermal stimulation, where the hand of a volunteer is undergone to a cold stress (whether with water or air), which reduces temporarily its temperature values. During the test (previous rest time, stimulation, and thermal recovery time), thermographic images of the hand are acquired until this reaches temperature values close to those values before the stimulation. The most straightforward way to analyze hand thermoregulation is to perform a manual segmentation of regions of interest (ROI) into the images and to plot the average temperature values in these ROI against time [6]. It is expected that significant differences in the temperature recovery curve, in contrast to that obtained for healthy subjects, serve as an indicator of the potential presence of a pathological condition.

The evidence suggests that thermoregulation should be analyzed in acral body regions such as the fingertips [7–9], because of the presence of some structures that converge in the microvasculature of the digits, such as terminal arterioles, capillary loops, postcapillary venules, and arteriovenous anastomoses. The latter are directly involved in thermoregulation [10]. There exist published papers that analyze the hand thermoregulation specifically the fingertips. One of the first related works is [11] and uses mathematical morphology to segment the palm and the fingers. Based on this first work, in [12] is proposed a methodology to automatically segment the fingertips. For this last work, some parameters involved in the segmentation are assigned arbitrarily. In general, none of these works include a comparison against a ground truth and in both, authors provide few information about the tuning and specifications of structuring elements involved.

In this work, morphological information of the hand is used to modify a recently proposed algorithm which identifies and labels the five fingers of the hand, from infrared thermography images of the dorsal region. Presented modification allows to delimit the area of each finger that encloses its distal phalange, since it is an acral region and according to the evidence, is an area of great importance for analyzing the hand thermoregulation.

3   Materials and Methods

The sequences of images used in this work came from the database proposed and created in [6]. This database consists of sequences of thermal images with the hand thermal recovery of 17 healthy subjects. The Table 1 summarizes the protocol used to acquire the images analyzed in this work.

The apparent reflected temperature was determined according to the method proposed by the standard ASTM - E1862 [13]. The working distance was set by considering the field-of-view of the thermal camera used, and the expression proposed in Ref. [5]. These two parameters and the ambient temperature, were measured before each acquisition by using a calibrated hygro-thermometer.

Table 1. Acquisition protocol summary. More information is provided in Ref. [6].

Camera details	Model	FLIR A655sc
	Spectral range	7.5–14 μm
	Image size	480 × 640 pixels
Acquisition details	Ambient temperature	[26.8–28 °C]
	Relative humidity	47.18 ± 2.16%
	Apparent reflected temperature	Equal to ambient temperature
	Working distance	0.58 m
	Emissivity	0.97
	Sampling frequency	1 fps
Acquisition duration	Basal temperature acquisition time	120 s
	Cold stimulus	30 s
	Recovery acquisition time	600 s

After an acclimatization period of 900 s, approximately 750 frames are acquired to record: the basal temperature of the volunteer before the cold stimulation, the time period where the hand is stimulated with water at 15 °C, and the thermal recovering of the hand once the cold-stimulation is removed.

3.1 Fingertip Segmentation

Expectation-Maximization algorithm [14] was used to threshold the images used in this work. The algorithm for segmenting the distal phalanges relies on the method proposed by Zapata et al. [15]. In such work, the fingertips are localized in a time-series created with pixels in the hand contour. Then, a distance-based criterion is used to label the fingertips. The method locates eleven pixels-of-interest in the hand geometry: a pixel at each fingertip and four pixels at inter-digital regions of the finger pairs: thumb-index, index-middle, middle-annular, and annular-little. The algorithm also finds two pixels useful to demarcate the outer contour of the thumb and little fingers. With these pixels located, the following process is applied to each finger:

1. Find the midpoint in the line segment delimiting the base of each finger.
2. Draw a new line segment to join the midpoint found in step 1, with their corresponding fingertip.

3. On the line segment found in step 2, locate a pixel whose distance to the fingertip agrees with the percentage proposed in [16] for the length of the distal phalanx.
4. From the fingertip, the pixels in the hand contour are considered, advancing towards the inner and outer sides of the finger. For each pixel in both directions, a straight line is drawn to the pixel found in point 3. In each direction, the slope of the generated straight line is evaluated and compared to the slope of the straight line delimiting the base of the finger.
5. When the straight lines are approximately parallel to the line that delimits the base of the correspondent finger, such pixels are selected.
6. Localization of pixels before and after the analyzed fingertip can lead to finding a single discontinuous line for delimiting the distal phalanx. For this reason, the algorithm concludes after drawing a single straight line that connects the two pixels found in step 5.

3.2   Normalized Probabilistic Rand Index (NPR)

To assess the effectiveness of the proposed method, three humans were asked to segment the distal phalanges from the analyzed hand images. Such segmentations were used as the ground truth.

To compare the algorithm and the humans segmentation, the Normalized Probabilistic Rand index (NPR) is used [17]. This index gives the proportion of pairs of pixels correctly classified in the algorithm-based segmented image and the ground truth, allowing to average multiple ground truth segmentations (three in our case), thus taking into account the variation of human perception. Its important to note that the NPR measure allows to compare the performed segmentation against those made by humans since it is highly correlated with human-made segmentations [18].

3.3   Experimental Framework

The steps involved at experimental framework are depicted in Fig. 1. For every image sequence, we compute the image representing the mean of the first five frames in order to avoid unwanted noise due to subject movement. The proposed algorithm and three humans segment the fingertips from such mean frame. The NPR measure is used to compare the segmentations carried out by the proposed algorithm and humans. Calculation of the NPR measure is computationally expensive because it estimates the pairs of pixels correctly segmented, therefore, it evaluates all possible pairs, images are resized to 15%, 30%, 50% and 70%, leading to images of sizes 72×96, 144×192, 240×320 and 336×448, respectively.

Table 2 shows an example of the resizing operation for one segmentation mean frame. The pseudo-color image at the top is the mean image computed with the first five frames in the image sequence. Then, both human and algorithm fingertip segmentations are depicted for every used resize ratio. Visually, it can be seen how similar are human and algorithm segmentations. Moreover, is clear how fingertip segmentation is not compromised when the resizing is performed.

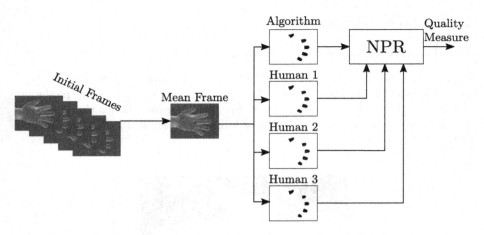

Fig. 1. Experimental scheme to evaluate the proposed algorithm.

Table 2. Top: Original mean Frame. Bottom: Algorithm and human segmentations resized.

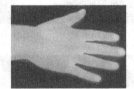

4   Results and Discussion

4.1   Distal Phalanges Segmentation

Figure 2(a) shows a thermal image, (b) their thresholding using the Expectation-Maximization algorithm and (c) the segmentation of the distal phalanges achieved with the method proposed in Sect. 3.1. Figure 2 shows that the algorithm effectively segments the region including the nailfolds and in general, the distal phalanges.

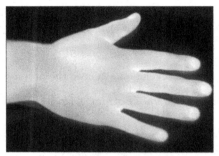

(a) Thermal example image.

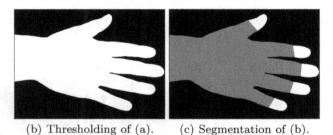

(b) Thresholding of (a).   (c) Segmentation of (b).

Fig. 2. In (a) a termographic image (pseudocolor) used in this work, and (b), the result of segmenting the distal phalanges with the proposed method.

Frequently, the Otsu's method is used as a strategy of global thresholding for thermal images. This is a plausible strategy when thermal contrast in the analyzed thermal images is not a problem. In this work, we found that Otsu's method was not a suitable thresholding method. As a possible cause of this situation, we observed a diminished contrast along the edges between the hands and the background of the considered images.

The surface where the hands rested during the acquisition was built on acrylic. We found that such surface stored heat which was visible as false hot-spots in the thermal images in presence of fingers movement. The expectation-maximization algorithm showed a better performance than Otsu's thresholding method under these conditions.

The segmentations made by human and the proposed method are similar, as shown in Table 2. However, the overall quantitative performance of such segmentations was measured with the NPR index. Results are summarized in Fig. 3. The boxplots show how the image resizing needed to accelerate NPR time evaluation, does not seriously affect the general results. Furthermore, NPR outstanding results above 0.965 gives evidence of how similar is the segmentation achieved by means of our proposed algorithm with the three ground truth, i.e. the human-based segmentations.

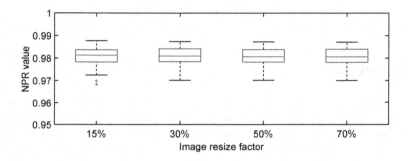

Fig. 3. NPR values for used resize factors.

It is clear that the proposed segmentation is simple. This work is pertinent because it presents a tool that could be used in the automatic analysis of the hand thermoregulation. In addition to a proper setup for acquiring the thermal images, a dedicated image processing technique is necessary for analyzing the thermal recovery exactly at the fingertips. This is a transversal issue in applications where the hand is analyzed, not only in the infrared spectrum [19].

Consider for example the thermal recovering profiles presented in Fig. 4. This figure shows the mean temperatures calculated on eight image sequences, but in two different regions-of-interest, the annular finger and its distal phalange.

The part 4(a) shows the mean temperature evolution of the annular fingers. Part 4(b) shows the mean temperature evolution measured only at fingertips. The solid lines in the figures show the mean temperature and the gray regions it standard deviation. An clear result after comparing the recovering profiles is that complete fingers exhibit higher temperature variations than the fingertips.

Both regions exhibit a similar characteristic curve with a minimum point when the cold stimulation occurs and a progressive temperature rise starting when this stimulation finishes. A less evident result is that temperatures at the fingertips are comparatively higher than temperatures measured in the complete fingers.

The utility of the differences found is that analysis can be focused on more compact regions-of-interest which. Future research should analyze the impact of these subtle temperature variations in characterizing the hand thermoregulation.

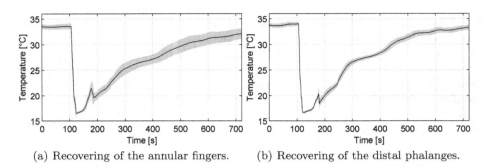

(a) Recovering of the annular fingers.   (b) Recovering of the distal phalanges.

Fig. 4. Mean thermal recovering profiles calculated in eight image sequences.

5   Conclusions

An algorithm for distal phalanges segmentation from thermographic images of the hand was presented. Attained segmentation results are compared against human-based segmentation through the normalized probabilistic rand index, which uses human segmentations as ground truth and averages comparison results. Achieved results over 0.965 demonstrate the effectiveness of the proposed algorithm when segmenting fingertips.

As a future work, it will be necessary characterizing the thermal recovery profiles. Considering the trajectories described by these profiles, the strategies could be based on curve fitting schemes. Also, it will be necessary to explore machine learning techniques for analyzing the extracted patterns against those exhibited by pathological subjects.

Acknowledgements. This work is supported by Colciencias grant 564-2015.

References

1. Park, K.S., Park, K.I., Kim, J.W., Yun, Y.J., Kim, S.H., Lee, C.H., Park, J.W., Lee, J.M.: Efficacy and safety of korean red ginseng for cold hypersensitivity in the hands and feet: a randomized, double-blind, placebo-controlled trial. J. Ethnopharmacol. **158**, 25–32 (2014)
2. Packham, T.L., Fok, D., Frederiksen, K., Thabane, L., Buckley, N.: Reliability of infrared thermometric measurements of skin temperature in the hand. J. Hand Ther. **25**(4), 358–362 (2012)
3. Pauling, J.D., Shipley, J.A., Hart, D.J., McGrogan, A., McHugh, N.J.: Use of laser speckle contrast imaging to assess digital microvascular function in primary raynaud phenomenon and systemic sclerosis: a comparison using the raynaud condition score diary. J. Rheumatol. **42**(7), 1163–1168 (2015)
4. Lim, M.J., Kwon, S.R., Jung, K.H., Joo, K., Park, S.G., Park, W.: Digital thermography of the fingers and toes in raynaud's phenomenon. J. Korean Med. Sci. **29**(4), 502–506 (2014)

5. Roldan, K.E., Piedrahita, M.A.O., Benitez, H.D.: Spatial-temporal features of thermal images for carpal tunnel syndrome detection. In: IS&T/SPIE Electronic Imaging, International Society for Optics and Photonics, pp. 90190E–90190E (2014)
6. Ramírez, L., Jiménez, K., Correa, A., Giraldo, J., Fandiño Toro, H.: Protocolo de adquisición de imágenes diagnósticas por termografía infrarroja. Medicina Laboratorio 21, 161–178 (2015)
7. Niehof, S.P., Huygen, F.J., van der Weerd, R.W., Westra, M., Zijlstra, F.J.: Thermography imaging during static and controlled thermoregulation in complex regional pain syndrome type 1: diagnostic value and involvement of the central sympathetic system. Biomed. Eng. Online 5(1), 30 (2006)
8. Kistler, A., Mariauzouls, C., von Berlepsch, K.: Fingertip temperature as an indicator for sympathetic responses. Int. J. Psychophysiol. 29(1), 35–41 (1998)
9. Romeijn, N., Verweij, I.M., Koeleman, A., Mooij, A., Steimke, R., Virkkala, J., van der Werf, Y., Van Someren, E.J.: Cold hands, warm feet: sleep deprivation disrupts thermoregulation and its association with vigilance. Sleep 35(12), 1673–1683 (2012)
10. Ruch, D.S., Vallee, J., Li, Z., Smith, B.P., Holden, M., Koman, L.A.: The acute effect of peripheral nerve transection on digital thermoregulatory function. J. Hand Surg. 28(3), 481–488 (2003)
11. Blank, M., Kargel, C.: Infrared imaging to measure temperature changes of the extremities caused by cigarette smoke and nicotine gums. In: 2006 Proceedings of the IEEE Instrumentation and Measurement Technology Conference, IMTC 2006, pp. 794–799. IEEE (2006)
12. Zhang, H.D., He, Y., Wang, X., Shao, H.W., Mu, L.Z., Zhang, J.: Dynamic infrared imaging for analysis of fingertip temperature after cold water stimulation and neurothermal modeling study. Comput. Biol.Med. 40(7), 650–656 (2010)
13. Bauer, E., De Freitas, V.P., Mustelier, N., Barreira, E., de Freitas, S.S.: Infrared thermography-evaluation of the results reproducibility. Struct. Surv. 33(1), 20–35 (2015)
14. Moon, T.K.: The expectation-maximization algorithm. IEEE Signal Process. Mag. 13(6), 47–60 (1996)
15. Zapata, N., Orrego, S., Ramírez, L., Castro, A., Fandiño Toro, H.: Processing of thermal images oriented to the automatic analysis of hand thermoregulation. In: IFMBE Proceedings (2016)
16. Ertas, I.H., Hocaoglu, E., Patoglu, V.: Assiston-finger: An under-actuated finger exoskeleton for robot-assisted tendon therapy. Robotica 32(08), 1363–1382 (2014)
17. Unnikrishnan, R., Pantofaru, C., Hebert, M.: A measure for objective evaluation of image segmentation algorithms. In: 2005 IEEE Computer Society Conference on Computer Vision and Pattern Recognition-Workshops, CVPR Workshops, p. 34. IEEE (2005)
18. Yang, A.Y., Wright, J., Ma, Y., Sastry, S.S.: Unsupervised segmentation of natural images via lossy data compression. omput. Vis. Image Underst. 110(2), 212–225 (2008)
19. Arroyave-Giraldo, M., Restrepo-Martíne, A., Vargas-Bonilla, F.: Incidencia de la segmentación en la obtención de región de interés en imágenes de palma de la mano. Tecno Lógicas, vol. 27, pp. 119–138 (2011)

Data Mining Applications

Listen to This: Music Recommendation Based on One-Class Support Vector Machine

Fabio A. Yepes, Vivian F. López$^{(\boxtimes)}$, Javier Pérez-Marcos, Ana B. Gil, and Gabriel Villarrubia

Departamento de informática y automática, Universidad de Salamanca, Salamanca, Spain
{andresyv,vivian,jpmarco,abg,gvg}@usal.es,
http://diaweb.usal.es

Abstract. The streaming services are here to stay. In recent years we have witnessed their consolidation and success, which is manifested in their exponential growth, while the sale of songs/albums in physical or digital format has declined. An important part of these services are recommendation systems, which facilitate the exploration of content to users. This article proposes a content-based approach, using the One-Class Support Vector Machine classification algorithm as an anomaly detector. The aim is to generate a playlist that adapts to the user's tastes, incorporating the novelties of new releases. The model is capable of detecting elements that belong to the profile of the user's tastes with great accuracy, facilitating the implementation of an Android mobile application that scans and detects changes in user preferences. This will make it possible not only to manage the playlist that has been recommended, but also periodically to incorporate new songs to the profile from the list of new music.

Keywords: Recommender system · Collaborative filtering
Content-based · Streaming · Spotify · Data mining
One-Class Support Vector Machine

1 Introduction

We find ourselves in the "streaming era", as it is known, due the consolidation of this type of content-on-demand services from the year 2015 to the present. The success of streaming content is undeniable. In 2016, despite the fall in the sale of CDs and any songs in physical and digital format, the figures generated by the direct listening of music are impressive.

© Springer International Publishing AG, part of Springer Nature 2018
F. J. de Cos Juez et al. (Eds.): HAIS 2018, LNAI 10870, pp. 467–478, 2018.
https://doi.org/10.1007/978-3-319-92639-1_39

It is increasingly common to see people resorting to streaming services, such as *Spotify*,[1] *Tidal*[2] and *PandoraMusic*[3] for legal use of multimedia, unlike a few years ago when there was a greater tendency to download illegal content through torrents or link aggregators. The report written by the New York firm BuzzAngle [7], reflects the behaviour of the American public in its musical consumption. The figures indicate that in 2016, sales of albums in any format fell by 15.6% (11.7% in physical format and 19.4% in digital support) and independent songs by 24.8%, but consumption of general music grew by 4.9%. Streaming grew 82.6% and reached a record of 250 thousand million reproductions.

To facilitate this type of service, it is very common to use Recommendation Systems (RS), which generate a profile of the users' tastes and, based on that, offer personalized content which may be of interest to each user, making it easier to scan the content since exploring the entire catalogue in search of something that we like would be totally unfeasible; such is the case, for example, of Spotify. The implementation of these systems is crucial in streaming platforms and poses a challenge when it comes to satisfying users. But the way in which users consume multimedia content is also of great interest. The year 2016 was the year in which, for the first time in history, the mobile phone surpassed the computer to access the internet, and frequently people leave their computer aside, since they have smartphones that can meet their daily needs on the network. In addition, Spain is a leader in the consumption of devices such as tablets, eBooks or Smart TVs in comparison to countries such as the United Kingdom, Germany, Argentina and Brazil. Seventy-eight point seven percent of the entire population between 16 and 74 years of age connects regularly to the internet, with 67.5% (the highest growth) doing so to consume movies, videos and music [2].

Taking into account the above, this work aims to create a music RS based on content, which makes it possible to generate a list of songs that will suit the user's taste, employing a One-Class Support Vector Machine. This will facilitate the implementation of an application for Android mobile devices to track and detect changes in user preferences and benefit from the RS, making it possible not only to create, delete, reproduce and export the playlist that has been recommended to the user, but also to periodically incorporate new songs to the user profile from the list of new features.

2   Recommender Systems

RSs are a set of powerful tools and techniques for analysing large amounts of data, especially product and user information, to subsequently provide relevant suggestions based on data mining approaches [9]. Examples of these systems are

---

[1] spotify.com.

[2] tidal.com.

[3] play.google.com/store/apps/details?id=com.pandora.android&hl=es&rdid= com.spandora.android.

those that provide suggestions for books on *Amazon*[4], movies on *Netflix*[5] music on Spotify, and social networks such as Twitter[6] in their section "to follow".

The design of these systems depends on the domain and the characteristics of the data that can be extracted from the available information. Spotify, for example, takes into account the sequence of songs that users listen to or like in their interaction with the system. To make the recommendation of a playlist in your Spotify's Discover Weekly, it fixes on a song that the user has heard, or on others that users have added to a playlist in which the song was included. If they match the musical vector of the user and if the user has never heard it, a playlist of music that may be of their liking is generated. This is just one way to do it, because RSs differ in how they analyse data sources and create relationships between users and elements that can be used to identify taste patterns.

As indicated in [1], we can find RS with the closest neighbour approach, among which is collaborative filtering [11] and its two types: user-based and item-based. The advantage of these systems is that they are easy to implement and very accurate. However, they have their limitations, such as the cold start problem [6], which fails to recommend items to new users for which there is no information available in the system. In addition, there are personalized RSs [8], such as those based on content [5], context-based ones and, finally, those based on models such as the Support Vector Machines (SVM), classification techniques, K-Nearest Neighbours (KNN), factorization of matrices and RS hybrids. The latter consist of combining several types of RS to build a more robust system.

2.1   Content-Based Filtering

This RS uses knowledge about the elements present in the system to show those that are similar to those that the user has liked previously. Unlike the collaborative approach, the properties of the items and user prereferences are used here to generate the recommendation. The recommendation takes into account the properties of the items and the user profiles. For example, if a user liked the song *Angie - The Rolling Stones*, the system will learn their preferences and recommend related songs, such as *Paint it Black*, which belong to the same group. In the construction of a content-based RS, three main steps are included [9]:

1. A user profile is generated containing their preferences regarding the properties of the products.
2. Content information for products is generated.
3. The model to produce recommendations to the active user is generated and predicts a list of items that the user may like.

These systems improve their behaviour regarding the cold start problem, as they present the collaborative filtering when new users appear. They are also often used in hybrid approaches [8] which seek to overcome the disadvantages

---

[4] amazon.com.

[5] netflix.com.

[6] twitter.com.

of some types of systems with the advantages of others and make the system more robust. For example, if we combine an RS of collaborative filtering with a content-based RS, we can eliminate the cold-start problem that collaborative filtering systems suffer from and that content-based RSs can solve. The latter use characteristics of resources to obtain recommendations. Most resources contain in their description high-level information (title, categories, descriptions or metadata) or low-level features (images, audio, ratings) that help solve the problem. This is their main advantage in the recommendations compared to the other approaches, because they are much more difficult to implement.

2.2   Recommendation Systems Based on Models

So far we have talked about approaches that involve similarity calculations between users or products to determine the affinity of the user with the product. These types of approaches have their own limitations. Since in order to make similarity calculations the information has to be loaded into memory, these approaches are called memory-based models. They are quite slow in scenarios that require a response in real time when the amount of data is very large. Another problem is that when the similarity weights are calculated, they are not learned automatically, giving rise to the cold start problem already mentioned.

To overcome the above limitations and improve the performance of the RSs, more advanced techniques have been applied [9], such as probability models, machine-learning models (supervised and non-supervised learning) and matrix approaches, like factoring. In this type of system a model based on the learned weights is generated automatically thanks to the available historical data [9]. The new product predictions will be made using the weights learned and then the final results will be classified in a specific order before making the recommendations. The simplest methods for classification are those based on the similarity between resources depending on their characteristics. These approaches recommend resource labels similar to those evaluated by using similarity coefficients, such as the cosine coefficient, on the feature vectors. Within this classification are the SVM [4], used in this work as part of the recommendation algorithm.

SVM is a supervised learning algorithm that can be used for both regression and classification, although it is mainly applied as a classifier [10]. In this algorithm, each data element is represented as a point in an n-dimensional space (with n equal to the number of properties that the element has), the value of each characteristic being the value of a particular coordinate. Then, the classification is done by finding the best hyperplane that clearly differentiates the two classes (Fig. 1).

The best hyperplane is the one that maximizes the margin between the two classes, the one whose distance is greater with each class. The hyperplane is learned by training the algorithm with information that uses procedures that maximize that margin. In practice, the real data is chaotic and cannot be separated perfectly in a hyperplane. The restriction of maximizing the margin of the line that separates the classes must be relaxed. This is usually called a soft margin classifier. This change allows some points in the training data to cross the

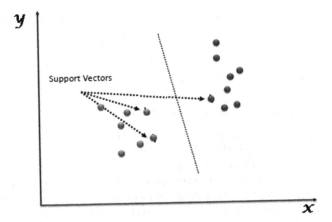

Fig. 1. The best hyperplane that clearly differentiates the two classes.

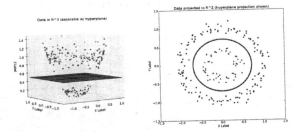

Fig. 2. Intelligent mapping using SVM.

line of separation. To solve it, an adjustment parameter called C is introduced, which defines the number of violations.

The previous example (Fig. 1) is quite easy since they are clearly linearly separable data. In practice, this is not always the case. To solve these cases, the SVMs perform an intelligent mapping (Fig. 2) of the space in a higher dimension, making classification easier.

In practice, SVMs are implemented using a *kernel* [3]. The smart mapping performed by the SVM entails a high computational cost: many new dimensions appear, which generate a greater calculation. Doing this for each vector in the set can be quite expensive. To solve it, kernel functions are used that allow operating in a space of implicit characteristics of high dimensionality without ever calculating the coordinates of the data in that space, but simply compiling the internal products among all the pairs of data in the space.

Usually when the problem is linearly separable, a linear kernel is used. For non-linearly separable problems, the polynomial kernel or Radial Basis Function (RBF) is used, the latter being the most popular.

The RBF kernel of two samples x y x', represented as feature vectors in some input space, is defined in [3], as shown in Eq. (1):

$$K_{RBF} = \exp[-\gamma||x - x'||^2] \tag{1}$$

where γ is a parameter that establishes kernel propagation.

2.3   One-Class Support Vector Machines

Traditionally, machine learning classifiers seek to classify the elements into different classes, distinguishing them from each other. SVM, as a supervised learning method, requires examples of tag-two training as belonging to one of the two classes. The problem arises in scenarios where a lot of information is available for one class, but in the other it is scarce or non-existent. In these cases, the SVM can be used as anomaly or novelty detectors (Fig. 3), using them as classifiers of a class [10]. Then the model is trained with data from a single class that is what would be considered normal. Inferring the properties of normal cases, and from these properties, it can be predicted that examples are different.

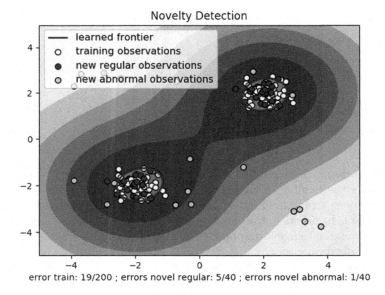

error train: 19/200 ; errors novel regular: 5/40 ; errors novel abnormal: 1/40

Fig. 3. One-Class SVM (it has been extracted from scikit-learn.org).

3   Case Study: Validation of the Proposal

In order to implement the RS, it is necessary to have information about songs that can be used in a content-based RS. *Spotify API*[7] provides endpoints with

---

[7] developer.spotify.com/web-api/.

which you can obtain 9 properties of each song, which are very useful for generating a user profile.

3.1   Data Set

The data to train the algorithm was obtained from *Spotify's API*, which provides a service of analysis of existing songs in the platform, offering values that are quite interesting for content-based filtering. As shown in Fig. 4, each of these properties is a different indicator, for example *acousticness* is an indicator of

Get audio features for a track

```
$ curl -X GET "https://api.spotify.com/v1/audio-features/06AKE
BrKUckW0KREUWRnvT" -H "Authorization: Bearer {your access toke
n}"
```

```json
{
    "danceability" : 0.735,
    "energy" : 0.578,
    "key" : 5,
    "loudness" : -11.840,
    "mode" : 0,
    "speechiness" : 0.0461,
    "acousticness" : 0.514,
    "instrumentalness" : 0.0902,
    "liveness" : 0.159,
    "valence" : 0.624,
    "tempo" : 98.002,
    "type" : "audio_features",
    "id" : "06AKEBrKUckW0KREUWRnvT",
    "uri" : "spotify:track:06AKEBrKUckW0KREUWRnvT",
    "track_href" : "https://api.spotify.com/v1/tracks/06AKEBrKUckW
    "analysis_url" : "https://api.spotify.com/v1/audio-analysis/06
    "duration_ms" : 255349,
    "time_signature" : 4
}
```

Fig. 4. Track Features obtained from Spotify's API.

whether the song is acoustic (0 <= *acousticness* <= 1); *valence*, is an indicator of positivism of the song, like *acousticness*, (0 <= *valence* <= 1). Songs with high *valence* are more positive (cheerful, euphoric) while songs with low *valence* are more negative (sad, depressing, angry). The rest of the properties can be consulted in *Spotify's Web API*. These properties are very useful for implementing machine learning algorithms such as SVM. It is only necessary to label each song with like or dislike for the training process. Songs that the user likes can be easily obtained through their interplay with the system, either by using their playback history or the songs they listen to most each day. The only disadvantage is that it may find songs that the user does not like, since there is no reference in the repository that provides this information. To solve this, it was decided to use SVM as a classifier of only one class, as an anomaly detector.

3.2 Training

To train the algorithm, we take the top 100 of the songs most listened to by each user on Spotify, since we can assume that these are the songs that he or she likes the most. Then we created a user profile based on it and extracted the *track_id* of the songs with the information on their properties and put them in a training matrix (Table 1). The rows of the matrix correspond to the songs and the columns to the properties. Each row is labelled with a 1 (like), a column is added indicating the class to which each song belongs, (class 1), since they are the songs that we consider will adapt to the profile of the user's tastes. Note that we are working with a single-class SVM as an anomaly detector and, by default, the algorithm takes the training data as belonging to the class that is considered normal. If we were working with an SVM of binary classification, we would have to label each row with its corresponding class.

Table 1. Training matrix.

Song	Danceability	Energy	Key	Loudness	...	Class
track1	0.735	0.538	5	−11840	...	1
track2	0.224	0.435	3	−9586	...	1
track3	0.115	0.458	6	635	...	1
...	...	...	...	...	...	...

As indicated above, the recommendation is carried out through an SVM of a class as a detector of new features. As can be seen, there are certain properties of the tracks whose values include ranges greater than those of the rest. To solve this, the library has a parameter in the training function, *normalize*, to normalize the values so that they all have the same rank. It also makes it possible to configure other parameters, such as the number *kFold* for cross validation, *kernelType* to choose the type of kernel to be used, *gamma* to choose one or several values required for the kernels (polynomials, sigmoids and RBF).*nu*, a parameter whose function is similar to parameter C in standard SVMs and *reduces* to indicate whether you want to use Principal Component Analysis (PCA) reduction to reduce the dimensions of the dataset. These parameters are added to the training function in JSON[8]. The configuration parameters of the algorithm are shown below:

```
{
    "svmOptions":{
        "svmType": "ONE_CLASS",
        "kernelType": "RBF",
        "gamma": [0.001,0.01,0.5],
        "kFold": 4,
        "nu": [0.01,0.125,0.5,1],
        "reduce": true,
        "normalize": true
    }
}
```

With these datasets, the model is trained to recognize elements that belong to the class that fits the user's taste profile and excludes the rest of the songs that are not to his or her liking. It thus obtains the songs whose properties are very similar to those of the profile, with which the recommended list for the user is obtained.

3.3 Classification

Once the training is finished, the model should be able to identify tracks that do not belong to the normal class (anomaly), as well as obtain new songs to recommend that are related to the profiles used. For this, we decided to extract from Spotify the latest albums released and then 3 songs from each of them. As in training, a list is created with the *track_id* of each song and its characteristics are obtained. Finally, a matrix similar to the training matrix is created (Table 1) but without the class column. Each row of the matrix is passed through the algorithm for its classification. For each of them, a prediction is generated: 1 if it is the kind of song that the user likes or 0 otherwise.

Finally, the *track_id* of the songs whose properties have been tagged with a 1 are placed in a playlist and sent to the mobile application in JSON format, where they will be formatted and adapted so that the user can interact with them. Then, these new songs are sent to the application so they can be shown to the user (Fig. 5). Once the user enters their credentials correctly, the main menu will appear, where they can navigate in three directions:

– Player: Play music with the Spotify SDK.

---

[8] json.org.

- History: Access the most listened to songs.
- Playlists: Find the generated playlists.

With this last option the user can generate new playlists, based on the music recommended for the user with the RS. In addition, the system will allow you to consult the songs that comprise it, including extra information, and in general manage all the playlists.

Fig. 5. Playlists recommended.

3.4   Experimental Results and Evaluation

To evaluate the system, tests were carried out with 11 different users who use Spotify's music transmission service with great frequency. Eleven episodes were created over a period of 4 days, with which between 200 and 400 songs that were adapted to the taste of each one could be obtained. The results of the classification after the training with the indicated parameters were quite good. An accuracy that oscillates between 0.86 and 0.92 was detected, which means that the model is able to detect elements belonging to the user's taste profile with an accuracy of $\approx 89\%$

Regarding the feedback from the users, in most cases they really liked the recommended music and only in a few cases did the recommended songs have no relation with the songs associated with the profile, showing a quite normal behaviour of the system.

The system was implemented using a client-server architecture, the client being a mobile application for Android devices. On the server is the RS, which is responsible for executing the recommendation algorithm with the data sent by the application. The recommendation process is carried out by means of a request of the application through *REpresentational State Transfer (REST)*[9] services, with which the server itself returns a list of songs that adapts to the user profile.

The code was implemented in *JavaScript language* in a *Nodejs*[10] environment with the help of the *Expressjs* and *Node-svm* modules. We also used *Scikit-Learn library*[11] for application of the machine learning algorithms. This library makes it possible to use the SVM algorithm and its variants, such as the One-Class SVM, providing functions to train the algorithm and perform the classifications once the model is created. The database system used was *MongoDB*[12].

4 Conclusions

The objective of this work was to develop a music RS based on content by using information easily collectable from music platforms. The results obtained are satisfactory because the proposal shown makes it possible to generate a list of songs that adapts to the user's tastes, incorporating the novelties of new launches. The system developed fulfils its purpose: with the use of SVM as novelty detectors it is capable of detecting the songs that belong to a certain profile with an accuracy of 89%, solving to a large extent the cold start problems.

As future lines of work, we intend to add a dimension of context to the recommender according to the place of interest or topic, which could lead to an improvement in the recommendations, since it would make it possible, for example, to recommend energetic music when the user is doing sports and also songs that reflect their mood in general. Social labelling that would help discover music within a network of friends who may have the same taste profiles could also be added.

Acknowledgments. This paper has been partially supported by: La convocatoria en concurrencia no competitiva para la concesión de subvenciones para la realización de proyectos de I+D de las PYMES de la Agencia de Innovación, Financiación e Internacionalización Empresarial de Castilla y León (ADE). Research Project: *Mejora de habilidades operativas ante riesgos emergentes en entornos inteligentes de producción: aplicación a las redes eléctricas inteligentes.*

---

[9] en.wikipedia.org/wiki/Representational_state_transfer.
[10] nodejs.org/en/.
[11] scikit-learn.org/.
[12] en.wikipedia.org/wiki/MongoDB.

References

1. Ricci, F., Rokach, L., Shapira, B., Kantor, P.: Recommender Systems Handbook, 1st edn. Springer, US (2010)
2. Lantigua I.F. El móvil supera por primera vez al ordenador para acceder a internet, Abril 2016. (posted 4-Abril-2016)
3. Brownlee, J.: Support vector machines for machine learning, Abril 2016. (posted 20-Abril-2016)
4. Illig, J., Hotho, A., Jäschke, R., Stumme, G.: A comparison of content-based tag recommendations in folksonomy systems. In: Wolff, K.E., Palchunov, D.E., Zagoruiko, N.G., Andelfinger, U. (eds.) KONT/KPP -2007. LNCS (LNAI), vol. 6581, pp. 136–149. Springer, Heidelberg (2011). https://doi.org/10.1007/978-3-642-22140-8_9
5. Adams, J.M., Bennett, P.N., Tomasic, A.: Combining personalized agents to improve content-based recommendations. Master's thesis, Language Technologies Institute, Carnegie Mellon University, Pittsburgh (2007)
6. Claypool, M., Gokhale, A., Miranda, T., Murnikov, P., Netes, D., Sartin, M.: Combining content-based and collaborative filters in an online newspaper. In: Proceedings of ACM SIGIR Workshop on Recommender Systems (1999)
7. BuzzAngle Music. U. s. music industry report, Enero 2017. (posted 3-Enero-2017)
8. Burke, R.: Hybrid web recommender systems. In: Brusilovsky, P., Kobsa, A., Nejdl, W. (eds.) The Adaptive Web. LNCS, vol. 4321, pp. 377–408. Springer, Heidelberg (2007). https://doi.org/10.1007/978-3-540-72079-9_12
9. Gorakala, S.: Building Recommendation Engines, 1st edn. Packt Publishing, Birmingham (2016)
10. Ray, S.: Understanding support vector machine algorithm from examples (along with code), Octubre 2015. (13-Septiembre-2017)
11. Sarwar, K.G., Konstan, J., Riedl, J.: Item-based collaborative filtering recommendation algorithms. ACM, Hong Kong, 1–5 May 2001

Improving Forecasting Using Information Fusion in Local Agricultural Markets

Washington R. Padilla[1], Jesús García[2], and José M. Molina[2(✉)]

[1] Research Group Ideia Geoca, Salesian Polytechnic University
of Quito-Ecuador Engineer Systems, Quito, Ecuador
wpadillaa@ups.edu.ec
[2] Applied Artificial Intelligence Group, Carlos III University, Madrid, Spain
jgherrer@inf.uc3m.es, molina@ia.uc3m.es

Abstract. This research explores the capacity of Information Fusion to extract knowledge about associations among agricultural products, which allows prediction for future consumption in local markets in the Andean region of Ecuador. This commercial activity is performed using Alternative Marketing Circuits (CIALCO), seeking to establish a direct relationship between producer and consumer prices, and promote buying and selling among family groups. In the results we see that, information fusion from heterogenous data sources that are spatially located allows to establish best association rules among data sources (several products on several local markets) to infer significant improvement in time forecasting and spatial prediction accuracy for the future sales of agricultural products.

Keywords: Data Fusion · Alternative circuits of commercialization
Associations mining · Predictive analysis

1 Introduction

In today's digital world, information is the key factor to take decisions. Ubiquitous electronic sources such as sensors and video provide a steady stream of data, while text-based data from databases, Internet, email, chat and VOIP, and social media is growing exponentially. The ability to make sense of data by fusing it into new knowledge would provide clear advantages in making decisions. Therefore, multi-source Data Fusion (DF) is a very important area of current research (Liggins et al. 2008)

Fusion systems aim to integrate sensor data and information in databases, knowledge bases, contextual information, user mission, etc. (Llinas et al. 2004), in order to describe situations. In a sense, the goal of information fusion is to attain a global view of the scenarios to take the best decision.

The ability to fuse digital data into useful information is hampered by the fact that the inputs, whether device-derived or text-based, are generated in different formats, some of them unstructured. Whether the information is unstructured by nature or the fact that information exchange (metadata) standards are lacking (or not adhered to) all

© Springer International Publishing AG, part of Springer Nature 2018
F. J. de Cos Juez et al. (Eds.): HAIS 2018, LNAI 10870, pp. 479–489, 2018.
https://doi.org/10.1007/978-3-319-92639-1_40

of this hinders automatic processing by computers. Furthermore, the data to be fused may be inaccurate, incomplete, ambiguous, or contradictory; it may be false or corrupted by hostile measures. Moreover, much information may be hard to formalize, i.e., imaging information. Consequently, information exchange is often overly complex. In many instances, there is a lack of communication between the various information sources, simply because there is no mechanism to support this exchange (Hall et al. 2017).

The key aspect in modern DF applications is the appropriate integration of all types of information or knowledge: observational data, knowledge models (a priori or inductively learned), and contextual information. Each of these categories has a distinctive nature and potential support to the result of the fusion process.

- Observational Data: Observational data are the fundamental data about the dynamic scenario, as collected from some observational capability (sensors of any type). These data are about the observable entities in the world that are of interest.
- Contextual Information: Context and the elements of what could be called Contextual Information (Snidaro et al. 2014) could be defined as "the set of circumstances surrounding a task that are potentially of relevance to its completion." Because of its task-relevance; fusion or estimating/inferring task implies the development of a best-possible estimate taking into account this lateral knowledge (Gómez-Romero et al. 2015). We can see the context as background, i.e., not the specific entity, event, or behavior of prime interest but that information which is influential to the formation of a best estimate of these items.
- Learned Knowledge: In those cases where a priori knowledge for DF process development cannot be formed, one possibility is to try and excise the knowledge through online machine learning processes operating on the observational and other data. These are procedural and algorithmic methods for discovering relationships among and behaviors of entities of interest (Gómez-Romero et al. 2011). There is a trade-off involved in trying to develop fully-automated algorithmic DF processes for complex problems where the insertion of human intelligence at some point in the process may be a much more judicious choice.

This research is aimed at finding patterns in the behavior of consumption of agricultural products, using Information Fusion of spatially distributed information of several products and learned knowledge about relationships among these products to predict future situations. In this work, we apply Artificial Intelligence techniques over the market fused data, to extract association rules for improving the prediction on the behavior of consumption of agricultural products to establish better policies to potentiate local operations.

The final objective is increasing the incomes of small farmers in the Andean region of Ecuador, preventing their migration to large population centers. The CIALCO acronym comes from using the first two letters of the words Alternative Circuit marketing (in Spanish 'Circuito Alternativo de Comercialización'). There are several types of circuits, this study is limited to information of groups involved in circuits of fair type, defined as specific places where agricultural producers meet periodically to conduct its business (Padilla and García 2016).

The analysis is based on information from 2014, provided by the General Coordination Network Marketing Ministry of Agriculture, and Livestock of Ecuador. It contains the weekly performance of sales of agricultural products made by small farmers located in Ecuador's central highlands specifically the provinces of Tungurahua and Chimborazo. The country Ecuador is crossed by the Equatorial line, i.e., its territory is located both north and south from latitude zero. The provinces of Tungurahua and Chimborazo are in the south and central region, and the information has been collected on the sale of agricultural products in these alternative marketing circuits, which seek to establish a direct relationship between producers and consumers.

The available data contains information about the number and volume of sales of products such as vegetables, legumes, meat, dairy, fruits, tubers and processed products, finding an average of 1,200 items per month divided on a weekly basis.

The document is organized as follows, in the second part there is an exhibition of different works that coincide with the research line proposed here, in the third section the methodology used is detailed, processes carried out to merge the information and obtain a group of products using the Apriori algorithm for association rules, in the fourth part the activities are detailed to find the future estimates for the tomato product and the prediction of the set of association products, to finalize it presents, as a result, an improvement significant with the sales estimate using the associated products.

2 Association Rules and Prediction

2.1 Association Rules

An data mining it is very important to look for causal relations between variables to predict changes in some variables based on knowledge of other ones. An association rule of the form $\{A\} \Rightarrow \{B\}$ can be interpreted as: "if A appears then B also will appear", and aims to identify relationships not explicit between categorical attributes. Among the existing algorithms for association rule discovery, *Apriori* algorithm (Zulfikar et al. 2016) is the most representative technique for this task., it searches for trends based on the performance parameters mentioned above, based on prior knowledge or "a priori" frequent sets. It is summarized below:

Step 1: Generate all item sets L with a single element; this set is used to form a new set with two, three or more elements
all possible pairs which are taken Sup equals minsup
Step 2: For every frequent item set L' found:
    For each subset J, of L'
        Determine all association rules of the form:
        If L'-J→J
        Select those rules whose confidence is greater or equal
    than minconf
Repeat Step 1, including next element into L

There are extensions of Apriori algorithm considering sequences of objects ordered in time as items, being the searched result the associations in form of sequences of items. Some cases derived from Apriori are GSP for spatio-time associations (Patil et al. 2016; Chang et al. 2005) algorithms. As a different alternative, the paradigm of "temporal" association rules (rules applicable within temporal frames), also based on Apriori schema, consider temporal support (Schlüter and Conrad 2011), or even meta-rules describing how relationships vary in time (Spiliopoulou and Roddick 2000). A series of studies conducted in various fields of science try to use the rules of association as a criterion to establish future estimates, so we can see some jobs such as (Asadifar and Kahani 2017) analyse the stock of a supermarket, or in (Mane and Ghorpade 2016) predict admission decisions by students. Finally, works like (Kumar et al. 2016) have explored relationships between association rules and a fuzzy classification.

2.2   Time Series

As complementary to temporal data association, time series analysis builds hypotheses about cause-effect relationships that can be expressed in terms of forecasting. In simple statistical methods, the hypothesis that X causes Y implies a correlation between X and Y, which in turn implies that Y can be predicted from X. More generally, the hypothesis on the effect of X on Y is tested to see if the scores on Y can be predicted more accurately from a model that includes X that of a model that excludes X. There is abundant bibliography in time-series regression, both with classical models and application of advanced machine learning techniques (neural networks, support vector regression, etc.) (Theophano Mitsa 2010).

2.3   Spatial Estimation Algorithms

The use of spatial data analysis and regression is very usual with geo-located data in multiple domains. In (Won and Ray 2004) an explanation of the mathematical development of kriging and cokriging based on substitution models within the framework of optimization is made (Chen et al. 2016), they basically propose to improve the construction of the variogram using information of magnitude and direction applied to data of the National Network of the Geomagnetic Observatories of China. As mentioned in (Celemín 2009), "in the geographical space everything is related to everything, but the closest spaces are more related to each other".

The theoretical approaches made in these articles in the part of the variogram and cokriging have been considered for the development of this research

This work uses different geo-located data sources as basis for information fusion applying the spatial prediction paradigm, but also searches for the group of products that have the highest ratio of marketing associativity (output of Apriori association rules), determined the one with the highest consumption, for the improvement in estimation processes of future commercial behaviour using the association set.

3   Information Fusion for Knowledge Extraction

The global market is composed by several local markets geographically dispersed. Each local market can measure the information of sales for each product in the market. Then, each market is a knowledge source generating:

1. Information related to local market: position, number of products, number of sellers, quantity of population that use the local market, etc...
2. For each product, information related with the level of sales for this product.

The system to be developed use the fusion of this information send for each market to extract high-level knowledge to be applied for prediction of future sales. Figure 1 sketches the global process for improving predictions. First, system integrates the information from every knowledge source to generate a global information with all available sources. Using this information, system searches for the best association rules to be used later to improve predictions of sales (learned model). The result of first task is the integration of local information to obtain a global view of the sales in the region, the second task is extract knowledge from this information about relevant relationships among products, and the third step uses this knowledge to improve the prediction output. In summary, the global fusion process system needs time, at the beginning, to gather some representative information to be interpreted by the machine learning procedure, and then the developed model is applied to generate improved predictions as result of fusion of all observations and learned knowledge.

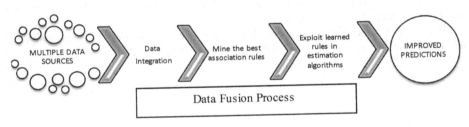

Fig. 1. Global process description

The complete process is divided in three main steps:

3.1   Methodology

1. **Integration of different sources of local data (Obtain the global information)**
    1.1   Receive the data from local market
        1.1.1   Delete records without information
        1.1.2   Standardize value of similar records
        1.1.3.   Establish units of measurement valid
        1.1.4.   Generate the set of products for the study
    1.2.   Generate the final record (database) from local information ready for pattern search

2. **Obtain the best rules of Association**
   2.1. Discredit the archive of products sold
   2.2. Apply *Apriori* algorithm to obtain the best rules of Association
   2.3. Establish the set of associated products
3. **Information fusion using prediction for future sales**
   3.1. Make predictions using temporal series of past sales of one product in one local market using temporal series forecast
   3.2 Make predictions using information of sales of one product in every local market using spatial estimation
   3.3. Compare error metrics of single-product prediction vs. the predictions using learning knowledge

3.2 Receiving Data from Local Markets

This process begins by validating the data originally provided, deletes non-significant information, and standardizes product names and units of measure.

The information provided includes data collected weekly in fairs of CIALCO type, in the provinces of Napo, Chimborazo and Tungurahua belonging to the central area of the Andes in Ecuador.

The record contains the products that are part of the marketing, the value of sales, date and fair to which the transaction belongs. The data sheet contains all the recorded transactions, organized in packages named "canastas" (baskets), each one representing a sale of certain products, containing the products present in each purchase, and implicitly also contains the spatial geo-localization of the operation (the location of fair) and the time stamp (date) of operation.

If an agricultural product is part of the purchase transaction, an 's' is assigned, otherwise 'no' for all months of 2014 as can be seen in Fig. 2.

	5/10/2014	5/10/2014	5/10/2014	5/10/2014	5/10/2014	5/10/2014	5/10/2014	5/10/2014	5/10/2
ACELGA	s	s	no	no	s	no	s	no	s
Ajo	no	no	no	s	no	s	s	s	no
Arveja	no	s	no	no	s	s	s	no	no
Babacos	no	no	no	no	no	no	no	s	no
BROCOLI	s	s	no	s	no	s	s	s	s
CEBOLLA BLA	no	no	s	s	s	no	s	s	s
CEBOLLA PAI	s	no	s	no	s	no	no	no	s
Choclo	s	s	s	no	s	s	s	no	no
COL	s	no	s	s	s	s	s	no	s
COLES VERDE	no	no	no	no	no	no	no	no	no
COLIFLOR	no	no	no	s	no	s	no	no	no
Espinaca	no	no	no	no	no	no	no	no	no
Frejol	no	s	no	s	s	s	no	no	no
FRUTILLA	no	no	no	s	no	no	s	s	no
HABAS	no	no	no	s	no	s	s	s	no
Hierbas	s	no	no	no	s	no	no	no	no
LECHUGA	s	no	no	s	s	s	s	no	s
Melloco	no	s	no	no	no	no	s	no	no
Nabo	no	no	no	no	no	no	no	no	no
PAPAS	no	s	s	s	s	s	s	no	s
Paquetes de	no	no	no	no	no	no	no	no	no
Pepinillo	no	no	s	s	no	no	no	no	s
Pepino	no	no	no	no	no	no	s	no	no
Pimiento	no	s	s	s	s	s	s	s	no
Rabano	no	no	s	s	s	no	no	s	no
Remolacha	no	no	no	s	no	no	s	no	s

| ◄ ► ... | Febrero | Marzo | Abril | Mayo | Junio | Julio | Agosto | Septiembre | **Octubre** |

Fig. 2. Transactions with products contained in each sale ("canasta")

The initial information has been subject to a data cleansing process in which homogenization is done, mainly in the names of products, values for unit sales and elimination of products that do not have relevant information for this study.

To find association rules information must be quantized, so we can identify whether an agricultural product is part of the procurement process. To optimize the process of searching for the best association rules, the 'no' symbol has been replaced by '?' symbol in the first place. The reason for this operation was to focus the search only in "positive" rules, those relating the presence of products and avoid the search of those rules relating absence of products. The main reason for this decision was that transactions in the dataset select sparse products in the table, so most of values would be negative.

3.3 Search for Best Association Rules

The data file contained 549 pre-processed transactions containing subsets of the 31 elements acquired. In the search for association rules, the first condition was that the value of **support** is higher or equal to 0.4, implying the number of times that appears in the database must be greater than 220 to ensures this constraint.

Among the elements that make up the subset of data L1 and L2, we obtain the subset of data that satisfy the minimum confidence value equal to 0.8, as can be verified in Fig. 3. For example, the white onion appears 338 times in L1, while the intersection between white onion and kidney tomato appears in L2 293 times, whose confidence for this example is 293/338 = 0.86. As a result, it is obtained that the kidney tomato is found in the five resulting association rules on the consequent side.

Fig. 3. Best association rules

4 Improving Predictions as Result of Information Fusion

The data used correspond to the commercialization of the input variable (kidney tomato) in the month of July 2014, corresponding to the fairs located in provinces of Tungurahua and Chimborazo, as shown in Fig. 4.

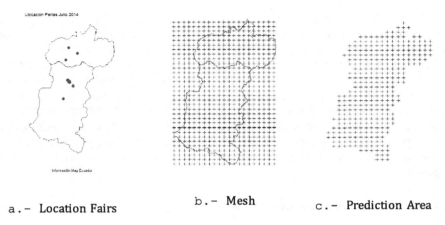

a. - Location Fairs   b. - Mesh   c. - Prediction Area

Fig. 4. Location of Fairs and Prediction Mesh

In the first place, the available data from different located sources (data registered in fairs) were fused to build the interpolation model. Based on the ordinary kriging method, considered the best unbiased linear estimator type, the results shown in Fig. 5 can be appreciated. The values found in the interpolation vary especially in two foci on which the predictions were generated. The values closest to the points of information are more influenced than those located further.

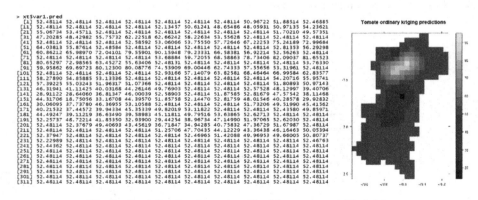

Fig. 5. Spatial estimation using (single) tomato variable kriging

As a second step, to carry out a prediction with multiple variables, the spatial prediction model was extended as result of interrelation with the products found from the association rules.

The associated products with the highest incidence in the process were identified, the five products resulting from A priori algorithm are {tomato, brocoli, white onion, tree tomato, carrot}. Then, we proceed to estimate the spatial distribution of this variable considering the associated products, as shown in Fig. 6.

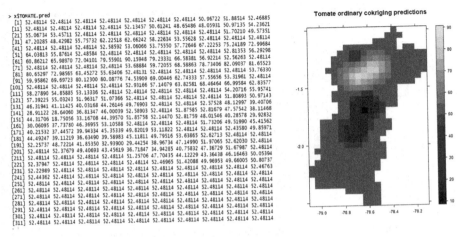

```
> x$TOMATE.pred
  [1] 52.48114 52.48114 52.48114 52.48114 52.48114 52.48114 50.96722 51.88514 52.46885
 [11] 52.48114 52.48114 52.48114 52.48114 52.13457 50.61241 48.65486 48.05931 50.97135 54.23621
 [21] 55.06734 53.45711 52.48114 52.48114 52.48114 52.48114 52.48114 52.48114 51.70210 49.57351
 [31] 47.20285 48.42982 55.75732 62.22518 62.66242 58.22634 53.55628 52.48114 52.48114 52.48114
 [41] 52.48114 52.48114 52.48114 52.58592 53.06066 53.75550 57.72646 67.22253 75.24189 72.99684
 [51] 64.03813 55.87614 52.48584 52.48114 52.48114 52.48114 52.48114 52.48114 52.81353 56.29298
 [61] 60.86212 65.98970 72.04101 79.55901 90.15948 79.23331 66.58381 56.92214 52.56263 52.48114
 [71] 52.48114 52.48114 52.48114 52.48114 53.68884 59.72055 68.58863 78.73406 82.09037 81.65523
 [81] 80.65297 72.98565 63.45272 55.63406 52.48131 52.48114 52.48114 52.48114 53.76330
 [91] 59.95862 69.69723 80.12300 80.08776 74.53909 69.00446 62.74333 57.55656 53.31961 52.48114
[101] 52.48114 52.48114 52.48114 52.48114 52.93166 57.14079 63.82581 68.46464 66.99584 62.83577
[111] 58.27890 54.85885 53.13336 52.48114 52.48114 52.48114 52.48114 54.20716 55.95741
[121] 57.39225 55.03243 51.96317 51.07366 52.48114 52.48114 52.48114 51.80893 50.97143
[131] 46.31941 41.11425 40.03168 44.26146 49.76903 52.48114 52.48114 52.57528 48.12997 39.40706
[141] 28.91122 28.64060 36.81347 46.00039 52.58903 52.48114 51.87585 52.81679 47.57542 38.11468
[151] 44.31706 18.75056 33.16708 44.39570 51.85758 52.14470 52.81759 48.01546 40.28578 29.92832
[161] 30.06095 37.73780 46.36955 53.10588 52.48114 52.48114 51.73206 49.51990 45.41562
[171] 40.21532 37.44572 39.94334 45.35339 49.82019 53.11822 52.48114 52.48114 52.43580 49.85971
[181] 44.49247 39.11219 36.63490 39.58983 45.11811 49.79516 53.63865 52.62713 52.48114 52.48114
[191] 52.25737 48.72214 41.85350 32.93900 29.44254 38.96734 47.14990 51.97065 52.62030 52.48114
[201] 52.48114 52.37679 49.40693 43.45619 36.71847 34.94285 40.75832 47.36729 51.67987 52.48114
[211] 52.48114 52.48114 52.48114 52.48114 51.25706 47.70435 44.12229 43.36438 46.16463 50.05394
[221] 52.37947 52.48114 52.48114 52.48114 52.48114 52.46965 51.42088 49.96953 49.66005 50.80737
[231] 52.22989 52.48114 52.48114 52.48114 52.48114 52.48114 52.48114 52.48114 52.48114 52.46763
[241] 52.44362 52.48114 52.48114 52.48114 52.48114 52.48114 52.48114 52.48114 52.48114 52.48114
[251] 52.48114 52.48114 52.48114 52.48114 52.48114 52.48114 52.48114 52.48114 52.48114 52.48114
[261] 52.48114 52.48114 52.48114 52.48114 52.48114 52.48114 52.48114 52.48114 52.48114 52.48114
[271] 52.48114 52.48114 52.48114 52.48114 52.48114 52.48114 52.48114 52.48114 52.48114 52.48114
[281] 52.48114 52.48114 52.48114 52.48114 52.48114 52.48114 52.48114 52.48114 52.48114 52.48114
[291] 52.48114 52.48114 52.48114 52.48114 52.48114 52.48114 52.48114 52.48114 52.48114 52.48114
[301] 52.48114 52.48114 52.48114 52.48114 52.48114 52.48114 52.48114 52.48114 52.48114 52.48114
[311] 52.48114 52.48114 52.48114 52.48114 52.48114 52.48114 52.48114 52.48114
```

Fig. 6. Multivariable prediction

To perform the validation of the prediction data, the procedure known as "leave-one-out" cross-validation (LOOCV) was applied. As can be seen in Fig. 7, the average of the residuals has a value of 0.7.

```
> summary(cvtom)
object of class SpatialPointsDataFrame
coordinates:
          min       max
x -78.723827 -78.551033
y  -1.888003  -1.257143
Is projected: NA
proj4string : [NA]
Number of points: 14
Data attributes:
  var1.pred          var1.var          observed          residual          zscore             fold
 Min.   :16.05    Min.   :  23.76    Min.   : 6.90    Min.   :-51.0621    Min.   :-7.9858    Min.   : 1.00
 1st Qu.:23.75    1st Qu.:  69.43    1st Qu.:20.00    1st Qu.:-19.6668    1st Qu.:-0.7889    1st Qu.: 4.25
 Median :36.58    Median : 142.48    Median :31.50    Median :  4.4379    Median : 0.2507    Median : 7.50
 Mean   :38.93    Mean   : 823.78    Mean   :39.65    Mean   :  0.7189    Mean   :-0.1132    Mean   : 7.50
 3rd Qu.:56.27    3rd Qu.:1993.55    3rd Qu.:57.75    3rd Qu.: 21.3920    3rd Qu.: 0.9124    3rd Qu.:10.75
 Max.   :62.74    Max.   :2496.56    Max.   :92.00    Max.   : 44.1241    Max.   : 9.0169    Max.   :14.00
```

Fig. 7. Cross validation for tomato sales prediction

Figure 8 shows the result of subtracting the residual values between the prediction of the tomato variable and the residual values of the multivariate prediction.

The values of the multivariate prediction are lower in eight locations, with positive values of the differences, and negatives in the remaining six locations.

In summary, the method has two stages. The first one established a consumption prediction using geostatistical techniques such as the variogram to find a continuous function to interpolate the estimated values of product sales in points with no available observations. Based on the functions of kriging, the sales values of products are established according to their spatial location and the influence of their close neighbours. To finalize this phase, a cross validation is carried out using the Leave-one-out cross-validation evaluation, which allows verifying the error of estimated future values.

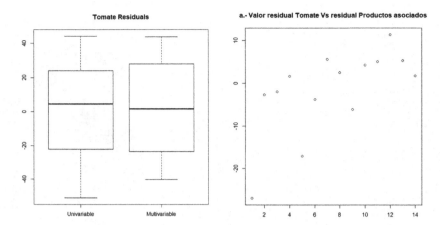

Fig. 8. Comparison of cross-validation results in spatial predictions

In the second stage, the Apriori algorithm was used to search the products that have the greatest associativity with the selected variable: tomato, broccoli, tree tomato, white onion and carrot. Based on this multivariable set, the prediction values are calculated using the same procedure described in the first stage.

5 Conclusions

This work has proposed a methodology to improve accuracy of predictions in time and space based on association rules, applied to the domain of agricultural local markets. The research has been carried out on marketing data of agricultural products in the central zone of the Ecuadorian Andes. Different data mining techniques have been applied with the objective of establishing scenarios that facilitate the generation of policies to improve the income of the farmer families.

Acknowledgements. This work was supported in part by Project MINECO TEC2017-88048-C2-2-R and by Commercial Coordination Network, Ministry of Agriculture, Livestock, Aquaculture and Fisheries Ecuador.

References

Asadifar, S., Kahani, M.: Semantic association rule mining: a new approach for stock market prediction. In: 2017 2nd Conference on Swarm Intelligence and Evolutionary Computation (CSIEC). pp. 106–111 (2017)

Celemín, J.P.: Autocorrelación espacial e indicadores locales de asociación espacial: Importancia, estructura y aplicación. Rev. Univ. Geogr. **18**, 11–31 (2009)

Chang, C.-C., Li, Y.-C., Lee, J.-S.: An efficient algorithm for incremental mining of association rules. In: 15th International Workshop on Research Issues in Data Engineering: Stream Data Mining and Applications (RIDE-SDMA 2005), pp. 3–10 (2005)

Chen, D., Liu, D., Li, Y., et al.: Improve spatiotemporal kriging with magnitude and direction information in variogram construction. Chin. J. Electron. **25**, 527–532 (2016). https://doi.org/10.1049/cje.2016.05.019

Gómez-Romero, J., Patricio, M.A., García, J., Molina, J.M.: Ontology-based context representation and reasoning for object tracking and scene interpretation in video. Expert Syst. Appl. **38**, 7494–7510 (2011). https://doi.org/10.1016/j.eswa.2010.12.118

Gómez-Romero, J., Serrano, M.A., García, J., et al.: Context-based multi-level information fusion for harbor surveillance. Inf. Fusion **21**, 173–186 (2015). https://doi.org/10.1016/j.inffus.2014.01.011

Hall, D., Chong, C.Y., Llinas, J., Liggins, M.: Distributed Data Fusion for Network-Centric Operations. CRC Press, Boca Raton (2017)

Kumar, P.S.V.V.S.R., Maddireddi, L.R.D.P., Lakshmi, V.A., Dirisala, J.N.K.: Novel fuzzy classification approaches based on optimisation of association rules. In: 2016 2nd International Conference on Applied and Theoretical Computing and Communication Technology (iCATccT), pp. 1–5 (2016)

Liggins, M., Hall, D., Llinas, J.: Handbook of Multisensor Data Fusion: Theory and Practice, 2nd edn. CRC Press, Boca Raton (2008)

Llinas, J., Bowman, C., Rogova, G., et al.: Revisiting the JDL data fusion model II. Space and Naval Warfare Systems Command, San Diego, CA (2004)

Mane, R.V., Ghorpade, V.R.: Predicting student admission decisions by association rule mining with pattern growth approach. In: 2016 International Conference on Electrical, Electronics, Communication, Computer and Optimization Techniques (ICEECCOT), pp. 202–207 (2016)

Padilla, W.R., García, H.J.: CIALCO: alternative marketing channels. In: Bajo, J., et al. (eds.) PAAMS 2016. CCIS, vol. 616, pp. 313–321. Springer, Cham (2016). https://doi.org/10.1007/978-3-319-39387-2_26

Patil, S.D., Deshmukh, R.R., Kirange, D.K.: Adaptive Apriori Algorithm for frequent itemset mining. In: 2016 International Conference System Modeling Advancement in Research Trends (SMART), pp. 7–13 (2016)

Schlüter, T., Conrad, S.: About the analysis of time series with temporal association rule mining. In: 2011 IEEE Symposium on Computational Intelligence and Data Mining (CIDM), pp. 325–332 (2011)

Snidaro, L., Garcia, J., Corchado Rodríguez, J.: Guest Editorial: Context-based Information Fusion. Information Fusion, vol. 21 (2014)

Spiliopoulou, M., Roddick, J.F.: Higher order mining: modelling and mining the results of knowledge discovery (2000)

Mitsa, T.: Temporal Data Mining. CRC Press (2010). https://www.crcpress.com/Temporal-Data-Mining/Mitsa/p/book/9781420089769

Won, K.S., Ray, T.: Performance of kriging and cokriging based surrogate models within the unified framework for surrogate assisted optimization. In: Proceedings of the 2004 Congress on Evolutionary Computation (IEEE Cat. No. 04TH8753), vol. 2, pp. 1577–1585 (2004)

Zulfikar, W.B., Wahana, A., Uriawan, W., Lukman, N.: Implementation of association rules with apriori algorithm for increasing the quality of promotion. In: 2016 4th International Conference on Cyber and IT Service Management, pp. 1–5 (2016)

Chord Progressions Selection Based on Song Audio Features

Noelia Rico and Irene Díaz[✉]

Computer Science Department at University of Oviedo, Oviedo, Spain
noripa@gmail.com, sirene@uniovi.es

Abstract. A chord progression is an essential building block in music. In the field of music theory is usually assumed that these progressions influence the mood, emotion, genre or other critical aspects of the songs, and also in the perception that they will cause on the listener. Therefore, it is natural to think that musical and audio features of a track should be related to its chord progressions. Choosing carefully these progressions when it comes the time of creating a new song, is a fundamental aspect depending on the feelings we want to evoke on the listener. Also, two songs can be considered alike or classified into the same emotions or genres if they use the same chord progressions. Many music classification studies are presented nowadays, but none of them take into account chord progressions, probably due to the lack of this kind of data. In this paper, classification algorithms are used to illustrate the influence of the songs' features when it comes to pick up chord progressions to create a new song.

Keywords: Music information retrieval · Chord progressions
Data mining · Machine learning · Classification

1 Introduction

It is generally accepted that music makes us feel a wide range of emotions. The mechanisms through which music evokes feelings and impacts in our brain is a vast field of research with a lot of unanswered questions.

On many occasions, the songs can be stirring even though they do not have any lyrics, and this is why music is considered an international language. In these cases, the songs' emotions are purely related to their structure [1]. Structural features of a song include rhythm, musical nuances, melody and the chords over which these melodies are settled. The same song can convey entirely different emotions after changing its chord progressions. Thus, listeners' feelings after hearing the modified song could be completely opposite. Those changes might also modify other aspects of the song such as its energy or popularity.

In the field of music regarding chord progressions, previous work has only focused on the analysis of music scores (most of the time using a MIDI file) in order to automatize the process of generating new music. Some of the previous

© Springer International Publishing AG, part of Springer Nature 2018
F. J. de Cos Juez et al. (Eds.): HAIS 2018, LNAI 10870, pp. 490–501, 2018.
https://doi.org/10.1007/978-3-319-92639-1_41

research in this discipline is based on the identification of the chords and the progressions themselves [2]. Others, center their efforts in using machine learning techniques to generate music chords progression, based on songs that are used to build the model and taking into account only their musical score [3–5]. Related work in this field of study combines musical features with the song lyrics [6] to classify the songs or use specific musical features to divide the songs into emotions [7].

However, there has been little discussion on how the audio features and the impact of the song on the listener are correlated with the chord progressions found in the song. One question that needs to be asked is how the chosen progressions for the composition of each song influence some musical aspects not directly related to the music score, such as the energy that the songs transmit or their future popularity.

This paper proposes a methodology to find the audio features of a track affecting their chord progressions. These audio features can be described as the musical features of the song in addition to some other attributes about its impact on the listener.

This paper is divided into five sections. Section 2 describes the dataset, explaining its creation and describing its features in detail. Section 3 gives an overview of the methods used to build the models. In Sect. 4 the results of the machine learning process are analyzed. Finally, the last section discusses the evaluation and possible application of the classification process and future work is proposed.

2   Datasets Creation

One of the most complex tasks in this work is getting the data set. For this purpose, data from Hook Theory [8] and Spotify [9] websites are considered. The information about a song provided by these two different sources is supposed to be related one to each other. Thus, the two datasets obtained are joined in order to perform a better prediction. Then, this section first details how to obtain information from Hook Theory and Spotify websites and how to put them together.

2.1   Hook Theory

Hook Theory [8] is a popular website in the music field that allows their users to analyze songs regarding chord progressions and share their analysis. It is designed to be used by people who are learning how to compose new music, integrating an interactive interface for people who do not know anything about music theory. For this purpose, this website includes a section showing the most popular chord progressions in music of all styles, ages and genres. Those "most popular progressions" are grouped into categories, to make easier the learning process for the amateur composers.

Using the API [10] of the website we have gathered all the songs that contain a specific song progression. The chord progressions categories that have been used for this work are shown in Table 1. The second column describes the name used on the website to identify each progression and the first one indicates the group where the progression is wrapped, regarding how challenging is to use it when composing a song. The third column represents the chord progression itself and the fourth column shows the name that has been given to each progression in this project, and therefore the one which is used to identify the variable representing the progression in the final dataset.

The third column represents the chord progression as the succession of musical chords expressed by Roman numerals, as is common in Classical music theory. Following the standard, the capital letters are used to express whether the chord is major or minor. The progressions with lower difficulty are built with chords corresponding to basic degrees in the scale (I, V, iii...). Whilst difficulty increases, the chords evolve to more complex constructions, and therefore the appearance of indexes becomes popular to indicate the inversion of the chord.

Table 1. Attributes obtained from the Hook Theory API

Difficulty	Name	Chord progression	Dataset variable
Beginner	Most popular progression	$I - V - vi - IV$	most.popular
	Pachelbel's progression	$I - V - vi - iii$	pachelbel
	Effective in all genres	$vi - V - IV - V$	effective.all.genres
	Those magic changes	$I - vi - IV - V$	magic.changes
	Gaining popularity	$I - IV - vi - V$	gaining.popularity
	Timeless	$I - V - IV - V$	timeless
Intermediate I	The cadential $^6_4$	$I^6_4 - V$	cadential64
	Stepwise bass down	$I - V^6 - vi$	stepwise.bass.down
	Stepwise bass up	$I - ii^7 - I^6$	stepwise.bass.up
	The newcomer	$I - iii^6_4 - vi$	newcomer
	I^6 as a precadence	$IV - I^6 - V$	I.as.precadence
	Simple yet powerful	$IV - I^6 - ii$	simple.yet.powerful
Intermediate II	Chord pleased the Lord	$V - V^6/vi - vi$	chord.pleased.Lord
	Expanding with V/vi	$I - V/vi$	expanding.V.vi
	vi^4_2 as a passing chord	$vi - vi^4_2 - IV$	vi42.passing.chord
	Secondary dominant	$V^7/IV - IV$	secondary.dominant
	Applied vii°	$vii°/vi - vi$	applied.vii
Advanced	Cadencing in style	$IV - iv - I$	candencing.in.style
	A mixolydian cadence	$\flat VII - IV - I$	mixolydian.cadence
	Using $\flat$ VI to set up a V	$\flat VI - V$	bVI.to.V
	Cadencing via $\flat$ VII	$\flat VII - I$	cadencing.bVII

Once all the songs have been collected, the dataset is build using a row to represent each song. Songs are identified by two columns, which are title and artist, and the dataset contains 21 more columns, each one associated to a different chord progression. If the song of the row i contains the progression j, the $cell_{ij}$ has the value 1, otherwise its value will be 0. The final dataset extracted from Hook Theory website [8] contains 1990 songs characterized by 23 attributes.

2.2 Spotify

The Spotify API fetches data [9] of all the available songs in the Spotify music catalog. To construct the second dataset based on Spotify, the songs contained in HookTheory dataset are considered. For each song in the hooktheory dataset, a search with the Spotify API is made using the primary key of the songs (artist and title). If the track is found, their features are extracted and the song is added to this new dataset. Among all the data available in the Spotify API, only the attributes that have been considered useful for this research will be included. Some of these features are strictly pulled out of the song's musical attributes like tempo or key, but others are calculated from its audio features like valence, or based on their reproductions on Spotify like popularity. These features are used as predictors of the final dataset. A thorough description of the attributes can be found in Table 2.

2.3 Features and Class of the Final Dataset

Using the data obtained from both Spotify and Hook Theory websites, different features characterizing 1990 songs have been collected. Features extracted from Spotify (see 2) represent the attributes associated with each song, while features obtained from Hook Theory website represent the class to learn. Thus, Features extracted from Spotify will be used to predict each chord progression as a binary problem (whether the song contains a chord progression or not).

Note that 21 different datasets will be created, each one linked with a different chord prediction Table 1. To sum up, 21 binary classification problems will be solved, taking into account the Spotify features as predictor variables. All the problems are independent one to each other, but they have the same structure. For each problem, a different binary class variable will be considered. The aim of each problem is to classify the songs into two possible categories, 0 or 1. The value will be 1 if the song uses the chord considered as the class variable for that concrete problem and 0 otherwise.

3 Learning Procedure

3.1 Methods

The classification will be made using R, and more precisely the caret [11] package. Six different algorithms will be used to build the models. This will allow us to make a comparison about the performance and the final results:

Table 2. Features fetched from the Spotify API

Feature	Type	Description
Artist	chr	Used as id of the song together with title
Title	chr	Used as id of the song together with artist
Popularity	int	The value will be between 0 and 100 expressing the popularity of the track, with 100 being the most popular. Songs that are being played a lot now will have a higher popularity than songs that were played a lot in the past
Explicit	Factor	Two possible levels: TRUE/FALSE. Whether or not the track has explicit lyrics
Danceability	num	Value between 0.0 and 1.0 describing how suitable a track is for dancing based on a combination of musical elements (tempo, rhythm stability, beat strength, overall regularity). A value of 0.0 is least danceable and 1.0 is most danceable
Energy	num	Value between 0.0 to 1.0 that represents a perceptual measure of intensity and activity. Perceptual features contributing to this attribute include dynamic range, perceived loudness, timbre, onset rate, and general entropy
Key	Factor	12 possible levels. The key track is in. Integers map to pitches using standard Pitch Class notation
Loudness	num	The overall loudness of a track in decibels (dB). Loudness values are averaged across the entire track. Values typical range between -60 and $0\,db$
Speechiness	num	Value between 0.0 to 1.0. Detects the presence of spoken words in a track. The more exclusively speech-like the recording the closer to 1.0 the attribute value
Acousticness	num	A confidence measure from 0.0 to 1.0 of whether the track is acoustic. 1.0 represents high confidence the track is acoustic
Instrumentalness	num	Value between 0.0 and 1.0. The closer the instrumental ness value is to 1.0, the greater likelihood that the track contains no vocal content. Vocals like "Ooh" and "aah" are treated as instrumental in this context
Valence	num	A measure from 0.0 to 1.0 describing the musical positiveness conveyed by a track. Tracks with high valence sound more positive while tracks with low valence sound more negative
Tempo	num	The overall estimated tempo of a track in beats per minute (BPM). In musical terminology, tempo is the speed or pace of a given piece and derives directly from the average beat duration
duration_ms	int	The duration of the track in milliseconds
time_signature	int	An estimated overall time signature of a track. The time signature (meter) is a notational convention to specify how many beats are in each bar (or measure)
Mode	Factor	Mode indicate the modality (major or minor) of a track, the type of scale from which its melodic content is derived. Major is represented by 1 and minor is 0

- **rpart**: PART tree
- **glm**: Generalized Linear Model
- **nb**: Naive Bayes
- **svmLinear**: Support Vector Machines with Linear Kernel
- **svmRadial**: Support Vector Machines with Radial Basis Function Kernel
- **ranger**: Random Forest

The performance and utility of the above-detailed Machine Learning methods in different domains [12–16] allows us to use them for predicting chord progressions. Different methods have been selected to cover different learning paradigms.

Method **glm** [17] fits generalized linear models, specified by giving a symbolic description of the linear predictor and a description of the error distribution. It is suitable for binary classification problems.

Tree modeling is performed through **rpart** [18] and **ranger** [19] (upgrade of the classical **randomForest**). **rpart** carries out a CART (Classification and regression trees) modeling. First of all, it grows the (classification in this case) tree until one of the possible stop conditions are reached: the number of observations in a node is under a threshold, the minimum cost complexity factor cannot be reached when attempting to do a split or the tree has reached the maximum depth allowed. Once the tree has been grown, the method examines the results and if over fitting is detected, the tree is pruned to find an optimal point between over fitting and under fitting. These methods split the data based on how the creation of sub-nodes increases the homogeneity of those resultant sub-nodes. Random forests technique is based on trees too but it improves predictive accuracy by generating not one but a large number of bootstrapped trees using random samples of variables for each iteration, classifying a case using each tree in this newest created forest, and deciding the final predicted outcome by combining the results across all of the trees.

Naive Bayes is based on the Bayes Theorem and although all the predictors are assumed to be independent within each class label, it gives excellent results for many real problems. The predictor variables are handled by assuming that they follow a Gaussian distribution, given the class label. The outcome variable is predicted using the probabilities of each attribute when **nb** [20] is applied. Support Vector Machines [21] performs classification looking for the hyperplane that maximizes the margin between the classes? closest points. When a linear separator cannot be found, data points are projected into a higher dimensional space where the data become linearly separable. This projection is made by kernel techniques. Two different methods with linear **svmLinear** and non-linear **svmRadial** kernel have been chosen

The above introduced methods have been applied considering the following settings:

- Method configuration: Default parameter settings provided by caret package in order to perform a first approach of how the datasets behave with each algorithm.
- Validation: Stratified split into training (80%) and test (20%). The training set will be used to build the model and the test set to evaluate it.

– Training: 10 fold cross validation with 3 repetitions has been used for training
the models. In this 10 fold cross validation, all the samples in the dataset were
grouped in 10 different sets, using one of them for testing while keeping the
remaining for training. This process is repeated three times.

Note that most of the datasets are imbalanced (30%(1)/70%(0)). Thus, different sampling strategies are tested and the resulting datasets represent the input
to each one of the learning methods. For each of the six algorithms, four different
models have been built according to a different sampling strategy (over, down,
ROSE and SMOTE [11]). For each algorithm, the model yielding the best results on
the test set will be selected as the best model. The evaluation of this previously
unseen instances of the test set let us compare the real performance of the final
models. Algorithm 1 shows the whole procedure.

algorithms = {glm, rpart, randomForest, nb, svmLinear, svmRadial};
foreach *dataset* **do**
> **foreach** *alg in algorithms* **do**
>> ; /* train the models   */
>> modOver = train(alg, data = train, sampling = over);
>> modDown = train(alg, data = train, sampling = down);
>> modROSE = train(alg, data = train, sampling = ROSE);
>> modSMOTE = train(alg, data = train, sampling = SMOTE);
>>
>> ; /* predict the class: test set   */
>> predOver = predict(modOver, data = test);
>> predDown = predict(modDown, data = test);
>> predROSE = predict(modROSE, data = test);
>> predSMOTE = predict(modSMOTE, data = test);
>>
>> ; /* evaluate the models   */
>> resOver = logLoss(confMatrix(test$class, predOver));
>> resDown = logLoss(confMatrix(test$class, predDown));
>> resROSE = logLoss(confMatrix(test$class, predROSE));
>> resSMOTE = logLoss(confMatrix(test$class, predSMOTE));
>>
>> best = maxMicroF1(resOver, resDown, resROSE, resSMOTE);
> **end**
end

Algorithm 1. Build and select the best model for each dataset

3.2   Metrics

As explained before, some of the datasets' class are imbalanced. In addition to
Accuracy (A), the micro averaged version of the well known metrics Precision
(P), Recall (R) and F_1 [11] is selected together with logLoss [22]. Considering
that TP is the number of examples correctly classified as the positive class, TN
is the number of examples correctly classified as the negative class, FP is the
number of examples wrongly classified as the positive class and FN the number

of examples wrongly classified as the negative class, A, P, R and F_1 are defined as follows.

$$A = \frac{TP + TN}{TP + TN + FP + FN} \qquad P = \frac{TP}{TP + FP}$$

$$R = \frac{TP}{TP + FN} \qquad F_1 = 2 \times \frac{P \times R}{P + R}$$

LogLoss quantifies the accuracy of a classifier by penalizing wrong classifications. Minimizing the Log Loss is equivalent to maximizing the accuracy of the classifier, but there is a subtle difference. In order to calculate the Log Loss, the classifier must assign a probability (p_i) to each class rather than simply yielding the most likely class. It is defined below.

$$LogLoss = -\frac{1}{N} \sum_{i=1}^{N} [y_i log p_i + (1 - y_i) log(1 - p_i)]$$

Table 3. F_1 for each model on test set

Progression	rpart	glm	nb	svmL	svmR	rf
effective.all.genres	0.74825	0.69578	0.73521	0.69849	0.74934	**0.79193**
gaining.popularity	0.84074	0.72467	**0.83956**	0.75179	0.82217	0.80793
magic.changes	**0.86839**	0.76379	0.29982	0.72214	0.82843	0.86310
most.popular	0.67680	0.65479	0.65490	0.67097	**0.69075**	0.68522
pachelbel	0.74137	0.86923	0.83064	0.87494	**0.90270**	0.75444
timeless	0.73474	0.75117	0.41014	0.68525	**0.82318**	0.56898
cadential64	0.83682	0.81701	0.62892	0.82006	0.88455	**0.89908**
I.as.precadence	**0.95234**	0.79190	0.83517	0.75889	0.81397	0.76771
newcomer	0.77865	0.81385	0.68669	0.82314	0.74944	**0.91111**
simple.yet.powerful	**0.94063**	0.76104	0.51293	0.80395	0.91206	0.92766
stepwise.bass.down	0.71599	0.69285	0.76457	0.70821	0.72704	**0.78109**
stepwise.bass.up	0.48695	0.84957	0.63792	0.89845	**0.92008**	0.82191
applied.vii	0.82918	0.80775	0.81376	0.83822	**0.88389**	0.74348
chord.pleased.Lord	**0.88399**	0.85250	0.41050	0.82230	0.78972	0.63735
expanding.V.vi	0.81187	0.70590	**0.84821**	0.72716	0.81105	0.80814
secondary.dominant	0.90437	0.88067	0.85573	0.88628	**0.94234**	0.91330
vi42.passing.chord	0.89668	0.82953	0.86953	0.83557	**0.91233**	0.82343
bVI.to.V	0.84705	0.85576	0.81376	0.83221	**0.90814**	0.66306
cadencing.bVII	0.80866	0.73626	**0.83309**	0.75970	0.82873	0.82798
candencing.in.style	0.84614	0.78441	**0.93802**	0.82860	0.88073	0.93221
mixolydian.cadence	0.90957	0.79328	0.49135	0.76125	0.84047	**0.90957**

4 Results

In this section, the aforementioned experiments are analyzed. Table 3 shows the F_1 obtained for each binary problem and each different machine learning method. Table 4 contains the LogLoss value for the experiments whose F_1 is shown in Table 3.

Table 4. LogLoss of each model on the test set

Progression	rpart	glm	nb	svmL	svmR	rF
effective.all.genres	2.29080	0.61709	1.06604	0.64638	0.54090	**0.50819**
gaining.popularity	2.29586	0.59268	0.68104	0.55032	0.56741	**0.53052**
magic.changes	0.45711	0.88835	5.69869	0.58640	0.45529	**0.41536**
most.popular	0.63065	0.61723	0.92931	0.61045	0.61680	**0.59226**
pachelbel	0.73712	0.57187	2.20434	0.49506	**0.42239**	0.66887
timeless	1.95744	0.85849	4.27938	0.64156	**0.45389**	0.74188
cadential64	0.55121	0.66113	3.76041	0.51237	**0.36370**	0.39110
I.as.precadence	0.76120	0.63053	1.47563	**0.56608**	0.61571	0.62741
newcomer	0.67570	1.00073	2.20718	0.49264	0.80781	**0.39696**
simple.yet.powerful	1.38791	0.73773	2.11598	0.60028	**0.32238**	0.39084
stepwise.bass.down	2.03425	0.64715	0.96973	0.64085	0.61982	**0.60725**
stepwise.bass.up	1.14947	0.50541	1.76185	0.51010	**0.34447**	0.60753
applied.vii	0.69399	0.61890	1.22321	0.58839	**0.40091**	0.66001
chord.pleased.Lord	**0.39493**	2.57653	10.74308	0.46906	0.85578	1.33383
expanding.V.vi	0.55673	0.66775	**0.47183**	0.65301	0.49104	0.59195
secondary.dominant	0.36368	0.39952	0.77317	0.35223	**0.17554**	0.26221
vi42.passing.chord	0.66081	0.62432	0.88796	0.50365	**0.34007**	0.58048
bVI.to.V	1.78251	0.51041	3.98525	0.47484	**0.32722**	0.72091
cadencing.bVII	0.78476	0.60006	0.72928	0.56762	**0.49911**	0.45051
candencing.in.style	0.58210	0.61831	0.53075	0.44616	0.38625	**0.38411**
mixolydian.cadence	1.59762	0.63714	3.21393	0.53388	0.52423	**0.40109**

Note that F_1 is often high. However, the highest F_1 is obtained with the combination of SMOTE resampling method and svmRadial. In particular, the results with highest F_1 and lowest logLoss are obtained when model is trained with svmRadial, being this the best for the 38% of the progressions, followed by the method ranger, which achieves the best results for the 33% of the progressions. We recommend the use of svmRadial because it has lower computational cost, and therefore it is quicker building the model. Table 5 shows the most important variables used for building the models. It should be noticed how the *valence*

Table 5. Most important variables of each model

Progression	First	Second	Third	Fourth
effective.all.genres	duration_ms	acousticness	energy	instrumentalness
gaining.popularity	duration_ms	loudness	valence	speechiness
magic.changes	tempo	acousticness	danceability	valence
most.popular	loudness	duration_ms	valence	popularity
pachelbel	key8	valence	mode1	instrumentalness
timeless	mode1	valence	acousticness	key1
cadential64	acousticness	speechiness	valence	duration_ms
I.as.precadence	valence	acousticness	danceability	duration_ms
newcomer	acousticness	mode1	valence	danceability
simple.yet.powerful	valence	danceability	popularity	instrumentalness
stepwise.bass.down	valence	popularity	acousticness	duration_ms
stepwise.bass.up	acousticness	energy	loudness	key10
applied.vii	acousticness	instrumentalness	energy	valence
chord.pleased.Lord	loudness	speechiness	popularity	energy
expanding.V.vi	speechiness	valence	loudness	instrumentalness
secondary.dominant	energy	loudness	acousticness	valence
vi42.passing.chord	key6	acousticness	loudness	valence
bVI.to.V	loudness	valence	energy	danceability
cadencing.bVII	popularity	loudness	instrumentalness	acousticness
candencing.in.style	acousticness	loudness	popularity	instrumentalness
mixolydian.cadence	loudness	instrumentalness	acousticness	valence

appears in the top four important variables for most of the models. Other features such as *danceability, valence, loudness* or *popularity* together with other strictly musical related like *instrumentalness* or *key*, stand out from the list of important variables used to build the models.

The evidence from this study points towards the idea that information gathered from Spotify could be useful for helping us decide which chord progressions should be used when a new song is being created.

5   Conclusions and Future Work

This paper presents a method to improve knowledge about how useful chord progressions can be in other music classification problems, which aims to classify songs into emotions or genres.

The method presented in this work has many interesting applications in the composition of new music. If done manually, the work of analyzing all the chord progressions in a music score is not an easy task. It requires time and a great knowledge of music theory. Automating this process is not trivial either, it has a high computing cost, and the results are not always satisfactory.

These results represent an excellent initial step toward a new field of application. The principal advantages are the creation of new songs based on the features of similar and already existing songs, without the need to analyze their structure and taking into account not only their musical features but aspects of the tracks as their energy or valence.

We are currently working on the multi-label version of this problem, that can address the issue of creating a new track based on a bunch of songs we want it to be similar to. Given as input existing songs, the classification model will return the list of chord progressions that should and should not be used to create the new composition.

Further work will look into the lyrics of the songs, in order to perform a sentiment analysis process to explore the correlation between the Spotify musical features, the chord progressions and the emotions the song intends to cause on the listener, based on the vocabulary they are using.

Acknowledgments. This research has been funded by the Spanish MINECO project TIN2017-87600-P.

References

1. Juslin, P.N., Sloboda, J.: Handbook of Music and Emotion: Theory, Research, Applications. Oxford University Press, Oxford (2011)
2. Cho, Y.H., Lim, H., Kim, D.W., Lee, I.K.: Music emotion recognition using chord progressions. In: 2016 IEEE International Conference on Systems, Man, and Cybernetics (SMC), pp. 002588–002593. IEEE (2016)
3. Stamatatos, E., Widmer, G.: Automatic identification of music performers with learning ensembles. Artif. Intell. **165**(1), 37–56 (2005)
4. Arutyunov, V., Averkin, A.: Genetic algorithms for music variation on genom platform. Procedia Comput. Sci. **120**, 317–324 (2017). 9th International Conference on Theory and Application of Soft Computing, Computing with Words and Perception, ICSCCW 2017, Budapest, Hungary, 22–23 August 2017
5. Costa, Y.M., Oliveira, L.S., Silla, C.N.: An evaluation of convolutional neural networks for music classification using spectrograms. Appl. Soft Comput. **52**, 28–38 (2017)
6. Hu, X., Downie, J.S.: Improving mood classification in music digital libraries by combining lyrics and audio. In: Proceedings of the 10th Annual Joint Conference on Digital Libraries, JCDL 2010, pp. 159–168. ACM, New York (2010)
7. Gómez, L.M., Cáceres, M.N.: Applying data mining for sentiment analysis in music. In: De la Prieta, F., Vale, Z., Antunes, L., Pinto, T., Campbell, A.T., Julián, V., Neves, A.J.R., Moreno, M.N. (eds.) PAAMS 2017. AISC, vol. 619, pp. 198–205. Springer, Cham (2018). https://doi.org/10.1007/978-3-319-61578-3_20
8. Famous-Chord-Progressions, January 2018. https://www.hooktheory.com/theorytab/common-chord-progressions
9. Spotify, January 2018. https://developer.spotify.com/web-api/get-audio-features/
10. HookTheory-API, January 2018. https://www.hooktheory.com/api/trends/docs
11. Kuhn, M.: Building predictive models in R using the caret package. J. Stat. Softw. **28**(5), 1–26 (2008)

12. Villar, J.R., Chira, C., Sedano, J., González, S., Trejo, J.M.: A hybrid intelligent recognition system for the early detection of strokes. Integr. Comput. Aided Eng. **22**(3), 215–227 (2015)
13. Herrero, Á., Sedano, J., Baruque, B., Quintián, H., Corchado, E. (eds.): SOCO 2015. AISC, vol. 368. Springer, Berlin (2015). https://doi.org/10.1007/978-3-319-19719-7
14. Troiano, L., Rodríguez-Muñiz, L.J., Ranilla, J., Díaz, I.: Interpretability of fuzzy association rules as means of discovering threats to privacy. Int. J. Comput. Math. **89**(3), 325–333 (2012)
15. Gil-Pita, R., Ayllón, D., Ranilla, J., Llerena-Aguilar, C., Díaz, I.: A computationally efficient sound environment classifier for hearing aids. IEEE Trans. Biomed. Eng. **62**(10), 2358–2368 (2015)
16. Montañés, E., Quevedo, J.R., Díaz, I., Ranilla, J.: Collaborative tag recommendation system based on logistic regression. In: Proceedings of ECML PKDD (The European Conference on Machine Learning and Principles and Practice of Knowledge Discovery in Databases) Discovery Challenge 2009, Bled, Slovenia, 7 September 2009 (2009)
17. Chambers, J.M.: Statistical Models in S. CRC Press, Inc., Boca Raton (1991)
18. Breiman, L., Friedman, J., Olshen, R., Stone, C.: Classification and Regression Trees. Wadsworth and Brooks, Monterey (1984)
19. Wright, M.N., Ziegler, A.: ranger: a fast implementation of random forests for high dimensional data in C++ and R. J. Stat. Softw. **77**(1), 1–17 (2017)
20. Rish, I.: An empirical study of the naive bayes classifier. In: IJCAI 2001 Workshop on Empirical Methods in Artificial Intelligence, vol. 3, pp. 41–46. IBM, New York (2001)
21. Cortes, C., Vapnik, V.: Support-vector networks. Mach. Learn. **20**(3), 273–297 (1995)
22. Masnadi-shirazi, H., Vasconcelos, N.: On the design of loss functions for classification: theory, robustness to outliers, and savageboost. In: Koller, D., Schuurmans, D., Bengio, Y., Bottou, L. (eds.) Advances in Neural Information Processing Systems 21, pp. 1049–1056. Curran Associates, Inc. (2009)

LearnSec: A Framework for Full Text Analysis

Carlos Gonçalves[1,3]([✉]), E. L. Iglesias[1], L. Borrajo[1], Rui Camacho[2,3],
A. Seara Vieira[1], and Célia Talma Gonçalves[4,5]

[1] Higher Technical School of Computer Engineering, University of Vigo,
Campus Universitario As Lagoas s/n, 32004 Ourense, Spain
{coliveira,eva,lborrajo,adrseara,}@uvigo.es
[2] FEUP-U.Porto, Rua Dr. Roberto Frias s/n, 4200-465 Porto, Portugal
rcamacho@fe.up.pt
[3] LIAAD/INESC TEC, Porto, Portugal
[4] ISCAP-P.Porto, Rua Jaime Lopes Amorim, s/n,
4465-004 S. Mamede de Infesta, Portugal
celia@iscap.ipp.pt
[5] LIACC, U.Porto, Porto, Portugal

Abstract. Large corpus of scientific research papers have been available for a long time. However, most of those corpus store only the title and the abstract of the paper. For some domains this information may not be enough to achieve high performance in text mining tasks. This problem has been recently reduced by the growing availability of full text scientific research papers. A full text version provides more detailed information but, on the other hand, a large amount of data needs to be processed. *A priori*, it is difficult to know if the extra work of the full text analysis has a significant impact in the performance of text mining tasks, or if the effect depends on the scientific domain or the specific corpus under analysis.

The goal of this paper is to show a framework for full text analysis, called LearnSec, which incorporates domain specific knowledge and information about the content of the document sections to improve the classification process with propositional and relational learning.

To demonstrate the usefulness of the tool, we process a scientific corpus based on OSHUMED, generating an attribute/value dataset in Weka format and a First Order Logic dataset in Inductive Logic Programming (ILP) format. Results show a successful assessment of the framework.

Keywords: Full text analyses · Text preprocessing · Text mining
Use of background knowledge · Inductive Logic Programming

1 Introduction

The amount of biomedical scientific articles stored in public resources continues to grow. Automatic retrieving and classification of articles has become a very

© Springer International Publishing AG, part of Springer Nature 2018
F. J. de Cos Juez et al. (Eds.): HAIS 2018, LNAI 10870, pp. 502–513, 2018.
https://doi.org/10.1007/978-3-319-92639-1_42

interesting and hot research topic. Thus, in order to manage this volume of documents, the use of sophisticated computer tools to preprocess the documents must be considered.

The goal of this paper is to show a framework for full text analysis, called LearnSec, which incorporates domain specific knowledge and information about the content of the document sections to improve the classification process with propositional learning and relational learning.

To demonstrate the usefulness of the tool, we process a scientific corpus based on OSHUMED [1,2], generating an attribute/value dataset in Weka format for relational learning, and a First Order Logic dataset in Inductive Logic Programming (ILP) format for propositional learning.

This paper is organized as follows. Section 2 introduces ILP together with its advantages and drawbacks. Section 3 presents the general architecture of the framework. Three case studies illustrating the use of the framework are described in Sect. 4 and the results are shown in Sect. 5. Finally, in Sect. 6 we draw the conclusions of the work.

2 ILP for Text Mining

Inductive Logic Programming (ILP) is a research area at the intersection of Machine Learning and Logic Programming [3]. Its objective is to learn logical programs from examples and knowledge in the domain.

From Logic Programming it inherits the representation scheme for both data and models -a subset of First Order Logic. It is a supervised machine learning method. ILP addresses the problem of inducing hypotheses as predicate definitions from examples and background knowledge. According to [3], an ILP learner requires an initial theory (background knowledge) and evidences (examples), to induce a hypothesis that, together with the background knowledge, *explains* some properties of the evidences. In traditional ILP, the evidences come in two forms: positive and negative. Positive examples are instances of the concept to learn, whereas negative examples are not. Negative examples are used to avoid over-generalization.

Using First Order Logic to encode both data and the background knowledge, structured data can be easily handled in ILP. The hypothesis is also encoded in First Order Logic and can therefore represent highly complex models. Traditional ILP systems transform the induction process into a search over a very large space (sometimes infinite) which may cause efficiency problems when dealing with complex problems. To address this problem, the user may constrain the language of the hypothesis and use a set of parameters for that purpose. ILP systems like Aleph (used in our framework) have a highly powerful expressive language to constrain the hypothesis language.

The expressive representational language of First Order Logic gives ILP advantages over propositional learners like SVM, K-Nearest Neighbors, decision trees, etc. ILP can easily handle data with structure. Most of ILP constructed models are comprehensible. It is easy to provide information that the experts find relevant for the construction of models. Induced models combine harmoniously relational information with numerical computations.

Due to the above mentioned features, ILP has relevant applications to problems in complex domains like the natural language and the molecular computational biology [3].

In [4] an overview of inducing part-of-speech taggers using Inductive Logic Programming is presented. It also describes how Progol [5] was applied to the induction of rules based on constraint grammar for several languages (English, Hungarian, Slovene, Swedish).

3   Framework Architecture

In this section we provide detailed information about the full text analysis tool LearnSec. The architecture and its components are presented and described. In order to facilitate their comprehension and clarity, the component diagram has been split across three figures (Figs. 1, 2 and 3).

To explain the process of analysis of the full text document corpus, we have used the OHSUMED corpus. OHSUMED text collection, compiled by William Hersh [1], is a set of 348, 566 references from Medline, the on-line medical information database, consisting of titles and/or abstracts from 270 medical journals over a five-year period (1987–1991). Each document is tagged with one or more categories (from 23 disease categories).

As the OSHUMED documents only have title and abstract, we get the full text papers from the NCBI PubMed Central (PMC). In addition, for the classification and selection of relevant papers we use the MeSH classes.

Data Acquisition and Storing

From PubMed we get the papers in XML format and store them in a local SQL database. Papers are then filtered by the MeSH classes and the SQL database is created (see Fig. 1).

During this first stage of acquiring and storing, a XML parser was applied to remove unnecessary HTML/XML tags and to identify the document sections. All documents of the corpus have sections according to the following common structure: Title, Abstract, Introduction, Methods (Materials and Methods, Methods, Experimental Procedures), Results (Results, Discussion, Results and Discussion), and Conclusions.

Text Mining Preprocessing

The preprocessing of documents is divided in four steps: *traditional* text preprocessing, context awareness, domain independent feature construction, and domain dependent feature construction. The main operations of text preprocessing are presented in Fig. 2.

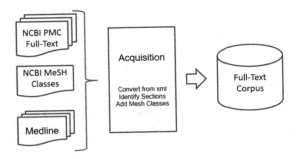

Fig. 1. LearnSec architecture (Part I). XML documents are converted to plain texts, broken down into sections and stored in a database. MeSH classes are associated with each document.

The so called *traditional* preprocessing performs the following operations in documents: Named Entity Recognition (NER) identification, stopwords removal, synonyms replacement, word validation using dictionaries and ontologies, stemming, and filtering out highly frequent words and highly infrequent words. To associate context to the words we use n-grams (bi-grams and tri-grams).

As first set of features to the machine learning algorithms, documents are represented as a set of attributes related to the frequency of words in documents, the well-known *bag-of-words* representation. The value of an attribute is given by the standard Term Frequency-Inverse Document Frequency (*TF-IDF*) [6]. In addition, we calculate the term frequency on each section of the document.

This representation is suitable for the use of propositional classifiers in text classification, since they require an "attribute/value" representation, as mentioned in Sect. 2.

Propositional learners, like Support Vector Machines (SVM) or decision trees, have a very restrictive representation language for the examples (attribute/value). In contrast with the relational learners, the attribute/value representation makes it very hard for the user to provide useful background/domain information. We try to mitigate this situation by encoding in an attribute/value scheme domain information available in ontologies and taxonomies. We take advantage of *is_a* and *part_of* domain relations to create extra attributes. This process it is called the text enrichment step.

The text enrichment step is different from the method for propositional datasets. For relational datasets we just encode all the information in the ontology or taxonomy in Prolog. A propositional learner like Aleph [7] can upload and use the whole ontology and taxonomy.

Generating Weka and ILP Datasets

The last phase of the framework is the generation of the datasets (see Fig. 3). We distinguish the generation of Weka datasets from the Aleph (ILP) datasets.

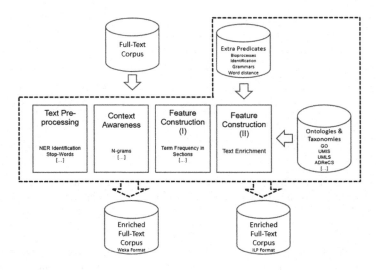

Fig. 2. LearnSec architecture (Part II)

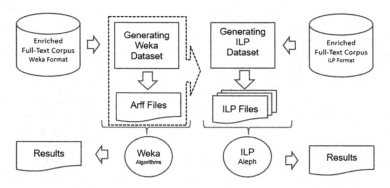

Fig. 3. LearnSec architecture (Part III). Generating Weka in ILP datasets.

For Weka datasets the framework just generates Attribute Relation File Format (ARFF) files with the attributes defined and explained previously.

ILP datasets may use the term frequencies and n-grams plus the available information concerning ontologies and taxonomies. However, the usual number of attributes is overwhelming for ILP. We have implemented a filtering procedure to reduce their number. We compute the document/section term frequencies separately for positive and negative examples and for each term. Then, again for each term, we compute the absolute value of the difference between the value on the positive and on the negative examples. We sort by descendant order and establish a threshold below which the terms are discarded. With this procedure we choose the most discriminative terms and reduce substantially the number of attributes.

Actually, an ILP system can use any information the expert finds useful for the induction of the models. As may be seen in Fig. 3, the framework supports a repository of ontologies and taxonomies that can help in the Weka generalization process and can be fully used by an ILP systems.

Furthermore, the framework has another repository with predicates specially encoded for specific domains or datasets. These predicates may identify distances between word and sentences, do Part-Of-Speech Tagging (POST), be entire grammars if syntactic information is required, etc.

4 Case Studies

To evaluate the proposed architecture, we build a corpus based on OSHUMED. The initial document collection includes $50,216$ abstracts from the year 1991. Table 1 contains information about the datasets.

The corpus was implemented following the Moschitti study [2]. Classes C01 to C26 totalize $18,907$ documents. For testing purposes we have used only the classes C04, C06, C14, C20, and C23.

Each line of the Table 1 shows the class name, its description, the number of documents in the class and the percentage and number of documents of the class that will be used as non-relevant documents when assembling a dataset for other class as target concept.

The documents used as non-relevant for each class are obtained through a process that randomly selects documents from the other classes using the percentage indicated in the Table 1, given by

$$\left\lceil \frac{n_c * 100}{n} \right\rceil \tag{1}$$

where n_c represents the number of documents of the class c, and n denotes the total number of documents in the corpus.

For example, for the class C04, the number of documents is $5,598$ and the number of documents that will be used as non-relevant for the other classes is $\lceil (5598/18907) * 100 \rceil = 30$ documents.

4.1 Full Text Corpus Construction

The first step was to obtain a reliable, balanced and preclassified full text corpus. As OHSUMED only contains title and abstract of the documents, we have downloaded a full text corpus available at NCBI[1] in XML format ($459,009$ documents in total). To obtain the Mesh terms we have downloaded, from NCBI[2], the 2017 MeSH trees.

To associate the available full text documents with the Mesh trees, we used the methodology indicated in a previous study [8]. The association of MeSH term

---

[1] ftp://ftp.ncbi.nlm.nih.gov/pub/pmc/manuscript/.

[2] https://www.nlm.nih.gov/mesh/download_mesh.html.

Table 1. Number of relevant documents of the categories of Oshumed, and percentage and number of documents to be used as non-relevant

Class	Description	Relevant	Used as non-relevant	
		#	%	#
C01	Bacterial Infections and Mycoses	862	5%	39
C02	Virus Diseases	1574	8%	131
C03	Parasitic Diseases	198	1%	2
C04	**Neoplasms**	**5598**	**30%**	**1657**
C05	Musculoskeletal Diseases	614	3%	20
C06	**Digestive System Diseases**	**1693**	**9%**	**152**
C07	Stomatognathic Diseases	245	1%	3
C08	Respiratory Tract Diseases	1036	5%	57
C09	Otorhinolaryngologic Diseases	244	1%	3
C10	Nervous System Diseases	4272	23%	965
C11	Eye Diseases	747	4%	30
C12	Urologic and Male Genital Diseases	1396	7%	103
C13	Female Genital Diseases, Pregnancy Complications	1511	8%	121
C14	**Cardiovascular Diseases**	**2614**	**14%**	**361**
C15	Hemic and Lymphatic Diseases	568	3%	17
C16	Neonatal Diseases and Abnormalities	896	5%	42
C17	Skin and Connective Tissue Diseases	1366	7%	99
C18	Nutritional and Metabolic Diseases	1516	8%	122
C19	Endocrine Diseases	914	5%	44
C20	**Immunologic Diseases**	**1744**	**9%**	**161**
C21	Disorders of Environmental Origin	1	0%	0
C22	Animal Diseases	175	1%	2
C23	**Pathological Conditions, Signs and Symptoms**	**7377**	**39%**	**2878**
C24	Occupational Diseases	29	0%	0
C25	Chemically-Induced Disorders	669	4%	27
C26	Wounds and Injuries	611	3%	18

to documents was done exploring the Medline 2010 database that contains only the title and abstract of the documents. To transform the texts into datasets, a series of preprocessing steps were required (see [9] for information detailed).

Processing the XML Files

As described in Sect. 3, we parse the XML and the sections of the documents and insert the results into a local SQL database.

Table 2 shows examples of term frequencies on each section for a specific document (pmid identifier 18459944), available on PubMed Central repository[3].

Table 2. Example of term frequencies by section, for some terms of the document with pmid identificator 18459944. Divisors represent the total number of occurrences of each term on each section for all documents of the corpus

Term	Title	Abstract	Introduction	Methods	Results	Conclusions
angiotonin	0/39	1/126	0/604	2/287	20/1150	0/143
collagen	0/23	0/435	0/2391	2/2734	0/12180	0/682
diabet	0/311	4/2005	5/7121	0/8009	42/28910	0/2994
fibroid	0/8	0/69	0/266	0/231	6/956	0/144
hypertens	1/158	7/961	6/3674	6/3985	164/15284	0/1258
insulin	0/153	0/1337	0/6247	4/5251	6/23356	0/1271
kidnei	1/120	2/788	4/3375	28/3853	58/14030	0/751
methanol	0/259	0/1689	0/8785	4/4135	0/21290	0/1686
pathogenesi	0/114	0/943	0/3997	0/361	2/6096	0/563

Tokenization, Special Characters Removal and case folding
After removing HTML/XML tags, punctuation, digits and some special characters such as (",",",",".", "!", "?", "/", "[", "]"), tokenization is applied. Characters like "+" and "-" are not removed because they might be important in some biology domains and if they are removed, the term may lose sense (for example:"blood-lead"). The encoding of characters was also considered, using the standard text encoding format *utf8*, to avoid and resolve unicode problems. In the tokenization process, we also remove punctuations marks and HTML/XML tags [10]. All terms were converted to the lowercase format (case folding).

Stopwords Removal
We have used a list of 659 stopwords (words that are meaningless such as articles, conjunctions and prepositions) to be identified and removed from the article. With this step we also reduce significantly the number of attributes of the dataset.

Synonyms Handling and Dictionary Validation
Handling synonyms allows to significantly reduce again the number of attributes in the datasets without changing the semantic of words.
    We have used *WordNet* [11][4] and Gene Ontology [12][5] to check technical terms.

---

[3] https://www.ncbi.nlm.nih.gov/pubmed/18459944.
[4] http://wordnet.princeton.edu/.
[5] http://www.geneontology.org/.

For the purpose of our work, we consider that a term is valid if it appears in a dictionary. We have gathered several dictionaries for the common English terms (such as ISPELL[6] and the already mentioned *WordNet*). Since we are dealing with a biomedical corpus, we also have used the *BioLexicon* dictionary [13], the *Hosford Medical Terms* Dictionary [14] and Gene Ontology, totaling 1186273 terms.

Stemming
Stemming is the process of removing inflectional affixes of words reducing the words to their root. We have implemented the Porter Stemmer algorithm [15], and have applied it to all the words of the corpus, normalizing several variants into the same form.

4.2   Generating WEKA Datasets

To illustrate how the framework can be used to generate and run a Weka [16] dataset, we have used C04 from Table 2 encoded in ARFF format. We have used in the experiments the Weka implementation of Support Vector Machine (SVM) and the Naive Bayes algorithm.

To perform feature selection we have used Information Gain of Weka [17], called Info Gain Attribute Eval [18]. A cut of threshold of 0.01 was used.

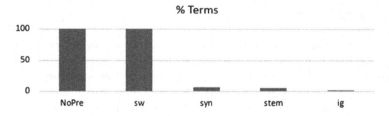

Fig. 4. Percentage of reduction of terms after applying the pre-processing techniques. *sw* stands for stopWords removal. *syn* Synonyms Handling and Word Validation. *stem* Stemming. *ig* stands for Information Gain.

Figure 4 shows the percentage of final terms based on the pre-processing techniques. In C04 dataset the initial number of terms is 458406, corresponding to the value 100%. The techniques were applied in a sequence. The application of stopwords removal produces a reduction of 0, 38%. Synonyms handling and word validation reduce the number of terms in 94, 23%. The Synonym handling and word validation has an high impact in reducing the number of terms because a term is consider if either have a synonym or exist in the several dictionaries used in this study (regular English and biomedicine domain dictionaries), otherwise the term is discarded. Stemming reduces the number of terms in 94, 83%. Finally,

---

[6] http://www.lasr.cs.ucla.edu/geoff/ispell.html.

Information Gain has an impact of 99, 67% reduction. The cumulative reduction results in a final number of 1525 terms for the C04 title and abstract dataset. The results obtained using the SVM and Naive Bayes algorithms are shown in the Table 3.

4.3 Generating ILP Datasets

We have also tested the framework on the use of ILP with the same C04 dataset as of Weka. We started by using all the final features given to Weka. As described in Sect. 3, even using "traditional" feature selection techniques, the number of attributes is still very large for ILP. We have applied the filtering procedure described in Sect. 3 using different cut off values. The cut off threshold of 10 terms produced the best results. As background knowledge to ILP we used the filtered terms plus ≤ and ≥ relations. Aleph ILP system was used in the experiments. As shown in Table 3, the results obtained by ILP are very close to those obtained by Weka, but using a considerably less amount of information. This is a very promising result for full text analysis.

5 Experimental Results

The experiments were done using a dataset of full text for the class C04-Neoplasms, and the original dataset (with only title and abstract). The results obtained are shown in Table 3. The results presented used a cross validation of 10 fold in the propositional learners and 3 fold in the ILP experiments.

Table 3. Results of application of classifiers on the corpus created by LearnSec

		SVM		Bayes		ILP	
Full text	F-measure	0,946	(0,007)	0,749	(0,011)	0,928	(0,002)
	Kappa	0,893	(0,015)	0,523	(0,020)	0,855	(0,004)
Title & Abs	F-measure	0,941	(0,007)	0,841	(0,011)	–	(–)
	Kappa	0,882	(0,014)	0,685	(0,021)	–	(–)

In the classification process of full texts, using the propositional learning, we achieved a 94, 6% in terms of the F-measure metric, and 89, 3% for the Kappa value, representing a "very good agreement", since the values are between 0.8 and 1.00 (according to the interpretation of [19]).

For the relational learner (Aleph) we also obtained very good results. A value of 92, 8% in terms of the F-measure metric and a Kappa value of 85, 5%. The kappa value is also very good which proves the good results achieved for both types of algorithms.

In the classification process of documents with title and abstract only propositional learning was applied and the results were worse. With this results, we decided not to make the relational learning process.

6   Conclusions

In this paper we offer a framework for full text analysis, called LearnSec, which incorporates domain specific knowledge and information about the content of the document sections to improve the classification process with propositional and relational learning.

The framework includes a set of functionalities for the generation of text corpora originally stored in XML format. The framework presents a set of database operations that improves the efficiency of the process of dataset generation. The framework includes preprocessing techniques and weighting of terms based on the document sections.

LearnSec creates both propositional learners *attribute/value* datasets and the adequate background knowledge for relational learners such as ILP systems. Besides the traditional *bag-of-words* approach to text classification, the developed framework incorporates domain specific background knowledge which helps in the classification process tasks for both the propositional and relational learners.

We have shown a successful first assessment of the framework in several case studies: a full text analysis with two propositional and one relational learner, using a generated full text corpus, and an analysis with the propositional learning using a corpus where the documents only have title and abstract.

Note that the experiments made are not an exhaustive comparative study neither between propositional learners and ILP nor between title and and abstract and full text documents. We only are demonstrating the utility of LearnSec and assessing the effectiveness of different preprocessing procedures.

References

1. Hersh, W., Buckley, C., Leone, T.J., Hickam, D.: OHSUMED: an interactive retrieval evaluation and new large test collection for research. In: Croft, B.W., van Rijsbergen, C.J. (eds.) SIGIR 1994. Springer, London (1994). https://doi.org/10.1007/978-1-4471-2099-5_20

2. Moschitti, A., Basili, R.: Complex linguistic features for text classification: a comprehensive study. In: McDonald, S., Tait, J. (eds.) ECIR 2004. LNCS, vol. 2997, pp. 181–196. Springer, Heidelberg (2004). https://doi.org/10.1007/978-3-540-24752-4_14

3. Muggleton, S., De Raedt, L.: Inductive Logic Programming: theory and methods. J. Logic Program. **19/20**, 629–679 (1994)

4. Eineborg, M., Lindberg, N.: ILP in Part-of-Speech Tagging — an overview. In: Cussens, J., Džeroski, S. (eds.) LLL 1999. LNCS (LNAI), vol. 1925, pp. 157–169. Springer, Heidelberg (2000). https://doi.org/10.1007/3-540-40030-3_10

5. Muggleton, S.: Inverse entailment and progol. New Gener. Comput. **1**(3–4), 245–286 (1995). Special issue on Inductive Logic Programming

6. Zhou, W., Smalheiser, N.R., Yu, C.: A tutorial on information retrieval, basics terms and concepts. J. Biomed. Discov. Collab. **1**, 2 (2006)

7. Srinivasan, A.: The aleph manual (2001)

8. Gonçalves, C.T., Camacho, R., Oliveira, E.: BioTextRetriever: a tool to retrieve relevant papers. Int. J. Knowl. Discov. Bioinform. **2**, 21–36 (2011). IGI Publishing

9. Gonçalves, C.A., Gonçalves, C.T., Camacho, R., Oliveira, E.: The Impact of pre-processing in classifying MEDLINE documents. In: Proceedings of the 10th International Workshop on Pattern Recognition in Information Systems (PRIS2010), Funchal, Madeira, pp. 53–61 (2010)

10. Aprile, A., Castellano, M., Mastronardi, G., Tarricone, G.: A web text mining flexible architecture. Int. J. Comput. Sci. Eng. (2007)

11. Oram, P.: WordNet: an electronical lexical database. Appl. Psycholinguist. **22**, 131–134 (1998). Cambridge University Press

12. Ashburner, M., Ball, C.A., Blake, J.A., Botstein, D., Butler, H., Sherlock, G.: Gene ontology: tool for the unification of biology. Nat. Genet. **25**, 25–29 (2000)

13. Rebholz-Schuhmann, D., Pezik, P., Lee, V., Kim, J.J., Del Gratta, R., Sasaki, Y., McNaught, J., Montemagni, S., Monachini, M., Calzolari, N., Ananiadou, S.: BioLexicon: towards a reference terminological resource in the biomedical domain. In: Proceedings of the 16th Annual International Conference on Intelligent Systems for Molecular Biology (2008)

14. The Hosford Medical Terms Dictionary v3.0 (2004)

15. Porter, M.F.: An algorithm for suffix stripping. In: Readings in Information Retrieval, pp. 313–316. Morgan Kaufmann Publishers Inc. (1997)

16. Witten, I.H., Eibe, F., Trigg, L., Hall, M., Holmes, G., Cunningham, S.J.: WEKA: practical machine learning tools and techniques with Java implementations. In: Proceedings of the ICONIP/ANZIIS/ANNES99 Future Directions for Intelligent Systems and Information Sciences, pp. 192–196. Morgan Kaufmann (1999)

17. Borase, P.N., Kinariwala, S.A.: Image Re-ranking using Information Gain and relative consistency through multi-graph learning. Int. J. Comput. Appl. **147**, 29–32 (2016). Foundation of Computer Science, NY, USA

18. Hall, M.: Correlation-based feature selection for machine learning. Ph.D. thesis, University of Waikato (1999)

19. Landis, J.R., Koch, G.G.: The measurement of observer agreement for categorical data. Biometrics **33**, 159–174 (1977)

A Mood Analysis on Youtube Comments and a Method for Improved Social Spam Detection

Enaitz Ezpeleta[(✉)], Mikel Iturbe, Iñaki Garitano, Iñaki Velez de Mendizabal, and Urko Zurutuza

Electronics and Computing Department, Mondragon University, Goiru Kalea, 2, 20500 Arrasate-Mondragón, Spain
{eezpeleta,miturbe,igaritano,ivelez,uzurutuza}@mondragon.edu

Abstract. In the same manner that Online Social Networks (OSN) usage increases, non-legitimate campaigns over these types of web services are growing. This is the reason why significant number of users are affected by social spam every day and therefore, their privacy is threatened. To deal with this issue in this study we focus on mood analysis, among all content-based analysis techniques. We demonstrate that using this technique social spam filtering results are improved. First, the best spam filtering classifiers are identified using a labeled dataset consisting of Youtube comments, including spam. Then, a new dataset is created adding the mood feature to each comment, and the best classifiers are applied to it. A comparison between obtained results with and without mood information shows that this feature can help to improve social spam filtering results: the best accuracy is improved in two different datasets, and the number of false positives is reduced 13.76% and 11.41% on average. Moreover, the results are validated carrying out the same experiment but using a different dataset.

Keywords: Spam · Social spam · Mood analysis
Online Social Networks · Youtube

1 Introduction

In recent years, Online Social Networks (OSNs) have extensively expanded around the world. The amount of users per each OSN platform shows the importance of these communication channels in our society: Facebook reached 1.4 billion daily active users on average as of December 2017[1]; Youtube has counted over a billion users in 2017[2]; and Twitter has 330 million monthly active users as of June 30, 2017[3].

---

[1] http://newsroom.fb.com/company-info/.

[2] https://www.youtube.com/yt/about/press/.

[3] https://www.statista.com/statistics/282087/number-of-monthly-active-twitter-users/.

© Springer International Publishing AG, part of Springer Nature 2018
F. J. de Cos Juez et al. (Eds.): HAIS 2018, LNAI 10870, pp. 514–525, 2018.
https://doi.org/10.1007/978-3-319-92639-1_43

The sudden increase in users gives malicious organizations the possibility to reach a vast amount of people easily. Authors in [11] demonstrate that these sites are now a major delivery platform targeted for spam. They analyze spam in several OSNs, and they quantify spam campaigns delivered from accounts in OSNs.

To deal with this problem, authors in [8] prove that content-based techniques can help to improve spam filtering results. They perform sentiment analysis on email messages in order to enrich the original dataset, and they obtain improved accuracy. Following a similar procedure, in this study we use another content-based technique, the mood analysis of the messages, to improve social media spam filtering results.

First, several spam filtering classifiers and different settings are applied to a Youtube comment dataset in order to identify the best ten filtering methods. After that, a mood analyzer is applied to each comment with the purpose of creating a new dataset adding this feature to the original dataset. Once the enhanced dataset is created, the previously selected ten classifiers are applied to the dataset with the mood feature. Finally, a comparison and an analysis of the results is carried out.

The remainder of this paper is organized as follows. Section 2 describes the previous work conducted in the area of social media spam filtering techniques. Section 3 describes the process of the aforementioned experiments, regarding Bayesian spam filtering and spam filtering using the mood of the texts. In Sect. 4, the obtained results are described, and finally, we summarize our findings and give conclusions in Sect. 5.

2   Related Work

OSN-related spam is an active research field [2] that has received wide attention from the scientific community. Stringhini et al. [22] demonstrated that it is possible to identify spammer accounts in large OSNs in an automatic manner and later block them, applying their approach in Facebook, Twitter and MySpace. Moreover, Wang et al. [25] proposed a spam detection system that is able to analyze OSNs in search of spam. Similarly, Egele et al. [7] presented COMPA, a tool to detect compromised OSN accounts based on anomalous user behavior. In a similar note, Gao et al. [10] employed classification and clustering to detect OSN spam campaigns in Twitter and Facebook. Ezpeleta et al. [9] showed that personalizing spam messages using publicly available OSN profile information lead to a significantly higher success rate than conventional, non-personalized spam.

In the field of OSN spam, the case of Twitter-based spam has been particularly well studied. Kwak et al. [13] identify this OSN as a useful tool for information diffusion. Therefore, it can be stated that Twitter is an attractive platform to perform spam campaigns. Yang et al. [26] described the dynamics of criminal accounts in Twitter and how they interact between them, and to other non-criminal accounts. Additionally, Song et al. [21] showed how to detect spammers based on the measurement of relation features, such as the distance and connectivity between receiver and recipient, instead of focusing on Twitter account features, as these account features are more prone to spammer manipulation.

Even if studied not as much as email, Twitter or Facebook, Youtube spam has also been an object of study. Chaudhary and Sureka [6] mined video descriptions, along with temporal and popularity based features, to detect spam videos on Youtube. O'Callaghan et al. [15] use network motif profiling to identify recurring Youtube spam campaigns, by characterizing Youtube users as motifs. By identifying users with distinctive motifs, they where able to label users in spamming campaigns.

However, even if numerous novel spam detection techniques have been published [20,24,27], OSN spam messages remain an open problem yet to be solved [11].

In this direction, content-based analysis, where text is analyzed to infer its meaning or purpose using different techniques such as Sentiment Analysis (SA), stands as a promising procedure for improving spam detection in OSNs [8,19]. The main objective of SA resides in the identification of the positive or negative nature of a document [17]. In order to reach this objective, it is possible to use a supervised learning approach, with three previously defined classes (positive, negative and neutral) [18] or a unsupervised one, where opinion words or phrases are the dominating indicators for sentiment classification [23].

It has been already demonstrated that SA helps in spam detection in different cases, such as social spammer detection [12], short informal messages [3], Twitter spam [19], email spam [8] and fraud detection [14].

As authors presented in [5], sentiment extraction from text can be used to predict mood, which can be used to prevent or mitigate security threats. Defined as "a temporary state of mind or feeling"[4], mood was used by Bollen et al. [4] in their analysis of Twitter feeds.

In this paper, we go beyond the State of the Art by using mood analysis for improved spam filtering, focusing on a popular OSN, the Youtube video service. To the best of our knowledge, no previous research has focused on using this approach for improved spam detection, let alone in the case of analyzing Youtube comments.

3   Design and Implementation

Having an original dataset, the process followed in this study is divided in two main parts, depicted on Fig. 1.

1. First, several classifiers are applied to a dataset consisting of social media messages (spam and ham) in order to identify and select the best ten social spam filtering classifiers. In the same step, the best 10 results are also extracted.

---

[4] https://en.oxforddictionaries.com/definition/mood.

2. Second, the mood of each message is added to the original dataset to create a new dataset. During the mood analysis, a descriptive experiment is carried out. In the next phase, the best ten classifiers selected in the previous step are applied to the created dataset in order to compare the results.

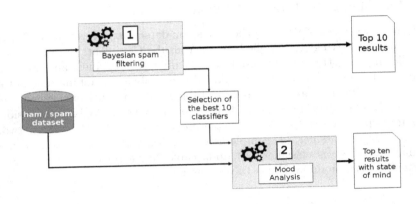

Fig. 1. Improving social spam detection using the mood of the comments.

To validate the algorithms and the obtained results the 10-fold cross-validation technique is used, and the results are analyzed in terms of the number of false positives and the accuracy. Accuracy is the percentage of testing set examples correctly classified by the classifier. Legitimate messages classified as spam are considered false positives. In order to validate these results, the same process is followed using another dataset.

3.1 Datasets

During this work two publicly available datasets are used:

- *Youtube Comments Dataset*[5]: Presented in [16]. This dataset contains multilingual 6,431,471 comments from a popular social media website, Youtube[6]. Among all the comments, 481,334 are marked as spam.
  In order to use similar number of texts messages to the experiments presented in [8] we created a new subset consisting of 1,000 spam and 3,000 ham, i.e. legitimate, comments. Those texts have been selected randomly and only taking into account comments written in English.
- *YouTube Spam Collection Dataset*[7]: Published by Alberto et al. [1]. Composed by 1,956 real messages divided in five subsets. The comments were extracted

---

[5] http://mlg.ucd.ie/yt/.
[6] www.youtube.com.
[7] https://archive.ics.uci.edu/ml/datasets/YouTube+Spam+Collection.

from five out of the ten most popular videos on the collection period. It consists of 1,005 spam and 951 ham texts. During this study, we use this dataset to validate the results of the previous dataset, repeating the experimental workflow.

3.2 Identifying the Best Social Spam Classifiers

With the objective of identifying the best spam detectors, several spam classifiers using different settings are applied to the Youtube Comments dataset.

Following the strategy presented in [8], 7 different classifiers and 56 settings combinations per each classifier are applied (we apply 392 combinations in total) The best ten results are presented in Table 1. During this experiment seven different classifiers have been used: (1) Large-scale Bayesian logistic regression for text categorization, (2) discriminative parameter learning for Bayesian networks, (3) Naive Bayes classifier, (4) complement class Naive Bayes classifier, (5) multinominal Naive Bayes classifier, (6) updateable Naive Bayes classifier, and (7) updateable multi-nominal Naive Bayes classifier.

Table 1. Results of the best ten classifiers

#	Spam classifier	TP	TN	FP	FN	Acc
1	NBM.c.stwv.go.ngtok	389	2911	89	611	82.50
2	NBMU.c.stwv.go.ngtok	389	2911	89	611	82.50
3	NBM.stwv.go.ngtok	370	2929	71	630	82.48
4	NBMU.stwv.go.ngtok	370	2929	71	630	82.48
5	NBM.c.stwv.go.ngtok.stemmer	379	2919	81	621	82.45
6	NBMU.c.stwv.go.ngtok.stemmer	379	2919	81	621	82.45
7	NBM.stwv.go.ngtok.stemmer	358	2936	64	642	82.35
8	NBMU.stwv.go.ngtok.stemmer	358	2936	64	642	82.35
9	CNB.stwv.go.ngtok	417	2875	125	583	82.30
10	CNB.stwv.go.ngtok.stemmer	400	2891	109	600	82.28

Nomenclatures and acronyms used in Table 1 and also throughout the paper are explained in Table 2.

Once the best classifiers and the best results are identified using the Youtube Comments dataset, a mood analysis of each message is carried out.

3.3 Mood Analysis

In order to analyze the mood of the youtubers' comments, each text is analyzed and a new feature (mood) is added to the original dataset. In this way a new dataset is created, and the best ten classifiers identified in the first phase are applied to it. This process is presented in Fig. 2.

Table 2. Nomenclatures

	Meaning		Meaning
CNB	Complement Naive Bayes	.stwv	String to Word Vector
NBM	Naive Bayes Multinomial	.go	General options
NBMU	Naive Bayes Multinomial Updatable	.wtok	Word Tokenizer
.c	idft F, tft F, outwc T [a]	.ngtok	NGram Tokenizer 1-3
.i.c	idft T, tft F, outwc T [a]	.stemmer	Stemmer
.i.t.c	idft T, tft T, outwc T [a]	.igain	Attribute selection using InfoGainAttributeEval

[a] idft means Inverse Document Frequency (IDF) Transformation; tft means Term Frequency score (TF) Transformation; outwc counts the words occurrences.

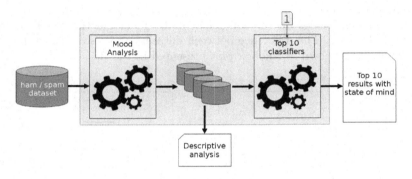

Fig. 2. Mood analysis.

To extract the mood of the writer, we use a publicly available machine learning Software as a Service (SaaS). This tool is hosted in *uClassify*[8], and taking into account the reached accuracy (96%) among all the possibilities we selected Mood classifier developed by Mattias Östmar.

As the author explains, this function determines the state of mind of the writer (upset or happy). On the extreme side there are angry, hateful writers while on the other extreme there are joyful and loving writers. The accuracy of the presented analyzer was measured using 10-fold cross validation by the author, and a 96% was reached.

The web service returns a float within the range [0.0, 1.0] specifying the happiness of each text. Using this value, it is possible to calculate also the upset level. Consequently, only one feature (mood) per each comment is added to the original dataset, and a new dataset is created.

---

[8] https://uclassify.com.

4   Experimental Results

In order to achieve the objective of this study, several experiments are carried out using the previously mentioned Youtube comments dataset. The results of these tests are presented in this section.

4.1   Descriptive Analysis

First, we perform a descriptive experiment of the two publicly available datasets in terms of the mood of the comments. We start by applying the mood analyzer to the dataset, then we continue by extracting statistics about the distribution and finally, we add a new feature, mood, to the original dataset.

As a result of this analysis, we find out that the state of mind of the spam texts in Youtube differs depending on the data collection strategy. On the one hand, Youtube Comments Dataset was created by crawling the comments from 6,407 different videos. On the other hand, researchers used only 5 out of the 10 most popular videos on the collection period to create the YouTube Spam Collection Dataset. The difference between datasets and also between ham and spam message is shown in the box plot presented in Fig. 3.

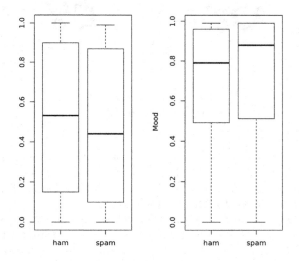

Fig. 3. Mood analysis of the dataset. (left-hand side) Youtube Comments Dataset and (right-hand side) YouTube Spam Collection Dataset.

This means that there is a difference between spam and ham social comments in terms of mood, so so the discriminative nature of this feature can aid in improving social spam filtering.

4.2   Predictive Experiment

To analyze the influence of the mood analysis in social spam filtering a predictive experiment is carried out.

We apply the best ten classifiers identified in the Youtube Comments Dataset and we compare the results with and without mood feature. The comparison between different results is presented in Table 3.

Table 3. Comparison between the best ten classifiers with and without mood. Using the first dataset.

Name	Normal		Mood		FP reduction (%)
	FP	Acc	FP	Acc	
NBM.c.stwv.go.ngtok	89	82.50	**77**	**82.53**	13.48
NBMU.c.stwv.go.ngtok	89	82.50	**70**	**82.58**	21.35
NBM.stwv.go.ngtok	71	**82.48**	63	82.43	11.27
NBMU.stwv.go.ngtok	71	**82.48**	59	82.43	16.90
NBM.c.stwv.go.ngtok.stemmer	81	82.45	**73**	**82.45**	9.88
NBMU.c.stwv.go.ngtok.stemmer	81	82.45	**68**	**82.48**	16.05
NBM.stwv.go.ngtok.stemmer	64	82.35	**58**	**82.38**	9.38
NBMU.stwv.go.ngtok.stemmer	64	82.35	**53**	**82.35**	17.19
CNB.stwv.go.ngtok	125	82.30	**110**	**82.43**	12.00
CNB.stwv.go.ngtok.stemmer	109	**82.28**	98	82.20	10.09
				avg:	_13.76_

As it is possible to see in Table 3, the comparison has been done taking into account the accuracy and the number of false positives. Results show that in almost all the top classifiers the accuracy is improved with the added mood analysis, and the best accuracy is obtained reaching an 82.58%. Moreover, the number of false positives is reduced in every cases between 9.38% and 21.35%.

In order to validate the results obtained with the Youtube Comments Dataset, the same experiments are carried out using the YouTube Spam Collection Dataset. The obtained results are shown in Table 4.

Using the validation dataset results are also improved adding mood feature. In this case the best accuracy is improved from 93.97% to 94.38%, and the number of false positives is reduced 11.41% on average.

Furthermore, to compare statistical behavior of the best classifier in both datasets, the area under the curve is analyzed. To create this ROC curve the specificity and sensitivity of the classifiers are taken into account. Figure 4 shows that the ROC area using mood analysis (0.756 and 0.943) is larger than without using it (0.753 and 0.939) in both datasets.

Table 4. Comparison between the best ten classifiers with and without mood. Using the validation dataset.

Name	Normal		Mood		FP reduction (%)
	FP	**Acc**	**FP**	**Acc**	
CNB.stwv.go.ngtok	85	93.97	**76**	**94.38**	10.59
CNB.stwv.go.ngtok.stemmer	89	93.87	**85**	**94.02**	4.49
NBM.stwv.go.ngtok	113	92.69	**80**	**94.17**	29.20
NBMU.stwv.go.ngtok	**113**	**92.69**	116	92.54	-2.65
NBM.stwv.go.ngtok.stemmer	119	92.38	**86**	**93.97**	27.73
NBMU.stwv.go.ngtok.stemmer	**119**	**92.38**	123	92.23	-3.36
NBM.c.stwv.go.ngtok	127	92.13	**96**	**93.66**	24.41
NBMU.c.stwv.go.ngtok	127	92.13	**127**	**92.13**	0.00
NBM.c.stwv.go.ngtok.stemmer	135	91.72	**101**	**93.35**	25.19
NBMU.c.stwv.go.ngtok.stemmer	**135**	**91.72**	137	91.62	-1.48
				avg:	*11.41*

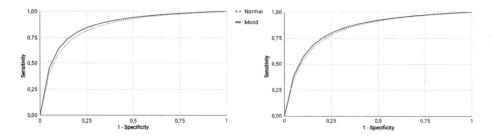

Fig. 4. ROC Curve of the best classifier with and without mood. (left-hand side) Youtube Comments Dataset and (right-hand side) YouTube Spam Collection Dataset.

5   Conclusions

This paper presents a new social spam filtering method. We provide means to validate our hypothesis that it is possible to improve current social spam filtering results extracting the mood of the texts.

First, using a social spam dataset, different experiments are carried out with and without mood feature. Next, we compare the obtained results, and we demonstrate that mood analysis can help to improve social spam filtering results.

Results show that the best accuracy obtained with the original dataset is improved from 82.50% to 82.58% using the Youtube Comments Dataset, and from 93.97% to 94.38% using the validation dataset. Despite the difference in the percentage does not seem to be relevant, if we take into account the amount of Youtube comments and the daily active users in this website, the improvement in absolute comment number is significant. Additionally, the number of false positives is reduced, on average 13.76% and 11.47%. This means that mood analysis is capable to highlight differences between spam and legitimate social comments. As descriptive analysis shows, the mood feature adds a distinctive feature for comments in each type of video (more positive or more upset). This variation helps classifiers to filter spam comments, and to improve the results.

Acknowledgments. This work has been developed by the intelligent systems for industrial systems group supported by the Department of Education, Language policy and Culture of the Basque Government. This work was partially supported by the project Semantic Knowledge Integration for Content-Based Spam Filtering (TIN2017-84658-C2-2-R) from the Spanish Ministry of Economy, Industry and Competitiveness (SMEIC), State Research Agency (SRA) and the European Regional Development Fund (ERDF).

We thank Mattias Östmar for the valuable tools developed and published. And we thank Jon Kâgström (Founder of uClassify(https://www.uclassify.com)) for the opportunity to use their API for research purposes.

Iñaki Garitano is partially supported by the INCIBE grant "INCIBEC-2015-02495" corresponding to the "Ayudas para la Excelencia de los Equipos de Investigación avanzada en ciberseguridad".

References

1. Alberto, T.C., Lochter, J.V., Almeida, T.A.: TubeSpam: Comment spam filtering on YouTube. In: 2015 IEEE 14th International Conference on Machine Learning and Applications (ICMLA), pp. 138–143, December 2015
2. Almaatouq, A., Shmueli, E., Nouh, M., Alabdulkareem, A., Singh, V.K., Alsaleh, M., Alarifi, A., Alfaris, A., Pentland, A.S.: If it looks like a spammer and behaves like a spammer, it must be a spammer: analysis and detection of microblogging spam accounts. Int. J. Inf. Secur. **15**(5), 475–491 (2016). https://doi.org/10.1007/s10207-016-0321-5
3. Arif, M.H., Li, J., Iqbal, M., Liu, K.: Sentiment analysis and spam detection in short informal text using learning classifier systems. Soft Comput. July 2017. https://doi.org/10.1007/s00500-017-2729-x
4. Bollen, J., Mao, H., Zeng, X.: Twitter mood predicts the stock market. J. Comput. Sci. **2**(1), 1–8 (2011)
5. Chandramouli, R.: Emerging social media threats: technology and policy perspectives. In: 2011 Second Worldwide Cybersecurity Summit (WCS), pp. 1–4, June 2011
6. Chaudhary, V., Sureka, A.: Contextual feature based one-class classifier approach for detecting video response spam on YouTube. In: 2013 Eleventh Annual Conference on Privacy, Security and Trust, pp. 195–204, July 2013
7. Egele, M., Stringhini, G., Kruegel, C., Vigna, G.: COMPA: detecting compromised accounts on social networks. In: NDSS. The Internet Society (2013). http://dblp.uni-trier.de/db/conf/ndss/ndss2013.html#EgeleSKV13
8. Ezpeleta, E., Zurutuza, U., Gómez Hidalgo, J.M.: Does sentiment analysis help in Bayesian spam filtering? In: Martínez-Álvarez, F., Troncoso, A., Quintián, H., Corchado, E. (eds.) HAIS 2016. LNCS (LNAI), vol. 9648, pp. 79–90. Springer, Cham (2016). https://doi.org/10.1007/978-3-319-32034-2_7
9. Ezpeleta, E., Zurutuza, U., Hidalgo, J.M.G.: A study of the personalization of spam content using facebook public information. Logic J. IGPL **25**(1), 30–41 (2017). https://doi.org/10.1093/jigpal/jzw040

10. Gao, H., Chen, Y., Lee, K., Palsetia, D., Choudhary, A.N.: Towards online spam filtering in social networks. In: NDSS. The Internet Society (2012). http://dblp. uni-trier.de/db/conf/ndss/ndss2012.html#GaoCLPC12

11. Gao, H., Hu, J., Wilson, C., Li, Z., Chen, Y., Zhao, B.Y.: Detecting and characterizing social spam campaigns. In: Proceedings of the 17th ACM Conference on Computer and Communications Security, CCS 2010, pp. 681–683. ACM, New York (2010). http://doi.acm.org/10.1145/1866307.1866396

12. Hu, X., Tang, J., Gao, H., Liu, H.: Social spammer detection with sentiment information. In: Proceedings of the 2014 IEEE International Conference on Data Mining, ICDM 2014, pp. 180–189. IEEE Computer Society, Washington, DC (2014). http://dx.doi.org/10.1109/ICDM.2014.141

13. Kwak, H., Lee, C., Park, H., Moon, S.: What is Twitter, a social network or a news media? In: Proceedings of the 19th International Conference on World Wide Web, pp. 591–600. ACM (2010)

14. Mahajan, S., Rana, V.: Spam detection on social network through sentiment analysis. Adv. Comput. Sci. Technol. **10**(8), 2225–2231 (2017)

15. O'Callaghan, D., Harrigan, M., Carthy, J., Cunningham, P.: Network analysis of recurring YouTube spam campaigns. arXiv preprint arXiv:1201.3783 (2012)

16. O'Callaghan, D., Harrigan, M., Carthy, J., Cunningham, P.: Network analysis of recurring YouTube spam campaigns. CoRR abs/1201.3783 (2012). http://arxiv. org/abs/1201.3783

17. Pang, B., Lee, L.: Opinion mining and sentiment analysis. Found. Trends Inf. Retrieval **2**(1–2), 1–135 (2008)

18. Pang, B., Lee, L., Vaithyanathan, S.: Thumbs up?: sentiment classification using machine learning techniques. In: Proceedings of the ACL-02 Conference on Empirical Methods in Natural Language Processing, EMNLP 2002, vol. 10, pp. 79–86. Association for Computational Linguistics, Stroudsburg (2002). http://dx.doi.org/ 10.3115/1118693.1118704

19. Perveen, N., Missen, M.M.S., Rasool, Q., Akhtar, N.: Sentiment based Twitter spam detection. Int. J. Adv. Comput. Sci. Appl. (IJACSA) **7**(7), 568–573 (2016)

20. Shehnepoor, S., Salehi, M., Farahbakhsh, R., Crespi, N.: NetSpam: a network-based spam detection framework for reviews in online social media. IEEE Trans. Inf. Forensics Secur. **12**(7), 1585–1595 (2017)

21. Song, J., Lee, S., Kim, J.: Spam filtering in Twitter using sender-receiver relationship. In: Sommer, R., Balzarotti, D., Maier, G. (eds.) RAID 2011. LNCS, vol. 6961, pp. 301–317. Springer, Heidelberg (2011). https://doi.org/10.1007/978-3-642-23644-0_16

22. Stringhini, G., Kruegel, C., Vigna, G.: Detecting spammers on social networks. In: Proceedings of the 26th Annual Computer Security Applications Conference, ACSAC 2010, pp. 1–9. ACM, New York (2010). http://doi.acm.org/10.1145/ 1920261.1920263

23. Turney, P.D.: Thumbs up or thumbs down?: semantic orientation applied to unsupervised classification of reviews. In: Proceedings of the 40th Annual Meeting on Association for Computational Linguistics, ACL 2002, pp. 417–424. Association for Computational Linguistics, Stroudsburg (2002). http://dx.doi.org/10. 3115/1073083.1073153

24. Wang, A.H.: Don't follow me: spam detection in Twitter. In: Proceedings of the 2010 International Conference on Security and Cryptography (SECRYPT), pp. 1–10. IEEE (2010)

25. Wang, D., Irani, D., Pu, C.: A social-spam detection framework. In: Proceedings of the 8th Annual Collaboration, Electronic messaging, Anti-Abuse and Spam Conference, pp. 46–54. ACM (2011)
26. Yang, C., Harkreader, R., Zhang, J., Shin, S., Gu, G.: Analyzing spammers' social networks for fun and profit: a case study of cyber criminal ecosystem on Twitter. In: Proceedings of the 21st International Conference on World Wide Web, pp. 71–80. ACM (2012)
27. Zheng, X., Zeng, Z., Chen, Z., Yu, Y., Rong, C.: Detecting spammers on social networks. Neurocomputing **159**, 27–34 (2015). http://www.sciencedirect.com/science/article/pii/S0925231215002106

An Improved Comfort Biased Smart Home Load Manager for Grid Connected Homes Under Direct Load Control

Chukwuka G. Monyei[✉] and Serestina Viriri

School of Mathematics, Statistics and Computer Science,
University of KwaZulu-Natal, Westville Campus, Durban, South Africa
chiejinamonyei@gmail.com, viriris@ukzn.ac.za

Abstract. This paper presents an improved comfort biased smart home load manager (iCBSHLM) for grid connected residential houses. The proposed algorithm discriminates household loads into class 1 (air-conditioner, heating) and class 2 loads (dishwasher, cloth washer and cloth dryer) and achieves electricity consumption reduction and electricity cost reduction of up to 2.9% and 7.5% respectively using dynamic pricing (Price1) over time of use pricing (Price0), while ensuring that indoor temperature is kept within the user pre-scribed range and without any violation. iCBSHLM advances existing home energy management systems (HEMs) by ensuring that vulnerable household residents (especially the elderly) can still benefit from smart grid initiatives like HEMs without any discomfort. Furthermore, this research presents a simplistic model for heating, ventilation and cooling (HVAC) loads using capacitor charging/discharging behaviour.

Keywords: iCBSHLM · Indoor temperature · Vulnerable
Consumer preference · Electricity cost reduction · Dynamic pricing
Smart grid

1 Introduction

The problem of energy (electricity) access is still a major problem for over 800 million people in sub-Sahara Africa (SSA) and South Asia. In Nigeria for example, over 80 million people are still without access to grid electricity. Beyond energy access is the problem of energy mobility. Energy mobility which is the ability of households to transit from one energy level to a higher one is being hampered for most households especially in SSA. This is due to rising electricity costs which further leads to energy poverty and consequently economic poverty. According to [1, 2], declining electricity consumption has been observed for Nigeria and South Africa respectively. The novel approach adopted in computing electricity consumption, utilizes state/provincial date on using actual energy consumed and actual/estimated connected houses.

However, the increasing tariffs by the utility companies is necessary to recoup investments in expanding the supply capacity and transmission network to meet with growing demands. In [2] for instance, grid capacity expansion within a period is estimated to be over 500% of the estimated demand increase within that same period.

© Springer International Publishing AG, part of Springer Nature 2018
F. J. de Cos Juez et al. (Eds.): HAIS 2018, LNAI 10870, pp. 526–536, 2018.
https://doi.org/10.1007/978-3-319-92639-1_44

Considering the limitation of technical solutions and the inadequacy of supply side measures in mitigating the growing electricity demand/supply imbalance, solutions that target the behaviour of consumers are being exploited. These solutions seek to ensure that compromises from every electricity participant (without adversely affecting any participant or violating grid constraints) guarantee optimality in the electricity grid operation.

Demand side management (DSM) is becoming a prominent technique currently receiving acclaimed attention across the world for its ability to both influence consumer electricity behaviour and ensure demand/supply balance while providing some reprieve to the electricity utility companies (mostly from expanding electricity supply capacity). According to [3], the concept of DSM connotes a relationship between the supply and utilization side for mutual benefit. This definition is further elaborated by [4] to be a set of flexible and interconnected programs that permit the end users of electricity (consumers) a greater role in altering their consumption pattern/profile by shifting their electricity usage from peak to off-peak periods. Generally, DSM programs are either energy efficiency (EE) based or demand response (DR) based [3, 4]. While the EE based DSM programs aim at curtailing electricity usage through the use of more energy efficient devices (use of energy saving bulbs instead of incandescent bulbs) and techniques (double glazed windows, insulation, sealing) or a compromise in comfort derived from certain electrical appliances (like reduced thermostat settings and reduced light brightness settings), the DR based programs aim at soliciting from electricity users a 'positive'[1] response due to increase in electricity rate or some other incentives [3, 4].

The basis for the exploitation of DR-DSM is the home energy management system (HEMS) which plays a crucial role in advancing the smart grid concept. According to [5], smart HEMS are an essential component in ensuring that the demand-side management aspect of the smart grid is successful. Smart HEMS are further defined by [5] to be optimal systems that provide energy management services for the efficient monitoring and management of electricity generation, storage and its consumption in smart houses. This perspective is further reified by [6] who posit that smart HEMS has formidable applications in the generation, transmission and distribution systems of electricity networks.

The advent of solar home systems (SHSs) has led to increased discussions and research on HEMS due to the increasing need to match supply with demand. Furthermore, because of the variable and stochastic nature of weather elements, HEMS are proving to be a viable platform for ensuring that SHSs are well utilized to guarantee consumer comfort and satisfaction. For instance, in [7], an energy flow management algorithm was presented for a grid-connected photovoltaic (PV) system that incorporated battery storage while in [8], a HEMS that integrated a learning prediction algorithm for forecasting power production of a house's solar PV plant and its power consumption across a time span and based on neural-network was designed. The effect of sending feedback on previous energy consumption to households was also evaluated by comparing consumption drop/increase across a time frame in [9] where a 3.4% and

---

[1] By positive we imply an inverse relationship in which electricity users reduce their usage of electricity based on increasing electricity rate.

5.4% drop in energy (electricity) consumption for average and higher electricity consuming households was observed during the winter. Furthermore, in [9], there was a significant 11.4% drop in the usage of space heating for the higher electricity consuming households. Data error impact on HEMs was studied in [10] while [11] presented a conceptual distributed integrated energy management (diEM) system for residential buildings. The aim of [11] is to minimize operational energy cost for households through load shifting to maximize renewable energy power produced. A life cycle assessment was conducted by [12] where the environmental impact of HEMs in terms of their potential benefits and detrimental impacts was evaluated. A negative energy payback time was computed for home automation devices due to the energy consumption of smart plugs. Foresee^{TM} was presented by [13] as a user-centred HEMs for optimizing its operations to achieve efficiency and utility cost savings. Abushnaf et al. in [14] made extensive arguments on the ability of HEMs to optimize residential building energy use especially in tackling the problems of green-house gas emission and energy wastage. Further reading on HEMs can be found in [6, 15].

In most of the literature surveyed, indoor comfort with respect to user specification has been vaguely researched with most literature overlooking the issue of sensitivity in temperature variations while trying to schedule the operations of heating, ventilation and cooling (HVAC) loads. Furthermore, load reduction has been achieved most times at the expense of some discomfort (though minimal) to the user. For instance, in [16], the proposed algorithm for the HEMs had variations in temperature as high as 1.5 °C with load reduction of about 5.7%. Considering the importance of adequate heating especially for elderly persons and the fact that by 2025 there will be over 1.2 billion people aged 60 and over worldwide [17], the need arises for smarter home energy management systems that can achieve significant cost savings while still maintaining indoor temperature strictly within the permissible range specified by the occupants. This system must thus advance the existing smart HEMs available and be interoperable with existing smart grid infrastructure while ensuring that its initial investments payoff within an appropriate time frame.

We thus present an improved comfort biased smart home load manager (iCBSHLM) for the optimal scheduling of demand response (DR) loads with a hierarchical biased nature. The proposed iCBSHLM contributes to the limited literature on HEMS design with user preference [13]. Two classes of loads are generally handled by iCBSHLM – class 1 and class 2 loads. The class 1 loads refer to the loads that primarily affect the ambient temperature or environment of a confined space. Examples of class 1 loads include air-conditioners (ACs) and heaters, refrigerator etc. The class 2 loads refer to other loads that can be scheduled or have their operation interrupted without significantly affecting users comfort. Examples of class 2 loads include dishwashers, cloth washers, cloth dryers etc. In deploying iCBSHLM, class 1 loads are expected to be fully on direct load control (DLC)^2 by the utility with user override provided while the class two loads can be on full of semi DLC. The full DLC assumes the utility

---

^{2} Direct load control (DLC) is defined as a system in which electricity users give the utility control over the operation of an equipment. The utility can in the case of peak demand or faults trigger such loads on/off without notifying the user. However, the user maintains an override over such device.

oversees the start and end time of operation of the participating demand response (DR) loads while the semi DLC assumes a case where the utility has operational control over the utilisation of a device within a strict time frame offered by the household. In all cases, the incentive for participation in DR is reduced electricity cost.

2 Methods

In modelling iCBSHLM, we utilise the concept of the charging and discharging of capacitors to represent the heating and cooling of a room. We further show by dimensions that the two cases are dimensionally equivalent thus providing a valid framework for the application of iCBSHLM. Consider the circuit shown in Fig. 1 [18]. The circuit represents the charging of a capacitor (C) through a resistor (R). We can thus define the charging rate of the capacitor as shown in Eq. (1).

$$V_c = V_s(1 - e^{\frac{t}{RC}}) \tag{1}$$

Where V_c is the voltage across the capacitor, V_s is the voltage from the battery source, t is the time elapsed since battery source is turned "ON", R is the resistance (Ω) and C is the capacitance (F). The value RC is the time constant and determines the charging rate of the capacitor. It is generally established that the capacitor charges fully by $5 \times RC$. Dimensionally, resistance $R = \frac{V}{A} = ML^2t^{-3}I^{-2}$ while capacitance $C = \frac{Q}{V} = M^{-1}L^{-2}t^4I^2$ which implies that $RC = t$, where M, L, t and I are the dimensional symbols for mass, length, time and current.

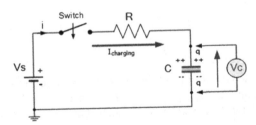

Fig. 1. The capacitor charging circuit [18].

In creating an analogy for the heating of a confined space, we relate the heat source of temperature T_s to the V_s, thermal resistance of air R_λ to R and thermal capacitance of air C_λ to C. We thus define R_λ and C_λ (where C_λ is the thermal conductivity) as shown in Eqs. (2)–(3) respectively.

$$R_\lambda = \frac{temperature}{J/S} = kg^{-1}m^{-2}s^3K \tag{2}$$

$$C_\lambda = \frac{heat}{temperature} = kgm^2s^{-2}K^{-1} \tag{3}$$

The time constant τ for this scenario is defined as $\tau = R_\lambda C_\lambda$ which is dimensionally equivalent with RC. We thus define the heating of the confined space as shown in Eq. (4).

$$T_c = T_s(1 - e^{-\frac{t}{R_\lambda C_\lambda}}) \tag{4}$$

Where T_c is the temperature of the confined space and T_s is the temperature of the heat source. However, if we assume that heat in the confined space depletes with time, we can also model the depletion/decay of heat in the confined space to the discharging of a capacitor. Consider Fig. 2 [19], which presents the discharging of a capacitor. If all values are assumed to have their initial meanings, then the discharging of the capacitor is shown in Eq. (5) while Eq. (6) presents the depletion/decay of heat in the confined space.

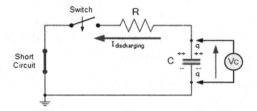

Fig. 2. The capacitor discharging circuit [19].

$$V_c = V_s e^{-\frac{t}{RC}} \tag{5}$$

$$T_c = T_s e^{-\frac{t}{R_\lambda C_\lambda}} \tag{6}$$

Given a user defined comfort level (in terms of temperature) to be T_{user} such that the user can tolerate variations $T_{min} \le T_{user} \le T_{max}$, then we seek to optimize the duration of operation of the heat source $R_\lambda C_\lambda$ such that the variation of temperature is within the limit $T_{min} \le T_{user} \le T_{max}$. Figure 3 presents a description of the optimization objective while Algorithm 1 presents the description of iCBSHLM.

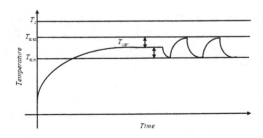

Fig. 3. Description of optimization objective

Algorithm 1. Improved Comfort Biased Smart Home Load Manager

Input: $T_{user\_defined}, t_{DW}^{start}, t_{DW}^{end}, t_{CW}^{start}, t_{CW}^{end}, t_{CD}^{start}, t_{CD}^{end}, \text{Price0}, \text{Price1}$

Output: $T_{profile}, t_{profile}$

Perform class 1 optimisation:

1. Generate open matrix $gen_k = [1, 3500]$ for house k under consideration.

2. Fill gen_k using random bit lengths of $1's$ and $0's$ such that
   $gen_k = [\{11111\}, \{111111111\}, \{00000\} \{111\}...]$

   Perform targeted mutation on gen_k

   - Scan gen_k for all zero bits and randomly mutate 10% such that:
     If $gen_k(i) == 0$ *and* $deefac == 1$ *then*
         Force mutate gen_k such that $gen_k(1, i-x : i+x) = 1$
     Endif
   - Scan gen_k for all $0 \rightarrow 1$ transition points and randomly force mutate 10% such that:
     If $gen_k(1, i : i+1) = \{0,1\}$ *and* $deefac == 1$ *then*
         Force mutate gen_k such that $gen_k(1, i) = 1$
     Endif

   Evaluate equivalent temperature values

   - Scan gen_k:
     If $gen_k(i) == 0$ *then*
         AC is "OFF"
     $$T_{profile}(i+1) = temp_{init} e^{-\frac{t}{r_\lambda \times c_\lambda}}$$
     Elseif $gen_k(i) == 1$ *then*
         AC is "ON"
     $$T_{profile}(i+1) = temp_{init} + temp_{diff}(1 - e^{-\frac{t}{r_\lambda \times c_\lambda}})$$
     Endif

   Check for violations

   - Scan $T_{profile}$ such that:
     If $any(T_{profile}(.) < T_{user\_defined})$ *then*
         Discard $T_{profile}$
     Else
         Compute cost
     Endif

   Compute and update cost and $T_{profile}$

       If $(\text{Price1}(T_{profile}) < \text{Price0}(T_{profile}))$ *then*
           Update cost and $T_{profile}$
       Endif

Where *deefac* is a decision factor randomly generated and used in moderating mutation exercises, i is an index position in gen_k, x is a carefully selected variable for adjusting the range of bits to be forced mutated and $T_{profile}$ contains the temperature values across the time horizon.

Perform class 2 optimisation:
This is done using the modified genetic algorithm (MGA) proposed in [2].

---

3   Results and Discussion

A single room apartment of an elderly person is considered. Winter season is assumed while 3 different scenarios are run with iCBSHLM. Table 1 presents the associated statistics for the house including loads considered, time range for optimisation and user defined temperature settings for all scenarios. Two pricing schemes are considered – Price0 (time of use pricing) and Price1 (dynamic pricing).

Table 1. Associated statistics for house under consideration

Device	Rating (W)	Operation time	Duration
Heater	1440	9AM–8PM (+1 day)	23 h
Dish washer	1200	9AM–5PM	1.75 h
Cloth washer	500	9AM–5PM	1.25 h
Cloth dryer	1000	7PM–11PM	1.75 h

Scenario	$T_{user\_defined}$	Allowance	Operation range
Scenario 1	26 °C	+0.5 °C	26 °C–26.5 °C
Scenario 2	27 °C	+0.5 °C	27 °C–27.5 °C
Scenario 3	28 °C	+0.5 °C	28 °C–28.5 °C

Figure 4 presents the pricing profile for Price0 and Price1 while the profile of allocation for class 1 and class 2 loads and for the various temperature scenarios (26 °C, 27 °C and 28 °C) are presented in the Figs. 5, 6 and 7 respectively. The summary of the results is shown in Table 2. It is observed from Table 2 that iCBSHLM achieves the strict temperature operating range of the household under consideration with average temperature for 26 °C, 27 °C and 28 °C exceeding their ideal cases by 0.42%, 0.47% and 0.25% which all fall within the 1.92%, 1.85% and 1.79% specified by the user for 26 °C, 27 °C and 28 °C respectively. Further observed from Table 2 is the fact that the application of iCBSHLM and Price1 achieves a reduction in electricity consumption by 2.9%, 2.8% and 2.9% for 26 °C, 27 °C and 28 °C while reduction in electricity cost was 7.2%, 7.3% and 7.5% for 26 °C, 27 °C and 28 °C respectively.

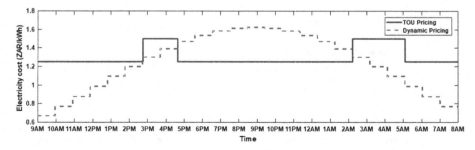

Fig. 4. Pricing profile for Price0 (TOU) and Price1 (dynamic pricing)

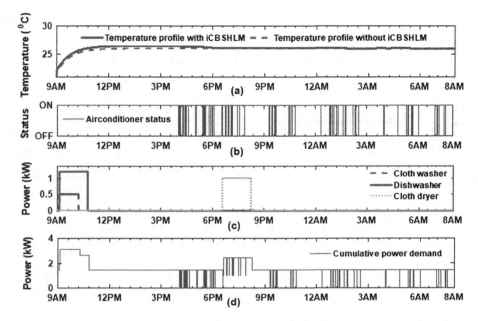

Fig. 5. Profile for (a) room temperature at 26 °C (b) air-conditioner status (c) cloth washer, cloth dryer and dishwasher power dispatch and (d) cumulative power demand

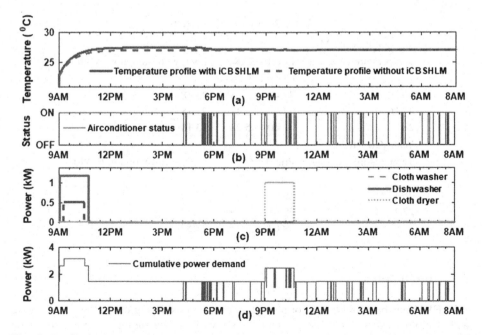

Fig. 6. Profile for (a) room temperature at 27 °C (b) air-conditioner status (c) cloth washer, cloth dryer and dishwasher power dispatch and (d) cumulative power demand

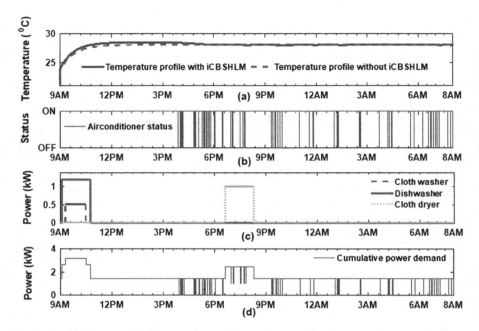

Fig. 7. Profile for (a) room temperature at 28 °C (b) air-conditioner status (c) cloth washer, cloth dryer and dishwasher power dispatch and (d) cumulative power demand

Table 2. Summary of results

Without iCBSHLM			
Temperature (Average)	Price0 (ZAR)		Energy consumption (kWh)
	Class 1	Class 2	
26 °C	42.40	5.28	23.18
27 °C	42.40	5.28	23.18
28 °C	42.40	5.28	23.18
With iCBSHLM			
Temperature (Average)	Price1 (ZAR)		Energy consumption (kWh)
	Class 1	Class 2	
26.11 °C	39.78	4.48	22.52
27.13 °C	39.71	4.49	22.54
28.07 °C	39.64	4.47	22.52

4   Conclusion

iCBSHLM has been presented for a single scenario considering one HVAC load along with 1 cloth washer, 1 cloth dryer and 1 dish washer. The savings accrued show the impact of consumer preferences on potential savings. Considering the fact that flexibility in the allocation of loads have been greatly constrained by the consumer preference, a greater factor capable of influencing savings would be the real time pricing trigger. However, the authors aver that aggregating more participants or incorporating more household loads – refrigerator, geysers etc. creates the opportunity for greater savings. Furthermore, the adopted modelling representation of HVAC loads using the charging/discharging of capacitors is quite flexible and easily adaptable to larger spaces once R_λ and C_λ can be determined.

Acknowledgement. The authors acknowledge the financial assistance of the National Research Foundation (NRF) and The World Academy of Sciences (TWAS) through the DST-NRF-TWAS doctoral fellowship towards this research. Opinions expressed and conclusions arrived at, are those of the authors and are not necessarily to be attributed to the NRF.

References

1. Monyei, C.G., et al.: Nigeria's energy poverty: insights and implications for smart policies and framework towards a smart Nigeria electricity network. Renew. Sustain. Energy Rev. **81**, 1582–1601 (2018)
2. Monyei, C.G., Adewumi, A.O.: Demand side management potentials for mitigating energy poverty in South Africa. Energy Policy **111**, 298–311 (2017)
3. Sharifi, R., Fathi, S.H., Vahidinasab, V.: A review on demand-side tools in electricity market. Renew. Sustain. Energy Rev. **72**, 565–572 (2017)
4. Setlhaolo, D.: Optimal management of household load under demand response, in Electrical, Electronic and Computer Engineering. University of Pretoria, Pretoria (2016)
5. Zhou, B., et al.: Smart home energy management systems: concept, configurations, and scheduling strategies. Renew. Sustain. Energy Rev. **61**, 30–40 (2016)
6. Liu, Y., et al.: Review of smart home energy management systems. Energy Procedia **104**, 504–508 (2016)
7. Chekired, F., et al.: An energy flow management algorithm for a photovoltaic solar home. Energy Procedia **111**, 934–943 (2017)
8. Ciabattoni, L., et al.: Design of a home energy management system by online neural networks. IFAC Proc. **46**(11), 677–682 (2013)
9. Iwafune, Y., et al.: Energy-saving effect of automatic home energy report utilizing home energy management system data in Japan. Energy **125**, 382–392 (2017)
10. Choi, D.-H., Xie, L.: A framework for sensitivity analysis of data errors on home energy management system. Energy **117**, 166–175 (2016)
11. Honold, J., et al.: Distributed integrated energy management systems in residential buildings. Appl. Therm. Eng. **114**, 1468–1475 (2017)
12. Louis, J., Calo, A., Leiviska, K., Pongracz, E.: Environmental impacts and benefits of smart home automation: life cycle assessment of home energy management system. In: International Federation of Automatic Control, pp. 880–885. Elsevier (2015)
13. Jin, X., et al.: Foresee: a user-centric home energy management system for energy efficiency and demand response. Appl. Energy **205**, 1583–1595 (2017)
14. Abushnaf, J., Rassau, A., Gornisiewicz, W.: Impact of dynamic energy pricing schemes on a novel multi-user home energy management system. Electric Power Syst. Res. **125**, 124–132 (2015)
15. Beaudin, M., Zareipour, H.: Home energy management systems: a review of modelling and complexity. Renew. Sustain. Energy Rev. **45**, 318–335 (2015)

16. Shakeri, M., et al.: Implementation of a novel home energy management system (HEMS) architecture with solar photovoltaic system as supplementary source. Renew. Energy **125**, 108–120 (2018)
17. Yang, J., Nam, I., Sohn, J.-R.: The influence of seasonal characteristics in elderly thermal comfort in Korea. Energy Build. **128**, 583–591 (2016)
18. RC Charging Circuit. https://www.electronics-tutorials.ws/rc/rc_1.html. Accessed 25 Feb 2018
19. RC Discharging Circuit. https://www.electronics-tutorials.ws/rc/rc_2.html. Accessed 25 Feb 2018

Remifentanil Dose Prediction for Patients During General Anesthesia

Esteban Jove[1,2]([✉]), Jose M. Gonzalez-Cava[2], José-Luis Casteleiro-Roca[1,2], Héctor Quintián[1], Juan Albino Méndez-Pérez[2], José Luis Calvo-Rolle[1], Francisco Javier de Cos Juez[3], Ana León[4], María Martín[4], and José Reboso[4]

[1] Department of Industrial Engineering, University of A Coruña,
Avda. 19 de febrero s/n, Ferrol, A Coruña 15495, Spain
esteban.jove@udc.es
[2] Department of Computer Science and System Engineering, Universidad de La Laguna, Avda. Astrof. Francisco Sánchez s/n, S/C de Tenerife 38200, Spain
jamendez@ull.edu.es
[3] Department of Mining Exploitation, University of Oviedo,
Calle San Francisco, 1, Oviedo 33004, Spain
[4] Hospital Universitario de Canarias, Tenerife, Spain

Abstract. In the anesthesia field there are some challenges, such as achieving new methods to control, and, of course, for reducing the pain suffered for the patients during surgeries. The first steps in this field were focused on obtaining representative measurements for pain measurement. Nowadays, one of the most promiser index is the ANI (Antinociception Index). This research works deals the model for the remifentanil dose prediction for patients undergoing general anesthesia. To do that, a hybrid model based on intelligent techniques is implemented. The model was trained using Support Vector Regression (SVR) and Artificial Neural Networks (ANN) algorithms. Results were validated with a real dataset of patients. It was possible to check the really successful model performance.

Keywords: EMG · ANI · MLP · SVR

1 Introduction

The application of automatic control techniques to the anesthesia field has outperformed manual administration of drugs [1]. The anesthetic process involves the control of three main variables: hypnosis, analgesia and muscular blockade. Different automatic controllers have been proposed in order to adapt the drug titration regarding the real needs of patients to control the hypnosis level [2,3]. However, further research is required in order to propose reliable strategies to deal with the control of analgesia. The main problem relies on the absence of a feedback variable capable of quantifying the analgesic state of patients during surgery.

© Springer International Publishing AG, part of Springer Nature 2018
F. J. de Cos Juez et al. (Eds.): HAIS 2018, LNAI 10870, pp. 537–546, 2018.
https://doi.org/10.1007/978-3-319-92639-1_45

Different variables and monitors have been proposed as measures that could be correlated to the analgesic state of patients [4–6]. Nevertheless, the reliability of these monitors has not been widely studied. Among the different alternatives, the Analgesia Nociception Index (ANI) developed by Mdoloris Medical Systems, has shown good results in clinical practice [7,8]. The ANI monitor is a non-invasive system that computes an index obtained from the Autonomic Nervous System (ANS) through the electrocardiogram (ECG). This index ranges from 0 to 100 in order to quantify the parasympathetic activity in patients undergoing surgery. A value between the 50–70 range is supposed to ensure an adequate analgesia. Values under 50 involves the possibility of future hemodynamic reactions, while values over 70 indicates the possibility to decrease the opioids administration without any risk.

The work presented in [9] used intelligent techniques with the aim of predicting the ANI level taking into account the Remifentanil drug rate administered to the patient, the EMG (ElectroMyoGram) and the previous values of ANI. In a similar way, other work [10] predicts the EMG and BIS (Bispectral Index). However, the present work deals the opposite problem and in this case, the proper Remifentanil infusion rate is predicted for the next five seconds from the previous values of ANI, EMG and Remifentanil. In a similar way, in [11], fuzzy algorithms are used in the drug supply decision-making process.

The proposed predictive model can be obtained considering different regression methods. The most commonly used methods are based on Multiple Regression Analysis (MRA) techniques. In [12–14], MRA techniques are used to address problems in applications of different fields. In some cases, the nature of the system to model can lead to wrong performance [13,15,16], and for this reason, Soft Computing techniques are employed [17–21]. Their use improves significantly the performance of the predictive model [22–27].

This work develops a model to predict the Remifentanil infusion rate using the ANI and EMG signals. To achieve this objective, Artificial Neural Networks and Support Vector Regression were employed. Then, the performance of the model is evaluated using the Mean Squared Error (MSE).

This paper is organized following the next structure. After this section, a brief case of study is presented. Then, the model approach and the used techniques used to obtain the model are shown. The results section shows the best configuration achieved by the proposed hybrid model. Finally, the conclusions and future works are presented.

2   Case of Study

Data to train and test the model were obtained from fifteen patients undergoing cholecystectomy surgery at the Hospital Universitario de Canarias (HUC). All patients received an informative document about the study and an informed consent was signed. A Total IntraVenous Anesthesia (TIVA) was performed. After the induction phase, propofol (hypnotic drug) and remifentanil (analgesic drug) were manually delivered according to the anesthesiologist criteria. Two

Graseby 3500 pumps were used. The Bispectral Index (BIS) monitor was considered to guide propofol titration, with a target value of 50. Remifentanil dose was adjusted in steps of 0.05 mg/kg/min regarding autonomic reactions and the presence of surgical events. During the surgery, the Analgesia Nociception Index information was registered in parallel to the process so that the physician was not able to see the information provided. As a result, their decisions were based only on traditional clinical parameters. Not only ANI, but also EMG signal and Remifentanil rate were automatically registered with a sample time of 5 s using a laptop via RS232 interfaces. The studied problem could be represented as shown in Fig. 1.

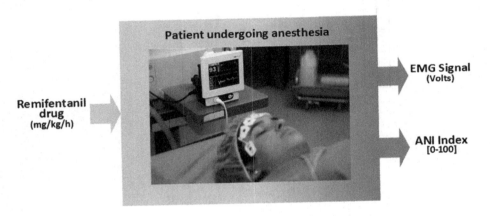

Fig. 1. Case of study. Input/Output representation

3   Model Approach

The model approach follows the structure presented on Fig. 2. Three signals are considered as inputs: Remifentanil, EMG and ANI. With the aim of representing the dynamics of the system, the five previous measures of each signal are taken into consideration. Hence, the model has a total amount of fifteen inputs. The model output is the predicted value of the Remifentanil infusion rate.

The model is implemented through the process described in Fig. 3. The dataset was divided in two groups, a group used to train the model and a group used to check its performance. The K-Fold Cross-Validation algorithm shown in Fig. 4 was employed in order to ensure the best model performance.

3.1   Dataset Obtaining and Description

The dataset used on this research consist on the registered Remifentanil infusion rate, EMG and ANI signals from fifteen different patients undergoing general anesthesia. The signal monitoring is performed with a sample time of 5 s.

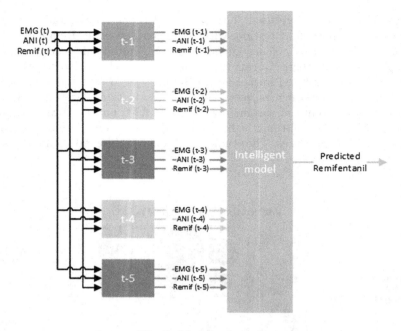

Fig. 2. Model approach

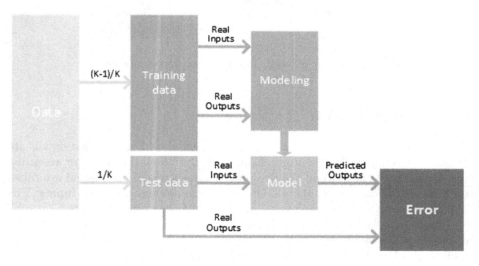

Fig. 3. Modeling process

Before applying the intelligent regression techniques to the dataset, an initial preprocessing of the raw data is conducted on the three monitored signals. This preprocessing consisted on filtering the data using a low pass filter in order to avoid the undesired noise. This research focuses only on the

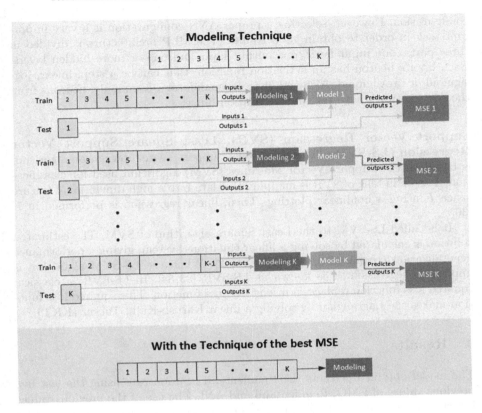

Fig. 4. K-Fold training process

anesthesia maintenance phase. Hence, induction and recovery phases were not considered. After the preprocessing, and taking into account only the maintenance phase, the dataset is composed of 17064 samples.

3.2 Used Techniques

With the aim of obtaining the best prediction model, different intelligent techniques were applied and tested in this work. The model performance is assessed using an MSE criteria with 10 k-Fold Cross Validation. This validation method leads to a more general measurement than the Hold-Out method. The number of patients taken into account to achieve the model is thirteen, and its performance was checked with two more patients.

Artificial Neural Networks (ANN). Multi-layer Perceptron (MLP). The Multi-Layer Perceptron is the most used feedforward Artificial Neural Network (ANN) [28,29] because of different factors, such as its robustness and simple configuration. However, selecting a proper ANN configuration is a very important task in order to obtain good results. The MLP architecture is divided in three parts, one input layer, one output layer and one or more hidden layers. Each layer's neuron has an activation function, that can be a step, linear, log-sigmoid or tan-sigmoid. Although it is not mandatory, commonly, neurons from the same network have the same activation function.

Support Vector Regression (SVR), Least Square Support Vector Regression (LS-SVR). The well known Support Vector Regression algorithm is based on the Support Vector Machines (SVM) algorithm used for classification. The main task of SVR is mapping the data into a high-dimensional feature space F using a nonlinear plotting. Then, linear regression is performed in F [30].

It is called LS-SVM to the Least Square algorithm of SVM. The estimated solution is calculated by solving a linear equations system, giving a performance generalization compared to SVM [21,31,32]. When the LS-SVM algorithm is applied to regression, it is known as LS-SVR [33,34]. In LS-SVR, a classical squared loss function replaces a insensitive loss function. This squared loss function makes the Lagrangian by solving a linear Karush-Kuhn-Tucker (KKT).

4   Results

The model obtained predicts the Remifentanil infusion rate using the last five previous values of EMG, Remifentanil and ANI. The use of the previous values incorporates to the model the dynamics of the system. The use of more previous values was tested, concluding that taking into account more than the last five values does not improve the model performance. Similarly, reducing the number of previous values led to worse results.

The artificial neural network algorithm was trained using different configurations: the number of neurons of the unique hidden layer was tested from 1 to 8. The hidden layer neurons have a tan-sigmoid activation function while the output layer neuron had a linear activation function. The training algorithm was Levenberg-Marquardt, the learning algorithm was gradient descent and the mean squared error was set as performance function.

The LS-SVR algorithm was trained with the KULeuven-ESAT-SCD autotuning algorithm implemented with the Matlab toolbox. The regression type was set to 'Function Estimation' and the model kernel was configured as Radial Basis Function (RBF). The cost criteria is 'leaveoneoutlssvm', the performance function is 'mse' and the optimization function is set as 'simplex'.

The best MSE, NMSE (Normalized Mean Squared Error) and MAE (Mean Absolute Error) for each regression technique is presented in Table 1. Increasing the number of neurons in the hidden layer for the ANN technique does not lead to better results. In this case, the best configuration is obtained with only one neuron in the hidden layer.

Table 1. MSE, NMSE and MAE for each regression algorithm

	MSE	NMSE	MAE
ANN-1	**8.6708e−5**	**0.0082**	**0.0012**
ANN-2	8.7155e−5	0.0083	0.0012
ANN-3	9.2794e−5	0.0088	0.0012
ANN-4	8.9512e−5	0.0085	0.0012
ANN-5	1.1584e−4	0.0110	0.0013
ANN-6	6.3378e−4	0.0601	0.0015
ANN-7	0.0135	1.2828	0.0029
ANN-8	0.0015	0.1457	0.0018
LS-SVR	9.0380e−5	0.0086	0.0014

After selecting the best configuration, the final model is trained with whole amount of data from the thirteen patients (without K-Fold Cross Validation). This final model is validated using the data from the two patients left. Hence, two complete surgeries are tested obtaining successfully results. In Fig. 5 and 6 the predictions of the remifentanil (red line) is compared with its real value (blue dashed) on both testing patients. It can be observed that, in both cases, the prediction is similar to the real value.

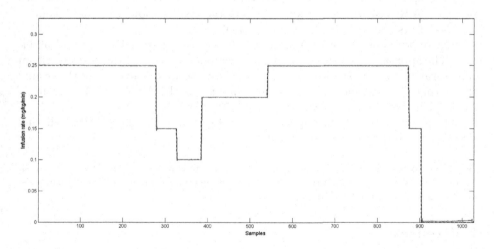

Fig. 5. All the surgery Remifentanil signal for the first testing patient. The real (blue dashed) and predicted (red line) signals are as close that in the global figure is difficult to see the two signals. (Color figure online)

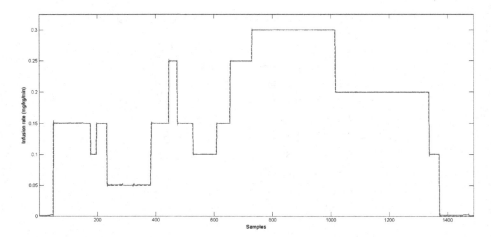

Fig. 6. All the surgery Remifentanil signal for the second testing patient. The real (blue dashed) and predicted (red line) signals are as close that in the global figure is difficult to see the two signals. (Color figure online)

5   Conclusions

It is important to emphasize that quite satisfactory results have been obtained with the approach proposed in this research. The main goal of the work was to predict the Remifentanil infusion rate, and the results show that the output of the model fits the real signal with very small error.

This model was obtained from a real dataset using 10 K-Fold Cross Validation. The approach is based on intelligent techniques, selecting the best algorithm configuration to train the final model. After some tests, the analysis of the results shows that the best model regression techniques is the ANN with 1 neurons in the hidden layer, that achieves less than 0.5% of predicted error.

This technique could be applied to several systems with the aim of predicting signals or to improve some specifications.

Acknowledgments. Jose M. Gonzalez-Cava's research was supported by the Spanish Ministry of Education, Culture and Sport (www.mecd.gob.es), under the "Formación de Profesorado" grant FPU15/03347.

References

1. Chang, J.J., Syafiie, S., Kamil, R., Lim, T.A.: Automation of anaesthesia: a review on multivariable control. J. Clin. Monit. Comput. **29**(2), 231–239 (2015)
2. Mendez, J.A., Marrero, A., Reboso, J.A., Leon, A.: Adaptive fuzzy predictive controller for anesthesia delivery. Control Eng. Pract. **46**, 1–9 (2016)
3. Marrero, A., Méndez, J.A., Reboso, J.A., Martín, I., Calvo, J.L.: Adaptive fuzzy modeling of the hypnotic process in anesthesia. J. Clin. Monit. Comput. **31**(2), 319–330 (2017)

4. Casteleiro-Roca, J., Calvo-Rolle, J., Meizoso-Lopez, M., Piñon-Pazos, A., Rodriguez-Gómez, B.: New approach for the QCM sensors characterization. Sens. Actuators, A **207**, 1–9 (2014)
5. Crespo-Ramos, M.J., Machón-González, I., López-García, H., Calvo-Rolle, J.L.: Detection of locally relevant variables using SOM-NG algorithm. Eng. Appl. Artif. Intell. **26**(8), 1992–2000 (2013)
6. Cowen, R., Stasiowska, M.K., Laycock, H., Bantel, C.: Assessing pain objectively: the use of physiological markers. Anaesthesia **70**(7), 828–847 (2015)
7. Ledowski, T.: Analgesia-nociception index. Br. J. Anaesth. **112**(5), 937 (2014)
8. Jeanne, M., Clément, C., De Jonckheere, J., Logier, R., Tavernier, B.: Variations of the analgesia nociception index during general anaesthesia for laparoscopic abdominal surgery. J. Clin. Monit. Comput. **26**(4), 289–294 (2012)
9. Jove, E., Gonzalez-Cava, J.M., Casteleiro-Roca, J.L., Pérez, J.A.M., Calvo-Rolle, J.L., de Cos Juez, F.J.: An intelligent model to predict ANI in patients undergoing general anesthesia. In: Pérez García, H., Alfonso-Cendón, J., Sánchez González, L., Quintián, H., Corchado, E. (eds.) SOCO/CISIS/ICEUTE -2017. AISC, vol. 649, pp. 492–501. Springer, Cham (2018). https://doi.org/10.1007/978-3-319-67180-2_48
10. Casteleiro-Roca, J.L., Pérez, J.A.M., Piñón-Pazos, A.J., Calvo-Rolle, J.L., Corchado, E.: Modeling the electromyogram (EMG) of patients undergoing anesthesia during surgery. In: Herrero, Á., Sedano, J., Baruque, B., Quintián, H., Corchado, E. (eds.) 10th International Conference on Soft Computing Models in Industrial and Environmental Applications. AISC, vol. 368, pp. 273–283. Springer, Cham (2015). https://doi.org/10.1007/978-3-319-19719-7_24
11. Gonzalez-Cava, J.M., Reboso, J.A., Casteleiro-Roca, J.L., Calvo-Rolle, J.L., Méndez Pérez, J.A.: A novel fuzzy algorithm to introduce new variables in the drug supply decision-making process in medicine. In: Complexity 2018 (2018)
12. Ghanghermeh, A., Roshan, G., Orosa, J.A., Calvo-Rolle, J.L., Costa, A.M.: New climatic indicators for improving urban sprawl: a case study of Tehran city. Entropy **15**(3), 999–1013 (2013)
13. Calvo-Rolle, J.L., Quintian-Pardo, H., Corchado, E., del Carmen Meizoso-López, M., García, R.F.: Simplified method based on an intelligent model to obtain the extinction angle of the current for a single-phase half wave controlled rectifier with resistive and inductive load. J. Appl. Logic **13**(1), 37–47 (2015)
14. Calvo-Rolle, J.L., Fontenla-Romero, O., Pérez-Sánchez, B., Guijarro-Berdinas, B.: Adaptive inverse control using an online learning algorithm for neural networks. Informatica **25**(3), 401–414 (2014)
15. Casteleiro-Roca, J.L., Calvo-Rolle, J.L., Meizoso-López, M.C., Piñón-Pazos, A., Rodríguez-Gómez, B.A.: Bio-inspired model of ground temperature behavior on the horizontal geothermal exchanger of an installation based on a heat pump. Neurocomputing **150**, 90–98 (2015)
16. Machón-González, I., López-García, H., Calvo-Rolle, J.L.: A hybrid batch SOM-NG algorithm. In: The 2010 International Joint Conference on Neural Networks (IJCNN), pp. 1–5. IEEE (2010)
17. Alaiz Moretón, H., Calvo Rolle, J., García, I., Alonso Alvarez, A.: Formalization and practical implementation of a conceptual model for PID controller tuning. Asian J. Control **13**(6), 773–784 (2011)
18. Rolle, J., Gonzalez, I., Garcia, H.: Neuro-robust controller for non-linear systems. DYNA **86**(3), 308–317 (2011)

19. Jove, E., Aláiz-Moretón, H., Casteleiro-Roca, J.L., Corchado, E., Calvo-Rolle, J.L.: Modeling of bicomponent mixing system used in the manufacture of wind generator blades. In: Corchado, E., Lozano, J.A., Quintián, H., Yin, H. (eds.) IDEAL 2014. LNCS, vol. 8669, pp. 275–285. Springer, Cham (2014). https://doi.org/10.1007/978-3-319-10840-7_34

20. Casteleiro-Roca, J.L., Jove, E., Sánchez-Lasheras, F., Méndez-Pérez, J.A., Calvo-Rolle, J.L., de Cos Juez, F.J.: Power cell SOC modelling for intelligent virtual sensor implementation. J. Sens. **2017**, 10 (2017)

21. Casteleiro-Roca, J.L., Calvo-Rolle, J.L., Méndez Pérez, J.A., Roqueñí Gutiérrez, N., de Cos Juez, F.J.: Hybrid intelligent system to perform fault detection on BIS sensor during surgeries. Sensors **17**(1), 179 (2017)

22. Gonzalez-Cava, J.M., et al.: A machine learning based system for analgesic drug delivery. In: Pérez García, H., Alfonso-Cendón, J., Sánchez González, L., Quintián, H., Corchado, E. (eds.) SOCO/CISIS/ICEUTE -2017. AISC, vol. 649, pp. 461–470. Springer, Cham (2018). https://doi.org/10.1007/978-3-319-67180-2_45

23. García, R.F., Rolle, J.L.C., Gomez, M.R., Catoira, A.D.: Expert condition monitoring on hydrostatic self-levitating bearings. Expert Syst. Appl. **40**(8), 2975–2984 (2013)

24. Calvo-Rolle, J.L., Casteleiro-Roca, J.L., Quintián, H., del Carmen Meizoso-Lopez, M.: A hybrid intelligent system for PID controller using in a steel rolling process. Expert Syst. Appl. **40**(13), 5188–5196 (2013)

25. García, R.F., Rolle, J.L.C., Castelo, J.P., Gomez, M.R.: On the monitoring task of solar thermal fluid transfer systems using NN based models and rule based techniques. Eng. Appl. Artif. Intell. **27**, 129–136 (2014)

26. Quintián, H., Calvo-Rolle, J.L., Corchado, E.: A hybrid regression system based on local models for solar energy prediction. Informatica **25**(2), 265–282 (2014)

27. Quintian Pardo, H., Calvo Rolle, J.L., Fontenla Romero, O.: Application of a low cost commercial robot in tasks of tracking of objects. DYNA **79**(175), 24–33 (2012)

28. Wasserman, P.: Advanced Methods in Neural Computing, 1st edn. Wiley, New York (1993)

29. Zeng, Z., Wang, J.: Advances in Neural Network Research and Applications, 1st edn. Springer, Heidelberg (2010). https://doi.org/10.1007/978-3-642-12990-2

30. Vapnik, V.: The Nature of Statistical Learning Theory. Springer, New York (1995). https://doi.org/10.1007/978-1-4757-3264-1

31. Kaski, S., Sinkkonen, J., Klami, A.: Discriminative clustering. Neurocomputing **69**(1–3), 18–41 (2005)

32. Fernández-Serantes, L.A., Estrada Vázquez, R., Casteleiro-Roca, J.L., Calvo-Rolle, J.L., Corchado, E.: Hybrid intelligent model to predict the SOC of a LFP power cell type. In: Polycarpou, M., de Carvalho, A.C.P.L.F., Pan, J.-S., Woźniak, M., Quintian, H., Corchado, E. (eds.) HAIS 2014. LNCS (LNAI), vol. 8480, pp. 561–572. Springer, Cham (2014). https://doi.org/10.1007/978-3-319-07617-1_49

33. Li, Y., Shao, X., Cai, W.: A consensus least squares support vector regression (LS-SVR) for analysis of near-infrared spectra of plant samples. Talanta **72**(1), 217–222 (2007)

34. Casteleiro-Roca, J.L., Quintián, H., Calvo-Rolle, J.L., Corchado, E., del Carmen Meizoso-López, M., Piñón-Pazos, A.: An intelligent fault detection system for a heat pump installation based on a geothermal heat exchanger. J. Appl. Logic **17**, 36–47 (2016)

Classification of Prostate Cancer Patients and Healthy Individuals by Means of a Hybrid Algorithm Combing SVM and Evolutionary Algorithms

Juan Enrique Sánchez Lasheras[1], Fernando Sánchez Lasheras[2(✉)],
Carmen González Donquiles[3,4,5], Adonina Tardón[3,4],
Gemma Castaño Vynals[6,7,8,9], Beatriz Pérez Gómez[6,10,11],
Camilo Palazuelos[12], Dolors Sala[13,14],
and Francisco Javier de Cos Juez[15]

[1] Anesthesiology and Resuscitation Service, Hospital Carmen y Severo Ochoa, Cangas del Narcea, Spain
[2] Department of Mathematics, University of Oviedo, Oviedo, Spain
sanchezfernando@uniovi.es
[3] Centro Investigación Biomédica en Red Epidemiología y Salud Pública (CIBERESP), Madrid, Spain
[4] Universitary Institute of Oncology of Asturias (IUOPA), University of Oviedo, Oviedo, Spain
[5] Research Group of Gene-Environment-Health Interactions, Biomedicine Institute (IBIOMED), University of Leon, León, Spain
[6] Consortium for Biomedical Research in Epidemiology & Public Health (CIBERESP), Carlos III Institute of Health, Madrid, Spain
[7] ISGlobal, Centre for Research in Environmental Epidemiology (CREAL), Barcelona, Spain
[8] Universitat Pompeu Fabra (UPF), Barcelona, Spain
[9] IMIM (Hospital del Mar Medical Research Institute), Barcelona, Spain
[10] Cancer Epidemiology Unit, National Centre for Epidemiology, Carlos III Institute of Health, Madrid, Spain
[11] Cancer Epidemiology Research Group, Oncology and Hematology Area, IIS Puerta de Hierro (IDIPHIM), Madrid, Spain
[12] Universidad de Cantabria – IDIVAL, Santander, Spain
[13] Valencia Cancer and Public Health Area, FISABIO - Public Health, Valencia, Spain
[14] General Directorate Public Health, Valencian Community, Valencia, Spain
[15] Department of Mines Prospecting and Exploitation, University of Oviedo, Oviedo, Spain

Abstract. This research presents a new hybrid algorithm able to select a set of features that makes it possible to classify healthy individuals and those affected by prostate cancer.

In this research the feature selection is performed with the help of evolutionary algorithms. This kind of algorithms, have proven in previous researches their ability for obtaining solutions for optimization problems in very different

© Springer International Publishing AG, part of Springer Nature 2018
F. J. de Cos Juez et al. (Eds.): HAIS 2018, LNAI 10870, pp. 547–557, 2018.
https://doi.org/10.1007/978-3-319-92639-1_46

fields. In this study, a hybrid algorithm based on evolutionary methods and support vector machine is developed for the selection of optimal feature subsets for the classification of data sets. The results of the algorithm using a reduced data set demonstrates the performance of the method when compared with non-hybrid methodologies.

Keywords: Support vector machines · Genetic algorithms · Prostate cancer
Single nucleotide polymorphism

1   Introduction

Prostate cancer (PCa) caused 300.000 deaths during 2012, and 1.1 million cases of this tumuor were diagnosed worldwide. In UE-28, it was the first cancer among men and the third in mortality, with more than 360,000 cases and 72,000 deaths. In Spain, 30,000 new cases were diagnosed and 5,000 deaths were estimated to have been due to this tumor [1]. In our country, an increase in incidence has been observed since the widespread use of prostate-specific antigen (PSA), while the mortality rate has been falling by 3.6% annually since 1998 [2].

Despite its importance, only three risk factors for PCa are firmly established: age, race and family history, none of which are modifiable [3]. Regarding modifiable environment factors, especially in those linked to lifestyle (alcohol and tobacco consumption, diet, and obesity) very heterogeneous results have been observed and their link with PCa is not well established, so it is also not known if the high adherence would reduce their incidence [4, 5].

It is also known that there is variability in the prognostic and clinical range of PCa, ranging from chronic indolent disease to aggressive, systemic and fulminant malignancy, and as a general rule, these lifestyle-related environmental factors behave differently depending on the aggressiveness and prognosis of the tumour. They behave as protective factors or have a neutral or weak effect in less aggressive cases, but are a relevant risk factor in the most aggressive tumours, in terms of recurrence and mortality of PCa. In turn, this same prognostic variability within the same disease range (stage TNM, PSA levels and Gleason or D'Amico score) is very relevant, and may be related to individual genetic factors and/or their interaction with environmental factors [6].

For this reason, it is necessary to apply novel methodologies that help to compute the analysis of genetic and environmental interactions, including intelligent information management systems based on advanced data extraction and automatic learning. These methodologies are capable of treating a large amount of information in a fast and effective way, with the objective of getting closer to an increasingly personalized and precise medicine to better predict the risk of disease [7].

The objective of this work will be the development of a model that allows us to classify individuals in prostate cases and controls based on the information of the most relevant environmental variables selected by it.

2 Methods

2.1 Support Vector Machines

Support Vector Machines (SVM) is a class of statistical models which have their origin in the research of Vapnik [8]. Although nowadays there are models feasible for regression, it is a technique that was originally developed for classification [9] and it is employed for this purpose in the present research.

Let us suppose the problem, in which by means of two variables we want to separate samples of two classes that are completely separable. If this information is represented in a two-dimensional space, we would find an infinite number of lines able to separate the data. In order to assess the performance of the different lines that would separate the data perfectly, Vapnik [8] defined a metric called marging. In the case of the problem with two categories, the maximum margin classifier creates a decision value $D(x)$ that classifies samples so that if $D(x) > 0$, we would predict a sample to be of, for example, class a, and otherwise that the sample corresponds to class b.

The maximum margin classifier can be written in terms of each data point in the sample with the help of the following equation [9, 10]:

$$D(u) = \beta_0 + \sum_{j=1}^{P} \beta_j u_j = \beta_0 + \sum_{i=1}^{n} y_i \alpha_i x_i' u$$

Where:

u	is an unknown sample.
β_0	is the independent term of the hyperplane equation.
β_j	are coefficients of the variables of the hyperplane.
u_j	are the components of vector u.
y_i	are the components of vector is the y component of the i-th data point in the sample.
α_i	is the set of nonzero values are the points that fell on the boundary of the margin.
$x_i' u$	is the dot product, where x_i is the x component of the i-th data point in the sample.

The equation presented above is only a function of the training set samples that are closest to the boundary and that are predicted with the least amount of certainty. As the prediction equation is supported only by these data points, the maximum margin classifier is usually called the support vector machine.

For those cases in which classes are not completely separable, Cortes and Vapnik [8] develop extensions to the early maximum margin classifier in order to accommodate this situation. This new formulation penalizes those points that are on the wrong side of the boundary.

The SVM equation was extended to non-linear models using a linear cross product [10, 11]:

$$D(\mu) = \beta_0 + \sum_{i=1}^{n} y_i \alpha_i x_i^1 \mu = \beta_0 + \sum_{i=1}^{n} y_i \alpha_i K(x_i, u)$$

Where $K(., .)$ is a kernel function of the two vectors. Some of the most common and well-known kernel functions are polynomial, radial basis function and hyperbolic tangent.

It must also be remarked that when the samples are not separable, a penalty term that approximates the total training error is considered. This is performed to avoid the reduction training error resulting in poor generalization error and overfitting. One of the most employed regularization terms is LASSO (Least Absolute Shrinkage and Selection Operator for parameters estimation) [12]. It has been adapted to SVM [13] by means of the following expression:

$$Pen_\gamma = \gamma \cdot \sum_{j=1}^{n} |u_j|$$

Where γ represents a tuning parameter.

The use of different kinds of kernels, makes it possible to produce extremely flexible decision boundaries. In the case of the present research, four well-known different kernels have been employed: linear, polynomial, radial basis function and tangent. These kernels have been used by the authors [14–17] in previous research.

2.2 Genetic Algorithms

Genetic Algorithms (GA) are a kind of adaptive methos generally employed for problems related to the search and optimization of parameters. The methodology of GA is based on the principles of natural reproduction and survival.

In order to find a solution, the problem starts with an initial set of individuals called population that, generally speaking, and in the case of the present research also is generated at random. Those individuals are evaluated using the rules of natural reproduction in search of a solution.

The origins of evolutive programming can be found in the work of Fogel, Evans and Walsh, published in 1966 [18]. We should also consider that GA, as they are known nowadays, were first presented by John Holland in his book called Adaptation in Natural and Artificial Systems [19]. It must be also remarked that the ability of GA goes beyond the capability of estimating parameters in a Physical Model.

In a GA, the population initially created is processed with the help of these main operators, which mimic biological processes [20].

- Selection: this mechanism behaves in the same way as the natural selection in nature. The best-performing individuals are most likely to transmit their information (genes) to the next generation.
- Crossover: produces a combination of different individuals of the same generation in a similar way to how natural reproduction does so.
- The mutation operation is employed to randomly change (flip) the value of single bits within an individual.

Although mutation is a very interesting mechanism for improving the variations of solutions, it must be employed with low probability values [20] in order to avoid the interruption of convergence process.

After these three mechanisms have been applied to the initial population, a new population will have been formed and the next generation is ready for iteration. This selection process continues until a fixed number of generations have elapsed, or until some form of convergence criterion has been met.

2.3   Study Population

MCC-Spain is a multicase-control study performed between September 2008 and December of 2013, based on population from Spain. As a part of the study, a database was created. The aim of this database was to assess of the influence of environmental exposures and their interaction with genetic factors in the most common tumors in Spain [21]. Therefore, the database includes cases of different kinds of cancer (colorectal, breast, gastric, prostate and chronic lymphocytic leukaemia). The total amount of individuals in the study was 10,106 subjects.

For the present research, the data subset was generated by taking data from MCC-Spain relating to the prostate cases and controls, specifically, the database designed is comprised of 1,112 (42.7%) subjects that have prostate cancer and 1,493 (57.3%) controls. A total of 93 different variables for each individual were employed, one of which was age, with an average value of 66.33 and a standard deviation of 8.09 years. Also, weight, height and educational level of the patients and if they were smokers or have smoked in past was recorded. In those that suffered from prostate cancer, the aggressiveness of the tumor was assessed by means of the Gleason Score. All the individuals enrolled in the study took a food frequency questionnaire that allowed us to know their daily intake of white meat, read meat, organ meat, cured meat, white fish, sugar, coffee, eggs, milk, soy, fruits, vegetables, bread rice etc. also, taking into account their answers to the questionnaire and the consumption of the different kinds of food, it was also possible to estimate the daily intake in terms of fiber, proteins, fats, ethanol, calcium, vitamin D, etc. Please note that although in the MCC-Spain genetic variables are measured, they are not included in the present study.

2.4   The Proposed Algorithm

The algorithm proposed in this research can be considered as a hybrid algorithm [22] and combines two well-known methodologies: GA and SVM. The use of GA is two-fold: on the one hand, they are responsible for the selection of the variables that take part in the SVM models and on the other hand, the optimization of parameters was also performed with the help of GA.

The flowchart of the algorithm is presented in Fig. 1. First of all, the data set is randomly split, using 80% of the information for training and the other 20% for validation. In the next step, the population of the GA, which is employed in order to determine which variables will be used in the SVM models is initialized. Each member of the population or in other words, each one of variables subset is employed for training a SVM model. Please note that each member of the GA population is formed for a string of '1s' and '0s' and there is a univocal relationship of each input variable with each one of this numbers. It means that when the number is '0' the variable is not taken into account for the SVM model while if the variable is '1', it is taken into account for the referred model. In order to form the individuals of the next generation, the crossover mechanism is applied cutting the bits strings of two members of the population in one point and combining them. The mutation mechanism is applied, given a low probability value to all the bits of the string for mutation (changing from '0' to '1' or from '1' to '0').

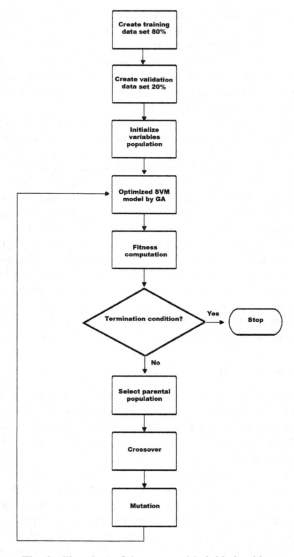

Fig. 1. Flowchart of the proposed hybrid algorithm.

Using as input information the selected variables and the training data set, the kernel and SVM parameters are tuned with the help of a new GA.

The performance of each SVM model is evaluated with the help of the validation data set. In the case of the present research, the fitness function calculates the area under the ROC (Receiver Operating Characteristic) curve. The termination condition of the GA is that no improvement in the area under the ROC is achieved after 100 cycles. The process of the flowchart of the figure was repeated 1,000 times using different training and validation data sets.

3   Results and Discussion

The application of the SVM classification model to the validation data sets gave us the results that are shown in Table 1. Please note that these values correspond to the best model, which in the case of the present research is considered to be the one with the highest values of specificity and sensitivity.

Table 1. SVM model performance. Point estimates and 95% CI.

Performance indicator	Point estimate	95% CI
Apparent prevalence	0.59	(0.54, 0.63)
True prevalence	0.59	(0.54, 0.63)
Sensitivity	0.91	(0.87, 0.94)
Specificity	0.87	(0.81, 0.91)
Positive likelihood ratio	6.80	(4.71, 9.81)
Negative likelihood ratio	0.10	(0.07, 0.15)

Please note that the selection of SVM model parameters and kernel is also performed by a GA. The quality of models built by machine learning algorithms mostly depends on the right tuning of parameters. In the case of this research, the assessment of the results with the help of validation data sets help us to avoid the consequences of overfitting.

In the case of the present research, the best results were employed using radial basis function as the kernel. In this kernel the G parameter that controls the scale was tuned together with the cost value. The optimum G value was 0.0057 and the cost value 235.

The developed GA is able to classify healthy individuals and PCa cases using only 16 variables: height, Body Mass Index, age, and the intake of artificial sweeteners, sodas, bread, sugar, glucoses, 'junk' food, dairy desserts, tomatoes, tinned fish, red and processed meat, proteins, soy milk and vitamin D. Although this is one of the strengths of the algorithm, in order to give a complete overview of the proposed method, limitations should also be taken into account. In the case of the results obtained, the model can be tuned if the association among variables is studied. From the existing evidence, age is the only marker which is a risk factor and is selected by this algorithm.

Most variables are correlated with obesity, so with BMI too: intake of sodas, bread, junk food or dairy products. All of them have a high quantity of glucoses or sugar. Several studies analysed the hypothesis that the decrease of carbohydrates in the diet delays PCa growth. However, not every publication has shown a positive statistical relationship between carbohydrates and PCa, and in some cases they have even been observed as protective factors.

Until now all the selected variables referred to different components of the diet. Closely related to this is the BMI variable. The observational studies found a statistically significant relationship between people with high BMI that present a specific predisposition and an increase in the rate of mortality from different cancer types, among them PCa [23].

Regarding dairy desserts, a two-fold relation with PCa can be found. On the one hand, they are sugar-rich, so they are associated to a high BMI. On the other hand, they also have a high calcium concentration. There are some evidence about diets with high intake of calcium-rich foods, which have been linked to the increased risk of prostate cancer, so high calcium concentrations down-regulate the formation of an active form of vitamin D which could play an important role in the carcinogenesis of PCa by inhibiting cell proliferation [24].

The intake of sweeteners has aroused great controversy concerning its effect in relation to cancer. Gallu et al. [25] showed that there does exist a relationship between sweeteners and some common types of cancer exists, including PCa.

Another product selected by this algorithm is soy milk, which is considered a protective factor due to a reduction in the inflammation that it produces [26]. In regard to vegetables, a protective role in the development of PCa has been found in some epidemiological studies, although not every publication has reached this conclusion [27]. Tomatoes have a high concentration of lycopene which is considered to be a protective factor for the development of several cancer types (particularly prostate) [28]. High intake of tomatoes or high levels of serum lycopene were associated to low PSA, decreased risk of PCa and reduction of urological symptoms.

Variables related to meat intake and the total proteins have a close relation. In human diet, proteins come mostly from meats, fish and dairy. Several studies have shown the potential benefit of chicken and fish intake over red meat, which is associated to a greater risk of suffering PCa [28]. Omega-3 (ω-3) fatty acids are a group of long chain polyunsaturated fatty acids, present in large proportion in fish. Giovanucci and Rimn found an inverse relationship between fish origin ω-3 and PCa, although it was not statistically significant.

Regarding non-modifiable factors, age and height can be mentioned. A rising incidence of microscopic foci of prostate cancer is found in men with increasing age. Results of autopsy studies have shown that almost 30% of men over the age of 50 have histological evidence of prostate cancer [29]. In regard to height, there is evidence of an effect of genetically-elevated height on prostate cancer risk [30].

Finally, another issue that should be taken into account is that there is not only one kind PCa among all the individuals considered in the database: some of the PCa are really aggressive while others are not. In the case of the non-aggressive cancers, high values of BMI are a protection against risk factors whereas they are a risk factor for the aggressive kinds of cancers. In other words, obesity does not seem to increase the overall risk of prostate cancer [23]. However, some studies [31] have found that obese men have a lower risk of a low-grade (less dangerous) form of the disease, but an increased risk of more aggressive prostate cancer [32] The reasons for this are not clear.

4 Conclusions

In the construction of a SVM model, the selection of the right kernel and optimum parameter value are really important tasks. In this paper, a new algorithm for the selection of variables was proposed. The experimental results demonstrate that the algorithm can be used for selecting a model of an SVM for any kernel.

As an overall conclusion of this research, it must be highlighted the need of a multidisciplinary work is necessary in order to overcome the possible technical limitations so as to in turn avoid including in the model those variables that are highly correlated in the model and working only with the minimum amount of variable able to represent all the information contained in the data set.

References

1. Sánchez, M.J., Payer, T., De Angelis, R., Larrañaga, N., Capocaccia, R., Martinez, C.: CIBERESP working group. Cancer incidence and mortality in Spain: estimates and projections for the period 1981-2012. Ann. Oncol. 21(Suppl. 3), iii30–iii36 (2010)
2. Larrañaga, N., Galceran, J., Ardanaz, E., Franch, P., Navarro, C., Sánchez, M.J., Pastor-Barriuso, R.: Prostate cancer working group. Prostate cancer incidence trends in Spain before and during the prostate-specific antigen era: impact on mortality. Ann. Oncol. 21(Suppl. 3), iii83–iii89 (2010). https://doi.org/10.1093/annonc/mdq087
3. Discacciati, A., Wolk, A.: Lifestyle and dietary factors in prostate cancer prevention. Recent Results Cancer Res. 202, 27–37 (2014). https://doi.org/10.1007/978-3-642-45195-9_3
4. Cuzick, J., Thorat, M.A., Andriole, G., Brawley, O.W., Brown, P.H., Culig, Z., Eeles, R.A., Ford, L.G., Hamdy, F.C., Holmberg, L., Ilic, D., Key, T.J., La Vecchia, C., Lilja, H., Marberger, M., Meyskens, F.L., Minasian, L.M., Parker, C., Parnes, H.L., Perner, S., Rittenhouse, H., Schalken, J., Schmid, H.P., Schmitz-Dräger, B.J., Schröder, F.H., Stenzl, A., Tombal, B., Wilt, T.J., Wolk, A.: Prevention and early detection of prostate cancer. Lancet Oncol. 15(11), e484–e492 (2014). https://doi.org/10.1016/S1470-2045(14)70211-6
5. Er, V., Lane, J.A., Martin, R.M., Emmett, P., Gilbert, R., Avery, K.N., Walsh, E., Donovan, J.L., Neal, D.E., Hamdy, F.C., Jeffreys, M.: Adherence to dietary and lifestyle recommendations and prostate cancer risk in the prostate testing for cancer and treatment (ProtecT) trial. Cancer Epidemiol. Biomark. Prev. 23(10), 2066–2077 (2014)
6. Al Olama, A.A., Kote-Jarai, Z., Berndt, S.I., Conti, D.V., Schumacher, F., Han, Y., et al.: A meta-analysis of 87,040 individuals identifies 23 new susceptibility loci for prostate cancer. Nat. Genet. 46(10), 1103–1109 (2014)
7. Ghasemi, M., Nabipour, I., Omrani, A., Alipour, Z., Assadi, M.: Precision medicine and molecular imaging: new targeted approaches toward cancer therapeutic and diagnosis. Am. J. Nucl. Med. Mol. Imaging 6(6), 310–327 (2016)
8. Vapnik, V.: The Nature of Statistical Learning Theory. Springer, New York (2010)

9. Artime Ríos, E.M., Seguí Crespo, M.M., Suarez Sánchez, A., Suárez Gómez, S.L., Sánchez Lasheras, F.: Genetic algorithm based on support vector machines for computer vision syndrome classification. In: Pérez García, H., Alfonso-Cendón, J., Sánchez González, L., Quintián, H., Corchado, E. (eds.) SOCO/CISIS/ICEUTE 2017. AISC, vol. 649, pp. 381–390. Springer, Cham (2018). https://doi.org/10.1007/978-3-319-67180-2_37

10. Boser, B., Guyon, I., Vapnik, V.: A training algorithm for optimal margin classifiers. In: Proceedings of the Fith Annual Workshop on Computation Learning Theory, pp. 144–152 (1992)

11. Nieto, P.J.G., Lasheras, F.S., García-Gonzalo, E., de Cos Juez, F.J.: PM_{10} concentration forecasting in the metropolitan area of Oviedo (Northern Spain) using models based on SVM, MLP, VARMA and ARIMA: a case study. Sci. Total Environ. **621**, 753–761 (2018)

12. Tibshirani, R.: Regression shrinkage and selection via the lasso. J. R. Stat. Soc. B **58**, 267–288 (1996)

13. Bradley, P.S., Mangasarian, O.L., Shavlik, J.: Feature selection via concave minimization and support vector machines. In: Proceedings of the Fifteenth International Conference on Machine Learning (ICML 1998), pp. 82–90. Morgan Kaufmann, San Francisco (1998)

14. Rosado, P., Lequerica-Fernández, P., Villallaín, L., Peña, I., Sánchez Lasheras, F.: Survival model in oral squamous cell carcinoma based on clinicopathological parameters, molecular markers and support vector machines. Expert Syst. Appl. **40**(12), 4770–4776 (2013)

15. Vilán, J.A.V., Fernández, J.R.A., Nieto, P.J.G., Lasheras, F.S., de Cos Juez, F.J.: Support vector machines and multilayer perceptron networks used to evaluate the cyanotoxins presence from experimental cyanobacteria concentrations in the Trasona reservoir (Northern Spain). Water Resour. Manag. **27**(9), 3457–3476 (2013)

16. Sánchez, A.S., Fernández, P.R., Lasheras, F.S., de Cos Juez, F.J., Nieto, P.J.G.: Prediction of work-related accidents according to working conditions using support vector machines. Appl. Math. Comput. **218**(7), 3539–3552 (2011)

17. Álvarez Antón, J.C., Nieto, P.J.G., de Cos Juez, F.J., Lasheras, F.S., Vega, M.G.: Battery state-of-charge estimator using the SVM technique. Appl. Math. Model. **37**(9), 6244–6253 (2013)

18. Fogel, L., Evans, M., Walsh, M.: Artificial Intelligence through Simulated Evolution. Wiley, New York (1966). Evolutionary Programming

19. Holland, J.: Adaptation in Natural and Artificial Systems. University of Michigan Press, Ann Arbor (1975)

20. Galán, C.O., Lasheras, F.S., de Cos Juez, F.J., Sánchez, A.B.: Missing data imputation of questionnaires by means of genetic algorithms with different fitness functions. J. Comput. Appl. Math. **311**, 704–717 (2017)

21. Castaño-Vinyals, G., Aragonés, N., Pérez-Gómez, B., Martín, V., Llorca, J., Moreno, V., Altzibar, J.M., Ardanaz, E., De Sanjosé, S., Jiménez-Moleón, J.J., et al.: Population-based multicase-control study in common tumors in Spain (MCC-Spain): rationale and study design. Gac. Sanit. **29**(4), 308–315 (2015)

22. Nieto, P.J.G., García-Gonzalo, E., Lasheras, F.S., de Cos Juez, F.J.: Hybrid PSO–SVM-based method for forecasting of the remaining useful life for aircraft engines and evaluation of its reliability. Reliab. Eng. Syst. Saf. **138**, 219–231 (2015)

23. Allott, E.H., Masko, E.M., Freedland, S.J.: Obesity and prostate cancer: weighing the evidence. Eur. Urol. **63**(5), 800–809 (2013). https://doi.org/10.1016/j.eururo.2012.11.013. Epub 2012 Nov 15. Review

24. Rodriguez, C., McCullough, M.L., Mondul, A.M., Jacobs, E.J., Fakhrabadi- Shokoohi, D., Giovannucci, E.L., et al.: Calcium, dairy products, and risk of prostate cancer in a prospective cohort of United States men. Cancer Epidemiol. Biomark. Prev. **12**, 597–603 (2003)

25. Gallus, S., Scotti, L., Negri, E., Talamini, R., Franceschi, S., Montella, M., Giacosa, A., Dal Maso, L., La Vecchia, C.: Artificial sweeteners and cancer risk in a network of case-control studies. Ann. Oncol. **18**(1), 40–44 (2007)
26. Xiudong, X., Ying, W., Xiaoli, L., Ying, L., Jianzhong, Z.: Soymilk residue (okara) as a natural immobilization carrier for Lactobacillus plantarum cells enhances soymilk fermentation, glucosidic isoflavone bioconversion, and cell survival under simulated gastric and intestinal conditions. PeerJ **4**, e2701 (2016)
27. Diallo, A., Deschasaux, M., Galan, P., et al.: Associations between fruit, vegetable and legume intakes and prostate cancer risk: results from the prospective Supplementation en Vitamines et Mineraux Anti-oxydants (SU.VI.MAX) cohort. Br. J. Nutr. **115**, 1579–1585 (2016)
28. Lin, P.H., Aronson, W., Freedland, S.J.: Nutrition, dietary interventions and prostate cancer: the latest evidence. BMC Med. **13**, 3 (2015)
29. Stangelberger, A., Waldert, M., Djavan, B.: Prostate cancer in elderly men. Rev. Urol. **10**(2), 111–119 (2008)
30. Davies, N.M., Gaunt, T.R., Lewis, S.J., Holly, J., Donovan, J.L., Hamdy, F.C., Kemp, J.P., Eeles, R., Easton, D., Kote-Jarai, Z., Al Olama, A.A., Benlloch, S., Muir, K., Giles, G.G., Wiklund, F., Gronberg, H., Haiman, C.A., Schleutker, J., Nordestgaard, B.G., Travis, R.C., Neal, D., Pashayan, N., Khaw, K.T., Stanford, J.L., Blot, W.J., Thibodeau, S., Maier, C., Kibel, A.S., Cybulski, C., Cannon-Albright, L., Brenner, H., Park, J., Kaneva, R., Batra, J., Teixeira, M.R., Pandha, H., PRACTICAL consortium, Lathrop, M., Smith, G.D., Martin, R. M.: The effects of height and BMI on prostate cancer incidence and mortality: a Mendelian randomization study in 20,848 cases and 20,214 controls from the PRACTICAL consortium. Cancer Causes Control **26**(11), 1603–1616 (2015)
31. Calle, E.E., Kaaks, R.: Overweight, obesity and cancer: epidemiological evidence and proposed mechanisms. Nat. Rev. Cancer **4**, 579–591 (2004)
32. Keto, C.J., Aronson, W.J., Terris, M.K., Presti, J.C., Kane, C.J., Amling, C.L., Freedland, S. J.: Obesity is associated with castration-resistant disease and metastasis in men treated with androgen deprivation therapy after radical prostatectomy: results from the SEARCH database. BJU Int. **110**(4), 492–498 (2012)

Hybrid Intelligent Applications

A Hybrid Deep Learning System of CNN and LRCN to Detect Cyberbullying from SNS Comments

Seok-Jun Bu and Sung-Bae Cho[✉]

Department of Computer Science, Yonsei University, Seoul, South Korea
{sjbuhan, sbcho}@yonsei.ac.kr

Abstract. The cyberbullying is becoming a significant social issue, in proportion to the proliferation of Social Network Service (SNS). The cyberbullying commentaries can be categorized into syntactic and semantic subsets. In this paper, we propose an ensemble method of the two deep learning models: One is character-level CNN which captures low-level syntactic information from the sequence of characters and is robust to noise using the transfer learning. The other is word-level LRCN which captures high-level semantic information from the sequence of words, complementing the CNN model. Empirical results show that the performance of the ensemble method is significantly enhanced, outperforming the state-of-the-art methods for detecting cyberbullying comment. The model is analyzed by t-SNE algorithm to investigate the mutually cooperative relations between syntactic and semantic models.

1 Introduction

Though the original purpose of Social Network Service (SNS) was to help people to communicate, SNS provides a rich medium exposed to verbal violence [1]. The term "cyberbullying" defined as the repeated injurious use of harassing, insulting or attacking to someone, is becoming severe social issues [2]. According to statistics, close to 25% of parents reported whose child had been involved in a cyberbullying incident [3], and 43% of teenagers in the USA alone were subjected to cyberbullying at some point [4]. Classifying task whether the SNS comment is cyberbullying can be a partial solution to prevent such social problems since the cyberbullying on SNS is mainly in the form of comments.

Detecting cyberbullying using traditional machine learning methods seems insufficient to identify and handle a cyberbullying, since these conventional methods are monotonous to model the variation of natural language. Complicated metaphors, polysemy, sarcasm and neologism is a well-known technical barrier in the field of Natural Language Processing (NLP).

In this paper, we propose an ensemble model based on deep learning approach to classify if the comments belong to cyberbullying. One of the components implemented to model the syntactics of the cyberbullying comments is character-level convolutional neural network (CNN) with knowledge-transfer. The CNN is the most successful model among the deep learning architectures and robust to signal-level noise via convolution and pooling operations [5]. The other component is implemented to model the semantics of the cyberbullying comments using word-embedding and long-term recurrent convolutional networks (LRCN). The spatial and temporal features of cyberbullying comments are modeled with convolution-pooling operations and LSTM cells, after the word vectors are extracted. The word-embedding algorithm that captures the continuous vector representations based on frequency is used to map each word to semantic space [6]. Our ensemble model achieved the best classification accuracy compared to other machine learning classifiers including deep learning classifiers.

2   Related Works

Sentimental analysis, one of the primary tasks in NLP, is about classifying the polarity of document or sentence. The problem of identifying comments whether it is belonged to cyberbullying is also the subset of sentiment analysis.

For early work in sentiment analysis, Turney et al. applied PMI-IR algorithm to estimate the semantic orientation of movie review dataset [7]. The algorithm compares semantic differences between words: it calculates semantic orientation using predefined good/bad word vocabulary [8]. Though the results showed the PMI-IR algorithm outperforms human-produced baselines, still there is a limitation that person manually selects a feature for classification.

From the mid-2000s, term frequency-inverse document frequency (TF-IDF) solidified as a measurement of a feature extracted from the text. TF-IDF is a statistical method of measuring how important a word is in the document. Yun-tao et al. improved TF-IDF approach using confidence, support, and characteristic words to enhance the performance in sentiment analysis [9]. Forman et al. also developed TF-IDF method using bi-normal separation (BNS), which shows substantially better performance in ranking words for feature selection filtering [10]. Though the TF-IDF measurement is useful to represent documents, it is not enough to implement in comment classification since TF-IDF approach ignores the order of words, so that information loss occurs.

To deal with metaphors, polysemy, sarcasm and neologism in the commentary, we considered ordering of words. LeCun et al. regarded the language as a signal with no difference from any other kind, and by applying CNN on the sequence of the characters, they achieved state-of-the-art performance in text classification task [5]. Motivated by [5], we enhance the noise robustness by using knowledge-transfer. Table 1 shows the summary of conventional methods.

Table 1. Related works for sentiment analysis and text classification

Author	Method	Description
Turney [7]	PMI-IR	Compare the semantic orientation using predefined keywords
Pang [8]	NB, SVM	Extract features using predefined bag-of-words
Yun [9]	TF-IDF	Improved TF-IDF approach using characteristic words
Forman [10]	TF-IDF	Improved TF-IDF approach using BNS
Zhang [5]	CNN	Treat text as a kind of raw signal at character level

3   The Proposed System

In this section, we present the hybrid deep learning architecture and the two main components: Character-level CNN and word-level LRCN. Each of deep neural networks models the syntactics and semantics from natural language to classify the cyberbullying comments and has complementary relation.

3.1   Hybrid Deep Networks

We categorize cyberbullying comments into syntactic and semantic, inspired by the way of dividing the field of NLP [11]. The syntactic cyberbullying is the kind of violence based on the particular slang, which is relatively easy to detect using domain knowledge. The semantic cyberbullying, however, is difficult to find hidden meanings since it consists of complex metaphors.

The syntactic and semantic information of cyberbullying is inherent in the sequence of characters or words. On the one hand, the syntactics of cyberbullying consists of particular slangs and neologism modeled as low-level signals from the character sequence. The neologism of slangs, for example, using the word 'fu<k' instead of 'fuck' is the noise of original signal. On the other hand, the semantics of cyberbullying modeled as high-level signals from the sequence of word vectors. We use word-embedding algorithm to extract vector representation of each word based on its frequency, to keep the statistical correlation between words [12].

We use CNN and long short-term memory (LSTM) for learning the features from the sequence of characters and word vectors, respectively. The two deep learning models are combined through an ensemble technique, while maintaining complementary relationships from syntactic and semantic aspects of language. Figure 1 shows the proposed hybrid architecture of two deep learning classifiers and an ensemble method.

Each model outputs the continuous value $\hat{y}$ through sigmoid activation function σ as follows:

$$\hat{y}^l = w_i^{l-1}\left(\sigma\left(p_i^{l-1}\right) + b_i\right) \tag{1}$$

where l is the depth of the model, w_i^{l-1} is the weight between the i th node from the $l-1$ th layer and output node from last layer, p^{l-1} is flattened feature vectors from previous layers and b_i is the bias term.

The outputs from the mth model $\hat{y}_m$, which has a value between 0 to 1, is more likely as cyberbullying if the value is close to 1. We calculate the arithmetic mean of the outputs from each of M models after converted each of the outputs in the log-scale score s to emphasize the role of each model:

$$s = \frac{1}{M}\sum_m \log(\hat{y}_m + 1) \tag{2}$$

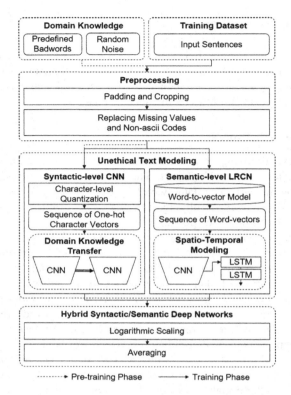

Fig. 1. Hybrid architecture of deep neural networks which captures character-level and word-level features from comments

3.2   Character-Level CNN and Knowledge Transfer

CNN is one of the successful deep learning architectures and achieved the best performance in image and pattern classification tasks. Reducing the spectral variations and modeling spectral correlations with local connectivity [13], convolution and pooling operations are known to be robust to noise. Hence, CNN suits for modeling the neologisms of cyberbullying comments.

We encode each character from comments by replacing each alphabet 1-of-m predefined integers. 26 alphabets, 10 numbers and 35 other characters including whitespace character are encoded as shown in Fig. 2. Since the predefined maximum length of comment is 225, we have cropped out exceeding characters. Any characters that are not included in predefined 71 characters are replaced with whitespace character. The dimensionality of input vector for CNN is $(N, 225, 71)$ after the encoding process.

abcdefghijklmnopqrstuvwxyz012345
6789-,;.!?:'"/\|_@#$%^&*~+-=<>()[]{}

Fig. 2. 71 characters to be encoded as integer, including 26 alphabets, 10 numbers and 35 other characters including whitespace character

Given input vector x_i, the output c_{xy}^l from the l th convolutional layer of our CNN performs the convolution operation with the output of last layer y^{l-1}, using $m \times m$ sized filter w:

$$c_{xy}^l = \sum_{a=0}^{m-1} \sum_{b=0}^{m-1} w_{ab} y_{(x+a)(y+b)}^{l-1} \tag{3}$$

The summary statistics of nearby outputs is derived from c^{l-1} by max-pooling operation [14]. The output p_{xy}^l from the l th pooling layer performs the max-pooling operations under $k \times k$ sized area with pooling stride τ:

$$p_{xy}^l = \max c_{xy \times \tau}^{l-1} \tag{4}$$

The character-level CNN is implemented with six couples of convolution-pooling modules, with 32-32-64-128-256-512 of 2×2 convolution filter and pooling size, to verify the modeling performance of various neologisms of natural languages with practical complexity. The 512-64-1 sized fully-connected layer is stacked over convolution-pooling modules to classify the cyberbullying.

Since the main purpose of character-level CNN is capturing the syntactic feature of cyberbullying, it can be expected that classification performance can be enhanced by pre-training of the domain knowledge such as known slangs. Knowledge-transfer, as known as transfer-learning, has main advantage in reusing previously learned feature distribution [15].

The pseudo-badword dataset is generated for constructing a classifier which has a strong resistance to noise. The 550 instances of Google-bad-words, a list of bad words banned by Google, were randomly inserted after being mapped random integer between 0 to 70 instead of each character. The word from Google-bad-words was chosen to keep the distributed features closed to cyberbullying domain. As a result, 200 thousands of uniformly distributed pseudo-badword dataset is generated. The virtual data generated to ensure the noise robustness of the CNN is pre-trained before the real dataset is learned.

3.3   Word-Level LRCN and Word-Embedding

Character-level CNN explained above gets mostly syntactic information rather than semantic one. To learn semantics, we train the word-level LRCN after embedding the word as a vector using skip-gram. The spatial and temporal features from the sequence of word vectors are modeled using convolution-pooling modules and LSTM [16].

In skip-gram approach, the word-embedding algorithm generates several input-output pairs of words extracted by windows whose size is C. The pairs are consisted of one input word w and $2 * C$ surrounding words as context. The set of context words given w is $C(w)$ and D is the set of all words and parameter θ_w is defined as (5) and optimized as (6):

$$p(c|w; \theta_w) = \frac{e^{v_c \cdot v_w}}{\sum_{c' \in C} e^{v_c' \cdot v_w}} \tag{5}$$

$$argmax_{\theta_w} \sum_{(w,c) \in D} \log p(c|w) = \sum_{(w,c) \in D} \left(\log e^{v_c \cdot v_w} - \log \sum_{c'} e^{v_{c'} \cdot v_w} \right) \tag{6}$$

Skip-gram model updates its hidden matrix with predicting context words given a target word that is represented as one-hot encoded vector product weight matrix [17]. We apply the LSTM layers to model the temporal features from the sequence of word vectors [18], after applying the convolution-pooling operations to model the spatial features. LSTM makes a mapping of temporal relation between an input sequence $(x_1, \ldots, x_T)$ and an output sequence $(y_1, \ldots, y_T)$ by calculating iteratively the unit activations using the following equations from $t = 1$ to T:

$$
\begin{aligned}
i_t &= \sigma(w_{ix}x_t + w_{im}m_{t-1} + w_{ic}C_{t-1} + b_i) \\
f_t &= \sigma(w_{fx}x_t + w_{fm}m_{t-1} + w_{fc}C_{t-1} + b_f) \\
c_t &= f_t \odot c_{t-1} + i_t \odot g(w_{cx}x_t + w_{cm}m_{t-1} + b_f) \\
o_t &= \sigma(w_{ox}x_t + w_{om}m_{t-1} + w_{oc}C_t + b_o) \\
m_t &= o_t \odot h(c_t) \\
y_t &= \pi(w_{ym}m_t + b_y)
\end{aligned}
\tag{7}
$$

where w denotes weight matrix, b is biased vector, and σ is the sigmoid function. Parameter i is input gate, o is the output gate, and f is the forget gate. C is an activation vector of a single LSTM cell, $\odot$ is the element-wise product of the vectors. Parameter g and h are activation functions of cell input and cell output, respectively, and π is the overall output activation function.

4   Experiments

In this section, we explain the dataset and detailed implementation of our architecture and experimental results. We also show how hybrid architecture performs in accordance with different hyper-parameters. For the fairness of evaluation, 10-fold cross validation was conducted.

4.1   Dataset and Experimental Setup

To verify the classification performance of cyberbullying comments, we used the dataset released in September 2012 by Kaggle, the popular data science competition portal. The dataset contains in total 8,815 comments which are binary labeled as 0 (neutral) and

1 (cyberbullying). Only a total of 2,818 comments are labeled as cyberbullying, which reveals the class imbalance problem in cyberbullying classification task.

We adjusted the depth of each network experimentally to avoid the over-fitting or degradation problem using training dataset [19]. At the top of the transfer CNN, 512-128-2 sized fully-connected layer is stacked to classify whether the input vector contains the encoded slangs. The fully-connected layer is removed before we overwrite the weights with training dataset using knowledge-transfer methods.

The proposed hybrid architecture of character-level CNN and word-level LRCN is implemented with Tensorflow that is a very efficient platform for matrix multiplication using GPU [20]. We used four NVIDIA GTX 1080 servers to facilitate a large number of CNN and LRCN prototypes.

4.2 Results

To verify and compare the proposed hybrid architecture with V. Sharma et al., which performed best in the Kaggle cyberbullying classification competition held in 2012, we calculated the ROC curve and AUC in Fig. 3. Compared to 0.8424, which was the highest AUC record at competition, the word-level LRCN alone outperforms the state-of-the-art methods with 0.8639. Moreover, the proposed hybrid classifier recorded 0.8854. Classification accuracy of the proposed hybrid classifier is 87.22% which is the best accuracy among conventional machine learning methods including deep learning methods. For the fairness of evaluation, we conducted 10-fold cross validation and visualized the box-plot in Fig. 4. The single CNN and LRCN achieved 85.41% and 85.73% respectively, which is not a significant improvement compared to random forest methods. However, the improvement of 2% accuracy shows the complementary characteristics of character-level CNN and word-level LRCN.

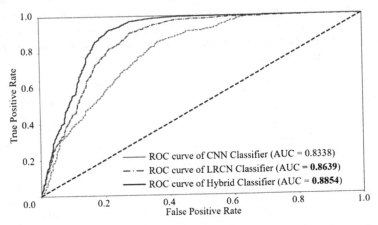

Fig. 3. The comparison of ROC-curve and AUC with the deep learning methods

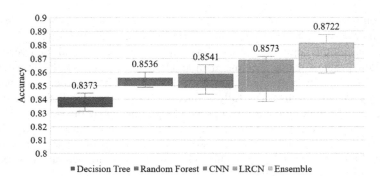

Fig. 4. The comparison of 10-CV classification accuracy with conventional machine learning methods

Table 2. The confusion matrix analysis of hybrid deep networks

Actual	Predicted		
		Cyberbullying	Neutral
	Cyberbullying	**245**	168
	Neutral	101	**897**

For more accurate analysis of the model, confusion matrix was analyzed in Table 2. The sensitivity and precision of the hybrid model are 0.5932 and 0.7081. The sensitivity, which seems to have declined compared to precision, has been caused by class imbalance problem.

In Fig. 5, the activation function values before the output layer of the character-level CNN and word-level LRCN are mapped in 2-dimensional space using the t-SNE dimension reduction algorithm [21]. Each point refers to one comment. On the right side of the figure, the comments with slangs and syntactics are clustered by character-level CNN. The comments with metaphors and semantics are clustered in the bottom of the figure by word-level LRCN.

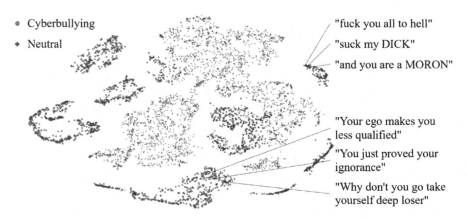

Fig. 5. Visualization of activation values of CNN and LRCN using t-SNE algorithm

The case analysis of CNN and LRCN are shown in Table 3. Each model has its own advantages depending on the implementation method. The character-level CNN, which captures the low-level syntactics from sequence of characters, showed advantages in classification based on syntactics and disadvantages on semantics. On the contrary, the word-level LRCN, which captures the high-level semantics from sequence of words, showed advantages on semantics. The case that is misclassified in both models is also hard to judge by the human, and can be thought of as the noise of the dataset.

Table 3. Case analysis of CNN and LRCN: Advantages and misclassified cases

	Class	Commentary	CNN score	LRCN score
CNN Advantages	1	You are just wonderful and stupid	0.9795	0.2203
	0	Would probably be cheaper just to legally change you name	0.0032	0.6326
LRCN Advantages	1	Why did your parents wish you were adopted	0.4415	0.8978
	1	If it were your mom sister and wife it would be only one person	0.1833	0.6870
Misclassified	1	You're a real bore ya know it	0.1428	0.3398
	1	We will bury you in November parasites	0.0092	0.0363

In order to enhance CNN, we used the knowledge-transfer method. To verify if the knowledge-transfer method is used properly, we applied the method to the convolution-pooling modules and compared the performance in Fig. 6. T6 is applied the knowledge-transfer method to all 6 convolution-pooling modules, T1 is applied to one module and naïve is not applied. The best performance was achieved when knowledge-transfer methods were applied to all modules.

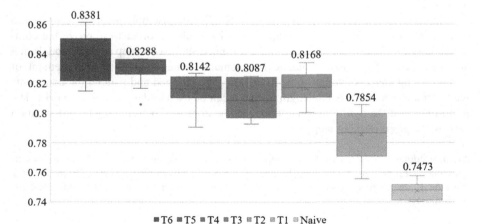

Fig. 6. CNN performance per number of knowledge-transferred modules

The LRCN model has many hyper-parameters, such as the embedding dimension and the highest frequency word used for embedding, in a serially connected with skip-gram model. In order to verify if the LRCN has been properly trained, we conducted grid-manner repetitive experiments in Fig. 7. Experiments were performed on the training data and achieved the best performance with 400-dimensional 3,500 most common words.

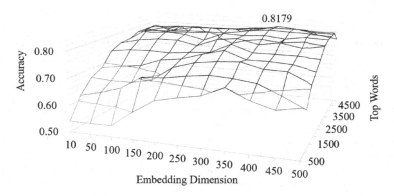

Fig. 7. Grid-manner repetitive experiments of LRCN hyper-parameters

5   Conclusions

In this paper, we propose a hybrid architecture of character-level CNN and word-level LRCN, which model syntactics and semantics of cyberbullying comments. Experimental results have shown that the proposed hybrid deep networks outperformed other machine learning methods including the state-of-the-art methods. The performance of each deep network is enhanced with knowledge-transfer and word-embedding methods.

Future work will include a further study of efficient ensemble method, compared to current method of averaging the outputs in a log-scale. In order to improve the complementary characteristics of each model, it is necessary to develop several models that are more subdivided than syntactic and semantic categories. The improvement of preprocessing steps is another issue to explore. We cropped and padded the comments with variable length to fix the length so that it can be used in deep neural networks. The performance of the hybrid deep networks can be improved by reducing the information loss in the preprocessing step.

Acknowledgements. This work was supported by Institute for Information & Communications Technology Promotion (IITP) grant funded by the Korea government (MSIT) (2016-0-00562, Emotional Intelligence Technology to Infer Human Emotion and Carry on Dialogue Accordingly).

References

1. Reynolds, K., Kontostathis, A., Edwards, L.: Using machine learning to detect cyberbullying. In: 2011 10th International Conference on Machine Learning and Applications and Workshops (ICMLA), vol. 2, pp. 241–244 (2011)
2. Olweus, D.: Bullying at School: What We Know and What We Can Do. Blackwell Publishing Google Scholar, Malden (1995)
3. Patchin, J.W., Hinduja, S.: Bullies move beyond the schoolyard a preliminary look at cyberbullying. Youth Violence Juv. Justice **4**, 148–169 (2006)
4. Ybarra, M.: Trends in technology-based sexual and non-sexual aggression over time and linkages to nontechnology aggression. National Summit on Interpersonal Violence and Abuse Across the Lifespan: Forging a Shared Agenda (2010)
5. Zhang, X., Zhao, J., LeCun, Y.: Character-level convolutional networks for text classification. In: Advances in Neural Information Processing Systems, pp. 649–657 (2015)
6. Mikolov, T., Chen, K., Corrado, G., Dean, J.: Efficient estimation of word representations in vector space. arXiv Preprint arXiv:1301.3781 (2013)
7. Turney, P.D.: Thumbs up of thumbs down?: semantic orientation applied to unsupervised classification of reviews. In: Proceedings of the 40th Annual Meeting on Association for Computational Linguistics, pp. 417–424 (2002)
8. Pang, B., Lee, L., Vaithyanathan, S.: Thumbs up?: sentiment classification using machine learning techniques. In: Proceedings of the ACL-02 Conference on Empirical Methods in Natural Language Processing, vol. 10, pp. 79–86 (2002)
9. Yun-tao, Z., Ling, G., Yong-cheng, W.: An improved TF-IDF approach for text classification. J. Zhejiang Univ. Sci. A **6**, 49–55 (2005)
10. Forman, G.: BNS feature scaling: an improved representation over tf-idf for SVM text classification. In: Proceedings of the 17th ACM Conference on Information and Knowledge Management, pp. 263–270 (2008)
11. Collobert, R., Weston, J.: A unified architecture for natural language processing: deep neural networks with multitask learning. In: Proceedings of the 25th International Conference on Machine Learning, pp. 160–167 (2008)
12. Mikolov, T., Sutskever, I., Chen, K., Corrado, G.S., Dean, J.: Distributed representations of words and phrases and their compositionality. In: Advances in Neural Information Processing Systems, pp. 3111–3119 (2013)
13. Sainath, T.N., Mohamed, A.R., Kingsbury, B., Ramabhadran, B.: Deep convolutional neural networks for LVCSR. In: 2013 IEEE International Conference on Acoustics, Speech and Signal Processing (ICASSP), pp. 8614–8618 (2013)
14. Bu, S.J., Cho, S.B.: A hybrid system of deep learning and learning classifier system for database intrusion detection. In: International Conference on Hybrid Artificial Intelligence Systems, pp. 615–625 (2017)
15. Bengio, Y.: Deep learning of representations for unsupervised and transfer learning. In: Proceedings of the ICML Workshop on Unsupervised and Transfer Learning, pp. 17–36 (2012)
16. Donahue, J., Hendricks, L., Guadarrama, S., Rohrbach, M., Venugopalan, S., Saenko, K., Darrel, T.: Long-term recurrent convolutional networks for visual recognition and description. In: Proceedings of the IEEE Conference on Computer Vision and Pattern Recognition, pp. 2625–2634 (2015)
17. Goldberg, Y., Levy, O.: Word2vec Explained: Deriving Mikolov et al.'s Negative-sampling Word-embedding Method. arXiv preprint arXiv:1402.3722 (2014)

18. Sainath, T.N., Vinyals, O., Senior, A., Sak, H.: Convolutional, long short-term memory, fully connected deep neural networks. In: 2015 IEEE International Conference on Acoustics, Speech and Signal Processing (ICASSP), pp. 4580–4584 (2015)
19. He, K., Zhang, X., Ren, S., Sun, J.: Deep residual learning for image recognition. In: Proceedings of the IEEE Conference on Computer Vision and Pattern Recognition, pp. 770–778 (2016)
20. Abadi, M., Barham, P., Chen, J., Chen, Z., Davis, A., Dean, J., Kudlur, M.: Tensorflow: a system for large-scale machine learning. In: Proceedings of the 12th USENIX Symposium on Operating Systems Design and Implementation (2016)
21. Maaten, L.V.D., Hinton, G.: Visualizing Data using t-SNE. J. Mach. Learn. Res. **9**, 2579–2605 (2008)

Taxonomy-Based Detection of User Emotions for Advanced Artificial Intelligent Applications

Alfredo Cuzzocrea[1,2] and Giovanni Pilato[2(✉)]

[1] University of Trieste, Trieste, Italy
alfredo.cuzzocrea@dia.units.it
[2] ICAR-CNR, Rende, Italy
giovanni.pilato@cnr.it

Abstract. Catching the attention of a new acquaintance and empathize with her can improve the social skills of a robot. For this reason, we illustrate here the first step towards a system which can be used by a social robot in order to "break the ice" between a robot and a new acquaintance. After a training phase, the robot acquires a sub-symbolic coding of the main concepts being expressed in tweets about the IAB Tier-1 categories. Then this knowledge is used to catch the new acquaintance interests, which let arouse in her a joyful sentiment. The analysis process is done alongside a general small talk, and once the process is finished, the robot can propose to talk about something that catches the attention of the user, hopefully letting arise in him a mix of feelings which involve surprise and joy, triggering, therefore, an engagement between the user and the social robot.

1 Introduction

Engagement is one of the most basic and important phases in interactions between human beings. In the last years there has been a growing interest about this topic throughout the human-machine-interaction (HMI) and related fields [6]. Researchers have highlighted that engagement is a very complex phenomenon, including both cognitive and affective components: it should involve attention and enjoyment [3,11].

We refer to this term as the "starting or intention to start an interaction". In particular, we focus our attention on the fact that, in making new acquaintances, the first impression is very important, and finding as soon as possible common interests to talk about, allows starting an empathetic interaction between two persons, with all that this implies.

In order to trigger both attention and enjoyment, given these premises, it would be useful to design a social robotic system which tries to find the topic of interest of a new acquaintance while attempting to understand what might raise a sentiment of joy attempting to catch an empathetic attention of the user.

© Springer International Publishing AG, part of Springer Nature 2018
F. J. de Cos Juez et al. (Eds.): HAIS 2018, LNAI 10870, pp. 573–585, 2018.
https://doi.org/10.1007/978-3-319-92639-1_48

As a matter of fact, the knowledge of the topics of interest and the "joyful" subjects for the user can lead the first stages of a conversational interaction that allows the robot to facilitate the engagement of a friendly interaction, instead of a classical trivial interaction between a robot and an human user.

To reach this goal, the robot can access the social network data of the new acquaintance trying to coarsely profile her/his interests, catching useful information to engage a possibly interesting conversation for the user.

Social networks represent a great place, maybe the best, to gather information about people's opinions, since they are generally used to express personal thoughts and to discuss with other people about specific subjects [8,26]. These opinions are really useful to understand and classify the emotion of an event, a product, a person, etc. and analyze his trend [9,17,18].

In this paper we illustrate the design of a system which can be used by a social robot in order to "break the ice" between the robot and a new acquaintance.

First of all, the robot acquires a knowledge about the construction of prototypes describing each entry of the IAB Taxonomy.

The system needs a training phase where fundamental concepts, induced by a data driven construction of a conceptual space by using the Latent Semantic Analysis (LSA) procedure and a set of topics derived by the Latent Dirichlet Allocation (LDA) methodology, representing the Tier1 categories of the IAB v2.0 taxonomy are mapped in a semantic space.

A set of tweets is therefore retrieved for each word describing each entry of the IAB Taxonomy. A set of words describing the conceptual axes of two "conceptual spaces" induced from each set of tweets associated to a single IAB entry is built. Each conceptual axis is therefore described by a specific "bag of words" which constitute the axis description. Each axis is therefore coded as a vector in a semantic space built through LSA, and it is associated to the specific IAB entry. At the end of the procedure, each entry of the IAB taxonomy is associated to set of vectors, associated to the labels of each fundamental axis of the category, in the built semantic space.

On the other hand, a system which is able to detect a pattern of basic Eckman emotions [20], given a text, is trained too.

Once the system is trained, during a general conversation with a new acquaintance, the robot asks for the user Twitter ID, and while the conversation continues, it retrieves the most recent tweets of the user.

Each tweet is then encoded as a vector in a semantic space. The semantic similarity between each tweet and each vector representing each entry of the IAB taxonomy is computed, and the highest value of similarity is retained.

The above procedure allows to associate a tweet of the user to a pattern of IAB categories; furthermore, for each tweet a vector of Eckman fundamental emotions is computed. This leads to a selection of the Tier1 categories of the IAB taxonomy which are of interest of the user and that let arise in the user a specific emotion. In our case we have chosen to select the "joy" emotion, which is the most desirable when a person meets for the first time another human being.

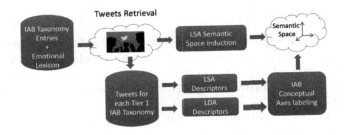

Fig. 1. Training process

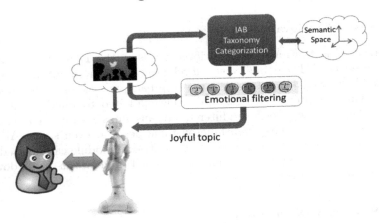

Fig. 2. Joyful-topic-detection process

The goal is to engage a conversation somehow polarizing it on topics that catch the attention of the user, trying to establish an empathetic relationship. Under some extensions, this approach has relationships with *adaptive metaphors*, like those developed in other scientific contexts (e.g., [5]).

The remaining of the paper is organized as follows: the next section illustrates the different modules of the system; Sect. 3 describes the experimental setup, and in Sect. 4 conclusions are given.

2   The System

The proposed system is composed of a set of modules interacting in order to catch the attention of the user. The modularity of the proposed architecture makes it suitable to be implemented on top of Cloud infrastructure (e.g., [7]). The system has a training phase, shown in Fig. 1, where a semantic space S is induced from Twitter data and a joyful-topic-detection process, illustrated in Fig. 2, which exploits the Twitter ID of the user in order to retrieve her posts and trying to catch the interests of the user that somehow let arise a "joy" emotion.

2.1   The IAB Taxonomy

The IAB (Interactive Advertising Bureau) Tech Lab Content Taxonomy is a concise taxonomy which is also an international standard to map contextual business categories [13,14]. The latest release of the taxonomy, namely version 2.0, has been released on November 2017 and it accounts 698 entries distributed over 29 Tier-1 classes.

This taxonomy is particularly suited for being used by companies in the market, it is standardized and industry-neutral. These characteristics can be effectively exploited for profiling an user interests.

2.2   Tweets Retrieval Module

The dataset object of analysis is retrieved by using the Twitter APIs with the default access level. The default access level gives a random sample of the streaming of publicly available tweets. For our approach, we use only the tweet text content, which is preprocessed before being exploited to build a data-driven conceptual space. Stop-words are filtered out, and links are removed before processing the text since they often hide off-topic posts or even spam. Abnormal sequences of characters were discarded. The retrieval module can be used either for retrieving tweets satisfying a query composed of keywords or to download the last tweets of a given twitter user ID.

2.3   LSA-based Descriptors

The Latent Semantic Analysis (LSA) technique is a well-known methodology that is capable of giving a coarse sub-symbolic encoding of word semantics [16] and of simulating several human cognitive phenomena [15]. The LSA procedure is based on a term-document occurrence matrix $\mathbf{A}$, whose generic element represents the number of times a term is present in a document. Let K be the rank of $\mathbf{A}$. The factorization named Singular Value Decomposition (SVD) holds for the matrix $\mathbf{A}$:

$$\mathbf{A} = \mathbf{U}\Sigma\mathbf{V}^T \tag{1}$$

Let R be an integer > 0 with $R < N$, and let $\mathbf{U}_R$ be the $M \times R$ matrix obtained from $\mathbf{U}$ by suppressing the last $N - R$ columns, let Σ_R be the matrix obtained from Σ by suppressing the last $N - R$ rows and the last $N - R$ columns; let $\mathbf{V}_R$ be the $N \times R$ matrix obtained from $\mathbf{V}$ by suppressing the last $N - R$ columns. Then:

$$\mathbf{A}_R = \mathbf{U}_R\Sigma_R\mathbf{V}_R^T \tag{2}$$

$\mathbf{A}_R$ is a $M \times N$ matrix of rank R, and it is the best rank R approximation of $\mathbf{A}$ (among the $M \times N$ matrices) with respect to the Frobenius metric. The i-th row of the matrix $\mathbf{U}_R$ may be considered as representative of the i-th word. The columns of the $\mathbf{U}_R$ matrix represent the R independent dimensions of the

$\Re^{\Re}$ space S. Each j-th dimension is weighted by the corresponding value σ_j of Σ_R. Furthermore, each j-th dimension can be tagged by considering the words having the highest module values of u_{ij}. This makes it possible to interpret the space S as a "conceptual" space, according to the procedure illustrated in [1,21].

2.4 LDA-Based Descriptors

In the last years a Bayesian probabilistic model of text corpora, namely the Latent Dirichlet Allocation (LDA), has been proposed with the aim of finding topics in documents [2] by associating a set of words to each topic, obtaining a rough representation of a textual corpus.

One on the main advantages of LDA, like LSA, is the fact that the approach is completely unsupervised The only thing required by LDA to setup a priori is the number N of topics to extract. Latent topics are discovered through the identification of sets of words in the corpus that often occur together within documents. LDA is based on a generative process according to these two steps:

- For each topic $n = 1, 2, \cdots, N$, $\phi^{(n)} \sim Dirichlet(\beta)$ is a discrete probability distribution over a fixed Vocabulary constituting the $n-th$ topic distribution, and β is a hyperparameter for the symmetric Dirichlet distribution.
- For each document d_k of the document corpus, $\theta_{d_k} \sim Dirichlet(\alpha)$, which is a symmetric Dirichlet distribution for the specific document d_k over the available topics is computed. θ_{d_k} is a low dimensional coding of d_k in the topic space. For each word w_i belonging to the d_k document, $z_i \sim Discrete(\theta_{d_j})$ and $w_i \sim Discrete(\phi^{(z_i)})$ are being computed, where z_i is the topic index for w_i

The above process leads to the following distribution

$$p(\mathbf{w}, \mathbf{z}, \theta, \phi | \alpha, \beta) = p(\phi|\beta)p(\theta|\alpha)p(\mathbf{z}|\theta)p(\mathbf{w}|\phi_z) \tag{3}$$

where $\mathbf{z}, \theta, \phi$ are the latent variables of interest. In LDA the posterior inference is given by:

$$p(\theta, \phi, \mathbf{z} | \mathbf{w}, \alpha, \beta) = \frac{p(\theta, \phi, \mathbf{z} | \mathbf{w}, \alpha, \beta)}{p(\mathbf{w}|\alpha, \beta)} \tag{4}$$

which represents the learning of the latent variables given the observed data. The above formula is usually computed through variational inference and Gibbs sampling, as reported in literature [2,10,25].

2.5 Emotion Detection Module

This module deals with the detection of emotions in tweets. For the emotional labeling of tweets, we have considered the six Ekman basic emotions: *anger, disgust, fear, joy, sadness* and *surprise* [12], exploiting an emotions lexicon obtained from the Word-Net Affect Lexicon, as described in [22,23] and adopting a procedure that has been illustrated in [20], which we briefly recap below.

The methodology is based on LSA and starts from the fact that any text d can be mapped into a Data Driven "conceptual" space in the sense illustrated above, by computing a vector $\mathbf{d}$ whose i-th component is the number of times the i-th word of the vocabulary, corresponding to the i-th row of $\mathbf{U}_R$, appears in d. This leads to the mapping of the text as:

$$\mathbf{d}_R = \mathbf{d}^T \mathbf{U}_R \Sigma_R^{-1} \tag{5}$$

The emotional lexicon has been split into six lists, each one associated to one of the basic Ekman emotions {$anger$, $disgust$, $fear$, joy, $sadness$, $surprise$}. Fixed an emotion e, a set of 300 artificial sentences has been built by using five randomly selected words belonging to the list related to e. This procedure has been done for each list associated with a fundamental Ekman emotion, leading to a set of 1800 artificial sentences. Furthermore, all the 1542 words of the lexicon have been considered. Each one of the 3342 (i.e. 1542 + 1800) b texts associated with an emotion e has been mapped into the data driven "conceptual space" induced by TSVD according to the transformation in Eq. (2). The above procedure leads to a cloud of 3342 (i.e. 1542 + 1800) vectors that have been used to map a tweet from the conceptual space to the emotional space. In particular, we have six sets $E_{anger}, E_{disgust}, \cdots, E_{surprise}$ of vectors constituting the sub-symbolic coding of the words belonging to the lexicon for a particular emotion together with their artifact sentences. The generic vector belonging to one of the sets will be denoted in the following as $\mathbf{b}_i^{(e)}$ where $e \in \{$ "$anger$", "$disgust$", "$fear$", "joy", "$sadness$", "$surprise$" $\}$ and i is the index that identifies the i-th $\mathbf{b}_i^{(e)}$ in the e set. Specifically, $\mathbf{b}_i^{(e)}$ is computed as:

$$\mathbf{b}_i^{(e)} = \mathbf{b}^T \mathbf{U}_R \Sigma_R^{-1} \tag{6}$$

where $\mathbf{b}$ is, time by time, the vector computed starting from one of the 3342 textual artifacts b according to the procedure illustrated at the beginning of this section.

Analogously, any textual content t of a tweet can be mapped into the Data Driven "conceptual" space by computing a vector $\mathbf{t}$ whose i-th component in the number of times the i-th word of the vocabulary, corresponding to the i-th row of $\mathbf{U}_R$, appears in t. This leads to the mapping of the tweet as:

$$\mathbf{t}_R = \mathbf{t}^T \mathbf{U}_R \Sigma_R^{-1} \tag{7}$$

Once the tweet t is mapped into the "conceptual" space as a vector $\mathbf{t}_R$, it is possible to compute its emotional fingerprint by exploiting the vectors $\mathbf{b}_i^{(e)}$, which act as "beacons" for the vector $\mathbf{t}_R$, helping in finding its position inside the conceptual space.

In particular, fixed $\mathbf{t}_R$, for each set E_e it is computed the weight:

$$w_e = \max cos(\mathbf{t}_R, \mathbf{b}_i^{(e)}) \tag{8}$$

once all the six w_e weights are computed, the vector $\mathbf{f}_t$, associated to the vector $\mathbf{t}_R$, and by consequence to the tweet t, is calculated as:

$$\mathbf{f}_t = \left[\frac{w_{(anger)}}{\sqrt{\sum_e w_e^2}}, \frac{w_{(fear)}}{\sqrt{\sum_e w_e^2}}, \cdots, \frac{w_{(surprise)}}{\sqrt{\sum_e w_e^2}} \right] \tag{9}$$

The vector $\mathbf{f}_t$ finally constitutes the *emotional fingerprint* of the tweet t in the emotional space. The emotional space is therefore a six-dimensional hypersphere where all tweets can be mapped and grouped. We call the fingerprint $\mathbf{f}_t$ "emoxel", analogously as the *knoxel* in the conceptual space paradigm [4].

2.6   Conversational Engine

The conversational engine exploits a speech-recognition module which makes use of the Google speech recognition APIs; after that the speech-to-text task is performed, the recognized string is sent to a dialogue manager. A set of question-answer rules are set-up into the conversation engine in order to start a conversation that leads to the detection of the user interests by transparently invoking the most adequate procedures which analyze the social network posts of the new acquaintance.

The conversational agent engine allows for a natural human-robot interaction. The conversational module is based on a Rivescript engine, which is a simple scripting language for realizing chatbots and other conversational entities.

We have chosen this kind of engine because of the following interesting features: it is plain text, line-based scripting language, simple to learn, quick to type, and easy to read and maintain [19]. The syntax required to build a Rivescript "knowledge base" is very simple: Question-Answers pairs are encoded in plain text; it is easy to write a set of rules that can be combined to build effective conversational agents; its core library is focused on rendering responses, and it is straightforward to make custom modules and scripts; last but not least is an Open Source tool released under the MIT license [24].

The choice of such an engine allows us to easily connect it to other kind of robots or other kind of services. As a matter of fact, the conversational engine is invoked through a REST service and the answer is delivered to the user after its processing.

A Rivescript knowledge base is made up of *Triggers/Replies* pairs. *Triggers* are identified by a "+" sign, while *Replies* are denoted by a "−" sign.

For example:

```
+ hi
- Hello there, my name is SocialRobot,
  please could you tell me your twitter ID?
```

the above pair makes it possible that whenever the user says "Hi", the conversational engine replies with "Hello there, my name is SocialRobot, please could you tell me your twitter ID?".

At the beginning of the conversation, a specific Rivescript *Topic* is activated. Topics are logical groupings of triggers. When the conversation is bound in a topic, what the user says can only match triggers that belong to the activated topic [19].

The topic is aimed at entertain a general conversation while the robot peeks the tweets of the user trying to roughly identify the subjects that interest the user and those that specifically trigger joyful emotions. Once the predominant subject has been identified, the robot activates another Rivescript Topic which is of particular interest for the user, trying to establish an empathetic engagement with the new acquaintance.

As an example, let us say that the system finds that, among the different higher level categories of the IAB taxonomy, the user is particularly interested in the "Automobiles" topic and that some of his tweets show the "joy" emotion for that topic, the conversation will be switched to the "Automobile" Topic and specific sentences will be said by the robot in order to catch the user attention and empathy, like "Great! with my superpowers I can see that you like automobiles. I like the *brand* automobiles! Which one do you prefer?"

3 Experimental Setup

For training the system, a set of tweets has been retrieved in order to build a semantic space S where to map tweets. In particular, we have retrieved and downloaded 254235 unique tweets by using as keywords the name of the 704 items of the IAB taxonomy, grouped for the 30 Tier 1 categories C_i (*Automotive, Books and Literature, Business and Finance, Careers, Education, Events and Attractions, Family and Relationships, Fine Art, Food and Drink, Healthy Living, Hobbies and Interests, Home and Garden, Medical Health, Movies, Music and Audio, News and Politics, Personal Finance, Pets, Pop Culture, Real Estate, Religion and Spirituality, Science, Shopping, Sports, Style and Fashion, Technology and Computing, Television, Television, Travel, Video Gaming*), with an average of 8474 tweets for category.

We have added to the previous dataset of tweets a dataset of 75078 other tweets that has been used in [20] for computing the emotional fingerprints of Twitter posts. This led to a dataset of 329313 tweets used to build a semantic space, based on LSA, of 300 dimensions, where both user tweets and conceptual labels can be mapped in order to catch an idea of the user interest which make her somehow joyful.

The tweets of each Tier 1 category have been "synthesized" through both LSA and LDA procedures in order to obtain a set of conceptual descriptors of each category. At present, we have chosen to have only a coarse description of each category C_i. In particular, for each category we have used the tweets which have been retrieved by using the name of the items belonging to each specific category through the Twitter APIs. Subsequently, we have induced both an LSA micro space of 20 dimensions and a set of 20 LDA based topic descriptors. The axes of the LSA i-th microspace have been labeled with a set of 10 words, and the same number of words have been used for the LDA topic. This allows to code each Tier1 category with a set of 40 "artificial sentences" constituted by the descriptors of the LSA based conceptual axes of the microspace and 20 random topics induced by the LDA procedure.

These 40 artificial sentences have been mapped in the semantic space S, all together representing the main conceptual descriptors of the C_i Tier1 Category taken into account. The emotional fingerprints of the user tweets have been obtained through the procedure described in [20].

Once the sub-symbolic conceptual representations of the categories for all the 30 Tier1 categories has been created, the system is ready to catch some relevant information about the interests of the new acquaintance.

The system asks the user for her twitter ID and retrieves a sample of her tweets that are available for that account, according to the Twitter API policy and limitations.

Once the tweets have been retrieved, another LSA microspace and another set of LDA topics is computed by using only the tweets of the user. This allows to obtain the main conceptual descriptors of the user tweets. The procedure is analogous to that one used for computing the conceptual descriptors of the i-th Tier 1 category, leading to a set of 40 "artificial sentences" describing the main conceptual topics of the user, which are also mapped into the S semantic space.

At the same time, all the tweets of the user are labeled with the emotional fingerprint procedure in order to extract those that express a "joy" emotion, i.e. whose joy component of the emoxel is above a given threshold $\theta_j oy$ experimentally determined. Those tweets are saved in a set J.

Fixed a Tier 1 category C_i, for every one of the 40 C_i descriptors and the 40 user tweets descriptors, their similarity value is computed and the maximum value obtained is retained. The average value α_i of the 40 obtained maximum values is then computed.

The procedure is repeated for all the conceptual representations C_i of the 30 Tier-1 categories. Once this operation is completed for all the categories, we retain the five topics that are the best candidates as being joyful for the user, i.e. the five categories whose conceptual sub-symbolic description is closer to the user tweets conceptual sub-symbolic description.

After that the five candidate topics have been identified, all the tweets of the user belonging to J are mapped in the S space and compared with the conceptual descriptors of the five Tier1 candidate categories.

For every one of the five selected C_i categories we then calculate the coefficient β_i, which represents the number of conceptual axes of the i-th IAB Tier1 category which are closer than a given threshold θ_β to a tweet belonging to the set J.

Furthermore, we compute also the coefficient γ_i as the average value of the cosine between all the joyful tweets that are closer to each conceptual axis of the i-th IAB Tier1 category according to an experimental threshold θ_γ.

Each IAB Tier1 category is then ranked according to this formula:

$$s_i = \alpha_i \cdot \beta_i \cdot \gamma_i \tag{10}$$

The i-th category with the highest s_i value triggers the topic to be activated in the conversational engine in order to catch the user attention, trying to establish an empathetic relationship.

As a case of study, we report here two examples: one regarding Donald Trump and the other one regarding Barack Obama. The results are illustrated in the following subsections:

3.1   Barack Obama Profile

We have retrieved a sample of 3241 tweets from 12 Aug 2014 to 15 Jan 2018 from the *BarackObama* tweet account. The methodology identified 134 tweets strongly characterizing an emotion (i.e. tweets with an emotional component larger than the value 0.8): in particular the system detected 11 tweets with an "anger" emotional content, 107 for joy and 34 for sadness.

In Table 1 we have reported the top IAB Tier 1 categories regarding Obama according to the formula 10. The main category on which the robot will trigger the attention will be that one related to the "Medical Health" IAB Tier1 Category; therefore the robot will say:

```
Hi Obama, I see that you are particularly interested into to the
Medical Health topic.  I am concerned to this topic too.
Would you talk about it?
```

Table 1. Barack Obama top IAB Tier 1 detected categories

Class	α	β	γ	Score s
Medical Health	0.409	28	0.585	6.698
Family and Relationship	0.386	20	0.500	3.856
Business and Finance	0.412	18	0.437	3.239
Careers	0.388	13	0.359	1.810
Real Estate	0.377	10	0.322	1.215

3.2   Donald Trump Profile

We have retrieved a sample of 3234 tweets from 15 Sept 2016 to 14 Nov 2017 from the *realDonaldTrump* tweet account. The methodology identified 148 tweets strongly characterizing an emotion (i.e. tweets with an emotional component larger than the value 0.8): in particular the system detected 10 tweets with an "anger" emotional content, 5 for the "fear" emotion, 141 for joy and 66 for sadness.

In Table 2 we have reported the top IAB Tier 1 categories regarding Donald Trump according to the formula 10. The main category on which the robot will trigger the attention will be that related to the "News and Politics" IAB Tier1 Category Table 1; therefore the robot will say:

Hi Donald, I see that you are particularly interested to
the News and Politics topic. This topic interests me very much.
Is there something in particular that concerns you?

Table 2. Donald Trump top IAB Tier 1 detected categories

Class	α	β	γ	Score s
News and Politics	0.437	30	0.533	6.986
Business and Finance	0.444	27	0.480	5.754
Pop Culture	0.363	26	0.518	4.889
Family and Relationships	0.361	24	0.491	4.251
Real Estate	0.404	12	0.362	1.753

4 Conclusions and Future Works

We have presented a preliminary work on a system that tries to catch the attention of a new acquaintance with the aim of establishing a first engagement with the user.

The system uses both LSA and LDA descriptors, as well as an emotion detection module to reach this goal. A conversational engine guides the initial process and continues with t the conversation.

Many issues have to be enhanced, starting from a more fine grained classification which should be also fast and reliable, the selection of specific entities that can catch in a more effective manner the attention of the user, as well as the automatic generation of conversational statements starting from the user tweets.

References

1. Agostaro, F., Augello, A., Pilato, G., Vassallo, G., Gaglio, S.: A conversational agent based on a conceptual interpretation of a data driven semantic space. In: Bandini, S., Manzoni, S. (eds.) AI*IA 2005. LNCS (LNAI), vol. 3673, pp. 381–392. Springer, Heidelberg (2005). https://doi.org/10.1007/11558590_39
2. Blei, D., Ng, A., Jordan, M.: Latent Dirichlet allocation. J. Mach. Learn. Res. **3**, 993–1022 (2003)
3. Brethes, L., Menezes, P., Lerasle, F., Hayet, J.: Face tracking and hand gesture recognition for human-robot interaction. In: IEEE International Conference on Robotics and Automation, vol 2, pp 1901–1906. IEEE (2004)
4. Chella, A., Frixione, M., Gaglio, S.: A cognitive architecture for robot self consciousness. Artif. Intell. Med. **44**(2), 147–154 (2008)
5. Cannataro, M., Cuzzocrea, A., Pugliese, A.: A probabilistic approach to model adaptive hypermedia systems. In: 1st International Workshop on Web Dynamics, in conjunction on ICDT 2001 (2001)

6. Corrigan, L.J., Peters, C., Küster, D., Castellano, G.: Engagement perception and generation for social robots and virtual agents. In: Esposito, A., Jain, L.C. (eds.) Toward Robotic Socially Believable Behaving Systems - Volume I. ISRL, vol. 105, pp. 29–51. Springer, Cham (2016). https://doi.org/10.1007/978-3-319-31056-5_4

7. Cuzzocrea, A., Fortino, G., Rana, O.: Managing Data and processes in cloud-enabled large-scale sensor networks: state-of-the-art and future research directions. In: 13th IEEE/ACM International Symposium on Cluster, Cloud, and Grid Computing, CCGrid 2013, pp. 583–588 (2013)

8. D'Avanzo, E., Pilato, G.: Mining social network users opinions' to aid buyers' shopping decisions. Comput. Hum. Behav. **51**, 1284–1294 (2014)

9. D'Avanzo E., Pilato G., Lytras M.D.: Using Twitter sentiment and emotions analysis of Google trends for decisions making. Program **51**(3), 322–350 (2017)

10. Darling, W.M.: A theoretical and practical implementation tutorial on topic modeling and gibbs sampling. In: Proceedings of the 49th Annual Meeting of the Association for Computational Linguistics: Human Language Technologies, pp. 642–647, 1 Dec 2011

11. Delaherche, E., Dumas, G., Nadel, J., Chetouani, M.: Automatic measure of imitation during social interaction: a behavioral and hyperscanning-EEG benchmark. Pattern Recognit. Lett. **66**, 118–126 (2015)

12. Ekman, P., Friesen, W.V.: Constants across cultures in the face and emotion. J. Pers. Soc. Psychol. **17**, 124 (1971)

13. Interactive Advertising Bureau (IAB) Contextual Taxonomy. http://www.iab.net/. Retrieved December 2017

14. Kanagasabai, R., Veeramani, A., Ngan, L.D., Yap, G.E., Decraene, J., Nash, A.S.: Using semantic technologies to mine customer insights in telecom industry. In: International Semantic Web Conference (Industry Track) (2014)

15. Landauer, T.K., Dumais, S.T.: A solution to Plato's problem: the latent semantic analysis theory of acquisition, induction, and representation of knowledge. Psychol. Rev. **104**(2), 211–223 (1990)

16. Landauer, T.K., Foltz, P.W., Laham, D.: An introduction to latent semantic analysis. Discourse Process. **25**, 259–284 (1998)

17. Liu, B.: Sentiment analysis and subjectivity. In: Indurkhya, N., Damerau, F.J. (eds.) Handbook of Natural Language Processing, pp. 627–665. CRC Press (2010)

18. Pang, B., Lee, L., Vaithyanathan, S.: Thumbs up? sentiment classification using machine learning techniques. In: Proceedings of the ACL-02 Conference on Empirical Methods in Natural Language Processing, vol. 10, pp. 79–86. Association for Computational Linguistics (2002)

19. Petherbridge, N.: Artifical Intelligence Scripting Language. Rivescript.com

20. Pilato, G., D'Avanzo, E.: Data-driven social mood analysis through the conceptualization of emotional fingerprints. Procedia Comput. Sci. **123**, 360–365 (2018)

21. Santilli, S., Nota, L., Pilato, G.: The use of latent semantic analysis in the positive psychology: a comparison with Twitter posts. In: 2017 IEEE 11th International Conference on Semantic Computing (ICSC), pp. 494–498. IEEE (2017)

22. Strapparava, C., Mihalcea, R.: Semeval-2007 task 14: affective text. In: Proceedings of the 4th International Workshop on Semantic Evaluations, pp. 70–74. Association for Computational Linguistics (2007)

23. Strapparava, C., Mihalcea, R.: Learning to identify emotions in text. In: SAC 2008 Proceedings of the 2008 ACM Symposium on Applied Computing (2008)

24. Siddharth, G., Borkar, D., De Mello, C., Patil, S.: An E-Commerce Website based Chatbot. Proc. (IJCSIT) Int. J. Comput. Sci. Inf. Technol. **6**(2), 1483–1485 (2015)
25. Teh, Y.W., Newman, D., Welling, M.: A collapsed variational Bayesian inference algorithm for latent Dirichlet allocation. NIPS **6**, 1378–1385 (2006)
26. Terrana, D., Augello, A., Pilato, G.: Facebook users relationships analysis based on sentiment classification. In: Proceedings of 2014 IEEE International Conference on Semantic Computing (ICSC), pp. 290–296 (2014)

Prediction of the Energy Demand of a Hotel Using an Artificial Intelligence-Based Model

José-Luis Casteleiro-Roca[1,2], José Francisco Gómez-González[3],
José Luis Calvo-Rolle[1], Esteban Jove[1,2(✉)], Héctor Quintián[1],
Juan Francisco Acosta Martín[3], Sara Gonzalez Perez[3],
Benjamin Gonzalez Diaz[3], Francisco Calero-Garcia[1,2,3,4],
and Juan Albino Méndez-Perez[2]

[1] Department of Industrial Engineering, University of A Coruña, A Coruña, Spain
esteban.jove@udc.es
[2] Department of Computer Science and System Engineering,
Universidad de La Laguna, La Laguna, Spain
jamendez@ull.edu.es
[3] Department of Industrial Engineering, Universidad de La Laguna Engineering,
La Laguna, Spain
[4] Departamento de Economia, Contabilidad y Finanzas,
Universidad de La Laguna, La Laguna, Spain

Abstract. The growth of the hotel industry in the world, is a reality that increasingly needs a greater use of energy resources, and their optimal management. Of all the available energy resources, renewable energies can give greater economic efficiency and lower environmental impact. To manage these resources it is important the availability of energy prediction models. This allows managing the demand for power and the available energy resources, to obtain maximum efficiency and stability, with the consequent economic savings. This paper focuses in the use of Artificial Intelligence methods for energy prediction in luxury hotels. As a case of study, the energy performance data used were taken from the hotel complex The Ritz-Carlton, Abama, located in the South of the island of Tenerife, in the Canary Islands, Spain. This is a high complexity infrastructure with many services that require a lot of energy, such as restaurants, kitchens, swimming pools, vehicle fleet, etc., which make the hotel a good study model for other resorts. The model developed for the artificial intelligence system is based on a hybrid topology with artificial neural networks. In this paper, the daily power demand prediction using information of last 24 h is presented. This prediction allows the development of appropriate actions to optimize energy management.

Keywords: Artificial intelligence · Artificial neural network · Hotel Tourism

© Springer International Publishing AG, part of Springer Nature 2018
F. J. de Cos Juez et al. (Eds.): HAIS 2018, LNAI 10870, pp. 586–596, 2018.
https://doi.org/10.1007/978-3-319-92639-1_49

1   Introduction

The tourism sector as an industrial activity is considered of great importance from several points of view: economic, energetic, ecological, cultural, socio-political, etc. The hotel sector plays a capital role in the tourism development and energy demand profile, affecting the entire region where it is located. All this leads us to the fact that any scenario we can pose for the study of energy consumption and the incorporation of renewable energies in the tourism sector should be mainly focused on the study of the hotel sector.

So far, some tools have been defined for energy forecasting and management with the purpose of reducing the energy consumption but with disparate results. It has been seen that indicators of the intensity of energy use, such as the energy consumed per useful area during a given time, are not good indicators for the management of energy in hotels [1]. Others have tried to correlate the consumption of electricity with several factors, such as the number of rooms, number of workers, guests, etc., but they have not found indexes that serve all hotels [2–4].

In last years, artificial intelligence (AI) based on predictive models have been applied to forecast the energy demand and generation [5–9]. The AI predictive models embedded in smart meters analyze and suggest actions in seconds, shifting the energy demand, reducing cost and improving the grid efficiency, without changing consumer requirements.

In the forecasting energy consumption field to contribute to design demand-side management strategies. In [10] a model based on an Artificial Neural Network (ANN) to forecast building energy consumption is proposed. Recently, Muralitharan et al. [11] compared different Neural Network for the energy demand prediction in smart grid. In the same sense, other authors [12,13], used fuzzy logic to calculate the load curve of a residential consumer. All this shows the great interest that awakened by having an adequate model of load demand prediction, nowadays. This would allow a better demand management, costs reduction, improvement of the grid efficiency and increasing the penetration of renewable energy sources.

This paper aims at developing a machine learning methodology to predict the short-term electricity demand of a hotel, at least, 12 h in advance, based on an ANN, which will be trained with previous consumption information. The proposed hybrid model is able to predict energy consumption and would be helpful for decision making for an optima energy management. To develop this work, data of a luxury 5 star hotel and resorts from Tenerife, Canary Islands, Spain, is used.

This paper is organized as follows: in Section 'Case of study' the location of the hotel and its energy demand profile is described; in Section 'Model approach' the predictive model is defined; and finally in sections 'Results' and 'Conclusions and future works', the comparison between the real energy demand and the predicted demand is compared and analyzed.

2   Case of Study

This case of study is based on a luxury 5 start hotel called The Ritz-Carlton Abama, in the south of Tenerife, Canary Islands (Spain) located on the Atlantic Ocean (28.100°N 15.400°W).

In Canary Islands, the tourism sector represented 34.3% of gross domestic product (GDP) in 2016 and generates 39.7% of employment [14]. In the island of Tenerife, foreign tourists staying on the island recorded an increase of +11% in 2016 and the average duration of the stays of the tourism hosted was 7.54 days. Until July 2016, 68% of the lodged did so in hotels, compared to 32% that were extra-hoteliers. Therefore, one can intuit the key importance that the hotel sector plays in tourism development. According to the number of hotel places in Tenerife, the 74% of places have 4 or more stars (data obtained from information from the Canary Islands Statistics Institute).

The daily power demand of hospitality sector and a single hotel (our case of study) is shown in Fig. 1. From the figure, it could be notice that the demand profile of the hotel, our case of study, is very similar to the hospitality sector. The data corresponding to the hotel under study, have been provided by those responsible for the hotel.

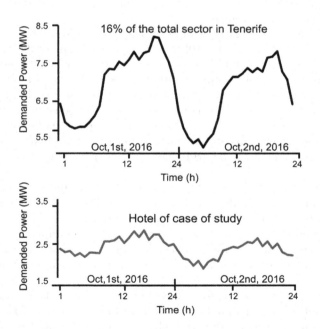

Fig. 1. Power demand profile in hospitality sector. Elaborated with information provided by ENDESA and The Ritz-Carlton Abama hotel.

2.1   Description of the Hotel

The hotel has 461 rooms, on the one hand, distributed in a main building or citadel (313 rooms), and on the other, in four residential streets or villas (148 rooms). It also has 10 restaurants and a 1,500 m² meeting room. The common and exterior areas of the complex consist of 8 pools with a total water volume of almost 3,300 m³. In addition, the entire hotel zone is distributed in almost 95,000 m², so there is a fleet of electric vehicles of 116 units, for the movement of guests and for the provision of services. In addition to the horizontal displacement, the hotel has 10 floors, so it has 36 elevators both for the movement of guests, as for cleaning services, room service and maintenance. The need for air-conditioning in common areas and rooms extends almost all year round. On the other hand, the guests are guaranteed an average temperature of 26 °C in all the pools, so that the energy demand increases in those months in which the temperature of the environment is lower than 26 °C.

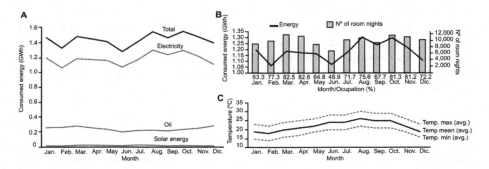

Fig. 2. Energy consumed and occupancy in 2017 in the hotel under study. (A) total energy consumed with different energy sources. (B) Electrical energy consumed and occupancy in 2017 in the hotel under study. (C) The average exterior temperatures (maximum, average and minimum) throughout the year 2017 are taken from the nearest meteorological station, that is, at the Tenerife South Airport (GCTS) (Source www.wunderground.com).

The types of sources of energy used in the hotel are: electricity from the power grid, gas-oil, and thermal solar. The energy from the power grid is used for: air conditioning and ventilation, kitchens and restaurants, pumps, elevators, lighting and electric vehicles. The gas-oil is used for: hot water, heating indoor SPA pool, and vehicles. Finally, the thermal solar energy is used for the hot sanitary water (HSW). The Fig. 2A shows the average energy consumed in the hotel during 2017 for the different sources. We can notice that the main consumed energy comes from the electrical energy from the power grid.

Regarding the consumption of electrical energy, Fig. 2B shows the occupancy percent of the hotel and the same time, the consumed electrical energy during 2017. As expected, the highest the demand, the highest the electric power demand. In addition, we are able to notice also that although the occupancy percent is very similar around the month of March and October, the consumed electrical energy is higher in this second, likely, due to the higher temperature in that period (Fig. 2C). From these data, we can estimate that on average a room night spends 118 KWh of electrical energy (or 144 KWh in total taking account the rest of energy sources).

An adequate energy management in a hotel implies the characterization of the state of energy efficiency and environmental impact of the hotel. The global potential for decreasing consumption, energy costs and environmental impacts of the hotel must be determined [15, 16]. With this aim, different basic actions can be taken into consideration, such as automated lighting or engine shutdown policies. Due to the complexity involved in the energy management of a hotel, it is very important to have a centralized computing center with an energy Building Management System (BMS) or Building Management and Control Systems (BMCS) [17–20].

In order to plan the energy management actions that will be carried out in the hotel in an automated way according to the expected energy demand, it will be necessary to have an intelligent system with a predictive model that indicates with sufficient anticipation the behavior of the power demand. In this sense, models developed with artificial intelligence can provide the information about the future energy profile based on the available information and historical data to the BMS system.

3   Hybrid Model Approach

The aim of this research is to predict the power consumption of the hotel under study to optimize its energy operation. The hybrid model approach proposed to achieve this goal is presented in the Fig. 3. In this scheme there are 96 models to predict the consumption registered during a day. The input for all the models, are the power consumption the day before the predicted one.

As the data acquisition system used in the hotel records the power with a sample rate of 15 min, there are 96 measures that represents the consumption over a day. Each model uses the previous 96 measures to predict the power consumption the next day; then, with the 96 models, it is possible to predict the power consumption for all the next day (each model predict the power consumption at an specific instant). As the personal for the hotel needs time to analyze the predicted future consumption, the models are trained with data from 20:00 to 19:45, to predict 24 h of power consumption starting next day at 08:00 (Fig. 4).

The used dataset represents data from 376 days, and it is based on the mentioned power measure recording system.

A Multi-Layer Perceptron (MLP) is the most known feedforward Artificial Neural Network due to its simple configuration and its robustness [21]. Despite of that fact, the MLP architecture must be carefully chosen in order to achieve satisfactory results. MLP is made of one input layer, one or more hidden layers and one output layer. The layers have neurons with an activation function.

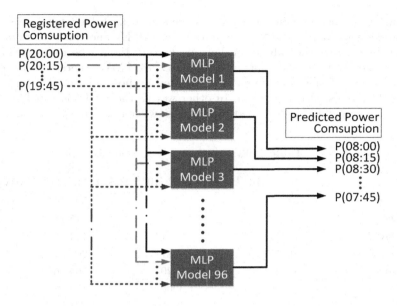

Fig. 3. Hybrid model approach.

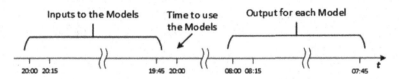

Fig. 4. Time distribution of the used variables.

In a typical configuration, all neurons in a layer have the same activation function, but this is not a restriction. This function could be step, linear, log-sigmoid or tan-sigmoid. Its successful performance has been proven in many different applications [22–31].

4   Results

The model was obtained using the previous power consumption as inputs, and several future power consumption are predicted. To validate the model, data of eight days were separate from the initial dataset; the rest of data (368 days) is used to train the models.

The configuration for all the 96 internal models are the same; they are MLP with one hidden layer with a fix number of neurons and tan-sigmoid as the activation function. In the output layer, as the MLP is used for regression, it is selected the linear activation function. The optimization algorithm used was Levenberg-Marquardt; gradient descent was used to finish the training phase, and the performance function was set to mean squared error.

The results of the validation process are shown in Table 1 where the Mean Absolute Error (MAE), the Mean Square Error (MSE), and the Maximum Error (Max.), for each of the validation days are shown. From where, the average of MAE of the eight-day data set, is around 26×10^{-3} MW, therefore, if the contracted power of the hotel (1.84 MW) is used as power base, the normalized averaged MAE is 0.0141 (1.41%). In other words, the model is capable of predicting the power demanded in advance of 24 h with an accuracy of around 98%.

Table 1. Errors achieve by the model.

	MAE (10^{-3} MW)	MSE (10^{-3} MW^2)	Max.(10^{-3} MW)
Day 1	57.391	14.214	597.34
Day 2	23.342	2.7553	215.12
Day 3	23.618	2.9911	224.68
Day 4	19.649	2.0681	203.61
Day 5	10.874	7.9320	164.66
Day 6	18.274	1.5006	174.38
Day 7	39.412	5.6905	221.09
Day 8	17.507	1.4570	152.10
Mean	26.258	3.9337	244.12
Normalized mean	1.43%	0.12%	13.24%

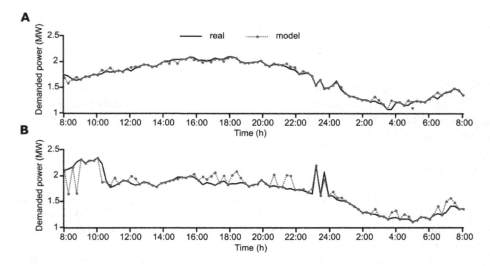

Fig. 5. Predicted power demand by the model. The best (A) and the worst (B) validation days - Real consumption in black continuous line, and predicted consumption in gray dashed line.

Two examples of the expected demand profile are shown in the Fig. 5. In this case, the best (Fig. 5A) and worst (Fig. 5B) case of the power demand prediction of the eight-day set used to test the model is presented. If we compare the two real demand profiles, it can be seen that there is a different behavior in both. Presumably, the profile of Fig. 5A is more common inside the daily profile set of the year selected to the training, so the model could better adjust this profile. The profile shown in Fig. 5B is a day with a very high demand early in the morning, that is unusual in the regular hotel demand profile. The model accuracy in this case is not as good as in Fig. 5A, however the prediction is able to include the main trend for the next 24 h. This suggests the need to incorporate extra information about the hotel's activity and weather data, in order to be able to pick up in the model those causes that could generate a different demand profile at a certain moment.

5   Conclusions and Future Works

An energy demand model was proposed for a touristic resort using artificial intelligence techniques. Its structure is composed of 96 sub-models based on Artificial Neural Network in charge of the prediction of energy demand in the resort. The final aim of the research is to create a complete hybrid model with an optimize one for each predicted power. In this paper, to achieve our first result in this field, all the used models have the same number of neurons in the hidden layer of each ANN model. To select the neuron's number, some tests had been made for different points, the conclusion was to use 5 neurons for this initial work with the aim of improving the global model in a future. As each model predicts only the power consumption in a specific instant, the 96 models used internally are different between each others.

The model trained with real data was able to predict the energy demand for the next 24 h. Results obtained were satisfactory when compared to the real demand. The Mean Absolute Error achieved was 0.0263 MW, but this value must be considered according to the installation high power demand. The value of the Mean Absolute Percentage Error is 1.43% with an averaged maximum error of 13.24% attest the accuracy of the proposed model.

This new model, based on Artificial Neural Network, could be a good solution for the prediction of residential complex power demand. In particular, it is a valuable tool to improve the energy management in a hotel, reduce energy consumption and strengthen environmental sustainability.

Concerning the prediction methods, different alternatives can be found in the literature such as autoregressive integrated moving average model, artificial neural-network model, multiple linear regression, etc. In the present work, we have chosen a Multi-Layer Perceptron (MLP) based model which have given similar results than Support Vector Machine in others works. In particular, the control of electrical energy in the hotel under study is currently based on a prediction model based in a linear correlation by least squares, where consumption (Energy daily prediction EDP) is directly proportional to the Cooling Degree

Days (CDD), calculated by BizEE Software Limited and occupation. The new proposal presented here offers a great potential to improve the accuracy of the existing prediction method in the hotel.

In future works, new improvements in the model with the inclusion of additional input variables that collects the state of the hotel (e.g. occupancy), weather conditions and planned hotel activity will be studied. It would be interesting to design a method to choose the best time window for the model input to improve its performance. Also, alternative model proposals for each output can be evaluated to improve the global accuracy obtained.

Acknowledgments. This study was supported by CajaCanarias Foundation with the project PR705752 (GreenTourist, 2016TUR17) and The Ritz-Carlton Abama Hotel in Tenerife, Spain.

References

1. Deng, S.M., Burnett, J.: Study of energy performance of hotel buildings in Hong Kong. Energ. Build. **31**(1), 7–12 (2000)
2. Papamarcou, M., Kalogirou, S.: Financial appraisal of a combined heat and power system for a hotel in Cyprus. Energ. Convers. Manag. **42**(6), 689–708 (2001)
3. Priyadarsini, R., Xuchao, W., Eang, L.S.: A study on energy performance of hotel buildings in Singapore. Energ. Build. **41**(12), 1319–1324 (2009)
4. Cabello Eras, J.J., Sousa Santos, V., Sagastume Gutiérrez, A., Guerra Plasencia, M.Á., Haeseldonckx, D., Vandecasteele, C.: Tools to improve forecasting and control of the electricity consumption in hotels. J. Clean. Prod. **137**, 803–812 (2016)
5. Suganthi, L., Samuel, A.A.: Energy models for demand forecasting: a review. Renew. Sustain. Energ. Rev. **16**, 1223–1240 (2012)
6. Singh, A.K., Khatoon, S.: An overview of electricity demand forecasting techniques. In: National Conference on Emerging Trends in Electrical, Instrumentation & Communication Engineering (2013)
7. Shao, Z., Chao, F., Yang, S.L., Zhou, K.L.: A review of the decomposition methodology for extracting and identifying the fluctuation characteristics in electricity demand forecasting (2017)
8. Khosravani, H., Castilla, M., Berenguel, M., Ruano, A., Ferreira, P.: A comparison of energy consumption prediction models based on neural networks of a bioclimatic building. Energies **9**, 57 (2016)
9. Torres, J.M., Aguilar, R., Aguilar, R.M., Zúñiga, K.V.: Deep learning to predict the generation of a wind farm. J. Renew. Sustain. Energ. **10**, 013305 (2018)
10. Neto, A.H., Fiorelli, F.A.S.: Comparison between detailed model simulation and artificial neural network for forecasting building energy consumption. Energ. Build. **40**, 2169–2176 (2008)
11. Muralitharan, K., Sakthivel, R., Vishnuvarthan, R.: Neural network based optimization approach for energy demand prediction in smart grid. Neurocomputing **273**, 199–208 (2018)
12. Zúñiga, K.V., Castilla, I., Aguilar, R.M.: Using fuzzy logic to model the behavior of residential electrical utility customers. Appl. Energ. **115**, 384–393 (2014)
13. Abreu, T., Alves, U.N., Minussi, C.R., Lotufo, A.D.P., Lopes, M.L.M.: Residential electric load curve profile based on fuzzy systems. In: 2015 IEEE PES Innovative Smart Grid Technologies Latin America (ISGT LATAM), pp. 591–596. IEEE, October 2015

14. EXCELTUR: EXCELTUR, Alliance for Excellency in Tourism (2017). http://www.exceltur.org/exceltur-in-english/
15. Hotel Energy Solutions: Energy Efficiency Series: Key Energy efficiency solutions for SME hotels. Hotel Energy Solutions project publications (2011)
16. Proyect, E.: Guía de Eficiencia Energética para instalaciones hoteleras en Canarias. Technical report, Instituto Tecnológico de Canarias, S.A (2009)
17. Brickfield, P., Mahling, D., Noyes, M., Weaver, D.: Automatic energy management and energy consumption reduction, especially in commercial and multi-building systems. US Patent App. 11/889, 513, 24 July 2008
18. Levermore, G.J.: Building energy management systems : applications to low energy HVAC and natural ventilation control. E & FN Spon (2000)
19. Siemens.com Global Website: Hotel Building Management Systems - Hospitality - Siemens (2018). http://w3.siemens.com/market-specific/global/en/hospitality/hotels-resorts-casinos/hotel-energy-efficiency/Building-management-systems/Pages/Building-management-systems.aspx
20. Honeywell International Inc: Honeywell Building Solutions – BMS – Commercial Buildings Distributed Control Systems (2018). https://buildingsolutions.honeywell.com/en-US/Pages/default.aspx
21. Zeng, Z., Wang, J.: Advances in neural network research and applications. Lecture Notes in Electrical Engineering. Springer Publishing Company, Heidelberg (2010). https://doi.org/10.1007/978-3-642-12990-2
22. Casteleiro-Roca, J.L., Jove, E., Sánchez-Lasheras, F., Méndez-Pérez, J.A., Calvo-Rolle, J.L., de Cos Juez, F.J.: Power cell SOC modelling for intelligent virtual sensor implementation. J. Sens. **2017**, 12 (2017)
23. Fernández-Serantes, L.A., Estrada Vázquez, R., Casteleiro-Roca, J.L., Calvo-Rolle, J.L., Corchado, E.: Hybrid intelligent model to predict the SOC of a LFP power cell type. In: Polycarpou, M., de Carvalho, A.C.P.L.F., Pan, J.-S., Woźniak, M., Quintian, H., Corchado, E. (eds.) HAIS 2014. LNCS (LNAI), vol. 8480, pp. 561–572. Springer, Cham (2014). https://doi.org/10.1007/978-3-319-07617-1_49
24. Casteleiro-Roca, J.L., Calvo-Rolle, J.L., Méndez Pérez, J.A., Roqueñí Gutiérrez, N., de Cos Juez, F.J.: Hybrid intelligent system to perform fault detection on bis sensor during surgeries. Sensors **17**(1), 179 (2017)
25. Gonzalez-Cava, J.M., Reboso, J.A., Casteleiro-Roca, J.L., Calvo-Rolle, J.L., Méndez Pérez, J.A.: A novel fuzzy algorithm to introduce new variables in the drug supply decision-making process in medicine. Complexity **2018**, 15 (2018)
26. Jove, E., Gonzalez-Cava, J.M., Casteleiro-Roca, J.L., Pérez, J.A.M., Calvo-Rolle, J.L., de Cos Juez, F.J.: An intelligent model to predict ANI in patients undergoing general anesthesia. In: Pérez García, H., Alfonso-Cendón, J., Sánchez González, L., Quintián, H., Corchado, E. (eds.) SOCO/CISIS/ICEUTE -2017. AISC, vol. 649, pp. 492–501. Springer, Cham (2018). https://doi.org/10.1007/978-3-319-67180-2_48
27. Jove, E., Blanco-Rodríguez, P., Casteleiro-Roca, J.L., Moreno-Arboleda, J., López-Vázquez, J.A., de Cos Juez, F.J., Calvo-Rolle, J.L.: Attempts Prediction by Missing Data Imputation in Engineering Degree. In: Pérez García, H., Alfonso-Cendón, J., Sánchez González, L., Quintián, H., Corchado, E. (eds.) SOCO/CISIS/ICEUTE -2017. AISC, vol. 649, pp. 167–176. Springer, Cham (2018). https://doi.org/10.1007/978-3-319-67180-2_16
28. Quintián, H., Corchado, E.: Beta hebbian learning as a new method for exploratory projection pursuit. Int. J. Neural Syst. **27**(6), 1–16 (2017)

29. Casteleiro-Roca, J.L., Quintián, H., Calvo-Rolle, J.L., Corchado, E., del Carmen Meizoso-López, M., Piñón-Pazos, A.: An intelligent fault detection system for a heat pump installation based on a geothermal heat exchanger. J. Appl. Logic **17**, 36–47 (2016)

30. Calvo-Rolle, J.L., Quintian-Pardo, H., Corchado, E., del Carmen Meizoso-López, M., García, R.F.: Simplified method based on an intelligent model to obtain the extinction angle of the current for a single-phase half wave controlled rectifier with resistive and inductive load. J. Appl. Logic **13**(1), 37–47 (2015)

31. Machón González, I.J., López García, H., Calvo Rolle, J.L.: Neuro-robust controller for non-linear systems (controlador neurorobusto para sistemas no lineales). Dyna (2011)

A Hybrid Algorithm for the Prediction of Computer Vision Syndrome in Health Personnel Based on Trees and Evolutionary Algorithms

Eva María Artime Ríos[1], Fernando Sánchez Lasheras[2],
Ana Suárez Sánchez[3(✉)], Francisco J. Iglesias-Rodríguez[3],
and María del Mar Seguí Crespo[4]

[1] Hospital Universitario Central de Asturias, Oviedo, Spain
[2] Department of Mathematics, University of Oviedo, Oviedo, Spain
[3] Department of Business Administration, University of Oviedo, Oviedo, Spain
suarezana@uniovi.es
[4] Department of Optics, Pharmacology and Anatomy, University of Alicante,
Alicante, Spain

Abstract. In the last decades, the use of video display terminals in workplaces has become more and more common. Despite their remarkable advantages, they imply a series of risks for the health of the workers, as they can be responsible for ocular and visual disorders.

In this research, certain problems associated to prolonged computer use classified under the name of Computer Vision Syndrome are studied with the help of a hybrid algorithm based on regression trees and genetic algorithms. The importance of the different symptoms on the Computer Vision Syndrome is evaluated.

Also, the proposed algorithm is tested in order to know its performance as a prediction model that can determine how prone an individual is to suffering from Computer Vision Syndrome.

Keywords: Hybrid algorithms · Genetic algorithms · Regression tree
Computer vision syndrome

1 Introduction

The introduction of visual display terminals (VDTs) brought new forms of work, management and organization into the world of work, implicating transformations and changes inside the enterprise organization as well as several risks for the health of workers as they can be responsible of ocular and visual disorders. The term "computer vision syndrome (CVS)" is defined as a group of eye and vision-related problems that result from prolonged use of VDTs in the workplace. The most common symptoms associated with CVS are eyestrain, blurred vision, dry eyes and headaches, among others [1]. VDT's use during long periods of time relates to visual intense efforts [2, 3]

© Springer International Publishing AG, part of Springer Nature 2018
F. J. de Cos Juez et al. (Eds.): HAIS 2018, LNAI 10870, pp. 597–608, 2018.
https://doi.org/10.1007/978-3-319-92639-1_50

as well as to changes in the ocular surface and in the condition of the tear film [4, 5]. In addition, during computer work the frequency of blinking decreased, with a consequent increase in the evaporation of the tear film that compromises the good condition of the ocular surface. In Spain, there are two instruments that have recently been developed and validated to measure this syndrome [6, 7].

CVS has been widely studied in the last decade, sometimes under a different name, as asthenopia, visual fatigue, eyestrain, visual strain or ocular symptoms. Different studies have reported diverse prevalence rates of CVS ranging from less than 20% to more than 80% [8–12]. Limitations in comparisons of studies in different populations include differences in the characteristics of the sample, the methodology and the instruments used for data collection. The probability of developing CVS is related to individual and work-related factors, increasing with the daily duration of VDT use (especially when it is longer than 4 h), a reduced number of breaks during the work shift and with work-related prolonged years of use. It can affect any race and is more frequent in women. Several environmental factors were pointed as possible causes of eye symptoms, including indoor air quality, high room temperature, low relative room humidity, lighting conditions, glare or reflections, screen quality or design or the work station [13, 14]. The most common occupations of the studied samples in the revised bibliography include office workers, bank employees, high-tech workers, graphical editors and call centers' operators.

Especially since the introduction of the Electronic Health Records (EHR) in the National Health System, healthcare workers have become users of video display terminals (VDT). However, according to the bibliographical review, until now, no studies have been identified that evaluate the effects on the visual health caused by the exposure to VDT in health personnel. Only two studies carried out in Turkey included a sample of workers from two hospitals who used computers. The first one [15] included secretaries, computer operators and hospital data management system users. The second one [5] did not specify whether they were health personnel or not.

Decision trees learning is one of the simplest, yet most popular methods for inductive inference and optimization. It is able to obtain the best possible values of decision variables based on the selected objective function. In recent times, evolutionary algorithms have been extensively used in various fields for finding near-optimal solutions.

The aim of the present work was to develop a hybrid algorithm based on regression trees and genetic algorithms, and then, to test this model in a sample of health workers in order to know its performance as a model to predict the score achieved by an individual in the Computer Vision Syndrome Questionnaire.

2 Methods

2.1 Genetic Algorithms

Genetic algorithms (GA) are a kind of heuristic search that can be applied to a vast number of optimization problems. The foundations of these algorithms are the evolution of a set of individuals that constitute a set of candidate solutions to an optimization problem [16]. The origin of this technique is the works of Holland in the decade of 1970 [17]. A family of operators called genetic operators produce new solutions in the chosen representation and allow the walk in the solution space.

The mechanic of a GA starts with the creation of an initial population. The most common approach and the one used in the present research is creating a random initial population in order to cover all the solution space [18]. Only in those cases in which it is possible to take advantage of expert knowledge, this is the preferred solution. Please also note that as in our case the GA applied to a variable selection problem, each member of the population is a bit string where 1 means that the variable is going to take part in the model, and 0 that not. This approach has already been used in previous researches [19] for very different kinds of problems.

After setting the initial population, the main generational loop of the GA generates new offsprings candidate solutions by means of crossover and mutation operators until the population is complete.

The crossover operator combines the [20] genetic material of two or more solutions. In the case of the present research, only two different solutions are combined in order to create a new offspring. The mutation operator changes an individual by making random changes in it using all the possible values that are present in the solution space. The use of the mutation operator makes possible the creation of individuals that has a part of information that is completely different from the one of the individuals of the previous generation. In some cases, it allows us to reach an optimum solution faster. In spite of these and in order to guarantee the convergence of the method, it is recommendable to maintain the mutation probability in a relatively low value.

When a new population is created with the help of the mutation and crossover operators, it must be evaluated with the help of the objective function. The objective function measures the quality of the solution the GA has generated. In order to allow the convergence of the algorithm, the best offspring solutions are selected and they become the parents of the next generation of solutions with the help of the crossover and mutation algorithms.

This is another operator that can be used in a GA and that is also employed in the present research. It is called elitism. By means of this operator, some of the best solutions are transferred with no changes from one generation to the next.

Finally, it must be taken into account that the performance of a GA is related to the selection of its parameters and that this selection is linked to the nature of the problem under study. For the present research, and taking into account not only the literature but also our previous experiences, the probability values applied for crossover were those from 0.5 to 1 in steps of 0.1, while the mutation probability employed values of 0.1, 0.2 and 0.3. Finally, and as it was already stated before, relatively smell values of elitism were employed: 0.01, 0.05 and 0.1 as in many other applications in the literature [21].

2.2 Regression Trees

In machine learning, regression analysis seeks to estimate the relationships between an output variable and a set of independent input variables. In those cases, in which researchers are interested in knowing about the variables relationship a methodology such as the regression trees would be considered of great interest [22].

The regression trees use a tree-like structure built by means of an iterative process. In a regression tree the internal nodes are labelled with a set of mutually exclusive conditions on the input variables $(x_1, x_2, ..., x_n)$.

For each condition there is an associated edge that leads to a unique child node. All terminal nodes are linked to an estimation of the output variable (y). For each individual, a prediction is performed using the regression tree and assigning it to any of the leaf nodes. In this work, the standard binary regression tree is considered.

2.3   The Algorithm

In the present research, the proposed algorithm combines the capabilities of genetic algorithm and regression trees in order to determinate the importance of those variables that would have influence on the Computer Vision Syndrome.

As the database employed in this study has a large amount of variables, the use of GA is really useful in order to reduce the number of variables required to a minimum number. As shown in Fig. 1, the algorithm starts with the random selection of the information corresponding to the 80% of the individuals that answered the survey. They will be used as training data set, while the other 20% will be employed as validation set.

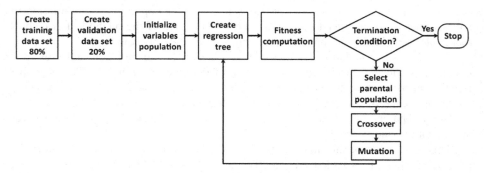

Fig. 1.   Algorithm diagram.

The next step consists on the creation of an initial population of individuals formed by 0 and 1, in which 0 represents those variables that are not going to be included in the model and 1 those that yes. Afterwards, regression trees are created and their performance evaluated by means of the root-mean-square error (RMSE). Until the termination condition is reached (RMSE not improved in more than 0.01% in the last 100 generations), many generations are created in the research of the optimum variables subset.

One of the main drawbacks of the convergence of this method is that the fitness function is not valuated on those individuals that are part of the training data set, if not in those that form the validation data set. Although this makes the convergence lower, it issues the generalizability of the result obtained.

Please note that in this work, and due to the total amount of individuals available instead of a k-fold cross-validation method, we have preferred to employ 80% of individuals for training and 20% for validation, but the algorithm was repeated 1,000 times. In other word, it was tested using 1,000 different subsets for training and validation.

3 Results

3.1 Case of Study

We perform an observational cross-sectional epidemiological study, based on the completion of two self-administered questionnaires, on health personnel of the Monte Naranco Hospital of Oviedo (HMN) and the Central University Hospital of Asturias (HUCA), Spain. The first hospital is specialized in geriatrics and palliative care and it started using the EHR in 2007, and the second one is the referral hospital for the Health Service of the Principality of Asturias and it started using the EHR in 2014.

The study included health personnel that were using VDT at work of the following occupational categories: physicians and surgeons, residents, nurses, advanced practice nurses (APNs) in training and auxiliary nurses.

Of a total of 668 potentially eligible workers, 539 questionnaires were finally obtained (the response rate was 80.69%). Among those, 196 workers were excluded, and the final sample that took part in the study was 343 workers (202 HUCA and 141 HMN) from 47 different services or hospitalization units. The reasons of exclusion in 66 workers was their suffering: dry eye (35), amblyopia (6), squint (1), conjunctivitis (22), corneal ulcers (5), non-surgically controlled cataracts (5), glaucoma (1), blepharitis (2), keratitis (1), uveitis (1), vitreous disorders (8) and retinal disease (4). Also 116 were excluded because they did not use a computer for their job, 13 were excluded because they had occupational categories out of the sampling criteria, and one was excluded because of the lack of information on their seniority.

All the participants signed an informed consent form, in which data confidentiality was guaranteed during the entire process. The study was authorized by the Health Authority and approved both by the Research Ethics Committee of the Principality of Asturias and by the Ethics Committee of the University of Alicante, Spain (coordinator of the study), in accordance with the tenets of the Declaration of Helsinki.

3.2 Data Collection

The survey was carried out from January to October 2017 using self-administered questionnaires. The "Anamnesis and History of Exposure Questionnaire" was specifically developed for this study, to gather information about gender, age, ophthalmic or/and contact lenses use, history of eye diseases and treatment, previous eye surgeries, occupational categories (physicians and surgeons included residents, nurses included advanced practice nurses (APNs) in training and auxiliary nurses), work schedule (included morning shifts, evening shifts, rotating shifts without nights, rotating shifts including nights and mornings with on call shifts), services or hospitalization units of work, seniority, information about the ease of use of the software application and daily VDT usage at and outside work.

The "Computer Vision Syndrome Questionnaire (CVS-Q)", designed and validated by Seguí et al. in 2015 [6], was used to measure perceived ocular and visual symptoms during or immediately following computer work. This questionnaire evaluates the frequency (never, occasionally or often/always) and the intensity (moderate or intense) of 16 ocular and visual symptoms: burning, itching, feeling of foreign body, tearing, excessive blinking, eye redness, eye pain, heavy eyelids, dryness, blurred vision, double vision, difficulty focusing for near vision, increased sensitivity to light, colored halos around objects, feeling that sight is worsening, and headache. Individuals with a score of 6 or more on the questionnaire are classified as symptomatic (suffering CVS).

3.3   Implementation of the Algorithm

After refining the database obtained, a total of 255 initial variables were used to implement the algorithm. Figure 2 shows the evolution of the RMSE value by the number of iterations in one of the 1,000 repetitions of the algorithm. The average best RMSE value obtained in the referred repetitions was of 3.779221. From a total of 255 input variables included in the database, 8 variables took part is more than the 75% of the models and only 30 in more than the 34%. Table 1 shows the variables that take part in most of the regression trees obtained.

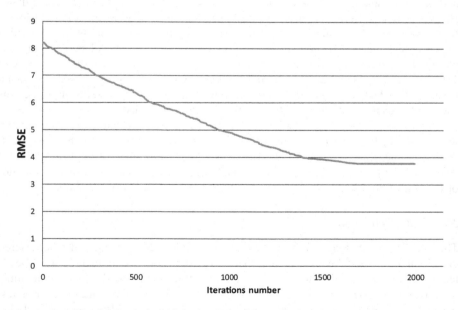

Fig. 2. Evolution of the RMSE value by the iterations number.

Table 1. Percentage of regression trees in which each variable is included.

Variables	Percentage of models
occupational_seniority	97.20%
h_day_VDT_workplace	96.90%
service_seniority	96.30%
past_conjunctivitis	88.50%
current_eye_treatment	79.80%
shifts_incl_nights	74.60%
refractive_surgery	74.20%
VDT_duration_ >2 years	74.10%
eye_surgery	69.40%
opthalmic_wearers	64.10%
Acute_palliative_geriatrics	60.70%
h_day_VDT_outside_work	60.60%
h_day_VDT_workplace&outside_work	60.10%
mornings with on call shifts	57.20%
sterilization_service	53.40%
contact_lenses_wearers	50.30%
surgery_service	48.60%
past_ocular_herpes	46.10%
Age	42.40%
blood_bank_servive	41.10%
pathological_anatomy_service	40.90%
endocrine_service	40.70%
morning shifts	40.50%
gender	40.20%
traumatology_service	39.90%
nephrology_service	39.70%
easy_software_application	39.60%
past_keratitis	39.30%
anesthesia_service	34.60%
evening shifts	34.20%

Figure 3 shows one of the regression trees obtained, where some of the most frequent variables are included. If this regression tree is employed to calculate the CVS score, the RMSE obtained is 3.740692. The performance of the model is evaluated in Fig. 4 by means of the cumulative distribution function of the difference in CVS score of real values and predictions. Please note how in 50% of the individuals the difference in absolute value of their real CVS score and the forecast is below 3 points.

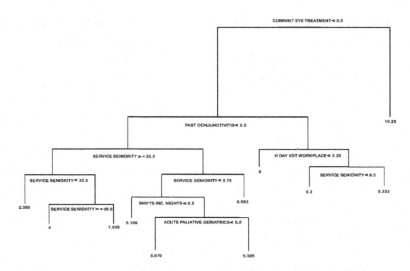

Fig. 3. Example of regression tree obtained with the proposed algorithm.

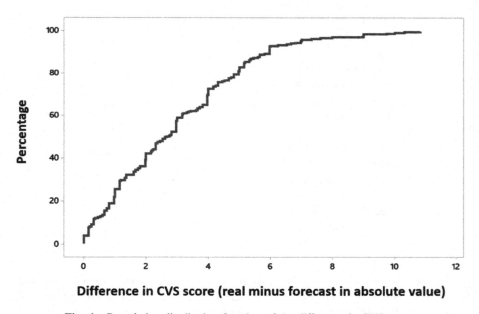

Difference in CVS score (real minus forecast in absolute value)

Fig. 4. Cumulative distribution function of the difference in CVS score.

In order to know how large is the difference of the real CVS values when compared whit their forecasts, Table 2 shows the average differences in absolute value of the real CVS scores and the forecasts. As it can be observed in this table, the minimum differences correspond to the CVS scores of 5, 6 and 7.

Table 2. Average differences in absolute value of the real CVS value and the forecast.

CVS value	Avg. difference	N
0	5.224	19
1	4.356	24
2	3.855	16
3	3.558	17
4	2.351	38
5	1.587	34
6	1.21	27
7	0.991	27
8	1.671	31
9	2.313	20
10	2.499	24
11	3.27	21
12	**3.899**	**14**
13	5.647	8
14	5.204	8
15	6.127	3
16	9.386	7
17	8.816	2
19	9.667	1
20	9.75	1
All	**2.976**	**343**

4 Discussion and Conclusions

There are many methods that can be chosen to analyze regression problems. Tree classification techniques, are well-known for producing accurate predictions based on a relative low amount of if-then conditions, that is the reason why they have been chosen for the present research. Also, it must be considered that the interpretation of results summarized in a tree is very simple. Tree methods are nonparametric and nonlinear. Therefore, there is no implicit assumption that the underlying relationships between the predictor variables and the dependent variable are linear, follow some specific non-linear link function or that they are even monotonic in nature.

The algorithm proposed in the present research, hybridizes regression trees and GA with the aim of finding better results than the regression tree. The GA are employed to select those variables that would take part in the regression trees models. As it is well-known, regression trees also select features by construction. So, before the implementation of the present algorithm, regression trees were employed, and their performance assessed in this case, the best RMSE value achieved was 4.804772 what is a value a 27% larger that the one obtained with the proposed hybrid algorithm. In other words, the proposed algorithm, although has a larger computational cost, allow us to achieve a better result.

The number of hours spent at work with a computer and the seniority in the service predicted the risk of CVS. The amount of hours working daily on a computer (both at and outside the workplace) was found to be related to visual discomfort [23, 24]. Also prolonged VDT use at work was connected with an increased risk of CVS [9, 10] and dry eye disease in workers using a VDT longer than 8 h per day (OR = 1.94; 95% CI, 1.22 − 3.09) [25]. The longer the seniority on the service, the higher the prevalence of CVS. Ranasinghe et al. [9] also observed that the prevalence of CVS significantly increased ($p < 0.01$) with the number of years working in a job using a computer.

We also observe that CVS is higher in workers who have had conjunctivitis in the past, are using eye drops for dryness or another eye treatment. An extrinsic cause of evaporative dry eye could be some diseases as allergic conjunctivitis, so in that workers, the presence of dryness could be higher. The presence of dryness or dry eye may play a significant role in the etiology of CVS. However, the benefit of dry eye therapies, which have been proposed to minimize symptoms of CVS, is unproven [13].

Regarding work schedule, working with shifts including nights and working mornings with on call shifts was associated with the prevalence of CVS. Previous studies [26–28] suggest an association between rotating night shift work and several diseases, including cardiovascular disease, cancer, diabetes, hypertension, chronic fatigue, overweight and obesity, sleeping problems and early spontaneous pregnancy loss.

Scores of CVS are higher in workers with a history of eye surgery, especially refractive eye surgery. Laser surgery may change the shape of the cornea and interfere with the relationship between the eyelids and ocular surface that disturbs normal blinking patterns [29].

The model predicts higher scores of CVS in ophthalmic and contact lenses wearers. Many studies have concluded that the prevalence of ocular and visual symptoms increases in daily computer users with a regular use of contact lenses [10, 13, 14], but none have obtained a relationship between CVS and being an ophthalmic wearer.

Health workers that work in a geriatric service tend to achieve higher CVS scores. This can be explained because, in our study, all the workers in that service belonged to the Monte Naranco Hospital. The EHR was introduced earlier in this hospital, so its workers have been exposed to VDT much longer.

Nevertheless, these results should be interpreted with caution given the limitations of our study. First, it was a cross-sectional design, and we cannot be sure that the cause (use of PVD) precedes the effect (CVS). Second, we did not include ophthalmic examinations that inform us of the workers' refractive state. Not including neck and shoulder pain as a symptom of CVS was also a limitation, as it has been considered an extra-ocular symptom of CVS [13, 14]. Finally, we did not take into consideration the use of mobile devices at and outside work, which could be a confounding factor.

Despite these limitations, the use of a validated questionnaire to measure CVS is a particular strength in this work, and this is the first study that develops a prediction model of the score achieved by an individual in the CVS-Q. Finally, we would like to remark that the main contribution of the present research is a novel hybrid methodology able to determine the most important variables in a prediction problem.

References

1. American Optometric Association. Computer Vision Syndrome, 16 Jan 2018. https://www. aoa.org/patients-and-public/caring-for-your-vision/protecting-your-vision/computer-vision-syndrome
2. Scheiman, M.: Accommodative and binocular vision disorders associated with video display terminals: diagnosis and management issues. J. Am. Optom. Assoc. **67**(9), 531–539 (1996)
3. Bergqvist, U.O., Knave, B.G.: Eye discomfort and work with visual display terminals. Scand. J. Work Environ. Health **20**(1), 27–33 (1994)
4. Fenga, C., Aragona, P., Di Nola, C., Spinella, R.: Comparison of ocular surface disease index and tear osmolarity as markers of ocular surface dysfunction in video terminal display workers. Am. J. Ophthalmol. **158**(1), 41–48 (2014)
5. Ünlü, C., Güney, E., Akçay, B., Akçali, G., Erdoğan, G., Bayramlar, H.: Comparison of ocular-surface disease index questionnaire, tearfilm break-up time, and Schirmer tests for the evaluation of the tearfilm in computer users with and without dry-eye symptomatology. Clin. Ophthalmol. **6**, 1303–1306 (2012)
6. Seguí, M.M., Cabrero-García, J., Crespo, A., Verdú, J., Ronda, E.: A reliable and valid questionnaire was developed to measure computer vision syndrome at the workplace. J. Clin. Epidemiol. **68**(6), 662–673 (2015)
7. González-Pérez, M., Susi, R., Antona, B., Barrio, A., González, E.: The Computer-Vision Symptom Scale (CVSS17): development and initial validation. Invest. Ophthalmol. Vis. Sci. **55**(7), 4504–4511 (2014)
8. Ye, Z., Honda, S., Abe, Y., et al.: Influence of work duration or physical symptoms on mental health among Japanese visual display users. Ind. Health **45**(2), 328–333 (2007)
9. Ranasinghe, P., Wathurapatha, W.S., Perera, Y.S., et al.: Computer vision syndrome among computer office workers in a developing country: an evaluation of prevalence and risk factors. BMC Res. Notes **9**(1), 150 (2016)
10. Tauste, A., Ronda, E., Molina, M.J., Seguí, M.: Effect of contact lens use on computer vision syndrome. Ophthalmic Physiol. Opt. **36**(2), 112–119 (2016)
11. Sa, E.C., Ferreira, M.: Junior, L. E. Rocha. Risk factors for computer visual syndrome (CVS) among operators of two call centers in São Paulo, Brazil. Work **41**(Supplementry 1), 3568–3574 (2012)
12. Sen, A., Richardson, S.: A study of computer-related upper limb discomfort and computer vision syndrome. J. Human Ergol. **36**(2), 45–50 (2007)
13. Rosenfield, M.: Computer vision syndrome: a review of ocular causes and potential treatments. Ophthalmic Physiol. Opt. **31**(5), 502–515 (2011)
14. Parihar, J.K.S., Jain, V.K., Chaturvedi, P., Kaushik, J., Jain, G., Parihar, A.K.S.: Computer and visual display terminals (VDT) vision syndrome (CVDTS). Med. J. Armed Forces India **72**(3), 270–276 (2016)
15. Yazici, A., Sari, E.S., Sahin, G.: Change in tear film characteristics in visual display terminal users. Eur. J. Ophthalmol. **25**(2), 85–89 (2015)
16. García Nieto, P.J., Álvarez Fernández, J.R., de Cos Juez, F.J., Sánchez Lasheras, F., Díaz Muñiz, C.: Hybrid modelling based on support vector regression with genetic algorithms in forecasting the cyanotoxins presence in the Trasona reservoir (Northern Spain). Environ. Res. **122**, 1–10 (2013)
17. Holland, J.H.: Adaptation in Natural and Artificial Systems: An Introductory Analysis with Applications to Biology, Control, and Artificial Intelligence. University of Michigan Press, Ann Arbor, MI (1975)

18. Artime Ríos, E.M., Seguí Crespo, M.d.M, Suarez Sánchez, A., Suárez Gómez, S.L., Sánchez Lasheras, F.: Genetic algorithm based on support vector machines for computer vision syndrome classification. In: Pérez García, H., Alfonso-Cendón, J., Sánchez González, L., Quintián, H., Corchado, E. (eds.) SOCO/CISIS/ICEUTE -2017. AISC, vol. 649, pp. 381–390. Springer, Cham (2018). https://doi.org/10.1007/978-3-319-67180-2_37

19. Ordóñez Galán, C., Sánchez Lasheras, F., de Cos Juez, F.J., Bernardo Sánchez, A.: Missing data imputation of questionnaires by means of genetic algorithms with different fitness functions. J. Comput. Appl. Math. **311**, 704–717 (2017)

20. Sánchez Lasheras, F., Suárez Gómez, S.L., Riesgo García, M.V., Krzemień, A., Suárez Sánchez, V.: Time series and artificial intelligence with a genetic algorithm hybrid approach for rare earth price prediction. In: ITISE 2017, Granada Spain, September 2017

21. Alonso Fernández, J.R., Díaz Muñiz, C., García Nieto, P.J., de Cos Juez, F.J., Sánchez Lasheras, F., Roqueñí, M.N.: Forecasting the cyanotoxins presence in fresh waters: A new model based on genetic algorithms combined with the MARS technique. Ecol. Eng. **53**, 68–78 (2013)

22. Sánchez Lasheras, F., García Nieto, P.J., de Cos Juez, F.J., Mayo Bayón, R., González Suárez, V.M.: A hybrid PCA-CART-MARS-based prognostic approach of the remaining useful life for aircraft engines. Sensors **15**, 7062–7083 (2015)

23. Robertson, M.M., Huang, Y.H., Larson, N.: The relationship among computer work, environmental design, and musculoskeletal and visual discomfort: examining the moderating role of supervisory relations and co-worker support. Int. Arch. Occup. Environ. Health **89**(1), 7–22 (2016)

24. Portello, J.K., Rosenfield, M., Bababekova, Y., Estrada, J.M., Leon, A.: Computer-related visual symptoms in office workers. Ophthalmic Physiol. Opt. **32**(5), 375–382 (2012)

25. Uchino, M., Yokoi, N., Uchino, Y., et al.: Prevalence of dry eye disease and its risk factors in visual display terminal users: the Osaka study. Am. J. Ophthalmol. **156**(4), 759–766 (2013)

26. Ramin, C., Devore, E.E., Wang, W., Pierre-Paul, J., Wegrzyn, L.R., Schernhammer, E.S.: Night shift work at specific age ranges and chronic disease risk factors. Occup. Environ. Med. **72**(2), 100–107 (2015)

27. Gu, F., Han, J., Laden, F., et al.: Total and cause-specific mortality of U.S. nurses working rotating night shifts. Am. J. Prev. Med. **48**(3), 241–252 (2015)

28. Stocker, L.J., Macklon, N.S., Cheong, Y.C., Bewley, S.J.: Influence of shift work on early reproductive outcomes: a systematic review and meta-analysis. Obstet. Gynecol. **124**(1), 99–110 (2014)

29. Shtein, R.M.: Post-LASIK dry eye. Exper. Rev. Ophthalmol. **6**(5), 575–582 (2011)

An Algorithm Based on Satellite Observations to Quality Control Ground Solar Sensors: Analysis of Spanish Meteorological Networks

Ruben Urraca, Javier Antonanzas, Andres Sanz-Garcia, Alvaro Aldama, and Francisco Javier Martinez-de-Pison[✉]

EDMANS Group, Department of Mechanical Engineering, University of La Rioja, 26004 Logroño, Spain
{ruben.urraca,fjmartin}@unirioja.es, edmans@dim.unirioja.es

Abstract. We present a hybrid quality control (QC) for identifying defects in ground sensors of solar radiation. The method combines a window function that flags potential defects in radiation time series with a visual decision support system that eases the detection of false alarms and the identification of the causes of the defects. The core of the algorithm is the window function that filters out groups of daily records where the errors of several radiation products, mainly satellite-based models, are greater than the typical values for that product, region and time of the year.

The QC method was tested in 748 Spanish ground stations finding different operational errors such as shading or soiling, and some equipment errors related to the deficiencies of silicon-based photodiode pyranometers. The majority of these errors cannot be detected by traditional QC methods based on physical or statistical limits, and hence produce problems in most of the applications that require solar radiation data. Besides, these results manifest the low-quality of Spanish networks such as SIAR, Meteocat, Euskalmet and SOS Rioja, which show defects in more than a 50% of the stations and should be consequently avoided.

Keywords: Solar radiation · Pyranometer · Satellite-based model
Quality control

1 Introduction

Solar radiation data is essential in many disciplines such as environmental sciences, energy production [2] and climate analysis [11]. The variable most widely used is global horizontal irradiance (G), which is the total amount of downwelling shortwave irradiance reaching the Earth's surface over a horizontal plane. Thermopile pyranometers and silicon-based photodiodes are the two types of outdoor sensors for measuring G. Thermopiles typically achieve the highest quality

© Springer International Publishing AG, part of Springer Nature 2018
F. J. de Cos Juez et al. (Eds.): HAIS 2018, LNAI 10870, pp. 609–621, 2018.
https://doi.org/10.1007/978-3-319-92639-1_51

and they are based on the thermoelectric effect: a blackened surface absorbs the incoming radiation creating a thermal gradient that is measured with a thermocouple. Photodiodes are made by small silicon cells and are thus based on the photovoltaic effect. They are a low-cost option requiring fewer maintenance and having faster response times, but they have higher uncertainty than thermopiles because their responsivity is limited by the spectral response of silicon.

Ground measurements obtained with pyranometers are the most accurate source of G data [12], but they still contain different types of errors that can be broadly divided into operational and equipment errors [5,15]. Operational errors are related to the particular operation conditions in the station, such as the station location and maintenance procedures. Some examples include shading by nearby objects, the accumulation of dust over the senor or electrical shutdowns. On the contrary, equipment errors are related to the intrinsic limitations of the instruments and inadequate calibration procedures, being typically more severe in photodiodes than in thermopile pyranometers.

Several quality control (QC) methods have been developed to detect errors in ground records. The majority of QC methods are simple range tests that establish the most probable physical or statistical limits for G values to discard samples out of these ranges [15]. Here, the most widely used method is the BSRN QC [7]. Alternative methods include interpolation between nearby stations [5], graphical analysis [8], analysis of the symmetry of irradiance profiles [3] and coherence tests of the different irradiance components [7]. All these methods are only able to detect large errors whereas most common defects, such as shading and soiling, introduce small deviations in G. The detection of these small but long-lasting errors is not straightforward because filters cannot be too restrictive due to the wide range of physically possible irradiance values.

We recently presented a new QC algorithm specially tailored for detecting small errors in ground records [13]. The method flags those samples in which the deviations between several radiation estimations and the ground records are out of the typical range for that region and time of the year. We assume that if the deviations of several independent radiation models are out of the typical ranges, the most likely cause is an error on the ground record. Even though the quality of estimations is not as high as that of ground records, the algorithm exploits the advances on the stability of solar radiation modeling techniques such as satellite-based products [12] and reanalysis. Besides, filtering out samples in terms of deviations instead of in terms of G produces a more restrictive filter, enabling the detection of low-magnitude defects.

The first goal of this study is to validate the QC algorithm with a heterogeneous dataset comprised by all Spanish monitoring stations that measure G (748 stations), including several regional networks where the probability of finding errors is higher. The second objective is to guide potential users by deriving some statistics about the quality of the monitoring networks available in Spain. In addition, we include an enhanced visual decision support tool for the analysis of flagged samples.

2   Methods

2.1   Radiation Data

Ground records were retrieved from all Spanish stations that measure G from 2005 to 2013 at the highest temporal resolution available at no cost. This results in a dataset comprised by 748 stations and 9 networks (Fig. 1): six dedicated meteorological networks (BSRN, AEMET, Meteo Navarra, Meteocat, Euskalmet, MeteoGalica), two networks for agricultural purposes (SIAR and SIAR Rioja) and one network for emergency situations (SOS Rioja). The networks can be also categorized based on their spatial coverage in worldwide networks (BSRN), national networks (AEMET and SIAR), while the remaining ones are regional networks. Most meteorological networks included thermopile pyranometers (285 stations) whereas photodiodes are the common sensor of agricultural networks (386 sensors). Thermopile pyranometers were classified from highest to lowest quality according to ISO 9060:1990 [4] in (i) Secondary Standard, (ii) First Class and (iii) Second Class. The description of the sensor was not provided in 77 stations.

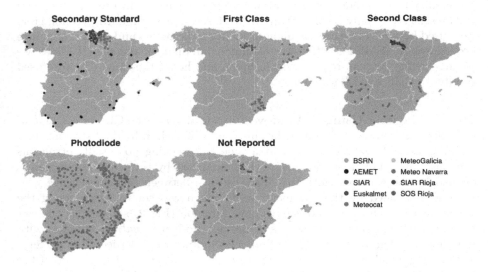

Fig. 1. Location and type of pyranometer installed in the stations used in this study.

The QC algorithm uses daily values of global horizontal irradiance (G_d). All night values (sun elevation $<0°$) were initially set to 0. In stations with 1-min resolution, 15-min means were calculated (5 valid values required) and subsequently averaged to obtain the hourly means (all four 15-min values required). In stations with time resolutions from 5-min to 30-min, hourly means were directly obtained by averaging the original data (all original values required). Daily values were finally obtained by averaging hourly values if at least 20 h values were available.

2.2   Description of the Quality Control (QC) Algorithm

Step 1: Calculation of the Confidence Intervals (CIs). The first step of the QC algorithm is to find the characteristic values for the daily deviations (δ_d) between each radiation product p and the ground records. This is done by calculating the confidence intervals (CIs) for the monthly bias of each product (temporal averaging) over an spatial region g with uniform irradiance conditions (spatial averaging). These CIs are defined as the median absolute deviation of this bias ($MAD^p_{m',g}$) around the median ($\widehat{Bias}$). They include a tuning parameter (n) to adjust the restriction level of the QC procedure (1).

$$CI^p_{m',g} = \widehat{Bias}^p_{m',g} \pm n \times MAD^p_{m',g} \quad m' \in (Jan, ..., Dec), g \in (g_1, ..., g_n), \ p \in (p_1, ..., p_n)$$
(1)

where m' are the different months of the year, p the radiation products used and g the spatial regions defined. The use of median and MAD statistics along with the spatio-temporal averaging of the bias increase the robustness of the CIs. The CIs were calculated only with high-quality ground records in order to reduce the probability of including operational and equipment errors in the CIs. Thus, in the present study only records from AEMET secondary standard pyranometers were used because these are the highest quality radiometers and the maintenance procedures of AEMET are the strictest among all Spanish networks. Besides, we did not define any sub-regions within Spain and hence the same CIs were used to filter out all Spanish stations.

Step 2: Flagging Using a Window Function. Once the CIs are calculated, a window function goes through the time series of each individual stations analyzing groups of consecutive days at a time and flagging potentially erroneous samples. The number of consecutive days analyzed by the window function is defined by the window width (w). The distance between two consecutive windows is controlled by the parameter *step*, which was set to 5 days along the experiments. Each analysis of the window function (Fig. 2) starts with the calculation of the number of available samples per product (d_valid). Products with less than 20% samples available within the window are discarded. The percentage of samples above the upper limit (d_over) or below the lower limit of the CIs (d_under) are subsequently calculated. If at least one of the products covers the 80% of the window days, the average of d_over and d_under are computed. These thresholds were set experimentally to ensure that all products used have sufficient amount of samples, and that at least one of these products covers most of the window width. Finally, daily records within the window are flagged if more than 80% of the samples are either over or under the CIs. If estimations from all the products present the same type of unusual deviation (above or below the CIs), we assume that the most likely cause will be a defect in the ground records.

Three independent radiation products were used in this study: two satellite-based models, SARAH-1 [9] and CLARA-A1 [6], and one reanalysis, ERA-Interim [1]. The two most important tuning parameters of the QC algorithm

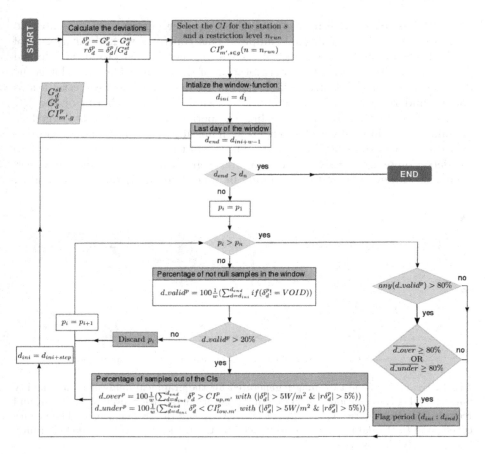

Fig. 2. Flowchart for one run of the window function.

are w and n. The best configuration was found by varying w within (5, 10, 15, 20, 30, 40, 60, 90, 120) and n from 0.2 to 3.5 in intervals of 0.1. Results were analyzed in terms of the Precision-Recall curve, which plots the precision (2) against the recall (3).

$$Precision = \frac{TP}{TP + FP} \tag{2}$$

$$Recall = \frac{TP}{TP + FN} \tag{3}$$

where TP stands for true positives, FP false positives and FN false negatives. The analysis of the PR curves was performed with the dataset of European stations described in [13] and it revealed that the best configuration consisted on running the window function two times. One run was to look for short-lived defects ($n = 2.4$, $w = 20$ days), and the second was to look for permanent low-magnitude deviations ($n = 0.4$, $w = 90$ days).

Step 3: Visual Decision Support System. Two graphs are generated to facilitate the analysis of flagged samples. This contrasts with the majority of available QC methods for solar radiation, which generally just produce numerical flags and leave to the user the interpretation of those flags. The first plot is the daily deviation plot (Fig. 3A), which depicts the deviations between estimations and ground records. The plot includes a visual flag for each run of the QC algorithm (yellow and orange), shading the periods of daily records flagged. It includes two additional flags for periods with missing data (grey) and for samples that do not pass the BSRN QC (red). The use of the BRSN range tests enables the detection of errors that are masked after aggregating to the daily values, e.g. time lags. However, these tests can only be used if sub-daily data are available.

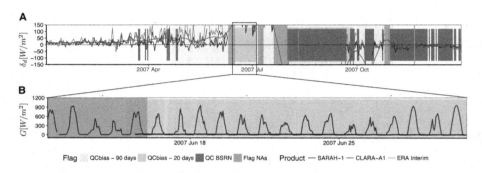

Fig. 3. Example of the two images generated for the graphical analysis of the quality flags. **(A)** Daily deviation between estimations from radiation products and ground records. **(B)** Hourly irradiance values of SARAH-1 and the ground sensor. The images correspond to the data recorded during 2007 by the Euskalmet station C064 (Zarautz, Camping). (Color figure online)

The second plot is the hourly irradiance profiles of measured and estimated data overlapped (Fig. 3B). It is only generated for stations with sub-daily time resolution data and it requires at least one product with hourly time resolution, e.g. SARAH-1. Whereas the first plot could be sufficient for detecting false alarms, the second plot provides valuable information for identifying the causes of the defects.

Software. The QC algorithm was implemented in R programming language using the `tidyverse` [14] collection of packages: `dplyr` and `tidyr` for data manipulation, `lubridate` to work with time series and `ggplot2` to create the plots of the visual decision support system.

3 Results and Discussion

3.1 Setting up the QC Algorithm

Results obtained with each combination of w and n represent one point in the PR space (Fig. 4). Two types of PR curves were calculated. The first one (Fig. 4A) considers that each sample of the PR curve is one daily record of a specific station. Even though this is the straightforward analysis of the output of the QC method, it is not the most practical approach. The algorithm rarely finds the exact number of days with defects because it flags all the daily records within a window. This is especially evident at the edges of periods with errors and with low radiation values (winter months or high latitude locations). Moreover, most of these misadjustments are corrected by visual inspecting the flagged samples, so this first set of PR curves do not show the real performance of the QC method. As a consequence, the second set of PR curves was generated considering that each sample corresponds to one ground station (Fig. 4B). These curves illustrate whether the QC method is able to detect the presence of a defect in a ground station, regardless it finds the exact daily records where the error occurs.

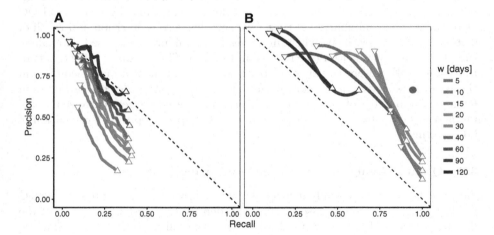

Fig. 4. Precision-Recall (PR) curves obtained for the different combinations of n (tuning parameter of the CIs) and w (window width). **(A)** One sample corresponds to one daily record of the station. **(B)** One sample corresponds to one station. The variable n goes from 0.2 (up-pointing triangle) to 3.5 (down-pointing triangle) in 0.1 intervals. The red dot represents the results obtained with the chosen configuration based on two runs of the window function. (Color figure online)

The PR curves show that using wider CIs by increasing n leads to a greater recall (more defects detected) but to a smaller precision (more false alarms). The same pattern is observed for decreasing values of w, reducing the number of days analyzed by the window function. With both parameters, more restrictive conditions (small n, small w) lead to larger number of defects identified at the expense of a larger amount of false alarms. In principle, the best configuration should be an intermediate solution that balances the number of true positives and false alarms, somewhat around $w = 30$ and $n = 1.5$. Nonetheless, the selection of the best configuration is also affected by the different characteristics of the defects present in ground sensors. Short-lived defects, such as electronic shutdowns or equipment failures, typically last from few hours to few days but the magnitude of the deviations created is usually large. On the contrary, long-lived defects introduce small deviations that can even become permanent, such is the case of shading by surrounding objects. Hence, the type of defects detected with narrow windows ($w < 20$ days) are not the same as those found with wide ones ($w > 30$ days), so the use of an intermediate solution is not sufficient to detect all types of defects present in ground sensors.

We found that the best configuration of the QC algorithm was obtained with two runs of the window function. One run looking for short-lived defects ($w = 20$ days, $n = 2.4$), using a wide CIs (high n) in order to reduce the number of false alarms. And another run looking for almost permanent defects ($w = 90$ days, $n = 0.4$), using a more restrictive CIs (small n) in order to detect low-magnitude semi-permanent defects. The use of a window function along with the trade off between w and n enables the detection of defects not found by traditional QC algorithms. The combination of these two runs leads to a precision and recall of 0.66 and 0.92, respectively, which improves the configurations based on a single run of the window function (see red dot in Fig. 4B). The parameters for this two-run configuration were tuned prioritizing the attainment of a high recall. From the users perspective, it is more useful to find all existing defects rather than having a low number of false alarms. This is even more clear here because the QC method incorporates a visual inspection tool that speeds up the identification of false alarms.

3.2   Quality Analysis of Spanish Monitoring Stations

Samples flagged were visually inspected using the two plots generated by the QC method (Fig. 3) to detect false alarms and identify the most likely cause of the deviation observed. Errors were classified into the following categories: shading by nearby objects (shading), accumulation or dust over the sensor (soiling), time lags, diurnal periods with irradiance equal to 0 (diurnal G = 0), incorrect leveling of the sensor (leveling), large errors due to major equipment failures (large errors) and errors of unknown cause (unknown cause).

The QC algorithm detected errors in 310 out of 748 stations (Fig. 5), whereas the BSRN QC, which is the most common QC procedure for solar data [10], only found time lags (49 stations) and some isolated cases of leveling issues and large errors. The majority of the defects were found in SIAR, which is also the largest network, with 225 defects (47% of SIAR stations). SIAR is an agricultural network created by the Spanish Ministry for irrigation planning. Most SIAR stations were installed in agricultural areas such as Ebro and Guadalquivir Valleys or the Mediterranean Coast. In some cases the exact placement of the sensor was even

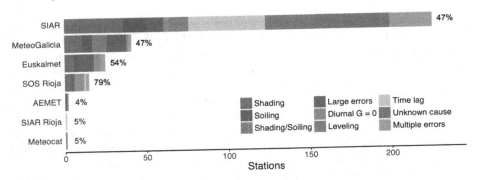

Fig. 5. Number of stations with errors and types of defect detected by the QC method in the different networks. The numeric values represent the percentage of stations with defects in each network. No errors were found in BSRN and Meteo Navarra. The "multiple errors" category represents the stations with more than one type of defect.

influenced by the proximity of other government facilities in order to facilitate the maintenance of the sensors. By contrast, pyranometers must be installed in locations with an obstacle-free horizon and far from potential sources of contamination such as industrial areas, airports or busy roads. This inadequate location selection explains the high amount of shading defects found (36 stations). Moreover, other variables such as temperature and precipitation are more frequently used for agricultural purposes than incoming solar radiation. This little use of G data, along with the poor maintenance of the stations, may also explain the presence of defects such as large errors, time lags or diurnal $G = 0$ in SIAR.

MeteoGalicia and Euskalmet are similar networks with a substantial amount of defects as well. They are the regional meteorological agencies of Galicia and Euskadi, respectively, providing G data with a high time resolution (10 min). Euskalmet records are obtained with high-quality secondary standard pyranometers, whereas MeteoGalicia uses different types of sensors including a large number of first class pyranometers (Fig. 1). However, the number of defects found in both cases is too high for a meteorological network with high quality equipment (54% for Euskalmet and 47% for MeteoGalicia). The most common defect is large errors, which could be partly explained by the high time resolution provided by both networks. Large errors are usually short-lived defects that get masked when aggregating the data to hourly or daily values. Nonetheless, some of the defects identified, such as long nocturnal periods with physically impossible values (Euskalmet), evidence the lack of quality checks in both meteorological agencies. Shading and soiling are other common defects in both networks that questions the maintenance routines of these networks as well. As a consequence, despite the fact that high quality is a priori expected from meteorological agencies, G records from these two networks should generally be avoided.

Ground records from SOS Rioja present the worst quality overall, with the presence of defects in 79% of the stations (15 out of 19). The most common defect is diurnal periods with G equal 0, which is some cases extend the whole year indicating a null maintenance in either the network or the acquisition system. Shading is another frequent defect in SOS Rioja (4 stations), but compared to other networks the shades are visible around solar noon. This excludes the possibility of shades being caused by obstacles in the horizon, such as mountains, trees or buildings. SOS Rioja sensors are installed in lattice towers, so the most likely scenario is that the shades are being caused the own structure. This evidences an inadequate planning during the installation of the equipment. In addition, the lack of maintenance and quality checks ruins the quality of the sensors (first class thermopile pyranometers), proving that the acquisition of high-quality equipment does not guarantee collecting high-quality records.

The networks with the highest quality are AEMET, SIAR Rioja, Meteo Navarra, Meteocat and BSRN, with only 4 defects among all these networks. The good quality of BSRN and AEMET was expected. BSRN is considered the highest quality radiation network worldwide. It has even one dedicated researcher at each station revising the sensors and checking the consistency of the data. AEMET is the Spanish national meteorology agency and it also includes high quality sensors with elaborated maintenance procedures. Besides, in both networks the pyranometers are always ventilated reducing the accumulation or snow, dust and humidity over the dome of the sensors. The use of BSRN and AEMET data should be therefore preferred in applications that require a small uncertainty of solar radiation data. We conclude that data from Meteocat, Meteo Navarra and SIAR Rioja have enough quality for being used for regional studies in Cataluña, Navarra and La Rioja, respectively.

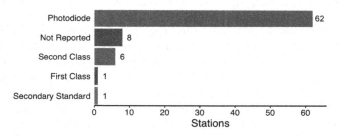

Fig. 6. Total number of stations in which defects were found but the cause of the error could not be identified categorized by the type of sensor used.

All errors identified were operational errors related to maintenance routines, the location of the sensor and QC of the data. However, there are 78 stations in which the presence of an error was evident but it was not possible to identify the exact cause of the defect. The classification of these errors by the type of pyranometer of the station (Fig. 6) reveals that the majority of these defects appear in low-quality sensors: photodiodes (62 sensors) and second class thermopiles (6 sensors). Besides, the 8 stations without information about the pyranometer belong to SIAR, where the majority of the sensors are photodiodes as well. The cause of the error was unknown only for two high quality sensors, one AEMET secondary standard and one Meteocat first class pyranometers. Both networks only provide daily data without cost, which prevents creating the hourly irradiance plot (plot B) for the visual analysis of flagged samples. Hence, the most likely cause here is the existence of an unidentified low-magnitude operational error. In the case of second class sensors, and specially in photodiodes, the most likely cause of these deviations is the presence of equipment errors.

Compared to thermopiles, photodiodes are more affected by cosine and temperature errors, an besides, they have a limited spectral response because they are made by silicon detectors. Silicon has a spectral response within 350–1100 nm that includes only about the 70–75% of total shortwave incoming radiation. Hence, the calibration constant does not account for non-linear variations of the solar spectrum out of the bandwidths covered by silicon. This occurs with changes in aerosol or water vapor concentrations and with variations of sun elevation that modify the main atmospheric scattering process. As a consequence, photodiodes need to be carefully calibrated against thermopile instruments. Independent correction factors for cosine errors, temperature dependence and spectral response are required to obtain field accuracies within the ranges specified by the manufacturer. These corrections should be applied individually for each location taking into consideration the particular conditions of each place and sensor. Therefore, the use of the same correction factors for all SIAR photodiodes, or even the lack of correction factors, may be the cause of the deviations observed. The maintenance procedures of SIAR network are also questioned after finding a large number of operational errors in SIAR stations. Therefore, these small deviations may be also caused by undetected operational errors. It is not easy to identify which cause leads to the deviations observed in each photodiode. Further work is required to gain a better understanding of the limitations of photodiodes, analyzing the half-hourly measurements provided by SIAR stations. Nonetheless, it is clear there are significant differences in terms of quality between SIAR photodiodes and thermopile sensors. Overall, our QC method was not only able to detect operational errors but also some equipment errors, which are the most difficult to detect due to the low-magnitude deviations introduced.

4   Conclusions

A hybrid QC algorithm for solar radiation data, which is based on the analysis of the deviations between satellite-based models and ground records, was validated using 748 Spanish ground monitoring stations that measure global horizontal irradiance. The results reveal that the QC algorithm can detect operational and equipment errors that are rarely found by conventional QC methods, such as the BSRN tests. Besides, this study manifests the low-quality of some of Spanish networks such as SIAR, MeteoGalicia, Euskalmet and SOS Rioja. These networks present defects in 50% or more of the stations. Most of these defects are operational errors related to an inadequate placement of the sensor, a lack of maintenance and a lack of quality control of the data, but the method was also able to identify potential equipment errors in silicon-based photodiode pyranometers. We conclude that data from these networks should be generally avoided in applications requiring solar radiation data.

Acknowledgments. R. Urraca and J. Antonanzas would like to acknowledge the fellowship FPI-UR-2014 granted by University of La Rioja. A.S.G. is funded by Academy of Finland under the FINSKIN Project. A.S.G. also wants to express his gratitude to Instituto de Estudios Riojanos (IER). This work used the Beronia cluster (University of La Rioja), which is supported by FEDER-MINECO grant number UNLR-094E-2C-225.

References

1. Dee, D.P., Uppala, S.M., Simmons, A.J., Berrisford, P., Poli, P., Kobayashi, S., Andrae, U., Balmaseda, M.A., Balsamo, G., Bauer, P., Bechtold, P., Beljaars, A.C.M., van de Berg, L., Bidlot, J., Bormann, N., Delsol, C., Dragani, R., Fuentes, M., Geer, A.J., Haimberger, L., Healy, S.B., Hersbach, H., Hólm, E.V., Isaksen, L., Kållberg, P., Köhler, M., Matricardi, M., McNally, A.P., Monge-Sanz, B.M., Morcrette, J.J., Park, B.K., Peubey, C., de Rosnay, P., Tavolato, C., Thépaut, J.N., Vitart, F.: The ERA-Interim reanalysis: configuration and performance of the data assimilation system. Q. J. R. Meteorol. Soc. **137**(656), 553–97 (2011). http://dx.doi.org/

2. Huld, T., Salis, E., Pozza, A., Herrmann, W., Müllejans, H.: Photovoltaic energy rating data sets for Europe. Solar Energy **133**, 349–362 (2016). http://dx.doi.org/10.1016/j.solener.2016.03.071

3. Ineichen, P.: Solar radiation resource in Geneva: measurements, modeling, data quality control, format and accessibility. 333.7–333.9 (2013). http://archive-ouverte.unige.ch/unige:29599. ID: unige:29599

4. ISO: ISO 9060:1990: specification and classification of instruments for measuring hemispherical solar and direct solar radiation, Geneva, Switzerland (1990)

5. Journée, M., Bertrand, C.: Quality control of solar radiation data within the RMIB solar measurements network. Solar Energy **85**, 72–86 (2011). http://dx.doi.org/10.1016/j.solener.2010.10.021

6. Karlsson, K., Riihelä, A., Müller, R., Meirink, J., Sedlar, J., Stengel, M., Lockhoff, M., Trentmann, J., Kaspar, F., Hollmann, R., Wolters, E.: CLARA-A1: CM SAF cLouds, Albedo and Radiation Dataset from AVHRR Data - Edition 1 - Monthly Means/Daily Means/Pentad Means/Monthly Histograms (2012). http://dx.doi.org/10.5676/EUM_SAF_CM/CLARA_AVHRR/V001

7. Long, C.N., Dutton, E.G.: BSRN Global Network Recommended QC tests, V2.0. BSRN Technical report (2002). http://ezksun3.ethz.ch/bsrn/admin/dokus/qualitycheck.pdf

8. Moreno-Tejada, S., Ramírez-Santigosa, L., Silva-Pérez, M.A.: A proposed methodology for quick assessment of timestamp and quality control results of solar radiation data. Renew. Energy **78**, 531–537 (2015). https://doi.org/10.1016/j.renene.2015.01.031

9. Müller, R., Pfeifroth, U., Träger-Chatterjee, C., Cremer, R., Trentmann, J., Hollmann, R.: Surface Solar Radiation Data Set - Heliosat (SARAH) - Edition 1. Satellite Application Facility on Climate Monitoring (CM SAF) (2015). http://dx.doi.org/10.5676/EUM_SAF_CM/SARAH/V001

10. Roesch, A., Wild, M., Ohmura, A., Dutton, E.G., Long, C.N., Zhang, T.: Assessment of BSRN radiation records for the computation of monthly means. Atmos. Meas. Tech. **4**, 339–354 (2011). http://dx.doi.org/10.5194/amt-4-339-2011

11. Sanchez-Lorenzo, A., Enriquez-Alonso, A., Wild, M., Trentmann, J., Vicente-Serrano, S., Sanchez-Romero, A., Posselt, R., Hakuba, M.: Trends in downward surface solar radiation from satellite and ground observations over Europe 1983–2010. Remote Sens. Environ. **189**, 108–117 (2017). https://doi.org/10.1016/j.rse.2016.11.018

12. Sengupta, M., Habte, A., Gueymard, C., Wilbert, S., Renné, D.: Best practices handbook for the collection and use of solar resource data for solar energy applications, 2nd edn. NREL Technical report (2017). https://www.nrel.gov/docs/fy18osti/68886.pdf

13. Urraca, R., Gracia Amillo, A., Huld, T., Martinez-de Pison, F.J., Trentmann, J., Lindfors, A., Riihelä, A., Sanz Garcia, A.: Quality control of solar radiation data with satellite-based products. Solar Energy **158**, 49–62 (2017). http://dx.doi.org/10.1016/j.solener.2017.09.032

14. Wickham, H.: tidyverse: Easily Install and Load 'Tidyverse' Packages (2017). r package version 1.1.1. https://CRAN.R-project.org/package=tidyverse

15. Younes, S., Claywell, R., Muneer, T.: Quality control of solar radiation data: present status and proposed new approaches. Energy **30**, 1533–1549 (2005). http://dx.doi.org/10.1016/j.energy.2004.04.031

Predicting Global Irradiance Combining Forecasting Models Through Machine Learning

J. Huertas-Tato[1]([⊠]), R. Aler[1], F. J. Rodríguez-Benítez[2], C. Arbizu-Barrena[2], D. Pozo-Vázquez[2], and I. M. Galván[1]

[1] Computer Science Departament, Universidad Carlos III de Madrid, Madrid, Spain
jahuerta@inf.uc3m.es
[2] Department of Physics, Universidad de Jaén, Jaén, Spain

Abstract. Predicting solar irradiance is an active research problem, with many physical models having being designed to accurately predict Global Horizontal Irradiance. However, some of the models are better at short time horizons, while others are more accurate for medium and long horizons. The aim of this research is to automatically combine the predictions of four different models (Smart Persistence, Satellite, Cloud Index Advection and Diffusion, and Solar Weather Research and Forecasting) by means of a state-of-the-art machine learning method (Extreme Gradient Boosting). With this purpose, the four models are used as inputs to the machine learning model, so that the output is an improved Global Irradiance forecast. A 2-year dataset of predictions and measures at one radiometric station in Seville has been gathered to validate the method proposed. Three approaches are studied: a general model, a model for each horizon, and models for groups of horizons. Experimental results show that the machine learning combination of predictors is, on average, more accurate than the predictors themselves.

Keywords: Global irradiance forecasting · Machine learning
Combining forecasting models

1 Introduction

A key issue to increase the competitiveness of the solar energy and to increase their share in the electric systems is the improvement of the reliability of the solar energy forecasts. In the last years, a wide range of forecasting methodologies has been developed, with very different characteristics, such as the spatial and temporal resolution or their forecasting horizon [1].

Machine Learning [2] has played an important role on improving solar energy forecasting [3,4]. Nevertheless, there is still room for improvement. In this regard, there have been some efforts to combine different sources of information (observations, camera, satellite, ...) and take advantage of the possible synergies.

© Springer International Publishing AG, part of Springer Nature 2018
F. J. de Cos Juez et al. (Eds.): HAIS 2018, LNAI 10870, pp. 622–633, 2018.
https://doi.org/10.1007/978-3-319-92639-1_52

For example, real measures and camera have been combined using an artificial neural network optimized through a genetic algorithm [5]. [6] proposes a combination of NAM (North American Mesoscale Model) with cloudiness information obtained from satellite images. This model improves spatial resolution of the NAM, while improving intra-day and 1-day predictions. A system based in extreme learning machines optimized through evolutionary computation (coral reef algorithm) combines direct measures, radiosondes and NWP to obtain the daily prediction of solar irradiance [7]. In [8], a combined prediction of cloud cover derived from a sky-camera and satellite offers a forecast of up to three hours. In a similar way, a combination of cloudiness estimation from satellite and the NWP from European Center for Medium range Weather Forecasting (ECMWF) is proposed [9]. Further results [10] show that the combination of statistical models and NWP is able to reduce the forecasting error at one hour horizons. In [11] a machine learning blending of irradiance forecasts using a Random Forest is used. This approach combines three models: the NAM model, the SREF (Short Range Ensemble Forecast) model and the GFS model. A recent work combines satellite-derived, ground data, solar radiation, and total cloud cover to improve solar radiation forecasting for horizons between 1 h and 6 h [12].

Similar combination approaches have also been applied for wind energy. [13] combines different predictions at several horizons using an adaptive weighting, dependent on the error yielded by the past sources.

The novelty of this research is to use machine learning to combine Global Horizontal Irradiance (GHI) forecasts obtained from four sources, in order to output an improved GHI forecast. The sources are: Smart Persistence [14], Satellite [15], Cloud Index Advection and Diffusion (CIADCast) [16], and Solar Weather Research and Forecasting (WRF-Solar) [17]. These perform differently under different situations and forecasting horizons. The aim of our approach is to use machine learning to combine them automatically, so as to take advantage of their synergies and to improve the performance of sources used separately.

In this work, several prediction horizons have been tested from 15 to 360 min, in steps of 15 min. Three different approaches are proposed for integrating the four sources: general, horizon-individual and horizon-group. The approaches are different ways to treat the horizon information either by using a single general model valid for all horizons, by making specialized models for each horizon (horizon-individual), or a compromise between both (horizon-group).

The machine learning method chosen for combining them is extreme gradient boosting [18]. Gradient boosting has been used before in solar forecasting. For instance, [19,20] use meteorological variables to predict GHI. However, the aim of our work is different, because our inputs are not meteorological variables but GHI forecasts themselves.

The structure of the paper is the following: first, the predictors used as inputs of machine learning method are presented in Sect. 2. In Sect. 3, the dataset used to the evaluation is described. Section 4 explains the different machine learning approaches to be studied in this work. In Sect. 5, the experimental methodology is explained and the experimental results are presented. The final conclusions of this research and future lines of work are presented in Sect. 6.

2 Description of the Predictors

This section describes the four forecasting models (or predictors) that will be combined by a machine learning model.

2.1 Smart Persistence

This model is computed with the actual measured irradiance I_0 and corrected with the variation of the clear-sky (cs) irradiances I_{cs} from the initial time to a future time t. The relation between actual irradiance and cs irradiance at a certain time 0 is kept constant and multiplied by the clear-sky irradiances in future t. The European Solar Radiation Atlas [14] cs model is used (Eq. 1).

$$I(t) = \frac{I_0}{I_{cs}(0)} * I_{cs}(t) \tag{1}$$

2.2 Satellite-Based Model

In this method, satellite images are first processed to derive the so-called cloud index images, an intermediate step to retrieve the clear-sky index images and then the solar radiation maps [21]. Secondly, a statistical comparison of various consecutive cloud index images allows deriving the cloud motion vector field. In this case OpenPIV is used (http://www.openpiv.net/openpiv-python/). The discrete cloud motion vector field is transformed into a continuous flow computing the streamlines, i.e., a family of curves tangent to this wind field [15]. The streamline passing through the station location is used to obtain the future cloud index values, then the clear-sky index values and, finally, the GHI forecast

2.3 CIADCast

The CIADCast model [16] for short-term solar radiation forecasting is based on the advection and diffusion of cloud index estimates derived from satellite using the Weather Research and Forecasting [22] NWP models. Cloud index values are inserted in the WRF cell which corresponds to the cloud top height provided by the EUMETSAT product. Then, WRF is used to advect and diffuse the cloud index values as dynamical tracers both horizontally and vertically.

After the model run, the sum of each column of cloud index values is computed to obtain again a two-dimensional cloud index map. The cloud index values at the station location are used finally to derive the GHI forecast, similarly to the satellite-based model. CIADCast was run with the standard WRF model version 3.7.1, configured with 37 vertical levels and three nested domains of 27, 9 and 3 km spatial resolution. The cloud index maps were ingested in the inner domain, which has similar resolution to the satellite images. 18-h simulations were run discarding the first 6 simulated hours as spin-up.

2.4   WRF-Solar

NWP uses mathematical models based on physical principles of the atmosphere and oceans to predict the weather based on current weather conditions. WRF-Solar [17] is a particular physical configuration of the WRF numerical weather prediction model version 3.6 devised for solar energy applications. It has improved parameterizations for the interactions of solar radiation with clouds and aerosols. The model configuration used here consisted of two nested domains with 9 and 3 km spatial resolution and 37 vertical levels. As with CIADCast, simulations with 18 h of forecasting horizon and 6 h of spin-up were run.

3   Data

The evaluation is conducted at one radiometric station in Seville (southern Spain) where GHI (the total amount of shortwave radiation received from above by a surface horizontal to the ground) has been measured. GHI has been acquired with a Kipp & Zonen CMP6 pyranometer with a 1 min sample rate. The maintenance of radiometric stations follows World Meteorological Organization recommendations and the quality control of the data is applied following Long and Dutton [23]. The observations cover from March 2015 to March 2017.

To ensure the quality of the data, a preprocessing of the dataset has been made. Only predictions taken when the zenith is less than 75° are included in the dataset, because the hours selected by this filter are the most relevant to global irradiance in the day. Forecasts of GHI up to 6 h ahead, with a time step of 15 min, are obtained based on four different models: Smart Persistence, Satellite-based, CIADCast, and WRF-Solar. An example of the structure of the dataset is shown in Table 1. There are four different numerical inputs (four predictors), a target (measure column). On average, each horizon contains 2400 instances,

Table 1. Dataset example

Date	Hour	Horizon	SmartPers	Satellite	CIADCast	WRF-Solar	Measure
2015-03-03	10:15	15	299.72	620.9	230.3	226.29	283.1
2015-03-03	10:15	30	427.23	649.3	240.42	283.04	254.55
2015-03-03	10:15	45	627.45	674.16	249.23	326.29	303.18
...	...	...	...	...	...	...	...
2015-03-03	10:15	360	71.80	312.25	119.01	31.23	280.22
2015-03-03	10:30	15	417.26	649.01	295.54	296.93	254.55
2015-03-03	10:30	30	666.79	673.98	306.37	401.20	303.18
2015-03-03	10:30	45	636.63	619.67	445.28	599.42	347.69
...	...	...	...	...	...	...	...
2015-03-03	10:30	360	471.29	556.22	298.87	411.25	546.71
...	...	...	...	...	...	...	...

although the distribution of the number of instances per horizon is not uniform (short horizons have more instances than long horizons).

4   Methods

The approach to predict GHI is to combine a set of n predictors by means of machine learning models, which the aim to improve the final prediction for every forecast horizon. Therefore, the GHI can be described by Eq. 2, where machine learning model f is used to combine several predictors P_i.:

$$ghi = f(P_1, P_2, ..., P_n) \tag{2}$$

In this work there are four predictors available which are used as inputs for the machine learning algorithm. The predictors combined in this work have been described in detail in Sect. 2. The machine learning method for finding f is the extreme gradient boosting tree ensemble (*xgbtree*) [18]. This decision has been taken after comparing preliminary results with random forests and support vector machines. *Xgbtree* displayed a good performance, while at the same time it is a very fast and efficient implementation. In any case, other methods could have been used within the schema proposed in this article.

Given that each predictor performance depends on the horizon, three different approaches for dealing with horizons have been studied: general, horizon-individual, and horizon-group. They are described in the following subsections.

4.1   General Model

The first approach constructs a model that minimizes error for all horizons considered together. This is achieved by combining data from the different horizons (i.e. excluding the horizon column in Table 1), and training a single model f from the joint dataset. Equation 3 shows how model f can be used for forecasting GHI at time t for horizon h. It can be seen that f is common for all horizons.

$$ghi(t + h) = f(P_1(t, h), P_2(t, h), ..., P_n(t, h)) \tag{3}$$

4.2   Horizon-Individual Model

The second approach builds a different machine learning model f_h for each horizon h. Each f_h is trained using data from each horizon h only. The end result will be a set of 25 machine learning models specialized in predicting GHI for every single horizon. There is a model for horizon 15, able to predict GHI at 15 min forward in time, another model for horizon 30, able to predict GHI at 30 min forward in time, and so on. Equation 4 shows how to use models f_h for making GHI forecasts at time t for each horizon h.

$$ghi(t + h) = \begin{cases} f_{15}(P_1(t, 15), ..P_n(t, 15)), & h \equiv 15 \\ f_{30}(P_1(t, 30), ..P_n(t, 30)), & h \equiv 30 \\ ... \\ f_{360}(P_1(t, 360), ..P_n(t, 360)), & h \equiv 360 \end{cases} \tag{4}$$

4.3   Horizon-Group Model

This last approach builds a set of models, this time by using groups of horizons instead of individual horizons (as in the previous approach). Given that some predictors work better for close horizons (Smart Persistence and Satellite) and others for medium or long horizons (WRF-Solar), the aim is to identify groups of horizons for which some predictors are better than others. The advantage over the horizon-individual approach is that now each group of horizons have more data for training. The end result will be g machine learning models, each one specialized in predicting GHI for a horizon group, where g is the number of groups. Each group model is trained using data from horizons belonging to that group only. This model is represented by Eq. 5, where the p_i's represent the partition points in the horizon range and the $15 \leq h \leq p_1$, $p_1 \leq h \leq p_2$, ..., $p_{g-1} \leq h \leq 360$ are the g horizon groups.

$$ghi(t + h) = \begin{cases} f_1(P_1(t,h),..P_n(t,h)), & 15 \leq h \leq p_1 \\ f_2(P_1(t,h),..P_n(t,h)), & p_1 \leq h \leq p_2 \\ ... \\ f_g(P_1(t,h),..P_n(t,h)), & p_{g-1} \leq h \leq 360 \end{cases} \tag{5}$$

After visual analysis and taking into account the performance of the forecasting models for the different horizons, three groups ($g = 3$) have been used, although larger values could be considered at the expense of computational cost and diminishing the number of data for each group.

In order to decide the actual location of p_1 and p_2, a greedy search has been implemented. Starting from some initial values for p_1 and p_2, all combinations of neighboring points are explored. The set of neighbors of (p_1, p_2) is considered to be $(p_1 \pm 0, 15, 30, p_2 \pm 0, 15, 30)$. Table 2 shows those neighboring points. For each partition explored, three different models are obtained (one per group), each one trained with data from each horizon group and evaluated on validation sets. Out of all the neighbors, the four combinations with lowest errors are kept.

The reason for keeping more than one (p_1, p_2) combination is to avoid falling into local minima. The combination with lowest error, and still unvisited by the algorithm, is the next explored combination of points. p_1 and p_2 are then updated to those locations that minimize the average validation error. This process is repeated until there is an empty list of combinations or the 4 best possible combinations have already been explored. At the end of the search, the algorithm chooses the partition (i.e. combination of p_1 and p_2) with the best error found throughout all the search.

The method is detailed in Algorithm 1. Line 1 creates a table (*VisitedTable*) that stores information about all combinations (p_1, p_2) explored by the algorithm. That is, *Accuracy* is the average validation error for each particular combination; *visited* informs whether this combination has already been explored; and *lastVisit* marks one of the pairs as the one that should be selected for expanding the neighbors.

Table 2. Neighbors for any p_1 and p_2

p1	p2	p1	p2
$p_1 - 30$	$p_2 - 30$	$p_1 - 15$	$p_2 - 15$
$p_1 - 30$	$p_2 - 15$	...	...
$p_1 - 30$	p_2	p_1	$p_2 - 15$
$p_1 - 30$	$p_2 + 15$	p_1	$p_2 + 15$
$p_1 - 30$	$p_2 + 30$	...	...
$p_1 + 15$	$p_2 - 30$	$p_1 + 30$	$p_2 + 30$

Loop in lines 2–17 runs while it is possible to find combinations that improve the error. In line 3 the last state is retrieved from *VisitedTable* (i.e. *lastVisit* == TRUE). After being retrieved, *lastVisit* will be set to FALSE. In line 4, the (p_1, p_2) combination is retrieved from *LastState* and all possible neighbors are calculated in line 5 (*AllNeighbours*(p_1, p_2)). The loop in lines 6–12 checks every pair of points from the list of neighbors (see Table 2) previously expanded (*NewPointList*). If the pair has already been visited, it can be extracted out of the *VisitedTable*. Otherwise, its performance is computed (*Accuracy*(p_1, p_2)) in line 10. At the end of the loop, *PairErrors* contains the performance of all neighbors. Line 13 selects the best four combinations (*BestErrors*), out of which the best unvisited pair is finally selected (line 14). This best pair is marked with *lastVisit* = TRUE, so that it will be selected in the next iteration for computing neighbors. All information regarding new explored pairs and their respective errors is included into *VisitedTable* (line 15). Exploration will continue, as far as at least one of the four best pairs was unvisited (*BestError* not empty, line 18). Once the termination condition is satisfied, the best pair (p_1, p_2) from *VisitedTable* is returned.

Algorithm 1. Horizon-group greedy search process

```
 1: VisitedTable ← Table(p₁, p₂, Accuracy(p₁, p₂), visited, lastVisit)
 2: while continue do
 3:     LastState ← lastVisit in VisitedTable is TRUE
 4:     p₁, p₂ ← p₁ and p₂ in LastState
 5:     NewPointList ← AllNeighbours(p₁, p₂)
 6:     for each PointPair in NewPointList do
 7:         if p₁, p₂ exists in VisitedTable then
 8:             PairError ← error in VisitedTable
 9:         else
10:             PairError ← Accuracy(p₁, p₂)
11:         end if
12:     end for
13:     BestErrors ← select 4 lowest from PairErrors
14:     BestError ← select lowest and !visited from BestErrors. Mark this pair with lastVisit ←
               TRUE
15:     update VisitedTable with NewPointList and PairErrors
16:     continue ← !(BestError is empty)
17: end while
18: BestSeparation ← min error from VisitedTable
```

5   Experimentation

The aim of the experimentation is to compare the skill of each method presented here to predict GHI at each horizon from $h = 15$ to $h = 360$. There are seven different methods to compare, four of them are the predictors (WRF-Solar, CIADCast, Smart Persistence and Satellite) and the other three are the different machine learning approaches proposed in this work (General, Horizon-individual and Horizon-group).

5.1   Methodology

The dataset is divided into two subsets: training and test. The former is made up of the 21 first days of each month (3 weeks of data), the latter is for the test set, that will be used to evaluate the trained model. The training set itself is divided into a model-training set and a validation set. The first one contains the 14 first days of the month and it is used for training the models. The validation set is used for hyper-parameter tuning and to guide the search process for the horizon-group approach and to select the best horizon groups (see Sect. 4.3). The metric used for comparison purposes is the normalized root mean square error, which is calculated in Eq. 6.

$$nRMSE = \frac{\sqrt{\sum (x_i - o_i)^2/N}}{\sum (o_i)/N} \tag{6}$$

where x_i is a prediction, o_i is an observation and N is the number of samples. nRMSE is calculated for each horizon. The global nRMSE is the mean of all horizon nRMSE values.

5.2   Results

In Table 3 the global test nRMSE for each model is shown. The first four rows refer to the predictors. The fifth row displays the error that would be obtained if for each horizon, the best predictor would be selected (called optimal selection in Table 3). Given that this selection is done using the test set, it could not be applied in practice. It is provided only for comparison purposes with the machine learning approaches. It can be seen that all machine learning approaches have better global error than any of the predictors or even their optimal selection. On average, WRF-Solar is the most reliable predictor (0.2603 global nRMSE), with Smart Persistence being the second best (0.2837 nRMSE). Observing the machine learning blending approaches, the most accurate prediction method is the horizon-group approach with a global nRMSE of 0.227. After applying Algorithm 1, horizon groups are 15–60, 75–270 and 285–360.

Table 4 shows the nRMSE broken down by horizon (from $h = 15$ to $h = 360$) for each of the different predictors and approaches. This information is also displayed in Figs. 1 and 2. Figure 1 compares the predictors. Under an hour, CIADCast is the best predictor available, then Smart Persistence is best until

Table 3. Global nRMSE

Method	nRMSE	Method	nRMSE
Smart persistence	0.2837	General	0.2291
Satellite	0.3117	Horizon-individual	0.2312
CIADCast	0.3039	**Horizon-group**	**0.227**
WRF-Solar	0.2603		
Optimal selection	0.2543		

90 min, when WRF-Solar starts being the best model from that point onwards. There are a couple of times where WRF-Solar is worse, at 105 and 150 min. Interestingly, for WRF-Solar, the nRMSE decreases as h increases, although after 285 min the error starts increasing again.

Table 4. nRMSE by horizon

h=	15	30	45	60	75	90	105	120	135	150	165	180
General	0.197	0.205	0.211	0.217	0.226	0.229	0.22	0.217	0.219	0.215	0.216	0.225
h-individual	0.208	0.21	0.213	0.221	0.234	0.229	0.22	0.217	0.225	0.223	0.22	0.227
h-group	0.203	0.208	0.209	0.212	0.228	0.23	0.222	0.218	0.218	0.217	0.216	0.226
CIADCast	0.229	0.244	0.252	0.263	0.283	0.287	0.281	0.29	0.297	0.282	0.273	0.270
Satellite	0.229	0.248	0.255	0.26	0.288	0.298	0.291	0.28	0.268	0.267	0.275	0.289
SmartPer	0.245	0.252	0.255	0.259	0.274	0.28	0.265	0.271	0.281	0.281	0.276	0.293
WRFSolar	0.284	0.272	0.279	0.277	0.284	0.274	0.269	0.264	0.262	0.268	0.266	0.265
h=	195	210	225	240	255	270	285	300	315	33	345	360
General	0.22	0.221	0.22	0.233	0.236	0.234	0.232	0.246	0.261	0.275	0.264	0.260
h-individual	0.226	0.227	0.223	0.226	0.241	0.234	0.225	0.241	0.264	0.271	0.256	0.269
h-group	0.22	0.224	0.217	0.231	0.234	0.23	0.227	0.244	0.256	0.257	0.255	0.248
CIADCast	0.264	0.291	0.301	0.312	0.356	0.348	0.337	0.353	0.358	0.372	0.37	0.381
Satellite	0.298	0.306	0.311	0.346	0.332	0.345	0.348	0.353	0.382	0.391	0.422	0.4
SmartPer	0.288	0.279	0.275	0.294	0.296	0.288	0.275	0.304	0.318	0.333	0.314	0.313
WRFSolar	0.263	0.253	0.251	0.237	0.229	0.242	0.231	0.237	0.254	0.26	0.264	0.263

In Fig. 2 the machine learning approaches (General, Horizon-individual and Horizon-group) are compared to the optimal selection of predictors mentioned above. All machine learning combination of predictors are better than the original predictors up to 255 min. At that point, it is difficult to observe a difference respect to the predictors (WRF-Solar being the most accurate one at those horizons, as shown in Fig. 1). The machine learning approaches show minor differences. First, the horizon-individual model is consistently worse during the early horizons, while both the general and horizon-group models are similar in their performance. However when the 255 min horizon is reached, they become

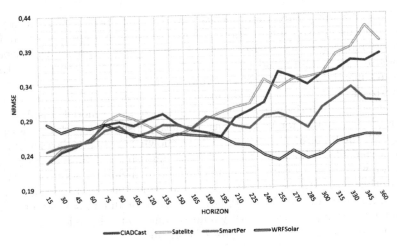

Fig. 1. Predictor performance at different horizons.

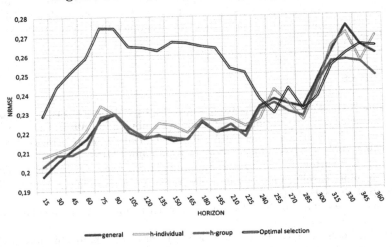

Fig. 2. Performance of the three machine learning approaches at different horizons.

harder to differentiate and behave similarly. At long horizons, starting at the 330 horizon, the horizon-group approach outperforms the other approaches and predictors. All machine learning models consistently increase in nRMSE as the horizon increases.

6   Conclusions

In this paper machine learning methods have been tested in order to combine GHI forecasting models (Smart Persistence, Satellite, CIADCast and WRF-Solar) as inputs to Xgboost, with the aim of improving predictions in horizons from 15 to 360 min. Three approaches have been studied: a general approach that disregards horizon information, a horizon-individual approach that builds a Xgboost model for each horizon, and a horizon-group approach that separates horizons in three groups and builds a Xgboost model for each group. Experimental results show a great accuracy improvement over the predictors on short time horizons, and equivalent performance to the best predictor on further horizons. The general, horizon-individual and horizon-group models display similar performance, although for far horizons, the latter displays a better performance.

Overall, the final results are satisfactory, showing that there is a lot of margin for improvement in the field of solar forecasting using machine learning. In the future, we would like to improve results further by including additional features, such as other predictors or historical solar radiation data. It can be also interesting training different models for different seasons or different weather regimes. Another interesting research work would be automatically selecting subsets of predictors for each horizons or groups of horizons.

Acknowledgments. The authors are supported by the Spanish Ministry of Economy and Competitiveness, projects ENE2014-56126-C2-1-R and ENE2014-56126-C2-2-R and FEDER funds. Some of the authors are also funded by the Junta de Andalucía (research group TEP-220).

References

1. Inman, R.H., Pedro, H.T.C., Coimbra, C.F.M.: Solar forecasting methods for renewable energy integration. Prog. Energy Combust. Sci. **39**(6), 535–576 (2013)
2. Witten, I.H., Frank, E., Hall, M.A., Pal, C.J.: Data Mining: Practical Machine Learning Tools and Techniques. Morgan Kaufmann (2016)
3. Voyant, C., Notton, G., Kalogirou, S., Nivet, M.-L., Paoli, C., Motte, F., Fouilloy, A.: Machine learning methods for solar radiation forecasting: a review. Renew. Energy **105**, 569–582 (2017)
4. Terren-Serrano, G.: Machine learning approach to forecast global solar radiation time series (2016)
5. Chu, Y., Pedro, H.T.C., Coimbra, C.F.M.: Hybrid intra-hour DNI forecasts with sky image processing enhanced by stochastic learning. Sol. Energy **98**, 592–603 (2013)
6. Mathiesen, P., Collier, C., Kleissl, J.: A high-resolution, cloud-assimilating numerical weather prediction model for solar irradiance forecasting. Sol. Energy **92**, 47–61 (2013)
7. Salcedo-Sanz, S., Casanova-Mateo, C., Pastor-Sánchez, A., Sánchez-Girón, M.: Daily global solar radiation prediction based on a hybrid coral reefs optimization-extreme learning machine approach. Sol. Energy **105**, 91–98 (2014)
8. Alonso, J., Batlles, F.J.: Short and medium-term cloudiness forecasting using remote sensing techniques and sky camera imagery. Energy **73**, 890–897 (2014)
9. Lorenz, E., Kühnert, J., Heinemann, D.: Short term forecasting of solar irradiance by combining satellite data and numerical weather predictions. In: Proceedings of 27th European Photovoltaic Solar Energy Conference, Valencia, Spain, pp. 4401–440 (2012)

10. Huang, J., Korolkiewicz, M., Agrawal, M., Boland, J.: Forecasting solar radiation on an hourly time scale using a coupled autoregressive and dynamical system (cards) model. Sol. Energy **87**, 136–149 (2013)

11. Lu, S., Hwang, Y., Khabibrakhmanov, I., Marianno, F.J., Shao, X., Zhang, J., Hodge, B.M., Hamann, H.F.: Machine learning based multi-physical-model blending for enhancing renewable energy forecast - improvement via situation dependent error correction. In: European Control Conference (ECC), pp. 283–290 (2015)

12. Mazorra Aguiar, L., Pereira, B., Lauret, P., Díaz, F., David, M.: Combining solar irradiance measurements, satellite-derived data and a numerical weather prediction model to improve intra-day solar forecasting. Renew. Energy **97**, 599–610 (2016)

13. Sánchez, I.: Adaptive combination of forecasts with application to wind energy. Int. J. Forecast. **24**(4), 679–693 (2008)

14. Rigollier, C., Bauer, O., Wald, L.: On the clear sky model of the ESRA-European Solar Radiation Atlas-with respect to the heliosat method. Sol. Energy **68**(1), 33–48 (2000)

15. Nonnenmacher, L., Coimbra, C.F.M.: Streamline-based method for intra-day solar forecasting through remote sensing. Sol. Energy **108**, 447–459 (2014)

16. Arbizu-Barrena, C., Ruiz-Arias, J.A., Rodríguez-Benítez, F.J., Pozo-Vázquez, D., Tovar-Pescador, J.: Short-term solar radiation forecasting by advecting and diffusing MSG cloud index. Sol. Energy **155**, 1092–1103 (2017)

17. Jimenez, P.A., Hacker, J.P., Dudhia, J., Haupt, S.E., Ruiz-Arias, J.A., Gueymard, C.A., Thompson, G., Eidhammer, T., Deng, A.: WRF-solar: description and clear-sky assessment of an augmented NWP model for solar power prediction. Bull. Am. Meteorol. Soc. **97**(7), 1249–1264 (2015)

18. Chen, T., Guestrin, C.: XGBoost: a scalable tree boosting system. In: Proceedings of the 22nd ACM SIGKDD International Conference on Knowledge Discovery and Data Mining, KDD 2016, pp. 785–794. ACM (2016)

19. Urraca, R., Antonanzas, J., Antonanzas-Torres, F., Martinez-de-Pison, F.J.: Estimation of daily global horizontal irradiation using extreme gradient boosting machines. In: Graña, M., López-Guede, J.M., Etxaniz, O., Herrero, Á., Quintián, H., Corchado, E. (eds.) ICEUTE/SOCO/CISIS-2016. AISC, vol. 527, pp. 105–113. Springer, Cham (2017). https://doi.org/10.1007/978-3-319-47364-2_11

20. Fan, J., Wang, X., Lifeng, W., Zhou, H., Zhang, F., Xiang, Y., Xianghui, L., Xiang, Y.: Comparison of support vector machine and extreme gradient boosting for predicting daily global solar radiation using temperature and precipitation in humid subtropical climates: a case study in China. Energy Conv. Manag. **164**, 102–111 (2018)

21. Rigollier, C., Lefèvre, M., Wald, L.: The method heliosat-2 for deriving shortwave solar radiation from satellite images. Sol. Energy **77**(2), 159–169 (2004)

22. Skamarock William, C., Joseph, B.K., Jimy, D., David, O.G., Dale, M.B., Michael, G.D., Huang, X.Y., Wang, W., Jordan, G.P.: A description of the advanced research WRF version 3. NCAR technical note, 126 (2008)

23. Long, C.N., Dutton, E.G.: BSRN global network recommended QC tests, v2. x. (2010)

A Hybrid Algorithm for the Assessment of the Influence of Risk Factors in the Development of Upper Limb Musculoskeletal Disorders

Nélida M. Busto Serrano[1], Paulino J. García Nieto[2],
Ana Suárez Sánchez[3]([✉]), Fernando Sánchez Lasheras[2],
and Pedro Riesgo Fernández[3]

[1] Occupational Health and Safety, Oviedo, Spain
[2] Department of Mathematics, University of Oviedo, Oviedo, Spain
[3] Department of Business Administration, University of Oviedo, Oviedo, Spain
suarezana@uniovi.es

Abstract. A hybrid model based on genetic algorithms, classification trees and multivariate adaptive regression splines is applied to identify the risk factors that have the strongest influence on the development of an upper limb musculoskeletal disorder using the data of the Spanish Seventh National Survey on Working Conditions. The study is performed among a sample of workers from the extractive and manufacturing industry sector, where upper limb have been the most frequently reported disorders during 2016.

The considered variables are connected to employment conditions, physical conditions at workplace, safety conditions, workstation design and ergonomics, psychosocial and organizational factors, Health and Safety management and health damages. These variables are either continuous, Liker scale or binary. The chosen output variable is built taking into consideration the presence or absence of three conditions: the existence of upper limb pain, the perception of a work-related nature and the requirement of medical care in relation with it. The results show that WMSD have a multifactorial origin and the categories that include the most relevant variables are: ergonomics and psychosocial factors, workplace conditions and workers' individual characteristics.

Keywords: Hybrid algorithms · Genetic algorithms · Classification tree
Multivariate adaptive regression splines (MARS)
Spanish Seventh National Survey on Working Conditions
Work-related musculoskeletal disorders (WMSD) · Upper limb pain

1 Introduction

Musculoskeletal disorders are injuries and illnesses that affect muscles, nerves, tendons, ligaments, joints or spinal discs that could be caused or aggravated by various hazards or risk factors in the workplace [1].

© Springer International Publishing AG, part of Springer Nature 2018
F. J. de Cos Juez et al. (Eds.): HAIS 2018, LNAI 10870, pp. 634–646, 2018.
https://doi.org/10.1007/978-3-319-92639-1_53

During last years, several studies [2, 3] have focused on the prevalence of musculoskeletal disorders on working population. Companies and public organizations are concerned about this increasing problem that has a serious impact on sick leaves, working productivity, employee's health care costs and other factors that affect the cost-benefit balance. Therefore, throughout the world, many attempts have been carried out to calculate the economic impact of musculoskeletal disorders for enterprises and society [4].

Analyzing latest years, several intervention programs with different results can be found in this area. Some of them were centered in ergonomic assessment and workers training [5]. Other works focused in psychosocial factors [6], or in lifestyle and environmental changes, considering risk factors such as dietary intake and physical activity [7]. Anyway, the key to success for this kind of programs is a correct diagnosis of the risk factors and the causes behind the Work-related Musculoskeletal Disorders (WMSD) to be able to prioritize the most damaging circumstances with regard to intervention. General epidemiological prediction can be a clear option for those who want to define policies and global acting strategies, as is the case of public organizations. On the other hand, enterprises might prefer to know the most frequent causes related with a specific kind of disorder with a high incidence rate in their sector or among a working population typology, and so be able to prevent it.

In this direction, there are some studies [8–10] that focus on ergonomics to explain the reasons that cause some specific WMSD, such as the relation between postures and epicondylitis, or wrist tendinosis, and loads lifting and low-back pain. Other works study psychosocial causes as work stressors, stress and work organization factors among manufacturing workers or burnout syndrome in relation to musculoskeletal complaints among hospital nurses. Also there are studies that look for the causes in the working conditions, such as the impact of vibrations exposure and musculoskeletal disorders in the furniture industry, or physical risk factors among manufacturing workers, or among office workers. Finally, there are authors that consider the workers' personal characteristics as the key factor. Such individual characteristics include factors such as gender, age, demography and lifestyle and origins.

The objective of the present research is to identify the factors with the strongest influence on the appearance of upper-limb musculoskeletal disorders among workers from the manufacturing and extractive industry sector in Spain.

This paper is divided into four parts. This first section is an introduction that synthetizes the state of the art. Section 2 describes the methods, including the case of study and data collection and the development of the algorithms. Finally, Sect. 3 discusses the main results and Sect. 4 summarizes the conclusions of the work.

2 Methods

2.1 Case of Study

The study employs data from the Seventh National Survey on Working Conditions (VII Encuesta Nacional de Condiciones de Trabajo) [11]. This survey was developed by the Spanish National Institute for Safety and Hygiene at Work (INSHT) and it was published in 2007. The results of the survey are presented in the following groups: employment conditions and kind of work, working physical environment, environmental conditions, safety conditions, physical workload, workstation design, psychosocial risk factors, working time, safety-specialized union representatives, preventive activities and health damages.

The chosen output variable was included in the survey's item number 54, which records the worker's perceived health status. Specifically, upper limb pain was selected due to the significant reporting ratio of this kind of injuries by extractive and manufacturing industry employers during 2016 [12]. The chosen disorders include pain in the shoulder, arm, elbow and forearm (excluding wrist, hand and finger pain).

The survey item was recorded including three different levels of importance: having pain, perceiving this pain as work-related and requiring medical care in relation to this disorder. To exploit all the information available, a variable combining the three levels (UppLimb) was constructed and used as the model output. Thus, the relation between each item and the existence and seriousness of an upper-limb musculoskeletal disorder could be studied.

2.2   Data Collection

The population consisted in 18,518,444 workers employed by Spanish enterprises belonging to the whole range of economic activities. The methodology included personal interviews to respond to a questionnaire consisting of 78 items conducted in the homes of 11,054 workers. Data collection was carried out between December 2006 and April 2007. The sampling procedure was multistage, stratified cluster sampling, with a random selection of primary and secondary sampling units, and workers were selected by random routes. For a confidence level of 95.5% and P = Q the error for the overall sample was ±0.95% [13]. Although the survey's initial sample included the whole range of economic activities, this research focuses on extractive and manufacturing industry workers, so the final sample consisted of 1.269 workers from these sectors.

2.3   Genetic Algorithms

A genetic algorithm (GA) is a metaheuristic algorithm inspired by the process of natural selection that belongs to the larger class of evolutionary algorithms (EA). In a genetic algorithm, a *population* of *candidate solutions* (called individuals, creatures, or *phenotypes*) to an optimization problem is evolved toward better solutions. Each candidate solution has a set of properties (its *chromosomes* or *genotype*) which can be mutated and altered. Traditionally, solutions are represented in binary as strings of 0s and 1s, but other encodings are also possible [14, 15].

The evolution usually starts from a population of randomly generated individuals, and is an *iterative process*, with the population in each iteration called a generation. In each generation, the fitness of every individual in the population is evaluated. The fitness is usually the value of the *objective function* in the optimization problem being solved. The more fit individuals are stochastically selected from the current population, and each individual's genome is modified (*recombined* and possibly randomly mutated) to form a new generation. The new generation of candidate solutions is then used in the next iteration of the *algorithm*. Commonly, the algorithm terminates when either a maximum number of generations has been produced, or a satisfactory fitness level has been reached for the population.

A typical genetic algorithm requires:

1. a *genetic representation* of the solution domain;
2. a *fitness function* to evaluate the solution domain.

A standard representation of each candidate solution is as an *array of bits* [16]. Once the genetic representation and the fitness function are defined, a GA proceeds to initialize a population of solutions and then to improve it through repetitive application of the mutation, crossover, inversion and selection operators.

- *Initialization*

The population size depends on the nature of the problem, but typically contains several hundreds or thousands of possible solutions. Often, the initial population is generated randomly, allowing the entire range of possible solutions (the search space).

- *Selection*

During each successive generation, a portion of the existing population is selected to breed a new generation. Individual solutions are selected through a fitness-based process, where *fitter* solutions (as measured by a *fitness function*) are typically more likely to be selected. The next step is to generate a second generation population of solutions from those selected through a combination of *genetic operators*: *crossover* (also called *recombination*), and *mutation*.

For each new solution to be produced, a pair of *parent* solutions is selected for breeding from the pool selected previously. By producing a *child* solution using the above methods of crossover and mutation, a new solution is created which typically shares many of the characteristics of its *parents*. New parents are selected for each new child, and the process continues until a new population of solutions of appropriate size is generated. Although reproduction methods that are based on the use of two parents are more *biology inspired*, some research [17] suggests that more than two *parents* generate higher quality chromosomes. These processes ultimately result in the next generation population of chromosomes that differs from the initial generation.

Figure 1 shows the flowchart of a GA algorithm. Although crossover and mutation are known as the main genetic operators, it is possible to use other operators such as regrouping, colonization-extinction, or migration in genetic algorithms [18]. It is worth tuning parameters such as the *mutation* probability, *crossover* probability and population size to find reasonable settings for the problem class being worked on. A very small mutation rate may lead to *genetic drift* (which is *non-ergodic* in nature). A recombination rate that is too high may lead to premature convergence of the genetic algorithm. A mutation rate that is too high may lead to loss of good solutions, unless *elitist selection* is employed. This generational process is repeated until a termination condition has been reached. In this research, the termination criterion employed for the GA is when the result does not improve after 100 generations.

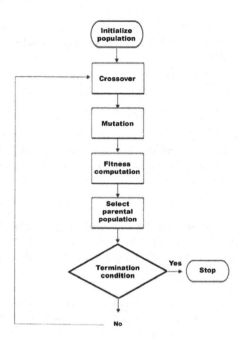

Fig. 1. Algorithm diagram.

2.4   Classification and Regression Trees (CARTs)

In the present research, those variables selected by the GA are employed as input data for CART. The tree is built by the following process: first the single variable is found which best splits the data into two groups ('best' will be defined later). The data is separated, and then this process is applied separately to each sub-group, and so on recursively until the subgroups either reach a minimum size or until no improvement can be made [19]. The resultant model is, with a certainty, too complex, and the question arises as it does with all stepwise procedures of when to stop. The second stage of the procedure consists of using cross-validation to trim back the full tree. Normally, the full tree has a specific number of terminal regions. A cross validated estimate of risk was computed for a nested set of sub trees. The final model was that sub tree with the lowest estimate of risk. In the case of this work, tree construction ends using a predefined stopping criterion, considering a minimum number (0.5%) of training instances assigned to each leaf node of the tree.

2.5   Multivariate Adaptive Regression Splines (MARS)

Multivariate adaptive regression splines (MARS) is a multivariate nonparametric classification/regression procedure [20]. Its main objective is the prediction of the values of a continuous dependent variable, $y(n \times 1)$, from a set of independent input variables, $X(n \times p)$. The MARS model can be written as:

$$y = f(\mathbf{X}) + \mathbf{e} \tag{6}$$

so that f is a weighted sum of basis functions that rely on $\mathbf{X}$ and $\mathbf{e}$ is the error vector whose dimension is $(n \times 1)$. MARS model does not demand any a priori hypotheses about the underlying functional relationship among dependent and independent variables. Otherwise, this relation is established from an ensemble of coefficients and piecewise polynomials of degree q (basis functions), completely driven from the regression data $(\mathbf{X}, \mathbf{y})$. The MARS regression model is built by fitting basis functions to different intervals of the independent variables. Indeed, MARS uses two-sided truncated power functions as spline basis functions, given by the equations [15]:

$$[-(x - t)]_+^q = \begin{cases} (t - x)^q & \text{if } x < t \\ 0 & \text{otherwise} \end{cases} \tag{7}$$

$$[+(x - t)]_+^q = \begin{cases} (t - x)^q & \text{if } x \geq t \\ 0 & \text{otherwise} \end{cases} \tag{8}$$

so that $q\ (\geq 0)$ is the power to which the splines are lifted and which defines the degree of evenness of the resulting function estimate. Note that when $q = 1$, only simple linear splines are evaluated.

The MARS model of a dependent variable $\mathbf{y}$ with M basis functions (terms) can be expressed as:

$$\hat{\mathbf{y}} = \hat{f}_M(\mathbf{x}) = c_0 + \sum_{m=1}^{M} c_m B_m(\mathbf{x}) \tag{9}$$

being $\hat{\mathbf{y}}$ the dependent variable prognosticated by the MARS model, c_0 a constant, $B_m(\mathbf{x})$ the m-th basis function and c_m the coefficient of the m-th basis functions. Specifically, both variables introduced into the model and the knot positions for each individual variable must be optimized. Furthermore, MARS uses the generalized cross-validation (GCV) [20] to define the basis functions included in the model.

2.6   The Algorithm

The algorithm developed for the assessment of the influence of the risk factors in the development of upper limb musculoskeletal disorders, combines GA, classification trees and MARS following the flowchart presented in Fig. 1.

First, training and validation data sets are created using the k-fold cross-validation methodology for $k = 5$. Also, an initial population of variables is created. This population is formed for a total of 50 individual. Each of them is a string of 209 bits, one for each variable. When the bit has value 1 it means that the variable it represents is introduced in the model, while a value of 0 means that the variable is not considered for the model. Figure 2(a) represents two members of a population with 209 bits. The recombination mechanism of the individuals in to create an offspring can be seen in Fig. 2(b). Finally, in that new individual the mutation operator is applied as it can be observed in Fig. 2(c). Using all those active variables, a classification tree is created. In the classification tree not all the variables are employed. Afterwards, and using as input variables those not employed by the classification tree, a MARS model is trained to improve the classification of each one of the terminal nodes of the tree.

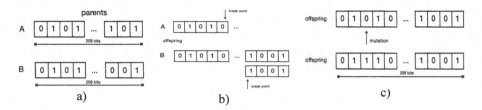

Fig. 2. (a) Example of two members of the population, (b) Recombination of two individuals in an offspring, (c) Mutation mechanism.

The reason why these three algorithms are combined is as follows: first of all, the use of GA allows us to employ a different number of variables in order to introduce in the models those variables that are better in order to predict the development of upper limb musculoskeletal disorders. Afterwards, a classification tree is performed, and it allows to classify individuals taking into account as input variables their answers to questionnaire and assigning them to the available leafs. Finally, different MARS models are developed for each one of the leafs of the regression tree. As a model is performed for each leaf, it is able to perform a most accurate prediction, taking into account those workers classified in each subset.

3   Results and Discussion

3.1   Results

The classification tree obtained is shown in Fig. 3. It employs variables P51, P28_7, P28_1, P26_6_Glp, P26_9_Pr and P55_17. The tree has seven final nodes each one with a number of cases that goes from 7 to 410. Table 1 shows the overall classification ability of the algorithm that makes use of a regression tree for the prediction of the UppLimb values of the individuals in each group. Forecasted values have been rounded to the nearest integer. Columns represents real classification values, while rows are forecasted values. As it can be observed, a total of 758 individuals with a real value of 0 are classified by the model in the same category, while another 148 with a real value of 0 are considered to be 1 and the other 8 with a real value of 0 are considered like 2 by the model. In this case, 75.14% of the individual are assigned to the right UppLimb value; this value can be compared with a benchmark technique MARS, that is only able to classify in the right UppLimb value 61.78% of individuals.

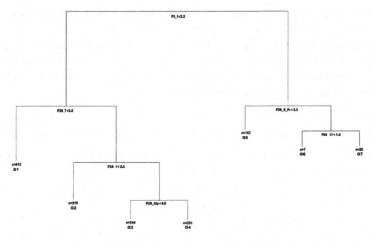

Fig. 3. Classification tree obtained.

Table 1. Overall classification matrix.

		Real values			
		0	1	2	3
Forecasts	0	758	17	8	5
	1	148	23	27	24
	2	8	2	46	60
	3	0	0	4	89

Figure 4 shows the boxplots of the values obtained by the regression model when compared with the real UppLimb values of the two leafs with a largest number of individuals, G1 with 410 and G4 with 244. As it can be observed there is a clear relationship of real UppLimb values when compared with the forecasts obtained. Please note that forecasts are continuous values.

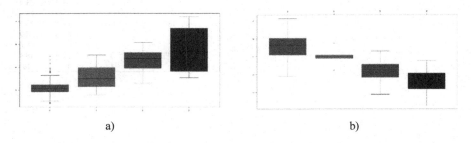

a)   b)

Fig. 4. Forecasted UppLim values versus real values in (a) Group G1 and (b) Group G3.

Table 2 shows those variables that take part in any of the MARS models and the number of models in which they take part. As it can be observed the variable that takes part in most models is P14_antig_puesto with five, followed by P26_16_Carg, P28_2, P30_4, P30_7, P55_1. Variables importance for each of the MARS models was also calculated using the nsubset, RSS and GCV criteria. Finally, Table 3 shows the results of variable importance for group 1 and group 3.

Table 2. Variables that appear in any MARS model-number of models where they appear.

Variable	# of leafs	Variable	# of leafs	Variable	# of leafs
P14_antig_puesto	5	P58_SEXO	3	P55_18	2
P26_16_Carg	4	P17_Mano,Brazo	2	P55_5	2
P28_2	4	P27_16_PF	2	P55_8	2
P30_4	4	P29_1_Nuca	2	P56	2
P30_7	4	P30_2	2	P26_2_PMN	1
P55_1	4	P31_1	2	P27_8_Minad	1
P27_1_Aguj	3	P31_4	2	P30_5	1
P30_3	3	P35_todoh	2	P30_6	1
P31_3	3	P38_1	2	P36_5_EqMT	1
P31_6	3	P43_1_Ruid	2	P47_3_Casc	1
P32_1	3	P50_5	2	P52	1
P38_2	3	P53_15_Noaten	2	P55_6	1
P43_2_AmbT	3	P55_12	2	P55_7	1

Table 3. Variables importance in the MARS model for (a) Group G1 and (b) Group G3.

Variable	nsubsets	GCV	RSS
P28_1	6	100	100
P55_1	5	83.8	75.6
P28_2	4	50.1	40.5
P30_7	3	45.2	32
P26_16_Carg	2	34.5	21.1
P43_1_Ruid	1	18.6	9.8

a)

Variable	nsubsets	GCV	RSS
P32_1	19	100	100
P14_antig_puesto	18	90.3	93.9
P30_3	16	85.1	87.5
P31_6	16	85.1	87.5
P26_16_Carg	16	85.1	87.5
P55_5	15	87.8	86.3
P30_4	12	73.2	74.4
P43_2_AmbT	12	73.2	74.4
P31_4	11	79.4	73.7
P55_18	11	79.4	73.7
P31_3	6	44.8	48.2
P28_2	5	43.7	44.6
P27_16_PF	3	52.4	48.4
P55_8	1	13.6	18

b)

3.2 Discussion

According to the model, there is a relation between suffering upper-limb pain and four different types of factors, as summarized in Table 4.

Table 4. Factors affecting the development of upper-limb pain.

Factors	Classification tree variables		MARS variables (most frequent)	
Individual	P51	Worker's perceived general health status in the last twelve months	P14_antig_puesto	Job seniority
	P55_17	Worker's worry about the risk of suffering a work-related disease		
Ergonomics	P28_7	Exposure to hand and arm repetitive movements	P26_16_Carg	Overexertion due to manual handling tasks
	P28_1	Exposure to painful and tiring postures	P28_2	Exposure to prolonged standing
Workplace conditions	P26_6_Glp	Struck-by hazard		
	P26_9_Pr	Flying objects hazard		
Psychosocial			P30_4	Multitasking demands at work
			P30_7	Carrying out monotonous tasks
			P55_1	Worker's worry about their job autonomy

The worker's perceived general health status, their worry about suffering a disease and their job seniority can be considered as individual factors. The presence of the first two suggests that the worker's lifestyle has an impact on the probability of suffering a musculoskeletal disorder. Indeed, that is one of the main beliefs behind the so-called "workers' behavior modifying programs" that some enterprises carry out to reduce absenteeism, making an effort to control health indicators such as body-mass index, physical activity, sleeping quality, tobacco, alcohol and other drugs consumption, etc. In the case job seniority, it could be an evidence for the negative impact of an aging workforce. That endorses one of the reasons that inspire the EU Occupational Safety and Health (OSH) Strategic Framework 2014–2020 [21] that includes among its objectives addressing the aging of the European workforce.

Secondly, there are some variables related with poor ergonomics, such as the exposure to hand and arm repetitive movements, painful and tiring postures, manual handling and prolonged standing. That is the most predictable result considering that a reasonable consequence of all kind of upper-limb musculoskeletal overload at work is the appearance of some kind of disorder or injury. Anyway, these results confirm the generally accepted principle that avoiding all kind of physical efforts at work helps in controlling absenteeism due to WMSD.

Regarding workplace conditions, the presence of struck-by and flying objects hazards is consistent with the frequent coincidence in industrial environments of hazardous jobs and poor ergonomics. In this sense, the coexistence of mechanical hazards and WMSD could be perceived as a symptom or consequence rather than a cause.

Finally, the inclusion of multitasking, monotonous tasks and worker's worry about job autonomy in the MARS models implies that psychosocial factors play an important role in the development of WMSD. This is consistent with several works suggesting that WMSD could be interpreted as workers' psychosomatic response in relation with psychological demands of the work.

Despite the fact that all the obtained relevant variables are somehow interrelated, the results show how ergonomics play the most important role in the studied disorder. Attending to the factors that can be controlled by the employer (ergonomics, workplace conditions and psychosocial factors), a clear conclusion is that dealing with the WMSD cannot be done partially, but through a comprehensive intervention program to act on the three identified groups. In other words, there is no other option than applying an integrated view of the Health and Safety management where every discipline is included. In this sense, it is important to emphasize the real meaning of the multidisciplinary nature of Health and Safety management, where the involvement of sanitary, engineering and executive professionals is necessary.

On the other hand, it is also important to note the significant absence of some variables among the selected twelve, such as all items related with specific risk assessment activities. That seems to have two possible explanations:

- Specific assessment activities, as those included in ergonomics and industrial hygiene programs, are not effectively carried out in Spanish enterprises, so there are not improvement plans that prevent WMSD.
- Either the specific assessment activities carried out by the Spanish enterprises are not well designed or the subsequent intervention programs are wrongly implemented and do not reach their objectives.

4   Conclusions

WMSD have a multifactorial origin that includes causes related with ergonomics, psychosocial factors, workplace conditions and workers' individual characteristics. Poor ergonomics seems to be the factor with the strongest influence on the development of WMSD. Psychosocial factors can provoke psychosomatic effects among workers generating or aggravating musculoskeletal injuries.

Even though individual characteristics have effects on WMSD, most of their causes lay within the employers' management scope of application. That means the employers hold the key to avoiding WMSD and improving the health status of their workforce.

Every intervention program aimed at reducing WMSD should start from an integrating perspective that includes activities from each health and safety management discipline: technical, sanitary and organizational.

References

1. Arab, R.: MSD prevention. Int. J. Sci. Eng. Res. **5**(5), 1067–1069 (2014)
2. Hildebrandt, V.H.: Back pain in the working population: prevalence rates in Dutch trades and professions. Ergonomics **38**, 1283–1298 (1995)
3. Morken, T., Moen, B., Riise, T., Bergum, O., Bua, L., Hauge, S.H., Holien, S., Langedrag, A., Olson, H.O., Pedersen, S., Saue, I.L., Seljebø, G.M., Thoppil, V.: Prevalence of musculoskeletal symptoms among aluminium workers. Occup. Med. **50**, 414–421 (2000)
4. Hanson, M.A., Burton, K., Kendall, N.A.S., Lancaster, R.J., Pilkington, A.: The costs and benefits of active case management and rehabilitation for musculoskeletal disorders (RR 493). Health and Safety Executive Research Report. HSE, Sudbury (2006)
5. Hoe, V.C.W., Urquhart, D.M., Kelsall, H.L., Sim, M.R.: Ergonomic design and training for preventing work-related musculoskeletal disorders of the upper limb and neck in adults. Cochrane Database Syst. Rev. **8** (2012) Art. No. CD008570
6. Choobineh, A., Motamedzade, M., Kazemi, M., Moghimbeigi, A., Pahlavian, A.H.: The impact of ergonomics intervention on psychosocial factors and musculoskeletal symptoms among office workers. Int. J. Ind. Ergon. **41**, 671–676 (2011)
7. Engbers, L.H., van Poppel, M.N., Chin, M.J., Paw, A., van Mechelen, W.: Worksite health promotion programs with environmental changes: a systematic review. Am. J. Prev. Med. **29**, 61–70 (2005)
8. Eatough, E.M., Way, J.D., Chang, C.H.: Understanding the link between psychosocial work stressors and work-related musculoskeletal complaints. Appl. Ergon. **43**, 554–563 (2012)
9. Luttmann, A., Schmidt, K.H., Jager, M.: Working conditions, muscular activity and complaints of office workers. Int. J. Ind. Ergon. **40**, 549–559 (2010)
10. Govindu, N.K., Babski-Reeves, K.: Effects of personal, psychosocial and occupational factors on low back pain severity in workers. Int. J. Ind. Ergon. **44**, 335–341 (2014)
11. Instituto Nacional de Seguridad e Higiene en el Trabajo: VI Encuesta Nacional de Condiciones de Trabajo (ENCT 200). Ministerio de Trabajo y Asuntos Sociales, Madrid (2007)
12. Secretaria De Estado De La Seguridad Social-Dirección General De Ordenación De La Seguridad Social. Observatorio De Enfermedades Profesionales (CEPROSS) Y De Enfermedades Causadas O Agravadas Por El Trabajo (PANOTRATSS). Informe Anual 2016 (2017)
13. Suárez Sánchez, A., Iglesias-Rodriguez, F.J., Riesgo, P., de Cos Juez, F.: Applying the K-nearest neighbor technique to the classification of workers according to their risk of suffering musculoskeletal disorders. Int. J. Ind. Ergon. **52**, 92–99 (2015)
14. Galán, C.O., Sánchez Lasheras, F., de Cos Juez, F.J., Bernardo Sánchez, A.: Missing data imputation of questionnaires by means of genetic algorithms with different fitness functions. J. Comput. Appl. Math. **311**, 704–717 (2017)
15. García Nieto, P.J., Alonso Fernández, J.R., de Cos Juez, F.J., Sánchez Lasheras, F., Díaz Muñíz, C.: Hybrid modelling based on support vector regression with genetic algorithms in forecasting the cyanotoxins presence in the Trasona reservoir (Northern Spain). Environ. Res. **122**, 1–10 (2013)

16. Michalewicz, Z.: Genetic Algorithms + Data Structures = Evolution Programs. Springer, Heidelberg (1998). https://doi.org/10.1007/978-3-662-03315-9
17. Alonso Fernández, J.R., Díaz Muñiz, C., García Nieto, P.J., de Cos Juez, F.J., Lasheras, F.S., Roqueñí, M.N.: Forecasting the cyanotoxins presence in fresh waters: a new model based on genetic algorithms combined with the MARS technique. Ecol. Eng. **53**, 68–78 (2013)
18. Sánchez Lasheras, F., García Nieto, P.J., de Cos Juez, F.J., Vilán Vilán, J.A.: Evolutionary support vector regression algorithm applied to the prediction of the thickness of the chromium layer in a hard chromium plating process. Appl. Math. Comput. **227**, 164–170 (2014)
19. Hastie, T., Tibshirani, R., Friedman, J.H.: The Elements of Statistical Learning. Springer, New York (2003). https://doi.org/10.1007/978-0-387-84858-7
20. Friedman, J.H.: Multivariate adaptive regression splines. Ann. Stat. **19**, 1–141 (1991)
21. European Commission: Communication from the Commission to the European Parliament, the Council, the European Economic and Social Committee and the Committee of the Regions on an EU Strategic Framework on Health and Safety at Work 2014–2020, Brussels (2014). http://eur-lex.europa.eu/legal-content/EN/TXT/PDF/?uri=CELEX:52014DC0332

Evolutionary Computation on Road Safety

Bruno Fernandes[1] (ID), Henrique Vicente[1,2] (ID), Jorge Ribeiro[3] (ID),
Cesar Analide[1] (ID), and José Neves[1(✉)] (ID)

[1] Centro ALGORITMI, University of Minho, Braga, Portugal
bruno.fmf.8@gmail.com, {analide,jneves}@di.uminho.pt
[2] Department of Chemistry, Évora Chemistry Centre,
University of Évora, Évora, Portugal
hvicente@uevora.pt
[3] School of Technology and Management, ARC4DigiT – Applied Research
Center for Digital Transformation, Polytechnic Institute of Viana do Castelo,
Viana do Castelo, Portugal
jribeiro@estg.ipvc.pt

Abstract. This study examines the psychological research that focuses on road safety in *Smart Cities* as proposed by the *Vulnerable Road Users* (*VRUs*) sphere. It takes into account qualities such as VRUs' personal information, their habits, environmental measurements and things data. With the goal of seeing *VRUs* as active and proactive actors with differentiated feelings and behaviours, we are committed to integrating the social factors that characterize each *VRU* into our social machinery. As a result, we will focus on the development of a *VRU Social Machine* to assess *VRUs'* behaviour in order to improve road safety. The formal background will be to use Logic Programming to define its architecture based on a *Deep Learning* approach to *Knowledge Representation* and *Reasoning*, complemented with an *Evolutionary* approach to *Computing*.

Keywords: Artificial Intelligence · Smart Cities · Vulnerable Road Users
Internet of People · Knowledge Representation and Reasoning
Evolutionary Computation

1 Introduction

The *VRU Social Machine* (*VRUSM*) may be defined as logical based computer system with a focus on *Artificial Intelligence* (*AI*) based methodologies for problem solving. The computing architecture is conceived in terms of a Logical Programming approach to model building and is grounded on a Genetic Programming attitude to computing (Fig. 1). The *Integrated Development Environment* (*IDE*), the algorithms and the data structures were taken from [1]. Further details are presented in Sect. 4.

When the output layer has more than one neuron, it means that there are different theories that model the universe of discourse (Fig. 2), which are ranked according to a *Quality-of-Information's* metric, as it is presented in Sect. 2.

© Springer International Publishing AG, part of Springer Nature 2018
F. J. de Cos Juez et al. (Eds.): HAIS 2018, LNAI 10870, pp. 647–657, 2018.
https://doi.org/10.1007/978-3-319-92639-1_54

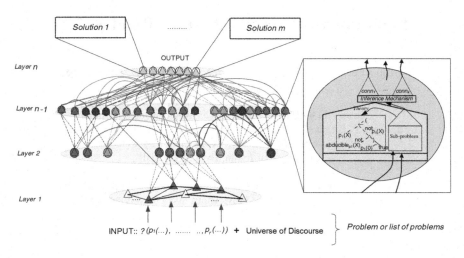

Fig. 1. A schematic representation of an intellect.

$\{$

$\quad \neg\, p \leftarrow not\ p, not\ exception_p$

$\quad p \leftarrow p_1, \cdots, p_n, not\ q_1, \cdots, not\ q_m$

$\quad ?\,(p_1, \cdots, p_n, not\ q_1, \cdots, not\ q_m)\ \ (n, m \geq 0)$

$\quad exception_{p_1}, \quad \cdots \quad , exception_{p_j}\ \ (0 \leq j \leq k), \quad being\ k\ an\ integer\ number$

$\}\, :: scoring_{value}$

Fig. 2. A generic result or solution.

Once a symbolic neuron implements a theorem proving process, it enables *VRUSMs* to benefit from shared information from distinct systems related to safety content and *VRUs* activities, thereby reducing the complexity of system development and maintenance while enhancing personalization, contextualization and interaction. By 2050, some 70% of the world's population will be living in cities [2]. How will we handle this? Indeed, road safety has become a major issue for both car manufacturers and governments, with a focus on reducing traffic accidents and improving the flow of vehicles on the road [3]. However, some issues do yet have to be deciphered, namely those with a strong impact on people outside the vehicle. Hence, one major issue consists on how to enhance the safety of those more vulnerable on the road, known as *Vulnerable Road Users* (*VRUs*). These users have been defined as non-motorized players, i.e., pedestrians, cyclists whose vulnerability stems from multiple directions, such as lack of external protection, physical, motor or visual impairment, or age [4–6]. Focusing on the *VRU* problem, people are seen as the key player who, along with things like vehicles or road infrastructures, spawns the *Internet of People* (*IoP*), an ecosystem in which things and people communicate, sense and understand each other and the world, and act on such data and knowledge with the aim to improve their quality of life. With people as an active, reactive and proactive player, social factors that characterize each *VRU* should also be taken into account [7, 8]. It is now possible to feature what we call a *VRUSM* whose architecture is envisaged as an intellect [9]. Consequently, the next section will describe one's route from *Deep Learning (DL)* to *Knowledge Representation and Reasoning (KRR)*, followed by a case study focused on the *VRU* problem and based on a previously conceived IoP architecture [9]. Afterwards, the evolving system is depicted and explained. Ultimately, conclusions are gathered and directions for future work are outlined.

2 A Deep Learning Route to Knowledge Representation and Reasoning

Knowledge Representation and Reasoning (KRR) aims at the understanding of the information's complexity and the associated inference mechanisms [10]. Indeed, automated reasoning capabilities enables a system to *fill in the blanks* when one is dealing with incomplete information, where data gaps are common, i.e., although *KRR* has been grounded on a symbolic logic in vector spaces, as it will be shown below, the fundamentals and the attributes of the logical functions go from discrete to continuous, allowing for the representation or handling of unknown, vague, or even self-contradictory information/knowledge. Such fact stands for the key distinction of one's approach with relation to *de facto* actual vision, otherwise it would be only symbolic logic in vector spaces, where the data items would remain essentially discrete, and therefore no added value would be attained. In this study, a data item is to be understood as find something smaller inside when taking anything apart, i.e., it is mostly formed from different elements, namely the *Interval Ends* where their values may be situated, the *Quality-of-Information (QoI)* they carry, and the *Degree-of-Confidence (DoC)* put on the fact that their values are inside the intervals just referred to above [11]. These are just three of over an endless element's number. Undeniably, one can make virtually anything one may think of by joining different elements together or, in other words, viz.

- What happens when one splits a data item? The broken pieces become data item for another element, a process that may be endless; and
- Can a data item be broken down? Basically, it is the smallest possible part of an element that still remains the element.

This makes one's route from *Deep Learning (DL)* to *KRR*. Therefore, the proposed approach to this issue, put in terms of the logical programs that elicit the universe of discourse, will be set as productions of the type, viz.

$$predicate_{1 \le i \le n} - \bigcap_{1 \le j \le m} clause_j(([A_{x_1}, B_{x_1}](QoI_{x_1}, DoC_{x_1})),$$
$$\cdots, ([A_{x_m}, B_{x_m}](QoI_{x_m}, DoC_{x_m}))) :: QoI_j :: DoC_j \tag{1}$$

that engender one's view to *DL*. *n*, ∩, *m* and A_{x_m}, B_{x_m} stand for the cardinality of the predicates' set, conjunction, predicate's extension, and the interval ends where the predicates attributes values may be situated, respectively. The metrics $[A_{x_m}, B_{x_m}]$, *QoI* and *DoC* show the way to data item dissection [11, 12], i.e., a data item is to be understood as the data's atomic structure. It consists of identifying not only all the sub items that are thought to make up an data item, but also to investigate the rules that oversee them, i.e., how $[A_{x_m}, B_{x_m}]$, QoI_{x_m}, and DoC_{x_m} are kept together and how much added value is created.

3   Case Study

A *Smart City* is a framework that consists predominantly of *Information and Communication Technologies* (*ICTs*) to develop, deploy and promote sustainable extension of practices to meet the growing challenges of urbanization [13]. A *Smart City* should promote its ability to reason upon the knowledge acquired through data gathered by sensorization, with focus on improving the quality of life at urban centers, considering sustainability and safety principles. Here, sustainability it is to be understood in terms of social, economic and environmental matters [5]. To achieve the *Smart City's* goal, one must add the human being to the *Internet of Things*, empowering the *IoP*, a dynamic global network, an ecosystem, where things and people communicate and understand each other; where everyone and everything can sense the other and the world, and act on such knowledge and information. Indeed, in [9] we had the opportunity to postulate a set of properties we envisage as essential for a successful implementation of the *IoP*. Among those properties, it is the novelty of including the Citizen Sensor as an active, reactive and proactive element that enables the *IoP*. Under this setting, wearable devices, which empower citizens with sensing capabilities, will allow people to join the *IoP* in a passive, non-intrusive, manner.

3.1   Data Collection

On the one hand, one's approach to the problem referred to above focuses on a thorough sensing of the environment, things and *VRUs*, with insights from the *Ambient Intelligence* (*AmI*) field being of the utmost importance, i.e., one is working with environments that are sensitive and responsive to the presence of human beings. On the other hand, special attention should be given to sensing not only the physical properties of the *VRU*, but also their feelings and emotions or even their own soul. Things such as vehicles, bikes or roadside infrastructures are required to sense the environment and to sense the thing itself, providing relevant data such as current position, velocity or direction. It is also imperative to focus on the social perspective of the problem, an environment comprising both humans and technology, interacting and producing actions and information that would not be possible to extract without having both parties present [8]. Therefore, regarding the factors that influence injury risk, the focus will be on the policies, viz.

- Economic strategies;
- The lifestyle of the individual and the conditions under which an individual lives and works; and
- The social and communal networks between individuals, such as the stability of social connections and their social participation.

3.2 Feature Extraction

The data used in this study will be given in terms of the extensions of the relations/tables depicted in Fig. 3, viz.

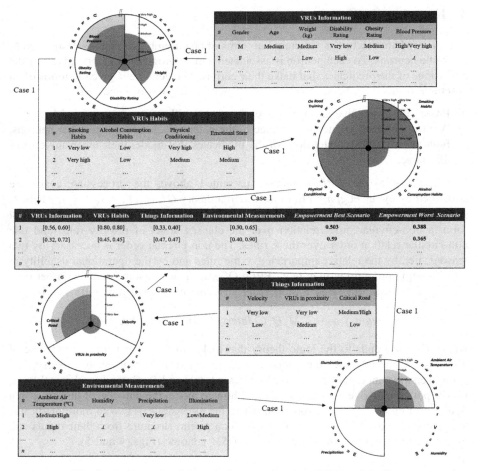

Fig. 3. A fragment of the *IoP* that supports the *VRUSocialMachine*.

- *Environmental Measurements* with attributes *Temperature, Humidity, Precipitation* and *Illumination*;
- *VRUs Information* with attributes as *Gender, Age, Weight, Disability Rating, Obesity Rating* and *Blood Pressure*;
- *VRU Habit's* with attributes *Smoking, Alcohol Consumption, Physical Exercise* and *Emotional State*; and
- *Things Information's* with attributes *Velocity, VRUs in Proximity*, and *Critical Road*.

Data on *VRU* movements such as *Position, Direction, Speed* are not considered. Last but not least, the qualitative attribute's values presented in the relations or tables depicted in Fig. 3 are set according to the scale *very low, low, average, high* and *very high*. For the sake of presentation all the functions' attributes have the same weight.

4 Evolving Systems

The *VRUSM* architecture is structured in terms of an ensemble of entities designated as symbolic neurons. To each neuron is associated a logic program or theory, given by the extensions of the predicates that make their corpus. The genome is given in terms of an ensemble of neurons, being each one coded with two types of genes, viz.

- Processing genes that specify how each neuron will assess its output; and
- A set of connection ones which specify the potential connections to other neurons, built in terms of the extensions of the predicates that model their inner universe of discourse.

According to the *KRR* formalism presented above, the processing genes are structured by the label $gpn_n(t)$ from the ordered theory $OT = (T, <, (S, \prec))$, where T, $<$, S and $\prec$ correspond, respectively, to the knowledge base of the gene in a clausal form, a non-circular order relation over the clauses, a set of priority rules and one non-circular relation order over those rules. The non-circular order is necessary by two reasons, i.e., by the relative importance of the rules and by the operational usability in which a logic program written (e.g., PROLOG) needs to set some concrete order over the set of rules. In this sense a processing gene can be described as follows, viz.

$$gpn(t) = <T, M, Q, C, Intervals\ Ends, QoI, DoC >$$

where T corresponds to the logic theory that make up the inner neuron's universe of discourse, M the inference mechanism, Q the question (or sub-problem) to solve, and C the scenarios under which Q is to be addressed. *QoI* and *DoC* stand for themselves. It is now possible to give a schematic view of how to model the universe of discourse in a dynamic or evolutionary environment, where the extensions of two or more predicates are projected into a fusion space that inherits a partial structure from their inputs and emerge with a structure of its own, the *VRUSM* genome (Figs. 4 and 5).

Fig. 4. The genome's archetype [1].

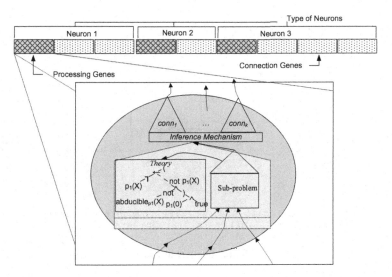

Fig. 5. The genome's scheme and its inner inference mechanisms.

The input of each neuron is given as a list of sub-problems to be solved according to the diverse scenarios referred to above (Fig. 1). Indeed, the evolutionary process to set the *VRUSocialMachine* (*VRUSM*) starts with an estimate representation of the universe of discourse in terms of its predicates' extensions and proceeds in order to optimize their attributes' metrics, i.e., the interval ends $[A_{x_j}, B_{x_j}]$, QoI_j, and DoC_j, once a problem to be solved is set as a theorem to be proved, i.e., the intellect's building is based on a continuous theorem proving process, which is depicted in Figs. 6, 7 and 8, viz.

A set with 270 records [8] was used to the *VRUSM's* analysis. Table 1 presents the *VRUSM's* confusion matrix [14], being the displayed values the average of 25 (twenty-five) runs. A perusal to Table 1 shows that the model accuracy was 92.6% (i.e., 250 instances correctly classified in 270). Based on confusion matrix it is possible to compute the *VRUSM's Sensitivity, Specificity, Positive Predictive Value* (*PPV*) and *Negative Predictive Value* (*NPV*) [15, 16]. *Sensitivity* measures the proportion of *True Positives* (*TP*) that are correctly identified as such, while *Specificity* measures the proportion of *True Negatives* (*TN*) that are correctly identified. *PPV* stands for the proportion of cases with positive values that were correctly diagnosed, while *NPV*

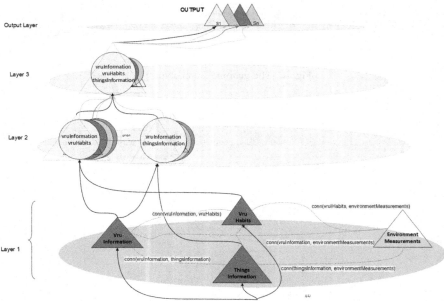

Input - ?(*vruInformation*(ID,A,B,C,D,E,F),*thingsInformation*(ID,G,H,I),
        vruHabits(ID,J,K,L,M)) + Initial Universe of Discourse

Fig. 6. Intellect at t = 0.

Input - ?(*thingsInformation*(ID,A,B,C),*vruHabits*(ID,D,E,F,G),
environment Measurements(ID,H,I,J,K)) + Initial Universe of Discourse

Fig. 7. Intellect at t = 1.

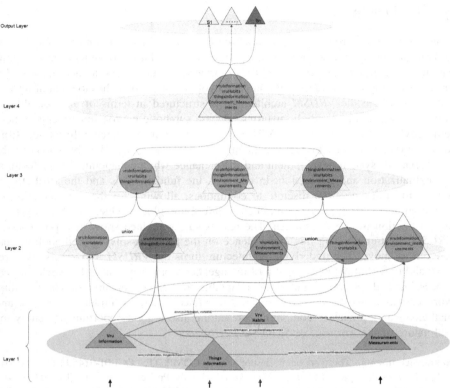

Input - ? (vruInformation (ID,A,B,C,D,E,F), thingsInformation(ID,G,H,I),
vruHabits(ID,J,K,L,M), environment_Measurements(ID,N,O,P,Q)) + Universe of
Discourse

Fig. 8. A view of the intellect's making and firming at present time.

Table 1. The *VRUSM's* confusion matrix.

Output	Model output	
	True (1)	False (0)
True (1)	TP = 139	FN = 8
False (0)	FP = 12	TN = 111

denotes the proportion of cases with negative values that were successfully labeled [15, 16]. The *Sensitivity* is 94.6% while the *Specificity* is 90.2%. *PPV*, in turns, is 92.1% whereas *NPV* is 93.3%. All the performance metrics mentioned above are higher than 90% and seem to suggest that the *VRUSM* performs well in advising the *VRUs* in their daily deals.

5   Conclusions

This work sets one route to apply *Machine Learning* and *Evolutionary Computation* tools to define suitable ways to develop *symbolic machines* and understand its application in the framework of *Smart Cities*. Indeed, in this work a methodology for problem solving grounded on symbolic and evolutionary approaches to computing was presented, where the *VRUSM* architecture is structured in terms of an ensemble of entities designated as symbolic neurons. Once a symbolic neuron implements a theorem proving process, it enables VRUSMs to benefit from shared information from distinct systems related to safety content and VRUs activities, thereby reducing the complexity of system development and maintenance while enhancing personalization, contextualization and interaction. In addition, the fundamentals and the attributes of logical functions go from discrete to continuous, allowing for the representation of unknown, vague or even self-contradictory information/knowledge, which stands for a key distinction of one's approach. The results so far attained show how promising a *VRUSM* is, strengthening one's confidence on the problem-solving methodology just referred to above. The underlying architecture turns the *VRUSM* into a versatile, creative and powerful computational tool to engender a practically infinite variety of data processing and analysis capabilities, adaptable to any conceivable situation. Future work encompasses the development of *Conscious Machines*, under a symbolic and mathematical approach to computing, presenting also a good opportunity to study the real nature of the *Artificial* or *Synthetic Intelligence*.

Acknowledgments. This work has been supported by COMPETE: POCI-01-0145-FEDER-007043 and FCT – *Fundação para a Ciência e Tecnologia* within the Project Scope: UID/CEC/00319/2013, being partially supported by a Portuguese doctoral grant, SFRH/BD/130125/2017, issued by FCT in Portugal.

References

1. Neves, J., Ribeiro, J., Pereira, P., Alves, V., Machado, J., Abelha, A., Novais, P., Analide, C., Santos, M., Fernandez-Delgado, M.: Evolutionary Intelligence in asphalt pavement modelling and quality-of-information. Progress Artif. Intell. J. 1(1), 119–135 (2012)
2. United Nations: World's Population Increasingly Urban with More Than Half Living in Urban Areas (2014). http://www.un.org/en/development/desa/news/population/world-urbanization-prospects-2014.html. Accessed 22 Jan 2018
3. Barba, C.T., Mateos, M.A., Soto, P.R., Mezher, A.M., Igartua, M.A.: Smart city for VANETs using warning messages, traffic statistics and intelligent traffic lights. In: Proceedings of the 2012 IEEE Intelligent Vehicles Symposium, pp. 902–907. IEEE Edition (2012)
4. Liebner, M., Klanner, F., Stiller, C.: Active safety for vulnerable road users based on smartphone position data. In: Proceedings of the 2013 IEEE Intelligent Vehicles Symposium (IV), pp. 256–261. IEEE Edition (2013)
5. Fernandes, B., Neves, J., Analide, C.: Road safety and vulnerable road users – internet of people insights. In: Proceedings of the 6th International Conference on Smart Cities and Green ICT Systems, vol. 1, pp. 311–316. Scitepress (2017)

6. European Parliament: Directive 2010/40/EU of the European Parliament and of the Council of 7 July 2010 on the framework for the deployment of Intelligent Transport Systems in the field of road transport and for interfaces with other modes of transport. Official Journal of European Union 53, L 207/1– L 207/13 (2010)
7. Hernández-Muñoz, J.M., Vercher, J.B., Muñoz, L., Galache, J.A., Presser, M., Hernández Gómez, L.A., Pettersson, J.: Smart cities at the forefront of the future internet. In: Domingue, J., et al. (eds.) FIA 2011. LNCS, vol. 6656, pp. 447–462. Springer, Heidelberg (2011). https://doi.org/10.1007/978-3-642-20898-0_32
8. Fernandes, B., Vicente, H., Neves, J., Analide, C.: Towards road safety – a social perception. In: Proceedings of the 2018 Intelligent Systems Conference (IntelliSys 2018), 8 pp. IEEE Edition (2018, to appear)
9. Fernandes, B., Neves, J., César, A.: Envisaging the internet of people an approach to the vulnerable road users problem. In: De Paz, J.F., Julián, V., Villarrubia, G., Marreiros, G., Novais, P. (eds.) ISAmI 2017. AISC, vol. 615, pp. 104–111. Springer, Cham (2017). https://doi.org/10.1007/978-3-319-61118-1_14
10. Neves, J.: A logic interpreter to handle time and negation in logic databases. In: Muller, R., Pottmyer, J. (eds.) Proceedings of the 1984 Annual Conference of the ACM on the 5th Generation Challenge, pp. 50–54. ACM, New York (1984)
11. Fernandes, F., Vicente, H., Abelha, A., Machado, J., Novais, P., Neves J.: Artificial neural networks in diabetes control. In: Proceedings of the 2015 Science and Information Conference (SAI 2015), pp. 362–370. IEEE Edition (2015)
12. Silva, A., Vicente, H., Abelha, A., Santos, M.F., Machado, J., Neves, J., Neves, J.: Length of stay in intensive care units – a case base evaluation. In: Fujita, H., Papadopoulos, G.A. (eds.) New Trends in Software Methodologies, Tools and Techniques, Frontiers in Artificial Intelligence and Applications, vol. 286, pp. 191–202. IOS Press, Amsterdam (2016)
13. Harrison, C., Eckman, B., Hamilton, R., Hartswick, P., Kalagnanam, J., Paraszczak, J., Williams, P.: Foundations for smarter cities. IBM J. Res. Dev. **54**(4), 1–16 (2010)
14. Kohavi, R., Provost, F.: Glossary of terms. Mach. Learn. **30**(2/3), 271–274 (1998)
15. Florkowski, C.M.: Sensitivity, specificity, receiver-operating characteristic (ROC) curves and likelihood ratios: communicating the performance of diagnostic tests. Clin. Biochem. Rev. **29**(Suppl. 1), S83–S87 (2008)
16. Vilhena, J., Vicente, H., Martins, M.R., Grañeda, J., Caldeira, F., Gusmão, R., Neves, J., Neves, J.: A case-based reasoning view of thrombophilia risk. J. Biomed. Inform. **62**, 265–275 (2016)

Memetic Modified Cuckoo Search Algorithm with ASSRS for the SSCF Problem in Self-Similar Fractal Image Reconstruction

Akemi Gálvez[1,2], Andrés Iglesias[1,2(✉)], Iztok Fister[3], Iztok Fister Jr.[3], Eneko Osaba[4], and Javier Del Ser[4,5,6]

[1] Toho University, 2-2-1 Miyama, Funabashi 274-8510, Japan
[2] University of Cantabria, Avenida de los Castros s/n, 39005 Santander, Spain
iglesias@unican.es
[3] University of Maribor, Smetanova, Maribor, Slovenia
[4] TECNALIA, Derio, Spain
[5] University of the Basque Country (UPV/EHU), Bilbao, Spain
[6] Basque Center for Applied Mathematics (BCAM), Bilbao, Spain
http://personales.unican.es/iglesias

Abstract. This paper proposes a new memetic approach to address the problem of obtaining the optimal set of individual *Self-Similar Contractive Functions* (SSCF) for the reconstruction of self-similar binary IFS fractal images, the so-called SSCF problem. This memetic approach is based on the hybridization of the modified cuckoo search method for global optimization with a new strategy for the Lévy flight step size (MMCS) and the adaptive step size random search (ASSRS) heuristics for local search. This new method is applied to some illustrative examples of self-similar fractal images with satisfactory graphical and numerical results. Our approach represents a substantial improvement with respect to a previous method based on the original cuckoo search algorithm for all contractive functions of the examples in this paper.

Keywords: Image reconstruction · Swarm intelligence
Cuckoo search algorithm · Fractal images · Iterated function systems
Contractive functions

1 Introduction

Fractals are one of the most interesting mathematical objects ever defined. They are also very popular in science due to their ability to describe many growing patterns and natural structures (branches of trees, river networks, coastlines, mountain ranges, and so on). Furthermore, fractals have also found remarkable applications in computer graphics, scientific visualization, image processing, dynamical systems, medicine, biology, arts, and other fields [1,2,8–10].

© Springer International Publishing AG, part of Springer Nature 2018
F. J. de Cos Juez et al. (Eds.): HAIS 2018, LNAI 10870, pp. 658–670, 2018.
https://doi.org/10.1007/978-3-319-92639-1_55

One of the most popular methods to obtain fractals images is the *Iterated Function Systems* (IFS), conceived by Hutchinson [11] and popularized by Barnsley in [1]. Roughly, an IFS consists of a finite system of contractive maps on a complete metric space. Any IFS has a unique non-empty compact fixed set $\mathcal{A}$ called the *attractor of the IFS*. The graphical representation of this attractor is (at least approximately) a self-similar fractal image. Conversely, each self-similar fractal image can be represented by an IFS. Obtaining the parameters of such IFS is called the *IFS inverse problem*. Basically, it consists of solving an image reconstruction problem: given a self-similar fractal image, determine the IFS whose attractor approximates such input image accurately. This IFS inverse problem is so difficult that only partial solutions have been reached so far. A very promising strategy is to split up the problem into two steps: firstly, obtain a suitable collection of individual self-similar contractive functions for the IFS, the so-called *SSCF problem*. The output of this step is then applied to compute the optimal solution for the general IFS inverse problem.

A previous paper addressed this first step by using the cuckoo search (CS) algorithm [13]. Although the method provided nice visual results, its accuracy was far from optimal, and can still be improved. Recently, the original CS has been improved and modified for better performance. In this sense, the present paper proposes a new hybrid scheme based on the CS and called *Memetic Modified Cuckoo Search* (MMCS). Our approach combines two techniques: firstly, we consider a variant proposed in [17] of the original cuckoo search algorithm for global optimization and called *Modified Cuckoo Search* (MCS). This variant is based on two important modifications: (1) the value of the Lévy flight step size is changed dynamically with the iterations; (2) the addition of information exchange between the eggs to speed up convergence to the optimum. In our approach, the Lévy flight step size is changed according to a new strategy proposed in this paper. This technique is hybridized with the *Adaptive Step Size Random Search* (ASSRS), a local search heuristics based on changing adaptively the radius of the hypersphere around the most promising solutions for higher accuracy and to escape from local optima.

The structure of this paper is as follows: Sect. 2 introduces the mathematical background about the iterated function systems and the SSCF problem. Then, Sect. 3 describes the original and the modified cuckoo search algorithms. Our proposed MMCS method is described in detail in Sect. 4, while the experimental results are briefly discussed in Sect. 5. The paper closes with the main conclusions and some ideas about future work in the field.

2   Mathematical Background

2.1   Iterated Function Systems

An *Iterated Function System* (IFS) is a finite set $\{\phi_i\}_{i=1,\ldots,\eta}$ of contractive maps $\phi_i : \Omega \longrightarrow \Omega$ defined on a complete metric space $\mathcal{M} = (\Omega, \Psi)$, where $\Omega \subset \mathbb{R}^n$ and Ψ is a distance on Ω. We refer to the IFS as $\mathcal{W} = \{\Omega; \phi_1, \ldots, \phi_\eta\}$. For visualization purposes, in this paper we consider that the metric space (Ω, Ψ) is

$\mathbb{R}^2$ along with the Euclidean distance d_2, which is a complete metric space. In this case, the affine transformations ϕ_κ are of the form:

$$\begin{bmatrix} \xi_1^* \\ \xi_2^* \end{bmatrix} = \phi_\kappa \begin{bmatrix} \xi_1 \\ \xi_2 \end{bmatrix} = \begin{bmatrix} \theta_{11}^\kappa & \theta_{12}^\kappa \\ \theta_{21}^\kappa & \theta_{22}^\kappa \end{bmatrix} \cdot \begin{bmatrix} \xi_1 \\ \xi_2 \end{bmatrix} + \begin{bmatrix} \sigma_1^\kappa \\ \sigma_2^\kappa \end{bmatrix} \qquad (1)$$

or equivalently: $\Phi_\kappa(\Xi) = \Theta_\kappa . \Xi + \Sigma_\kappa$ where Σ_κ is a translation vector and Θ_κ is a 2×2 matrix with eigenvalues $\lambda_1^\kappa, \lambda_2^\kappa$ such that $|\lambda_j^\kappa| < 1$. In fact, $\mu_\kappa = |det(\Theta_\kappa)| < 1$ meaning that ϕ_κ shrinks distances between points. Let us now define a transformation, Υ, in the set of compact subsets of Ω, $\mathcal{H}(\Omega)$, by

$$\Upsilon(\mathcal{S}) = \bigcup_{\kappa=1}^{\eta} \phi_\kappa(\mathcal{S}). \qquad (2)$$

If all the ϕ_κ are contractions, Υ is also a contraction in $\mathcal{H}(\Omega)$ with the induced Hausdorff metric [1,11]. Then, according to the fixed point theorem, Υ has a unique fixed point, $\Upsilon(\mathcal{A}) = \mathcal{A}$, called the *attractor of the IFS*.

Let us now consider a set of probabilities $\mathcal{P} = \{\omega_1, \ldots, \omega_\eta\}$, with $\sum_{\kappa=1}^{\eta} \omega_\kappa = 1$. There exists an efficient method, known as *probabilistic algorithm*, for the generation of the attractor of an IFS. Picking an initial point ξ_0, one of the mappings in the set $\{\phi_1, \ldots, \phi_\eta\}$ is chosen at random using the weights $\{\omega_1, \ldots, \omega_\eta\}$ and then applied to generate a new point; the same process is repeated again with the new point and so on. As a result, we obtain a sequence of points that converges to the fractal as the number of points increases. This set of points represents graphically the attractor of the IFS.

2.2   The Self-Similar Contractive Functions (SSCF) Problem

Suppose that we are given an initial self-similar fractal image $\mathcal{I}^\square$. The *Collage Theorem* says that it is possible to obtain an IFS $\mathcal{W}$ whose attractor has a graphical representation $\mathcal{I}^\blacksquare$ that approximates $\mathcal{I}^\square$ accurately according to a error function $\mathcal{E}$ between $\mathcal{I}^\square$ and $\mathcal{I}^\blacksquare$. Note that $\mathcal{I}^\blacksquare = \Upsilon(O^\square)$ for any image $O^\square$. Mathematically, this means that we have to solve the optimization problem:

$$\underset{\{\Theta_\kappa, \Sigma_\kappa, \omega_\kappa\}_{\kappa=1,\ldots,\eta}}{minimize} \quad \mathcal{E}\left(\mathcal{I}^\square, \Upsilon(O^\square)\right) \qquad (3)$$

which is a continuous constrained optimization problem, since all free variables in $\{\Theta_\kappa, \Sigma_\kappa, \omega_\kappa\}_\kappa$ are real-valued and must satisfy the condition that all ϕ_κ have to be contractive. It is also a multimodal problem, since there can be several global or local minima of the error function. So far only partial solutions have been reported, but the general problem remains unsolved.

A promising strategy to tackle this issue is to solve firstly the sub-problem of computing a suitable collection of self-similar contractive functions for the IFS (this is called the *SSCF problem*). However, even this SSCF problem is challenging because we do not have any information about the number of contractive functions and their parametric values. To overcome this limitation, a previous paper applied a given number of contractive maps ϕ_κ onto the original fractal image $\mathcal{I}^\square$ and compare the resulting images according to the error function $\mathcal{E}$ in order to obtain suitable values for the SSCF parameters [13]. With this strategy, the original problem (3) was transformed into the optimization problem:

$$\underset{\{\Theta_\kappa, \Sigma_\kappa, \omega_\kappa\}_{\kappa=1,\dots,\eta}}{minimize} \quad \mathcal{E}\left(\mathcal{I}^\square, \phi_\kappa(\mathcal{I}^\square)\right) \qquad\qquad (\kappa = 1, \dots, \eta) \qquad (4)$$

The cuckoo search algorithm was applied to solve this optimization problem [13]. Unfortunately, although the reconstructed figures looked nice visually, the accuracy was far from optimal in terms of the numerical similarity error rates. In this paper, we modify that CS-based method to improve those results.

3   The Cuckoo Search Algorithms

3.1   Original Cuckoo Search (CS)

The *cuckoo search* (CS) is a powerful metaheuristic algorithm originally proposed by Yang and Deb in 2009 [19]. Since then, it has been successfully applied to difficult optimization problems [4,12,18,20]. The algorithm is inspired by the obligate interspecific brood-parasitism of some cuckoo species that lay their eggs in the nests of host birds of other species to escape from the parental investment in raising their offspring and minimize the risk of egg loss to other species.

This interesting breeding behavioral pattern is the metaphor of the cuckoo search metaheuristic approach for solving optimization problems. In this algorithm, the eggs in the nest are interpreted as a pool of candidate solutions while the cuckoo egg represents a new coming solution. The ultimate goal of the method is to use these new (and potentially better) solutions associated with the parasitic cuckoo eggs to replace the current solution associated with the eggs in the nest. This replacement, carried out iteratively, will eventually lead to a very good solution of the problem. In addition to this representation scheme, the CS algorithm is also based on three idealized rules [19,20]:

1. Each cuckoo lays one egg at a time, and dumps it in a randomly chosen nest;
2. The best nests with high quality of eggs (solutions) will be carried over to the next generations;
3. The number of available host nests is fixed, and a host can discover an alien egg with a probability $p_a \in [0,1]$. For simplicity, this assumption can be approximated by a fraction p_a of the n nests being replaced by new nests (with new random solutions at new locations).

The basic steps of the CS algorithm are summarized in Table 1. It starts with an initial population of n host nests and it is performed iteratively. The initial values of the jth component of the ith nest are determined by the expression $x_i^j(0) = rand.(up_i^j - low_i^j) + low_i^j$, where up_i^j and low_i^j represent the upper and lower bounds of that jth component, respectively, and $rand$ represents a standard uniform random number on the interval $(0,1)$. With this choice, the

Table 1. Cuckoo search algorithm via Lévy flights as originally proposed in [19, 20].

Algorithm: Cuckoo Search via Lévy Flights

---

begin
    Objective function $f(\mathbf{x})$, $\mathbf{x} = (x_1, \ldots, x_D)^T$
    Generate initial population of n host nests $\mathbf{x}_i$ $(i = 1, 2, \ldots, n)$
    while $(t < MaxGeneration)$ or (stop criterion)
        Get a cuckoo (say, i) randomly by Lévy flights
        Evaluate its fitness F_i
        Choose a nest among n (say, j) randomly
        if $(F_i > F_j)$
            Replace j by the new solution
        end
        A fraction (p_a) of worse nests are abandoned and new ones
                are built via Lévy flights
        Keep the best solutions (or nests with quality solutions)
        Rank the solutions and find the current best
    end while
    Postprocess results and visualization
end

---

initial values are within the search space domain. These boundary conditions are also controlled in each iteration step. For each iteration t, a cuckoo egg i is selected randomly and new solutions $\mathbf{x}_i^{t+1}$ are generated by using the Lévy flight. The general equation for the Lévy flight is given by:

$$\mathbf{x}_i^{t+1} = \mathbf{x}_i^t + \alpha \oplus levy(\lambda) \qquad (5)$$

where $\alpha > 0$ indicates the step size (usually related to the scale of the problem) and $\oplus$ indicates the entry-wise multiplication. The second term of Eq. (5) is a transition probability modulated by the Lévy distribution as:

$$levy(\lambda) \sim t^{-\lambda}, \qquad (1 < \lambda \leq 3) \qquad (6)$$

which has an infinite variance with an infinite mean. The authors in [20] suggested to use the Mantegna's algorithm, which computes the factor:

$$\hat{\phi} = \left(\frac{\Gamma(1 + \hat{\beta}).sin\left(\frac{\pi.\hat{\beta}}{2}\right)}{\Gamma\left(\left(\frac{1+\hat{\beta}}{2}\right).\hat{\beta}.2^{\frac{\hat{\beta}-1}{2}}\right)} \right)^{\frac{1}{\hat{\beta}}} \qquad (7)$$

where Γ denotes the Gamma function and $\hat{\beta} = 3/2$ in [20]. This factor is used in Mantegna's algorithm to compute the step length as: $\varsigma = u/|v|^{\frac{1}{\hat{\beta}}}$, where u and v follow the normal distribution of zero mean and deviation σ_u^2 and σ_v^2, respectively, where σ_u obeys the Lévy distribution given by Eq. (7) and $\sigma_v = 1$. Then, the stepsize ζ is computed as $\zeta = 0.01\,\varsigma\,(\mathbf{x} - \mathbf{x}_{best})$. Finally, $\mathbf{x}$ is modified as: $\mathbf{x} \leftarrow \mathbf{x} + \zeta.\boldsymbol{\Delta}$ where $\boldsymbol{\Delta}$ is a random vector that follows the normal distribution $N(0,1)$. The CS evaluates the fitness of the new solution and compares it with the current one. In case that the new solution brings better fitness, it replaces the current one. On the other hand, a fraction of the worse nests are abandoned and replaced by new solutions to increase the exploration of the search space looking for more promising solutions. The rate of replacement is given by the probability p_a, a parameter of the model that has to be tuned for better performance. Moreover, for each iteration step, all current solutions are ranked according to their fitness and the best solution reached so far is stored as the vector $\mathbf{x}_{best}$.

3.2   Modified Cuckoo Search (MCS)

The *modified cuckoo search* (MCS) method [17] aims at improving the performance of the original CS described above through two important modifications:

1. the value of the Lévy flight step size, α, assumed constant in the CS, is decreased with the number of iterations. The reason is to promote local search as the individuals get closer to the solution, in a rather similar way to the inertia weight in PSO. In [17] an initial value of the Lévy step size $\alpha^0 = 1$ is chosen. At each generation t, the new step size is computed adaptively as:

$$\alpha^t = \frac{\alpha^0}{\sqrt{t}} \tag{8}$$

This modification is only applied on the set of nests to be abandoned.

2. the addition of information exchange between the eggs to speed up convergence to the optimum. In the original CS, the search relies on random walks so fast convergence is not guaranteed. In the MCS, some eggs with the best fitness are selected for a set of top eggs. For each of the top eggs, a second egg is chosen randomly and then a third egg is generated on the path from the top egg to the second one, at a distance given by the inverse of the golden ratio $\varphi = (1 + \sqrt{5})/2$, so that it gets closer to the top egg.

With these modifications, the MCS performs better than the CS for several examples, showing a higher convergence rate to the actual global minimum [17].

4   Proposed Approach

4.1   Memetic Modified Cuckoo Search (MSA-MCS)

To address the SSCF problem, a new hybrid CS scheme called *Memetic Modified Cuckoo Search* is proposed. Now, the exploration-exploitation trade-off is achieved through the combination of two techniques:

1. We adopt the MCS method for global optimization. However, instead of the adaptive method in Eq. (8), we consider a new strategy to modify α dynamically, given by:

$$\alpha^{t+1} = \alpha^t \, Exp \left[-2\pi \left(\frac{t-1}{\Lambda} \right) \right] \tag{9}$$

where Λ denotes the maximum number of iterations. The main difference between both strategies is that the values for α at early iterations are larger for Eq. (9), and the opposite for last iterations (i.e., Eq. (9) boosts a larger exploration at early stages and a larger exploitation at late stages).

2. This global-search technique is then hybridized with a local-search heuristics: the *Adaptive Step Size Random Search* technique [16]. It is based on the idea of changing adaptively the radius of the hypersphere around the most promising solutions for higher accuracy and to escape from local optima. Roughly, the method starts by sampling two points from a hypersphere surrounding the most promising solutions (using Marsaglia's technique [14]). These two points are sampled at different radius, the current one and a larger step in each iteration; the larger is accepted whenever it leads to an improved result. If neither of the two step values lead to improvement for several iterations in a row, smaller step sizes are taken, and the algorithm continues.

Of course, these new features introduce new control parameters in our method, that have also to be properly tuned. This issue will be discussed in Sect. 4.3.

4.2   Application to the SSCF Problem

Given a 2D self-similar binary fractal image $\mathcal{I}^\square$, we apply the MSA-MCS method to solve the SSCF problem. We consider an initial population of χ individuals $\{\mathcal{C}_i\}_{i=1,\dots,\chi}$, where each individual $\mathcal{C}_i = \{\mathcal{C}_i^\kappa\}_\kappa$ is a collection of η real-valued vectors $\mathcal{C}_\kappa^i$ of the free variables of Eq. (1), as:

$$\mathcal{C}_\kappa^i = (\theta_{1,1}^{\kappa,i}, \theta_{1,2}^{\kappa,i}, \theta_{2,1}^{\kappa,i}, \theta_{2,2}^{\kappa,i} | \sigma_1^{\kappa,i}, \sigma_2^{\kappa,i} | \omega_\kappa^i) \tag{10}$$

These individuals are initialized with uniform random values in $[-1, 1]$ for the variables in Θ_κ and Σ_κ, and in $[0, 1]$ for the ω_κ^i, such that $\sum_{\kappa=1}^\eta \omega_\kappa^i = 1$. After this initialization step, we compute the contractive factors μ_κ and reinitialize all functions ϕ_κ with $\mu_\kappa \geq 1$ to ensure that only contractive functions are included in the initial population. Before applying our method, we also need to define a suitable fitness function. Different metrics can be used for our problem. The most natural choice is the Hausdorff distance, but it is computationally expensive and inefficient for this problem. In this paper the Hamming distance is used instead: we consider the fractal images as binary bitmap images on a grid of pixels for a given resolution defined by a mesh size parameter, m_s. This yields matrices with 0s and 1s, where 1 means that the pixel is activated and 0 otherwise. Then, we count the number of mismatches between the original and the reconstructed matrices to determine the similarity error rate between both images. Dividing this value by the total number of active pixels in the image yields the *normalized similarity error rate* (NSER). This is the fitness function used in this paper.

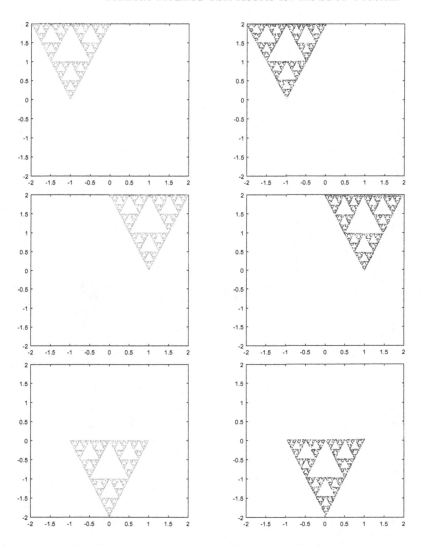

Fig. 1. Graphical results for the Sierpinsky gasket fractal: (left) original images of the three contractive functions; (right) reconstructed images with the MSA-MCS method. (Color figure online)

4.3   Parameter Tuning

The parameter tuning of metaheuristics is slow, difficult, and problem-dependent. Fortunately, the cuckoo search is specially advantageous in this regard, as it only depends on two parameters: the population size, χ, and the probability p_a. We carried out some numerical trials for different values of these parameters and found that $\chi = 40$ and $p_a = 0.25$ are very adequate for our problem.

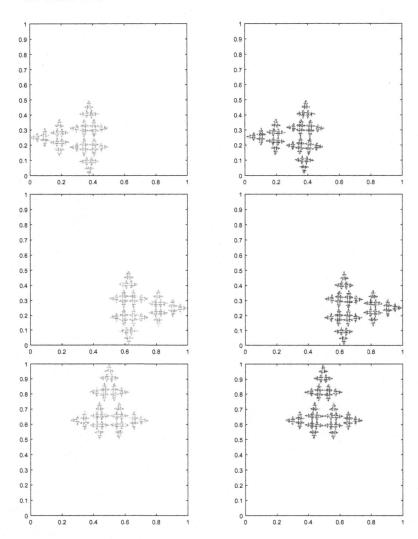

Fig. 2. Graphical results for the Christmas tree fractal: (left) original images of the three contractive functions; (right) reconstructed images with the MSA-MCS method. (Color figure online)

However, the MCS also requires three additional parameters: the initial step size for the Lévy flights, α^0, the number of nests to be abandoned, ρ, and the fraction of nests to make up the top nests, τ. Following some previous works, they have been set to $\alpha^0 = 1$, $\rho = 0.75$ and $\tau = 0.25$, respectively. Moreover, the method is executed for Λ iterations. In our simulations, we found that $\Lambda = 2500$ is enough to reach convergence in all cases. In addition to the control parameters for our method, we also need two more parameters related to the problem: the

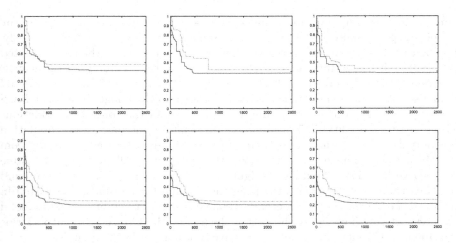

Fig. 3. Convergence diagram of the normalized similarity error rate for the three contractive functions (left to right) of the Sierpinsky gasket (top) and the Christmas tree (bottom) with the original CS algorithm (red dashed line) and our MSA-MCS method (blue solid line). (Color figure online)

number of contractive functions η and the mesh size, ν. In this work, they are set to $\eta = 3$ and $\nu = 40$, respectively. Unfortunately, we cannot analyze here how all our parameters affect the method performance because of limitations of space.

Table 2. Numerical results of the normalized similarity error rate for the three contractive functions of the examples in Figs. 1 and 2 with the original CS algorithm and our MSA-MCS method (see also Fig. 3 for their graphical representation).

	Sierpinsky gasket			Christmas tree		
	NSER(ϕ_1)	NSER(ϕ_2)	NSER(ϕ_3)	NSER(ϕ_1)	NSER(ϕ_2)	NSER(ϕ_3)
Best (CS):	0.4798	0.4178	0.4296	0.2445	0.2382	0.2547
Mean (CS):	0.4992	0.4333	0.4501	0.2603	0.2511	0.2769
Best (MSA-MCS):	0.4124	0.3805	0.3849	0.2014	0.2008	0.2113
Mean (MSA-MCS):	0.4403	0.4157	0.4195	0.2206	0.2257	0.2331

5   Experimental Results

All computations in this paper have been performed on a 2.6 GHz. Intel Core i7 processor with 16 GB. of RAM. The source code has been implemented by the authors in the native programming language of the popular scientific program *Matlab version 2015a* and using the numerical libraries for fractals in [3,5–7]. Our method has been applied to several examples of fractals with $\eta = 3$. Only two (already analyzed in [13]) are included here because of limitations of space: the Sierpinsky gasket and the Christmas tree, depicted in Figs. 1 and 2, respectively. The figures show the fractal images of the original (in red) and the reconstructed (in blue) contractive functions on the left and the right columns, respectively. The images correspond to the best value of the NSER fitness function selected from a set of 50 independent executions. As shown in the images, our MSA-MCS approach captures the structure and general shape of the contractive functions with high visual quality. This is a remarkable result because our initial population is totally random, meaning that their corresponding images are all very far from the given fractal image. Figure 3 shows the convergence diagram for the three contractive functions (from left to right) of the Sierpinsky gasket (top row) and the Christmas tree (bottom row) using the original CS method (as reported in [13]) and the new MSA-MCS method, displayed as red dashed lines and blue solid lines respectively. This figure shows that the new method MSA-MCS outperforms the previous CS method for all contractive functions of both examples.

The good visual appearance of the method in Figs. 1 and 2 and its graphical comparison with the CS method in Fig. 3 are all confirmed by our numerical results reported in Table 2. The table shows the best and the mean values of the normalized similarity error rate, $NSER(\phi_\kappa)$, for 50 independent runs. These results indicate that the new MSA-MCS method performs quite well. It also improves the previous results in [13] based on the original CS algorithm by a significant margin in all cases. For instance, we can see that the even the mean value of NSER for MSA-MCS is better that the best value of NSER with the original CS method. In other words, it is not a case of just an incremental improvement, but a significant one statistically.

6   Conclusions and Future Work

In this paper we address the problem to compute the optimal set of individual contractive functions for the reconstruction of self-similar binary fractal images. To this aim, we propose a memetic approach comprised of the modified CS method for global optimization with a new strategy for the Lévy flight step size (MMCS) and the ASSRS heuristics for local search. This approach is applied to some illustrative examples of fractal images with satisfactory results. This new method shows a significant improvement with respect to a previous approach based on the original CS for all functions in our benchmark.

In spite of these good results, there is still room for further improvement in the SSCF problem. We also wish to address the second step of the general IFS inverse problem for self-similar fractal images and its extension to the case of non self-similar fractals. We also plan to apply a very promising recent hybrid self-adaptive cuckoo search [15] to our problem as part of our future work.

Acknowledgements. This research is supported by the project PDE-GIR of the European Union's Horizon 2020 research and innovation programme under the Marie Sklodowska-Curie grant agreement No 778035, the Spanish Ministry of Economy and

Competitiveness under grant #TIN2017-89275-R of the Agencia Estatal de Investigación and European Funds FEDER (AEI/FEDER, UE), the project #JU12, of SODERCAN and European Funds FEDER (SODERCAN/FEDER UE) and the project EMAITEK of the Basque Government.

References

1. Barnsley, M.F.: Fractals Everywhere, 2nd edn. Academic Press, San Diego (1993)
2. Falconer, K.: Fractal Geometry: Mathematical Foundations and Applications, 2nd edn. Wiley, Chichester, England (2003)
3. Gálvez, A.: IFS Matlab generator: a computer tool for displaying IFS fractals. In: Proceedings of ICCSA 2009, pp. 132–142. IEEE CS Press, Los Alamitos, CA (2009)
4. Gálvez, A., Iglesias, A.: Cuckoo search with Lévy flights for weighted Bayesian energy functional optimization in global-support curve data fitting. Sci. World J. **2014**, 11 (2014). Article ID 138760
5. Gálvez, A., Iglesias, A., Takato, S.: Matlab-based KETpic add-on for generating and rendering IFS fractals. CCIS **56**, 334–341 (2009)
6. Gálvez, A., Iglesias, A., Takato, S.: KETpic Matlab binding for efficient handling of fractal images. Int. J. Future Gener. Comm. Networ. **3**(2), 1–14 (2010)
7. Gálvez, A., Kitahara, K., Kaneko, M.: *IFSGen4LaTeX* : Interactive graphical user interface for generation and visualization of iterated function systems in LaTeX. In: Hong, H., Yap, C. (eds.) ICMS 2014. LNCS, vol. 8592, pp. 554–561. Springer, Heidelberg (2014). https://doi.org/10.1007/978-3-662-44199-2_84
8. Gutiérrez, J.M., Iglesias, A.: A mathematica package for the analysis and control of chaos in nonlinear systems. Comput. Phys. **12**(6), 608–619 (1998)
9. Gutiérrez, J.M., Iglesias, A., Rodríguez, M.A.: A multifractal analysis of IFSP invariant measures with application to fractal image generation. Fractals **4**(1), 17–27 (1996)
10. Gutiérrez, J.M., Iglesias, A., Rodríguez, M.A., Burgos, J.D., Moreno, P.A.: Analyzing the multifractal structure of DNA nucleotide sequences. In: Chaos and Noise in Biology and Medicine, vol. 7, pp. 315–319. World Scientific, Singapore (1998)
11. Hutchinson, J.E.: Fractals and self similarity. Indiana Univ. Math. J. **30**(5), 713–747 (1981)
12. Iglesias, A., Gálvez, A.: Cuckoo search with Lévy flights for reconstruction of outline curves of computer fonts with rational Bézier curves. In: Proceedings of Congress on Evolutionary Computation, CEC 2016. IEEE CS Press, Los Alamitos, CA (2016)
13. Quirce, J., Gálvez, A., Iglesias, A.: Computing self-similar contractive functions for the IFS inverse problem through the cuckoo search algorithm. In: Del Ser, J. (ed.) ICHSA 2017. AISC, vol. 514, pp. 333–342. Springer, Singapore (2017). https://doi.org/10.1007/978-981-10-3728-3_33
14. Marsaglia, G.: Choosing a point from the surface of a sphere. Annal. Math. Stat. **43**(2), 645–646 (1972)
15. Mlakar, U., Fister Jr., I., Fister, I.: Hybrid self-adaptive cuckoo search optimization. Swarm Evol. Comput. **29**, 47–72 (2016)
16. Schumer, M.A., Steiglitz, K.: Adaptive step size random search. IEEE Trans. Autom. Control **13**(3), 270–276 (1968)
17. Walton, S., Hassan, O., Morgan, K., Brown, M.R.: Modified cuckoo search: a new gradient free optimisation algorithm. Chaos, Solitons Fractals **44**, 710–718 (2011)

18. Yang, X.-S.: Nature-Inspired Metaheuristic Algorithms, 2nd edn. Luniver Press, Frome, UK (2010)
19. Yang, X.S., Deb, S.: Cuckoo search via Lévy flights. In: Proceedings of World Congress on Nature & Biologically Inspired Computing (NaBIC), pp. 210–214. IEEE Press, New York (2009)
20. Yang, X.S., Deb, S.: Engineering optimization by cuckoo search. Int. J. Math. Model. Numer. Optim. 1(4), 330–343 (2010)

A Re-description Based Developmental Approach to the Generation of Value Functions for Cognitive Robots

A. Romero, F. Bellas[(⊠)], A. Prieto, and R. J. Duro

Integrated Group for Engineering Research, Universidade da Coruña,
A Coruña, Spain
{alejandro.romero.montero, francisco.bellas,
abprieto, richard}@udc.es
http://www.gii.udc.es

Abstract. Motivation is a fundamental topic when implementing cognitive architectures aimed at lifelong open-ended learning in autonomous robots. In particular, it is of paramount importance for these types of architectures to be able to establish goals that provide purpose to the robot's interaction with the world as well as to progressively learn value functions within its state space that allow reaching those goals whatever the starting point. This paper aims at exploring a developmental approach to the generation of high level neural network based value functions in complex continuous state spaces through a re-description process. This process starts by obtaining relatively simple Separable Utility Regions (SURs) which allow the system to consistently achieve goals, although not necessarily in the most efficient manner. The traces obtained by these SURs are then used to provide training data for a neural network based value function. Through a simple experiment with the Robobo robot, we show that this procedure can be more generalizable than attempting to directly obtain the value function through more traditional means.

Keywords: Cognitive Developmental Robotics · Motivation · Value function

1 Introduction

Natural cognitive architectures are the result of very long evolutionary processes of the species undergoing interactions of different lines of evolutionarily changing "hardware" (bodies, organs, sensors, etc.) with different sets of environments [1, 2]. As a consequence of these processes different mixtures of knowledge and structures present at birth (phylogenetically coded knowledge), together with capabilities for their modification in order to adapt to particular environments and situations (ontogenetically acquired knowledge) have been produced.

This paper is concerned with Cognitive Developmental Robotics (CDR) [3] in lifelong open ended learning. In general, CDR seeks to design robotic systems through the application of insights gained from the ontogenetic development of the cognitive capabilities of living organisms. In particular, from the progressive manner in which lower level competences are acquired and used as a scaffolding in order to support

F. J. de Cos Juez et al. (Eds.): HAIS 2018, LNAI 10870, pp. 671–683, 2018.
https://doi.org/10.1007/978-3-319-92639-1_56

higher level cognitive abilities. These can later be combined and reused to generate even higher level cognitive capacities through a process of representational re-description, as described from a psychological point of view by [4, 5]. These re-descriptions should permit going from low level continuous sensory-motor state spaces to higher level more symbolic and discrete state spaces that facilitate planning and other higher level functions.

1.1 Motivation in CDR

Motivation is a very important concept that is often sidelined when implementing intelligent robots, but it is key aspect in CDR, and it is the main topic of this work. Motivation is the driver that makes an organism perform some action [6, 7]. It can be related to simple individual needs such as maintaining the body in operational conditions (e.g. providing enough water or food for the body to function), to much more complex socially related cues (e.g. looking good in order to attract a partner). Motivation is a mechanism used by cognitive systems in order to choose actions (decide on resource assignment) based on the evaluation of states.

A lot has been written from a bio/psychological point of view on the different types of motivations and how they relate to each other. Already in 1943, Hull [8] talked about animal behavior being motivated by drives, conceived as temporal physiological deficits that the organism is led to reduce in order to achieve homeostatic equilibrium (hunger, thirst ...). Reducing these deficits was assimilated to obtaining utility (or reward, which is a term that has become central to Reinforcement Learning [9, 10]). From this point of view, all motivations are either physiological primary drives or secondary drives derived from primary ones through learning. Later, other authors reported on animal/human behaviors that did not seem to be reward driven. One of the first examples that were published, presented by Harlow and co-workers [11], is that of rhesus monkeys spending long periods of time trying to solve mechanical puzzles, which apparently did not cover any homeostatic need and, consequently, had no associated homeostatic drive.

In the following decades, many authors indirectly established that these activities did cover cognitive needs in terms of allowing the animals to receive an optimal level of stimulation (or of novelty of stimuli) [12], to reduce discrepancy (incongruity or dissonance) between their knowledge and their current perception [13, 14], to augment effectance, that is, the capacity to have effective interactions with their environments [15] and others. These needs are often really heterostatic, that is they may never be absolutely fulfilled, but they can be expressed in terms of drives.

Summarizing, any motivational structure for a cognitive system is based on establishing a set of basic innate drives. In the case of natural systems these innate drives (e.g. seek novelty, avoid pain, satisfy hunger....) have evolved over phylogenetic time following the path of greatest species survivability and provide the mechanisms to learn in ontogenetic time sets of derived or instrumental drives (get money...) so as to be able to operate more efficiently in the particular environments and circumstances the system faces. The main function of this set of drives is to allow the cognitive system to evaluate states, whether its current state or prospective states, so that decisions can be made on the actions to take and the resources to commit.

However, drives are application and environment independent and they only determine what is required. They do not provide any information on how to go about achieving it.

1.2   Goals and Expected Utility Models

As an organism acts in a given world, it may find regularities about many aspects within it. This knowledge, if appropriately managed, may help it be more efficient in its future interactions. Among these regularities, a very informative set is that of what perceptual space (state space) points led to the partial or total fulfilment of what drives. That is, what points provided utility. The main idea underlying this notion of utility is to be able to increase the likelihood of a given response to an external phenomenon that increases the satisfaction of drives provided that there exists a prior experience related to a similar phenomenon [16]. In other words, to establish some type of experience based mapping that allows the robot to decide on the actions that satisfy its needs.

This information is obviously vital for the system to be able to decide on actions that lead to the fulfilment of drives in an efficient manner (i.e. without having to randomly explore its state space). Here is where the concept of goal arises. A goal is just a state space point that provides utility (i.e. fulfils drives) and, consequently, a desired end-state [17]. Thus, the operation of any cognitive system can be stated in terms of establishing what goals should be active each moment of time, and being able to determine what actions lead to the active goals being fulfilled. In other words, a goal structure, and the capability of acquiring and autonomously extending it by learning new goals, establishing complex goal hierarchies, and relating them to different contexts they experience during their lifetimes, is a fundamental part of any cognitive architecture [18].

Reaching goals consistently involves not only knowing where they are in state space, but also finding efficient paths towards them whatever state the system starts from. Consequently, any cognitive architecture, in addition to goal search and goal representation, must provide for modeling the distribution of utility in its state space. These models can then be used in order to determine what paths to follow to reach the goals it has found from any starting perceptual point.

Directly modeling utility is often not the best solution as utility may be a very sparse function, leading to models where most areas of state space would present a utility value of zero, providing no information to the system about how to move. A more interesting approach is to model an expected utility, whether absolute, as in the case of *Value Functions (VF)* [14, 19–22], or relative, as in the case of *Separable Utility Regions (SUR)* [22]. In both cases, what is modeled for each point in state space is a measure of the expectation of reaching the goal when at that point. Modeling the expected utility is one of the most important aspects of motivational systems in CDR, and it is the specific topic of this paper.

The approach presented here addresses a staged production of expected utility models. On the one hand, VFs are often very hard to obtain directly in complex state spaces which are usually continuous, often very high dimensional (most relevant systems have a large number of sensors and effectors) and, in many cases, not too well behaved. Consequently, constructing appropriate VFs over the whole state space is in general no easy task, and often intractable. On the other hand, as indicated in [22], it is much easier to obtain local utility models based on sensorimotor contingencies that are structured into correlation chains where the correlation of sensor values to the direction of motion to reach a goal is established. These models are called SURs. Unfortunately, even though SURs are easier to produce, they are also much less precise and imply paths through state space that are Manhattan-like in terms of the sensors, that is, all motions in state space they induce are parallel to one of the axes.

Here, we explore the combination of both types of expected utility functions through a re-description process whereby the motivational system initially models expected utility using SURs, and when these become consistent, uses the traces obtained from reaching the goals using SURs in order to obtain a much more general representation in the form of artificial neural network based VFs. In other words, the approach allows cognitive systems to go from continuous and generally high dimensional perceptual spaces, to progressively higher level representations that allow for simpler and more generalized or abstract planning and decision processes.

2 Motivational Engine

The expected utility modelling we are describing has been implemented in the MotivEn motivational architecture [23, 24] developed within the EU DREAM research project [25]. In this context, we will use a simplified version of this motivational architecture, where we consider that the cognitive robot has only two types of drives. On the one hand, it will have one operational drive associated to the fulfilment of the task that will be controlled by the human user, that is, the utility will be provided as human feedback in an explicit way. On the other hand, a heterostatic exploratory cognitive drive will be provided that will focus on exploring the perceptual space in order to discover goals and to improve the expected utility model. Thus, this test version of MotivEn operates using these 2 simple rules:

1. The *exploratory* drive will be used to evaluate candidate perceptual states in those cases when the state cannot be predicted by the expected utility model. This can happen because there is no utility model, the goal has not been reached yet, or because the candidate state is out of the utility model domain.
2. The *human-feedback* drive will be used to evaluate candidate perceptual states if the state falls into the utility model domain.

2.1 Separable Utility Regions

Separable Utility Regions (SURs) [22] are a very coarse representation of variations of expected utility in state-space. They seek to correlate the appropriate direction of the motion of the system in state-space with the variation of the values of environmental sensors that point in that direction so that they can be used as guidance. Thus, the key idea behind the SUR approach is to determine the sensor tendencies each moment in time, choosing the strongest one and correlating it to the desired motion in state space. Moreover, the domain region where the correlation is applicable must be defined so that it can be assumed that within that region utility increases in the direction of the correlation with a particular sensor. This domain region is called the *certainty area*, and it is a key concept in our system.

A certainty function is a function that associates a certainty value with respect to a given utility function to points in state space. This function operates mainly as a density map based on the visited past states that are stored as traces in a trace buffer (*TB*). The certainty of a point is calculated based on its distance from points that have actually been explored and which are obtained from the *TB*. To create the certainty maps, the system must handle three types of traces:

- *Successful Traces (s-traces)*: created when a goal is reached while the robot is acting in an existing certainty area under the influence of its human-feedback drive.
- *Failed Traces (f-traces)*: created when a robot fails to reach a goal under the influence of the human-feedback drive in its associated certainty area.
- *Weak Traces (w-traces)*: created when a goal is reached executing actions guided by the exploratory drive.

A certainty map can be seen as a set of time changing regions under the influence of each type of trace (s, f or w-traces). The addition of s-traces expands the area of high certainty values (certainty area), while the addition of f-traces reduces it. The certainty area starts as a large area, covering most of the state space, and it gradually converges towards a range that is correlated with the variance of the available trace points. All the traces stored in memory are used to produce a sampling density model based on the distances between the samples in these traces and the points of the state space.

During the learning process, the certainty areas corresponding to different sensor tendencies are tuned using the traces that are being generated so that, after some trials, only the correct sensor will display positive certainty in the area corresponding to its correlation, while the others are going to return a value of zero due to the presence of failed traces. This way, the system will converge to the activation of the appropriate SURs on those regions where they are useful.

Under this approach, the robot can be in three situations according to its action selection criterion at that moment:

1. If the robot has previously selected a sensor tendency (a sensor tendency is active), the state evaluation uses this sensor tendency; else
2. If the robot has not selected a sensor tendency before (there is no active sensor tendency), but there is some state that for a sensor tendency presents a certainty value that is larger than zero, the robot selects that sensor tendency as active and evaluates all the states according to it; else
3. If the robot has not selected a sensor tendency before (there is no active sensor tendency) and there is no state for any sensor tendency with a certainty value larger than zero, the states are evaluated using the exploratory drive.

A tendency trace will be an ordered list of perceptual states $P(t)$ in a given instant of time for which a specific sensor value is increased/decreased with respect to the sensorial state of the previous instant $P(t + 1)$. Thus, after a goal achievement, a robot with n sensors, will create m tendency traces being $m \leq n$. As these tendency traces are created depending on the increase or decrease of a sensor value, both situations cannot be possible for the same sensor. In addition, some sensors could present a constant value, so no tendency trace would be created for them. These tendency traces would have different lengths depending on when the tendency is broken. Every time a new trace is obtained using the exploratory drive, the algorithm considers it may have an increasing or decreasing tendency and it goes through all the episodes of the trace checking whether this is so.

2.2   Sub-goal Identification

Certainty areas are limited to a state subspace that is close to the goal. This way, a large part of state space would be outside the certainty area assigned to the first expected utility model. To complete the human-feedback drive component of the motivational system, it is necessary to combine different certainty areas with their own evaluation structures in order to create a path from any point in state space to the goal. As a consequence, the concept of *sub-goal* arises formally in this new representation.

To see this in a clearer way, Fig. 1 contains a schematic representation of the sub-goal identification process in terms of the perceptual state space. It is also intended to demonstrate how this method will allow chaining simple structures to carry out more complex tasks. It corresponds to a situation where the main goal depends on sensors 1 and 2. As it can be observed, the first SUR certainty area (SUR 1) leads directly to the main goal and goes in the 2+ direction, that is, in its activation zone the expected utility is directly correlated with sensor 2. Therefore, increasing the value of sensor 2 will lead directly to the goal. In turn, arriving at this SUR certainty area is a sub-goal, and it can be seen how another SUR leads to it: SUR Region 2, which goes in direction 1+, that is, in its activation zone the expected utility is directly correlated to sensor 1 (increasing the value of sensor 1 will lead directly to the first sub-goal that leads to the main goal). Hence, the region of certainty of this last SUR will be also a sub-goal for new possible SURs.

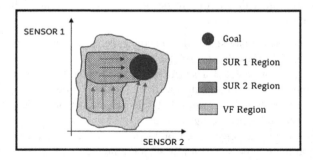

Fig. 1. Sub-goal identification and chaining process (Color figure online)

Therefore, in view of the representation of Fig. 1, it is possible to visualize how through the identification of sub-goals it is possible to link SURs, and how this can provide different ways to reach a goal as a function of the area of state space in which the robot is initially located.

2.3 Re-description Process

Once the SUR creation and concatenation towards the goal has been presented in the previous two sections, we can now introduce the re-description process that can be carried out using them. The idea behind this process is that, once MotivEn creates, at least, two SURs, which through their concatenation lead to the goal, the VF learning process can start, trying to generalize the response of these SURs. To this end, the following procedure, inspired on the one used in the MDB cognitive architecture [23], has been designed and tested:

- The VF is represented through a simple Feed Forward Artificial Neural Network, which has as many inputs as the selected SURs to be generalized (remember that SURs are one-dimensional), and one output, the expected utility.
- The VF parameters are adjusted using a Back-Propagation Algorithm, where the training set is made up of the trace buffers of the SURS to be generalized, which are combined into a single buffer. The traces are assigned a decreasing expected utility, which was not necessary for the SURs, following the classical scheme of eligibility traces used in Reinforcement Learning. Consequently, we have a set of perceptual states that are correlated in different sensors towards the goal state, and that the ANN must generalize.
- The Back-Propagation algorithm is executed for a predefined number of learning steps and, as a result, a VF is obtained.
- From this instant onwards, the VF is used in MotivEn to guide to robot towards the goal instead of using the SURs, and a certainty area associated to the VF arises.
- Each time the robot reaches the goal using the VF, a new trace is obtained, which is stored in the VF trace buffer implying the elimination of the oldest one. At this moment, a new training process is carried out, and the certainty area of the VF is updated (specifically, the re-training process is not carried out continuously, but every N traces, where N is a configurable parameter).

As we can see, VF learning is a dynamic process that is performed continuously as new traces are obtained. The relevant point here is that the VF learning starts from the traces obtained with the SUR methodology, that is, all the perceptual points follow a correlation towards the goal (utility), which simplifies the search space of the VF. Once learned, the VF is a re-description of the one-dimensional expected utility models created with the SURs, which generalizes their response. Figure 1 contains a simplified representation of this process, where we can see two SUR regions 1 and 2, that lead to the goal following one-dimensional paths in the state space. With the VF generalization, its certainty region is wider and it generalizes the other two, obtaining direct paths to the goal in more than one dimension (represented using green lines) in Fig. 1. In the following section, we will clarify this method with a robotic task.

3   Re-description Example with a Real Robot

The example is based on a collect-a-ball experiment. The final objective is to place an object in a predefined position without any prior knowledge. The experiment was carried out using a Robobo robot [26], which is based on a mobile base that carries a smartphone and provides all the high-level functionalities to the robot. Robobo is controlled by a simplified cognitive architecture where MotivEn is executed. Apart from the robot, the scenario consists on a table with a target area, marked in violet, a movable item, a red cylinder, and two beacons, one blue and the other green, as shown in Fig. 5. To simplify the setup, the target area is situated between both beacons and the Robobo always starts carrying the cylinder, so it does not have to find it previously. Also, when the robot reaches the target area carrying the cylinder, it is dropped there and the robot receives utility. This event triggers a reset of the scenario and the robot will be placed in a random place on the table again.

This setup works over a state space generated by two sensors: one provides the distance from the green beacon to the Robobo actuator ($db1$), and the other one the distance from the blue beacon to the Robobo actuator ($db2$). Thus, at each iteration, the system will perceive a sensorial state: $S(t) = (db1, db2)$. As for the actions ($A(t)$), Robobo moves freely on the table at a constant speed, so the action will change its orientation by an angle between $-90°$ and $90°$.

Initially, the robot has no idea where the goal is or how to reach it. Consequently, it will have to discover and associate it to a point in its perceptual space, and then learn to reach it from any part of the environment. In other words, it has to learn the expected utility model. As commented in the previous section, in this case MotivEn will first model the expected utility using SURs and, once learned, it will use the traces obtained by these SURs to create a more general representation in form of a neural network-based VF. Figure 2 shows the results obtained in this experiment using the re-description approach in a typical run. The first 10.000 iterations correspond to the learning process where the motivational system models the expected utility using SURs. The dashed pink vertical line marks the instant when the re-description process takes place and the motivational system starts using the ANN-based VF.

The figure displays how long it takes MotivEN to find the goal. In this case, each blue dot represents the iteration when the main target was reached by the robot (cylinder in the target area). The left Y axis represents how many iterations where necessary for the system to find the target in each episode, whereas the X axis represents the number of goal achievements at each point. The blue vertical lines indicate the creation of the different SURs. The green line indicates the iteration (time step counter) associated to each goal achieved, so the number of iterations is represented on the right Y axis. In this figure, it is possible to see how the time to reach the goal decreases gradually and that the creation and consolidation of SURs is associated with that reduction. This decrease is also reflected in the reduction of the slope of the green line. Thus, shortly after the creation of the fourth SUR, the system converges towards a stable performance in the pursuit of the goal. Moreover, after the change of the expected utility model used (re-description process), the efficiency of the system

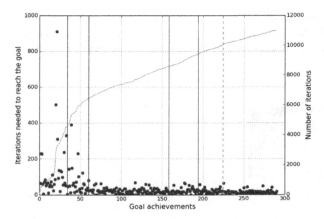

Fig. 2. Results obtained in a typical run of the robot experiment using the re-description approach. (Color figure online)

increases slightly, more utility is achieved in fewer iterations, which is reflected in a reduction of the green line slope.

To better understand how MotivEn generates and re-describes knowledge, that is, how it goes from the initial SURs to the final neural VF, Figs. 3 and 4 show the evolution of the expected utility models and the traces resulting from their use. The left part (top) of Fig. 3 shows some straight lines in real space and how they are reflected in state space (bottom) in order to give an idea of what the right part of the figure means. In this graph the two beacons are shown in state space. The right four graphs of Fig. 3 display a state space representation of the traces generated from the use of the different expected utility models. The initial point of each trace in state space is marked in green, whereas the final point, the point where the utility was obtained, is represented in red. Moreover, the grey zone represents a state space area that is unreachable for the robot due to its sensing: considering that both beacons are separated a distance 'd', the point of the state space in which the robot is located must fulfill that: $db1 + db2 \geq d$. Attending to the different figures of Fig. 3, it can be seen how the traces generated by the robot are evolving towards greater efficiency when it comes to reaching the goal. In the first steps, the traces obtained using SURs, Fig. 3(a) show how the system tended to move in straight lines reducing the distance to one of its sensors (as shown in the example in the left of Fig. 3, the sensors do not really conform a mathematical base and thus movement towards one sensor may induce a reduction in distance to the other, producing curves in the state space trajectories). The point is that it not able to reach the goal in a direct way, as it follows SURs. On the other hand, the following two graphs show the transition until obtaining fully efficient traces derived from the process of generalization of the expected utility through the re-description process. Thus, the traces shown in Fig. 3(b) already begin to separate from the main directions marked by the sensors, while Fig. 3(c) shows that the expected utility model is practically generalized, since most of the traces obtained go directly to the goal and are able to maintain the desired direction, although in certain initial positions the robot is not able to go in a totally direct way to the goal. Finally, Fig. 3(d) shows the resulting traces once the model has been correctly learned. It is possible to see how they take the robot directly to the goal whatever its initial position on the stage, or in this case, in the space of states is.

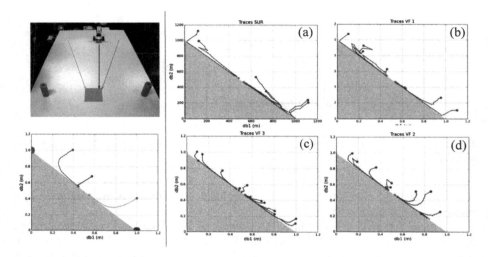

Fig. 3. Evolution of the traces (state space) generated from employing the different utility models. (Color figure online)

To understand the reason of this variation in the efficiency of the traces generated, it is necessary to pay attention to how the utility model has varied throughout the process of re-description. For that purpose, Fig. 4 shows a 3D representation of the VFs that were generated by the different traces over time. The X and Y axis of the different surfaces represent the normalized sensor values (distance) as inputs of the neural network, and the Z axis represents the value (arbitrary units), that is, the output of the VF. In this way, Fig. 4(a) corresponds to the VF generated from the traces obtained once the robot was able to reach goals consistently when using SURs (traces shown in Fig. 3(a)), which makes the new traces look like the ones shown in Fig. 3(b). In turn, Fig. 4(b) shows the result of training the neural network again with the traces obtained with the previous VF. In this case the traces generated using this VF as utility model are shown in Fig. 3(c). Finally, the surface shown in Fig. 4(c) shows the result of training the neural network with the previous traces. It can be seen that it is an almost concave surface which reaches its maximum value at the point of the state space corresponding to the goal. In turn, its geometry makes the new traces of the robot go directly to the goal and in a much more efficient way, as shown in Fig. 3(d).

Figure 5 shows the evolution of the trajectory followed by the real robot to reach the goal through the use of SURs and the use of the final neural network based VF, where the improvement in system efficiency when solving the task, derived from the re-description process, is clearly seen.

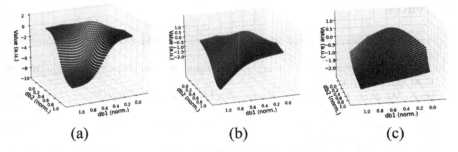

Fig. 4. 3D representation of the neural network based VFs generated.

(a) Representative trace using SURs   (b) Representative trace using VF

Fig. 5. Comparison between real robot traces before and after the re-description process. (Color figure online)

4   Conclusions

This paper has addressed the production of value functions within a robot motivational system through a two-step re-description process. This process involves first generating a SUR based representation of the utility model that is quite simple to obtain and only requires determining the correlations between the appropriate directions in state space and the values of the sensors. These correlations allow the robot to consistently reach the goal, albeit not in the most efficient or direct way. As a second step, the information derived from the traces obtained through the SUR model are used in order to efficiently train a neural network that progressively provides a much more efficient utility function representation.

The approach has been shown to be quite effective and we are now working on extending it to motivational problems involving very high dimensional state spaces.

Acknowledgements. This work was partially funded by the EU's H2020 research and innovation programme under grant agreement No. 640891 (DREAM project) and by the Xunta de Galicia and European Regional Development Funds under grant RedTEIC (ED341D R2016/012) and grant ED431C 2017/12.

References

1. Sweller, J.: Evolution of human cognitive architecture. In: The Psychology of Learning and Motivation (2003)
2. Scott, P.D., Markovitch, S.: Learning novel domains through curiosity and conjecture. In: IJCAI, pp. 669–674 (1989)
3. Asada, M., et al.: Cognitive developmental robotics: a survey. IEEE Trans. Auton. Ment. Dev. 1(1), 12–34 (2009)
4. Karmiloff-Smith, A.: Beyond modularity: a developmental perspective on cognitive science. Behav. Brain Sci. (1992)
5. Carassa, A., Tirassa, M.: Representational redescription and cognitive architectures, pp. 711–712 (1994)
6. Maslow, A.H.: A theory of human motivation. Psychol. Rev. 50(13), 370–396 (1943)
7. Maslow, A.H.: Human motivation. Hum. Motiv. (1987)
8. Hull, C.L.: Principles of Behavioir. Appleton-Century-Crofts, New York (1943)
9. Kober, J., Bagnell, J.A., Peters, J.: Reinforcement learning in robotics: a survey. Int. J. Robot. Res. 32(11), 1238–1274 (2013)
10. Sutton, R.S., Barto, A.G.: Introduction to reinforcement learning. Learning 4(1996), 1–5 (1998)
11. Harlow, H.F.: Learning and satiation of response in intrinsically motivated complex puzzle performance by monkeys. J. Comp. Physiol. Psychol. 43(4), 289–294 (1950)
12. Ryan, R., Deci, E.: Intrinsic and extrinsic motivations: classic definitions and new directions. Contemp. Educ. Psychol. 25(1), 54–67 (2000)
13. Kagan, J.: Motives and development. J. Pers. Soc. Psychol. 22(1), 51 (1972)
14. Oudeyer, P.-Y.: Intelligent adaptive curiosity: a source of self-development. Science 80 (2004)
15. Baldassarre, G., Mirolli, M.: Intrinsically motivated learning systems: an overview. In: Baldassarre, G., Mirolli, M. (eds.) Intrinsically Motivated Learning in Natural and Artificial Systems, pp. 1–14. Springer, Heidelberg (2013). https://doi.org/10.1007/978-3-642-32375-1_1
16. Friston, K.J., Tononi, G., Reeke, G.N., Sporns, O., Edelman, G.M.: Value-dependent selection in the brain: simulation in a synthetic neural model. Neuroscience 59, 229–243 (1994)
17. Rolf, M., Asada, M.: What are goals? And if so, how many? pp. 332–339 (2015)
18. Baldassarre, G., Stafford, T., Mirolli, M., Redgrave, P., Ryan, R., Barto, A.: Intrinsic motivations and open-ended development in animals, humans, and robots: an overview. Front. Psychol. 5, 1–8 (2014)
19. Huang, X., Weng, J.: Value system development for a robot. In: Proceedings of the IEEE International Conference on Neural Networks, vol. 4, pp. 2883–2888 (2004)
20. Huang, X., Weng, J.: Novelty and reinforcement learning in the value system of developmental robots. In: Proceedings of the Second International Workshop on Epigenetic Robotics, pp. 47–55 (2002)
21. Zhang, Y., Weng, J.: Action chaining by a developmental robot with a value system. In: Proceedings of the 2nd International Conference on Development and Learning, ICDL 2002 (2002)
22. Salgado, R., Prieto, A., Bellas, F., Duro, R.J.: Motivational engine for cognitive robotics in non-static tasks. In: Ferrández Vicente, J.M., Álvarez-Sánchez, J.R., de la Paz López, F., Toledo Moreo, J., Adeli, H. (eds.) IWINAC 2017. LNCS, vol. 10337, pp. 32–42. Springer, Cham (2017). https://doi.org/10.1007/978-3-319-59740-9_4

23. Salgado, R., Prieto, A., Bellas, F., Calvo-Varela, L., Duro, R.J.: Motivational engine with autonomous sub-goal identification for the Multilevel Darwinist Brain. Biol. Inspir. Cogn. Archit. **17**, 1–11 (2016)
24. Salgado, R., Prieto, A., Bellas, F., Calvo-Varela, L., Duro, R.J.: Neuroevolutionary motivational engine for autonomous robots. In: GECCO 2016 Companion - Proceedings of the 2016 Genetic and Evolutionary Computation Conference (2016)
25. EU DREAM H2020 Project. http://www.robotsthatdream.eu
26. Bellas, F., et al.: The Robobo project: bringing educational robotics closer to real-world applications. In: Lepuschitz, W., Merdan, M., Koppensteiner, G., Balogh, R., Obdržálek, D. (eds.) RiE 2017. AISC, vol. 630, pp. 226–237. Springer, Cham (2018). https://doi.org/10.1007/978-3-319-62875-2_20

A Hybrid Iterated Local Search for Solving a Particular Two-Stage Fixed-Charge Transportation Problem

Ovidiu Cosma[1], Petrica Pop[1(✉)], Matei Oliviu[2], and Ioana Zelina[1]

[1] Department of Mathematics and Computer Science, Technical University of Cluj-Napoca, North University Center at Baia Mare, Cluj-Napoca, Romania
{ovidiu.cosma,petrica.pop}@cunbm.utcluj.ro
[2] Department of Electronics and Computer Science, Technical University of Cluj-Napoca, North University Center at Baia Mare, Cluj-Napoca, Romania
oliviu.matei@holisun.com

Abstract. In the current paper we take a different approach to a particular capacitated two-stage fixed-charge transportation problem proposing an efficient hybrid Iterated Local Search (HILS) procedure as a means of solving the above-mentioned problem. Our approach is a heuristic one; it constructs an initial solution while using a local search procedure whose aim is to increase the exploration, namely a perturbation mechanism. For the purpose of diversifying the search, a neighborhood structure is used to hybridize it. The preliminary computational results that we achieved stand as proof to the fact that the solution we propose yields high-quality solutions within reasonable running-times.

1 Introduction

The current study approaches a certain transportation problem, i.e. the capacitated fixed-cost transportation problem (FCTP) which occurs in a two-stage supply chain network. When approaching this transportation problem, we aim to identify and select the distribution centers, link them to the manufacturer and meet the requirements of the customers at minimal costs. The principal attribute of the fixed-charge transportation problem refers to the fact that a fixed charge corresponds to each route connecting distribution centers to customers that may be opened. This charge is payable in addition to the variable transportation cost which is proportionally calculated in relation to the amount of goods shipped.

It was Geoffrion and Graves [1] who first introduced the two-stage transportation problem. Researchers have since then studied several variants of the problem and, using exact and heuristic algorithms, they have come up with several methods for solving these variants.

In our paper we consider the problem as it was defined by Molla et al. [2]. In their variant only one manufacturer is considered. Their approach was to describe an integer programming mathematical formulation of the problem in

© Springer International Publishing AG, part of Springer Nature 2018
F. J. de Cos Juez et al. (Eds.): HAIS 2018, LNAI 10870, pp. 684–693, 2018.
https://doi.org/10.1007/978-3-319-92639-1_57

question, to propose a spanning tree-based genetic algorithm with a Prüfer number representation, as well as an artificial immune algorithm meant for solving the problem. El-Sherbiny [9] approached this problem and came up with some comments regarding the mathematical formulation. Subsequently, Pintea et al. [3,5] proposed some classical approaches and came up with an improved hybrid algorithm in which they combined the Nearest Neighbor search heuristic with a local search procedure and thus solved the two-stage transportation problem with fixed costs. A new and improved hybrid heuristic approach was described by Pop et al. [7]; this was obtained through the combination of a genetic algorithm based on a hash table coding of the individuals with a powerful local search procedure.

There exists yet another version of this problem, and it takes into account the environmental impact by reducing the greenhouse gas emissions. This version was introduced by Santibanez-Gonzalez et al. [8] in order to deal with a practical application occurring in the public sector. Considering this variant of the problem, Pintea et al. [4] came up with a set of classical hybrid heuristic approaches and Pop et al. [6] suggested an efficient reverse distribution system for solving the problem.

As regards the structure of our paper, this is as follows: in Sect. 2, the considered particular two-stage fixed-cost transportation problem is defined; in Sect. 3, the developed hybrid Iterated Local Search algorithm is described, while in Sect. 4 we present the computational experiments and the achieved results. The paper ends with the summary of the obtained results and the presentation of the future research directions presented in the last section.

2 Definition of the Problem

In order to define the considered particular two-stage fixed-cost transportation problem we start by introducing the notations and symbols:

m	the number of distribution centers
n	the number of customers
i	distribution center identifier, $i \in \{1, ..., m\}$
j	customer identifier, $j \in \{1, ..., n\}$
$D[j]$	the demand of customer j
$SC[i]$	stocking capacity of DC i
$F1[i]$	the opening cost of the DC i
$F2[i, j]$	the fixed transportation charges for the link from DC i to customer j
$C1[i]$	unit cost of transportation from manufacturer to DC i
$C2[i, j]$	unit cost of transportation from DC i to customer j
$X1[i]$	the number of units transported from the manufacturer to DC i
$X2[i, j]$	the number of units transported from DC i to customer j
Z_{opt}	the optimal total cost of the distribution
Z	the total cost of the distribution
Z_{loc}	the local optimal solution
I_{max}	the total number of iterations
$In[j]$	the number of units supplied to customer j at a given moment.

Given a manufacturer, a set of m potential distribution centers (DC's) and a set of n customers satisfying the following properties:

- the manufacturer may ship to any distribution center at a transportation cost $C1[i]$, $i \in \{1, ..., m\}$,
- each DC may ship to any customer at a transportation cost $C2[i,j]$ from DC $i \in \{1, ..., m\}$ to customer $j \in \{1, ..., n\}$, plus a fixed-cost $F2[i,j]$ for operating the route,
- the opening costs for a potential DC i are denoted by $F1[i]$, $i \in \{1, ..., m\}$,
- each DCi, $i \in \{1, ..., m\}$ has $SC[i]$ units of stocking capacity and each customer j, $j \in \{1, ..., n\}$ has a demand $D[j]$,

the aim of the considered two-stage capacitated fixed-cost transportation problem is to determine the routes to be opened and corresponding shipment quantities on these routes, such that the customer demands are fulfilled, all shipment constraints are satisfied, and the total distribution costs are minimized.

An illustration of the particular investigated two-stage fixed-charge transportation problem is presented in the next figure.

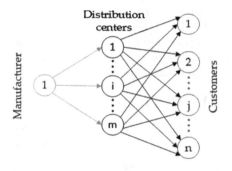

Fig. 1. Illustration of the particular two-stage fixed-charge transportation problem

3   Description of the Hybrid Iterated Local Search Heuristic

Our developed heuristic approach is a Hybrid Iterated Local Search (HILS) algorithm. We constructed initial solutions using a minimum cost procedure applied for each customer, combined with a neighborhood operator and hybridized with a powerful local search. When a solution is trapped in a local optimum, then we applied a perturbation mechanism that builds a new solution from scratch.

Algorithm 1 presents the pseudo-code of our HILS. The algorithm executes a fixed number I_{max} of iterations, established based on the number of customers. At each iteration, the customers are placed in a random order that has not been used in previous iterations (lines 6–8), using the Fischer-Yates shuffle algorithm.

Algorithm 1. HILS()

```
 1: procedure HILS()
 2:     choose Imax
 3:     Zopt ← ∞
 4:     iNo ← 0
 5:     while iNo<Imax do
 6:         repeat
 7:             shuffle customers
 8:         until order not used before
 9:         initialSolution()
10:         repeat
11:             Zloc ← Z
12:             Z ← localSearch()
13:             if Z <Zopt then
14:                 Zopt← Z
15:                 save solution
16:             end if
17:         until Z ≥ Zloc
18:         iNo ← iNo + 1
19:     end while
20: end procedure
```

The **initialSolution** procedure (line 9) constructs an initial solution, processing the customers in the order previously set. The initial solution is enhanced by a powerful local search procedure (line 12).

The **initialSolution** procedure constructs a solution in several steps. Every step is looking for a better supplying strategy for the customer j in the conditions created by the decisions made in the previous steps when distribution centers were opened and part of their storage capacity was consumed. As a consequence, the order in which the customers are processed is decisive for the final result. At the beginning of this procedure a reset of the distribution system is performed. All decisions made on the construction of the previous solution (lines 2 to 4) are invalidated and then it follows a loop where each client's request is satisfied (line 6). After ensuring the needs for each of the customers, a fast local search attempts to improve the partial solution generated so far (line 7).

The procedure **resolveCustomerDemand** seeks for supplying *amt* units to the customer j in one or more steps. For the selection of the distribution routes, the **findRoute** procedure (line 3) is called at each step, which determines the most advantageous supply route at the time of the call. The result is returned as a data structure $Route\{i, j, a\}$. This is a supply route for the customer j, from the distribution center i, with a units. The selected path is added to the solution by calling the **resolveCustomerDemand** procedure (line 6). Each route added to the distribution strategy in the **resolveCustomerDemand** procedure is also added to the external **addedRoutes** list so that it can be easily canceled later. This operation is used in the **efficientSearch** and **localSearch** procedures.

Algorithm 2. initialSolution

1: **procedure** INITIALSOLUTION
2:     $Out[i] \leftarrow X1[i] \leftarrow 0;\ i = \overline{1,m}$
3:     $In[j] \leftarrow 0;\ j = \overline{1,n}$
4:     $X2[i,j] \leftarrow 0;\ i = \overline{1,m},\ j = \overline{1,n}$
5:     **for** $j \leftarrow 1$ **to** n **do**
6:         resolveCustomerDemand(j,D[j])
7:         efficientSearch()
8:     **end for**
9: **end procedure**

Algorithm 3. resolveCustomerDemand

1: **procedure** RESOLVECUSTOMERDEMAND(j:int, amt:int)
2:     **while** amt > 0 **do**
3:         $rt \leftarrow$ findRoute (j,amt)
4:         addToRoute($rt, rt.a$)
5:         **add** rt **to** addedRoutes list
6:         amt $\leftarrow$ amt $- rt.a$
7:     **end while**
8: **end procedure**

The **findRoute** procedure looks for supplying routes and returns a best possible supply route with amt units to the client j. The returned route may support a number of units less than or equal to the one requested. The loop on lines 3–13 checks all the distribution centers. The distribution center i is taken into consideration only if its stocking capacity has not been fully utilized (lines 4–5). The quality of each possible route is estimated by calculating a unit cost that also takes into account any fixed costs incurred when opening new distribution routes. It is returned the route for which the unit cost calculated on line 7 is minimal. In the **findRoute** procedure $cost1(i,c)$ represents the transportation cost of c units from manufacturer to DCi and $cost2(i,j,c)$ represents the transportation cost of c units from DCi to customer j. The opening cost $F1[i]$ and the fixed-transportation charges $F2[i,j]$ are only included in the calculation of the two costs if the routes have not been previously opened.

The **addToRoute** procedure adds to the distribution strategy the amount amt on the route specified by the parameter r. In order to cancel a route, a negative amount is transmitted to the procedure via amt parameter.

The **efficientSearch** procedure tries to modify all routes that were originally booked. For each customer who has $In[j] > 0$, links to all distribution centers are checked. Each path linking the distribution center i and the customer j is evaluated as follows: if $X2[i,j] > 0$ then this quantity is taken out from the route: manufacturer - distribution center DCi - customer j. We look then for a better supply route with c units to the client j by calling the **resolveCustomerDemand** procedure (line 12). Since there is a relatively small number of routes for which $X2[i,j] > 0$, this procedure is fast because it performs a small

---

Algorithm 4. findRoute

1: **procedure** FINDROUTE(j:int, amt:int): Route
2:     best $\leftarrow \infty$
3:     **for** $i \leftarrow 1$ to m **do**
4:         $a \leftarrow SC\,[i] - $ Out $[i]$
5:         **if** $a > 0$ **then**
6:             $c \leftarrow \min(a, \text{amt})$
7:             $d \leftarrow \frac{\text{cost1}(i,c) + \text{cost2}(i,j,c)}{c}$
8:             **if** $d <$ best **then**
9:                 best $\leftarrow d$
10:                result $\leftarrow$ newRoute (i, j, c)
11:             **end if**
12:         **end if**
13:     **end for**
14: **end procedure**

Algorithm 5. addToRoute

1: **procedure** ADDTOROUTE(r:Route, amt:int)
2:     Out $[r.i] \leftarrow$ Out $[r.i] + $ amt
3:     In $[r.j] \leftarrow$ In $[r.j] + $ amt
4:     $X1\,[r.i] \leftarrow X1\,[r.i] + $ amt
5:     $X2\,[r.i, r.j] \leftarrow X2\,[r.i, r.j] + $ amt
6: **end procedure**

---

number of operations and can be called at each iteration of the loop on lines 5–8 of the **initialSolution** procedure. If the adjustment operations do not improve the solution, then they are abandoned, thus returning to the initial solution (line 16). This can be done efficiently with the **addedRoutes** list.

The **localSearch** procedure is called on line 9 of the **HILS** procedure in order to enhance the initial solution. In consequence, a number of routes from the distribution strategy will be canceled, after which we will replace them with better ones. Each try will work with a set of two routes. At the initialization of the procedure a list *DList* is created in which are added all the links between the distribution centers and the customers for which $X2[i, j] > 0$, as *Link*$\{i, j\}$ structures.

This procedure cancels the two complete routes $l1$ and $l2$ from manufacturer through the distribution centers to the customers and then a more preferred supply option is sought, under the new conditions.

4   Computational Results

In order to test the performance of our proposed hybrid Iterated Local Search algorithm, we conducted our computational experiments on two sets of instances containing in total 24 instances. The first set of instances contains 15 instances used in the computational experiments of Pintea and Pop [5] and Pop et al. [7].

Algorithm 6. efficientSearch

```
 1: procedure EFFICIENTSEARCH()
 2:     for j ← 1 to n do
 3:         if In[j] > 0 then
 4:             clear addedRoutes list
 5:             before ← Z
 6:             for i ← 1 to m do
 7:                 if X2[i, j] > 0 then
 8:                     c ← X2[i, j]
 9:                     rt ← newRoute(i, j, c)
10:                     add rt to addedRoutes list
11:                     addToRoute(rt, −c)
12:                     resolveCustomerDemand(j, c)
13:                 end if
14:             end for
15:             if Z > before then
16:                 restore initial solution
17:             end if
18:         end if
19:     end for
20: end procedure
```

The second set of instances contains 9 new randomly instances of larger sizes generated according to Molla et al. strategy [2]. All the instances used in our computational experiments are available at the address: https://sites.google.com/view/tstp-instances/.

Our algorithm was coded in Java 8 and we performed 10 independent runs for each instance on a Procesor Intel Core i5-4590 3.3GHz, 4GB RAM, Windows 10 Education 64 bit.

Table 1 presents the obtained computational results by our proposed heuristic hybrid ILS algorithm in comparison with the hybrid heuristic algorithm described by Pintea et al. [3], denoted by HA, the genetic algorithm and hybrid based genetic algorithm introduced by Pop et al. [7], denoted by GA and HGA.

Algorithm 7. localSearch()

```
 1: procedure LOCALSEARCH(): int
 2: build DList
 3: tc ← Z
 4:     for itNo ← 1 to totalIt do
 5:         randomly pick l_1, l_2 from DList
 6:         tc ← changeDistribution(l_1, l_2, tc)
 7:         tc ← changeDistribution(l_2, l_1, tc)
 8:     end for
 9:     return tc
10: end procedure
```

The first column of Table 1 gives the type of the instances followed by two columns that contain the number of distribution centers (m) and the number of customers (n). The next columns provide the best solution achieved by the hybrid heuristic algorithm described by Pintea et al. [3], the best and average solutions obtained by the genetic algorithm and hybrid based genetic algorithm introduced by Pop et al. [7] and the achieved results by our hybrid ILS heuristic algorithm: the best and average solutions and the average execution times in seconds necessary to obtain the provided solutions. The results written in bold represent cases for which the obtained solution is the best existing from the literature.

Table 1. Computational results achived by our proposed ILS compared to existing approaches

Type	m	n	HA [3]	GA [7]		HGA [7]		Our ILS algorithm		
				Best sol	Avg. sol	Best sol	Avg. sol	Best sol	Avg. sol	Exec. time
1	10	10	21980	20450	21430	20400	21320	**20395**	**20395**	0
2	10	10	12160	11240	11850	11220	11740	**11214**	**11214**	0
3	10	10	14000	14100	14620	14040	14520	**13999**	**13999**	0
1	10	20	36000	35400	36200	35380	35860	**35371**	**35371**	0.03
2	10	20	39660	37840	38470	**37800**	38250	**37800**	**37800**	0.97
3	10	20	36060	36000	36110	36000	36000	**35988**	**35988**	0.19
1	10	30	55660	52700	54880	52650	53700	**52643**	**52643**	0.06
2	10	30	55380	54650	55640	54540	54880	**54539**	**54539**	0.01
3	10	30	49860	48580	49470	48540	49240	**48535**	**48535**	5.87
1	15	15	26680	25420	27640	25420	26710	**25417**	**25417**	0
2	15	15	29100	28600	29230	28600	28940	**28598**	**28598**	0.03
3	15	15	29200	28840	29470	28750	29120	**28411**	**28411**	1.46
1	50	50	92400	91550	92410	91500	92104	**91455**	**91455**	71.85
2	50	50	116500	114660	117440	114150	115420	**114111**	**114114.2**	315.90
3	50	50	105000	105000	107400	105000	106480	**104904**	**104920.8**	571.61

Analyzing the computational results reported in Table 1, we can observe that our hybrid ILS approach has a better computational performance compared to the hybrid heuristic algorithm described by Pintea et al. [3], the genetic algorithm and hybrid based genetic algorithm introduced by Pop et al. [7]. In all the instances we were able to improve the existing best solution from the literature with one exception the instance containing 10 distribution centers and 20 customers where the solution delivered by our approach is the same with the one provided by the hybrid based genetic algorithm considered by Pop et al. [7].

In Table 2, we present the computational results achieved by our proposed hybrid heuristic ILS algorithm in the case of the 9 new randomly generated instances of larger sizes. The first column of Table 2 gives the type of the instances followed by two columns that contain the number of distribution centers (m) and the number of customers (n) and the next six columns contain the following results achieved by our hybrid ILS heuristic algorithm: the best solution, the average solution, the corresponding standard deviation and the best and average computational times necessary to obtain the provided solutions and the corresponding standard deviation. The execution times are reported in minutes (min).

Table 2. Computational results achieved by our proposed hybrid ILS in the case of large size instances

Type	m	n	Total cost			Execution time [min]		
			Best	Average	Std. deviation	Best	Average	Std. deviation
A	50	50	361171	361187.6	10.05	33.14	133.55	68.14
B	50	50	366870	366904.6	13.26	4.48	133.55	64.47
C	50	50	374285	374285	0	1.14	48.84	37.05
A	30	100	242099	242099	0	50.91	276.08	142.52
B	30	100	233279	233379.5	62.02	47.39	329.92	225.68
C	30	100	277087	277327.7	132.50	75.70	440.55	170.29
A	50	100	236916	237187.0	155.62	199.42	326.84	114.76
B	50	100	245628	245809.3	156.58	11.02	99.88	54.38
C	50	100	288768	289229.5	270.81	14.10	93.84	57.09

Analyzing our achieved results, we can conclude that our hybrid Iterated Local Search approach yields high-quality solutions within short running-times for the benchmark instances existing in the literature and within reasonable running-times in the case of the proposed new randomly generated instances of larger sizes.

5 Conclusions

This paper deals with a particular two-stage fixed-charge transportation problem, which models an important transportation application in a supply chain, from a given manufacturer to a number of customers through distribution centers.

For solving this optimization problem we developed a hybrid Iterated Local Search (HILS) heuristic algorithm. Our heuristic approach constructs an initial solution, uses a local search procedure to increase the exploration, a perturbation mechanism and it is hybridized with a neighborhood structure in order to diversify the search. The preliminary computational results show that our hybrid Iterated Local Search is robust and yields high-quality solutions within reasonable running-times.

In future, we plan to evaluate the generality and scalability of the proposed solution approach by testing it on larger instances.

References

1. Geoffrion, A.M., Graves, G.W.: Multicommodity distribution system design by Benders decomposition. Manag. Sci. **20**, 822–844 (1974)
2. Molla-Alizadeh-Zavardehi, S., Hajiaghaei-Kesteli, M., Tavakkoli-Moghaddam, R.: Solving a capacitated fixed-cost transportation problem by artificial immune and genetic algorithms with a Prüfer number representation. Expert Syst. Appl. **38**, 10462–10474 (2011)
3. Pintea, C.-M., Sitar, C.P., Hajdu-Macelaru, M., Petrica, P.: A hybrid classical approach to a fixed-charged transportation problem. In: Corchado, E., Snášel, V., Abraham, A., Woźniak, M., Graña, M., Cho, S.-B. (eds.) HAIS 2012. LNCS (LNAI), vol. 7208, pp. 557–566. Springer, Heidelberg (2012). https://doi.org/10.1007/978-3-642-28942-2_50
4. Pintea, C.-M., Pop, P.C., Hajdu-Măcelaru, M.: Classical hybrid approaches on a transportation problem with gas emissions constraints. Adv. Intell. Soft Comput. **188**, 449–458 (2013)
5. Pintea, C.M., Pop, P.C.: An improved hybrid algorithm for capacitated fixed-charge transportation problem. Log. J. IJPL **23**(3), 369–378 (2015)
6. Pop, P.C., Pintea, C.-M., Pop Sitar, C., Hajdu-Macelaru, M.: An efficient reverse distribution system for solving sustainable supply chain network design problem. J. Appl. Log. **13**(2), 105–113 (2015)
7. Pop, P.C., Matei, O., Pop Sitar, C., Zelina, I.: A hybrid based genetic algorithm for solving a capacitated fixed-charge transportation problem. Carpathian J. Math. **32**(2), 225–232 (2016)
8. Santibanez-Gonzalez, E., Del, R., Robson Mateus, G., Pacca Luna, H.: Solving a public sector sustainable supply chain problem: a Genetic Algorithm approach. In: Proceedings of International Conference of Artificial Intelligence (ICAI), Las Vegas, USA, pp. 507–512 (2011)
9. El-Sherbiny, M.M.: 'Solving a capacitated fixed-cost transportation problem by artificial immune and genetic algorithms with a Prüfer number representation" by Molla-Alizadeh-Zavardehi, S., et al. Expert Systems with Applications (2011). Expert Syst. Appl. **39**, 11321–11322 (2012)

Change Detection in Multidimensional Data Streams with Efficient Tensor Subspace Model

Bogusław Cyganek[✉]

AGH University of Science and Technology,
Al. Mickiewicza 30, 30-059 Kraków, Poland
cyganek@agh.edu.pl

Abstract. The paper presents a method for change detection in multidimensional streams of data based on a tensor model constructed from the Higher-Order Singular Value Decomposition of raw data tensors. The method was applied to the problem of video shot detection showing good accuracy and high speed of execution compared with other more time demanding tensor models. In this paper we show two efficient algorithms for tensor model construction and tensor model update from the stream of data.

Keywords: Tensor change detection · Video shot detection
Orthogonal tensor space · Higher-Order Singular Value Decomposition

1 Introduction

Signal change detection finds many applications. However, existing methods are highly specialized. That is, for different signals there are methods that operate only with specific features, such as color, texture, spare, etc. However, the problem gets even more difficult with growing number of types of signals, their dimensionality as well as amount and speed of incoming data. Thus, in the era of information coming in streams of multidimensional data there is a need for more general methods that can be applied to any type of signal. One of the approaches rely on tensor methods. Nevertheless, the price to pay is computational burden involved with this class of algorithms.

Specifically, signal change detection in video, called shot detection, is an active research topic [1–3, 7, 9–12, 20–22, 23, 24]. Recently, for this problem we proposed a solution based on a tensor method that relies on the best rank-R tensor decomposition [4]. However, although good results were obtained, the maximal speed of executions did not exceed 3–4 frames per second. In this paper we propose an improvement to this method in a form of a new tensor model based on the Higher-Order Singular Value Decomposition (HOSVD). Although HOSVD is usually treated as a simpler tensor approximation compared with the best rank-R, it fits well to the change detection problem, as will be shown. Also, in this paper we propose an efficient algorithms of the HOSVD model construction and an algorithm for the model update from the stream of incoming data. These are the main theoretical contributions of this paper, whereas experimental results show comparable accuracy and much faster execution than the previous tensor model.

© Springer International Publishing AG, part of Springer Nature 2018
F. J. de Cos Juez et al. (Eds.): HAIS 2018, LNAI 10870, pp. 694–705, 2018.
https://doi.org/10.1007/978-3-319-92639-1_58

2   Tensor Framework for Multidimensional Data Stream Analysis

Figure 1 depicts architecture of the multi-dimensional data stream analysis by tensor models. The input stream consists of potentially infinite series of D-dimensional tensors. From these a window of a fixed size W is selected from which a model tensor is computed, as will be discussed. Then, each incoming data tensor is checked to fit to the model. If it does, then the model is updated with that data. Otherwise the model is rebuilt starting at current data position, as will be discussed.

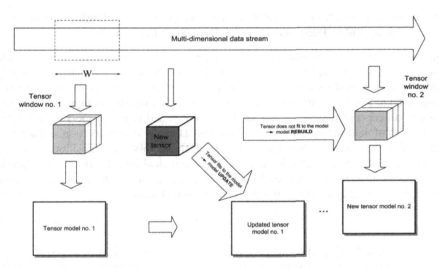

Fig. 1. Architecture of the multi-dimensional data stream analysis by tensor models. A window of size W is selected to build the tensor model. Further data are checked to fit to that model. If data fits the model, then the model is updated, otherwise the model is rebuilt starting at current data position.

In our previous publication we proposed a similar scheme, which was based on the model construction from the best rank-R tensor decomposition [4]. However, computation of this model is computationally demanding since the algorithm is iterative and also it requires P times eigenvalue decompositions. Thus, although our previous method produces good results, its operation time is long, reaching up to three color frames per second. In this paper we propose other, simpler tensor model, which is based on the HOSVD decomposition of the input tensors. Contrary to the best rank-R algorithm, HOSVD requires only one solution to the eigenvalue problem and, thanks to our representation, it is computed from a symmetric matrix of small size $W \times W$, as will be discussed. Thus, the HOSVD tensor model applied to multi-dimensional data stream analysis allows similar accuracy, offering a much faster performance, as will be discussed. We also show how the model can be built and efficiently updated with each incoming frame. These are the main contributions of this paper.

3   Construction of the Orthogonal Tensor Subspace Based Model

In processing multi-dimensional data, such as video or reflection seismic data, tensors offer many advantages compared to the classical vector based methods. In a tensor domain, each degree of freedom is represented by a separate index of a tensor. Another advantage is that tensor methods can work with any type of signal since no specific signal features are assumed. Here we introduce details to understand our tensor model. Further information on tensor processing can be found in literature, e.g. in [6, 17, 18].

3.1   Short Introduction to Tensor Analysis and the Tensor Higher-Order Singular Value Decomposition for Data Stream Analysis

Since the experimental results presented in this paper are related to the color video processing, i.e. each data in the stream can be represented as a 3D tensor, then the model is represented as a 4D tensor. Therefore, further analysis in this section is constrained to the 4D tensors, although it can be easily extended to higher dimensions. The key concepts of tensor analysis are presented related to tensor model construction for analysis of data streams. These are the HOSVD tensor decomposition and a way of constructing orthogonal tensor subspaces based on the this decomposition.

Although the tensor analysis deals with mathematical objects which fulfill precise transformation rules on a change of coordinate system, for our discussion a 4D tensor can be represented as a four-dimensional cube of real values, i.e.

$$\mathcal{T} \in \Re^{N_1 \times N_2 \times N_3 \times N_4}, \tag{1}$$

where N_j stands for a j-th dimension, for $1 \leq j \leq 4$. With no loss of information each tensor can be unanimously represented in a matrix representation. Such representation is known as a tensor flattening, which for a j-th dimension is defined as follows

$$\mathbf{T}_{(j)} \in \Re^{N_j \times \left(\ldots N_{j-1} N_{j+1} \ldots \right)}. \tag{2}$$

It is obtained from a tensor $\mathcal{T}$ by selecting its j-th dimension for a row dimension of $\mathbf{T}_{(j)}$, whereas a product of all other indices constitutes a column dimension of $\mathbf{T}_{(j)}$.

In further derivations we also use the concept of a k-th modal product of a tensor $\mathcal{T} \in \Re^{N_1 \times \ldots \times N_4}$ and a matrix $\mathbf{M} \in \Re^{Q \times N_k}$. The result of this product is a tensor $\mathcal{S} \in \Re^{N_1 \times \ldots N_{k-1} \times Q \times N_{k+1} \times \ldots N_4}$ with elements defined as follows

$$\mathcal{S}_{n_1 \ldots n_{k-1} q n_{k+1} \ldots n_4} = (\mathcal{T} \times_k \mathbf{M})_{n_1 \ldots n_{k-1} q n_{k+1} \ldots n_4} = \sum_{n_k=1}^{N_k} t_{n_1 \ldots n_{k-1} n_k n_{k+1} \ldots n_4} m_{q n_k}. \tag{3}$$

Equipped with the three tensor concepts (1)–(3), we can define the HOSVD decomposition of a tensor $\mathcal{T} \in \Re^{N_1 \times N_2 \times N_3 \times N_4}$, as follows [18, 19]

$$T = Z \times_1 S_1 \times_2 S_2 \times_3 S_3 \times_4 S_4, \tag{4}$$

where S_k stands for a unitary mode matrix of dimensions $N_k \times N_k$, and $Z \in \Re^{N_1 \times N_2 \times N_3 \times N_4}$ is a core tensor of the same dimensions as T. Further, it can be shown that Z satisfies the following conditions:

1. Two sub-tensors $Z_{n_k=a}$ and $Z_{n_k=b}$, obtained by fixing the n_k index to a, or b, are orthogonal, that is, the following holds

$$Z_{n_k=a} \cdot Z_{n_k=b} = 0, \tag{5}$$

  for all possible values of k for which $a \neq b$.
2. All sub-tensors can be ordered according to their Frobenius norms

$$\|Z_{n_k=1}\| \geq \|Z_{n_k=2}\| \geq \dots \geq \|Z_{n_k=N_P}\| \geq 0. \tag{6}$$

In the proposed stream processing framework the input tensor T is composed of a series of 3D frame-tensors F_w, for $1 \leq w \leq W$, i.e. color images, although in general case the tensor data can be of *any* dimension. That is, the input tensor is formed as follows

$$T = [F_1|F_2|\dots|F_W]. \tag{7}$$

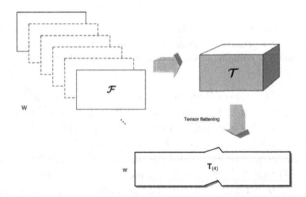

Fig. 2. The input stream consists of potentially infinite series of D-dimensional tensors. From these a window of a fixed size W is selected from which a model tensor of dimension $D + 1$ is constructed. For construction of the HOSVD model a tensor flattening alongside its last dimension is used. It is easy to observe that each data from the series constitutes one row in this flattening. The order of flattening is irrelevant if kept consistent among all tensors.

Construction of the tensor T is depicted in Fig. 2. T can be now decomposed as stated in (4). However, it can be easily noticed that (4) after simple re-arrangement can be written as follows

$$T = \underbrace{(\mathcal{Z} \times_1 \mathbf{S}_1 \times_2 \mathbf{S}_2 \times_3 \mathbf{S}_3)}_{\mathcal{B}_w} \times_4 \mathbf{S}_4. \tag{8}$$

where, it can be shown that $\mathcal{B}_w$, for $1 \leq w \leq W$, are *orthogonal tensors* thanks to the condition (5), i.e. their product gives 0. Equation (8) can be further written as follows

$$T = \sum_{w=1}^{W} \left(\mathcal{B}_w \times_4 \mathbf{s}_4^w \right). \tag{9}$$

where $\mathbf{s}_4^w$ denote columns of the unitary matrix $\mathbf{S}_4$. Because each tensor $\mathcal{B}_w$ is three-dimensional, then $\times_4$ denotes the outer product of each 3D tensor and a vector $\mathbf{s}_4^w$.

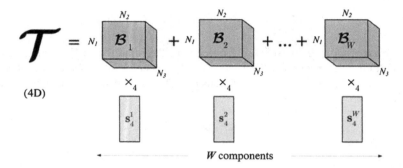

Fig. 3. Visualization of the OTS of a 4D model tensor spanned by a series of orthogonal 3D base tensors. The OTS represents a model of a window of W input data tensors. A distance of a tensor $\mathcal{F}$ to the model is computed as a projection of $\mathcal{F}$ onto the OTS.

This way, out of the tensor T the *orthogonal tensor subspace* (OTS) is constructed, as visualized in Fig. 3. In other words, this OTS constitutes a model. The reason of constructing such an OTS is its ability to represent the series of W frames, as well as to introduce a distance of a data tensor $\mathcal{F}$ to that space. This property will be used in checking a fitness degree of each frame to the model, as will be discussed.

For a given 3D tensor $\mathcal{F}$ (frame), its distance to the model represented by a series of base tensors $\{\mathcal{B}_w\}$ is computed as a sum of squared inner products, as follows [6]

$$R = \sum_{w=1}^{W} \left\langle \hat{\mathcal{B}}_w, \hat{\mathcal{F}} \right\rangle^2, \tag{10}$$

where "hat" denotes tensor normalization. The above formula will be used to assess model consistency, by computing R for all frames belonging to the model, as well as to assess if a new frame is consistent or not with the model, as will be discussed.

3.2 Model Construction - Efficient Computation of the Orthogonal Tensor Subspaces

Since efficient computation of the base tensors $\mathcal{B}_w$ from the input stream is essential for operation of the method, in this section we propose an efficient algorithm. Practically, $\mathcal{B}_w$ can be simply computed after rearrangement of Eq. (8), as follows

$$\mathcal{T} \times_4 \mathbf{S}_4^T = \mathcal{Z} \times_1 \mathbf{S}_1 \times_2 \mathbf{S}_2 \times_3 \mathbf{S}_3 = \mathcal{B}_w. \tag{11}$$

Thus, to compute $\mathcal{B}_w$ it is sufficient to compute only the mode matrix $\mathbf{S}_4$. It can be computed from the SVD decomposition of the flattened matrix $\mathbf{T}_{(4)}$ of $\mathcal{T}$, that is

$$\mathbf{T}_{(4)} = \mathbf{S}_4 \mathbf{V}_4 \mathbf{D}_4^T. \tag{12}$$

However, $\mathbf{T}_{(4)}$ is of large dimension, having a number of rows equal to W and the number of columns equal to the product of dimensions 1–3. That is, for color video processing, it is a total number of pixels in the input frames times three color channels. To overcome this problem, both sides of (12) can be multiplied by $\mathbf{T}_{(4)}^T$, as follows

$$\mathbf{T}_{(4)} \mathbf{T}_{(4)}^T = \mathbf{S}_4 \mathbf{V}_4^2 \mathbf{S}_4^T. \tag{13}$$

The above product $\mathbf{T}_{(4)} \mathbf{T}_{(4)}^T$ has dimensions of only $W \times W$ and moreover it is a symmetrical matrix for which an efficient fixed-point eigenvalue decomposition algorithm is used. This is the same algorithm which was used for computation of the best rank-R decomposition, described in our previous publication [5]. However, in the proposed model with HOSVD, eigen-decomposition it is computed only once and from smaller matrix. Thanks to this, computation of the base tensors can be much faster. Summarizing, the model build algorithm looks as follows.

Model build algorithm:

1. Fill the buffer with W input data and construct the tensor $\mathcal{T}$ as in (7);
2. Construct the flattened matrix $\mathbf{T}_{(4)}$ of a tensor $\mathcal{T}$;
3. Compute the product $\mathbf{T}_{(4)} \mathbf{T}_{(4)}^T$;
4. Compute the $\mathbf{S}_4$ as eigenvectors of the symmetric matrix $\mathbf{T}_{(4)} \mathbf{T}_{(4)}^T$ in (13);
5. Compute the bases $\mathcal{B}_w$ from (11);

3.3 Model Fitness Measure

As alluded to previously, the measure R in (10) can be used to tell a distance of a tensor $\mathcal{F}$ to the model represented by the basis $\{\mathcal{B}_w\}$. Values of R for the model frames, as well as for all other frames from the stream can be used for the statistical analysis of shot boundaries in the stream. However, instead of the absolute values of R, better results are obtained when the differences of ΔR are used for computations. That is, the following error function is used

$$\Delta R_i \equiv R_i - R_{i-1}. \tag{14}$$

For proper detection of the shots with slowly changing content, the following drift measure is proposed [4]

$$\|\Delta R_{\mathcal{F}} - \bar{R}_\Delta\| < a\,\sigma_\Delta + b. \tag{15}$$

where a is a multiplicative factor (3.0–4.0) and b is an additive component (0.2–2.5), $\bar{R}_\Delta$ and σ_Δ are the mean and standard deviation computed from the differences of fit values in (14) for the model frames from (7), as follows

$$\bar{R}_\Delta = \frac{1}{W} \sum_{w=1}^{W} R_{\Delta w}, \text{and } \sigma^2 = \frac{1}{W-1} \sum_{w=1}^{W} (R_{\Delta w} - \bar{R}_\Delta)^2. \tag{16}$$

3.4 Efficient Model Update Scheme

Each new tensor $\mathcal{F}$ is checked to fit to the model in accordance with (16). If it does not fit, the model has to be built from a new set of frames, starting at position of $\mathcal{F}$.

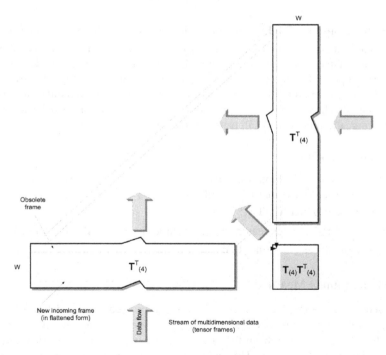

Fig. 4. Update scheme for the flattened version $\mathbf{T}_{(4)}$ of the model tensor. If a new frame fits to the model, the model is updated. This is done by simple insertion of the new row and obliteration of the oldest one in the $\mathbf{T}_{(4)}$. In the product matrix $\mathbf{T}_{(4)}\mathbf{T}_{(4)}^T$ all values, except for one row and one column, can be reused as shown in Fig. 5.

On the other hand, if $\mathcal{F}$ fits to the model, the model needs only to be updated. Figure 4 depicts the process of updating of the flattened version $\mathbf{T}_{(4)}$ of the model tensor. This is done by simple insertion of the new row and obliteration of the oldest one in the $\mathbf{T}_{(4)}$ tensor. In the product matrix $\mathbf{T}_{(4)}\mathbf{T}_{(4)}^T$ all values except one row and one column can be reused. The following steps describe the model update algorithm (see also Fig. 1).

Model update algorithm:

1. Shift data in $\mathbf{T}_{(4)}$ by one row up (Fig. 4); Fill the last row with $\mathcal{F}$;
2. Shift all data in the old $\mathbf{T}_{(4)}\mathbf{T}_{(4)}^T$ matrix by one row up and to the left (Fig. 5);
3. Fill up the last row and right column in $\mathbf{T}_{(4)}\mathbf{T}_{(4)}^T$ with a product $\mathcal{F}$ and all remaining (old) frames from $\mathbf{T}_{(4)}$;
4. Perform steps 4 and 5 of the model build algorithm;

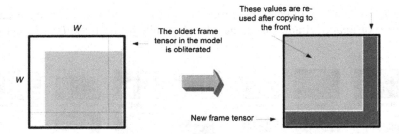

Fig. 5. Update scheme of the product matrix $\mathbf{T}_{(4)}\mathbf{T}_{(4)}^T$.

In the above model update algorithm the most time consuming is the step 3 which involves W-1 products of the tensor frames. The last step 4 is relatively fast and consumes the same amount of time as in the full model build step, since it requires solution of the eigenvalue problem of a matrix of size $W \times W$.

4   Experimental Results

The method was implemented and checked in the similar environment as in the previous publication [4]. That is, the method was entirely implemented in C++ in the Microsoft Visual 2017 IDE and with help of the *DeRecLib* library [8]. The experiments were run on a computer with the Intel® Xeon® E-1545 processor, operating at 2.9 GHz, with 64 GB RAM, and 64-bit version of Windows 10.

As already mentioned, the proposed method can work with any type of signal of any finite dimensions since no specific features are computed and signal is taken with no pre-processing. However, it is not easy to find suitable test streams with ground truth annotation. Therefore, and to compare the results with our previous work, for evaluation the VSUMM database with 50 color videos from the *Open Video Project*, endowed with the human annotated video shots summarizations, was used [2, 15]. This database contains color video sequences of resolution 352×240 pixels, 30 fps, with duration in the range of 1 to 4 min, encoded in the MPEG-1 [14].

For the qualitative evaluation, the parameters proposed by de Avila *et al.* [2, 15], called *Comparison of User Summaries* were used. These are $CUS_A = n_{AU}/n_U$ and $CUS_E = \sim n_{AU}/n_U$, where n_A denotes a number of *matching* keyframes from the automatic summary (*AS*) and the user annotated summary, $\sim n_{AU}$ is the complement of this set (i.e. the frames that were not matched), whereas n_U is a total number of keyframes from the user summary only (*US*). However, in other works the *precision P* and *recall R*, parameters are preferred, since they convey also information on the keyframes present in *AS* and not present in *US* or vice versa, as discussed by Mahmoud *et al.* [16, 21]. As a tradeoff of the two, in many works the *F* measure is also used [13, 21]. The latter are also used in our experiments. These are defined as follows

$$P = \frac{n_{AU}}{n_A}, R = \frac{n_{AU}}{n_U}, \text{and } F = 2\frac{PR}{P+R}, \quad (17)$$

where n_{AU} is a number of keyframes from the *AS* that match the ones from the *US*, n_A is a number of total keyframes from the *AS* only, while n_U from the *US*, respectively.

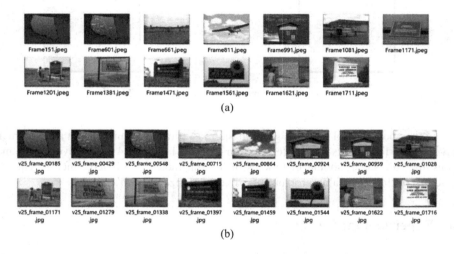

(a)

(b)

Fig. 6. User selected keyframes from the "A New Horizon" sequence (a). These are compared with the keyframes computed by our algorithm (b).

Figure 6a shows exemplary user selected keyframes from one test sequence "A New Horizon" from the Open Video Database [14]. These are compared with the keyframes computed by our algorithm shown in Fig. 6b. Based on such comparisons the average value of the accuracy parameter *F* was computed, as shown in Table 1.

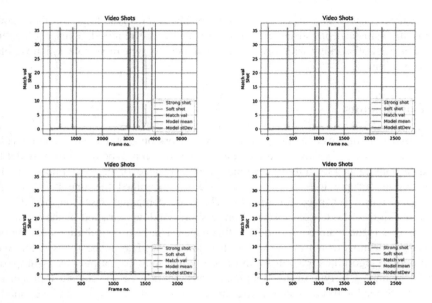

Fig. 7. Plots of the detected scene shots computed by our method in the video sequence no 64 to 67 from the Open Video Database [14], respectively.

Figure 7 show plots of the detected keyframes from the test sequences no 64 to 67, respectively. Large green bars show detected shots of a window W.

Table 1. Comparison of accuracy for different methods.

Method	OV [23]	DT [24]	STIMO [11]	VSUMM [2]	VSCAN [21]	Best rank-R [4]	**HOSVD** (this paper)
F	0.67	0.61	0.65	0.72	0.77	0.73	**0.73**

In Table 1 it is visible that the proposed method performs comparatively or better than other methods published in literature and attains the same results as the tensor method based on the best rank-R tensor decomposition.

Table 2. Average execution time of the tensor based methods for 352×240 color video.

Video type	Best rank-R [4]	HOSVD (this paper)
Processing time [frames/s]	3.2	**15.3**

On the other hand, Table 2 shows execution times of the two methods – the one presented in our previous paper [4] and the one proposed in this paper. The average execution time for the same accuracy are 5 times faster for the latter method. The important aspect of the proposed method is that, although it was tested on the color videos, it does not put any specific assumptions about the signal type. In other words, no specific statistical properties, nor specific features, are required. Thus, the method has a potential of working with any dimensional and any type of signals which we plan to test in a future.

5 Conclusions

In this paper an improved version of our previous work on shot detection in color video signals is presented. The proposed method assumes different type of a tensor model for the data coming in the multidimensional stream. Instead of the best rank-R model, proposed in our previous work, in this one we examined the one based on the orthogonal tensor subspace construction. These are computed based on the HOSVD decomposition of the tensors formed from a short window extracted of the input data stream. This is the main contribution of this paper. The next contributions are the two efficient algorithms of tensor model construction and update, respectively. Moreover, the proposed algorithms are general in the sense that they do not put any assumptions on dimensionality or type of the input data. Thus, neither specific statistical properties, nor features need to be computed. Therefore, the proposed framework can be applied to any type of multidimensional streams of data. Also, the proposed construction of the orthogonal tensor subspaces can be used in many other classification and clusterization frameworks suitable for processing of the multidimensional data.

In this paper properties of this method were experimentally verified in the problem of shot detection in streams of color video, due to availability of human annotated data, as well as to compare with our previous method. In this framework, the proposed HOSVD model allowed for the same accuracy as the previously proposed best rank-R method, achieving 5 times faster execution. This is due to much simpler computations of the former one thanks to the proposed fast algorithms of the model construction and update.

In the future we plan to test further properties of the proposed method, and especially with other multidimensional streams of data.

Acknowledgement. This work was supported by the Polish National Science Center NCN under the grant no. 2014/15/B/ST6/00609 as well as AGH Statutory Funds no. 11.11.230.017.

References

1. Asghar, M.N., Hussain, F., Manton, R.: Video indexing: a survey. Int. J. Comput. Inf. Technol. **03**(01), 148–169 (2014)
2. de Avila, S.E.F., Lopes, A.P.B., da Luz Jr., A., Araújo, A.A.: VSUMM: a mechanism designed to produce static video summaries and a novel evaluation method. Pattern Recogn. Lett. **32**, 56–68 (2011)
3. Cayllahua-Cahuina, E.J., Cámara-Chávez, G., Menotti, D.: A Static Video Summarization Approach with Automatic Shot Detection Using Color Histograms (2012)
4. Cyganek, B., Woźniak, M.: Tensor-based shot boundary detection in video streams. New Gener. Comput. **35**(4), 311–340 (2017)

5. Cyganek, B., Woźniak, M.: On robust computation of tensor classifiers based on the higher-order singular value decomposition. In: Silhavy, R., Senkerik, R., Oplatkova, Z.K., Silhavy, P., Prokopova, Z. (eds.) Software Engineering Perspectives and Application in Intelligent Systems. AISC, vol. 465, pp. 193–201. Springer, Cham (2016). https://doi.org/10. 1007/978-3-319-33622-0_18

6. Cyganek, B.: Object Detection and Recognition in Digital Images: Theory and Practice. Wiley, Hoboken (2013)

7. Cyganek, B.: Recognition of road signs with mixture of neural networks and arbitration modules. In: Wang, J., Yi, Z., Zurada, J.M., Lu, B.-L., Yin, H. (eds.) ISNN 2006. LNCS, vol. 3973, pp. 52–57. Springer, Heidelberg (2006). https://doi.org/10.1007/11760191_8

8. DeRecLib, 2013. http://www.wiley.com/go/cyganekobject

9. Del Fabro, M., Böszörmenyi, L.: State-of-the-art and future challenges in video scene detection: a survey. Multimedia Syst. **19**(5), 427–454 (2013)

10. Fu, Y., Guo, Y., Zhu, Y., Liu, F., Song, C., Zhou, Z.-H.: Multi-view video summarization. IEEE Trans. Multimedia **12**(7), 717–729 (2010)

11. Furini, M., Geraci, F., Montangero, M., Pellegrini, M.: STIMO: STIll and moving video storyboard for the web scenario. Multimedia Tools Appl. **46**(1), 47–69 (2010)

12. Gao, Y., Wang, W.-B., Yong, J.-H., Gu, H.-J.: Dynamic video summarization using two-level redundancy detection. Multimedia Tools Appl. **42**(2), 233–250 (2009)

13. Guan, G., Wang, Z., Yu, K., Mei, S., He, M., Feng, D.: Video summarization with global and local features. In: Proceedings of the 2012 IEEE International Conference on Multimedia and Expo Workshops, pp. 570–575. IEEE Computer Society, Washington, DC (2012)

14. https://open-video.org/

15. https://sites.google.com/site/vsummsite/home

16. https://sites.google.com/site/vscansite/home

17. Kolda, T.G., Bader, B.W.: Tensor decompositions and applications. SIAM Rev. **51**(3), 455–500 (2008)

18. de Lathauwer, L.: Signal Processing Based on Multilinear Algebra. Ph.D. dissertation, Katholieke Universiteit Leuven (1997)

19. de Lathauwer, L., de Moor, B., Vandewalle, J.: A multilinear singular value decomposition. SIAM J. Matrix Anal. Appl. **21**(4), 1253–1278 (2000)

20. Lee, H., Yu, J., Im, Y., Gil, J.-M., Park, D.: A unified scheme of shot boundary detection and anchor shot detection in news video story parsing. Multimedia Tools Appl. **51**, 1127–1145 (2011)

21. Mahmoud, K.M., Ismail, M.A., Ghanem, N.M.: VSCAN: an enhanced video summarization using density-based spatial clustering. In: Petrosino, A. (ed.) ICIAP 2013. LNCS, vol. 8156, pp. 733–742. Springer, Heidelberg (2013). https://doi.org/10.1007/978-3-642-41181-6_74

22. Medentzidou, P., Kotropoulos, C.: Video summarization based on shot boundary detection with penalized contrasts. In: IEEE 9th International Symposium on Image and Signal Processing and Analysis (ISPA), pp. 199–203 (2015)

23. DeMenthon, D., Kobla, V., Doermann, D.: Video summarization by curve simplification. In: Proceedings of the Sixth ACM International Conference on Multimedia, pp. 211–218. ACM (1998)

24. Mundur, P., Rao, Y., Yesha, Y.: Keyframe-based video summarization using delaunay clustering. Int. J. Dig. Libr. **6**(2), 219–232 (2006)

A View of the State of the Art of Dialogue Systems

Leire Ozaeta and Manuel Graña[✉]

Computational Intelligence Group, Department CCIA,
University of the Basque Country, Leioa, Spain
manuel.grana@ehu.es

Abstract. Dialogue systems are becoming central tools in human computer interface systems. New interaction systems, e.g. Siri, Echo and others, are proposed by the day, and new features are added to these systems at breathtaking pace. The conventional approaches based on traditional artificial intelligence techniques, such as ontologies and tree based search, have been superseded by machine learning approaches and, more recently, deep learning. In this paper we give a view of the current state of dialogue systems, describing the areas of application, as well as the current technical approaches and challenges. We propose two emerging domains of application of dialogue systems that may be highly influential in the near future: storytelling and therapeutic systems.

1 Introduction

The development of dialogue systems it's been a topic of remarkable interest since the very beginning of the Artificial Intelligence [18]. dialogue systems can be divided into goal-driven systems, such as technical support services, and goal-free systems, such as language learning tools or computer game characters [15]. There has been a long journey since the first conversational system, ELIZA, considered one of the most important chatbot dialogue systems in the history of the field [9], to the task-oriented personal assistants that are currently present in most cellphones or home controllers i.e.: Siri, Cortana, Alexa, Google Now/Home, etc. This spread of the dialogue systems is linked to the development of a wide range of data-driven machine learning methods have been shown to be effective for natural language processing [14] including the tremendous succeed for large vocabulary continuous speech recognition of Deep Neural Networks (DNNs), such as Convolutional Neural Networks (CNNs) and Long-Short Term Memory Recurrent Neural Networks (LSTMs) [13]. Until very recently, most deployed task-oriented dialogue systems used hand-crafted features for the state and action space representations, and require either a large annotated task-specific corpus or a large number of human subjects willing to interact with the unfinished system. This did not only made it expensive and time-consuming to deploy, it also limited its usage to a narrow domain. Conversational systems, however, have drawn inspiration from the use of neural networks in natural language modelling and machine translation tasks [15]. At this point, however, it

© Springer International Publishing AG, part of Springer Nature 2018
F. J. de Cos Juez et al. (Eds.): HAIS 2018, LNAI 10870, pp. 706–715, 2018.
https://doi.org/10.1007/978-3-319-92639-1_59

seems that a link between the two traditionally separated system development approaches can be achieved, where the task-oriented dialogue systems provide a more natural interaction where there is room for small talk and task-free dialogue. In the same fashion, latest improvements open the door for more complex areas to be approached with this dialogue systems such as therapeutic systems and robot interfaces [5].

Intended Contribution. The aim of the work in this article is to present the state-of-the-art of dialogue systems and some ideas about their future development and new fields of application. Tables 1 and 2 summarize the main references discussed in this paper. The contents of the paper are as follows: Sect. 2 discusses system architectures. Section 3 comments on the dialogue system categories. Section 4 discusses evaluation and training issues. Finally, Sect. 5 discusses some future challenges.

Table 1. Description of systems in the literature

Reference	Description
[16]	Developing a Hierarchical Encoder-Decoder Model
[15]	End-To-End dialogue Systems generated by Hierarchical Neural Networks
[7]	Speech Recognition with Deep Recurrent Neural Networks
[11]	Adversarial Learning for Neural dialogue Generation
[8]	Review of Machine Learning for dialogue State Tracking
[10]	Deep Reinforcement Learning for dialogue Generation
[21]	A Network-based End-to-End Trainable Task-oriented dialogue System
[20]	Speeding up adaptation of Spoken dialogue Systems by Recurrent Neural Networks shaping rewards
[22]	An Entropy Minimization Framework for Goal-Driven dialogue Management
[18]	End-to-end optimization of goal-driven and visually grounded dialogue systems
[2]	Learning End-to-End Goal-Oriented dialogue
[12]	Deep Sentence Embedding Using Long Short-Term Memory Networks: Analysis and Application to Information Retrieval
[4]	Deep Reinforcement Learning based Dialogue System
[17]	Adaptation of Spoken dialogue Systems using RL
[19]	On-line adaptation of Spoken dialogue Systems using Active Reward Learning

2   Architectures of Dialogue Systems

The traditional architecture for dialogue systems illustrated in Fig. 1 includes a series of system modules, each with specific functionality [14]:

- Speech Recognizer, in charge of providing the lexical units for the system extracting them from the voice signal,

Table 2. Main features of referenced works. EM Entropy Maximization, DRL Deep Reinforcement Learning, MN Memory Networks, RL Reinforcement Learning, CNN Convolutional Neural Network, LSTM Long Short Term Memory, RNN Recurrent Neural Network, DRNN Deep Recurrent Neural Network, HRNN Hierarchical Neural Network.

Reference	System type	Task-oriented?	Architecture	Module	Method
[16]	SEQ2SEQ	No	End-to-end	Dialogue generation	HRNN
[15]	SEQ2SEQ	No	End-to-end	Dialogue generation	HRNN
[7]	-	-	End-to-end	Speech recognition	DRNN
[11]	SEQ2SEQ	No	End-to-end	Response generation	RL
[8]	review	-	Traditional	State tracking	ALL
[10]	SEQ2SEQ	No	End-to-end	Dialogue generation	RL
[21]	SEQ2SEQ	Yes	End-to-end	Dialogue generation	NN
[20]		Yes	Traditional	Evaluation	SLTM
[22]	Inf. Retr	Yes	Traditional	Response generation	EM
[18]	-	Yes	End-to-end	Dialogue generation	DRL
[2]	-	Yes	End-to-end	Dialogue generation	MN
[12]	-	-	Traditional	Language interpreter	RNN with STLM
[4]	-	No	Traditional	Response generation	DRL
[17]	-	Yes	Traditional	State tracking	RL
[19]	SEQ2SEQ	Yes	-	Evaluation	RL

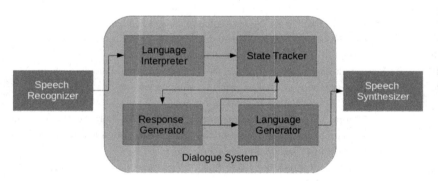

Fig. 1. Traditional dialogue System

- Language Interpreter, in charge of extracting meaning from the stream of lexical units by Natural Language Processing techniques,
- State Tracker, in charge of modelling the dialogue state and dynamics, it keeps track of the goal in task oriented systems and of the contextual information in task-free systems.
- Response Generator, produces the semantically grounded response to the current input,

- Language Generator, formulates the response in correct language constructs by Natural Language Generation techniques, and
- Speech Synthesizer generates a recognizable voice signal for the communication with the human side.

Obviously, the Speech Recognizer and Speech Synthesizer modules have meaning in voice-based dialogue systems. Text based dialogue systems do not need them. Each of these modules can be tackled with as an independent problem, hence they have been approached using different techniques. This variety is evident in the list of examples in Table 1. Speech Recognition advances due the renew interest in Neural Network are exemplified by [7], or the review in [1]. Deep Neural Networks have been also influential in Language Interpretation [12] and Response Generation [10,11,16]. State tracking has been addressed from many sides with a variety of techniques [8].

2.1 End-to-end Dialogue Systems

In recent years the so-called end-to-end dialogue system architectures have become popular and so, some modules, or even all of them, have been collapsed into a unique module of Response Generator, as illustrated in Fig. 2. Those systems, mostly based on neural networks, have shown promising results on several dialogue tasks [14]. The main difference between the classical approach and the end-to-end approach is the emphasis on data driven system construction. While the classical approach is much hand crafted and introduces a priori assumptions and design restrictions in all the modules, via specific computational models, the end-to-end approach assumes that the whole architecture can be induced from the data via learning algorithms [2,3]. This shift has been possible because of the success of Deep Learning Neural Network approaches [6]. There are two main categories of end-to-end dialogue approaches [14], on one hand those that search in a dataset of fixed possible responses, and on the other hand those that select the utterance that maximizes the posterior distribution over all possible utterances. The second approach allows more dynamic responses, as the response generation can be decomposed to the word level.

Fig. 2. End-to-end dialogue System [12]

3 Dialogue System Categories

There are two basic categories of dialogue systems:

- Task oriented systems: these systems have some specific goal that is to be achieved through the dialogue interaction. The assistant systems are designed to help search for specific information items. One consequence is that the iteration always reaches a termination state if the user achieves its goals.
- Conversational systems: these task-free systems have not a specific goal, so that iteration can evolve indefinitely, though it is expected that the iterations would produce some evolution of both the user and the cyber-side agent.

3.1 Task Oriented Dialogue Systems

These are the most useful applications of dialogue systems: personal assistants where the system needs to understand a request from the user and complete the related task within a limited number of dialogue turns. They are typically designed according to a structured ontology (or a database schema), which defines the domain that the system can talk about [19]. Getting the info is usually achieved using slot-filling, where a dialogue state is a set of slots to be filled during dialogue [2]. However, this system is inherently hard to scale to new domains as it has to be had all features and slots that might be needed manually encoded [2]. Task oriented systems are often designed to carry out some information retrieval dialogue [9], such as looking for the nearest restaurant, or to coordinate events, such as planning an appointment or a date.

This kind of systems have beneficed less of the end-to-end architecture and Machine Learning approaches, that do not make assumptions over the domain or dialogue state structure [2], because those methods cast the dialogue problem into one of supervised learning, predicting the distribution over possible next utterances given the discourse so far. The supervised learning framework does not account for the intrinsic planning problem that underlies dialogue, i.e. the sequential decision making process, which makes dialogue consistent over time [18].

3.2 Conversational Dialogue Systems

Conversational aka open dialogue systems try to produce meaningful and coherent responses in the framework of a dialogue history. They have applications ranging from technical support services, to language learning and entertainment, such as playing games with robots [5]. Approaches to build conversational architectures fall into two classes: rule-based systems and corpus-based systems [9]. The rule based systems correspond to the early attempts. Such as the famous ELIZA system, where rules were hand crafted following some a priori hints about the desired behaviour of the system. On the other hand, corpus-based approaches learn the system structure and parameters from the data in the corpus, making strong use of machine learning and other learning approaches, mining human-to-human conversations, or the human responses extracted from human-machine conversations [9]. Most either rule-based or corpus-based chatbots tend to do very little modelling of the conversational context. Instead they tend to focus on generating a single response turn that is appropriate given the user's immediately previous utterance. For this reason they are often called response generation systems [9]. Given the lack of precise goals, the conversational systems can be formulated as sequence-to-sequence transductors (SEQ2SEQ). However the SEQ2SEQ models tend to generate generic responses, which closes the conversation, or become stuck in an infinite loop of repetitive responses [10].

The most recent computational models used to build the conversational systems are generative models, such as the hierarchical recurrent encoder-decoder (HRED) [15,16], a kind of Recurrent Neural Networks (RNN) modelling the posterior of the next word in the sequence from the past context by using two contexts, that of the past words and that of the queries performed by the user. The encoder RNN maps each utterance to an utterance vector modelling the hidden state at both contexts, while the decoder RNN models the probability distribution of the utterances conditional to the hidden state. Utterance generation is achieved sampling the posterior probability density.

4   Evaluation and Training

System evaluation and training are closely related issues, because the quality measure used for evaluation may be used for training, and the resources employed for evaluation are closely related to the resources employed for training. Some approaches to evaluation use quality measures developed for machine translation systems, such as the bilingual evaluation understudy (BLEU), assuming that the dialogue process is akin to a translation process, between the system generated responses and the natural ones from humans. Other use the word perplexity measure [15] from probabilistic word modelling. This approach requires big corpora often unavailable for conversational dialogue systems, and scarce for task oriented systems. Most of the corpora available for dialogue system training and tuning come from very specific domains (e.g. chats about technical problems such as the Ubuntu IRC chats, or restaurant/movie picking) or were designed for other purposes such as automatic speech recognition system training [14].

Due to the lack of corpora containing precise desired responses for the supervised training of the systems, a natural trend is to resort to Reinforcement Learning (RL) approaches [11,18,20], which only require rewards at some point in time, such as the successful task achievement or some negative rewards when the task-free dialogue becomes senseless. The scientific community has turned towards the RL to train and evaluate the dialogue systems since it offers the possibility to treat dialogue design as an optimisation problem, and because RL-based systems can improve their performance over time with experience [4] following a life-long learning approach. However, training dialogue policies in an efficient, scalable and effective way across domains remains an unsolved problem as often requires significant time to explore the state-action space, which is a critical issue when the system is trained on-line with real users where learning costs are expensive [20].

Reinforcement learning approaches need some mechanism to generate the reward function values. The natural approach is to use human operators that provide rewards according to some quality criteria (i.e. easy of answering, coherence, informativeness, keyword retrieval) but in general it is difficult to extend the approach to wide open dialogue systems. A way to automate the process is to apply adversarial approaches [9] mimicking the Turing test of indistinguishability of the machine responses from the human responses. For example in [11] the authors use "a generator (a neural SEQ2SEQ model) that defines the probability of generating a dialogue sequence, and a discriminator analogous to the human evaluator in the Turing test that labels dialogues as human-generated or machine-generated. The generator is driven by the discriminator to generate utterances indistinguishable from human generated dialogues. In the end the human evaluation is the gold standard for all approaches, despite the high cost and inconvenience of having to deal with humans in the loop.

5   Future Applications and Challenges

We have tried to illustrate in Fig. 3 two emerging domains of application for dialogue systems which we have identified as Storytelling and Therapeutics systems. Storytelling is a hybridization of task and conversational systems with many applications in education and entertainment. The interaction is intended to reach the end of the plot, but it can wander along in the path, creating diverging paths that can be creative of new situations. The paradigm of telling a tale while allowing the audience to pose questions and/or ask the audience about their understanding of the current state of the plot and the personages, can be translated also to the teaching of formal concepts and personal training in specific topics in the academic curricula. The dialogue system is required to maintain unexpected paths of dialogue and to be able to answer about arbitrarily old states of the dialogue or even previous instances of the storytelling process. The system could be adjusted for various degrees of freedom relative to the story and alternative paths leading to the same conclusion of the story.

Therapeutic systems are focused on the user assuming that there is some kind of condition that needs to be reverted or alleviated, which can be pathological in the clinical sense or less dramatical. In the domain of education applications, children showing some aspect of the autistic spectrum can be more accessible to dialogue with anthropomorphic robots than with humans. In general, the therapeutic dialogue system needs to carry out the following tasks, which may or may not correspond to a specific module: diagnostic and evaluation of the user status, selection of treatment, application and assessment of the treatment effects.

Both kinds of innovative dialogue systems share the lack or, at best, the scarcity of the available data, because there are no corpuses covering these situations. The model free data drive approaches represented by Machine Learning and Deep Neural Networks may have some difficulties dealing with the need to explain to the medical staff the reasoning leading to some specific treatment and the assessment of the treatment outcomes. The lack of explicit state representation may be an issue when trying to follow divergent paths in storytelling or to sharing information with the medical staff. Therefore, new hybridization of the data-driven and the classical dialogue architectures may be required.

We will be involved in the development of storytelling systems for educational purposes, specifically the support of children with special needs in the framework of the CybSPEED european project, where we intend to embody these dialogue systems in the Nao anthropomorphic robot.

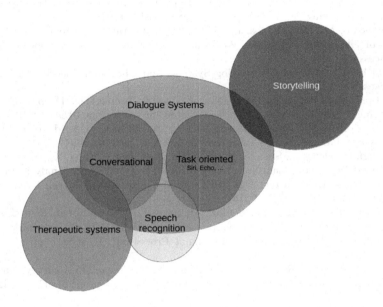

Fig. 3. Domains and challenges for dialogue systems

Acknowledgments. Leire Ozaeta has been supported by a Predoctoral grant from the Basque Government. This work has been partially supported by the EC through project CybSPEED funded by the MSCA-RISE grant agreement No 777720.

References

1. Baumann, T., Kennington, C., Hough, J., Schlangen, D.: Recognising conversational speech: what an incremental ASR should do for a dialogue system and how to get there. In: Jokinen, K., Wilcock, G. (eds.) Dialogues with Social Robots. LNEE, vol. 999, pp. 421–432. Springer, Singapore (2017). https://doi.org/10.1007/978-981-10-2585-3_35
2. Bordes, A., Boureau, Y.-L., Weston, J.: Learning end-to-end goal-oriented dialog. arXiv preprint arXiv:1605.07683 (2016)

3. Bowden, K.K., Oraby, S., Misra, A., Wu, J., Lukin, S.: Data-driven dialogue systems for social agents. arXiv preprint arXiv:1709.03190 (2017)
4. Cuayáhuitl, H.: *SimpleDS*: a simple deep reinforcement learning dialogue system. In: Jokinen, K., Wilcock, G. (eds.) Dialogues with Social Robots. LNEE, vol. 999, pp. 109–118. Springer, Singapore (2017). https://doi.org/10.1007/978-981-10-2585-3_8
5. Cuayahuitl, H., et al.: Deep reinforcement learning for conversational robots playing games (2017)
6. Goodfellow, I., Bengio, Y., Courville, A.: Deep Learning. MIT Press, Cambridge (2016)
7. Graves, A., Mohamed, A.R., Hinton, G.: Speech recognition with deep recurrent neural networks. In: 2013 IEEE International Conference on Acoustics, Speech and Signal Processing (ICASSP), pp. 6645–6649. IEEE (2013)
8. Henderson, M.: Machine learning for dialog state tracking: a review. In: Proceedings of The First International Workshop on Machine Learning in Spoken Language Processing (2015)
9. Jurafsky, D., Martin, J.H.: Dialog Systems and Chatbots. Speech and Language Processing, 3 (2014)
10. Li, J., Monroe, W., Ritter, A., Galley, M., Gao, J., Jurafsky, D.: Deep reinforcement learning for dialogue generation. arXiv preprint arXiv:1606.01541 (2016)
11. Li, J., Monroe, W., Shi, T., Jean, S., Ritter, A., Jurafsky, D.: Adversarial learning for neural dialogue generation. arXiv preprint arXiv:1701.06547 (2017)
12. Palangi, H., Deng, L., Shen, Y., Gao, J., He, X., Chen, J., Song, X., Ward, R.: Deep sentence embedding using long short-term memory networks: analysis and application to information retrieval. IEEE/ACM Trans. Audio Speech Lang. Process. (TASLP) **24**(4), 694–707 (2016)
13. Sainath, T.N., Vinyals, O., Senior, A., Sak, H.: Convolutional, long short-term memory, fully connected deep neural networks. In: 2015 IEEE International Conference on Acoustics, Speech and Signal Processing (ICASSP), pp. 4580–4584. IEEE (2015)
14. Serban, I.V., Lowe, R., Henderson, P., Charlin, L., Pineau, J.: A survey of available corpora for building data-driven dialogue systems. arXiv preprint arXiv:1512.05742 (2015)
15. Serban, I.V., Sordoni, A., Bengio, Y., Courville, A.C., Pineau, J.: Building end-to-end dialogue systems using generative hierarchical neural network models. In: AAAI, vol. 16, pp. 3776–3784 (2016)
16. Serban, I.V., Sordoni, A., Lowe, R., Charlin, L., Pineau, J., Courville, A.C., Bengio, Y.: A hierarchical latent variable encoder-decoder model for generating dialogues. In: AAAI, pp. 3295–3301 (2017)
17. Singh, S.P., Kearns, M.J., Litman, D.J., Walker, M.A.: Reinforcement learning for spoken dialogue systems. In: Advances in Neural Information Processing Systems, pp. 956–962 (2000)
18. Strub, F., De Vries, H., Mary, J., Piot, B., Courville, A., Pietquin, O.: End-to-end optimization of goal-driven and visually grounded dialogue systems. arXiv preprint arXiv:1703.05423 (2017)
19. Su, P.H., Gasic, M., Mrksic, N., Rojas-Barahona, L., Ultes, S., Vandyke, D., Wen, T.H., Young, S.: On-line active reward learning for policy optimisation in spoken dialogue systems. arXiv preprint arXiv:1605.07669 (2016)

20. Su, P.H., Vandyke, D., Gasic, M., Mrksic, N., Wen, T.H., Young, S.: Reward shaping with recurrent neural networks for speeding up on-line policy learning in spoken dialogue systems. arXiv preprint arXiv:1508.03391 (2015)
21. Wen, T.H., Vandyke, D., Mrksic, N., Gasic, M., Rojas-Barahona, L.M., Su, P.H., Ultes, S., Young, S.: A network-based end-to-end trainable task-oriented dialogue system. arXiv preprint arXiv:1604.04562 (2016)
22. Wu, J., Li, M., Lee, C.H.: An entropy minimization framework for goal-driven dialogue management. In: Sixteenth Annual Conference of the International Speech Communication Association (2015)

An Adaptive Approach for Index Tuning with Learning Classifier Systems on Hybrid Storage Environments

Wendel Góes Pedrozo[1,2(✉)], Júlio Cesar Nievola[1],
and Deborah Carvalho Ribeiro[1]

[1] PPGIa – Pontifícia Universidade Católica do Paraná, Curitiba, Brasil
{wendel, nievola}@ppgia.pucpr.br,
ribeiro.carvalho@pucpr.br
[2] COENC – Universidade Tecnológica Federal do Paraná, Apucarana, Brasil
wpedrozo@utfpr.edu.br

Abstract. Index tuning is an activity typically performed by database administrators (DBAs) and advisors tools to decrease the response times of commands submitted to a database management system (DBMS). With the introduction of solid state drive (SSD) storage, a new challenge has arisen for DBAs and tools because SSDs provide fast read operations and low random-access costs, and these new features must be considered to perform index tuning of the database. In this paper, we use a learning classifier system (LCS), which is a machine learning approach that combines learning by reinforcement and genetic algorithms and allows the updating and discovery of new rules to provide an efficient and flexible index tuning mechanism applicable for hybrid storage environments (HDD/SSD). The proposed approach, termed Index Tuning with Learning Classifier System (ITLCS), builds a rule-based mechanism designed to represent the knowledge of the system. Experimental results with the TPC-H benchmark showed that the ITLCS performed better than well-known advisor tools, indicating the feasibility of the proposed approach.

1 Introduction

Despite the significant efforts of database administrators (DBAs) and researchers, the performance tuning of relational database systems remains a challenge [10], mainly due to the way that data records are logically organized and accessed in a storage device. In this context, index tuning proves to be relevant because this technique can significantly improve data retrieval performance. Indexes correspond to logical and ordered structures that map the addresses indicating physical data storage locations; their use is optional but contributes significantly to increased performance [2]. Tuning is an action that aims to decrease the response time and/or increase the throughput of a certain application. In the indexing context, tuning actions include: selection of tables/columns that must have indexes, defragmentation indexes, and index exclusion.

Performing good tuning actions is not trivial because it involves several aspects, such as DBMS knowledge implementation, data characteristics and workloads [3], and

© Springer International Publishing AG, part of Springer Nature 2018
F. J. de Cos Juez et al. (Eds.): HAIS 2018, LNAI 10870, pp. 716–729, 2018.
https://doi.org/10.1007/978-3-319-92639-1_60

the I/O costs of secondary storage devices that can be of different types, including SSDs and HDDs. With the largest use of disks being flash memory SSDs, DBAs, in addition to considering the characteristics of HDDs, should also consider SSDs because many indexing actions can have greater advantages when using SSDs than using classic HDDs due to intrinsic aspects to SSDs, such as parallelism and fast reading.

The objective of this work is to design an index tuning mechanism applied to hybrid storage environments (HDD/SSD). The proposed solution uses a learning classifier system (LCS) to create and update knowledge represented by rules and generated to cover most situations in a DBMS, adjusting their performance by creating adjusting indexes. LCS is a hybrid approach that combines learning by reinforcement, evolutionary computation, and other heuristics to produce complex adaptive systems.

For the validation of the solution, a benchmark was used that contains workloads simulating real database system environments to demonstrate the advantages and relevance of the solution for the index tuning problem in hybrid storage environments.

2   Problem Formulation and Proposal

For decades, HDDs have been the dominant solution for secondary storage in DBMSs because they offer the highest cost/benefit ratio, albeit with lower performances than other non-volatile storage technologies, such as NVRAM memories [8]. Certain characteristics of flash-based SSDs render them superior to HDDs, such as (i) asymmetric reading/writing, which allows for reading more than 300 times faster than writing, (ii) an absence of mechanical latency, and (iii) parallelism in operations [8].

DBMSs have always had complex interactions with storage systems because there are various types of data/object layouts in a DBMS, such as indexes, tablespaces and temporary/cache storage. Thus, several types of I/O requests with various qualities of service requirements are employed [12]. DBMSs and database tool advisors typically treat storage systems as "black boxes" and do not consider the features of each device.

Storage systems have inevitably become hybrids with the heterogeneous use of storage devices. Consequently, solutions for tuning DBMSs that can differentiate the I/O characteristics of each storage device are relevant.

Among the main DBMS suppliers and research found in literature there are several proposals for database tools applied to index tuning [1, 3, 5, 10]. However, these solutions do not consider the differences in disk I/O costs. Considering that the dominant costs in a DBMS are the I/O costs from disks to memory [2], discerning these costs is essential. In addition, rules used by DBAs, which are used as a basis for the heuristic construction of many tools, consider only the magnetic disk (HDDs) costs.

Our proposal is to use an LCS, that is responsible for the automatic creation of indexes in the DBMS, reducing the execution time of the queries submitted. Thus, queries submitted to the DBMS can be sent as a message to the LCS for analysis and possible optimization. When users submit the same or similar queries again to DBMS, indexes previously created by the LCS can be used to decrease the response time.

When receiving the workload, the LCS identifies the disk type where the data are stored in order to differentiate the types of storage devices (HDD/SSD). Subsequently, through its population of LCS rules or classifiers (some intended for

HDDs and others for SSDs), classifiers are identified that can optimize the query by creating indexes.

3 LCS Applied to Index Tuning

First introduced by Holland in the mid-1970s [9], an LCS is an machine learning approach and consists of a population of classifiers represented by inference rules, following a format of type IF <condition> - THEN <action>. Unlike the Expert Systems, which have a set of static rules, an LCS allows the evolution and discovery of new rules, by the incorporation of a genetic algorithm (GA). LCS has been used successfully in areas such as data mining, autonomous navigation, and robotics. The main approaches are XCS [11], ACS [7], and UCS [6].

3.1 Genetic Algorithms

A GA is a meta-heuristics inspired by the theory of natural selection. It acts on a population of individuals, based on the fact that individuals with good genetic characteristics have a greater chance of survival and increasingly produce fit individuals, while less fit individuals tend to disappear [9].

The GAs differ from the exact optimization methods, mainly in three aspects: (i) work from a population of solutions, not with a single solution; (ii) it does not need knowledge derived from the problem, just a form of evaluation of the result (evaluation function); and (iii) it uses rules of probabilistic transitions, not deterministic rules.

GAs start from a random population, which can be seen as possible solutions of the problem. During the evolutionary process, a fitness value is assigned for each individual in the population, expressing their degree of adaptation to the environment. Through a selection mechanism it is determined which individuals will reproduce in order to generate descendants for the following generation, using genetic operators such as crossover and mutation [9]. The crossover performs the crossing between selected individuals (parents), forming new individuals (children), which potentially combines the best characteristics of individuals. The mutation changes randomly and with a small probability, some characteristic of the individual, allowing the genetic diversity of the population and the generation of new search points in the space of solutions.

In the implementation of a typical LCS, GA is used for the generation of high fitness classifiers, aiming at the substitution of low fitness classifiers present in the LCS [4].

3.2 Structure of the Classifiers

Similar to a production rule, classifiers contain antecedent and consequent parts. Furthermore, every classifier has a field representing its strength, defined primarily by the feedback value received for each action applied in the environment. Therefore, the greater the classifier strength (fitness) is, the greater its probability of acting in the environment and of perpetuating its characteristics in other classifiers in future generations, assuming that the aims and environment features remain the same.

In this work, each individual chromosome of the population has 16 genes in the antecedent (condition) and one gene in the consequent part (action).

Classifier Codification – Antecedent. Using the ternary alphabet {1, 0, #}, the size of the antecedent corresponds to the size of the message in the environment. The encoding process seeks to assign binary values, indicating that a particular condition is present or absent for any incoming input. The presence of the "#" symbol in the antecedent part means "do not care" can be interpreted as a "1" or as "0" and is related to the specificity of each classifier defined in Eq. 1. Table 1 presents the meaning of each position and was designed based on the criteria found in the literature and those used by DBAs to define whether a column should have an index. For example, position 2 is "1" if there is an index in that column and is "0" otherwise.

Table 1. Antecedent part (conditions)

Gene	Description
G1	Is the table stored on HDD or SSD?
G2	Is there any index in the column?
G3	Is there any aggregation function in the column?
G4	Is the column a foreign key?
G5	Is the column in a group by clause?
G6	Is the column in an order by clause?
G7	What is the type of data in the column?
G8	Which relational operator is used?
G9	If there is an index in the column, what is its degree of fragmentation?
G10	What is the degree of column selectivity?
G11	What is the percentage of null tuples for the column?
G12	What is the percentage of tuples returned in the query execution?
G13	What is the percentage of tuples updated by SQL updates when referencing the column?
G14	If there is any index, what is the height of the b-tree?
G15	What is the ratio of the index size to the shared buffer?
G16	What is the ratio for the shared buffer related to the size of the table?

Classifier Codification – Consequent. Using the binary alphabet {1, 0}, the field of action indicates the decision of the classifier, which in this work can be (i) rebuild, delete and create a B+ tree index or (ii) migrate an existing index from HDD to an SSD.

Message from the RDBMS to the LCS. In Fig. 1, a message received by ITLCS containing an SQL query is cited as an example. The columns present in the where clause (l_orderkey, l_partkey, and l_returnflag) were encoded in binary and used as input to the LCS. Columns with Max, Sum, and Avg SQL functions present in the select clause were also encoded in case they exist. The entries only have the antecedent part, whereas the classifiers have antecedent and consequent parts.

Message Received (workload)															

SELECT   l_orderkey, l_partkey, l_returnflag
FROM   lineitem
WHERE   l_orderkey >= 999 AND l_partkey=15
ORDER BY l_returnflag

Encoded Message															
Columns	G1	G2	G3	G4	G5	G6	G7	G8	G9	G10	G11	G12	G13	G14	G15	G16
l_orderkey	1	000	0	0	0	0	001	001	000	001	000	010	001	110	001	110
l_partkey	1	000	0	0	0	0	001	010	000	000	000	010	000	110	000	110
l_returnflag	1	000	0	0	0	1	010	000	000	101	011	010	011	110	011	110

Fig. 1. Message example and its encoding

3.3   Architecture Index Tuning with Learning Classifier Systems (ITLCS)

As shown in Fig. 2, the proposed architecture called ITLCS (Index Tuning with Learning Classifier Systems) interacts with the environment as follows:

(1) When the message detectors receive a message from the environment, this message is sent to the Rules and Messages Sub-system. Then, all classifiers attempt to combine the antecedent part with the message.

(2) If there are multiple classifiers that match the antecedent part with the message, these classifiers can compete, in order to find a winner that can act in the environment (Fig. 3). The winner is defined by the bid offered by each classifier, which can be calculated according to Eqs. 1 and 2, found in [18], using the coefficients $k0 = 0.1$, $k1 = 0.1$, $k2 = 0.083$, and $SPow = 3$, as defined in [18]. In these equations, the bid is calculated from the strength and specificity of each classifier to select the classifiers that have more "#" symbols in their antecedent part, i.e., more generic classifiers.

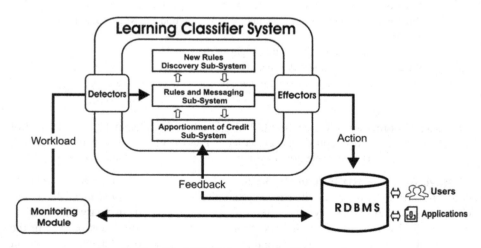

Fig. 2. Simplified ITLCS interaction with the environment

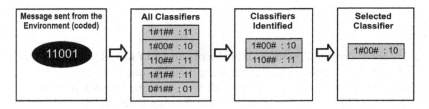

Fig. 3. Example of the input comparison process versus classifiers

$$Spec = \frac{\left(N - total_\#\right)}{N} \tag{1}$$

where

N: the size of the antecedent condition; and
$total_\#$: the number of # symbols in the antecedent.

$$Bid_t = k0 * \left(k1 + k2 * Spec^{SPow}\right) * S_t \tag{2}$$

where

Bid_t: "bid" at step t;
$k0$: bid coefficient 0, referring to classifier strength (positive constant less than 1);
$k1$: bid coefficient 1, referring to the non-specificity portion of the classifier (positive constant less than 1);
$k2$: bid coefficient 2, referring to the specificity portion of the classifier (positive constant less than 1);
$Spec$: specificity of the classifier (as defined in Eq. 1);
$SPow$: control parameter of the influence of the specificity on the bid value; and
S_t: strength of the classifier at step t.

(3) After the action of the winning classifier, the environment (DBMS in our work) provides feedback to the LCS in response to the action commanded by the executors. To calculate the feedback, a heuristic was developed, as presented in algorithm 1, based on the benefits of the classifier actions. The evaluation of benefit-based tuning actions has already been used in several studies, such as [3, 5, 13], and the basic premise of this evaluation is that the value attributed to the benefit should increase if cost reduction occurs with a query and should decrease if the opposite occurs.

Equation 3 demonstrates how the benefit provided by index ij in the execution of task tk is estimated. However, to use this metric as the feedback value of the classifiers, this value was converted to a percentage, according to Eq. 4.

$$B_{ij,tk} = \left(CS(tk) - CS(tk, ij)\right) \tag{3}$$

$$BP_{ij} = 1 - \left(\frac{CS(tk, ij)}{CS(tk)}\right) \tag{4}$$

where

$B_{ij,tk}$: benefit provided by index ij in the execution of task tk;
BP: percentage benefit related to reducing the SQL query cost;
$CS(tk)$: estimated execution cost for task tk without using index ij; and
$CS(tk, ij)$: estimated execution cost of task tk using index ij.

To define whether a particular classifier should be rewarded, the heuristic of benefits (algorithm 1) first checks whether the benefit is positive ($BP > 0$), and whether the index was used (idx_Used = true). If some of these conditions are not satisfied, a classifier will be punished, proportional to the performance worsening obtained.

Algorithm 1: Assignment of feedback using heuristics of benefits

	Input: list of applied actions and their classifiers
	Output: value of feedback to the classifier
1:	**for each** action in Action Set **do**
2:	**if** BP>0 and idx_Used=true **then**
3:	getReward (classifier);
4:	**else**
5:	getPunishment (classifier);
6:	**End**
7:	**End**

All classifiers suffer a decrease in strength at each iteration due to a demanded life tax. Thus, the rarely selected classifiers will have their strength decreased in every iteration, with a high probability of replacement by new classifiers.

(4) Upon receipt of feedback (reward or punishment), it is the responsibility of the credit sub-system to enter the value returned to the classifier's strength.

After feedback, the system will be able to receive a new message and returning to step 1 until there are no more messages. According to Fig. 4, at the end of each epoch, the LCS participates in an evolutionary process. It is the responsibility of the new rules discovery sub-system to apply intrinsic GA procedures [9], detailed in Sect. 3.1.

To decrease the GA run time, hypothetical index[1] are used through the extension installed in the DBMS [15]. Moreover, during the training and testing phases, queries need not be executed; due to the use of DBMS EXPLAIN <query> command, to estimate of queries costs. After the training and testing phases, the set of recommended indexes are physically created in the DBMS, and then the benchmark queries are executed several times, according to the parameters defined in Sect. 4.1.

---

[1] Hypothetical indexes are treated by the optimizer as if they existed physically in the DBMS.

```
┌─ BEGINNING OF THE ITLCS
│ ┌─ BEGIN OF ONE EPOCH
│ │   Step 1: Receive a DBMS workload (messages).
│ │   Step 2: Workload (messages) encoding.
│ │   Step 3: Select matched classifiers.
│ │   Step 4: Begin competition:
│ │               • Calculate each competitor's bid.
│ │               • Aim the winning classifier.
│ │               • Collect taxes from competitors and the winner.
│ │   Step 5: Act on the DBMS (environment).
│ │   Step 6: Receive DBMS's (environment) feedback.
│ │   Step 7: Reward or punishment the winning classifiers.
│ │   Step 8: Charge life tax from all classifiers.
│ │   Step 9: If it is not the end of the epoch, go to Step 1.
│ └─ END OF ONE EPOCH
│ ┌─ BEGIN OF THE DISCOVERY OF NEW RULES (run the GA)
│ │   Step 10: Begin a new generation of classifiers.
│ │   Step 11: Select classifiers for application of genetic operators.
│ │   Step 12: Apply crossover operator to generate children.
│ │   Step 13: Apply the mutation operator on a portion of children.
│ │   Step 14: Evaluate children, using Heuristics defined(Algorithm 1)
│ │   Step 15: If it is not the generation limit, go to Step 10.
│ └─ END OF THE DISCOVERY OF NEW RULES
│ ┌─ BEGIN OF REPLACEMENT OF CLASSIFIERS
│ │   Step 16: Select a predefined number of classifiers to replace.
│ │   Step 17: Replace the classifiers selected in Step 16 with
│ │             the children generated in step 14.
│ └─ END OF REPLACEMENT OF CLASSIFIERS
└─ END.
```

Fig. 4. Algorithm of the evolutionary process of the ITLCS

3.4 Initial Population and ITLCS Parameters

After the empirical evaluation, a population of 1,000 individuals was defined, with 500 classifiers destined for HDDs and the other half for SSDs. This is possible due to the first gene of the chromosome, which extracts information from the entries concerning the type of disk in which table/index were stored. The generation of the initial population is 50% random and 50% manual based on knowledge obtained in the literature and used by DBAs. After generating the initial population of classifiers, their strength values must be adjusted. Thus, a training set was created using the same database and containing similar queries to the TPC-H benchmark. The adjustment process consists of performing the steps described in the algorithm in Fig. 4 ten times. During the training and testing phases, the GA is applied to 10% of the classifier population, and for each phase of discovery of new rules (Fig. 4, steps 10–15), 20 generations are used. The selection method used in GA/LCS is roulette wheel [4]. The crossover type is random two-point, with 50% rate, elitism of 5 individuals and mutation of 5%.

4   Experimental Evaluation

In this section, the experimental results of the ITLCS proposed in this work are presented and compared to other methods that also perform index tuning. The goal is to verify the final set of indexes obtained after the execution of a benchmark to evaluate the performance and adaptation degree of the methods in hybrid environments. To reach these goals, all methods were analysed by executing the benchmark with the database stored on an SSD; after that, an HDD was also used for evaluation as well. The OS and other software, such as the DBMS itself, were installed only on the SSD. The strategy of changing only the local/device database simulates a hybrid storage where tables, indexes, and other objects can change storage devices.

4.1   Environment Configuration, Workload, and Scenarios

The experiments were performed using an HP server (2.5 GHz Intel Xeon quad-core, 8 GB of RAM) running on a GNU/Linux OS, 2.6 kernel (64-bit), and DBMS PostgreSQL 10.1 with the Hypopg extension [15]. ITLCS was developed using the Java SE platform. Because this work is oriented to hybrid storage environments, the server had four disks: two were HDDs (1 TB 6 GB/s SATA 3) and two were SSDs (500 GB 6 GB/s SATA 3). The OS and DBMS were installed on the SSD.

The adopted dataset and workload was OSDL DBT3 [16] - with 40-factor (40 GB). OSDL DBT3 is an open-source implementation of the TPC-H benchmark, adjusted for PostgreSQL, that provides 22 queries common to data warehouse environments.

To evaluate and compare the results obtained by the ITLCS with the other solutions, the following tools/methods were used:

- EDB: EnterpriseDB [14] is a commercial database advisor tool.
- POWA: PostgreSQL Workload Analyser [17] is an open-source advisor tool.
- NO-INDEX: Default setting, without any index tuning action.
- ITLCS: Solution for tuning indexes proposed in this work.

The two scenarios proposed for the evaluated methods are as follows:

- Scenario 1: Running the workload with the database and indexes stored on an HDD.
- Scenario 2: Running the workload with database and indexes stored on an SSD.

To estimate the average time of each benchmark query, the same query was performed seven times and the average of the runs were calculated, discarding the first two runs, due to the process of cache warm-up. After the seventh run, the DBMS and OS cache were cleared to avoid interference between one query and another.

4.2   Experimental Results

Table 2 presents a summary of the average results only the queries with the greatest execution times except for query 21, which extrapolated the DBMS default time (Scenario 1). Moreover, the recommended indexes for each query and the percentage of performance improvement or worsening in relation to NO-INDEX method are shown.

Table 2. Time and cost evaluation for Scenarios 1 (HDD) and 2 (SSD)

	Total execution time		Total query cost	
	HDD	SSD	HDD	SSD
IA-LCS	0 h 49 min	0 h 44 min	109,828,732.94	111,453,424.46
NO-INDEX	1 h 06 min	1 h 01 min	121,914,033.17	111,870,281.64
POWA	1 h 14 min	0 h 59 min	119,982,279.01	111,435,484.17
EDB	2 h 33 min	1 h 01 min	116,920,563.81	111,429,352.19

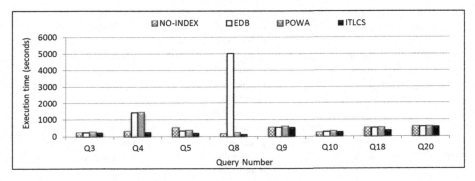

Fig. 5. Performance by query using HDD storage

Scenario 1. In this scenario, where only HDDs were used, the proposed ITLCS achieved the best overall performance, as shown in Table 2 and Fig. 5. Performance improved by 12% over the second-best time, i.e., the NO-INDEX method, which carried out the workload without the use of any secondary index.

Furthermore, Table 3 shows that although the ITLCS improved the performance of queries Q5, Q8, and Q18 by 52%, 8%, and 25%, respectively, the performance decreased for queries Q10 and Q20, thus reducing some of the overall gain. Performance degradation for the target query also occurred with the POWA and EDB, specifically queries Q4 and Q8, in which the deterioration was −372% and −2333%, respectively.

Considering the cost metrics, ITLCS obtained the best overall performance, achieving the lowest value (Table 2), followed by the EDB, POWA, and NO-INDEX.

Scenario 2. In this scenario, where only SSDs were used, the ITLCS also achieved the best overall performance, with a significant reduction of 27% in overall time compared to the NO-INDEX method, as shown in Table 2 and Fig. 6. Queries Q3, Q5, Q8 and Q18 were highlighted because they obtained an improvement of 28%, 84%, 8%, and 59% respectively. Unlike Scenario 1, the ITLCS in Scenario 2 created indexes for most queries. We also emphasize the quality of the proposed indexes, as shown in Table 3; except for query Q20, all queries exhibited increased performance.

Table 3. Set of suggested indexes for Scenarios 1 and 2

QN°	Disk type	EDB Column indexed	EDB Speed-up time	POWA Column indexed	POWA Speed-up time	ITLCS Column indexed	ITLCS Speed-up time
Q3	HDD	c_mktsegment	−2%	c_mktsegment	−2%	none	none
	SSD	c_mktsegment o_orderdate*	−6%	l_shipdate o_orderdate*	17%	l_shipdate	28%
Q4	HDD	o_orderdate	−372%	o_orderdate	−372%	none	none
	SSD	o_orderdate	4%	o_orderdate l_orderkey	7%	none	none
Q5	HDD	o_orderdate	32%	o_orderdate	32%	l_orderkey	52%
	SSD	o_orderdate	23%	o_orderdate o_orderkey	23%	l_orderkey	84%
Q8	HDD	o_orderdate l_partkety	−2333%	p_type o_orderdate	−16%	p_type	8%
	SSD	p_type l_partkey	24%	p_type o_orderkey* o_orderdate*	33%	p_type	34%
Q9	HDD	none	none	p_name*	−6%	none	none
	SSD	l_partkey*	−5%	p_name*	0%	none	none
Q10	HDD	o_orderdate	−21%	l_returnflag* o_orderdate	−23%	o_orderdate	−21%
	SSD	o_orderdate l_returnflag*	−19%	l_returnflag*	−19%	o_orderdate	0%
Q18	HDD	none	0%	none	0%	l_orderkey	25%
	SSD	none	0%	none	0%	l_orderkey	59%
Q20	HDD	s_name l_shipdate	1%	l_shipdate p_name*	0%	ps_partkey l_shipdate	0%
	SSD	s_name l_shipdate	−6%	l_shipdate p_name*	−11%	s_nationalkey l_shipdate	−6%

*Recommended index, although not used by the DBMS when executing the query.

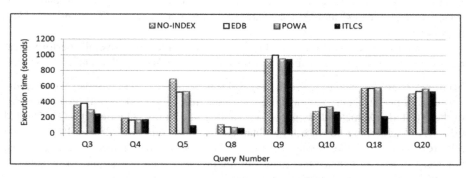

Fig. 6. Performance by query using SSD storage

In this scenario, although EDB and POWA improved performance for some queries, for the queries Q10 and Q20, significant losses occurred, reducing the gains obtained.

Analysis of Results from Scenarios 1 and 2. Considering only the queries that showed improved performances, the average gain obtained by the ITLCS was significant, namely, 27% and 51% for Scenarios 1 and 2, respectively. Another important point, as shown in Table 4, the ITLCS resulted in the fewest created indexes, which has the following direct advantages: (i) saving space, avoiding the creation of unused indexes, as occurs with the POWA method, and (ii) lower update index overhead for each insert/update performed in the source table.

Table 4. Summary recommendations for Scenarios 1 and 2

Description	Scenario 1			Scenario 2		
	ITLCS	POWA	EDB	ITLCS	POWA	EDB
Total recommended indexes	6	10	8	7	13	11
Queries without any recommendation	3	1	2	2	1	1
Total recommended but not used indexes	0	2	0	0	6	3

Table 5 shows the discovered rules with the greatest strength in each scenario, and the simplicity and comprehensibility of the developed rules model are verified.

Table 5. Best rules

Best rule for Scenario 1 (HDD)	Best rule for Scenario 2 (SSD)
IF (type storage)=(HDD) (is there an index in the column) = (no) (the type of data in the column) = (numeric) (the degree of column selectivity) = (5-10%) (the percentage of null tuples) < (5%) (the percentage of tuples returned) < (5%) THEN (create a b-tree index in the column)	IF (type storage) = (SSD) (is there an index in the column) = (no) (is there an aggregation function in column) = (yes) (type of data in the column) = (numeric) (the degree of column selectivity) = (10-20%) THEN (create a b-tree index in the column)

ITLCS Performance. Although approaches using evolutionary algorithms have the disadvantage of solution convergence time, in this work, this has been significantly reduced due to the use of hypothetical indexes [15] and to the storage of metadata frequently used by the LCS during the encoding phase of the inputs (see Table 1). The largest bottleneck in the process occurs in the first iteration of the training/adaptation phase, which required approximately 40% of the total time. The total times of the training and test phases are presented in Table 6.

Table 6. Performance of the ITLCS

Phase	HDD	SSD
Training phase	2 h 28 min	1 h 21 min
Test phase	0 h 39 min	0 h 27 min

5   Conclusion and Future Work

This work presented the methodological developments necessary for the use of LCS in the process of database indexes tuning, as well as the proposed architecture and integration of all components with the DBMS. The developed ITLCS utilized a rule-based approach designed to store the system knowledge and guide its performance in the environment, as well as allowing the discovery of new and understandable tuning rules in response to changes in the DBMS environment. We emphasize that the proposed approach allows its use in other relational DBMSs, requiring few adjustments.

Although the database advisor tools evaluated in this study provide positive improvements for some benchmark queries, the results were lower than those reported by the ITLCS. In general, database advisor tools tend to focus on optimizing record localization operations and often only at locations with the greatest margin of improvement. Approaches using evolutionary algorithms, as used in this paper, work in a search space where all possibilities are considered, and this contributes so that the final solution is not limited to local minima or maxima.

Future work should apply LCS to other actions, such as the creation of other types of indexes (hash, inverted index, and bitmap), recommendation of materialized views and the creation/updating of histograms. Another research direction is the evaluation this approach in scenarios with others disk layouts and other relational DBMSs.

Acknowledgements. This research was sponsored by grants from CAPES (Coordenação de Aperfeiçoamento de Pessoal de Nível Superior).

References

1. Schnaitter, K., Polyzotis, N.: Semi-automatic index tuning: keeping DBAs in the loop. VLDB Endowment **5**(5), 478–489 (2012)
2. Ramakrishnan, R., Gehrke, J.: Database Management Systems. McGraw-Hill, New York (2003)
3. Luhring, M., Sattler, K., Schmidt, K., Schallehn, E.: Autonomous management of soft indexes. In: International Conference on Data Engineering, ICDE 2007, Istanbul, Turkey, pp. 450–458 (2007)
4. Booker, L.B., Goldberg, D.E., Holland, J.H.: Classifier systems and genetic algorithms. J. Artif. Intell. **40**, 235–282 (1989)
5. Costa, R.L.C., Lifschitz, S.: Index self-tuning and agent based databases. In: Latin-American Conference on Informatics, CLEI 2002, Montevideu, Uruguay, 12 pp. (2002)
6. Bernado-Mansilla, E., Garrell, J.M.: Accuracy-based LCS. Models, analysis and applications to classification tasks. Evol. Comput. **11**(3), 209–238 (2003)
7. Butz, M.V., Goldberg, D.E., Stolzmann, W.: The anticipatory classifier system and genetic generalization. Nat. Comput. **1**, 427–467 (2002)
8. Ghodsnia, P., Bowman, I., Nica, A.: Parallel I/O aware query optimization. In: ACM Sigmod Conference on Management of Data, Snowbird, UT, USA, pp. 349–360 (2014)
9. Holland, J.H.: Adaptation in Natural and Artificial Systems: An Introductory Analysis with Applications to Biology, Control and Artificial Intelligence. The MIT Press, Ann Arbor (1992)

10. Jimenez, I., Sanchez, H., Tran, Q., Polyzotis, K.: Kaizen: a semi-automatic index advisor. In: ACM Sigmod Conference on Management of Data, Scottsdale, AR, USA, pp. 685–688 (2012)
11. Butz, M.V., Kovacs, T., Lanzi, P.L., Wilson, S.W.: Toward a theory of generalization and learning in XCS. IEEE Trans. Evol. Comput. **8**, 28–46 (2004)
12. Lee, S., Moon, B., Park, C., Kim, J., Kim. S.: A case for flash memory SSD in enterprise database applications. In: Sigmod, pp. 1075–1086 (2008)
13. Salles, M.V., Lifschitz, S.: Autonomic index management. In: International Conference on Autonomic Computing, ICAC 2005, Seattle, WA, USA, pp. 304–305 (2005)
14. EDB: Enterprise DB (2018). https://www.enterprisedb.com
15. HYPOPG: Hypothetical Indexes Support PostgreSQL (2018). http://dalibo.github.io/hypopg
16. OSDL DBT3 (2018). http://osdldbt.sourceforge.net/#dbt3
17. POWA: PostgreSQL Workload Analyzer (2018). http://powa.readthedocs.io
18. Richards, R.A.: Zeroth-order shape optimization utilizing learning classifier systems (1995)

Electrical Behavior Modeling of Solar Panels Using Extreme Learning Machines

Jose Manuel Lopez-Guede[✉], Jose Antonio Ramos-Hernanz, Julian Estevez, Asier Garmendia, Leyre Torre, and Manuel Graña

Computational Intelligence Group,
Basque Country University (UPV/EHU), Donostia-San Sebastian, Spain
jm.lopez@ehu.es

Abstract. Predicting the response of solar panels has a big potential impact on the economical viability of the insertion of alternative energy sources in our societies, diminishing the dependence on polluting fossil fuels. In this paper we approach the modeling of the electrical behavior of a commercial photovoltaic module Atersa A-55 using Extreme Learning Machines (ELMs). The training and validation data were extracted from the response of a real photovoltaic module installed at the Faculty of Engineering of Vitoria-Gasteiz (Basque Country University, Spain). The resulting predictive model has one input (V_{PV}) and one output (I_{PV}) variables. We achieve a Root Mean Squared Error (RMSE) of 0.026 in the electrical current measured in Amperes.

1 Introduction

In order to obtain the maximum performance of commercial photovoltaic modules their control algorithms carry out a maximum power point (MPP) tracking strategy. Tracking the MPP needs a model of the electrical behavior predicting the relation between the supplied voltage V_{PV} and the resulting output electrical current intensity I_{PV} of the photovoltaic module.

There are two main approaches to photovoltaic module electrical behavior modeling: the theoretical and the empirical approaches. Theoretical approaches use a characteristic equation [13] with different degrees of freedom leading to different models. These variations are due to diverse assumptions and simplifications of the modeling of the current generation process. These theoretical models are dependent of the accuracy of the estimation of their operational parameters, and the correct selection of the operational relations between some of their variables. On the other hand, empirical approaches learn the electrical behavior of the photovoltaic modules from their actual data, building predictive models that capture the behavior of the actual photovoltaic element [10,13].

In this paper we deal with the problem of accurate modeling of the electrical behavior of photovoltaic modules from their actual data using Artificial Neural Networks (ANNs), more specifically Extreme Learning Machines (ELMs). After following a detailed experimental setup using an Atersa A-55 photovoltaic module and performing a number of experiments, we have trained a predictive model

© Springer International Publishing AG, part of Springer Nature 2018
F. J. de Cos Juez et al. (Eds.): HAIS 2018, LNAI 10870, pp. 730–740, 2018.
https://doi.org/10.1007/978-3-319-92639-1_61

achieving a Root Mean Square Error (RMSE) of 0.026 in the measured output current in Ampere units.

The paper is structured as follows. Section 2 provides a background on photovoltaic module modeling, as well as on ANNs and ELMs. Section 3 states formally the objective of the paper. The detailed description of the experimental setup is given in Sect. 4, while Sect. 5 discusses the obtained results. Finally, our conclusions are given in Sect. 6.

2 Background

2.1 Theoretical Models of Photovoltaic Modules

The basic element of all photovoltaic modules is the theoretical photovoltaic cell, which can be modeled as a current source with an anti-parallel diode. When the cell is exposed to light, the direct electrical current generated varies linearly with the incoming solar radiation. Some authors have improved this basic model taking into account the effect of a shunt resistor and other one in series. The main magnitudes involved are the photo-generated electrical current (aka photocurrent) I_{PH}, the electrical current of the diode I_D, the series resistance R_S, and the shunt resistance R_{SH}.

A first attempt to characterize a photovoltaic cell is relating the current I_{PV} and the voltage V_{PV} provided by the photovoltaic cell through Eq. (1) [7], which is expanded in Eq. (2) [14]:

$$I_{PV} = I_{PH} - I_D - I_{SH}, \tag{1}$$

$$I_{PV} = I_{PH} - I_0 \left(e^{\frac{q(V_{PV}+I_{PV}R_S)}{aKT}} - 1 \right) - \frac{V_{PH} + I_{PH}R_S}{R_{SH}}, \tag{2}$$

where T is the cell temperature [°C], K is the Boltzmann's constant (1.38×10^{-23} [j/K]), a is the diode ideality factor, q is the charge of the electron (1.6×10^{-19} [C]), and I_0 is the saturation current of the diode [A]. Equation (2) involves also a number of manufacturing structural parameters of the photovoltaic cell, i.e., R_{SH}, R_S, a, I_D and I_{PH}. We can find in the literature theoretical models with 7 parameters (a_1, a_2, R_S, R_{SH}, I_{0_1}, I_{0_2} and I_{PH}) [3,11,12], with 5 parameters (a, R_S, R_{SH}, I_0 and I_{PH}) [2,15,16], 4 parameters [6], and even 3 parameters models [1,14].

In order to characterize a larger photovoltaic module it is possible to use the cell characteristic curves, such as IV, PV and PI curves, providing the relations between pairs of the most relevant magnitudes of a photovoltaic module operating at a given solar irradiation and ambient temperature.

As these are theoretical models and the enumerated parameters and the curves are the standard ones for a photovoltaic module commercial model, it is only possible to obtain approximate values when dealing with a specific photovoltaic module. The individual small errors for one photovoltaic cell could be relevant when dealing with large photovoltaic modules or large photovoltaic farms.

2.2 Artificial Neural Networks (ANNs)

Its is a matter of fact that the use of ANNs in this scope is boosted by their ability to model dynamic systems [17] of arbitrary complexity. There are a number of types of ANNs, each one with different characteristics that make them more suitable for solving specific problems. These bio-inspired computational devices have several general advantages, and among others, these are the most outstanding to our problem:

- Learning capabilities: If they are properly trained, they can learn complex non-linear mathematical models. There are several well known training algorithms and good and tested implementations of them. The main challenge concerning this issue is to choose appropriate inputs and outputs to the black box model and the internal structure.
- Generalization capabilities: If they are properly trained and the training examples cover a variety of different situations, the response of a neural network in unseen situations (i.e., with unseen inputs) will probably be acceptable and quite similar to the correct response. So it is said that they have the *generalization property*.
- Real time capabilities: Once they are trained, and due to their parallel internal structure, their response is always very fast. Their internal structure could be more or less complex, but in any case, all the internal operations that must be done are several multiplications and additions if it is a linear neural network. This fast response is independent of the complexity of the learned models, conferring them real time capabilities since they have an inherent parallel internal structure.

Due to these main properties and to other minor ones, ANNs have found a wide field of application in a number of areas [4,5].

2.3 Extreme Learning Machines (ELMs)

Extreme Learning Machines (ELMs) [8,9] are Single-Hidden Layer Feedforward Networks (SLFNs), trained as follows:

1. Generate randomly the hidden layer weights $\mathbf{W}$, computing the hidden layer output $\mathbf{H} = g(\mathbf{WX})$ for the given data inputs $\mathbf{X}$, where $g(x)$ denotes the hidden units activation function, which can be sigmoidal, Gaussian or even the identity.
2. Solve the linear problem $\mathbf{H}\boldsymbol{\beta} = \mathbf{Y}$, where $\mathbf{Y}$ are the outputs of the data sample, and $\boldsymbol{\beta}$ the weights from the hidden layer to the output layer, by the mean least squares approach. Therefore $\hat{\boldsymbol{\beta}} = \mathbf{H}^{\dagger}\mathbf{Y}$, where $\mathbf{H}^{\dagger}$ is the pseudo-inverse.

3   Motivation of the Modeling Approach

We aim to build an accurate electrical model of the Atersa A-55 photovoltaic module installed on the roof of the Faculty of Engineering of Vitoria-Gasteiz

(University of the Basque Country, Spain). The main requirement is that the model should be as accurate as possible compared to the collected data, but there are other requirements for that model design:

- The training speed should be high in order to be able to add new gathered data at any moment and get a new model without a high computational burden and without discarding the previous acquired knowledge.
- It is desirable a very small human intervention in the training.
- It is also desirable the model to be easy to adjust without excessive computational burden.
- It should be a useful model with a very fast response, in such a way that as it needs a very small amount of time to give an approximate response and it could be used in intensive simulations.

Therefore, we propose the ELM modeling approach as the most adequate for the task at hand, using only one input (V_{PV}) and one output (I_{PV}) variables.

4   Materials and Methods

Materials consisting of the actual physical photovoltaic modules and the data extracted from them are specified here. Methods consisting of the actual ELM training parameters tested are also specified.

4.1   Photovoltaic Modules Physical Settings

Photovoltaic modules Atersa A-A55 (637 × 527 × 35) are professional panels, not only for small systems but also for large installations. They are manufactured by a Spanish company, and as their specifications show in Table 1, they are composed of monocrystalline silicon cells that guarantee power production from dawn to dusk. The performance of solar cells is usually evaluated under the standard test condition (STC), where an average solar spectrum at AM 1.5 is used, the irradiance is normalized to 1,000 W/m^2, and the cell temperature is defined as 25 °C. In Fig. 1 we can see the placement of the real photovoltaic modules on the roof of the Faculty of Engineering of Vitoria-Gasteiz (Basque Country University, Spain).

4.2   Data Collection from the Photovoltaic Modules

On one hand, Fig. 2(a) shows the conceptual disposition of the measuring devices: the voltmeter is placed in parallel with the module and the amperemeter in series. Besides, there is a variable resistance to act as a variable load and obtain different pairs of voltage and current with the same irradiance and temperature. The variable resistance value is controlled according to our convenience, but the temperature and the irradiance depends on the climatological conditions. On the other hand, Fig. 2(b) shows the real devices that have been used to capture the data. The first device is the data logger Sineax CAM, and it was configured to

Fig. 1. Atersa A-55 solar panels

Table 1. Atersa A-55 photovoltaic module characteristics

Attribute	Value
Model	Atersa A-55
Cell type	Monocrystalline
Maximum Power [W]	55
Open Circuit Voltage Voc [V]	20,5
Short circuit Current Isc [A]	3,7
Voltage, max power Vmpp [V]	16,2
Current, max power Impp [A]	3,4
Number of cells in series	36
Temp. Coeff. of Isc [mA/°C]	1,66
Temp. Coeff. of Voc [mV/°C]	-84,08
Nominal operation cell temp. [°C]	47,5

generate records with the irradiance and temperature of the environment and the voltage and current supplied by the photovoltaic module under those environmental conditions. The second element is the multimeter TV809, which helps to isolate the data logging device from the photovoltaic module and converts voltage and current magnitudes to a predefined range. The third element is the irradiance sensor Si-420TC-T-K. It is placed outside close to the photovoltaic module and it is used to provide the irradiance and temperature conditions under which the module is working outside. Finally, the fourth element of the figure are current clamps Chauvin Arnous PAC12 used to measure direct currents provided by the module. After analyzing the theoretical arrangement of the measuring and logging elements, in Fig. 3 we can see the real elements that took part in the recording process.

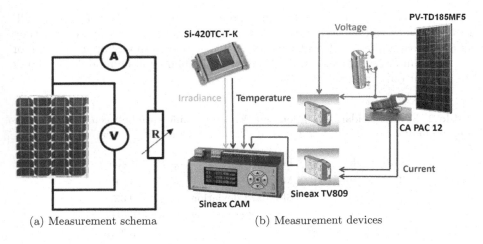

(a) Measurement schema   (b) Measurement devices

Fig. 2. Real photovoltaic module data measurement

Fig. 3. Real devices used during the data recording process

4.3   Data Basic Descriptors

In this subsection we analyze the characteristics of the data recorded during July 2014 through the basic descriptors that are in Table 2. For each one of the four recorded magnitudes, i.e., temperature, irradiance, voltage and current, we have calculated their mean, standard deviation and the relative difference between the maximum and minimum value of that magnitude. We note that the temperature and irradiance are quite stable values and probably they will not provide information for the modeling, since their standard deviation and the percentage difference between the maximum and minimum value are very small. However, both voltage and current provide meaningful information because they show a broader amplitude in their values. This circumstance support the aim of the paper modeling the electrical behavior of the photovoltaic module obtaining the current only from the voltage.

Table 2. Mean, Standard deviation and Percentage difference between the maximum and minimum value of the four recorded physical magnitudes

Magnitude	Descriptor	Value
Temperature [°C]	Mean	54.19
	Standard deviation	0.62
	Perc. difference max vs min	3.75%
Irradiance [W/m^2]	Mean	920.21
	Standard deviation	1.37
	Perc. difference max vs min	1.43%
Voltage [V]	Mean	13.27
	Standard deviation	5.93
	Perc. difference max vs min	6,177.22%
Current [A]	Mean	1.93
	Standard deviation	1.16
	Perc. difference max vs min	1,936.61%

4.4   ELM Training and Validation Process

As the first step we have fixed the structure of the neural network to train, choosing a regular ELM structure following the ideas explained in Subsect. 2.3 using the regression approach. There are only one input (V_{PV}) and one output (I_{PV}) in the ELM networks, as specified in Sect. 3, associating each target value with each input value. Regarding the internal structure of the ELMs, they have only one hidden layer. We have tested all possible hidden layer size in the range $h \in [1, 100]$. We carry out 10 repetitions of training and validation for each hidden layer size. We tested also diverse activation functions in the hidden layer, using for each ELM one element of the set $F = \{Sigmoidal, Sine, Hard limit, Triangular basis, Radial basis\}$.

The last issue to decide is how to use the gathered data in order to carry out the learning process. We have avoided to run any normalization or data preprocessing, using the data directly as were recorded. Both the input and target vectors have been divided into two sets using interleaved indices generating a partition where the 75% are used for training and the remaining 25% are used as a completely independent test of ELM generalization.

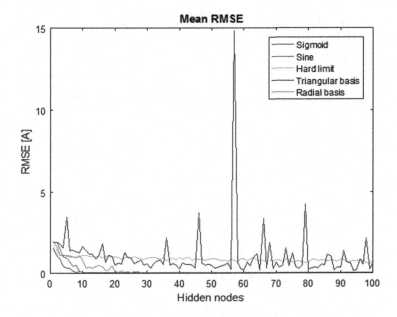

Fig. 4. Mean RMSE of the 10 trials for each combination

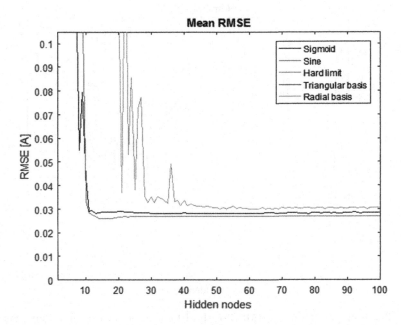

Fig. 5. Zoom of the mean RMSE of the 10 trials for each combination (Sigmoid, Sine and Radial basis activation functions)

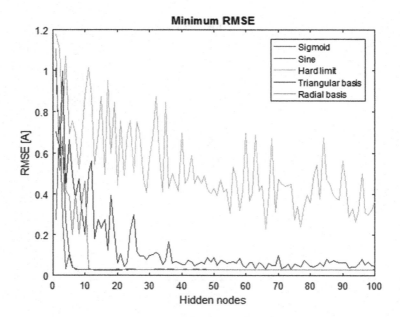

Fig. 6. Minimum RMSE of the 10 trials for each combination

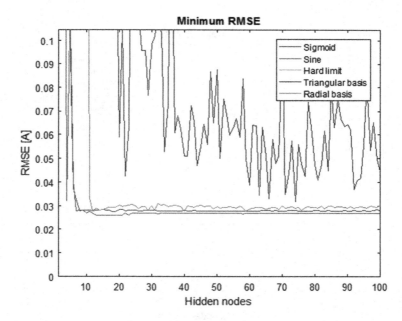

Fig. 7. Zoom of the minimum RMSE of the 10 trials for each combination (Sigmoid, Sine, Triangular basis and Radial basis activation functions)

5   Results

We recall that 10 repetitions of training and validation have been performed for each combination resulting of the cartesian product of the number of hidden nodes set H ($|H| = 100$) and the activation functions set F ($|F| = 5$), obtaining a total of 5,000 experimental settings. In Fig. 4 we show the mean accuracy of the 10 trials for each combination of the number of hidden neurons and activation function in terms of Root Mean Squared Error (RMSE). There we can see that there are no large differences regarding the number of hidden neurons used in the ELMs, excepting a peak value around 55 hidden neurons. However, there are clearly defined two groups of activation functions: on one hand there are the Triangular basis and Hard limit functions, while on the other hand the remaining ones with a very low RMSE. In order to get further insight into the behavior of this second group, Fig. 5 shows a zoom of Fig. 4, where we can see that the best mean activation function is the Sine one.

In order to analyze what is the best individual ELM, Fig. 6 shows the accuracy of the best ELM of the 10 trials for each combination of the number of neurons and activation function. In this case we notice that there is a dependence on the number of hidden neurons at least for two activation functions (Hard limit and Triangular basis). For the remaining ones, in Fig. 7 we can see that the results are very similar and above 10 hidden nodes, there is not difference regarding the number of neurons. The best ELM was configured with 16 hidden neurons and the Sine activation function, obtaining a test RMSE of 0.026 A. This RMSE is lower than the measurement tool (current clamps Chauvin Arnous PAC12).

6   Conclusions

This paper has faced the problem of modeling photovoltaic modules. We have started it reviewing previous works about modeling of photovoltaic cells with different level of accuracy. Next a brief background on photovoltaic elements, ANNs and ELMs has been given in Sect. 2. The formulation of the modeling problem has been stated in Sect. 3, while the experimental design to achieve it has been given in Sect. 4. Finally in Sect. 5 the obtained results have been discussed. In summary, this paper has been the first attempt to model the Atersa A-55 photovoltaic module using ELMs, obtaining a RMSE of 0.026 A, that is a very accurate value in this scope.

Acknowledgments. The research was supported by the Computational Intelligence Group of the Basque Country University (UPV/EHU) through Grant IT874-13 of Research Groups Call 2013–2017 (Basque Country Government).

References

1. Bandou, F., Arab, A.H., Belkaid, M.S., Logerais, P.-O., Riou, O., Charki, A.: Evaluation performance of photovoltaic modules after a long time operation in Saharan environment. Int. J. Hydrogen Energy **40**(39), 13839–13848 (2015)

2. Bastidas-Rodriguez, J.D., Petrone, G., Ramos-Paja, C.A., Spagnuolo, G.: A genetic algorithm for identifying the single diode model parameters of a photovoltaic panel. Math. Comput. Simul. **131**, 38–54 (2015)
3. Elbaset, A.A., Ali, H., Abd-El Sattar, M.: Novel seven-parameter model for photovoltaic modules. Solar Energy Mater. Solar Cells **130**, 442–455 (2014)
4. Fang, W., Quan, S.H., Xie, C.J., Tang, X.F., Wang, L.L., Huang, L.: Maximum power point tracking with dichotomy and gradient method for automobile exhaust thermoelectric generators. J. Electron. Mater. **45**(3), 1613–1624 (2016)
5. Gautam, A., Soh, Y.C.: Stabilizing model predictive control using parameter-dependent dynamic policy for nonlinear systems modeled with neural networks. J. Process Control **36**, 11–21 (2015)
6. Gonzaez-Longatt, F.: Model of photovoltaic in MatlabTM. In: 2do Congreso Iberoamericano de Estudiantes de Ingeniería Eléctrica, Electrónica y Computación (II CIBELEC 2005). Puerto la Cruz-Venezuela (2006)
7. Gow, J.A., Manning, C.D.: Development of a photovoltaic array model for use in power-electronics simulation studies. IEE Proc. Electric Power Appl. **146**(2), 193–200 (1999)
8. Huang, G.-B., Wang, D.H., Lan, Y.: Extreme learning machines: a survey. Int. J. Mach. Learn. Cybernet. **2**(2), 107–122 (2011)
9. Huang, G.-B., Zhu, Q.-Y., Siew, C.-K.: Extreme learning machine: theory and applications. Neurocomputing **70**(1–3), 489–501 (2006)
10. Karamirad, M., Omid, M., Alimardani, R., Mousazadeh, H., Heidari, S.N.: ANN based simulation and experimental verification of analytical four-and five-parameters models of pv modules. Simul. Model. Pract. Theory **34**, 86–98 (2013)
11. Miceli, R., Orioli, A., Di Gangi, A.: A procedure to calculate the i-v characteristics of thin-film photovoltaic modules using an explicit rational form. Appl. Energy **155**, 613–628 (2015)
12. Muhsen, D.H., Ghazali, A.B., Khatib, T., Abed, I.A.: Parameters extraction of double diode photovoltaic module model based on hybrid evolutionary algorithm. Energy Convers. Manag. **105**, 552–561 (2015)
13. Ramos Hernanz, J.A., Lopez Guede, J.M., Zamora Belver, I., Eguia Lopez, P., Zulueta, E., Barambones, O., Oterino Echavarri, F.: Modelling of a photovoltaic panel based on their actual measurements. Int. J. Tech. Phys. Prob. Eng. (IJTPE) **6**(4), 37–41 (2014)
14. Ramos-Hernanz, J.A., Campayo, J.J., Larranaga, J., Zulueta, E., Barambones, O., Motrico, J., Fernandez Gamiz, U., Zamora, I.: Two photovoltaic cell simulation models in Matlab/Simulink. Int. J. Tech. Phys. Prob. Eng. (IJTPE) **4**(1), 45–51 (2012)
15. De Soto, W., Klein, S.A., Beckman, W.A.: Improvement and validation of a model for photovoltaic array performance. Sol. Energy **80**(1), 78–88 (2006)
16. Villalva, M.G., Gazoli, J.R., Filho, E.R.: Modeling and circuit-based simulation of photovoltaic arrays. In: Power Electronics Conference, COBEP 2009, Brazilian, pp. 1244–1254 (2009)
17. Widrow, B., Lehr, M.A.: 30 years of adaptive neural networks: perceptron, madaline, and backpropagation. Proc. IEEE **78**(9), 1415–1442 (1990)

A Hybrid Clustering Approach
for Diagnosing Medical Diseases

Svetlana Simić[1], Zorana Banković[2], Dragan Simić[3($\boxtimes$)],
and Svetislav D. Simić[3]

[1] Faculty of Medicine, University of Novi Sad,
Hajduk Veljkova 1–9, 21000 Novi Sad, Serbia
svetlana.simic@mf.uns.ac.rs
[2] Frontiers Media SA, Pozuelo de Alarcón sn, Madrid, Spain
zbankovic@gmail.com
[3] Faculty of Technical Sciences, University of Novi Sad, Trg Dositeja
Obradovića 6, 21000 Novi Sad, Serbia
dsimic@eunet.rs, {dsimic,simicsvetislav}@uns.ac.rs

Abstract. Clustering is one of the most fundamental and essential data analysis tasks with broad applications. It has been studied extensively in various research fields, including data mining, machine learning, pattern recognition, and in scientific, engineering, social, economic, and biomedical data analysis. This paper is focused on a new strategy based on a hybrid model for combining fuzzy partition method and maximum likelihood estimates clustering algorithm for diagnosing medical diseases. The proposed hybrid system is first tested on well-known *Iris data set* and then on three data sets for diagnosing medical diseases from UCI data repository.

Keywords: Data clustering · Maximum likelihood estimates clustering
Number of clusters · Fuzzy partition method

1 Introduction

Clustering is one of the most fundamental and essential data analysis tasks with broad applications. It is a process in which a group of unlabeled patterns are partitioned into several sets so that similar patterns are assigned to the same cluster, and dissimilar patterns are assigned to different clusters. The purpose of clustering is to identify natural groupings of data from a large data set to produce a concise representation of a system's behavior.

The unsupervised nature of the problem implies that its structural characteristics are not known, except in case of domain knowledge available in advance. There are some goals for clustering algorithms: (1) estimate the optimal number of clusters, (2) determining good clusters and (3) doing so efficiently. One of the main difficulties for cluster analysis is estimating the optimal and correct number of clusters of different types of datasets.

© Springer International Publishing AG, part of Springer Nature 2018
F. J. de Cos Juez et al. (Eds.): HAIS 2018, LNAI 10870, pp. 741–752, 2018.
https://doi.org/10.1007/978-3-319-92639-1_62

Clustering techniques offer several advantages over manual grouping process. First, a clustering algorithm can apply a specified objective criterion consistently to form the groups. Second, a clustering algorithm can form the groups in a fraction of time necessary for manual grouping, particularly if long list of descriptors or features is associated with each object. The speed, reliability and consistency of clustering algorithm in organizing data represent an overwhelming reason to use it. Clustering can be roughly distinguished as:

- Hard clustering: each object belongs to specific cluster or not
- Soft clustering also named - *fuzzy clustering* - each object belongs to each cluster to a certain degree.

Cluster analysis, an important technology in data mining, is an effective method of analyzing and discovering useful information from numerous data. Modern medicine generates a great deal of information stored in the medical database. Extracting useful knowledge and making scientific decision for diagnosis and treatment of disease from the database increasingly becomes necessary. Medical field is primarily directed at patient care activity and only secondarily as research resource. The only justification for collecting medical data is to benefit the individual patient.

This paper presents hybrid clustering approach for diagnosing medical diseases combining fuzzy partition method and maximum likelihood estimates clustering algorithm. Also, this paper continuous the authors' previous research in clustering presented in [1–5].

The rest of the paper is organized in the following way: Sect. 2 provides an overview of the basic idea on clustering and classification. Section 3 presents modeling the fuzzy clustering approach and testing the proposed hybrid model with well-known *Iris Data Set*. The Preliminary experimental results are presented in Sect. 4. Section 5 provides conclusions and some points for future work.

2   Clustering, Classification and Related Work

Clustering and classification are basic scientific tools used to systematize knowledge and analyze the structure of phenomena. Clustering and classification are both fundamental tasks in data mining. Both refer to the process of partitioning a set of objects into groups as dissimilar as possible from one another. Unfortunately, although some basic distinctions in this process are recognized across discipline, common terminology is lacking, and the two terms are often used interchangeably.

The conventional distinction made between clustering and classification is the following. Clustering is a process of partitioning a set of items (or grouping individual items) into set of categories. Classification is a process of assigning a new item or observation to its proper place in an established set of categories [6]. In clustering, little or nothing is known about category structure, and the objective is to discover a structure that fits the observations.

Classification is used mostly as a supervised learning method, but on the other side clustering is used for unsupervised learning. The goal of clustering is descriptive, that of classification is predictive [7].

2.1 Clustering

Clustering groups data instances into subsets in such a manner that similar instances are grouped together, while different instances belong to different groups. The instances are thereby organized into an efficient representation that characterizes the population being sampled.

Formally, the clustering structure is represented as a set of subsets $C = C_1,..., C_k$ of S, such that: $S = \bigcup_{i=1}^{k} C_i$ and $C_i \cap C_j = 0$ for $i \neq j$. Consequently, any instance in S belongs to exactly one and only one subset.

Clustering of objects is as ancient as the human need for describing the salient characteristics of men and objects and identifying them with a type. Therefore, it embraces various scientific disciplines: from mathematics and statistics to biology and genetics, each of which uses different terms to describe the topologies formed using this analysis. From biological "taxonomies", to medical "syndromes" and genetic "genotypes" to manufacturing "group technology" — the problem is identical: forming categories of entities and assigning individuals to the proper groups within it [8].

Cluster analysis, an important technology in data mining, is an effective method of analyzing and discovering useful information from numerous data. Cluster algorithm groups the data into classes or clusters so that objects within a cluster have high similarity in comparison to one another but are very dissimilar to objects in other clusters.

General references regarding data clustering are presented in [9, 10]. A very good presentation of contemporary data mining clustering techniques can be found in the textbook [11].

2.2 Related Work in Clustering Medical Data

In the past decades, many approaches have been proposed to solve clustering problem in medical data to help physicians to make decision regarding patients illness and future treatments. It started as a clinical probability model and then proceeded to look at a wide range of techniques which: include logistic regression, neural networks, Bayesian networks, class probability trees, and hidden Markov models [12].

In [13] some very basic algorithms like *k-means*, *fuzzy c-means*, hierarchical clustering to come up with clusters are discussed as well as the use of R data mining tool. The results are tested on the datasets, namely Online News Popularity, Iris Data Set and from UCI data repository [14] and miRNA dataset [15] for medical data analysis.

In the paper, [16] is concerned with the ideas behind the design, implementation, testing and application of a novel swarm based intelligent system for medical data set analysis. The unique contribution of the research is in the implementation of a hybrid intelligent system Data Mining technique such as Bacteria Foraging Optimization Algorithm (BFOA) for solving novel practical problems. The detailed description of this technique and the illustrations of several applications solved by this novel technique are done.

Hypertension, dyslipidemia, diabetes and smoking are well-established risk factors for cardiovascular disease (CVD) and the damage caused by these factors is widespread across the developed world. The clustering of CVD risk factors is a serious threat for increasing medical expenses. The age-specific proportion and distribution of medical expenditure attributable to CVD risk factors, especially focused on the elderly, is thus indispensable for formulating public health policy given the extent of the ageing population in developed countries. Gamma regression models were applied to examine how the number of CVD risk factors affects mean medical expenditure. The four CVD risk factors are analyzed: hypertension, hypercholesterolaemia, high blood glucose and smoking [17].

Prognostic logistic regression models have been used in various ways for intensive-care medicine. These include the stratification of patients for therapeutic drug trials is presented in [18].

3 Modeling the Fuzzy Clustering Approach

Clustering algorithms can group given data set into clusters by different approaches: (1) *hard partitioning methods* – such as – *k*-means, *k*-medoids, *k*-medians, *k*-means++; (2) *fuzzy partitioning methods* – such as – *fuzzy c-means clustering* method, *fuzzy* Gustafson-Kessel clustering method, *fuzzy* Gath-Geva clustering method.

There are also finer distinctions possible, such as: (1) *strict partitioning clustering*: each object belongs to exactly one cluster; (2) *strict partitioning clustering with outliers*: objects can also belong to no cluster, and are considered outliers; (3) *overlapping clustering*: objects may belong to more than one cluster; (4) *hierarchical clustering*: objects that belong to a child cluster also belong to the parent cluster; (5) *subspace clustering*: while an overlapping clustering, within a uniquely defined subspace, clusters are not expected to overlap.

In this research *fuzzy maximum likelihood estimates with a direct distance norm* based on the fuzzy Gath-Geva clustering method is used [19]. *Fuzzy maximum likelihood estimates with a direct distance norm* belongs to fuzzy partitioning methods.

3.1 Optimal Number of Clusters

During every partitioning problem the number of subsets (called the clusters) must be given by the user before the calculation, but it is rarely known *a priori*. The useful technique named *Improved Covariance Estimation for Gustafson-Kessel Clustering* algorithm is employed in the extraction of the rules from data and proposed in this model and used to estimate the *optimal number of clusters*. It reduces the risk of over fitting when the number of training samples is low relative to the number of clusters. This is achieved by adding a scaled unity matrix to the calculated covariance matrix. The proposed algorithm is an extension of classical *Gustafson-Kessel* clustering algorithm which is well and in detail described in [20].

The proposed *Gustafson-Kessel* extension algorithm calculates seven different coefficients and compares them to estimate the optimal number of clusters. This method calculates: (1) Partition Coefficient (PC); (2) Classification Entropy (CE); (3) Partition Index (SC); (4) Separation Index (S); (6) Xie and Beni's Index (XB); Dunn's Index (DI); (7) Alternative Dunn Index (ADI). For estimate optimal number of clusters for *hard partitioning methods* it is recommended to calculate DI and ADI. For estimate optimal number of clusters for *fuzzy partitioning methods* there is no definitely scientific or experimental recommendation, but the usage of PC and CE parameters, as default, is proposed. The optimal number of clusters is at the maximum value of parameters.

3.2 Fuzzy Partition Method

The data set is typically an observation of some physical process. Each observation consists of n measured variables, grouped into an n-dimensional row vector $x_k = [x_{k1}, x_{k2},..., x_{kn}]^T$, $x_k \in R^n$. A set of N observations is denoted by $X = \{x_k \mid k = 1, 2,..., N\}$, and is represented as an $N \times n$ matrix, a data set. Since clusters can formally be viewed as subsets of the data set, the number of subsets (clusters) is denoted by c. Fuzzy partition can be seen as a generalization of hard partition, it allows μ_{ik} to attain real values in [0, 1]. A $N \times c$ matrix $U = [\mu_{ik}]$ represents the fuzzy partitions, its conditions are given by:

$$\mu_{ij} \in [0, 1], \ 1 \leq i \leq N, 1 \leq k \leq c \tag{1}$$

$$\sum_{k=1}^{c} \mu_{ik} = 1, \ 1 \leq i \leq N \tag{2}$$

$$0 < \sum_{i=1}^{N} \mu_{ik} < N, \ 1 \leq k \leq c \tag{3}$$

Let $X = [x_1, x_2,..., x_N]$ be a *finite set* and let $2 \leq c < N$ be an integer. The *fuzzy partitioning space* for X is the set

$$M_{fc} = \left\{ U \in \Re^{N \times c} \mid \mu_{ik} \in [0, 1], \forall i, k; \sum_{k=1}^{c} \mu_{ik} = 1, \forall i; 0 < \sum_{k=1}^{c} \mu_{ik} < N, \forall k \right\}. \tag{4}$$

The i-th column of U contains values of the *membership function* of the i-th fuzzy subset of X. The Eq. (2) constrains the sum of each column to 1, and thus the total membership of each x_k in X equals one. The distribution of memberships among the c fuzzy subsets is not constrained.

3.3 Fuzzy Maximum Likelihood Estimates Clustering Algorithm

The basic steps of the proposed hybrid algorithm for the fuzzy maximum likelihood estimates (FMLE) clustering algorithm employs a distance norm based on fuzzy maximum likelihood estimates proposed in [21], are summarized by the pseudo code shown in Algorithm 1.

Algorithm 1: *The algorithm for **Fuzzy Maximum Likelihood Estimates***

Begin

    Step 1:   --- *Initialization.*

                  $X; \; c; \; m > 1; \; \varepsilon > 0$

    Step 2:   --- *Calculate the cluster centers.*
                       Repeat for *l=1, 2, ...*

$$v_i^{(l)} = \frac{\sum\limits_{k=1}^{N} (\mu_{ik}^{(l-1)})^{w} x_k}{\sum\limits_{k=1}^{N} (\mu_{ik}^{(l-1)})^{w}}, \; 1 \le i \le c$$

    Step 3:   --- *Compute the distance measure* D_{ik}^{2}

$$F_i^{(l)} = \frac{\sum\limits_{k=1}^{N} (\mu_{ik}^{(l-1)})^{w} (x_k - v_i^{(l)})(x_k - v_i^{(l)})^{T}}{\sum\limits_{k=1}^{N} (\mu_{ik}^{(l-1)})^{w}}, \; 1 \le i \le c$$

                  --- *The distance function is chosen as*

$$D_{ik}^{2}(x_k, v_i) = \frac{(2\pi)^{(\frac{n}{2})} \sqrt{\det(F_i)}}{\alpha_i} \exp(\frac{1}{2}(x_k - v_i^{l})^{T} \; F_i^{-1}(x_k - v_i^{l}))$$

                  --- *with the **a priori** probability*

$$\alpha_i = \frac{1}{N} \sum\limits_{k=1}^{1} \mu_{ik}$$

    Step 4:   --- *Update the partition matrix*

$$\mu_{ik}^{(l)} = \frac{1}{\sum\limits_{j=1}^{c} (D_{ik}(x_k, v_i) / (D_{jk}(x_k, v_j))^{2/(m-1)}}, \; 1 \le i \le c, \; 1 \le k \le N$$

    Step 5:   *Until* $\| U^{(l)} - U^{(l-1)} \| < \varepsilon$

End.

In consistence with the theory, notice in previous subsection, *Fuzzy Partition Method*, there is a set of data *X* specify *c*, choose a *weighting exponent* $m > 1$ and a termination tolerance $\varepsilon > 0$. Initialize the partition matrix with a more robust method. It is important to mention that in *Step 3*. the distance to the cluster center (centroid) is calculated on the basis of the fuzzy covariance matrices of the cluster.

3.4   Testing the Hybrid Clustering Model – *Iris Data Set*

In order to test proposed hybrid fuzzy maximum likelihood estimates clustering algorithm employs with distance norm based on fuzzy maximum likelihood estimates, a well-known Iris data set, from UCI Machine Learning Repository is used [14]. It is the data set most often used in scientific work [13, 22]. Iris data set has 150 samples, 4 attributes, and 3 classes. A case in point is famed iris data which has been widely used as test data for clustering methods. Fifty specimens from each of three species of the flower are described on the basis of four measurements. One of the species exhibits distinctive values on two of the measurements and virtually any clustering method succeeds; however, the other two species overlap, and they provide a challenge.

Also, there are many discussions in the scientific papers in terms of how many classes are there in a data set. Some researchers report that Iris data set consists of 2, 3, or 4 classes. Many different techniques are used (Fig. 1).

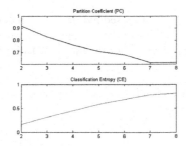

Fig. 1. Partition Coefficient (PC) and Classification Entropy (CE) for *Iris data set* to estimate the optimal number of clusters

The proposed *Gustafson-Kessel* extension algorithm calculates seven different coefficients to estimate the optimal number of clusters, and for *Iris data set* the calculated values for three clusters are: PC: 0.8276; CE: 0.3113; SC: 0.0977; S: 9.6888e-004; XB: 5.3214; DI: 0.0271; ADI: 0.0060. On the other hand, the same algorithm and the calculation is used for *Iris data set* and they are presented as: PC: 0.7617; CE: 0.4494; SC: 0.1308; S: 0.0011; XB: 3.9190; DI: 0.0247; ADI: 0.0044. The technique *Improved Covariance Estimation for Gustafson-Kessel Clustering* algorithm is employed in the extraction of the rules from data and estimation of the *optimal number of clusters* for *fuzzy partitioning methods*. This particular algorithm is used for calculating PC parameter (which should be as high as possible), which we used for well-known *Iris data set* to prove that the *optimal number of clusters* is three, as shown in Fig. 2. This experimental result is also supported by *Calinski-Harabasz criterion*, by using two measures, the *Calinski-Harabasz Index*, known as the *variance ratio criterion* and *Total within Sum of Squares,* for choosing the suitable *c*, number of clusters.

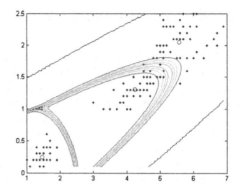

Fig. 2. Similarity of data for *Iris data set* calculated by hybrid fuzzy maximum likelihood estimates clustering algorithm and presented with three clusters based on only the most important features

It can be concluded that the proposed hybrid fuzzy maximum likelihood estimates clustering algorithm appropriate fit for well-known Iris data set and the algorithm can be tested, for example, with data set for diagnosing medical diseases (Table 1).

Table 1. The coordinates centroids for *Iris data set* for: (1) whole data set; (2) selected the most important features of the data set

Iris data set						
	Whole data set				Important features	
Centroid 1	5.9089	4.1893	2.7763	1.2919	4.2384	1.3065
Centroid 2	5.0060	3.4280	1.4620	0.2460	1.4620	0.2460
Centroid 3	6.5555	2.9513	5.4969	1.9953	5.5723	2.0488

4 Experimental Results

The proposed hybrid fuzzy maximum likelihood estimates clustering algorithm was further on, in our research, tested on medical data sets from UCI Machine Learning Repository presented Table 2.

Table 2. The UCI Machine Learning Repository data sets used in this research

Data set					
	Instances	Used instances	Attributes	Classes	Miss data
Echocardiogram	132	105	12	2	Yes
Hepatitis	155	145	19	2	Yes
Liver Disorders	345	345	7	2	No

4.1 Experimental Results *Echocardiogram Data Set*

In this research the missing data are removed and there are, therefore, 105 instances. The experimental results for PC = 0.5827 and CE = 0.6046 for *Echocardiogram Data Set* and estimated optimal number of clusters are presented in Fig. 3. Partition Coefficient is maximal for two clusters and it is used in FMLE algorithm.

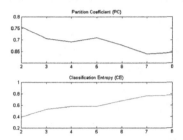

Fig. 3. Partition Coefficient (PC) and Classification Entropy (CE) for *Echocardiogram Data Set* to estimate the optimal number of clusters

When the most *important features* (1 and 3) are used after 69 iterations, the PC = 0.9857, CE = 0.0243, and Centroid 1: 27.61, 61.49; Centroid 2: 0.68, 68.05 are given. For all data sets, PC and CE are given in previous paragraph and the centroids are given by: Centroid 1: 25.36, 62.22, 10.48, 4.62 13.83 1.24; Centroid 2: 9.37 63.66 23.54 5.77 20.16 2.10. The comparison of the experimental results of classes given by hybrid fuzzy maximum likelihood estimates clustering algorithm and the original given classes shows the accuracy of 78%. The 82 instances out of 105 in data set have been correctly evaluated.

4.2 Experimental Results *Hepatitis Data Set*

In this research, also, the missing data have been removed and there are, therefore, 146, instead of 155, instances, with 19 attributes and two classes, as in original data set. The experimental results for PC = 0.8432 and CE = 0.2742 for *Hepatitis Data Set* and estimated optimal number of clusters are presented in Fig. 4. Partition Coefficient is maximal for two clusters and it is used in FMLE algorithm.

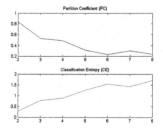

Fig. 4. Partition Coefficient (PC) and Classification Entropy (CE) for *Hepatitis Data Set* to estimate the optimal number of clusters

For all data sets, PC and CE are given in previous paragraph and the centroids are given by: Centroid 1: 1.32, 1.78, 2.75, 3.27; Centroid 2: 1.72, 1.81, 0.90, 3.98. The comparison of the experimental results of classes given by hybrid fuzzy maximum likelihood estimates clustering algorithm and the original given classes presents accuracy of 79%. The 116 instances of 146 in data set have correct evaluation. When the most *important features* (15 and 17) are used after 68 iterations, the PC = 0.9236, CE = 0.1412, and Centroid 1: 0.91, 3.95; Centroid 2: 3.08, 3.31 are given in Fig. 5.

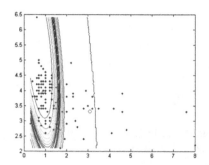

Fig. 5. Similarity of data for *Hepatitis data set* calculated by hybrid fuzzy maximum likelihood estimates clustering algorithm and presented with two clusters based on only the most important features

4.3   Experimental Results *Liver Disorders Data Set*

There are no missing data in Liver Disorders Data Set and therefore all 345 instances, with 7 attributes and two classes, as in original data set.

The experimental results for PC = 0.5675 and CE = 0.8208 for *Liver Disorders Data Set* and estimated optimal number of clusters and centroids are given by: Centroid 1: 68.03, 23.91, 21.19, 22.33, 2.54; Centroid 2: 74.60, 50.98, 35.49, 88.54, 6.06, are presented in Fig. 6. Partition Coefficient is maximal for two clusters and it is used in FMLE algorithm. When the most *important features* (1 and 2) are used after 228 iterations, the PC = 0.9446, CE = 0.0932, and Centroid 1: 90.04, 87.44; Centroid 2: 90.20, 58.89.

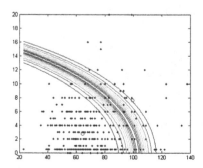

Fig. 6. Similarity of data for *Liver Disorder data set* calculated by hybrid FMLE estimates clustering algorithm and presented with two clusters based on only the most important features

The comparison of the experimental results of classes given by hybrid fuzzy maximum likelihood estimates clustering algorithm and the original given classes shows that 202 out of 345 data set instances have been correctly evaluated, which presents accuracy of 58%.

4.4 Discussion

The hybrid fuzzy maximum likelihood estimates clustering algorithm can detect clusters of varying shapes, sizes and densities. The cluster covariance matrix is used in conjunction with an "exponential" distance, and the clusters are not constrained in volume.

It can be concluded that the proposed hybrid fuzzy maximum likelihood estimates clustering algorithm appropriate fit for well-known *Iris data set*. The proposed algorithm includes the estimation of number of clusters tested with data set for diagnosing medical diseases. In two experimental data sets which are used from domain of diagnosis medical diseases both have very similar accuracy of 78%, but the third data set has accuracy of 58%.

5   Conclusion and Future Work

The aim of this paper is to propose the new hybrid strategy for diagnosing medical diseases. First, the algorithm is employed for the extraction of the rules from data and proposed model to estimate the *optimal number of clusters*. The new proposed hybrid approach is obtained by combining *fuzzy partition* method and *maximum likelihood estimates clustering* algorithm employs by a distance norm based on fuzzy maximum likelihood estimates.

Preliminary experimental results encourage the further research by the authors because three experimental data sets in domain of diagnosing medical diseases have accuracy: two of them have the accuracy of 78% and one has the accuracy of 58%. Our future research will focus on creating new hybrid model combined evolutionary techniques which will efficiently solve real-world data sets from the Clinical centre of Vojvodina in Serbia.

References

1. Simić, D., Ilin, V., Tanackov, I., Svirčević, V., Simić, S.: A hybrid analytic hierarchy process for clustering and ranking best location for logistics distribution center. In: Onieva, E., Santos, I., Osaba, E., Quintián, H., Corchado, E. (eds.) HAIS 2015. LNCS (LNAI), vol. 9121, pp. 477–488. Springer, Cham (2015). https://doi.org/10.1007/978-3-319-19644-2_40
2. Simić, D., Ilin, V., Svirčević, V., Simić, S.: A hybrid clustering and ranking method for best positioned logistics distribution centre in Balkan Peninsula. Logic J. IGPL **25**(6), 991–1005 (2017)
3. Simić, D., Svirčević, V., Sremac, S., Ilin, V., Simić, S.: An efficiency k-means data clustering in cotton textile imports. In: Burduk, R., Jackowski, K., Kurzyński, M., Woźniak, M., Żołnierek, A. (eds.) Proceedings of the 9th International Conference on Computer Recognition Systems CORES 2015. AISC, vol. 403, pp. 255–264. Springer, Cham (2016). https://doi.org/10.1007/978-3-319-26227-7_24

4. Simić, D., Jackowski, K., Jankowski, D., Simić, S.: Comparison of clustering methods in cotton textile industry. In: Jackowski, K., Burduk, R., Walkowiak, K., Woźniak, M., Yin, H. (eds.) IDEAL 2015. LNCS, vol. 9375, pp. 501–508. Springer, Cham (2015). https://doi.org/10.1007/978-3-319-24834-9_58
5. Krawczyk, B., Simić, D., Simić, S., Woźniak, M.: Automatic diagnosis of primary headaches by machine learning methods. Open Med. **8**(2), 157–165 (2013)
6. Anderberg, M.R.: Cluster Analysis for Applications. Academic Press, New York (1973)
7. Veyssieres, M.P., Plant, R.E.: Identification of vegetation state and transition domains in California's hardwood rangelands. University of California (1998). http://frap.fire.ca.gov/publications/state_and_trans2.pdf. Accessed 7 Feb 2018
8. Rokach, L., Maimon, O.: Clustering methods. In: Maimon, O., Rokach, L. (eds.) Data Mining and Knowledge Discovery Handbook, pp. 321–352. Springer, Boston (2005). https://doi.org/10.1007/0-387-25465-X_15
9. Hartigan, J.: Clustering Algorithms. Wiley, New York (1975)
10. Jain, A., Dubes, R.: Algorithms for Clustering Data. Prentice-Hall, Upper Saddle River (1988)
11. Han, J., Kamber, M.: Data Mining. Morgan Kaufmann Publishers, Boston (2001)
12. Dybowski, R., Roberts, S.: An anthology of probabilistic models for medical informatics. In: Husmeier, D., Dybowski, R., Roberts, R. (eds.) Probabilistic Modeling in Bioinformatics and Medical Informatics, pp. 297–349. Springer, London (2005). https://doi.org/10.1007/1-84628-119-9_10
13. Pamulaparty, L., Guru Rao, C.V., Sreenivasa, R.M.: Cluster analysis of medical research data using R. Glob. J. Comput. Sci. Technol. **16**(1), 16–22 (2016)
14. UCI Machine Learning Repository. http://archive.ics.uci.edu/ml/datasets.html. Accessed 28 Feb 2018
15. http://www.mirbase.org. Accessed 28 Feb 2018
16. Kalyani, P.: Medical data set analysis – a enhanced clustering approach. Int. J. Latest Res. Sci. Technol. **3**(1), 102–105 (2014)
17. Murakami, Y., Okamura, T., Nakamura, K., Miura, K., Ueshima, H.: The clustering of cardiovascular disease risk factors and their impacts on annual medical expenditure in Japan: community-based cost analysis using Gamma regression models. BMJ Open **3**(3) (2013). https://doi.org/10.1136/bmjopen-2012-002234
18. Knaus, W.A., Harrell, W.A., Fisher, C.J., Wagner, D.P., Opal, S.M., Sadoff, J.C., Draper, E.A., Walawander, C.A., Conboy, K., Grasela, T.H.: The clinical evaluation of new drugs for sepsis: a prospective study design based on survival analysis. J. Am. Med. Assoc. **270**(10), 1233–1241 (1993)
19. Gath, I., Geva, A.B.: Unsupervised optimal fuzzy clustering. IEEE Trans. Pattern Anal. Mach. Intell. **11**(7), 773–780 (1989)
20. Babuška, R., van der Veen, P.J., Kaymak, U.: Improved covariance estimation for Gustafson-Kessel clustering. In: IEEE International Conference on Fuzzy Systems, pp. 1081–1085 (2002)
21. Bezdek, J.C., Dunn, J.C.: Optimal fuzzy partitions: a heuristic for estimating the parameters in a mixture of normal distributions. IEEE Trans. Comput. **C-24**(8), 835–838 (1975)
22. Yao, H., Butz, C.J., Hamilton, H.J.: Causal discovery. In: Maimon, O., Rokach, L. (eds.) Data Mining and Knowledge Discovery Handbook, pp. 949–957. Springer, Boston (2005). https://doi.org/10.1007/0-387-25465-X_44

Author Index

Printed in the United States
By Bookmasters